W9-BVC-793

The authors have done an excellent job of updating the content with all the new developments in optical networking in this second edition. Optical networking is a rapidly changing technology that can support the enormous Internet traffic growth demands required by bandwidth intensive applications of the 21st century.
—Rao Lingampalli, Senior Technical Marketing Manager, Calient Networks

I really enjoy the bottoms-up approach taken by the authors to address fundamentals of optical components as the enablers, optical transmission system design and engineering as the building blocks, and network architecture and its management features that deliver applications to the network operators and services providers at the top of the food chain.
—Shoa-Kai Liu, Director of Advanced Technology, Worldcom

This book not only provides the fundamentals and details of photonics, but the pragmatic perspective presented enables the service provider, the equipment manufacturer, and the academician to view light from a real-life standpoint.
—Mathew Oommen, Vice President, Network Architecture, Williams Communications Group

This book functions as both an introduction to optical networking and as a text to reference again and again. Great for system designers as well as those marketing and selling those systems. Optical Networks provides theory and applications. While no text can be truly state-of-the-art in the fast moving area of optical networking, this one comes as close as possible.
—Alan Repech, System Architect, Cisco Systems Optical Transport

This book provides the most comprehensive coverage of both the theory and practice of optical networking. Its up-to-date coverage makes it an invaluable reference for both practitioners and researchers.
—Suresh Subramaniam, Assistant Professor, Department of Electrical and Computer Engineering, George Washington University

This book definitely provides the most comprehensive coverage of the technologies, architectures, and applications essential for the study of the optical networking field. This second edition is very well written and organized, and the new topics covered appropriately reflect many of the latest technology and standards developments within our industry.
—Mark R. Wilson, Ph.D., Adjunct Professor of Electrical Engineering, University of Pennsylvania; Senior Manager, Optical Network Architectures, Agere Systems

This book provides an excellent overview of the complex field of optical networking. I especially like how it ties the optical hardware functionality into the overall networking picture. Everybody who wants to be a player in the optical networking space should have this book within easy reach.
—Martin Zirngibl, Director, Photonics Network Research, Lucent Technologies, Bell Laboratories

Optical Networks

A Practical Perspective

Second Edition

The Morgan Kaufmann Series in Networking
Series Editor, David Clark, MIT

Optical Networks: A Practical Perspective, 2e
Rajiv Ramaswami and Kumar N. Sivarajan

Telecommunications Law in the Internet Age
Sharon K. Black

Internet QoS: Architectures and Mechanisms
Zheng Wang

TCP/IP Sockets in Java: Practical Guide for Programmers
Michael J. Donahoo and Kenneth L. Calvert

TCP/IP Sockets in C: Practical Guide for Programmers
Kenneth L. Calvert and Michael J. Donahoo

Multicast Communication: Protocols, Programming, and Applications
Ralph Wittmann and Martina Zitterbart

MPLS: Technology and Applications
Bruce Davie and Yakov Rekhter

High-Performance Communication Networks, 2e
Jean Walrand and Pravin Varaiya

Computer Networks: A Systems Approach, 2e
Larry L. Peterson and Bruce S. Davie

Internetworking Multimedia
Jon Crowcroft, Mark Handley, and Ian Wakeman

Understanding Networked Applications: A First Course
David G. Messerschmitt

Integrated Management of Networked Systems: Concepts, Architectures, and their Operational Application
Heinz-Gerd Hegering, Sebastian Abeck, and Bernhard Neumair

Virtual Private Networks: Making the Right Connection
Dennis Fowler

Networked Applications: A Guide to the New Computing Infrastructure
David G. Messerschmitt

Modern Cable Television Technology: Video, Voice, and Data Communications
Walter Ciciora, James Farmer, and David Large

Switching in IP Networks: IP Switching, Tag Switching, and Related Technologies
Bruce S. Davie, Paul Doolan, and Yakov Rekhter

Wide Area Network Design: Concepts and Tools for Optimization
Robert S. Cahn

Practical Computer Network Analysis and Design
James D. McCabe

Frame Relay Applications: Business and Technology Case Studies
James P. Cavanagh

For further information on these books and for a list of forthcoming titles,
please visit our Web site at *www.mkp.com*.

Optical Networks

A Practical Perspective

Second Edition

Rajiv Ramaswami

Kumar N. Sivarajan

MORGAN KAUFMANN PUBLISHERS

AN IMPRINT OF ACADEMIC PRESS

A Division of Harcourt, Inc.

SAN FRANCISCO   SAN DIEGO   NEW YORK   BOSTON
LONDON   SYDNEY   TOKYO

Editor   Rick Adams
Publishing Services Manager   Scott Norton
Senior Production Editor   Cheri Palmer
Assistant Acquisitions Editor   Karyn Johnson
Cover Design   Ross Carron Design
Cover Image   © Dominique Sarraute / Image Bank
Text Design   Windfall Software
Copyeditor   Ken DellaPenta
Proofreader   Jennifer McClain
Indexer   Steve Rath
Printer   Courier Corporation

This book has been author-typeset using LaTeX.

Designations used by companies to distinguish their products are often claimed as trademarks or registered trademarks. In all instances in which Morgan Kaufmann Publishers is aware of a claim, the product names appear in initial capital or all capital letters. Readers, however, should contact the appropriate companies for more complete information regarding trademarks and registration.

Morgan Kaufmann Publishers
340 Pine Street, Sixth Floor, San Francisco, CA 94104-3205, USA
http://www.mkp.com

ACADEMIC PRESS
A Division of Harcourt, Inc.
525 B Street, Suite 1900, San Diego, CA 92101-4495, USA
http://www.academicpress.com

Academic Press
Harcourt Place, 32 Jamestown Road, London, NW1 7BY, United Kingdom
http://www.academicpress.com

Library of Congress Control Number: 2001094371

ISBN 1-55860-655-6

This book is printed on acid-free paper.

To our parents.

Foreword

by Paul E. Green, Jr.

Director, Optical Network Technology (retired)

Tellabs, Inc.

If anyone needed evidence of the rapid pace of innovation in WDM optical telecommunications, a year-by-year perusal of the innovations reported in the journals and in the activities of start-up companies in the optical space would provide all needed persuasion. While the field is old—some fifteen years to be exact—it is also young in the sense that almost nothing in the technology has reached steady-state maturity. New optical bands are being opened up, unprecedented bitrates are being carried per wavelength over all-optical distances that were not thought possible not long ago, new forms of fiber are being introduced, and point-point links are now being organized by their owners into rings and other real network structures, using new optical switching and routing technologies.

Thus, it is a most welcome development to see that Ramaswami and Sivarajan have reorganized and updated their outstanding 1998 traversal of this fascinating and important field and have done so in a way that accurately portrays what is going on. The first edition reflected the authors' backgrounds, straddling the academic and industrial research community on one hand and the school of commercial hard knocks on the other, and it is still true that this new book provides a view of optical communication that is lacking in other frequently updated volumes in this field, whose authors lack the authority possessed by Ramaswami and Sivarajan.

At this juncture, there is a pause in the headlong rush to provide ever more bandwidth to everybody, and the reasons are not just economic. The bottleneck between the gigabit environment on everyone's desktop and the terabit environment of the telco and cable fiber infrastructure is a consequence of the failure of the available "last mile" communication technologies to seriously meet the challenge of providing enough bandwidth. Between now and the next edition of this book, one may hope that fiber last mile solutions will finally begin to proliferate, building on the technical ideas captured here.

There can be no doubt that fiber optic communication can only play an ever widening, if invisible, role in our lives, and once again, I cannot recommend the book highly enough to anyone interested in the technical underpinnings.

Foreword to the First Edition

by Paul E. Green, Jr.

Director, Optical Network Technology (retired)

Tellabs, Inc.

Not too many years ago, whenever one wanted to send messages effectively, there were really only two choices—send them by wire or send them by radio. This situation lasted for decades until the mid-1960s, when the fiber optics revolution began, quietly at first, and then with increasing force as people began to appreciate that sending pulses of light through tiny strands of glass wasn't so crazy after all. This revolution is now in full cry, with 4000 strand miles of fiber being installed per day, just in the United States alone. Fiber has been displacing wire in many applications, and gradually it is emerging as one of the two dominant Cinderella transmission technologies of today, wireless being the other. One of these (wireless) goes anywhere but doesn't do much when it gets there, whereas the other (fiber) will never go everywhere but does a great deal indeed wherever it reaches. From the earliest days of fiber communication, people realized that this simple glass medium has incredible amounts of untapped bandwidth capacity waiting to be mined, should the day come when we would actually need it, and should we be able to figure out how to tap it. That day has now come. The demand is here and so are the solutions.

This book describes a revolution within a revolution, the opening up of the capacity of the now-familiar optical fiber to carry more messages, handle a wider variety of transmission types, and provide improved reliabilities and ease of use. In many places where fiber has been installed simply as a better form of copper, even the gigabit capacities that result have not proved adequate to keep up with the demand. The inborn human voracity for more and more bandwidth, plus the growing realization that there are other flexibilities to be had by imaginative use of the fiber, have led people to explore all-optical networks, the subject of this book.

Such networks are those in which either wavelength division or time division is used in new ways to form entire network structures where the messages travel in purely optical form all the way from one user location to another.

When I attempted the same kind of book in 1993, nobody was quite sure whether optical networking would be a roaring success or disappear into the annals of "whatever happened to ..." stories of technology that had once sounded great on paper, but that had somehow never panned out in the real world. My book (*Fiber Optic Networks,* Prentice Hall) spent most of its pages talking about technology building blocks and lamenting their limitations since there was little to say about real networks, the architectural considerations underlying them, and what good they had ever done anybody.

In the last four years, optical networking has indeed really happened, essentially all of it based on wavelength division multiplexing, and with this book Ramaswami and Sivarajan, two of the principal architects of this success, have redressed the insufficiencies of earlier books such as mine. Today, hundreds of millions of dollars of wavelength division networking systems are being sold annually, major new businesses have been created that produce nothing but optical networks, and bandwidth bottlenecks are being relieved and proliferating protocol zoos tamed by this remarkably transparent new way of doing networking; what's more, there is a rich architectural understanding of where to go next. Network experts, fresh from the novelties of such excitements as the Web, now have still another wonderful toy shop to play in. The whole optical networking idea is endlessly fascinating in itself—based on a medium with thousands of gigabits of capacity yet so small as to be almost invisible, transmitters no larger than a grain of salt, amplifiers that amplify vast chunks of bandwidth purely as light, transmission designs that bypass 50 years of hard-won but complex coding, modulation and equalization insights, network architectures that subsume many functions usually done more clumsily in the lower layers of classical layered architectures—these are all fresh and interesting topics that await the reader of this book.

To understand this new networking revolution within a revolution, it is necessary to be led with a sure hand through territory that to many will be unfamiliar. The present authors, with their rare mixture of physics and network architecture expertise, are eminently qualified to serve as guides. After spending some time with this book, you will be more thoroughly conversant with all the important issues that today affect how optical networks are made, what their limitations and potentialities are, and how they fit in with more classical forms of communication networks based on electronic time division. Whether you are a computer network expert wondering how to use fiber to break the bandwidth bottlenecks that are limiting your system capabilities, a planner or implementer trying to future-proof your telephone network, a teacher planning a truly up-to-date communication engineering curriculum, a student looking for a fun lucrative career, or a midcareer person in need of a retread, this volume will provide the help you need.

The authors have captured what is going on and what is going to be going on in this field in a completely up-to-date treatment unavailable elsewhere. I learned a lot from reading it and expect that you will too.

Contents

Preface

Since the first edition of this book appeared in February 1998, we have witnessed a dramatic explosion in optical networking. Optical networking used to be confined to a fairly small community of researchers and engineers but is now of great interest to a broad audience including students; engineers in optical component, equipment, and service provider companies; network planners; investors; venture capitalists; and industry and investment analysts.

With the rapid pace in technological advances and the widespread deployment of optical networks over the past three years, the need for a second edition of this book became apparent. In this edition we have attempted to include the latest advances in optical networks and their underlying technologies. We have also tried to make the book more accessible to a broader community of people interested in learning about optical networking. With this in mind, we have rewritten several chapters, added a large amount of new material, and removed some material that is not as relevant to practical optical networks. We have also updated the references and added some new problems.

The major changes we've made are as follows: We have mostly rewritten the introduction to reflect the current understanding of optical networks, and we've added a section called "Transmission Basics" to introduce several terms commonly used in optical networking and wavelength division multiplexing (WDM) to the layperson.

In Chapter 2, we've added significant sections on dispersion management and solitons, along with a section describing the different fiber types now available.

In Chapter 3, we now cover electro-absorption modulated lasers, tunable lasers, Raman amplifiers, and L-band erbium-doped fiber amplifiers, and we have significantly expanded the section on optical switching to include the new types of switches using micro-electro-mechanical systems (MEMS) and other technologies.

In Chapter 4, we cover return-to-zero modulation and other newer modulation formats such as duobinary, as well as forward error correction, now widely used in high-bit-rate systems. Chapter 5 now includes expanded coverage of chromatic dispersion and polarization effects, which are important factors influencing the design of high-bit-rate long-haul systems.

The networking chapters of the book have been completely rewritten and expanded to reflect the significant progress made in this area. We have organized these chapters as follows: Chapter 6 now includes expanded coverage of SONET/SDH, ATM, and IP networks. Chapter 7 is devoted to architectural considerations underlying WDM network elements. Chapter 8 attempts to provide a unified view of the problems associated with network design and routing in optical networks. Chapter 9 provides significantly expanded coverage of network management and control. We have devoted Chapter 10 to network survivability, with a detailed discussion on optical layer protection. Chapter 11 covers access networks with a focus on emerging passive optical networks (PONs). Chapter 12 provides updated coverage of optical packet-switched networks. Finally, Chapter 13 focuses on deployment considerations and is intended to provide the reader with a broad understanding of how telecommunications networks are evolving. It includes a couple of detailed network planning case studies on a typical long-haul and metro network.

There is currently a great deal of standards activity in this field. We've added an appendix listing the relevant standards. We have also added another appendix listing the acronyms used in the book and moved some of the more advanced material on pulse propagation into an appendix.

While we have mostly added new material, we have also removed some chapters present in the first edition. We have eliminated the chapter on broadcast-and-select networks, as these networks are mostly of academic interest today. Likewise, we also removed the chapter describing optical networking testbeds as they are mostly of historical importance at this point. Interested readers can obtain a copy of these chapters on the Internet at *www.mkp.com/opticalnet2*.

Teaching and Learning from This Book

This book can be used as a textbook for graduate courses in electrical engineering or computer science. Much of the material in this book has been covered in courses taught by us. Chapters 2–5 cover components and transmission technology aspects of optical networking, and Chapters 6–13 deal with the networking aspects. To understand the networking issues, students will require a basic undergraduate-level knowledge of communication networks. We have tried to make the transmission-related chapters of the book accessible to networking professionals. For example, components are treated first in a simple qualitative manner from the viewpoint of a network designer, but their principle of operation is then explained in detail. Some prior knowledge of semiconductors and electromagnetics will be helpful in appreciating the detailed treatment in some of the sections.

Readers wishing to obtain a broad understanding of the major aspects of optical networking can read Chapters 1, 6, 7, and 13. Those interested in getting a basic appreciation of the underlying components and transmission technologies can read through Chapters 1–5, skipping the quantitative sections.

The book can be the basis for a graduate course in an electrical engineering or computer science curriculum. A networks-oriented course might emphasize network architectures and control and management, by focusing on Chapters 6–13, and skim over the technology portions of the book. Likewise, a course on optical transmission in an electrical engineering department might instead focus on Chapters 2–5 and omit the remaining chapters. Each chapter is accompanied by a number of problems, and instructors may obtain a solution manual by contacting the publisher at *mkp@mkp.com*.

Acknowledgments

We were fortunate to have an outstanding set of reviewers who made a significant effort in reading through the chapters in detail and providing us with many suggestions to improve the coverage and presentation of material. They have been invaluable in shaping this edition. Specifically, we would like to thank Paul Green, Goff Hill, David Hunter, Rao Lingampalli, Alan McGuire, Shawn O'Donnell, Walter Johnstone, Alan Repech, George Stewart, Suresh Subramaniam, Eric Verillow, and Martin Zirngibl. In addition, we would like to acknowledge Bijan Raahemi, Jim Refi, Krishna Thyagarajan, and Mark R. Wilson who provided inputs and comments on specific topics and pointed out some mistakes in the first edition. Mark R. Wilson was kind enough to provide us with several applications-oriented problems from his class, which we have included in this edition. We would also like to thank Amit Agarwal, Shyam Iyer, Ashutosh Kulshreshtha, and Sarath Kumar for the use of their mesh network design tool, Ashutosh Kulshreshtha for also computing the detailed mesh network design example, Tapan Kumar Nayak for computing the lightpath topology design example, Parthasarathi Palai for simulating the EDFA gain curves, and Rajeev Roy for verifying some of our results. As always, we take responsibility for any errors or omissions and would greatly appreciate hearing from you as you discover them. Please email your comments to *mkp@mkp.com*.

chapter 1

Introduction to Optical Networks

A S WE ENTER THE NEW MILLENNIUM, we are seeing dramatic changes in the telecommunications industry that have far-reaching implications for our lifestyles. There are many drivers for these changes. First and foremost is the continuing, relentless need for more capacity in the network. This demand is fueled by many factors. The tremendous growth of the Internet and the World Wide Web, both in terms of number of users as well as the amount of time and thus bandwidth taken by each user, is a major factor. A simple example of the latter phenomenon is the following: An average voice phone call lasts about 3 minutes; in contrast, users connecting to the Internet via dialup lines typically stay on for an average of 20 minutes. So an Internet call brings in about six times as much traffic into a network as a voice call.

Internet traffic has been doubling every four to six months, and this trend appears set to continue for a while. Meanwhile we are seeing the ongoing deployment of broadband access technologies such as digital subscriber line (DSL) and cable modems, which provide bandwidths per user on the order of 1 Mb/s, contrasted against the 28–56 kb/s available over dialup lines. The impact of such deployments is quite significant. A 10% increase in penetration of DSL among the 100 million total U.S. households will bring in another 1 Tb/s of traffic into the network, assuming that 10% of these users are on simultaneously.

At the same time, businesses today rely on high-speed networks to conduct their businesses. These networks are used to interconnect multiple locations within a company as well as between companies for business-to-business transactions. Large corporations that used to lease 1.5 Mb/s lines to interconnect their internal sites are commonly leasing 155 Mb/s connections today.

There is also a strong correlation between the increase in demand and the cost of bandwidth. Technological advances have succeeded in continuously reducing the cost of bandwidth. This reduced cost of bandwidth in turn spurs the development of a new set of applications that make use of more bandwidth and affects behavioral patterns. A simple example is that as phone calls get cheaper, people spend more time on the phone. This in turn drives the need for more bandwidth in the network. This positive feedback cycle shows no sign of abating in the near future.

Another factor causing major changes in the industry is the deregulation of the telephone industry and the breaking up of telephone monopolies. For several decades, the telecommunications business was controlled by service providers who were essentially monopolies. In fact this is still the case in many parts of the world. It is a well-known fact that monopolies impede rapid progress. Monopolistic companies can take their time adapting to changes and have no incentive to reduce costs and provide new services. Deregulation of these monopolies has stimulated competition in the marketplace, which in turn has resulted in lower costs to end users and faster deployment of new technologies and services. For example, since the long-distance market was deregulated in the United States in 1984, long-distance phone call rates have dropped by 1.8% annually. In contrast, in the monopoly-dominated local phone market, local phone call rates have increased by 3.6% annually. Deregulation has also resulted in creating a number of new startup service providers as well as startup companies providing equipment to these service providers. This is a big difference from the situation that existed until the mid- to late 1990s, where the telecommunications industry was dominated by a few large service providers and a few large equipment suppliers to these service providers.

There is also a significant change in the type of traffic that is increasingly dominating the network. Much of the new demand is being spurred by data, as opposed to traditional voice traffic—a trend that has already existed for quite a while. However, much of the network today is architected to efficiently support voice traffic, not data traffic. This change in traffic mix is causing service providers to reexamine the way they build their networks, the type of services they deliver, and even their entire business model, in many cases. We will study the impact of this later in this chapter and also in Chapter 13.

These factors have driven the development of high-capacity optical networks and their remarkably rapid transition from the research laboratories into commercial deployment. This book aims to cover optical network technologies, systems, and networking issues, as well as economic and other deployment considerations.

1.1   Telecommunications Network Architecture

Our focus in this book is primarily on the so-called *public* networks, which are networks operated by *service providers,* or *carriers,* as they are often called. Carriers use their network to provide a variety of services to their customers. Carriers used to be essentially telephone companies, but today there are many different breeds of carriers operating under different business models, many of whom do not even provide telephone service. In addition to the traditional carriers providing telephone and leased line services, today there are carriers who are dedicated to interconnecting Internet service providers (ISPs), carriers who are in the business of providing bulk bandwidth to other carriers, and even virtual carriers who provide services without owning any infrastructure.

In many cases, the carrier owns the facilities (for example, fiber links) and equipment deployed inside the network. Building fiber links requires right-of-way privileges. Not anybody can dig up streets! Fiber is deployed in many different ways today—buried underground, strung on overhead poles, and buried beside oil and gas pipelines and railroad tracks. In other cases, carriers may lease facilities from other carriers and in turn offer value-added services using these facilities. For example, a long-distance phone service provider may not own a network at all but rather simply buy bandwidth from another carrier and resell it to end users in smaller portions.

A *local-exchange* carrier (LEC) offers local services in metropolitan areas, and an *interexchange* carrier (IXC) offers long-distance services. This distinction is blurring rapidly as LECs expand into long distance and IXCs expand into local services. In order to understand this better, we need to step back and look at the history of deregulation in the telecommunications services industry. In the United States, before 1984, there was one phone company—AT&T. AT&T, along with the local Bell operating companies, which it owned, held a monopoly for both long-distance and local services. In 1984, with the passing of the telecommunications deregulation act, the overall entity was split into AT&T, which could offer only long-distance services, and a number of "baby" Bells, or regional Bell operating companies (RBOCs), which offered local services and were not allowed to offer long-distance services. Long-distance services were deregulated, and many other companies, such as MCI and Sprint, successfully entered the long-distance market. The baby Bells came to be known as the incumbent LECs (ILECs) and were still monopolies within their local regions. With all the consolidation that has happened in the industry, we are left with four RBOCs, Southwestern Bell Communications (SBC), Bell Atlantic (now Verizon), BellSouth, and U.S. West (now part of Qwest). In addition to the RBOCs, there are other competitive LECs (CLECs) that are less regulated and compete with the RBOCs to offer local services.

The terminology used above is prevalent mostly in North America. In Europe, we had a similar situation where the government-owned postal, telephone, and telegraph (PTT) companies held monopolies within their respective countries. Over the past decade, deregulation has set in, and we now have a number of new carriers in Europe offering both local and long-distance services.

In the rest of the book, we will take a more general approach and classify carriers as metro carriers or long-haul carriers. While the same carrier may offer metro and long-haul services, the networks used to deliver long-haul services are somewhat different from metro networks, and so it is useful to keep this distinction.

In contrast to public networks, *private* networks are networks owned and operated by corporations for their internal use. Many of these corporations in turn rely on capacity provided by public networks to implement their private networks, particularly if these networks cross public land where right-of-way permits are required to construct networks. Networks within buildings spanning at most a few kilometers are called *local-area networks* (LANs), those that span a campus or metropolitan area, typically tens to a few hundred kilometers, are called *metropolitan-area networks* (MANs), and networks that span even longer distances, ranging from several hundred to thousands of kilometers, are called *wide-area networks* (WANs). We will also see a similar type of classification used in public networks, which we study next.

Figure 1.1 shows an overview of a typical public fiber network architecture. The network is vast and complex, and different parts of the network may be owned and operated by different carriers. The nodes in the network are *central offices,* sometimes also called *points of presence* (POPs). (In some cases, POPs refer to "small" nodes and hubs refer to "large" nodes.) The links between the nodes consist of fiber pairs, and in many cases, multiple fiber pairs. Links in the long-haul network tend to be very expensive to construct. For this reason, the topology of many North American long-haul networks is fairly sparse. In Europe, the link lengths are shorter and the European long-haul network topologies tend to be denser. At the same time, it is imperative to provide alternate paths for traffic in case some of the links fail. These constraints have resulted in the widespread deployment of ring topologies, particularly in North America. Rings are sparse (only two links per node), but still provide an alternate path to reroute traffic. In many cases a meshed network is actually implemented in the form of interconnected ring networks.

At a high level, the network can be broken up into a *metropolitan* (or metro) network and a *long-haul* network. The metro network is the part of the network that lies within a large city or a region. The long-haul network interconnects cities or different regions. The metro network consists of a metro *access* network and a

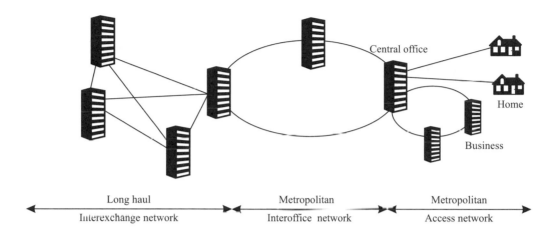

Figure 1.1   Different parts of a public network.

metro *interoffice* network. The access network extends from a central office out to individual businesses or homes (typically groups of homes rather than individual homes at this time). The access network's reach is typically a few kilometers, and it mostly collects traffic from customer locations into the carrier network. Thus most of the traffic in the access network is hubbed into the carrier's central office. The interoffice network connects groups of central offices within a city or region. This network typically spans a few kilometers to several tens of kilometers between offices. The long-haul network interconnects different cities or regions and spans hundreds to thousands of kilometers between central offices. In some cases, there is another part of the network that provides the handoff between the metro network and the long-haul network, particularly if these networks are operated by different carriers. In contrast to the access network, the traffic distribution in the metro interoffice and long-haul networks is meshed (or distributed). The distances indicated here are illustrative and vary widely based on the location of the network. For example, intercity distances in Europe are often only a few hundred kilometers, whereas intercity distances in North America can be as high as a few thousand kilometers.

The network shown in Figure 1.1 is a terrestrial network. Optical fiber is also extensively used in undersea networks. Undersea networks can range from a few hundred kilometers in distance to several thousands of kilometers for routes that cross the Atlantic and Pacific oceans.

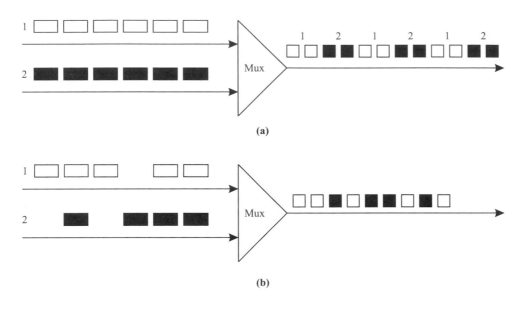

Figure 1.2   Different types of time division multiplexing: (a) fixed, (b) statistical.

1.2   Services, Circuit Switching, and Packet Switching

Many types of services are offered by carriers to their customers. In many cases, these are *connection-oriented* services, in that there is the notion of a connection between two or more parties across an underlying network. The differences lie in the bandwidth of the connection and the type of underlying network with which the connection is supported, which has a significant impact on the quality-of-service guarantees offered by the carriers to their customers. Networks can also provide *connectionless* service; we will discuss this later in this section.

There are two fundamental types of underlying network infrastructures based on how traffic is multiplexed and switched inside the network: *circuit-switched* and *packet-switched*. Figure 1.2 illustrates some of the differences in the type of multiplexing used in these cases.

A circuit-switched network provides circuit-switched connections to its customers. In circuit switching, a guaranteed amount of bandwidth is allocated to each connection and available to the connection all the time, once the connection is set up. The sum of the bandwidth of all the circuits, or connections, on a link must be less than the link bandwidth. The most common example of a circuit-switched network is the public-switched telephone network (PSTN), which provides a nailed-down connection to end users with a fixed amount of bandwidth (typically around 4 kHz) once the connection is established. This circuit is converted to a digital 64 kb/s circuit at the carrier central office. This network was designed to support voice streams and does a fine job for this application.

The circuit-switched services offered by carriers today include circuits at a variety of bit rates, ranging from 64 kb/s voice circuits all the way up to several Gb/s. These connections are typically leased by a carrier to its customers and remain nailed down for fairly long periods, ranging from several days to months to years as the bandwidth on the connection goes up. These services are also called *private line* services. The PSTN fits into this category with one important difference—in the PSTN, users dial up and establish connections between themselves, whereas with private line services, the carrier usually sets up the connection using a management system. This situation is changing, and we will no doubt see users dialing for higher-speed private lines in the future, particularly as the connection durations come down.

The problem with circuit switching is that it is not efficient at handling bursty data traffic. An example of a bursty traffic stream is traffic from a user typing on a keyboard. When the user is actively typing, bits are transmitted at more or less a steady rate. When the user pauses, there is no traffic. Another example is Web browsing. When a user is looking at a recently downloaded screen, there is almost no traffic. When she clicks on a hyperlink, a new page needs to be downloaded as soon as possible from the network. Thus a bursty stream requires a lot of bandwidth from the network whenever it is active and very little bandwidth when it is not active. It is usually characterized by an *average* bandwidth and a *peak* bandwidth, which correspond to the long-term average and the short-term burst rates, respectively. In a circuit-switched network, we would have to reserve sufficient bandwidth to deal with the peak rate, and this bandwidth would be unused a lot of the time.

Packet switching was invented to deal with the problem of tranporting bursty data traffic efficiently. In packet-switched networks, the data stream is broken up into small *packets* of data. These packets are multiplexed together with packets from other data streams inside the network. The packets are switched inside the network based on their destination. To facilitate this switching, a packet header is added to the payload in each packet. The header carries addressing information, for example, the destination address or the address of the next node in the path. The intermediate nodes read the header and determine where to switch the packet based on the information contained in the header. At the destination, packets belonging to a particular stream are received and the data stream is put back together. The predominant example of a packet-switched network is the Internet, which uses the Internet Protocol (IP) to route packets from their source to their destination.

Packet switching uses a technique called *statistical multiplexing* when multiplexing multiple bursty data streams together on a link. Since each data stream is bursty, it is likely that at any given time only some streams are active and others aren't. The probability that all streams are active simultaneously is quite small. Therefore the bandwidth required on the link can be made significantly smaller than the bandwidth that would be required if all streams were to be active simultaneously.

Statistical multiplexing improves the bandwidth utilization but leads to some other important effects. If more streams are active simultaneously than there is bandwidth available on the link, some packets will have to be *queued* or *buffered* until the link becomes free again. The delay experienced by a packet therefore depends on how many packets are queued up ahead of it. This causes the delay to be a random parameter. On occasion, the traffic may be so high that it causes the buffers to overflow. When this happens, some of the packets must be dropped from the network. Usually, a higher-layer transport protocol, such as the transmission control protocol (TCP) in the Internet, detects this and ensures that these packets are retransmitted. On top of this, a traditional packet-switched network does not even support the notion of a connection. Packets belonging to a connection are treated as independent entities, and different packets may take different routes through the network. This is the case with networks using IP. This type of *connectionless* service is called a *datagram* service. This leads to even more variations in the delays experienced by different packets and also forces the higher-layer transport protocol to resequence packets that arrive out of sequence at their destinations.

Thus, traditionally, such a packet-switched network provides what is called *best-effort* service. The network tries its best to get data from its source to its destination as quickly as possible, but offers no guarantees. This is indeed the case with much of the Internet today. Another example of this type of service is *frame relay*. Frame relay is a popular packet-switched service provided by carriers to interconnect corporate data networks. When a user signs up for frame relay service, she is promised a certain average bandwidth over time but allowed to have an instantaneous burst rate above this rate, however without any guarantees. In order to ensure that the network is not overloaded, the user data rate may be regulated at the input to the network so that the user does not exceed her committed average bandwidth over time. In other words, a user who is provided a committed rate of 64 kb/s may send data at 128 kb/s on occasion, and 32 kb/s at other times, but will not be allowed to exceed the average rate of 64 kb/s over a long period of time.

This best-effort service provided by packet-switched networks is fine for a number of applications, such as Web browsing and file transfers, which are not highly delay-sensitive applications. However, applications such as real-time video or voice calls cannot tolerate random packet delays. Therefore, there is a great deal of effort today to design packet-switched networks that can provide some guarantees on the *quality of service* that they offer. Examples of quality of service (QoS) may include certain guarantees on the maximum packet delay as well as the variation in the delay, and guarantees on providing a minimum average bandwidth for each connection. The asynchronous transfer mode (ATM) network is a consequence of this thinking. The Internet Protocol has also been enhanced to provide similar services. Most of these QoS efforts rely on the notion of having a connection-oriented layer. For example, in an IP network, multi-protocol label switching (MPLS) provides *virtual* circuits to support end-to-end traffic streams. A virtual circuit forces all packets belonging to that circuit to follow the same path through the network, allowing better allocation of resources in the network to meet certain quality-of-service guarantees, such as bounded delay for each packet. Unlike a real circuit-switched network, a virtual circuit does not provide a fixed guaranteed bandwidth along the path of the circuit due to the fact that statistical multiplexing is used to multiplex virtual circuits inside the network.

1.2.1 The Changing Services Landscape

The service model used by the carriers is changing rapidly as networks and technologies evolve and competition among carriers intensifies. The bandwidth delivered per connection is increasing, and it is becoming common to lease lines ranging in capacity from 155 Mb/s to 2.5 Gb/s and even 10 Gb/s. Note that in many cases, a carrier's customer is another carrier. The so-called carrier's carrier essentially delivers bandwidth in large quantities to interconnect other carriers' networks. Also, because of increased competition and customer demands, carriers now need to be able to deliver these connections rapidly in minutes to hours rather than days to months, once the bandwidth is requested. Moreover, rather than signing up for contracts that range from months to years, customers would like to sign up for much shorter durations. It is not unthinkable to have a situation where a user leases a large amount of bandwidth for a relatively short period of time, for example, to perform large backups at certain times of the day, or to handle special events, or to handle temporary surges in demands.

Another aspect of change has to do with the *availability* of these circuits, which is defined as the percentage of time the service is available to the user. Typically, carriers provide 99.999% availability, which corresponds to a downtime of less than five minutes per year. This in turn requires the network to be designed to provide very fast restoration of service in the event of failures such as fiber cuts, today in about 50 ms. While this will remain true for a subset of connections, there will be other connections carrying data that may be able to tolerate higher restoration times. There may be connections that may not need to be restored at all by the carrier, with the user dealing with rerouting traffic on these connections in the event of failures. Very fast restoration is usually accomplished by providing full redundancy—half the bandwidth in the network is reserved for this purpose. We will see in Chapter 10 that more sophisticated techniques can be used to improve the bandwidth efficiency but usually at the cost of slower restoration times. This realization is stimulating the development of service offerings that trade off restoration time against bandwidth efficiency in the network.

Thus carriers in the new world need to deploy networks that provide them with the flexibility to deliver bandwidth on demand *when* needed *where* needed, with the appropriate service attributes. The "where needed" is significant because carriers can rarely predict the location of future traffic demands. As a result it is difficult for them to plan and build networks optimized around specific assumptions on bandwidth demands.

At the same time, the mix of services offered by carriers is expanding. We talked about different circuit-switched and packet-switched services earlier. What is not commonly realized is that today these services are delivered over separate overlay networks, rather than a single network. Thus carriers need to operate and maintain multiple networks—a very expensive proposition over time. For most networks, the costs associated with operating the network over time (such as maintenance, provisioning of new connections, upgrades) far outweigh the initial cost of putting in the equipment to build the network. Carriers would thus like to migrate to maintaining a single network infrastructure that enables them to deliver multiple types of services.

1.3   Optical Networks

Optical networks offer the promise to solve many of the problems discussed above. In addition to providing enormous capacities in the network, an optical network provides a common infrastructure over which a variety of services can be delivered. These networks are also increasingly becoming capable of delivering bandwidth in a flexible manner where and when needed.

Optical fiber offers much higher bandwidth than copper cables and is less susceptible to various kinds of electromagnetic interferences and other undesirable effects. As a result, it is the preferred medium for transmission of data at anything more than a few tens of megabits per second over any distance more than a kilometer. It is also the preferred means of realizing short-distance (a few meters to hundreds of meters), high-speed (gigabits per second and above) interconnections inside large systems.

The latest statistics from the U. S. Federal Communications Commission [Kra99] indicate the ubiquity of fiber deployment. Optical fibers are widely deployed today in all kinds of telecommunications networks, except perhaps in residential access networks. Although fiber is provided to many businesses today, particularly in large cities, it has yet to reach individual homes, due to the huge cost of wiring the infrastructure and the questionable rate of return on this investment seen by the

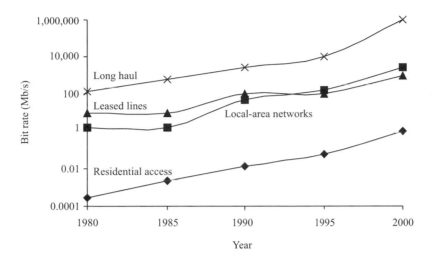

Figure 1.3 Bandwidth growth over time in different types of networks.

service providers. Before providing some more data, we introduce some terminology first. Each *route* in a network comprises many fiber *cables*. Each cable contains many *fibers*. For example, a 10-mile-long route using 3 fiber cables is said to have 10 route miles and 30 *sheath* (cable) miles. If each cable has 20 fibers, the same route is said to have 600 fiber miles. As of the end of 1998, the local-exchange carriers in the United States had deployed more than 355,000 sheath miles of fiber, containing more than 16 million fiber miles. More than 160,000 route miles of fiber had been deployed by the interexchange carriers in the United States, containing more than 3.6 million miles of optical fiber [Kra99].

Fiber transmission technology has evolved over the past few decades to offer higher and higher bit rates on a fiber over longer and longer distances. Figure 1.3 plots the growth in bandwidth over time of different types of networks, updated from a chart originally presented in [Fra93]. This tremendous growth in bandwidth is primarily due to the deployment of optical fiber communication systems.

Note from Figure 1.3 that the bandwidths into our homes are still limited by the bandwidth available on our phone line, which is made of twisted-pair copper cable. These lines are capable of carrying data at a few megabits per second using digital subscriber loop (DSL) technology, but voice-grade lines are limited at the central office to 4 kHz of bandwidth. (Note that in some cases we measure bandwidth in hertz and sometimes loosely use bit rates when talking about bandwidth. We will quantify the relationship between bit rate and bandwidth in Section 1.7.) The other alternative is the cable network, which again is capable of providing a few megabits per second to each subscriber on a shared basis using cable modem technology.

When we talk about optical networks, we are really talking about two generations of optical networks. In the first generation, optics was essentially used for transmission and simply to provide capacity. Optical fiber provided lower bit error rates and higher capacities than copper cables. All the switching and other intelligent network functions were handled by electronics. Examples of first-generation optical networks are SONET (synchronous optical network) and the essentially similar SDH (synchronous digital hierarchy) networks, which form the core of the telecommunications infrastructure in North America and in Europe and Asia, respectively, as well as a variety of enterprise networks such as ESCON (enterprise serial connection). We will study these first-generation networks in Chapter 6.

Today we are seeing the deployment of second-generation optical networks, where some of the routing, switching, and intelligence is moving into the *optical layer*. Before we discuss this new generation of networks, we will first look at the multiplexing techniques that provide the capacity needed to realize these networks.

1.3.1   Multiplexing Techniques

The need for multiplexing is driven by the fact that it is much more economical to transmit data at higher rates over a single fiber than it is to transmit at lower rates over multiple fibers, in most applications. There are fundamentally two ways of increasing the transmission capacity on a fiber, as shown in Figure 1.4. The first is to increase the bit rate. This requires higher-speed electronics. Many lower-speed data streams are multiplexed into a higher-speed stream at the transmission bit rate by means of electronic *time division multiplexing* (TDM). The multiplexer typically interleaves the lower-speed streams to obtain the higher-speed stream. For example, it could pick 1 byte of data from the first stream, the next byte from the second stream, and so on. As an example, 64 155 Mb/s streams may be multiplexed into a single 10 Gb/s stream. Today, the highest transmission rate in commercially available systems is around 10 Gb/s; 40 Gb/s TDM technology will be available soon. To push TDM technology beyond these rates, researchers are working on methods to perform the multiplexing and demultiplexing functions *optically*. This approach is called *optical time division multiplexing* (OTDM). Laboratory experiments have demonstrated the multiplexing/demultiplexing of several 10 Gb/s streams into/from a 250 Gb/s stream, although commercial implementation of OTDM is still several years away. We will study OTDM systems in Chapter 12. However, multiplexing and demultiplexing high-speed streams by itself is not sufficient to realize practical networks. We need to contend with the various impairments that arise as these very

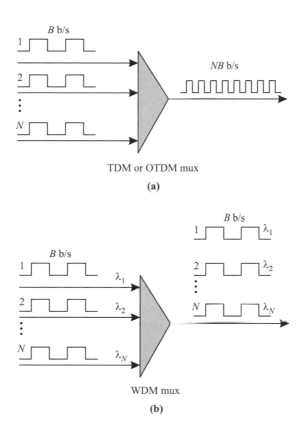

Figure 1.4   Different multiplexing techniques for increasing the transmission capacity on an optical fiber. (a) Electronic or optical time division multiplexing and (b) wavelength division multiplexing. Both multiplexing techniques take in N data streams, each of B b/s, and multiplex them into a single fiber with a total aggregate rate of NB b/s.

high-speed streams are transmitted over a fiber. As we will see in Chapters 5 and 13, the higher the bit rate, the more difficult it is to engineer around these impairments. However, similar bottlenecks have been encountered in the past, and people have always found ways to overcome them; so we can expect the transmission bit rates to continue to increase, although perhaps not at the breakneck pace of the past two decades.

Another way to increase the capacity is by a technique called *wavelength division multiplexing* (WDM). WDM is essentially the same as frequency division multiplexing (FDM), which has been used in radio systems for more than a century. For some reason, the term FDM is used widely in radio communication, but WDM is used in the context of optical communication, perhaps because FDM was studied first by communications engineers and WDM by physicists. The idea is to transmit data simultaneously at multiple carrier wavelengths (or, equivalently, frequencies or colors) over a fiber. To first order, these wavelengths do not interfere with each other provided they are kept sufficiently far apart. (There are some undesirable second-order effects where wavelengths do interfere with each other, and we will study these in Chapters 2 and 5.) Thus WDM provides *virtual fibers,* in that it makes a single fiber look like multiple "virtual" fibers, with each virtual fiber carrying a single data stream. WDM systems are widely deployed today in long-haul and undersea networks and are being deployed in metro networks as well.

WDM and TDM both provide ways to increase the transmission capacity and are complementary to each other. Therefore networks today use a combination of TDM and WDM. The question of what combination of TDM and WDM to use in systems is an important one facing carriers today. For example, suppose a carrier wants to install an 80 Gb/s link. Should we deploy 32 WDM channels at 2.5 Gb/s each or should we deploy 10 WDM channels at 8 Gb/s each? The answer depends on a number of factors, including the type and parameters of the fiber used in the link and the services that the carrier wishes to provide using that link. We will discuss this issue in Chapter 13. Using a combination of WDM and TDM, systems with transmission capacities of around 1 Tb/s over a single fiber are becoming commercially available, and no doubt we will see systems with higher capacities operating over longer distances emerge in the future.

1.3.2   Second-Generation Optical Networks

Optics is clearly the preferred means of transmission, and WDM transmission is now widely used in the network. In recent years, people have realized that optical networks are capable of providing more functions than just point-to-point transmission. Major advantages are to be gained by incorporating some of the switching and routing functions that were performed by electronics into the optical part of the network. For example, as data rates get higher and higher, it becomes more difficult for electronics to process data. Suppose the electronics must process data in blocks of 53 bytes each (this happens to be the block size in asynchronous transfer mode networks). In a 100 Mb/s data stream, we have 4.24 μs to process a block, whereas at 10 Gb/s, the same block must be processed within 42.4 ns. In first-generation networks, the electronics at a node must handle not only all the data intended for that node but also all the data that is being passed through that node on to other nodes in the network. If the latter data could be routed through in the optical domain, the burden on the underlying electronics at the node would be significantly reduced. This is one of the key drivers for second-generation optical networks.

Optical networks based on this paradigm are now being deployed. The architecture of such a network is shown in Figure 1.5. We call this network a *wavelength-routing* network. The network provides *lightpaths* to its users, such as SONET terminals or IP routers. Lightpaths are optical connections carried end to end from a source node to a destination node over a wavelength on each intermediate link. At intermediate nodes in the network, the lightpaths are routed and switched from one link to another link. In some cases, lightpaths may be converted from one wavelength to another wavelength as well along their route. Different lightpaths in a wavelength routing network can use the same wavelength as long as they do not share any common links. This allows the same wavelength to be reused spatially in different parts of the network. For example, Figure 1.5 shows six lightpaths. The lightpath between B and C, the lightpath between D and E, and one of the lightpaths between E and F do not share any links in the network and can therefore be set up using the same wavelength λ_1. At the same time, the lightpath between A and F shares a link with the lightpath between B and C and must therefore use a different wavelength. Likewise, the two lightpaths between E and F must be assigned different wavelengths. Note that these lightpaths all use the same wavelength on every link in their path. This is a constraint that we must deal with if we do not have *wavelength conversion* capabilities within the network. Suppose we had only two wavelengths available in the network and wanted to set up a new lightpath between nodes E and F. Without wavelength conversion, we would not be able to set up this lightpath. On the other hand, if the intermediate node X can perform wavelength conversion, then we can set up this lightpath using wavelength λ_2 on link EX and wavelength λ_1 on link XF.

The key network elements that enable optical networking are *optical line terminals* (OLTs), *optical add/drop multiplexers* (OADMs), and *optical crossconnects* (OXCs), as shown in Figure 1.5. An OLT multiplexes multiple wavelengths into a single fiber and demultiplexes a set of wavelengths on a single fiber into separate fibers. OLTs are used at the ends of a point-to-point WDM link. An OADM takes in signals at multiple wavelengths and selectively drops some of these wavelengths locally while letting others pass through. It also selectively adds wavelengths to the composite outbound signal. An OADM has two *line* ports where the composite WDM signals are present, and a number of *local* ports where individual wavelengths are dropped and added. An OXC essentially performs a similar function but at much larger sizes. OXCs have a large number of ports (ranging from a few tens to thousands) and are able to switch wavelengths from one input port to another. Both OADMs and OXCs may incorporate wavelength conversion capabilities. The detailed architecture of these networks will be discussed in Chapter 7.

Optical networks based on the architecture described above are already being deployed. OLTs have been widely deployed for point-to-point applications. OADMs

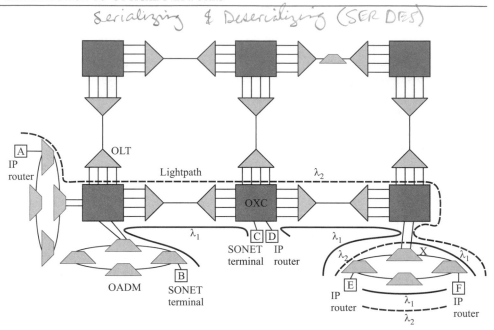

Figure 1.5 A WDM wavelength-routing network, showing optical line terminals (OLTs), optical add/drop multiplexers (OADMs), and optical crossconnects (OXCs). The network provides lightpaths to its users, which are typically IP routers or SONET terminals.

are now used in long-haul and metro networks. OXCs are beginning to be deployed first in long-haul networks because of the higher capacities in those networks.

1.4 The Optical Layer

Before delving into the details of the optical layer, we first introduce the notion of a layered network architecture. Networks are complicated entities with a variety of different functions being performed by different components of the network, with equipment from different vendors all interworking together. In order to simplify our view of the network, it is desirable to break up the functions of the network into different layers, as shown in Figure 1.6. This type of layered model was proposed by the International Standards Organization (ISO) in the early 1980s. Imagine the layers as being vertically stacked up. Each layer performs a certain set of functions

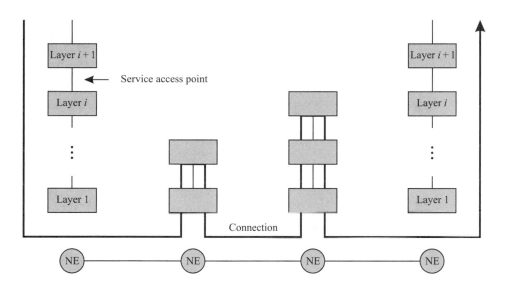

Figure 1.6   Layered hierarchy of a network showing the layers at each network element (NE).

and provides a certain set of services to the next higher layer. In turn, each layer expects the layer below it to deliver a certain set of services to it. The service interface between two adjacent layers is called a *service access point* (SAP), and there can be multiple SAPs between layers corresponding to different types of services offered.

In most cases, the network provides *connections* to the user. A connection is established between a source and a destination node. Setting up, taking down, and managing the state of a connection is the job of a separate network control and management entity (not shown in Figure 1.6), which may control each individual layer in the network. There are also examples where the network provides *connectionless* services to the user. These services are suitable for transmitting short messages across a network, without having to pay the overhead of setting up and taking down a connection for this purpose. We will confine the following discussion to the connection-oriented model.

Within a network element, data belonging to a connection flows between the layers. Each layer multiplexes a number of higher-layer connections and may add some additional overhead to data coming from the higher layer. Each intermediate network element along the path of a connection embodies a set of layers starting from the lowest layer up to a certain layer in the hierarchy.

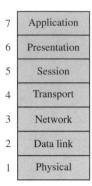

Figure 1.7   The classical layered hierarchy.

It is important to define the functions of each layer and the interfaces between layers. This is essential because it allows vendors to manufacture a variety of hardware and software products performing the functions of some, but not all, of the layers, and provide the appropriate interfaces to communicate with other products performing the functions of other layers.

There are many possible implementations and standards for each layer. A given layer may work together with a variety of lower or higher layers. Each of the different types of optical networks that we will study constitutes a layer. Each layer itself can in turn be broken up into several sublayers. As we study these networks, we will explore this layered hierarchy further.

Figure 1.7 shows a classical breakdown of the different layers in a network that was proposed by the International Standards Organization. The lowest layer in the hierarchy is the *physical layer*, which provides a "pipe" with a certain amount of bandwidth to the layer above it. The physical layer may be optical, wireless, or coaxial or twisted-pair cable. The next layer above is the *data link layer*, which is responsible for framing, multiplexing, and demultiplexing data sent over the physical layer. The framing protocol defines how data is transported over a physical link. Typically data is broken up into frames before being transmitted over a physical link. This is necessary to ensure reliable delivery of data across the link. The framing protocol provides clear delineation between frames, provides sufficient transitions in the signal so that it can be recovered at the other end, and usually includes additional overhead that enables link errors to be detected. Examples of data link protocols suitable for operation over point-to-point links include the *point-to-point protocol* (PPP) and the *high-level data link control* (HDLC) protocol. Included in the data link layer is the *media access control layer* (MAC), which coordinates the transmissions of different nodes when they all share common bandwidth, as is the case in many local-area networks, such as Ethernets and token rings.

Above the data link layer resides the *network layer*. The network layer usually provides *virtual circuits* or *datagram* services to the higher layer. A virtual circuit (VC) represents an end-to-end connection with a certain set of quality-of-service parameters associated with it, such as bandwidth and error rate. Data transmitted by the source over a VC is delivered in sequence at its destination. Datagrams, on the other hand, are short messages transmitted end to end, with no notion of a connection. The network layer performs the end-to-end routing function of taking a message at its source and delivering it to its destination. The predominant network layer today is IP, and the main network element in an IP network is an IP router. IP provides a way to route packets (or datagrams) end to end in a packet-switched network. IP includes statistical multiplexing of multiple packet streams and today also provides some simple and relatively slow and inefficient service restoration mechanisms. The Internet Protocol has been adapted to operate over a variety of data link and physical media, such as Ethernet, serial telephone lines, coaxial cable lines, and optical fiber lines.

The *transport layer* resides on top of the network layer and is responsible for ensuring the end-to-end, in-sequence, and error-free delivery of the transmitted messages. For example, the *transmission control protocol* (TCP) used in the Internet belongs to this layer. Above the transport layer reside other layers such as the *session*, *presentation*, and *application* layers, but we will not be concerned with these layers in this book.

Another important packet-switched layer is ATM. ATM provides a connection-oriented service (virtual circuits) and is capable of providing a variety of quality-of-service guarantees. Packets in ATM are called *cells* and are of fixed length (53 bytes). ATM is being used by many carriers as a vehicle to deliver reliable packet-switched services. More on this subject in Chapters 6 and 13.

This classical layered view of networks needs some embellishment to handle the variety of networks and protocols that are proliferating today. A more realistic layered model for today's networks would employ multiple protocol stacks residing one on top of the other. Each stack incorporates several sublayers, which may provide functions resembling traditional physical, link, and network layers. To provide a concrete example of this, consider an IP over SONET network shown in Figure 1.8. In this case, the IP network treats the SONET network as providing it with point-to-point links between IP routers. The SONET layer itself, however, internally routes and switches connections, and in a sense, incorporates its own link, physical, and network layers.

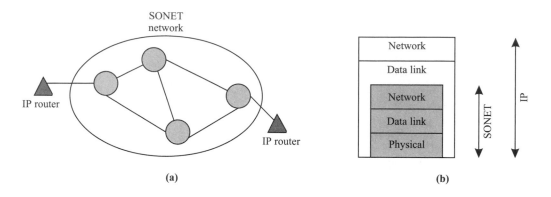

(a)   **(b)**

Figure 1.8   An IP over SONET network. (a) The network has IP switches with SONET adaptors that are connected to a SONET network. (b) The layered view of this network.

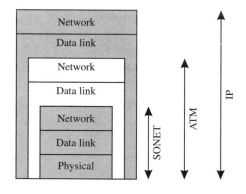

Figure 1.9   The layered view of an IP over ATM over SONET network.

Another example of this sort of layering arises in the context of an IP over ATM over SONET network. Some service providers are deploying an ATM network operating over a SONET infrastructure to provide services for IP users. In such a network, IP packets are converted to ATM cells at the periphery of the network. The ATM switches are connected through a SONET infrastructure. The layered view of such a network is shown in Figure 1.9. Again, the IP network treats the ATM network as its link layer, and the ATM network uses SONET as its link layer.

The introduction of second-generation optical networks adds yet another layer to the protocol hierarchy—the so-called optical layer. The optical layer is a *server* layer

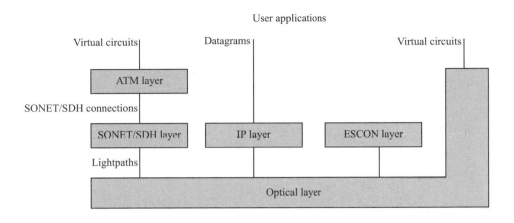

Figure 1.10 A layered view of a network consisting of a second-generation optical network layer that supports a variety of client layers above it.

that provides services to other *client* layers. This optical layer provides lightpaths to a variety of client layers, as shown in Figure 1.10. Examples of client layers residing above a second-generation optical network layer include IP, ATM, and SONET/SDH, as well as other possible protocols such as Gigabit Ethernet, ESCON (enterprise serial connection—a protocol used to interconnect computers to storage devices and other computers), or Fibre Channel (which performs the same function as ESCON, at higher speeds). As second-generation optical networks evolve, they may provide other services besides lightpaths, such as packet-switched virtual circuit or datagram services. These services may directly interface with user applications, as shown in Figure 1.10. Several other layer combinations are possible and not shown in the figure, such as IP over SONET over optical, and ATM over optical.

The client layers make use of the lightpaths provided by the optical layer. To a SONET network operating over the optical layer, the lightpaths are simply replacements for hardwired fiber connections between SONET terminals. As described earlier, a lightpath is a connection between two nodes in the network, and it is set up by assigning a dedicated wavelength to it on each link in its path. Note that individual wavelengths are likely to carry data at fairly high bit rates (a few gigabits per second), and this entire bandwidth is provided to the higher layer by a lightpath. Depending on the capabilities of the network, this lightpath could be set up or taken down in response to a request from the higher layer. This can be thought of as a *circuit-switched* service, akin to the service provided by today's telephone network: the network sets up or takes down calls in response to a request from the user. Alternatively, the network may provide only *permanent* lightpaths, which are set up at the time the network is deployed. This lightpath service can be used to support high-speed connections for a variety of overlying networks.

Optical networks today provide functions that might be thought of as falling primarily within the physical layer from the perspective of its users. However, the optical network itself incorporates several sublayers, which in turn correspond to the link and network layer functions in the classical layered view.

Before the emergence of the optical layer, SONET/SDH was the predominant transmission layer in the telecommunications network, and it is still the dominant layer in many parts of the network. We will study SONET/SDH in detail in Chapter 6. For convenience, we will use SONET terminology in the rest of this section. The SONET layer provides several key functions. It provides end-to-end, managed, circuit-switched connections. It provides an efficient mechanism for multiplexing lower-speed connections into higher-speed connections. For example, low-speed voice connections at 64 kb/s or private line 1.5 Mb/s connections can be multiplexed all the way up into 2.5 Gb/s or 10 Gb/s line rates for transport over the network. Moreover, at intermediate nodes, SONET provides an efficient way to extract individual low-speed streams from a high-speed stream, using an elegant multiplexing mechanism based on the use of pointers.

SONET also provides a high degree of network reliability and availability. Carriers expect their networks to provide 99.99% to 99.999% of availability. These numbers translate into an allowable network downtime of less than one hour per year and five minutes per year, respectively. SONET achieves this by incorporating sophisticated mechanisms for rapid service restoration in the event of failures in the network. This is a subject we will look at in Chapter 10.

Finally, SONET includes extensive overheads that allow operators to monitor and manage the network. Examples of these overheads include parity check bytes to determine whether frames are received in error or not, and connection identifiers that allow connections to be traced and verified across a complex network.

SONET network elements include line terminals, add/drop multiplexers (ADMs), regenerators, and digital crossconnects (DCSs). Line terminals multiplex and demultiplex traffic streams. ADMs are deployed in linear and ring network configurations. They provide an efficient way to drop part of the traffic at a node while allowing the remaining traffic to pass through. The ring topology allows traffic to be rerouted around failures in the network. Regenerators regenerate the SONET signal wherever needed. DCSs are deployed in larger nodes to switch a large number of traffic streams. Today's DCSs are capable of switching thousands of 45 Mb/s traffic streams.

The functions performed by the optical layer are in many ways analogous to those performed by the SONET layer. The optical layer multiplexes multiple lightpaths into a single fiber and allows individual lightpaths to be extracted efficiently from the composite multiplex signal at network nodes. It incorporates sophisticated service

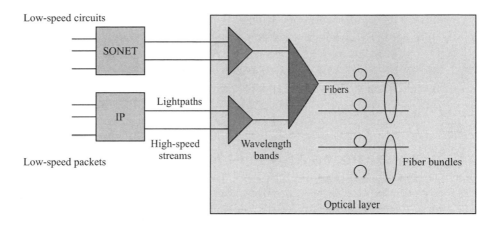

Figure 1.11   Example of a typical multiplexing layered hierarchy.

restoration techniques and management techniques as well. We will look at these techniques in Chapters 9 and 10.

Figure 1.11 shows a typical layered network hierarchy, highlighting the optical layer. The optical layer provides lightpaths that are used by SONET and IP network elements. The SONET layer multiplexes low-speed circuit-switched streams into higher-speed streams, which are then carried over lightpaths. The IP layer performs statistical multiplexing of packet-switched streams into higher-speed streams, which are also carried over lightpaths. Inside the optical layer itself is a multiplexing hierarchy. Multiple wavelengths or lightpaths are combined together into wavelength bands. Bands are combined together to produce a composite WDM signal on a fiber. The network itself may include multiple fibers and multiple fiber bundles, each of which carries a number of fibers.

So why have multiple layers in the network that perform similar functions? The answer is that this form of layering significantly reduces network equipment costs. Different layers are more efficient at performing functions at different bit rates. For example, the SONET layer can efficiently (that is, cost-effectively) switch and process traffic streams up to, say, 2.5 Gb/s today. However, it is very expensive to have this layer process 100 10 Gb/s streams coming in on a WDM link. The optical layer, on the other hand, is particularly efficient at processing traffic on a wavelength-by-wavelength basis, but not particularly good at processing traffic streams at lower granularities, for example, 155 Mb/s. Therefore, it makes sense to use the optical layer to process large amounts of bandwidth at a relatively coarse level and the SONET layer to process smaller amounts of bandwidth at a relatively finer level. This fundamental observation is the key driver to providing such functions in multiple layers, and we will study this in detail in Chapter 7.

A similar observation also holds for the service restoration function of these networks. Certain failures are better handled by the optical layer and certain others by the SONET layer or the IP layer. We will study this aspect in Chapter 10.

1.5   Transparency and All-Optical Networks

A major feature of the lightpath service provided by second-generation networks is that this type of service can be *transparent* to the actual data being sent over the lightpath once it is set up. For instance, a certain maximum and minimum bit rate might be specified, and the service may accept data at any bit rate and any protocol format within these limits. It may also be able to carry analog data.

Transparency in the network provides several advantages. An operator can provide a variety of different services using a single infrastructure. We can think of this as *service transparency*. Second, the infrastructure is future-proof in that if protocols or bit rates change, the equipment deployed in the network is still likely to be able to support the new protocols and/or bit rates without requiring a complete overhaul of the entire network. This allows new services to be deployed efficiently and rapidly, while allowing legacy services to be carried as well.

An example of a transparent network of this sort is the telephone network. Once a call is established in the telephone network, it provides 4 kHz of bandwidth over which a user can send a variety of different types of traffic such as voice, data, or fax. There is no question that transparency in the telephone network today has had a far-reaching impact on our lifestyles. Transparency has become a useful feature of second-generation optical networks as well.

Another term associated with transparent networks is the notion of an *all-optical network*. In an all-optical network, data is carried from its source to its destination in optical form, without undergoing any optical-to-electrical conversions along the way. In an ideal world, such a network would be *fully transparent*. However, all-optical networks are limited in their scope by several parameters of the physical layer, such as bandwidth and signal-to-noise ratios. For example, analog signals require much higher signal-to-noise ratios than digital signals. The actual requirements depend on the modulation format used as well as the bit rate. We will study these aspects in Chapter 5, where we will see that engineering the physical layer is a complex task with a variety of parameters to be taken into consideration. For this reason, it is very difficult to build and operate a network that can support analog as well as digital signals at arbitrary bit rates.

The other extreme is to build a network that handles essentially a single bit rate and protocol (say, 2.5 Gb/s SONET only). This would be a *nontransparent* network. In between is a *practical* network that handles digital signals at a range of bit rates up to a specified maximum. Most optical networks being deployed today fall into this category.

Although we talk about optical networks, they almost always include a fair amount of electronics. First, electronics plays a crucial role in performing the intelligent control and management functions within a network. However, even in the data path, in most cases, electronics is needed at the periphery of the network to adapt the signals entering the optical network. In many cases, the signal may not be able to remain in optical form all the way to its destination due to limitations imposed by the physical layer design and may have to be regenerated in between. In other cases, the signal may have to be converted from one wavelength to another wavelength. In all these situations, the signal is usually converted from optical form to electronic form and back again to optical form.

Having these electronic regenerators in the path of the signal reduces the transparency of that path. There are three types of electronic regeneration techniques for digital data. The standard one is called regeneration *with* retiming and reshaping, also known as 3R. Here the bit clock is extracted from the signal, and the signal is reclocked. This technique essentially produces a "fresh" copy of the signal at each regeneration step, allowing the signal to go through a very large number of regenerators. However, it eliminates transparency to bit rates and the framing protocols, since acquiring the clock usually requires knowledge of both of these. Some limited form of bit rate transparency is possible by making use of programmable clock recovery chips that can work at a set of bit rates that are multiples of one another. For example, chipsets that perform clock recovery at either 2.5 Gb/s or 622 Mb/s are commercially available today.

An implementation using regeneration of the optical signal *without* retiming, also called 2R, offers transparency to bit rates, without supporting analog data or different modulation formats [GJR96]. However, this approach limits the number of regeneration steps allowed, particularly at higher bit rates, over a few hundred megabits per second. The limitation is due to the jitter, which accumulates at each regeneration step.

The final form of electronic regeneration is 1R, where the signal is simply received and retransmitted without retiming or reshaping. This form of regeneration can handle analog data as well, but its performance is significantly poorer than the other two forms of regeneration. For this reason, the networks being deployed today use 2R or 3R electronic regeneration. Note, however, that optical amplifiers are widely used to amplify the signal in the optical domain, without converting the signal to the electrical domain. These can be thought of as 1R optical regenerators.

Table 1.1   Different types of transparency in an optical network.

Parameter	Transparency type		
	Fully transparent	Practical	Nontransparent
Analog/digital	Both	Digital	Digital
Bit rate	Arbitrary	Predetermined maximum	Fixed
Framing protocol	Arbitrary	Selected few	Single

Table 1.1 provides an overview of the different dimensions of transparency. At one end of the spectrum is a network that operates at a fixed bit rate and framing protocol, for example, SONET at 2.5 Gb/s. This would be truly an *opaque* network. In contrast, a fully transparent network would support analog and digital signals with arbitrary bit rates and framing protocols. As we argued earlier, however, such a network is not practical to engineer and build. Today, a practical alternative is to engineer the network to support a variety of digital signals up to a predetermined maximum bit rate and a specific set of framing protocols, such as SONET and Gigabit Ethernet. The network supports a variety of framing protocols either by making use of 2R regeneration inside the network or by providing specific 3R adaptation devices for each of the framing protocols. Such a network is shown in Figure 1.12. It can be viewed as consisting of islands of all-optical subnetworks with optical-to-electrical-to-optical conversion at their boundaries for the purposes of adaptation, regeneration, or wavelength conversion.

1.6   Optical Packet Switching

So far we have talked about optical networks that provide lightpaths. These networks are essentially circuit-switched. Researchers are also working on optical networks that can perform packet switching in the optical domain. Such a network would be able to offer *virtual circuit* services or *datagram* services, much like what is provided by ATM and IP networks. With a virtual circuit connection, the network offers what looks like a circuit-switched connection between two nodes. However, the bandwidth offered on the connection can be smaller than the full bandwidth available on a link or wavelength. For instance, individual connections in a future high-speed network may operate at 10 Gb/s, while transmission bit rates on a wavelength could be 100 Gb/s. Thus the network must incorporate some form of time division multiplexing to combine multiple connections onto the transmission bit rate. At these rates, it may be easier to do the multiplexing in the optical domain rather than in the electronic domain. This form of optical time domain multiplexing (OTDM) may

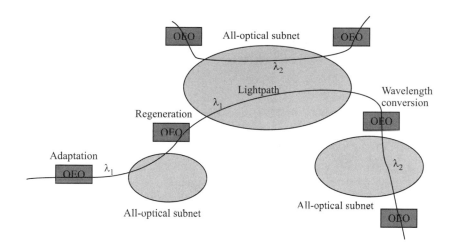

Figure 1.12   An optical network consisting of all-optical subnetworks interconnected by optical-to-electrical-to-optical (OEO) converters. OEO converters are used in the network for adapting external signals to the optical network, for regeneration, and for wavelength conversion.

be *fixed* or *statistical*. Those that perform statistical multiplexing are called optical packet-switched networks. For simplicity we will talk mostly about optical packet switching. Fixed OTDM can be thought of as a subset of optical packet switching where the multiplexing is fixed instead of statistical.

An optical packet-switching node is shown in Figure 1.13. The idea is to create packet-switching nodes with much higher capacities than can be envisioned with electronic packet switching. Such a node takes a packet coming in, reads its header, and switches it to the appropriate output port. The node may also impose a new header on the packet. It must also handle *contention* for output ports. If two packets coming in on different ports need to go out on the same output port, one of the packets must be buffered, or sent out on another port.

Ideally, all the functions inside the node would be performed in the optical domain, but in practice, certain functions, such as processing the header and controlling the switch, get relegated to the electronic domain. This is because of the very limited processing capabilities in the optical domain. The header itself could be sent at a lower bit rate than the data so that it can be processed electronically.

The mission of optical packet switching is to enable packet-switching capabilities at rates that cannot be contemplated using electronic packet switching. However, designers are handicapped by several limitations with respect to processing signals in the optical domain. One important factor is the lack of optical random access

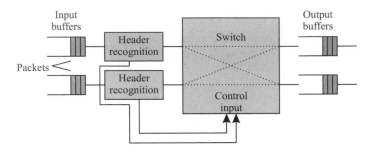

Figure 1.13   An optical packet-switching node. The node buffers the incoming packets, looks at the packet header, and routes the packets to an appropriate output port based on the information contained in the header.

memory for buffering. Optical buffers are realized by using a length of fiber and are just simple delay lines, not fully functional memories. Packet switches include a high amount of intelligent real-time software and dedicated hardware to control the network and provide quality-of-service guarantees, and these functions are difficult to perform in the optical domain. Another factor is the relatively primitive state of fast optical-switching technology, compared to electronics. For these reasons, optical packet switching is still in its infancy today in research laboratories. Chapter 12 covers all these aspects in detail.

1.7   Transmission Basics

In this section, we introduce and define the units for common parameters associated with optical communication systems.

1.7.1   Wavelengths, Frequencies, and Channel Spacing

When we talk about WDM signals, we will be talking about the wavelength, or frequency, of these signals. The wavelength λ and frequency f are related by the equation

$$c = f\lambda,$$

where c denotes the speed of light in free space, which is 3×10^8 m/s. We will reference all parameters to free space. The speed of light in fiber is actually somewhat lower (closer to 2×10^8 m/s), and the wavelengths are also correspondingly different.

To characterize a WDM signal, we can use either its frequency or wavelength interchangeably. Wavelength is measured in units of nanometers (nm) or micrometers (μm or microns). (1 nm $= 10^{-9}$ m, 1 μm $= 10^{-6}$ m.) The wavelengths of interest to optical fiber communication are centered around 0.8, 1.3, and 1.55 μm. These wavelengths lie in the infrared band, which is not visible to the human eye. Frequencies are measured in units of hertz (or cycles per second), more typically in megahertz (1 MHz $= 10^6$ Hz), gigahertz (1 GHz $= 10^9$ Hz), or terahertz (1 THz $= 10^{12}$ Hz). Using $c = 3 \times 10^8$ m/s, a wavelength of 1.55 μm would correspond to a frequency of approximately 193 THz, which is 193×10^{12} Hz.

Another parameter of interest is the channel spacing, which is the spacing between two wavelengths or frequencies in a WDM system. Again the channel spacing can be measured in units of wavelengths or frequencies. The relationship between the two can be obtained starting from the equation

$$f = \frac{c}{\lambda}.$$

Differentiating this equation around a center wavelength λ_0, we obtain the relationship between the frequency spacing Δf and the wavelength spacing $\Delta \lambda$ as

$$\Delta f = -\frac{c}{\lambda_0^2} \Delta \lambda.$$

This relationship is accurate as long as the wavelength (or frequency) spacing is small compared to the actual channel wavelength (or frequency), which is usually the case in optical communication systems. At a wavelength $\lambda_0 = 1550$ nm, a wavelength spacing of 0.8 nm corresponds to a frequency spacing of 100 GHz, a typical spacing in WDM systems.

Digital information signals in the time domain can be viewed as a periodic sequence of pulses, which are on or off, depending on whether the data is a 1 or a 0. The bit rate is simply the inverse of this period. These signals have an equivalent representation in the frequency domain, where the energy of the signal is spread across a set of frequencies. This representation is called the *power spectrum*, or simply *spectrum*. The signal *bandwidth* is a measure of the width of the spectrum of the signal. The bandwidth can also be measured either in the frequency domain or in the wavelength domain, but is mostly measured in units of frequency. Note that we have been using the term *bandwidth* rather loosely. The bandwidth and *bit rate* of a digital signal are related but not exactly the same. Bandwidth is usually specified in kilohertz or megahertz or gigahertz, whereas bit rate is specified in kilobits/second (kb/s), megabits/second (Mb/s), or gigabits/second (Gb/s). The relationship between the two depends on the type of modulation used. For instance, a phone line offers 4 kHz of bandwidth, but sophisticated modulation technology allows us to realize

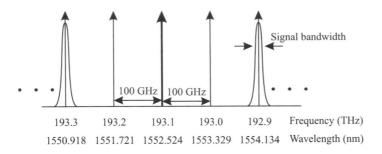

193.3	193.2	193.1	193.0	192.9	Frequency (THz)
1550.918	1551.721	1552.524	1553.329	1554.134	Wavelength (nm)

Figure 1.14 The 100 GHz ITU frequency grid based on a reference frequency of 193.1 THz. A 50 GHz grid has also been defined around the same reference frequency.

a bit rate of 56 kb/s over this phone line. This ratio of bit rate to available bandwidth is called *spectral efficiency*. Optical communication systems use rather simple modulation techniques that achieve a spectral efficiency of about 0.4 bits/s/Hz, and it is reasonable to assume therefore that a signal at a bit rate of 10 Gb/s uses up bandwidth of approximately 25 GHz. Note that the signal bandwidth needs to be sufficiently smaller than the channel spacing; otherwise we would have undesirable interference between adjacent channels and distortion of the signal itself.

1.7.2 Wavelength Standards

WDM systems today primarily use the 1.55 μm wavelength region for two reasons: the inherent loss in optical fiber is the lowest in that region, and excellent optical amplifiers are available in that region. We will discuss this in more detail in later chapters. The wavelengths and frequencies used in WDM systems have been standardized on a frequency grid by the International Telecommunications Union (ITU). It is an infinite grid centered at 193.1 THz, a segment of which is shown in Figure 1.14. The ITU decided to standardize the grid in the frequency domain based on equal channel spacings of 50 GHz or 100 GHz. Observe that if multiple channels are spaced apart equally in wavelength, they are not spaced apart exactly equally in frequency, and vice versa. The figure also shows the power spectrum of two channels 400 GHz apart in the grid populated by traffic-bearing signals, as indicated by the increased signal bandwidth on those channels.

The ITU grid only tells part of the story. Today, we are already starting to see systems using 25 GHz channel spacings. We are also seeing the use of several transmission bands. The early WDM systems used the so-called C-band, or conventional band (approximately 1530–1565 nm). The use of the L-band, or long wavelength band (approximately 1565–1625 nm), has become feasible recently with the development of optical amplifiers in this band. We will look at this and other bands in Section 1.8.

It has proven difficult to obtain agreement from the different WDM vendors and service providers on more concrete wavelength standards. As we will see in Chapters 2 and 5, designing WDM transmission systems is a complex endeavor, requiring trade-offs among many different parameters, including the specific wavelengths used in the system. Different WDM vendors use different methods for optimizing their system designs, and converging on a wavelength plan becomes difficult as a result. However, the ITU grid standard has helped accelerate the deployment of WDM systems because component vendors can build wavelength-selective parts to a specific grid, which helps significantly in inventory management and manufacturing.

1.7.3 Optical Power and Loss

In optical communication, it is quite common to use decibel units (dB) to measure power and signal levels, as opposed to conventional units. The reason for doing this is that powers vary over several orders of magnitude in a system, and this makes it easier to deal with a logarithmic rather than a linear scale. Moreover, by using such a scale, calculations that involve multiplication in the conventional domain become additive operations in the decibel domain. Decibel units are used to represent relative as well as absolute values.

To understand this system, let us consider an optical fiber link. Suppose we transmit a light signal with power P_t watts (W). In terms of dB units, we have

$$(P_t)_{\text{dBW}} = 10 \log(P_t)_{\text{W}}.$$

In many cases, it is more convenient to measure powers in milliwatts (mW), and we have an equivalent dBm value given as

$$(P_t)_{\text{dBm}} = 10 \log(P_t)_{\text{mW}}.$$

For example, a power of 1 mW corresponds to 0 dBm or −30 dBW. A power of 10 mW corresponds to 10 dBm or −20 dBW.

As the light signal propagates through the fiber, it is attenuated; that is, its power is decreased. At the end of the link, suppose the received power is P_r. The link loss γ is then defined as

$$\gamma = \frac{P_r}{P_t}.$$

In dB units, we would have

$$(\gamma)_{dB} = 10 \log \gamma = (P_r)_{dBm} - (P_t)_{dBm}.$$

Note that dB is used to indicate relative values, whereas dBm and dBW are used to indicate the absolute power value. As an example, if $P_t = 1$ mW and $P_r = 1$ μW, implying that $\gamma = 0.001$, we would have, equivalently,

$$(P_t)_{dBm} = 0 \text{ dBm or } -30 \text{ dBW},$$

$$(P_r)_{dBm} = -30 \text{ dBm or } -60 \text{ dBW},$$

and

$$(\gamma)_{dB} = -30 \text{ dB}.$$

In this context, a signal being attenuated by a factor of 1000 would equivalently undergo a 30 dB loss. A signal being amplified by a factor of 1000 would equivalently have a 30 dB gain.

We measure loss in optical fiber usually in units of dB/km. So, for example, a light signal traveling through 120 km of fiber with a loss of 0.25 dB/km would be attenuated by 30 dB.

1.8 Network Evolution

We conclude this chapter by outlining the trends and factors that have shaped the evolution of optical fiber transmission systems and networks. Figure 1.15 gives an overview. The history of optical fiber transmission has been all about how to transmit data at the highest capacity over the longest possible distance and is remarkable for its rapid progress. What is equally remarkable is the fact that researchers have successfully overcome numerous obstacles along this path, many of which when first discovered looked as though they would impede further increases in capacity and transmission distance. The net result of this is that capacity continues to grow in the network, while the cost per bit transmitted per kilometer continues to get lower and lower, to a point where it has become practical for carriers to price circuits independently of the distance.

We will introduce various types of fiber propagation impairments as well as optical components in this section. These will be covered in depth in Chapters 2, 3, and 5.

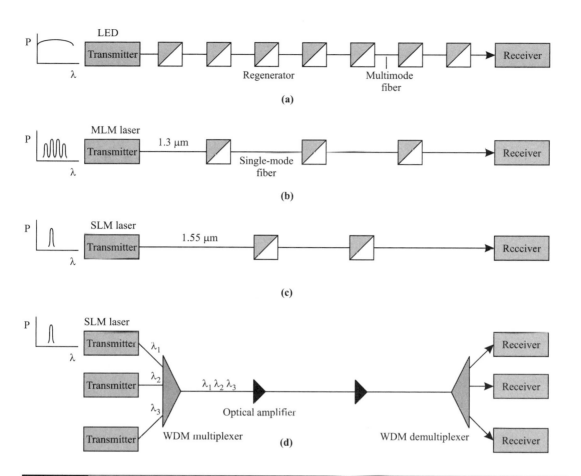

Figure 1.15   Evolution of optical fiber transmission systems. (a) An early system using LEDs over multimode fiber. (b) A system using MLM lasers over single-mode fiber in the 1.3 μm band to overcome intermodal dispersion in multimode fiber. (c) A later system using the 1.55 μm band for lower loss, and using SLM lasers to overcome chromatic dispersion limits. (d) A current-generation WDM system using multiple wavelengths at 1.55 μm and optical amplifiers instead of regenerators. The P-λ curves to the left of the transmitters indicate the power spectrum of the signal transmitted.

1.8.1   Early Days—Multimode Fiber

Early experiments in the mid-1960s demonstrated that information encoded in light signals could be transmitted over a glass fiber *waveguide*. A waveguide provides a medium that can *guide* the light signal, enabling it to stay focused for a reasonable distance without being scattered. This allows the signal to be received at the other end with sufficient strength so that the information can be decoded. These early experiments proved that optical transmission over fiber was feasible.

An optical fiber is a very thin cylindrical glass waveguide consisting of two parts: an inner *core* material and an outer *cladding* material. The core and cladding are designed so as to keep the light signals *guided* inside the fiber, allowing the light signal to be transmitted for reasonably long distances before the signal degrades in quality.

It was not until the invention of low-loss optical fiber in the early 1970s that optical fiber transmission systems really took off. This silica-based optical fiber has three low-loss windows in the 0.8, 1.3, and 1.55 μm infrared wavelength bands. The lowest loss is around 0.25 dB/km in the 1.55 μm band, and about 0.5 dB/km in the 1.3 μm band. These fibers enabled transmission of light signals over distances of several tens of kilometers before they needed to be *regenerated*. A regenerator converts the light signal into an electrical signal and retransmits a fresh copy of the data as a new light signal.

The early fibers were the so-called multimode fibers. Multimode fibers have core diameters of about 50 to 85 μm. This diameter is large compared to the operating wavelength of the light signal. A basic understanding of light propagation in these fibers can be obtained using the so-called geometrical optics model, illustrated in Figure 1.16. In this model, a light ray bounces back and forth in the core, being reflected at the core-cladding interface. The signal consists of multiple light rays, each of which potentially takes a different path through the fiber. Each of these different paths corresponds to a *propagation mode*. The length of the different paths is different, as seen in the figure. Each mode therefore travels with a slightly different speed compared to the other modes.

The other key devices needed for optical fiber transmission are light sources and receivers. Compact semiconductor lasers and light-emitting diodes (LEDs) provided practical light sources. These lasers and LEDs were simply turned on and off rapidly to transmit digital (binary) data. Semiconductor photodetectors enabled the conversion of the light signal back into the electrical domain.

The early telecommunication systems (late 1970s through the early 1980s) used multimode fibers along with LEDs or laser transmitters in the 0.8 and 1.3 μm wavelength bands. LEDs were relatively low-power devices that emitted light over a fairly wide spectrum of several nanometers to tens of nanometers. A laser provided higher output power than an LED and therefore allowed transmission over greater distances before regeneration. The early lasers were *multilongitudinal mode* (MLM) Fabry-Perot lasers. These MLM lasers emit light over a fairly wide spectrum of several nanometers to tens of nanometers. The actual spectrum consists of multiple spectral lines, which can be thought of as different longitudinal modes, hence the

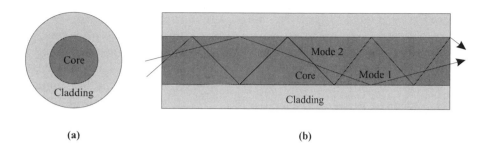

Figure 1.16 Geometrical optics model to illustrate the propagation of light in an optical fiber. (a) Cross section of an optical fiber. The fiber has an inner core and an outer cladding, with the core having a slightly higher refractive index than the cladding. (b) Longitudinal view. Light rays within the core hitting the core-cladding boundary are reflected back into the core by total internal reflection.

term MLM. Note that these longitudinal laser modes are different from the propagation modes inside the optical fiber! While both LEDs and MLM lasers emit light over a broad spectrum, the spectrum of an LED is continuous, whereas the spectrum of an MLM laser consists of many periodic lines.

These early systems had to have regenerators every few kilometers to regenerate the signal. Regenerators were expensive devices and continue to be expensive today, so it is highly desirable to maximize the distance between regenerators. In this case, the distance limitation was primarily due to a phenomenon known as *intermodal dispersion*. As we saw earlier, in a multimode fiber, the energy in a pulse travels in different modes, each with a different speed. At the end of the fiber, the different modes arrive at slightly different times, resulting in a smearing of the pulse. This smearing in general is called *dispersion*, and this specific form is called intermodal dispersion. Typically, these early systems operated at bit rates ranging from 32 to 140 Mb/s with regenerators every 10 km. Such systems are still used for low-cost computer interconnection at a few hundred megabits per second over a few kilometers.

1.8.2 Single-Mode Fiber

The next generation of systems deployed starting around 1984 used *single-mode* fiber as a means of eliminating intermodal dispersion, along with MLM Fabry-Perot lasers in the 1.3 μm wavelength band. Single-mode fiber has a relatively small core diameter of about 8 to 10 μm, which is a small multiple of the operating wavelength range of the light signal. This forces all the energy in a light signal to travel in the form of a single mode. Using single-mode fiber effectively eliminated intermodal dispersion and enabled a dramatic increase in the bit rates and distances possible between regenerators. These systems typically had regenerator spacings of about 40 km and operated at bit rates of a few hundred megabits per second. At this point, the distance between regenerators was limited primarily by the fiber loss.

The next step in this evolution in the late 1980s was to deploy systems in the 1.55 μm wavelength window to take advantage of the lower loss in this window, relative to the 1.3 μm window. This enabled longer spans between regenerators. At this point, another impairment, namely, *chromatic dispersion,* started becoming a limiting factor as far as increasing the bit rates was concerned. Chromatic dispersion is another form of dispersion in optical fiber (we looked at intermodal dispersion earlier). As we saw in Section 1.7, the energy in a light signal or pulse has a finite bandwidth. Even in a single-mode fiber, the different frequency components of a pulse propagate with different speeds. This is due to the fundamental physical properties of the glass. This effect again causes a smearing of the pulse at the output, just as with intermodal dispersion. The wider the spectrum of the pulse, the more the smearing due to chromatic dispersion. The chromatic dispersion in an optical fiber depends on the wavelength of the signal. It turns out that without any special effort, the standard silica-based optical fiber has essentially no chromatic dispersion in the 1.3 μm band, but has significant dispersion in the 1.55 μm band. Thus chromatic dispersion was not an issue in the earlier systems at 1.3 μm.

The high chromatic dispersion at 1.55 μm motivated the development of *dispersion-shifted fiber*. Dispersion-shifted fiber is carefully designed to have zero dispersion in the 1.55 μm wavelength window so that we need not worry about chromatic dispersion in this window. However, by this time there was already a large installed base of standard single-mode fiber deployed for which this solution could not be applied. Some carriers, particularly NTT in Japan and MCI (now part of Worldcom) in the United States, did deploy dispersion-shifted fiber.

At this time, researchers started looking for ways to overcome chromatic dispersion while still continuing to make use of standard fiber. The main technique that came into play was to reduce the width of the spectrum of the transmitted pulse. As we saw earlier, the wider the spectrum of the transmitted pulse, the greater the smearing due to chromatic dispersion. The bandwidth of the transmitted pulse is at least equal to its modulation bandwidth. On top of this, however, the bandwidth may be determined entirely by the width of the spectrum of the transmitter used. The MLM Fabry-Perot lasers, as we said earlier, emitted over a fairly wide spectrum of several nanometers (or, equivalently, hundreds of gigahertz), which is much larger than the modulation bandwidth of the signal itself. If we reduce the spectrum of the transmitted pulse to something close to its modulation bandwidth, the penalty due to chromatic dispersion is significantly reduced. This motivated the development of a laser source with a narrow spectral width—the *distributed-feedback* (DFB) laser. A DFB laser is an example of a *single-longitudinal mode* (SLM) laser. An SLM laser emits a narrow single-wavelength signal in a single spectral line, in contrast to MLM lasers whose spectrum consists of many spectral lines. This technological breakthrough spurred further increases in the bit rate to more than 1 Gb/s.

1.8.3   Optical Amplifiers and WDM

The next major milestone in the evolution of optical fiber transmission systems was the development of *erbium-doped fiber amplifiers* (EDFAs) in the late 1980s and early 1990s. The EDFA basically consists of a length of optical fiber, typically a few meters to tens of meters, doped with the rare earth element erbium. The erbium atoms in the fiber are *pumped* from their ground state to an excited state at a higher energy level using a pump source. An incoming signal photon triggers these atoms to come down to their ground state. In the process, each atom emits a photon. Thus incoming signal photons trigger the emission of additional photons, resulting in optical amplification. Due to a unique coincidence of nature, the difference in energy levels of the atomic states of erbium line up with the 1.5 μm low-loss window in the optical fiber. The pumping itself is done using a pump laser at a lower wavelength than the signal because photons with a lower wavelength have higher energies and energy can be transferred only from a photon of higher energy to that with a lower energy. The EDFA concept was invented in the 1960s but had to wait for the availability of reliable high-power semiconductor pump lasers in the late 1980s and early 1990s before becoming commercially viable.

EDFAs spurred the deployment of a completely new generation of systems. A major advantage of EDFAs is that they are capable of amplifying signals at many wavelengths simultaneously. This provided another way of increasing the system capacity: rather than increasing the bit rate, keep the bit rate the same and use more than one wavelength; that is, use wavelength division multiplexing. EDFAs were perhaps the single biggest catalyst aiding the deployment of WDM systems. The use of WDM and EDFAs dramatically brought down the cost of long-haul transmission systems and increased their capacity. At each regenerator location, a single optical amplifier could replace an entire array of expensive regenerators, one per fiber. This proved to be so compelling that almost every long-haul carrier has widely deployed amplified WDM systems today. Moreover WDM provided the ability to turn on capacity quickly, as opposed to the months to years it could take to deploy new fiber. WDM systems with EDFAs were deployed starting in the mid-1990s and are today achieving capacities over 1 Tb/s over a single fiber. At the same time, transmission bit rates on a single channel have risen to 10 Gb/s. Among the earliest

WDM systems deployed were AT&T's 4-wavelength long-haul system in 1995 and IBM's 20-wavelength MuxMaster metropolitan system in 1994.

With the advent of EDFAs, chromatic dispersion again reared its ugly head. Instead of regenerating the signal every 40 to 80 km, signals were now transmitted over much longer distances because of EDFAs, leading to significantly higher pulse smearing due to chromatic dispersion. Again, researchers found several techniques to deal with chromatic dispersion. The transmitted spectrum could be reduced further by using an external device to turn the laser on and off (called *external modulation*), instead of directly turning the laser on and off (called *direct modulation*). Using external modulators along with DFB lasers and EDFAs allowed systems to achieve distances of about 600 km at 2.5 Gb/s between regenerators over standard single-mode fiber at 1.55 μm. This number is substantially less at 10 Gb/s.

The next logical invention was to develop *chromatic dispersion compensation* techniques. A variety of chromatic dispersion compensators were developed to compensate for the dispersion introduced by the fiber, allowing the overall residual dispersion to be reduced to within manageable limits. These techniques have enabled commercial systems to achieve distances of several thousand kilometers between regenerators at bit rates as high as 10 Gb/s per channel.

At the same time, several other impairments that were second- or third-order effects earlier began to emerge as first-order effects. Today, this list includes nonlinear effects in fiber, the nonflat gain spectrum of EDFAs, and various polarization-related effects. There are several types of nonlinear effects that occur in optical fiber. One of them is called *four-wave mixing* (FWM). In FWM, three light signals at different wavelengths interact in the fiber to create a fourth light signal at a wavelength that may overlap with one of the light signals. As we can imagine, this signal interferes with the actual data that is being transmitted on that wavelength. It turns out paradoxically that the higher the chromatic dispersion, the lower the effect of fiber nonlinearities. Chromatic dispersion causes the light signals at different wavelengths to propagate at different speeds in the fiber. This in turn causes less overlap between these signals, as the signals go in and out of phase with each other, reducing the effect of the FWM nonlinearity.

The realization of this trade-off between chromatic dispersion and fiber nonlinearities stimulated the development of a variety of new types of single-mode fibers to manage the interaction between these two effects. These fibers are tailored to provide less chromatic dispersion than conventional fiber but, at the same time, reduce nonlinearities. We devote Chapter 5 to the study of these impairments and how they can be overcome; we discuss the origin of many of these effects in Chapter 2.

Today we are seeing the development of high-capacity amplified terabits/second WDM systems with hundreds of channels at 10 Gb/s, with channel spacings as low as 50 GHz, with distances between electrical regenerators extending to a few thousand

Table 1.2 Different wavelength bands in optical fiber. The ranges are approximate and have not yet been standardized.

Band	Descriptor	Wavelength range (nm)
O-band	Original	1260 to 1360
E-band	Extended	1360 to 1460
S-band	Short	1460 to 1530
C-band	Conventional	1530 to 1565
L-band	Long	1565 to 1625
U-band	Ultra-long	1625 to 1675

kilometers. Systems operating at 40 Gb/s channel rates are in the research laboratories, and no doubt we will see them become commercially available soon. Meanwhile, recent experiments have achieved terabit/second capacities and stretched the distance between regenerators to several thousand kilometers [Cai01, Bak01, VPM01], or achieved total capacities of over 10 Tb/s [Fuk01, Big01] over shorter distances.

Table 1.2 shows the different bands available for transmission in single-mode optical fiber. The early WDM systems used the C-band, primarily because that was where EDFAs existed. Today we have EDFAs that work in the L-band, which allow WDM systems to use both the C- and L-bands. We are also seeing the use of other types of amplification (such as Raman amplification, a topic that we will cover in Chapter 3) that complement EDFAs and hold the promise to open up other fiber bands such as the S-band and the U-band for WDM applications. Meanwhile, the development of new fiber types is also opening up a new window in the so-called E-band. This band was previously not feasible due to the high fiber loss in this wavelength range. New fibers have now been developed that reduce the loss in this range. However, there are still no good amplifiers in this band, so the E-band is useful mostly for short-distance applications.

1.8.4 Beyond Transmission Links to Networks

The late 1980s also witnessed the emergence of a variety of first-generation optical networks. In the data communications world, we saw the deployment of metropolitan-area networks, such as the 100 Mb/s fiber distributed data interface (FDDI), and networks to interconnect mainframe computers, such as the 200 Mb/s enterprise serial connection (ESCON). Today we are seeing the proliferation of storage networks using the 1 Gb/s Fibre Channel standard for similar applications. In the telecommunications world, standardization and mass deployment of SONET in North America and the similar SDH network in Europe and Japan began. All these networks are now widely deployed. Today it is common to have high-speed optical interfaces on a variety of other devices such as IP routers and ATM switches.

As these first-generation networks were being deployed in the late 1980s and early 1990s, people started thinking about innovative network architectures that would use fiber for more than just transmission. Most of the early experimental efforts were focused on optical networks for local-area network applications, but the high cost of the technology for these applications has hindered commercial viability of such networks. Research activity on optical packet-switched networks and local-area optical networks continues today. Meanwhile, wavelength-routing networks became a major focus area for several researchers in the early 1990s as people realized the benefits of having an optical layer. Optical add/drop multiplexers and crossconnects are now available as commercial products and are beginning to be introduced into telecommunications networks, stimulated by the fact that switching and routing high-capacity connections is much more economical at the optical layer than in the electrical layer. At the same time, the optical layer is evolving to provide additional functionality, including the ability to set up and take down lightpaths across the network in a dynamic fashion, and the ability to reroute lightpaths rapidly in case of a failure in the network. A combination of these factors is resulting in the introduction of intelligent optical ring and mesh networks, which provide lightpaths on demand and incorporate built-in restoration capabilities to deal with network failures.

There was also a major effort to promote the concept of *fiber to the home* (FTTH) and its many variants, such as *fiber to the curb* (FTTC), in the late 1980s and early 1990s. The problems with this concept were the high infrastructure cost and the questionable return on investment resulting from customers' reluctance to pay for a bevy of new services such as video to the home. However, telecommunications deregulation, coupled with the increasing demand for broadband services such as Internet access and video on demand, is accelerating the deployment of such networks by the major operators today. Both telecommunications carriers and cable operators are deploying fiber deeper into the access network and closer to the end user. Large businesses requiring very high capacities are being served by fiber-based SONET/SDH or Ethernet networks, while passive optical networks are emerging as possible candidates to provide high-speed services to homes and small businesses. This is the subject of Chapter 11.

Summary

We started this chapter by describing the changing face of the telecom industry—the large increase in traffic demands, the increase in data traffic relative to voice traffic, the deregulation of the telecom industry, the resulting emergence of a new set of carriers as well as equipment suppliers to these carriers, the need for new and flexible types of services, and an infrastructure to support all of these.

We described two generations of optical networks in this chapter: first-generation networks and second-generation networks. First-generation networks use optical fiber as a replacement for copper cable to get higher capacities. Second-generation networks provide circuit-switched lightpaths by routing and switching wavelengths inside the network. The key elements that enable this are optical line terminals (OLTs), optical add/drop multiplexers (OADMs), and optical crossconnects (OXCs). Optical packet switching may develop over time but faces several technological hurdles.

We saw that there were two complementary approaches to increasing transmission capacity: using more wavelengths on the fiber (WDM) and increasing the bit rate (TDM). We also traced the historical evolution of optical fiber transmission and networking. What is significant is that we are still far away from hitting the fundamental limits of capacity in optical fiber. While there are several roadblocks along the way, we will no doubt see the invention of new techniques that enable progressively higher and higher capacities, and the deployment of optical networks with increasing functionality.

Further Reading

The communications revolution is a topic that is receiving a lot of coverage across the board these days from the business press. A number of journal and magazine special issues have been focused on optical networks [GLM⁺00, CSH00, DYJ00, DL00, Alf99, HSS98, CHK⁺96, FGO⁺96, HD97, Bar96, NO94, KLHN93, CNW90, Pru89, Bra89].

Several conferences cover optical networks. The main ones are the Optical Fiber Communication Conference (OFC), Supercomm, and the National Fiber-Optic Engineers' Conference. Other conferences such as Next-Generation Networks (NGN), Networld-Interop, European Conference on Optical Communication (ECOC), IEEE Infocom, and the IEEE's International Conference on Communication (ICC) also cover optical networks. Archival journals such as the IEEE's *Journal of Lightwave Technology, Journal of Selected Areas in Communication, Journal of Quantum Electronics, Journal of Selected Topics in Quantum Electronics, Transactions on Networking,* and *Photonics Technology Letters,* and magazines such as the *IEEE Communications Magazine* and *Optical Networks Magazine* provide good coverage of this subject.

There are several excellent books devoted to fiber optic transmission and components, ranging from fairly basic [Hec98, ST91] to more advanced [KK97a, KK97b,

Agr97, Agr95, MK88, Lin89]. The 1993 book by Green [Gre93] provides specific coverage of WDM components, transmission, and networking aspects.

The historical evolution of transmission systems described here is also covered in a few other places in more detail. [Hec99] is an easily readable book devoted to the early history of fiber optics. [Wil00] is a special issue consisting of papers by many of the optical pioneers providing overviews and historical perspectives of various aspects of lasers, fiber optics, and other component and transmission technologies. [AKW00, Gla00, BKLW00] provide excellent, although Bell Labs–centric, overviews of the historical evolution of optical fiber technology and systems leading up to the current generation of WDM technology and systems. See also [MK88, Lin89].

Kao and Hockham [KH66] were the first to propose using low-loss glass fiber for optical communication. The processes used to fabricate low-loss fiber today were first reported in [KKM70] and refined in [Mac74]. [Sta83, CS83, MT83, Ish83] describe some of the early terrestrial optical fiber transmission systems. [RT84] describes one of the early undersea optical fiber transmission systems. See also [KM98] for a more recent overview.

Experiments reporting more than 1 Tb/s transmission over a single fiber were first reported at the Optical Fiber Communication Conference in 1996, and the numbers are being improved upon constantly. See, for example, [CT98, Ona96, Gna96, Mor96, Yan96]. Recent work on these frontiers has focused on (1) transmitting terabits-per-second aggregate traffic across transoceanic distances with individual channel data rates at 10 or 20 Gb/s [Cai01, Bak01, VPM01], or 40 Gb/s channel rates over shorter distances [Zhu01], or (2) obtaining over 10 Tb/s transmission capacity using 40 Gb/s channel rates over a few hundred kilometers [Fuk01, Big01].

Finally, we didn't cover standards in this chapter—but we will do so in Chapters 6, 9, and 10. The various standards bodies working on optical networking include the International Telecommunications Union (ITU), the American National Standards Institute (ANSI), the Optical Internetworking Forum (OIF), Internet Engineering Task Force (IETF), and Telcordia Technologies. Appendix C provides a list of relevant standards documents.

References

[Agr95]   G. P. Agrawal. *Nonlinear Fiber Optics*, 2nd edition. Academic Press, San Diego, CA, 1995.

[Agr97]   G. P. Agrawal. *Fiber-Optic Communication Systems*. John Wiley, New York, 1997.

[AKW00] R. C. Alferness, H. Kogelnik, and T. H. Wood. The evolution of optical systems: Optics everywhere. *Bell Labs Technical Journal*, 5(1):188–202, Jan.–March 2000.

[Alf99] R. Alferness, editor. *Bell Labs Technical Journal: Optical Networking*, volume 4, Jan.–Mar. 1999.

[Bak01] B. Bakhshi et al. 1 Tb/s (101 × 10 Gb/s) transmission over transpacific distance using 28 nm C-band EDFAs. In *OFC 2001 Technical Digest*, pages PD21/1–3, 2001.

[Bar96] R. A. Barry, editor. *IEEE Network: Special Issue on Optical Networks*, volume 10, Nov. 1996.

[Big01] S. Bigo et al. 10.2 Tb/s (256 × 42.7 Gbit/s PDM/WDM) transmission over 100 km TeraLight fiber with 1.28bit/s/Hz spectral efficiency. In *OFC 2001 Technical Digest*, pages PD25/1–3, 2001.

[BKLW00] W. F. Brinkman, T. L. Koch, D. V. Lang, and D. W. Wilt. The lasers behind the communications revolution. *Bell Labs Technical Journal*, 5(1):150–167, Jan.–March 2000.

[Bra89] C. A. Brackett, editor. *IEEE Communications Magazine: Special Issue on Lightwave Systems and Components*, volume 27, Oct. 1989.

[Cai01] J.-X. Cai et al. 2.4 Tb/s (120 × 20 Gb/s) transmission over transoceanic distance with optimum FEC overhead and 48% spectral efficiency. In *OFC 2001 Technical Digest*, pages PD20/1–3, 2001.

[CHK⁺96] R. L. Cruz, G. R. Hill, A. L. Kellner, R. Ramaswami, and G. H. Sasaki, editors. *IEEE JSAC/JLT Special Issue on Optical Networks*, volume 14, June 1996.

[CNW90] N. K. Cheung, G. Nosu, and G. Winzer, editors. *IEEE JSAC: Special Issue on Dense WDM Networks*, volume 8, Aug. 1990.

[CS83] J. S. Cook and O. I. Szentisi. North American field trials and early applications in telephony. *IEEE JSAC*, 1:393–397, 1983.

[CSH00] G. K. Chang, K. I. Sato, and D. K. Hunter, editors. *IEEE/OSA Journal of Lightwave Technology: Special Issue on Optical Networks*, volume 18, 2000.

[CT98] A. R. Chraplyvy and R. W. Tkach. Terabit/second transmission experiments. *IEEE Journal of Quantum Electronics*, 34(11):2103–2108, 1998.

[DL00] S. S. Dixit and P. J. Lin, editors. *IEEE Communications Magazine: Optical Networks Come of Age*, volume 38, Feb. 2000.

[DYJ00] S. S. Dixit and A. Yla-Jaaski, editors. *IEEE Communications Magazine: WDM Optical Networks: A Reality Check*, volume 38, Mar. 2000.

[FGO⁺96] M. Fujiwara, M. S. Goodman, M. J. O'Mahony, O. K. Tonguez, and A. E. Willner, editors. *IEEE/OSA JLT/JSAC Special Issue on Multiwavelength Optical Technology and Networks*, volume 14, June 1996.

[Fra93]   A. G. Fraser. Banquet speech. In *Proceedings of Workshop on High-Performance Communication Subsystems*, Williamsburg, VA, Sept. 1993.

[Fuk01]   K. Fukuchi et al. 10.92 Tb/s (273 × 40 Gb/s) triple-band/ultra-dense WDM optical-repeatered transmission experiment. In *OFC 2001 Technical Digest*, pages PD24/1–3, 2001.

[GJR96]   P. E. Green, F. J. Janniello, and R. Ramaswami. Multichannel protocol-transparent WDM distance extension using remodulation. *IEEE JSAC/JLT Special Issue on Optical Networks*, 14(6):962–967, June 1996.

[Gla00]   A. M. Glass et al. Advances in fiber optics. *Bell Labs Technical Journal*, 5(1):168–187, Jan.–March 2000.

[GLM+00] O. Gerstel, B. Li, A. McGuire, G. Rouskas, K. Sivalingam, and Z. Zhang, editors. *IEEE JSAC: Special Issue on Protocols and Architectures for Next-Generation Optical Networks*, Oct. 2000.

[Gna96]   A. H. Gnauck et al. One terabit/s transmission experiment. In *OFC'96 Technical Digest*, 1996. Postdeadline paper PD20.

[Gre93]   P. E. Green. *Fiber-Optic Networks*. Prentice Hall, Englewood Cliffs, NJ, 1993.

[HD97]   G. R. Hill and P. Demeester, editors. *IEEE Communications Magazine: Special Issue on Photonic Networks in Europe*, volume 35, April 1997.

[Hec98]   J. Hecht. *Understanding Fiber Optics*. Prentice Hall, Englewood Cliffs, NJ, 1998.

[Hec99]   J. Hecht. *City of Light: The Story of Fiber Optics*. Oxford University Press, New York, 1999.

[HSS98]   A. M. Hill, A. A. M. Saleh, and K. Sato, editors. *IEEE JSAC: Special Issue on High-Capacity Optical Transport Networks*, volume 16, Sept. 1998.

[Ish83]   H. Ishio. Japanese field trials and applications in telephony. *IEEE JSAC*, 1:404–412, 1983.

[KH66]   K. C. Kao and G. A. Hockham. Dielectric-fiber surface waveguides for optical frequencies. *Proceedings of IEE*, 133(3):1151–1158, July 1966.

[KK97a]   I. P. Kaminow and T. L. Koch, editors. *Optical Fiber Telecommunications IIIA*. Academic Press, San Diego, CA, 1997.

[KK97b]   I. P. Kaminow and T. L. Koch, editors. *Optical Fiber Telecommunications IIIB*. Academic Press, San Diego, CA, 1997.

[KKM70]   F. P. Kapron, D. B. Keck, and R. D. Maurer. Radiation losses in glass optical waveguides. *Applied Physics Letters*, 17(10):423–425, Nov. 1970.

[KLHN93] M. J. Karol, C. Lin, G. Hill, and K. Nosu, editors. *IEEE/OSA Journal of Lightwave Technology: Special Issue on Broadband Optical Networks*, May/June 1993.

[KM98]   F. W. Kerfoot and W. C. Marra. Undersea fiber optic networks: Past, present and future. *IEEE JSAC: Special Issue on High-Capacity Optical Transport Networks*, 16(7):1220–1225, Sept. 1998.

[Kra99]   J. M. Kraushaar. *Fiber Deployment Update: End of Year 1998*. Federal Communications Commission, Sept. 1999. Available from *http://www.fcc.gov*.

[Lin89]   C. Lin, editor. *Optoelectronic Technology and Lightwave Communications Systems*. Van Nostrand Reinhold, New York, 1989.

[Mac74]   J. B. MacChesney et al. Preparation of low-loss optical fibers using simultaneous vapor deposition and fusion. In *Proceedings of 10th International Congress on Glass*, volume 6, pages 40–44, Kyoto, Japan, 1974.

[MK88]   S. D. Miller and I. P. Kaminow, editors. *Optical Fiber Telecommunications II*. Academic Press, San Diego, CA, 1988.

[Mor96]   T. Morioka et al. 100 Gb/s × 10 channel OTDM/WDM transmission using a single supercontinuum WDM source. In *OFC'96 Technical Digest*, 1996. Postdeadline paper PD21.

[MT83]   A. Moncalvo and F. Tosco. European field trials and early applications in telephony. *IEEE JSAC*, 1:398–403, 1983.

[NO94]   K. Nosu and M. J. O'Mahony, editors. *IEEE Communications Magazine: Special Issue on Optically Multiplexed Networks*, volume 32, Dec. 1994.

[Ona96]   H. Onaka et al. 1.1 Tb/s WDM transmission over a 150 km 1.3 μm zero-dispersion single-mode fiber. In *OFC'96 Technical Digest*, 1996. Postdeadline paper PD19.

[Pru89]   P. R. Prucnal, editor. *IEEE Network: Special Issue on Optical Multiaccess Networks*, volume 3, March 1989.

[RT84]   P. K. Runge and P. R. Trischitta. The SL undersea lightwave system. *IEEE/OSA Journal on Lightwave Technology*, 2:744–753, 1984.

[ST91]   B. E. A. Saleh and M. C. Teich. *Fundamentals of Photonics*. Wiley, New York, 1991.

[Sta83]   J. R. Stauffer. FT3C—a lightwave system for metropolitan and intercity applications. *IEEE JSAC*, 1:413–419, 1983.

[VPM01]   G. Vareille, F. Pitel, and J. F. Marcerou. 3 Tb/s (300 × 11.6 Gbit/s) transmission over 7380 km using 28 nm C+L-band with 25 GHz channel spacing and NRZ format. In *OFC 2001 Technical Digest*, pages PD22/1–3, 2001.

[Wil00]   A. E. Willner, editor. *IEEE Journal of Selected Topics in Quantum Electronics: Millennium Issue*, volume 6, Nov./Dec. 2000.

[Yan96] Y. Yano et al. 2.6 Tb/s WDM transmission experiment using optical duobinary coding. In *Proceedings of European Conference on Optical Communication*, 1996. Postdeadline paper Th.B.3.1.

[Zhu01] B. Zhu et al. 3.08 Tb/s (77 × 42.7 Gb/s) transmission over 1200 km of non-zero dispersion-shifted fiber with 100-km spans using C- L-band distributed Raman amplification. In *OFC 2001 Technical Digest*, pages PD23/1–3, 2001.

part

I

Technology

chapter

Propagation of Signals in Optical Fiber

O PTICAL FIBER IS A REMARKABLE communication medium compared to other media such as copper or free space. An optical fiber provides low-loss transmission over an enormous frequency range of at least 25 THz—even higher with special fibers—which is orders of magnitude more than the bandwidth available in copper cables or any other transmission medium. For example, this bandwidth is sufficient to transmit hundreds of millions of phone calls simultaneously, or tens of millions of Web pages per second. The low-loss property allows signals to be transmitted over long distances at high speeds before they need to be amplified or regenerated. It is due to these two properties of low loss and high bandwidth that optical fiber communication systems are so widely used today.

As transmission systems evolved to longer distances and higher bit rates, *dispersion* became an important limiting factor. Dispersion refers to the phenomenon where different components of the signal travel at different velocities in the fiber. In particular, *chromatic* dispersion refers to the phenomenon where different frequency (or wavelength) components of the signal travel with different velocities in the fiber. In most situations, dispersion leads to broadening of pulses, and hence pulses corresponding to adjacent bits interfere with each other. In a communication system, this leads to the overlap of pulses representing adjacent bits. This phenomenon is called *Inter-Symbol Interference* (ISI). As systems evolved to larger numbers of wavelengths, and even higher bit rates and distances, *nonlinear effects* in the fiber began to present serious limitations. As we will see, there is a complex interplay of nonlinear effects with chromatic dispersion.

We start this chapter by discussing the basics of light propagation in optical fiber, starting from a simple geometrical optics model to the more general wave

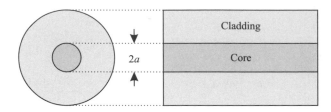

Figure 2.1   Cross section and longitudinal section of an optical fiber showing the core and cladding regions. *a* denotes the radius of the fiber core.

theory model based on solving Maxwell's equations. We then devote the rest of the chapter to understanding the basics of chromatic dispersion and fiber nonlinearities. Designing advanced systems optimized with respect to these parameters is treated in Chapter 5.

2.1   Light Propagation in Optical Fiber

An optical fiber consists of a cylindrical *core* surrounded by a *cladding*. The cross section of an optical fiber is shown in Figure 2.1. Both the core and the cladding are made primarily of silica (SiO_2), which has a refractive index of approximately 1.45. The *refractive index* of a material is the ratio of the speed of light in a vacuum to the speed of light in that material. During the manufacturing of the fiber, certain impurities (or dopants) are introduced in the core and/or the cladding so that the refractive index is slightly higher in the core than in the cladding. Materials such as germanium and phosphorous increase the refractive index of silica and are used as dopants for the core, whereas materials such as boron and fluorine that decrease the refractive index of silica are used as dopants for the cladding. As we will see, the resulting higher refractive index of the core enables light to be guided by the core, and thus propagate through the fiber.

2.1.1   Geometrical Optics Approach

We can obtain a simplified understanding of light propagation in optical fiber using the so-called *ray theory* or *geometrical optics* approach. This approach is valid when the fiber that is used has a core radius *a* that is much larger than the operating wavelength λ. Such fibers are termed *multimode*, and first-generation optical communication links were built using such fibers with *a* in the range of 25–100 μm and λ around 0.85 μm.

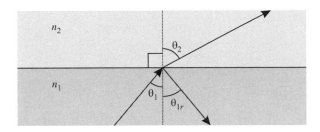

Figure 2.2 Reflection and refraction of light rays at the interface between two media.

In the geometrical optics approach, light can be thought of as consisting of a number of "rays" propagating in straight lines within a material (or medium) and getting reflected and/or refracted at the interfaces between two materials. Figure 2.2 shows the interface between two media of refractive index n_1 and n_2. A light ray from medium 1 is incident on the interface of medium 1 with medium 2. The *angle of incidence* is the angle between the *incident ray* and the normal to the interface between the two media and is denoted by θ_1. Part of the energy is reflected into medium 1 as a *reflected ray,* and the remainder (neglecting absorption) passes into medium 2 as a *refracted ray*. The *angle of reflection* θ_{1r} is the angle between the reflected ray and the normal to the interface; similarly, the *angle of refraction* θ_2 is the angle between the refracted ray and the normal. The laws of geometrical optics state that

$$\theta_{1r} = \theta_1$$

and

$$n_1 \sin \theta_1 = n_2 \sin \theta_2. \tag{2.1}$$

Equation (2.1) is known as *Snell's law*.

As the angle of incidence θ_1 increases, the angle of refraction θ_2 also increases. If $n_1 > n_2$, there comes a point when $\theta_2 = \pi/2$ radians. This happens when $\theta_1 = \sin^{-1} n_2/n_1$. For larger values of θ_1, there is no refracted ray, and all the energy from the incident ray is reflected. This phenomenon is called *total internal reflection*. The smallest angle of incidence for which we get total internal reflection is called the *critical angle* and equals $\sin^{-1} n_2/n_1$.

Simply stated, from the geometrical optics viewpoint, light propagates in optical fiber due to a series of total internal reflections that occur at the core-cladding interface. This is depicted in Figure 2.3. In this figure, the coupling of light from the medium outside (taken to be air with refractive index n_0) into the fiber is also shown.

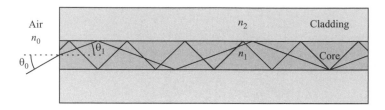

Figure 2.3   Propagation of light rays in optical fiber by total internal reflection.

It can be shown using Snell's law (see Problem 2.1) that only those light rays that are incident at an angle

$$\theta_0 < \theta_0^{\max} = \sin^{-1} \frac{\sqrt{n_1^2 - n_2^2}}{n_0} \tag{2.2}$$

at the air-core interface will undergo total internal reflection at the core-cladding interface and will thus propagate. Such rays are called *guided rays*, and $\theta_0^{\max}$ is called the *acceptance angle*. The refractive index difference $n_1 - n_2$ is usually small, and it is convenient to denote the fractional refractive index difference $(n_1 - n_2)/n_1$ by Δ. For small Δ, $\theta_0^{\max} \approx \sin^{-1} \frac{n_1 \sqrt{2\Delta}}{n_0}$. As an example, if $\Delta = 0.01$, which is a typical value for (multimode) fiber, and $n_1 = 1.5$, a typical value for silica, assuming we are coupling from air, so that $n_0 = 1$, we obtain $\theta_0^{\max} \approx 12°$.

Owing to the different lengths of the paths taken by different guided rays, the energy in a narrow (in time) pulse at the input of the fiber will be spread out over a larger time interval at the output of the fiber. A measure of this time spread, which is called *intermodal dispersion*, is obtained by taking the difference in time, δT, between the fastest and the slowest guided rays. We will see later that by suitably designing the fiber, intermodal dispersion can be significantly reduced (graded-index fiber) and even eliminated (single-mode fiber).

We now derive an approximate measure of the time spread due to intermodal dispersion. Consider a fiber of length L. The fastest guided ray is the one that travels along the center of the core and takes a time $T_f = Ln_1/c$ to traverse the fiber, c being the speed of light in a vacuum. The slowest guided ray is incident at the critical angle on the core-cladding interface, and it can be shown that it takes a time $T_s = Ln_1^2/cn_2$ to propagate through the fiber. Thus

$$\delta T = T_s - T_f = \frac{L}{c} \frac{n_1^2}{n_2} \Delta.$$

How large can δT be before it begins to matter? That depends on the bit rate used. A rough measure of the delay variation δT that can be tolerated at a bit rate of B b/s is half the bit period $1/2B$ s. Thus intermodal dispersion sets the following limit:

$$\delta T = \frac{L}{c} \frac{n_1^2}{n_2} \Delta < \frac{1}{2B}. \tag{2.3}$$

The capacity of an optical communication system is frequently measured in terms of the *bit rate–distance product*. If a system is capable of transmitting x Mb/s over a distance of y km, it is said to have a bit rate–distance product of xy (Mb/s)-km. The reason for doing this is that usually the same system is capable of transmitting x' Mb/s over y' km providing $x'y' < xy$; thus only the product of the bit rate and the distance is constrained. (This is true for simple systems that are limited by loss and/or intermodal dispersion, but is no longer true for systems that are limited by chromatic dispersion and nonlinear effects in the fiber.) From (2.3), the intermodal dispersion constrains the bit rate–distance product of an optical communication link to

$$BL < \frac{1}{2} \frac{n_2}{n_1^2} \frac{c}{\Delta}.$$

For example, if $\Delta = 0.01$ and $n_1 = 1.5 (\approx n_2)$, we get $BL < 10$ (Mb/s)-km. This limit is plotted in Figure 2.4.

Note that $\theta_0^{\max}$ increases with increasing Δ, which causes the limit on the bit rate–distance product to decrease with increasing Δ. The value of Δ is typically chosen to be less than 1% so as to minimize the effects of intermodal dispersion, and since $\theta_0^{\max}$ is consequently small, lenses or other suitable mechanisms are used to couple light into the fiber.

The fiber we have described is a *step-index* fiber since the variation of the refractive index along the fiber cross section can be represented as a function with a step at the core-cladding interface. However, almost all multimode fibers used today are *graded-index* fibers, and the refractive index decreases gradually, or continuously, from its maximum value at the center of the core to the value in the cladding at the core-cladding interface. This has the effect of reducing δT because the rays traversing the shortest path through the center of the core encounter the highest refractive index and travel slowest, whereas rays traversing longer paths encounter regions of lower refractive index and travel faster. For the optimum graded-index profile (which is very nearly a quadratic decrease of the refractive index in the core from its maximum value at the center to its value in the cladding), it can be shown that δT, the time

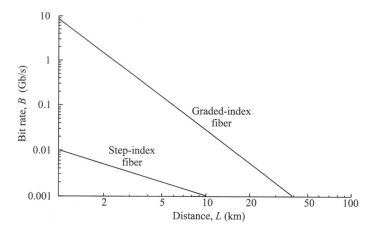

Figure 2.4   Limit on the bit rate–distance product due to intermodal dispersion in a step-index and a graded-index fiber. In both cases, $\Delta = 0.01$ and $n_1 = 1.5$.

difference between the fastest and slowest rays to travel a length L of the fiber, is given by

$$\delta T = \frac{L}{c} \frac{n_1 \Delta^2}{8}.$$

Assuming that the condition $\delta T < 1/2B$, where B is the bit rate, must be satisfied, we get the following limit on the bit rate–distance product of a communication system employing graded-index fiber:

$$BL < \frac{4c}{n_1 \Delta^2}.$$

For example, if $\Delta = 0.01$ and $n_1 = 1.5$, we get $BL < 8$ (Gb/s)-km. This limit is also plotted in Figure 2.4 along with the limit for step-index fiber. For instance, there are commercial systems operating at 200 Mb/s over a few kilometers using multimode fibers today.

Graded-index fibers significantly reduce the effects of intermodal dispersion. But in order to overcome intermodal dispersion completely, you must use fibers whose core radius is appreciably smaller and of the order of the operating wavelength. Such fibers are called *single-mode* fibers (the precise reason for this term will become clear later). Roughly speaking, the different paths that light rays can take in a multimode fiber can be termed as different *modes* in which light can propagate. In a single-mode fiber, there is only one mode in which light can propagate.

The physical reason for the confinement of the light within the fiber core can no longer be attributed to total internal reflection since this picture is invalid when the fiber core radius is comparable to the light wavelength, as is the case for single-mode fiber. The following physical explanation for the propagation of light in single-mode fiber is based on [Neu88]. In any medium with a constant refractive index, a narrow light beam tends to spread due to a phenomenon called *diffraction*. Thus in such a medium, the beam width will increase as light propagates. This effect can be counteracted by using an inhomogeneous medium in which the refractive index near the beam center is appropriately larger than the refractive index at the beam periphery. In such a medium, the beam center travels slightly slower than the beam periphery so that the medium effectively provides continuous focusing of the light to offset the spreading effect of diffraction. This allows the beam to be *guided* in the medium and go long distances with low loss, which would not be the case if the beam were allowed to spread out. A step-index optical fiber is an example of such an inhomogeneous medium since the refractive index in the core (beam center) is larger than that in the cladding (beam periphery).

In the following sections, we will provide a more quantitative description of the propagation of light in single-mode fibers using the wave theory approach. The wave theory is more general and is applicable for all values of the fiber radius. The ray theory is an approximation that holds when the optical wavelength is much smaller than the radius of the fiber core. Our objective is to gain a quantitative understanding of two phenomena that are important in the design of fiber optic communication systems: chromatic dispersion and fiber nonlinearities.

2.1.2 Wave Theory Approach

Light is an electromagnetic wave, and its propagation in any medium is governed by *Maxwell's equations*. These equations are stated in Appendix D. The propagation of light can be described by specifying the evolution of the associated electric and magnetic field vectors in space and time, denoted by $E(\mathbf{r}, t)$ and $H(\mathbf{r}, t)$, respectively. Here $\mathbf{r}$ denotes the position vector and t denotes time. Sometimes it will be more convenient to deal with the Fourier transforms of these vectors. The Fourier transform of E is defined as

$$\tilde{E}(\mathbf{r}, \omega) = \int_{-\infty}^{\infty} E(\mathbf{r}, t) \exp(i\omega t) \, dt. \tag{2.4}$$

The Fourier transform of H and other vectors that we will encounter later are defined similarly. Note that even when $E(\mathbf{r}, t)$ is real, $\tilde{E}(\mathbf{r}, \omega)$ can be complex. It turns out to be quite convenient, in many cases, to allow $E(\mathbf{r}, t)$ to be complex valued as well. But it is understood that we should consider only the real part of the solutions obtained.

The electrons in an atom are negatively charged, and the nucleus carries a positive charge. Thus when an electric field is applied to a material such as silica, the forces experienced by the nuclei and the electrons are in opposite directions. These forces result in the atoms being *polarized* or distorted. The *induced electric polarization* of the material, or *dielectric polarization,* can be described by a vector $\mathbf{P}$, which depends both on the material properties and the applied field. The dielectric polarization can be viewed as the response of the medium to the applied electric field. We will shortly discuss the relationship between $\mathbf{P}$ and $\mathbf{E}$ in detail. It is convenient to define another vector $\mathbf{D}$ called the *electric flux density,* which is simply related to the electric field $\mathbf{E}$ and dielectric polarization $\mathbf{P}$ by

$$\mathbf{D} = \epsilon_0 \mathbf{E} + \mathbf{P}, \tag{2.5}$$

where ϵ_0 is a constant called the *permittivity of vacuum.* The flux density in a vacuum is simply $\epsilon_0 \mathbf{E}$. The *magnetic polarization* $\mathbf{M}$ and the *magnetic flux density* $\mathbf{B}$ can be defined in an analogous fashion as

$$\mathbf{B} = \mu_0 (\mathbf{H} + \mathbf{M}). \tag{2.6}$$

However, since silica is a nonmagnetic material, $\mathbf{B} = \mu_0 \mathbf{H}$, where μ_0 is a constant called the *permeability of vacuum.* Maxwell's equations take into account the effect of material properties on the propagation of electromagnetic waves, since they not only involve $\mathbf{E}$ and $\mathbf{H}$ but also the flux densities $\mathbf{D}$ and the magnetic flux density $\mathbf{B}$.

The relationship between $\mathbf{P}$ and $\mathbf{E}$ in optical fiber due to the nature of silica is the origin of two important effects related to the propagation of light in fiber, namely, dispersion and nonlinearities. These two effects set limits on the performance of optical communication systems today. We will understand the origin of these effects in this chapter. Methods of dealing with these effects in optical communication systems will be discussed in Chapter 5.

The relationship between the vectors $\mathbf{P}$ and $\mathbf{E}$ depends on the nature of the medium. Next, we discuss five characteristics of a medium and their effect on the relationship between the dielectric polarization $\mathbf{P}$ in the medium and the applied electric field $\mathbf{E}$.

Locality of Response. In a medium whose response to the applied electric field is local, $\mathbf{P}(\mathbf{r})$ at $\mathbf{r} = \mathbf{r}_1$ depends only on $\mathbf{E}(\mathbf{r}_1)$. The values of $\mathbf{E}(\mathbf{r})$ for $\mathbf{r} \neq \mathbf{r}_1$ have no effect on $\mathbf{P}(\mathbf{r}_1)$. This property holds to a good degree of approximation for silica fibers in the 0.5–2 μm wavelength range that is of interest in optical communication systems.

Isotropy. An isotropic medium is one whose electromagnetic properties such as the refractive index are the same in all directions. In an isotropic medium, **E** and **P** are vectors with the same orientation. Silica is an isotropic medium, and a perfectly cylindrical optical fiber is isotropic in the transverse plane. However, this is not exactly true if the cylindrical symmetry of fiber is destroyed. A medium whose refractive indices along two different directions, for example, the x and y axes in an appropriate coordinate system, are different is said to *birefringent*. Birefringence can arise due to the geometry of the medium or due to the intrinsic property of the material. An optical fiber that does not possess cylindrical symmetry is therefore said to be geometrically birefringent. Birefringence of materials such as lithium niobate is exploited in designing certain components such as modulators, isolators, and tunable filters. We will discuss these components in Chapter 3. A bent fiber is also not an isotropic medium. Bending leads to additional loss, and we discuss this in Section 2.2.

Linearity. In a linear, isotropic medium,

$$\mathbf{P}(\mathbf{r}, t) = \epsilon_0 \int_{-\infty}^{t} \chi(\mathbf{r}, t - t') \mathbf{E}(\mathbf{r}, t') \, dt', \tag{2.7}$$

where χ is called the *susceptibility*, or more accurately, *linear susceptibility*, of the medium (silica). Thus the induced dielectric polarization is obtained by convolving the applied electric field with (ϵ_0 times) the susceptibility of the medium. If $\tilde{\mathbf{P}}$ and $\tilde{\chi}$ denote the Fourier transforms of **P** and χ, respectively, (2.7) can be written in terms of Fourier transforms as

$$\tilde{\mathbf{P}}(\mathbf{r}, \omega) = \epsilon_0 \tilde{\chi}(\mathbf{r}, \omega) \tilde{\mathbf{E}}(\mathbf{r}, \omega). \tag{2.8}$$

Electrical engineers will note that in this linear case, the dielectric polarization can be viewed as the output of a linear system with impulse response $\epsilon_0 \chi(\mathbf{r}, t)$, or transfer function $\epsilon_0 \tilde{\chi}(\mathbf{r}, \omega)$, and input $\mathbf{E}(\mathbf{r}, t)$ (or $\tilde{\mathbf{E}}(\mathbf{r}, \omega)$). It is important to note that the value of **P** at time t depends not only on the value of **E** at time t but also on the values of **E** before time t. Thus the response of the medium to the applied electric field is not instantaneous. (In other words, $\tilde{\chi}(\mathbf{r}, \omega)$ is not independent of ω.) This is the origin of an important type of dispersion known as *chromatic dispersion*, which sets a fundamental limit on the performance of optical communication systems. If the medium response is instantaneous so that the susceptibility (impulse response) is a Dirac delta function, its Fourier transform would be a constant, independent of ω, and chromatic dispersion would vanish. Thus the origin of chromatic dispersion lies in the delayed response of the dielectric polarization in the silica medium to the applied electric field.

This linear relationship between **P** and **E** does not hold exactly for silica but is a good approximation at moderate signal powers and bit rates. The effects of nonlinearities on the propagation of light will be discussed in Section 2.4.

Homogeneity. A homogeneous medium has the same electromagnetic properties at all points within it. In such a medium, χ, and hence $\tilde{\chi}$, are independent of the position vector **r**, and we can write $\chi(t)$ for $\chi(\mathbf{r}, t)$. Whereas silica is a homogeneous medium, optical fiber is not, since the refractive indices in the core and cladding are different. However, individually, the core and cladding regions in a step-index fiber are homogeneous. The core of a graded-index fiber is inhomogeneous. A discussion of the propagation of light in graded-index fiber is beyond the scope of this book.

Losslessness. Although silica fiber is certainly not lossless, the loss is negligible and can be assumed to be zero in the discussion of *propagation modes*. These modes would not change significantly if the nonzero loss of silica fiber were included in their derivation.

In this section, we assume that the core and the cladding regions of the silica fiber are *locally responsive, isotropic, linear, homogeneous,* and *lossless.* These assumptions are equivalent to assuming the appropriate properties for **P**, **E**, and χ in the fiber according to the preceding discussion.

Recall that the refractive index of a material n is the ratio of the speed of light in a vacuum to the speed of light in that material. It is related to the susceptibility as

$$n^2(\omega) = 1 + \tilde{\chi}(\omega). \tag{2.9}$$

Since the susceptibility $\tilde{\chi}$ is a function of the angular frequency ω, so is the refractive index. Hence we have written $n(\omega)$ for n in (2.9). This dependence of the refractive index on frequency is the origin of chromatic dispersion in optical fibers as we noted. For optical fibers, the value of $\tilde{\chi} \approx 1.25$, and the refractive index $n \approx 1.5$.

With these assumptions, starting from Maxwell's equations, it can be shown that the following wave equations hold for $\tilde{\mathbf{E}}$ and $\tilde{\mathbf{H}}$. These equations are derived in Appendix D.

$$\nabla^2 \tilde{\mathbf{E}} + \frac{\omega^2 n^2(\omega)}{c^2} \tilde{\mathbf{E}} = 0 \tag{2.10}$$

$$\nabla^2 \tilde{\mathbf{H}} + \frac{\omega^2 n^2(\omega)}{c^2} \tilde{\mathbf{H}} = 0. \tag{2.11}$$

Here ∇^2 denotes the Laplacian operator, which is given in Cartesian coordinates by $\frac{\partial^2}{\partial x^2} + \frac{\partial^2}{\partial y^2} + \frac{\partial^2}{\partial z^2}$. Thus the wave equations are second-order, linear, partial differential equations for the Fourier transforms of the electric and magnetic field vectors.

Note that each wave equation actually represents three equations—one for each component of the corresponding field vector.

Fiber Modes

The electric and magnetic field vectors in the core, $\tilde{\mathbf{E}}_{core}$ and $\tilde{\mathbf{H}}_{core}$, and the electric and magnetic field vectors in the cladding, $\tilde{\mathbf{E}}_{cladding}$ and $\tilde{\mathbf{H}}_{cladding}$, must satisfy the wave equations, (2.10) and (2.11), respectively. However, the solutions in the core and the cladding *are not independent;* they are related by boundary conditions on $\tilde{\mathbf{E}}$ and $\tilde{\mathbf{H}}$ at the core-cladding interface. Quite simply, every pair of solutions of these wave equations that satisfies these boundary conditions is a *fiber mode.*

Assume the direction of propagation of the electromagnetic wave (light) is z. Also assume that the fiber properties such as the core diameter and the core and cladding refractive indices are independent of z. Then it turns out that the z-dependence of the electric and magnetic fields of each fiber mode is of the form $e^{i\beta z}$. The quantity β is called the propagation constant of the mode. Each fiber mode has a different propagation constant β associated with it. (This is true for nondegenerate modes. We discuss degenerate modes in the context of polarization below.) The propagation constant is measured in units of radians per unit length. It determines the speed at which pulse energy in a mode propagates in the fiber. (Note that this concept of different propagation speeds for different modes has an analog in the geometrical optics approach. We can think of a "mode" as one possible path that a guided ray can take. Since the path lengths are different, the propagation speeds of the modes are different.) We will discuss this further in Section 2.3. The light energy propagating in the fiber will be divided among the modes supported by the fiber, and since the modes travel at different speeds in the fiber, the energy in a narrow pulse at the input of a length of fiber will be spread out at the output. Thus it is desirable to *design the fiber such that it supports only a single mode.* Such a fiber is called a *single-mode fiber,* and the mode that it supports is termed the *fundamental mode.* We had already come to a similar conclusion at the end of Section 2.1.1, but the wave theory approach enables us to get a clearer understanding of the concept of modes.

To better understand the notion of a propagation constant of a mode, consider the propagation of an electromagnetic wave in a homogeneous medium with refractive index n. Further assume that the wave is *monochromatic;* that is, all its energy is concentrated at a single angular frequency ω or free-space wavelength λ. In this case, the propagation constant is $\omega n/c = 2\pi n/\lambda$. The *wave number, k,* is defined by $k = 2\pi/\lambda$ and is simply the spatial frequency (in cycles per unit length). In terms of the wave number, the propagation constant is kn. Thus for a wave propagating purely in the core, the propagation constant is kn_1, and for a wave propagating only in the cladding, the propagation constant is kn_2. The fiber modes propagate partly in the cladding and partly in the core, and thus their propagation constants β satisfy $kn_2 < \beta < kn_1$. Instead of the propagation constant of a mode, we can consider its *effective index* $n_{\text{eff}} = \beta/k$. The effective index of a mode thus lies between the refractive indices of the cladding and the core. For a monochromatic wave in a single-mode fiber, the effective index is analogous to the refractive index: the speed at which the wave propagates is c/n_{eff}. We will discuss the propagation constant further in Section 2.3.

The solution of (2.10) and (2.11) is discussed in [Agr97, Jeu90]. We only state some important properties of the solution in the rest of this section.

The core radius a, the core refractive index n_1, and the cladding refractive index n_2 must satisfy the cutoff condition

$$ V \stackrel{\text{def}}{=} \frac{2\pi}{\lambda} a \sqrt{n_1^2 - n_2^2} < 2.405 \tag{2.12} $$

in order for a fiber to be *single moded at wavelength* λ. The smallest wavelength λ for which a given fiber is single moded is called the *cutoff wavelength* and denoted by λ_{cutoff}. Note that V decreases with a and $\Delta = (n_1 - n_2)/n_1$. Thus single-mode fibers tend to have small radii and small core-cladding refractive index differences. Typical values are $a = 4$ μm and $\Delta = 0.003$, giving a V value close to 2 at 1.55 μm. The calculation of the cutoff wavelength λ_{cutoff} for these parameters is left as an exercise (Problem 2.4).

Since the value of Δ is typically small, the refractive indices of the core and cladding are nearly equal, and the light energy is not strictly confined to the fiber core. In fact, a significant portion of the light energy can propagate in the fiber cladding. For this reason, the fiber modes are said to be *weakly guided*. For a given mode, for example, the fundamental mode, the proportion of light energy that propagates in the core depends on the wavelength. This gives rise to spreading of pulses through a phenomenon called *waveguide dispersion*, which we will discuss in Section 2.3.

A fiber with a large value of the V parameter is called a *multimode fiber* and supports several modes. For large V, the number of modes can be approximated by $V^2/2$. For multimode fibers, typical values are $a = 25$ μm and $\Delta = 0.005$, giving a V value of about 28 at 0.8 μm. Thus a typical multimode fiber supports a few hundred propagation modes.

The parameter V can be viewed as a normalized wave number since for a given fiber (fixed a, n_1, and n_2) it is proportional to the wave number. It is useful to know the propagation constant β of the fundamental mode supported by a fiber as a function of wavelength. This is needed to design components such as filters whose operation depends on coupling energy from one mode to another, as will become clear in Chapter 3. For example, such an expression can be used to calculate the velocity with which pulses at different wavelengths propagate in the fiber. The exact determination of β must be done numerically. But, analogous to the normalized wave number, we can define a normalized propagation constant (sometimes called a normalized effective index), b, by

$$b \overset{\text{def}}{=} \frac{\beta^2 - k^2 n_2^2}{k^2 n_1^2 - k^2 n_2^2} = \frac{n_{\text{eff}}^2 - n_2^2}{n_1^2 - n_2^2}.$$

This normalized propagation constant can be approximated with a relative error less than 0.2% by the equation

$$b(V) \approx (1.1428 - 0.9960/V)^2$$

for V in the interval $(1.5, 2.5)$; see [Neu88, p. 71] or [Jeu90, p. 25], where the result is attributed to [RN76]. This is the range of V that is of interest in the design of single-mode optical fibers.

Polarization

We defined a fiber mode as a solution of the wave equations that satisfies the boundary conditions at the core-cladding interface. Two linearly independent solutions of the wave equations exist for all λ, however large. Both these solutions correspond to the fundamental mode and have the same propagation constant. The other solutions exist only for $\lambda < \lambda_{\text{cutoff}}$.

Assume that the electric field $\tilde{\mathbf{E}}(\mathbf{r}, t)$ is written as $\tilde{\mathbf{E}}(\mathbf{r}, t) = \tilde{E}_x \hat{e}_x + \tilde{E}_y \hat{e}_y + \tilde{E}_z \hat{e}_z$, where $\hat{e}_x$, $\hat{e}_y$, and $\hat{e}_z$ are the unit vectors along the x, y, and z directions, respectively. Note that each of E_x, E_y, and E_z can depend, in general, on x, y, and z. We take the direction of propagation (fiber axis) as z and consider the two linearly independent solutions to (2.10) and (2.11) that correspond to the fundamental mode. It can be shown (see [Jeu90]) that one of these solutions has $\tilde{E}_x = 0$ but $\tilde{E}_y$, $\tilde{E}_z \neq 0$, whereas the other has $\tilde{E}_y = 0$ but $\tilde{E}_x$, $\tilde{E}_z \neq 0$. Since z is also the direction of propagation, E_z is called the *longitudinal* component. The other nonzero component, which is either E_x or E_y, is called the *transverse* component.

Before we discuss the electric field distributions of the fundamental mode further, we need to understand the concept of *polarization* of an electric field. Note that this is different from the dielectric polarization $\mathbf{P}$ discussed above. Since the electric field is a vector, for a time-varying electric field, both the magnitude and the direction can vary with time. A time-varying electric field is said to be *linearly polarized* if its direction is a constant, independent of time. If the electric field associated with an electromagnetic wave has no component along the direction of propagation of the wave, the electric field is said to be *transverse*. For the fundamental mode of a single-mode fiber, the magnitude of the longitudinal component (E_z) is much smaller than the magnitude of the transverse component (E_x or E_y). Thus the electric field associated with the fundamental mode can effectively be assumed to be a transverse field.

With this assumption, the two linearly independent solutions of the wave equations for the electric field are linearly polarized along the x and y directions. Since these two directions are perpendicular to each other, the two solutions are said to be *orthogonally polarized*. Since the wave equations are linear, any linear combination of these two linearly polarized fields is also a solution and thus a fundamental mode. The *state of polarization* (SOP) refers to the distribution of light energy among the two polarization modes. The reason the fiber is still termed *single mode* is that these two polarization modes are *degenerate;* that is, they have the same propagation constant, at least in an ideal, perfectly circularly symmetric fiber. Thus, though the energy of a pulse is divided between these two polarization modes, since they have the same propagation constant, it does not give rise to pulse spreading by the phenomenon of dispersion.

In practice, fibers are not perfectly circularly symmetric, and the two orthogonally polarized modes have slightly different propagation constants; that is, practical fibers are slightly birefringent. Since the light energy of a pulse propagating in a fiber will usually be split between these two modes, this birefringence gives rise to pulse spreading. This phenomenon is called *polarization-mode dispersion* (PMD). This is similar, in principle, to pulse spreading in the case of multimode fibers, but the effect is much weaker. We will study the effects of PMD on optical communication systems in Section 5.7.4.

PMD is illustrated in Figure 2.5. The assumption here is that the propagation constants of the two polarizations are constant throughout the length of the fiber. If the difference in propagation constants is denoted by $\Delta\beta$, then the time spread, or differential group delay (DGD), due to PMD after the pulse has propagated through a unit length of fiber is given by

$$\Delta\tau = \Delta\beta/\omega.$$

A typical value of the DGD is $\Delta\tau = 0.5$ ps/km, which suggests that after propagating through 100 km of fiber, the accumulated time spread will be 50 ps—comparable to the bit period of 100 ps for a 10 Gb/s system. This would effectively mean that 10 Gb/s transmission would not be feasible over any reasonable distances due to the effects of PMD.

However, the assumption of fixed propagation constants for each polarization mode is unrealistic for fibers of practical lengths since the fiber birefringence changes over the length of the fiber. (It also changes over time due to temperature and other environmental changes.) The net effect is that the PMD effects are not nearly as bad as indicated by this model since the time delays in different segments of the fiber vary randomly and tend to cancel each other. This results in an inverse dependence of the

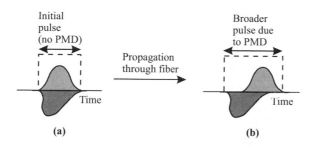

Figure 2.5   Illustration of pulse spreading due to PMD. The energy of the pulse is assumed to be split between the two orthogonally polarized modes, shown by horizontal and vertical pulses, in (a). Due to the fiber birefringence, one of these components travels slower than the other. Assuming the horizontal polarization component travels slower than the vertical one, the resulting relative positions of the horizontal and vertical pulses are shown in (b). The pulse has been broadened due to PMD since its energy is now spread over a larger time period.

DGD not on the link length, but on the square root of the link length. Typical values lie in the range 0.1–1 ps/$\sqrt{\text{km}}$. We undertake a quantitative discussion of the effects of PMD, and the system limitations imposed by it, in Section 5.7.4.

Many optical materials and components constructed using them respond differently to the different polarization components in the input light. Some components where these polarization effects are used include isolators, circulators, and acousto-optic tunable filters, which we will study in Chapter 3. The two polarization modes also see slightly different losses in many of these components. This dependence of the loss through a component on the state of polarization of the input light is termed the *polarization-dependent loss* (PDL) and is an important characteristic that has to be specified for most components.

Light Propagation in Dielectric Waveguides

A *dielectric* is a material whose conductivity is very small; silica is a dielectric material. Any dielectric region of higher refractive index placed in another dielectric of lower refractive index for the purpose of guiding (optical) waves can be called a *dielectric waveguide*. Thus an optical fiber is also a dielectric waveguide. However, the term is more often used to refer to a device where the guiding occurs in some region of a glass or dielectric slab. Examples of such devices include semiconductor amplifiers, semiconductor lasers, dielectric switches, multiplexers, and other integrated optic devices. In many applications, the guiding region has a rectangular cross section. In contrast, the guiding region of an optical fiber is its core, which has a circular cross section.

The propagation of light in waveguides can be analyzed in a fashion similar to that of propagation in optical fiber. In the ray theory approach, which is applicable when the dimensions of the guiding region are much larger than the wavelength, the guiding process is due to total internal reflection; light that is launched into the waveguide at one end is confined to the guiding region. When we use the wave theory approach, we again find that only certain distributions of the electromagnetic fields are supported or guided by the waveguide, and these are called the *modes* of the waveguide. Furthermore, the dimensions of the waveguide can be chosen so that the waveguide supports only a single mode, the *fundamental mode*, above a certain *cutoff wavelength*, just as in the case of optical fiber.

However, the modes of a rectangular waveguide are quite different from the fiber modes. For most rectangular waveguides, their width is much larger than their depth. For these waveguides, the modes can be classified into two groups: one for which the electric field is approximately transverse, called the *TE modes*, and the other for which the magnetic field is approximately transverse, called the *TM modes*. (The transverse approximation holds exactly if the waveguides have infinite width; such waveguides are called *slab waveguides*.) If the width of the waveguide is along the x direction (and much larger than the depth), the TE modes have an electric field that is approximately linearly polarized along the x direction. The same is true for the magnetic fields of TM modes.

The fundamental mode of a rectangular waveguide is a TE mode. But in some applications, for example, in the design of isolators and circulators (Section 3.2.1), the waveguide is designed to support two modes: the fundamental TE mode and the lowest-order TM mode. For most waveguides, for example, those made of silica, the propagation constants of the fundamental TE mode and lowest-order TM mode are very close to each other. The electric field vector of a light wave propagating in such a waveguide can be expressed as a linear combination of the TE and TM modes. In other words, the energy of the light wave is split between the TE and TM modes. The proportion of light energy in the two modes depends on the input excitation. This proportion also changes when gradual or abrupt discontinuities are present in the waveguide.

In some applications, for example, in the design of acousto-optic tunable filters (Section 3.3.9), it is desirable for the propagation constants of the fundamental TE mode and lowest-order TM mode to have a significant difference. This can be arranged by constructing the waveguide using a *birefringent material*, such as lithium niobate. For such a material, the refractive indices along different axes are quite different, resulting in the effective indices of the TE and TM modes being quite different.

$\underline{2.2}$ Loss and Bandwidth

Although we neglected the attenuation loss in the fiber in the derivation of propagation modes, its effect can be modeled easily as follows: the output power P_{out} at the end of a fiber of length L is related to the input power P_{in} by

$$P_{out} = P_{in}e^{-\alpha L}.$$

Here the parameter α represents the fiber attenuation. It is customary to express the loss in units of dB/km; thus a loss of α_{dB} dB/km means that the ratio P_{out}/P_{in} for $L = 1$ km satisfies

$$10\log_{10}\frac{P_{out}}{P_{in}} = -\alpha_{dB}$$

or

$$\alpha_{dB} = (10\log_{10}e)\alpha \approx 4.343\alpha.$$

The two main loss mechanisms in an optical fiber are *material absorption* and *Rayleigh scattering*. Material absorption includes absorption by silica as well as the impurities in the fiber. The material absorption of pure silica is negligible in the entire 0.8–1.6 μm band that is used for optical communication systems. The reduction of the loss due to material absorption by the impurities in silica has been very important in making optical fiber the remarkable communication medium that it is today. The loss has now been reduced to negligible levels at the wavelengths of interest for optical communication—so much so that the loss due to Rayleigh scattering is the dominant component in today's fibers in all the three wavelength bands used for optical communication: 0.8 μm, 1.3 μm, and 1.55 μm. Figure 2.6 shows the attenuation loss in silica as a function of wavelength. We see that the loss has local minima at these three wavelength bands with typical losses of 2.5, 0.4, and 0.25 dB/km. (In a typical optical communication system, a signal can undergo a loss of about 20–30 dB before it needs to be amplified or regenerated. At 0.25 dB/km, this corresponds to a distance of 80–120 km.) The attenuation peaks separating these bands are primarily due to absorption by the residual water vapor in the silica fiber.

The bandwidth can be measured either in terms of wavelength $\Delta\lambda$ or in terms of frequency Δf. These are related by the equation

$$\Delta f \approx \frac{c}{\lambda^2}\Delta\lambda.$$

This equation can be derived by differentiating the relation $f = c/\lambda$ with respect to λ. Consider the long wavelength 1.3 and 1.5 μm bands, which are the primary bands used today for optical communication. The usable bandwidth of optical fiber in

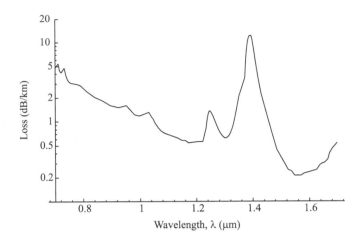

Figure 2.6   Attenuation loss in silica as a function of wavelength. (After [Agr97].)

these bands, which we can take as the bandwidth over which the loss in decibels per kilometer is within a factor of 2 of its minimum, is approximately 80 nm at 1.3 μm and 180 nm at 1.55 μm. In terms of optical frequency, these bandwidths correspond to about 35,000 GHz! This is an enormous amount of bandwidth indeed, considering that the bit rate needed for most user applications today is no more than a few tens of megabits per second.

The usable bandwidth of fiber in most of today's long-distance networks is limited by the bandwidth of the erbium-doped fiber amplifiers (see Section 3.4) that are widely deployed, rather than by the bandwidth of the silica fiber. Based on the availability of amplifiers, the low-loss band at 1.55 μm is divided into three regions, as shown in Figure 2.7. The middle band from 1530 to 1565 nm is the conventional or C-band where WDM systems have operated using conventional erbium-doped fiber amplifiers. The band from 1565 to 1625 nm, which consists of wavelengths longer than those in the C-band, is called the L-band and is today being used in high-capacity WDM systems, with the development of gain-shifted erbium-doped amplifiers (see Section 3.4) that provide amplification in this band. The band below 1530 nm, consisting of wavelengths shorter than those in the C-band, is called the S-band. Fiber Raman amplifiers (Section 3.4.4) provide amplification in this band.

Lucent introduced a new kind of single-mode optical fiber, called AllWave fiber, in 1998, which virtually eliminates the absorption peaks due to water vapor. This fiber has an even larger bandwidth and is expected to be deployed in metropolitan-area networks that do not use erbium-doped fiber amplifiers.

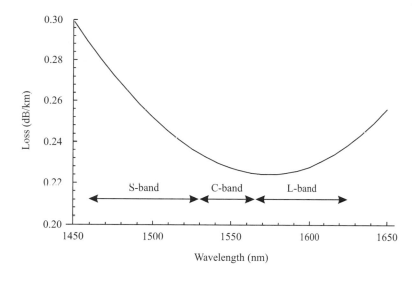

Figure 2.7   The three bands, S-band, C-band, and L-band, based on amplifier availability, within the low loss region around 1.55 μm in silica fiber. (After [Kan99].)

As we saw earlier in this section, the dominant loss mechanism in optical fiber is Rayleigh scattering. Rayleigh scattering arises because of fluctuations in the density of the medium (silica) at the microscopic level. We refer to [BW99] for a detailed description of the scattering mechanism. The loss due to Rayleigh scattering is a fundamental one and decreases with increasing wavelength. The loss coefficient α_R due to Rayleigh scattering at a wavelength λ can be written as $\alpha_R = A/\lambda^4$, where A is called the Rayleigh scattering coefficient. Note that the Rayleigh scattering loss decreases rapidly with increasing wavelength due to the λ^{-4} dependence. Glasses with substantially lower Rayleigh attenuation coefficients at 1.55 μm are not known. In order to reduce the fiber loss below the current best value of about 0.2 dB/km, one possibility is to operate at higher wavelengths, so as to reduce the loss due to Rayleigh scattering. However, at such higher wavelengths, the material absorption of silica is quite significant. It may be possible to use other materials such as fluorozirconate (ZiFr$_4$) in order to realize the low loss that is potentially possible by operating at these wavelengths [KK97, p. 69].

2.2.1   Bending Loss

Optical fibers need to be bent for various reasons both when deployed in the field and particularly within equipment. Bending leads to "leakage" of power out of the fiber core into the cladding, resulting in additional loss. A bend is characterized by the *bend radius*—the radius of curvature of the bend (radius of the circle whose arc approximates the bend). The "tighter" the bend, the smaller the bend radius and the larger the loss. The bend radius must be of the order of a few centimeters in order to keep the bending loss low. Also, the bending loss at 1550 nm is higher than at 1310 nm. The ITU-T standards specify that the additional loss at 1550 nm due to bending must be in the range 0.5–1 dB, depending on the fiber type, for 100 turns of fiber wound with a radius of 37.5 mm. Thus a bend with a radius of 4 cm results in a bending loss of < 0.01 dB. However, the loss increases rapidly as the bend radius is reduced, so that care must be taken to avoid sharp bends, especially within equipment.

2.3   Chromatic Dispersion

Dispersion is the name given to any effect wherein different components of the transmitted signal travel at different velocities in the fiber, arriving at different times at the receiver. We already discussed the phenomenon of intermodal dispersion in Section 2.1 and polarization-mode dispersion in Section 2.1.2. Our main goal in this section will be to understand the phenomenon of *chromatic dispersion* and the system limitations imposed by it. Other forms of dispersion and their effect on the design of the system are discussed in Section 5.7.

Chromatic dispersion is the term given to the phenomenon by which different *spectral* components of a pulse travel at different velocities. To understand the effect of chromatic dispersion, we must understand the significance of the propagation constant. We will restrict our discussion to single-mode fiber since in the case of multimode fiber, the effects of intermodal dispersion usually overshadow that of chromatic dispersion. So the propagation constant in our discussions will be that associated with the fundamental mode of the fiber.

Chromatic dispersion arises for two reasons. The first is that the refractive index of silica, the material used to make optical fiber, is frequency dependent. Thus different frequency components travel at different speeds in silica. This component of chromatic dispersion is termed *material dispersion*. Although this is the principal component of chromatic dispersion for most fibers, there is a second component, called *waveguide dispersion*. To understand the physical origin of waveguide dispersion, recall from Section 2.1.2 that the light energy of a mode propagates partly in the core and partly in the cladding. Also recall that the effective index of a mode lies between the refractive indices of the cladding and the core. The actual value of the effective index between these two limits depends on the proportion of power that is contained in the cladding and the core. If most of the power is contained in the core, the effective index is closer to the core refractive index; if most of it propagates in the cladding, the effective index is closer to the cladding refractive index. The power distribution of a mode between the core and cladding of the fiber is itself a function of the wavelength. More accurately, the longer the wavelength, the more power in the cladding. Thus, even in the absence of material dispersion—so that the refractive indices of the core and cladding are independent of wavelength—if the wavelength changes, this power distribution changes, causing the effective index or propagation constant of the mode to change. This is the physical explanation for waveguide dispersion.

A mathematical description of the propagation of pulses in the presence of chromatic dispersion is given in Appendix E. Here we just note that the shape of pulses propagating in optical fiber is not preserved, in general, due to the presence of chromatic dispersion. The key parameter governing the evolution of pulse shape is the second derivative $\beta_2 = d^2\beta/d\omega^2$ of the propagation constant β. β_2 is called the *group velocity dispersion parameter,* or simply the *GVD parameter.* The reason for this terminology is as follows. If $\beta_1 = d\beta/d\omega$, $1/\beta_1$ is the velocity with which a pulse propagates in optical fiber and is called the *group velocity.* The concept of group velocity is discussed in greater detail in Appendix E. Since β_2 is related to the rate of change of group velocity with frequency, chromatic dispersion is also called group velocity dispersion.

In the absence of chromatic dispersion, $\beta_2 = 0$, and in this ideal situation, all pulses would propagate without change in shape. In general, not only is $\beta_2 \neq 0$, it is also a function of the optical frequency or, equivalently, the optical wavelength. For most optical fibers, there is a so-called *zero-dispersion wavelength,* which is the wavelength at which the GVD parameter $\beta_2 = 0$. If $\beta_2 > 0$, the chromatic dispersion is said to be *normal.* When $\beta_2 < 0$, the chromatic dispersion is said to be *anomalous.*

2.3.1 Chirped Gaussian Pulses

We next discuss how a specific family of pulses changes shape as they propagate along a length of single-mode optical fiber. The pulses we consider are called *chirped Gaussian pulses.* An example is shown in Figure 2.8. The term *Gaussian* refers to the envelope of the launched pulse. *Chirped* means that the frequency of the launched pulse changes with time. Both aspects are illustrated in Figure 2.8, where the center frequency ω_0 has been greatly diminished for the purposes of illustration.

We consider chirped pulses for three reasons. First, the pulses emitted by semiconductor lasers when they are directly modulated are considerably chirped, and such transmitters are widely used in practice. As we will see in Chapter 5, this chirp has a significant effect on the design of optical communication systems. The second reason is that some nonlinear effects that we will study in Section 2.4 can cause otherwise

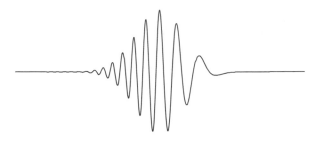

Figure 2.8   A (negatively) chirped Gaussian pulse. Here, and in all such figures, we show the shape of the pulse as a function of time.

unchirped pulses to acquire a chirp. It then becomes important to study the effect of chromatic dispersion on such pulses. The third reason is that the best transmission performance is achieved today by the use of Gaussian pulses that are deliberately chirped. (We will discuss these systems in Section 2.5.1 and in Chapter 5.)

Pulses with a Gaussian envelope are used in high-performance systems employing RZ modulation (see Section 4.1). For most other systems, the pulses used tend to be rectangular rather than Gaussian. However, the results we derive will be qualitatively valid for most pulse envelopes. In Appendix E, we describe mathematically how chirped Gaussian pulses propagate in optical fiber. The key result that we will use in subsequent discussions here is that after a pulse with initial width T_0 has propagated a distance z, its width T_z is given by

$$\frac{T_z}{T_0} = \sqrt{\left(1 + \frac{\kappa \beta_2 z}{T_0^2}\right)^2 + \left(\frac{\beta_2 z}{T_0^2}\right)^2}. \tag{2.13}$$

Here κ is called the *chirp factor* of the pulse and is proportional to the rate of change of the pulse frequency with time. (A related parameter, which depends on both the chirp and the pulse rise-time, is called the *source frequency chirp factor*, α, in the Telcordia SONET standard GR.253.)

Broadening of Chirped Gaussian Pulses

Figure 2.9 shows the pulse-broadening effect of chromatic dispersion graphically. In these figures, the center or carrier frequency of the pulse, ω_0, has deliberately been shown greatly diminished for the purposes of illustration. We assume β_2 is negative; this is true for standard single-mode fiber in the 1.55 μm band. Figure 2.9(a) shows an unchirped ($\kappa = 0$) Gaussian pulse, and Figure 2.9(b) shows the same pulse after

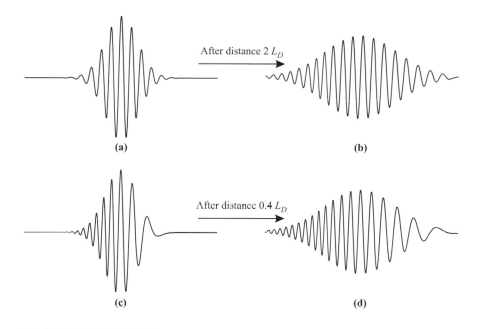

After distance $2 L_D$

(a)   (b)

After distance $0.4 L_D$

(c)   (d)

Figure 2.9   Illustration of the pulse-broadening effect of chromatic dispersion on unchirped and chirped Gaussian pulses (for $\beta_2 < 0$). (a) An unchirped Gaussian pulse at $z = 0$. (b) The pulse in (a) at $z = 2L_D$. (c) A chirped Gaussian pulse with $\kappa = -3$ at $z = 0$. (d) The pulse in (c) at $z = 0.4L_D$. For systems operating over standard single-mode fiber at 1.55 μm, $L_D \approx 1800$ km at 2.5 Gb/s, whereas $L_D \approx 115$ km at 10 Gb/s.

it has propagated a distance $2T_0^2/|\beta_2|$ along the fiber. Figure 2.9(c) shows a chirped Gaussian pulse with $\kappa = -3$, and Figure 2.9(d) shows the same pulse after it has propagated a distance of only $0.4T_0^2/|\beta_2|$ along the fiber. The amount of broadening can be seen to be about the same as that of the unchirped Gaussian pulse, but the distance traveled is only a fifth. This shows that the presence of chirp significantly exacerbates the pulse broadening due to chromatic dispersion (when the product $\kappa\beta_2$ is positive).

   The quantity $T_0^2/|\beta_2|$ is called the *dispersion length* and is denoted by L_D. It serves as a convenient normalizing measure for the distance z in discussing the effects of chromatic dispersion. For example, the effects of chromatic dispersion can be neglected if $z \ll L_D$ since in that case, from (2.13), $T_z/T_0 \approx 1$. It also has the interpretation that the width of an unchirped pulse at the $1/e$-intensity point increases by a factor of $\sqrt{2}$ after it has propagated a distance equal to the dispersion length.

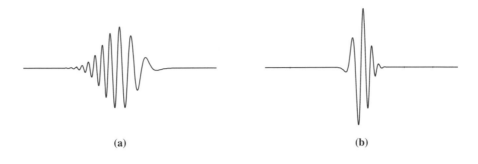

(a)   (b)

Figure 2.10   Illustration of the pulse compression effect of chromatic dispersion when $\kappa\beta_2 < 0$. (a) A chirped Gaussian pulse with $\kappa = -3$ at $z = 0$. (b) The pulse in (a) at $z = 0.4L_D$.

The dispersion length for a 2.5 Gb/s system operating over standard single-mode fiber at 1.55 μm is approximately 1800 km, assuming $T_0 = 0.2$ ns, which is half the bit interval. If the bit rate of the system is increased to 10 Gb/s with $T_0 = 0.05$ ns, again half bit interval, the dispersion length decreases to approximately 115 km. This indicates that the limitations on systems due to chromatic dispersion are much more severe at 10 Gb/s than at 2.5 Gb/s. We will discuss the system limitations of chromatic dispersion in Section 5.7.2. (The chromatic dispersion limit at 2.5 Gb/s is considerably shorter, about 600 km, than the dispersion length of 1800 km because NRZ pulses are used.)

For $\kappa = 0$ and $z = 2L_D$, (2.13) yields $T_z/T_0 = \sqrt{5} \approx 2.24$. For $\kappa = -3$ and $z = 0.4L_D$, (2.13) yields $T_z/T_0 = \sqrt{5} \approx 2.24$. Thus both pulses broaden to the same extent, and these values are in agreement with Figure 2.9.

An interesting phenomenon occurs when the product $\kappa\beta_2$ is negative. The pulse initially undergoes compression up to a certain distance and then undergoes broadening. This is illustrated in Figure 2.10. The pulse in Figure 2.10(a) is the same chirped Gaussian pulse shown in Figure 2.9(c) and has the chirp parameter $\kappa = -3$. But the sign of β_2 is now positive (which is the case, for example, in the lower portion of the 1.3 μm band), and the pulse, after it has propagated a distance $z = 0.4L_D$, is shown in Figure 2.10(b). The pulse has now undergone compression rather than broadening. This can also be seen from (2.13) since we now get $T_z/T_0 = 1/\sqrt{5} \approx 0.45$. However, as z increases further, the pulse will start to broaden quite rapidly. This can be seen from Figure 2.11, where we plot the pulse width evolution as a function of distance for different chirp parameters. (Also see Problem 2.11.) We will discuss this phenomenon further in Sections 2.4.5 and 2.4.6.

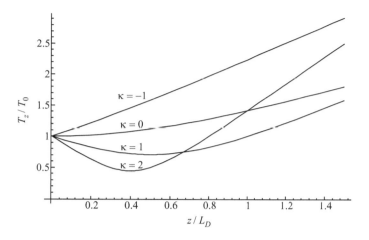

Figure 2.11 Evolution of pulse width as a function of distance (z/L_D) for chirped and unchirped pulses in the presence of chromatic dispersion. We assume $\beta_2 < 0$, which is the case for 1.55 μm systems operating over standard single-mode fiber. Note that for positive chirp the pulse width initially decreases but subsequently broadens more rapidly. For systems operating over standard single-mode fiber at 1.55 μm, $L_D \approx 1800$ km at 2.5 Gb/s, whereas $L_D \approx 115$ km at 10 Gb/s.

An intuitive explanation of pulse compression and broadening due to chromatic dispersion is as follows. For a negatively chirped pulse, the instantaneous frequency decreases with increasing time, as illustrated in Figures 2.9(c) and 2.10(a). When $\beta_2 > 0$, higher-frequency (components of) pulses travel faster than lower-frequency (components of) pulses, and vice versa. Thus, when $\beta_2 > 0$, the tail of the pulse, which has higher-frequency components, travels faster than the head of the pulse, which has lower-frequency components, resulting in pulse compression. This is the situation illustrated in Figure 2.10. When $\beta_2 < 0$, the situation is reversed: the tail of the pulse travels slower than the head of the pulse, and the pulse broadens. This is the situation illustrated in Figure 2.9(c) and (d).

The pulse compression phenomenon can be used to increase the transmission distance before chromatic dispersion becomes significant, if the sign of $\kappa\beta_2$ can be made negative. Since the output of directly modulated semiconductor lasers is negatively chirped, the fiber must have a positive β_2 for pulse compression to occur. While standard single-mode fiber cannot be used because it has negative β_2 in the 1.55 μm band, Corning's Metrocor fiber has positive β_2 in this band. This fiber has been designed specifically to take advantage of this pulse compression effect in the design of metropolitan systems.

A careful observation of Figure 2.9(b) shows that the unchirped Gaussian pulse acquires chirp when it has propagated some distance along the fiber. Furthermore, the acquired chirp is negative since the frequency of the pulse decreases with increasing time, t. The derivation of an expression for the acquired chirp is left as an exercise (Problem 2.9).

2.3.2   Controlling the Dispersion Profile

Group velocity dispersion is commonly expressed in terms of the *chromatic dispersion parameter D* that is related to β_2 as $D = -(2\pi c/\lambda^2)\beta_2$. The chromatic dispersion parameter is measured in units of ps/nm-km since it expresses the temporal spread (ps) per unit propagation distance (km), per unit pulse spectral width (nm). D can be written as $D = D_M + D_W$, where D_M is the material dispersion and D_W is the waveguide dispersion, both of which we have discussed earlier. Figure 2.12 shows D_M, D_W, and D for standard single-mode fiber. D_M increases monotonically with λ and equals 0 for $\lambda = 1.276\ \mu$m. On the other hand, D_W decreases monotonically with λ and is always negative. The total chromatic dispersion D is zero around $\lambda = 1.31\ \mu$m; thus the waveguide dispersion shifts the zero-dispersion wavelength

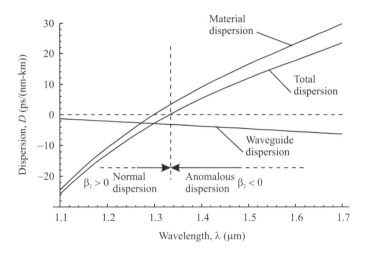

Figure 2.12   Material, waveguide, and total dispersion in standard single-mode optical fiber. Recall that chromatic dispersion is measured in units of ps/nm-km since it expresses the temporal spread (ps) per unit propagation distance (km), per unit pulse spectral width (nm). (After [Agr97].)

by a few tens of nanometers. Around the zero-dispersion wavelength, D may be approximated by a straight line whose slope is called the *chromatic dispersion slope* of the fiber.

For standard single-mode fiber, the chromatic dispersion effects are small in the 1.3 μm band, and systems operating in this wavelength range are loss limited. On the other hand, most optical communication systems operate in the 1.55 μm band today because of the low loss in this region and the well-developed erbium-doped fiber amplifier technology. But as we have already seen, optical communication systems in this band are chromatic dispersion limited. This limitation can be reduced if somehow the zero-dispersion wavelength were shifted to the 1.55 μm band.

We do not have much control over the material dispersion D_M though it can be varied slightly by doping the core and cladding regions of the fiber. However, we can vary the waveguide dispersion D_W considerably so as to shift the zero-dispersion wavelength into the 1.55 μm band. Fibers with this property are called *dispersion-shifted* fibers (DSF). Such fibers have a chromatic dispersion of at most 3.3 ps/nm-km in the 1.55 μm wavelength range and typically zero dispersion at 1550 nm. A large fraction of the installed base in Japan is DSF.

Recall that when $\beta_2 > 0$, the chromatic dispersion is said to be *normal*, and when $\beta_2 < 0$, the chromatic dispersion is said to be *anomalous*. Pulses in silica fiber experience normal chromatic dispersion below the zero-dispersion wavelength, which is around 1.3 μm for standard single-mode fiber. Pulses experience anomalous dispersion in the entire 1.55 μm band in standard single-mode fiber. For dispersion-shifted fiber, the dispersion zero lies in the 1.55 μm band. As a result, pulses in one part of the 1.55 μm band experience normal chromatic dispersion, and pulses in the other part of the band experience anomalous chromatic dispersion.

The waveguide dispersion can be varied by varying the *refractive index profile* of the fiber, that is, the variation of refractive index in the fiber core and cladding. A typical refractive index profile of a dispersion-shifted fiber is shown in Figure 2.13(b). Comparing this with the refractive index profile of a step-index fiber shown in Figure 2.13(a), we see that, in addition to a trapezoidal variation of the refractive index in the fiber core, there is step variation of the refractive index in the cladding. Such a variation leads to a single-mode fiber with a dispersion zero in the 1.55 μm band.

As we will see in Section 5.7.3, fibers with very large chromatic dispersions (but with the opposite sign) are used to compensate for the accumulated chromatic dispersion on a lengthy link. The refractive index profile of such a fiber is shown in Figure 2.13(c). The core radius of such a fiber is considerably smaller than that of standard single-mode fiber but has a higher refractive index. This leads to a large negative chromatic dispersion. This core is surrounded by a ring of lower refractive index, which is in turn surrounded by a ring of higher refractive index. Such a

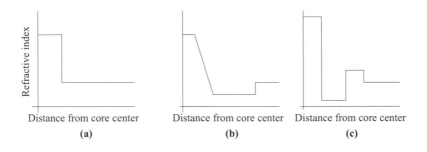

Figure 2.13   Typical refractive index profile of (a) step-index fiber, (b) dispersion-shifted fiber, and (c) dispersion-compensating fiber. (After [KK97, Chapter 4].)

variation leads to a negative chromatic dispersion slope, an important characteristic for chromatic dispersion compensation, as we will see in Section 5.7.3.

2.4   Nonlinear Effects

Our description of optical communication systems under the linearity assumption we made in Section 2.1.2 is adequate to understand the behavior of these systems when they are operated at moderate power (a few milliwatts) and at bit rates up to about 2.5 Gb/s. However, at higher bit rates such as 10 Gb/s and above and/or at higher transmitted powers, it is important to consider the effect of nonlinearities. In the case of WDM systems, nonlinear effects can become important even at moderate powers and bit rates.

There are two categories of nonlinear effects. The first arises due to the interaction of light waves with phonons (molecular vibrations) in the silica medium—one of several types of scattering effects, of which we have already met one, namely, Rayleigh scattering (Section 2.2). The two main effects in this category are *stimulated Brillouin scattering* (SBS) and *stimulated Raman scattering* (SRS).

The second set of nonlinear effects arises due to the dependence of the refractive index on the *intensity* of the applied electric field, which in turn is proportional to the square of the field amplitude. The most important nonlinear effects in this category are *self-phase modulation* (SPM) and *four-wave mixing* (FWM).

In scattering effects, energy gets transferred from one light wave to another wave at a longer wavelength (or lower energy). The lost energy is absorbed by the molecular vibrations, or phonons, in the medium. (The type of phonon involved is different for SBS and SRS.) This second wave is called the *Stokes wave*. The first wave can be thought of as being a "pump" wave that causes amplification of the

Stokes wave. As the pump propagates in the fiber, it loses power and the Stokes wave gains power. In the case of SBS, the pump wave is the signal wave, and the Stokes wave is the unwanted wave that is generated due to the scattering process. In the case of SRS, the pump wave is a high-power wave, and the Stokes wave is the signal wave that gets amplified at the expense of the pump wave.

In general, scattering effects are characterized by a gain coefficient g, measured in meters per watt, and spectral width Δf over which the gain is present. The gain coefficient is a measure of the strength of the nonlinear effect.

In the case of self-phase modulation, the transmitted pulses undergo chirping. This induced chirp factor becomes significant at high power levels. We have already seen in Section 2.3 that the pulse-broadening effects of chromatic dispersion can be enhanced in the presence of chirp. Thus the SPM-induced chirp can significantly increase the pulse spreading due to chromatic dispersion in these systems. For high-bit-rate systems, the SPM-induced chirp can significantly increase the pulse spreading due to chromatic dispersion even at moderate power levels. The precise effects of SPM are critically dependent not only on the sign of the GVD parameter β_2 but also on the length of the system.

In a WDM system with multiple channels, the induced chirp in one channel depends on the variation of the refractive index with the intensity on the other channels. This effect is called cross-phase modulation (CPM). When we discuss the induced chirp in a channel due to the variation of the refractive index with the intensity on the *same* channel, we call the effect SPM.

In the case of WDM systems, another important nonlinear effect is that of four-wave mixing. If the WDM system consists of frequencies $f_1, \ldots, f_n$, four-wave mixing gives rise to new signals at frequencies such as $2f_i - f_j$ and $f_i + f_j - f_k$. These signals appear as *crosstalk* to the existing signals in the system. These crosstalk effects are particularly severe when the channel spacing is tight. Reduced chromatic dispersion enhances the crosstalk induced by four-wave mixing. Thus systems using dispersion-shifted fibers are much more affected by four-wave mixing effects than systems using standard single-mode fiber.

We will devote the rest of this section to a detailed understanding of the various types of fiber nonlinearities.

2.4.1 Effective Length and Area

The nonlinear interaction depends on the transmission length and the cross-sectional area of the fiber. The longer the link length, the more the interaction and the worse the effect of the nonlinearity. However, as the signal propagates along the link, its power decreases because of fiber attenuation. Thus, most of the nonlinear effects occur early in the fiber span and diminish as the signal propagates.

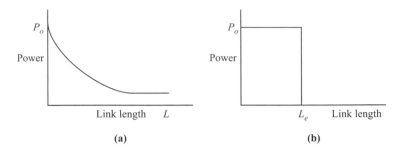

Figure 2.14 Effective transmission length calculation. (a) A typical distribution of the power along the length L of a link. The peak power is P_o. (b) A hypothetical uniform distribution of the power along a link up to the effective length L_e. This length L_e is chosen such that the area under the curve in (a) is equal to the area of the rectangle in (b).

Modeling this effect can be quite complicated, but in practice, a simple model that assumes that the power is constant over a certain *effective length* L_e has proved to be quite sufficient in understanding the effect of nonlinearities. Suppose P_o denotes the power transmitted into the fiber and $P(z) = P_o e^{-\alpha z}$ denotes the power at distance z along the link, with α being the fiber attenuation. Let L denote the actual link length. Then the effective length (see Figure 2.14) is defined as the length L_e such that

$$P_o L_e = \int_{z=0}^{L} P(z)dz.$$

This yields

$$L_e = \frac{1 - e^{-\alpha L}}{\alpha}.$$

Typically, $\alpha = 0.22$ dB/km at 1.55 μm wavelength, and for long links where $L \gg 1/\alpha$, we have $L_e \approx 20$ km.

In addition to the link length, the effect of a nonlinearity also grows with the intensity in the fiber. For a given power, the intensity is inversely proportional to the area of the core. Since the power is not uniformly distributed within the cross section of the fiber, it is convenient to use an *effective cross-sectional area* A_e (see Figure 2.15), related to the actual area A and the cross-sectional distribution of the fundamental mode $F(r, \theta)$, as

$$A_e = \frac{[\int_r \int_\theta |F(r, \theta)|^2 \, r dr d\theta]^2}{\int_r \int_\theta |F(r, \theta)|^4 \, r dr d\theta},$$

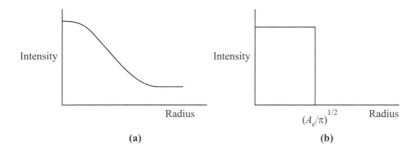

Figure 2.15   Effective cross-sectional area. (a) A typical distribution of the signal intensity along the radius of optical fiber. (b) A hypothetical intensity distribution, equivalent to that in (a) for many purposes, showing an intensity distribution that is nonzero only for an area A_e around the center of the fiber.

where r and θ denote the polar coordinates. The effective area, as defined above, has the significance that the dependence of most nonlinear effects can be expressed in terms of the effective area for the fundamental mode propagating in the given type of fiber. For example, the effective intensity of the pulse can be taken to be $I_e = P/A_e$, where P is the pulse power, in order to calculate the impact of certain nonlinear effects such as SPM, as we will see below. The effective area of SMF is around 85 μm^2 and that of DSF around 50 μm^2. The dispersion compensating fibers that we will study in Section 5.7.3 have even smaller effective areas and hence exhibit higher nonlinearities.

2.4.2   Stimulated Brillouin Scattering

In the case of SBS, the phonons involved in the scattering interaction are acoustic phonons, and the interaction occurs over a very narrow line width of $\Delta f_B = 20$ MHz at 1.55 μm. Also the Stokes and pump waves propagate in opposite directions. Thus SBS does not cause any interaction between different wavelengths, as long as the wavelength spacing is much greater than 20 MHz, which is typically the case. SBS can, however, create significant distortion within a single channel. SBS produces gain in the direction opposite to the direction of propagation of the signal, in other words, back toward the source. Thus it depletes the transmitted signal as well as generates a potentially strong signal back toward the transmitter, which must be shielded by an isolator. The SBS gain coefficient g_B is approximately 4×10^{-11} m/W, independent of the wavelength.

The intensities of the pump wave I_p and the Stokes wave I_s are related by the coupled-wave equations [Buc95]

$$\frac{dI_s}{dz} = -g_B I_p I_s + \alpha I_s, \tag{2.14}$$

and

$$\frac{dI_p}{dz} = -g_B I_p I_s - \alpha I_p. \tag{2.15}$$

The intensities are related to the powers as $P_s = A_e I_s$ and $P_p = A_e I_p$. For the case where the Stokes power is much smaller than the pump power, we can assume that the pump wave is not depleted. This amounts to neglecting the $-g_B I_p I_s$ term on the right-hand side of (2.15). With this assumption, (2.14) and (2.15) can be solved (see Problem 5.23) for a link of length L to yield

$$P_s(0) = P_s(L)e^{-\alpha L}e^{\frac{g_B P_p(0)L_e}{A_e}} \tag{2.16}$$

and

$$P_p(L) = P_p(0)e^{-\alpha L}. \tag{2.17}$$

Note that the output of the pump wave is at $z = L$, but the output of the Stokes wave is at $z = 0$ since the two waves are counterpropagating.

2.4.3 Stimulated Raman Scattering

If two or more signals at different wavelengths are injected into a fiber, SRS causes power to be transferred from the lower-wavelength channels to the higher-wavelength channels (see Figure 2.16). This coupling of energy from a lower-wavelength signal to a higher-wavelength signal is a fundamental effect that is also the basis of optical amplification and lasers. The energy of a photon at a wavelength λ is given by hc/λ, where h is Planck's constant (6.63×10^{-34} J s). Thus, a photon of lower wavelength has a higher energy. The transfer of energy from a signal of lower wavelength to a signal of higher wavelength corresponds to emission of photons of lower energy caused by photons of higher energy.

Unlike SBS, SRS is a broadband effect. Figure 2.17 shows its gain coefficient as a function of wavelength spacing. The peak gain coefficient g_R is approximately 6×10^{-14} m/W at 1.55 μm, which is much smaller than the gain coefficient for SBS. However, channels up to 15 THz (125 nm) apart will be coupled with SRS. Also SRS causes coupling in both the direction of propagation and the reverse direction.

We will study the system impact of SRS in Section 5.8.3. While SRS between channels in a WDM system is harmful to the system, we can also use SRS to provide

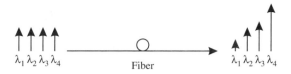

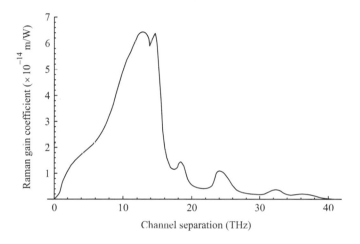

Figure 2.16   The effect of SRS. Power from lower-wavelength channels is transferred to the higher-wavelength channels.

Figure 2.17   SRS gain coefficient as a function of channel separation. (After [Agr97].)

amplification in the system, which benefits the overall system performance. We will discuss such amplifiers in Section 3.4.4.

2.4.4   Propagation in a Nonlinear Medium

In order to discuss the origin of SPM, CPM, and FWM in the following sections, we need to understand how the propagation of light waves is affected when we relax the linearity assumption we made in Section 2.1.2. This is the subject of this section. We will continue, however, to make the other assumptions of local responsivity, isotropy, homogeneity, and losslessness on the silica medium. The losslessness assumption can be removed by carrying out the remaining discussion using complex variables for the following fields and susceptibilities, as is done, for example, in [Agr95]. However, to keep the discussion simple, we use real variables for all the fields and neglect the effect of fiber loss.

For a linear medium, as we saw in Section 2.1.2, we have the relation shown in (2.8):

$$\tilde{\mathbf{P}}(\mathbf{r}, \omega) = \epsilon_0 \tilde{\chi}(\mathbf{r}, \omega) \tilde{\mathbf{E}}(\mathbf{r}, \omega)$$

between the Fourier transforms $\tilde{\mathbf{P}}$ and $\tilde{\mathbf{E}}$ of the induced dielectric polarization and applied electric field, respectively. Since we are considering nonlinearities in this section, it is no longer as convenient to work in the Fourier transform domain. By taking inverse Fourier transforms, this relation can be written in the time domain as (2.7):

$$\mathbf{P}_L(\mathbf{r}, t) = \epsilon_0 \int_{-\infty}^{t} \chi^{(1)}(t - t') \mathbf{E}(\mathbf{r}, t') \, dt', \tag{2.18}$$

where we have dropped the dependence of the susceptibility on $\mathbf{r}$ due to the homogeneity assumption, written $\mathbf{P}_L$ instead of $\mathbf{P}$ to emphasize the linearity assumption used in obtaining this relation, and used $\chi^{(1)}()$ instead of $\chi()$ for convenience in what follows.

In discussing the effect of nonlinearities, we will assume that the electric field of the fundamental mode is linearly polarized along the x direction. Recall from Section 2.1.2 that the electric field in a single-mode fiber is a linear combination of two modes, linearly polarized along the x and y directions. (Note that the term *polarization* here refers to the energy distribution of a propagation mode and is different from the dielectric polarization. The linearly polarized modes referred to here have no relation to the linear component of the dielectric polarization.) The following results can be generalized to this case, but the resulting expressions are significantly more complex. Hence we make the assumption of linearly polarized fields.

Because of the isotropy assumption, even in the presence of nonlinearities, the dielectric polarization is along the same direction as the electric field, which is the x direction, by assumption. Thus the vector functions $\mathbf{E}(\mathbf{r}, t)$ and $\mathbf{P}(\mathbf{r}, t)$ have only one component, which we will denote by the scalar functions $E(\mathbf{r}, t)$ and $P(\mathbf{r}, t)$, respectively. With this assumption, in the presence of nonlinearities, we show in Appendix F that we can write

$$P(\mathbf{r}, t) = \mathcal{P}_L(\mathbf{r}, t) + \mathcal{P}_{NL}(\mathbf{r}, t).$$

Here $\mathcal{P}_L(\mathbf{r}, t)$ is the *linear dielectric polarization* given by (2.18) with the vectors $\mathbf{P}_L(,)$ and $\mathbf{E}(,)$ replaced by the scalars $\mathcal{P}_L(,)$ and $E(,)$, respectively, due to the linear dielectric polarization assumption. The *nonlinear dielectric polarization* $\mathcal{P}_{NL}(\mathbf{r}, t)$ is given by

$$\mathcal{P}_{NL}(\mathbf{r}, t) = \epsilon_0 \chi^{(3)} E^3(\mathbf{r}, t), \tag{2.19}$$

where $\chi^{(3)}$ is called the *third-order nonlinear susceptibility* and is assumed to be a constant (independent of t). (With the assumption of linearly polarized modes, the dielectric polarization can be expanded in a power series in E with coefficients $\epsilon_0 \chi^{(i)}$, and the superscript i in $\chi^{(i)}$ refers to the power of the electric field in each term of such an expansion. Since $\chi^{(2)} = 0$ for silica, the dominant term in determining $\mathcal{P}_{NL}(\mathbf{r}, t)$ is not the E^2 term but the E^3 term.) Recall that the refractive index is related to the susceptibility by (2.9). Thus the nonlinear dielectric polarization causes the refractive index to become intensity dependent, which is the root cause of these nonlinear effects. We will use this equation (2.19) as the starting point in understanding three important nonlinear phenomena affecting the propagation of signals in optical fiber: *self-phase modulation* (SPM), *cross-phase modulation* (CPM), and *four-wave mixing* (FWM). For simplicity, we will assume that the signals used are *monochromatic plane waves*; that is, the electric field is of the form

$$E(\mathbf{r}, t) = E(z, t) = E \cos(\omega_0 t - \beta_0 z),$$

where E is a constant. The term *monochromatic* implies the electric field has a single frequency component, namely, ω_0, and the term *plane wave* indicates that the electric field is constant in the plane perpendicular to the direction of propagation, z. Hence we have also written $E(z, t)$ for $E(\mathbf{r}, t)$. In the case of wavelength division multiplexed (WDM) signals, we assume that the signal in each wavelength channel is a monochromatic plane wave. Thus if there are n wavelength channels at the angular frequencies $\omega_1, \ldots, \omega_n$, with the corresponding propagation constants $\beta_1, \ldots, \beta_n$, the electric field of the composite WDM signal is

$$E(\mathbf{r}, t) = E(z, t) = \sum_{i=1}^{n} E_i \cos(\omega_i t - \beta_i z).$$

(Since the signals on each WDM channel are not necessarily in phase, we should add an arbitrary phase ϕ_i to each of the sinusoids, but we omit this in order to keep the expressions simple.)

2.4.5 Self-Phase Modulation

SPM arises because the refractive index of the fiber has an intensity-dependent component. This nonlinear refractive index causes an induced phase shift that is proportional to the intensity of the pulse. Thus different parts of the pulse undergo different phase shifts, which gives rise to chirping of the pulses. Pulse chirping in turn enhances the pulse-broadening effects of chromatic dispersion. This chirping effect is proportional to the transmitted signal power so that SPM effects are more pronounced in systems using high transmitted powers. The SPM-induced chirp affects the pulse-broadening effects of chromatic dispersion and thus is important to consider for high-bit-rate systems that already have significant chromatic dispersion limitations. For systems operating at 10 Gb/s and above, or for lower-bit-rate systems that use high transmitted powers, SPM can significantly increase the pulse-broadening effects of chromatic dispersion.

In order to understand the effects of SPM, consider a single-channel system where the electric field is of the form

$$E(z, t) = E \cos(\omega_0 t - \beta_0 z).$$

In the presence of fiber nonlinearities, we want to find how this field evolves along the fiber. For the monochromatic plane wave we have assumed, this means finding the propagation constant β_0. Using (2.19), the nonlinear dielectric polarization is given by

$$
\begin{aligned}
\mathcal{P}_{NL}(\mathbf{r}, t) &= \epsilon_0 \chi^{(3)} E^3 \cos^3(\omega_0 t - \beta_0 z) \\
&= \epsilon_0 \chi^{(3)} E^3 \left(\frac{3}{4} \cos(\omega_0 t - \beta_0 z) + \frac{1}{4} \cos(3\omega_0 t - 3\beta_0 z) \right).
\end{aligned}
\tag{2.20}
$$

Thus the nonlinear dielectric polarization has a new frequency component at $3\omega_0$. The wave equation for the electric field (2.10) is derived assuming only the linear component of the dielectric polarization is present. In the presence of a nonlinear dielectric polarization component, it must be modified. We omit the details of how it should be modified but just remark that the solution of the modified equation will have, in general, electric fields at the new frequencies generated as a result of nonlinear dielectric polarization. Thus, in this case, the electric field will have a component at $3\omega_0$.

The fiber has a propagation constant at the angular frequency $3\omega_0$ of the generated field, which we will denote by $\beta(3\omega_0)$. From (2.20), the electric field generated as a result of nonlinear dielectric polarization at $3\omega_0$ has a propagation constant $3\beta_0$, where $\beta_0 = \beta(\omega_0)$ is the propagation constant at the angular frequency ω_0. In an ideal, dispersionless fiber, $\beta = \omega n/c$, where the refractive index n is a constant independent of ω so that $\beta(3\omega_0) = 3\beta(\omega_0)$. But in real fibers that have dispersion, n is not a constant, and $\beta(3\omega_0)$ will be very different from $3\beta(\omega_0)$. Because of this mismatch between the two propagation constants—which is usually described as a lack of *phase match*— the electric field component at $3\omega_0$ becomes negligible. This phase-matching condition will be important in our discussion of four-wave mixing in Section 2.4.8.

Neglecting the component at $3\omega_0$, the nonlinear dielectric polarization can be written as

$$P_{NL}(\mathbf{r}, t) = \left(\frac{3}{4}\epsilon_0 \chi^{(3)} E^2\right) E \cos(\omega_0 t - \beta_0 z). \tag{2.21}$$

When the wave equation (2.10) is modified to include the effect of nonlinear dielectric polarization and solved for β_0 with this expression for the nonlinear dielectric polarization, we get

$$\beta_0 = \frac{\omega_0}{c}\sqrt{1 + \tilde{\chi}^{(1)} + \frac{3}{4}\chi^{(3)} E^2}.$$

From (2.9), $n^2 = 1 + \tilde{\chi}^{(1)}$. Hence

$$\beta_0 = \frac{\omega_0 n}{c}\sqrt{1 + \frac{3}{4n^2}\chi^{(3)} E^2}.$$

Since $\chi^{(3)}$ is very small for silica fibers (as we will see), we can approximate this by

$$\beta_0 = \frac{\omega_0}{c}\left(n + \frac{3}{8n}\chi^{(3)} E^2\right). \tag{2.22}$$

Thus the electric field $E(z, t) = E \cos(\omega_0 t - \beta_0 z)$ is a sinusoid whose phase changes as $E^2 z$. This phenomenon is referred to as *self-phase modulation*. The *intensity* of the electric field corresponding to a plane wave with amplitude E is $I = \frac{1}{2}\epsilon_0 cn E^2$. Thus the phase change due to SPM is proportional to the intensity of the electric field. Note that this phase change increases as the propagation distance z increases. Since the relation between β and the refractive index n in the linear regime is $\beta = \omega n/c$, we can also interpret (2.22) as specifying an *intensity-dependent refractive index*

$$\hat{n}(E) = n + \bar{n} I \tag{2.23}$$

for the fiber, in the presence of nonlinearities. Here, $I = \frac{1}{2}\epsilon_0 cn|E|^2$ is the intensity of the field, and is measured in units of W/μm^2. The quantity $\bar{n} = \frac{2}{\epsilon_0 cn}\frac{3}{8n}\chi^{(3)}$ is called the *nonlinear index coefficient* and varies in the range 2.2–3.4×10^{-8} μm^2/W in silica fiber. We will assume the value 3.2×10^{-8} μm^2/W in the numerical examples we compute.

Pulses used in optical communication systems have finite temporal widths, and hence are not monochromatic. They are also not plane waves—that is, they have a transverse $((x, y)$-plane$)$ distribution of the electric field that is not constant but dictated by the geometry of the fiber. Nevertheless, the same qualitative effect of self-phase modulation holds for these pulses. In this section, we will give an intuitive explanation of the effect of SPM on pulses. A more quantitative explanation can be found in Sections 2.4.6 and E.2.

Because of SPM, the phase of the electric field contains a term that is proportional to the intensity of the electric field. However, because of their finite temporal extent, such pulses do not have a constant intensity for the electric field. Thus the phase shift undergone by different parts of the pulse is different. Note that the sign of the phase shift due to SPM is negative because of the minus sign in the expression for the phase, namely, $\omega_0 t - \beta_0 z$. The peak of the pulse undergoes the maximum phase shift in absolute value, and its leading and trailing edges undergo progressively smaller phase shifts. Since the frequency is the derivative of the phase, the trailing edges of the pulse undergo a negative frequency shift, and the leading edges a positive frequency shift. Since the chirp is proportional to the derivative of the frequency, this implies that the chirp factor κ is positive. Thus *SPM causes positive chirping* of pulses.

Because of the relatively small value of the nonlinear susceptibility $\chi^{(3)}$ in optical fiber, the effects of SPM become important only when high powers are used (since E^2 then becomes large). Since the SPM-induced chirp changes the chromatic dispersion effects, at the same power levels, it becomes important to consider SPM effects for shorter pulses (higher bit rates) that are already severely affected by chromatic dispersion. These two points must be kept in mind during the following discussion. We quantify the required powers and pulse durations in Section E.2.

The effect of this positive chirping depends on the sign of the GVD parameter β_2. Recall that when $\beta_2 > 0$, the chromatic dispersion is said to be normal, and when $\beta_2 < 0$, the chromatic dispersion is said to be *anomalous* (see Figure 2.12). We have seen in Section 2.3 that if the product $\kappa\beta_2 > 0$, the chirp significantly enhances the pulse-broadening effects of chromatic dispersion. Since the SPM-induced chirp is positive, *SPM causes enhanced, monotone, pulse broadening in the normal chromatic dispersion regime*. In the anomalous chromatic dispersion regime even the qualitative effect of SPM depends critically on the amount of chromatic dispersion present. When the effects of SPM and chromatic dispersion are nearly equal, but chromatic dispersion dominates, SPM can actually reduce the pulse-broadening effect of chromatic dispersion. This phenomenon can be understood from Figure 2.10, where we saw that a positively chirped pulse undergoes initial compression in the anomalous chromatic dispersion regime. The reason the pulse doesn't broaden considerably after this initial compression as described in Problem 2.11 is that the chirp factor is not constant for the entire pulse but dependent on the pulse amplitude (or intensity). This intensity dependence of the chirp factor is what leads to qualitatively different behaviors in the anomalous chromatic dispersion regime, depending on the amount of chromatic dispersion present. When the effects of chromatic dispersion and SPM are equal (we make this notion precise in Section E.2), the pulse remains stable, that is, doesn't broaden further, after undergoing some initial broadening. When the amount of chromatic dispersion is negligible, say, around the zero-dispersion wavelength, SPM leads to amplitude modulation of the pulse.

2.4.6 SPM-Induced Chirp for Gaussian Pulses

Consider an initially unchirped Gaussian pulse with envelope $U(0, \tau) = e^{-\tau^2/2}$. We have assumed a normalized envelope so that the pulse has unit peak amplitude and $1/e$-width $T_0 = 1$. For such a pulse, the parameter

$$L_{NL} - \frac{\lambda A_e}{2\pi \bar{n} P_0}$$

is called the nonlinear length. Here P_0 is the peak power of the pulse, assumed to be unity in this case. If the link length is comparable to, or greater than, the nonlinear length, the effect of the nonlinearity can be quite severe.

In the presence of SPM alone (neglecting chromatic dispersion), this pulse acquires a distance-dependent chirp. The initially unchirped pulse and the same pulse with an SPM-induced chirp after the pulse has propagated a distance $L = 5L_{NL}$ are shown in Figure 2.18. In this figure, the center frequency of the pulse is greatly diminished for the purposes of illustration.

Using (E.18), the SPM-induced phase change can be calculated to be $-(L/L_{NL})e^{-\tau^2}$. Using the definition of the instantaneous frequency and chirp factor from Section 2.3, we can calculate the instantaneous frequency of this pulse to be

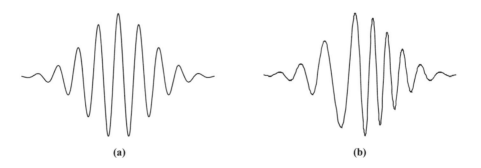

(a)   (b)

Figure 2.18 Illustration of the SPM-induced chirp. (a) An unchirped Gaussian pulse. (b) The pulse in (a) after it has propagated a distance $L = 5L_{NL}$ under the effect of SPM. (Dispersion has been neglected.)

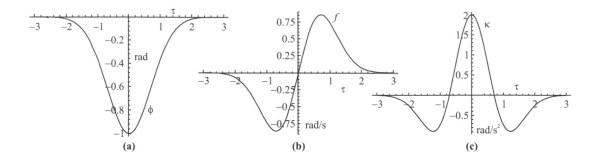

Figure 2.19 The phase (a), instantaneous frequency (b), and chirp (c) of an initially unchirped Gaussian pulse after it has propagated a distance $L = L_{NL}$.

$$\omega(\tau) = \omega_0 + \frac{2L}{L_{NL}}\tau e^{-\tau^2}$$

and the chirp factor of this pulse to be

$$\kappa_{\text{SPM}}(\tau) = \frac{2L}{L_{NL}}e^{-\tau^2}(1 - 2\tau^2). \tag{2.24}$$

Here ω_0 is the center frequency of the pulse. The SPM-induced phase change, the change, $\omega - \omega_0$, in the instantaneous frequency from the center frequency, and the chirp factor are plotted in Figure 2.19, for $L = L_{NL}$. Note that the SPM-induced chirp depends on τ. Near the center of the pulse when $\tau \approx 0$, $\kappa_{\text{SPM}} \approx 2L/L_{NL}$. The SPM-induced chirp is thus positive around the center of the pulse and is significant if L is comparable to L_{NL}. For example, if $L = L_{NL}$, the chirp factor at the pulse center is equal to 2.

The SPM-induced chirp appears to increase linearly with distance from (2.24). However, this is true only when losses are neglected. To take into account the effect of fiber loss, the expression (2.24) for the SPM-induced chirp should be modified by replacing L by the *effective length* L_e, given by

$$L_e \overset{\text{def}}{=} \frac{1 - e^{-\alpha L}}{\alpha} \tag{2.25}$$

and discussed in Section 2.4.1. Here α is the fiber loss discussed in Section 2.2. Note that $L_e < 1/\alpha$ and $L_e \to 1/\alpha$ for large L. Thus the SPM-induced chirp at the pulse center is bounded above by $2/L_{NL}\alpha$. At 1.55 μm, $\alpha \approx 0.22$ dB/km and $1/\alpha \approx 20$ km. Thus, regardless of the propagated distance L, the SPM-induced chirp is significant only if L_{NL} is comparable to 20 km. Since we calculated that the nonlinear length

$L_{NL} = 384$ km for a transmitted power of 1 mW, the SPM-induced effects can be neglected at these power levels. At a transmitted power level of 10 mW, $L_{NL} = 38$ km so that SPM effects cannot be neglected.

2.4.7   Cross-Phase Modulation

In WDM systems, the intensity-dependent nonlinear effects are enhanced since the combined signal from all the channels can be quite intense, even when individual channels are operated at moderate powers. Thus the intensity-dependent phase shift, and consequent chirping, induced by SPM alone is enhanced because of the intensities of the signals in the other channels. This effect is referred to as *cross-phase modulation* (CPM).

To understand the effects of CPM, it is sufficient to consider a WDM system with two channels. For such a system,

$$E(\mathbf{r}, t) = E_1 \cos(\omega_1 t - \beta_1 z) + E_2 \cos(\omega_2 t - \beta_2 z).$$

Using (2.19), the nonlinear dielectric polarization is given by

$$
\begin{aligned}
\mathcal{P}_{NL}(\mathbf{r}, t) &= \epsilon_0 \chi^{(3)} \left(E_1 \cos(\omega_1 t - \beta_1 z) + E_2 \cos(\omega_2 t - \beta_2 z) \right)^3 \\
&= \epsilon_0 \chi^{(3)} \left[\left(\frac{3E_1^3}{4} + \frac{3E_2^2 E_1}{2} \right) \cos(\omega_1 t - \beta_1 z) + \left(\frac{3E_2^3}{4} + \frac{3E_1^2 E_2}{2} \right) \cos(\omega_2 t - \beta_2 z) \right. \\
&\quad + \frac{3E_1^2 E_2}{4} \cos((2\omega_1 - \omega_2)t - (2\beta_1 - \beta_2)z) \\
&\quad + \frac{3E_2^2 E_1}{4} \cos((2\omega_2 - \omega_1)t - (2\beta_2 - \beta_1)z) \\
&\quad + \frac{3E_1^2 E_2}{4} \cos((2\omega_1 + \omega_2)t - (2\beta_1 + \beta_2)z) \\
&\quad + \frac{3E_2^2 E_1}{4} \cos((2\omega_2 + \omega_1)t - (2\beta_2 + \beta_1)z) \\
&\quad \left. + \frac{E_1^3}{4} \cos(3\omega_1 t - 3\beta_1 z) + \frac{E_2^3}{4} \cos(3\omega_1 t - 3\beta_1 z) \right].
\end{aligned}
\tag{2.26}
$$

The terms at $2\omega_1 + \omega_2$, $2\omega_2 + \omega_1$, $3\omega_1$, and $3\omega_2$ can be neglected since the phase-matching condition will not be satisfied for these terms owing to the presence of fiber chromatic dispersion. We will discuss the terms at $2\omega_1 - \omega_2$ and $2\omega_2 - \omega_1$ in

Section 2.4.8 when we consider four-wave mixing. The component of the nonlinear dielectric polarization at the frequency ω_1 is

$$\frac{3}{4}\epsilon_0 \chi^{(3)} \left(E_1^2 + 2E_2^2 \right) E_1 \cos(\omega_1 t - \beta_1 t). \tag{2.27}$$

When the wave equations (2.10) and (2.11) are modified to include the effect of nonlinear dielectric polarization and solved for the resulting electric field, this field has a sinusoidal component at ω_1 whose phase changes in proportion to $(E_1^2 + 2E_2^2)z$. The first term is due to SPM, whereas the effect of the second term is called *cross-phase modulation*. Note that if $E_1 = E_2$ so that the two fields have the same intensity, the effect of CPM appears to be twice as bad as that of SPM. Since the effect of CPM is qualitatively similar to that of SPM, we expect CPM to exacerbate the chirping and consequent pulse-spreading effects of SPM in WDM systems, which we discussed in Section 2.4.5.

In practice, the effect of CPM in WDM systems operating over standard single-mode fiber can be significantly reduced by increasing the wavelength spacing between the individual channels. Because of fiber chromatic dispersion, the propagation constants β_i of these channels then become sufficiently different so that the pulses corresponding to individual channels *walk away* from each other, rapidly. This happens as long as there is a small amount of chromatic dispersion (1–2 ps/nm-km) in the fiber, which is generally true except close to the zero-dispersion wavelength of the fiber. On account of this *pulse walk-off* phenomenon, the pulses, which were initially temporally coincident, cease to be so after propagating for some distance and cannot interact further. Thus the effect of CPM is reduced. For example, the effects of CPM are negligible in standard SMF operating in the 1550 nm band with 100 GHz channel spacings. In general, all nonlinear effects in optical fiber are weak and depend on long interaction lengths to build up to significant levels, so any mechanism that reduces the interaction length decreases the effect of the nonlinearity. Note, however, that in dispersion-shifted fiber, the pulses in different channels do not walk away from each other since they travel with approximately the same group velocities. Thus CPM can be a significant problem in high-speed (10 Gb/s and higher) WDM systems operating over dispersion-shifted fiber.

2.4.8   Four-Wave Mixing

In a WDM system using the angular frequencies $\omega_1, \ldots, \omega_n$, the intensity dependence of the refractive index not only induces phase shifts within a channel but also gives rise to signals at new frequencies such as $2\omega_i - \omega_j$ and $\omega_i + \omega_j - \omega_k$. This phenomenon is called *four-wave mixing*. In contrast to SPM and CPM, which are significant mainly for high-bit-rate systems, the four-wave mixing effect is independent of the bit rate but is critically dependent on the channel spacing and fiber chromatic dispersion. Decreasing the channel spacing increases the four-wave mixing effect, and so does decreasing the chromatic dispersion. Thus the effects of FWM must be considered even for moderate-bit-rate systems when the channels are closely spaced and/or dispersion-shifted fibers are used.

To understand the effects of four-wave mixing, consider a WDM signal that is the sum of n monochromatic plane waves. Thus the electric field of this signal can be written as

$$E(\mathbf{r}, t) = \sum_{i=1}^{n} E_i \cos(\omega_i t - \beta_i z).$$

Using (2.19), the nonlinear dielectric polarization is given by

$$
\begin{aligned}
\mathcal{P}_{NL}(\mathbf{r}, t) &= \epsilon_0 \chi^{(3)} \sum_{i=1}^{n} \sum_{j=1}^{n} \sum_{k=1}^{n} E_i \cos(\omega_i t - \beta_i z) E_j \cos(\omega_j t - \beta_j z) E_k \cos(\omega_k t - \beta_k z) \\
\end{aligned}
$$

$$
= \frac{3\epsilon_0 \chi^{(3)}}{4} \sum_{i=1}^{n} \left(E_i^2 + 2 \sum_{j \neq i} E_i E_j \right) E_i \cos(\omega_i t - \beta_i z) \tag{2.28}
$$

$$
+ \frac{\epsilon_0 \chi^{(3)}}{4} \sum_{i=1}^{n} E_i^3 \cos(3\omega_i t - 3\beta_i z) \tag{2.29}
$$

$$
+ \frac{3\epsilon_0 \chi^{(3)}}{4} \sum_{i=1}^{n} \sum_{j \neq i} E_i^2 E_j \cos((2\omega_i - \omega_j)t - (2\beta_i - \beta_j)z) \tag{2.30}
$$

$$
+ \frac{3\epsilon_0 \chi^{(3)}}{4} \sum_{i=1}^{n} \sum_{j \neq i} E_i^2 E_j \cos((2\omega_i + \omega_j)t - (2\beta_i + \beta_j)z) \tag{2.31}
$$

$$
+ \frac{6\epsilon_0 \chi^{(3)}}{4} \sum_{i=1}^{n} \sum_{j>i} \sum_{k>j} E_i E_j E_k
$$

$$
\Big(\cos((\omega_i + \omega_j + \omega_k)t - (\beta_i + \beta_j + \beta_k)z) \tag{2.32}
$$

$$
+ \cos((\omega_i + \omega_j - \omega_k)t - (\beta_i + \beta_j - \beta_k)z) \tag{2.33}
$$

$$
+ \cos((\omega_i - \omega_j + \omega_k)t - (\beta_i - \beta_j + \beta_k)z) \tag{2.34}
$$

$$
+ \cos((\omega_i - \omega_j - \omega_k)t - (\beta_i - \beta_j - \beta_k)z) \Big). \tag{2.35}
$$

Thus the nonlinear susceptibility of the fiber generates new fields (waves) at the frequencies $\omega_i \pm \omega_j \pm \omega_k$ (ω_i, ω_j, ω_k not necessarily distinct). This phenomenon is termed *four-wave mixing*. The reason for this term is that three *waves* with the frequencies ω_i, ω_j, and ω_k combine to generate a *fourth* wave at a frequency $\omega_i \pm \omega_j \pm \omega_k$. For equal frequency spacing, and certain choices of i, j, and k, the fourth wave contaminates ω_i. For example, for a frequency spacing $\Delta\omega$, taking ω_1, ω_2, and ω_k to be successive frequencies, that is, $\omega_2 = \omega_1 + \Delta\omega$ and $\omega_3 = \omega_1 + 2\Delta\omega$, we have $\omega_1 - \omega_2 + \omega_3 = \omega_2$, and $2\omega_2 - \omega_1 = \omega_3$.

The term (2.28) represents the effect of SPM and CPM that we have discussed in Sections 2.4.5 and 2.4.7. The terms (2.29), (2.31), and (2.32) can be neglected because of lack of phase matching. Under suitable circumstances, it is possible to approximately satisfy the phase-matching condition for the remaining terms, which are all of the form $\omega_i + \omega_j - \omega_k$, $i, j \neq k$ (ω_i, ω_j not necessarily distinct). For example, if the wavelengths in the WDM system are closely spaced, or are spaced near the dispersion zero of the fiber, then β is nearly constant over these frequencies and the phase-matching condition is nearly satisfied. When this is so, the power generated at these frequencies can be quite significant.

There is a compact way to express these four-wave mixing terms of the form $\omega_i + \omega_j - \omega_k$, $i, j \neq k$, that is frequently used in the literature. Define $\omega_{ijk} = \omega_i + \omega_j - \omega_k$ and the *degeneracy factor*

$$d_{ijk} = \begin{cases} 3, & i = j, \\ 6, & i \neq j. \end{cases}$$

Then the nonlinear dielectric polarization term at ω_{ijk} can be written as

$$\mathcal{P}_{ijk}(z, t) = \frac{\epsilon_0 \chi^{(3)}}{4} d_{ijk} E_i E_j E_k \cos((\omega_i + \omega_j - \omega_k)t - (\beta_i + \beta_j - \beta_k)z). \qquad (2.36)$$

If we assume that the optical signals propagate as plane waves over an effective cross-sectional area A_e within the fiber (see Figure 2.15) using (2.36), it can be shown that the power of the signal generated at the frequency ω_{ijk} after traversing a fiber length of L is

$$P_{ijk} = \left(\frac{\omega_{ijk} d_{ijk} \chi^{(3)}}{8 A_e n_{\text{eff}} c} \right)^2 P_i P_j P_k L^2,$$

where P_i, P_j, and P_k are the input powers at ω_i, ω_j, and ω_k. Note that the refractive index n is replaced by the effective index n_{eff} of the fundamental mode. In terms of the nonlinear refractive index $\bar{n}$, this can be written as

$$P_{ijk} = \left(\frac{\omega_{ijk} \bar{n} d_{ijk}}{3 c A_e} \right)^2 P_i P_j P_k L^2. \qquad (2.37)$$

We now consider a numerical example. We assume that each of the optical signals at ω_i, ω_j, and ω_k has a power of 1 mW and the effective cross-sectional area of the fiber is $A_e = 50~\mu m^2$. We also assume $\omega_i \neq \omega_j$ so that $d_{ijk} = 6$. Using $\bar{n} = 3.0 \times 10^{-8}~\mu m^2/W$, and taking the propagation distance $L = 20$ km, we calculate that the power P_{ijk} of the signal at the frequency ω_{ijk} generated by the four-wave mixing process is about 9.5 μW. Note that this is only about 20 dB below the signal power of 1 mW. In a WDM system, if another channel happens to be located at ω_{ljk}, the four-wave mixing process can produce significant degradation of that channel.

In practice, the signals generated by four-wave mixing have lower powers due to the lack of perfect phase matching and the attenuation of signals due to fiber loss. We will consider some numerical examples that include these effects in Chapter 5.

2.4.9   New Optical Fiber Types

Just as dispersion-shifted fibers were developed to reduce the pulse spreading due to chromatic dispersion in the 1.55 μm band, new fiber types have been developed to mitigate the effects of nonlinearities on optical communication systems. We discuss the salient characteristics of these new fibers in this section.

Nonzero-Dispersion Fiber

Although dispersion-shifted fiber overcomes the problems due to chromatic dispersion in the 1.55 μm wavelength window, unfortunately it is not suitable for use with WDM because of severe penalties due to four-wave mixing and other nonlinearities (see Section 5.8). As we shall see, these penalties are reduced if a little chromatic dispersion is present in the fiber because the different interacting waves then travel with different group velocities. This led to the development of nonzero-dispersion fibers (NZ-DSF). Such fibers have a chromatic dispersion between 1 and 6 ps/nm km, or between −1 and −6 ps/nm-km, in the 1.55 μm wavelength window. This reduces the penalties due to nonlinearities while retaining most of the advantages of DSF. This fiber is being used on many recently constructed long-haul routes in North America.

Examples include the LS fiber from Corning, which has a zero-dispersion wavelength of 1560 nm and a small chromatic dispersion of $0.092(\lambda - 1560)$ ps/nm-km in the 1550 nm wavelength window, and the TrueWave fiber from Lucent Technologies.

Since all NZ-DSFs are designed to have a small nonzero value of the dispersion in the C-band, their zero-dispersion wavelength lies outside the C-band but could lie in the L-band or in the S-band. In such cases, a large portion of the band around the zero-dispersion wavelength becomes unusable due to four-wave mixing. Alcatel's TeraLight fiber is an NZ-DSF with a zero-dispersion wavelength that lies below 1440 nm and is thus designed to be used in all three bands.

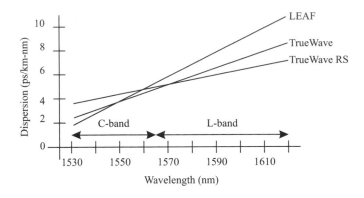

Figure 2.20   Dispersion profiles (slopes) of TrueWave fiber, TrueWave RS fiber, and LEAF.

As we shall see in Chapter 5, in addition to having a small value, it is important to have a small slope (versus wavelength) for the chromatic dispersion. Having a small slope reduces the spread in the accumulated chromatic dispersion among the different channels in a WDM system. If the spread is small, that is, the accumulated chromatic dispersion in different channels is close to being uniform, it may be possible to compensate the accumulated chromatic dispersion in all the channels with a single chromatic dispersion compensator (discussed in Chapter 5). This would be cheaper than using a chromatic dispersion compensator for each channel. The chromatic dispersion slopes of TrueWave fiber, TrueWave RS (reduced slope) fiber, and LEAF (which is discussed below) are shown in Figure 2.20. Lucent's TrueWave RS fiber has been designed to have a smaller value of the chromatic dispersion slope, about 0.05 ps/nm-km^2, compared to other NZ-DSFs, which have chromatic dispersion slopes in the range 0.07–0.11 ps/nm-km^2.

Large Effective Area Fiber

The effect of nonlinearities can be reduced by designing a fiber with a large effective area. We have seen that nonzero dispersion fibers have a small value of the chromatic dispersion in the $1.55~\mu$m band to minimize the effects of chromatic dispersion. Unfortunately such fibers also have a smaller effective area. Recently, an NZ-DSF with a large effective area—over $70~\mu$m^2—has been developed by both Corning (LEAF) and Lucent (TrueWave XL). This compares to about $50~\mu$m^2 for a typical NZ-DSF and $85~\mu$m^2 for SMF. These fibers thus achieve a better trade-off between chromatic dispersion and nonlinearities than normal NZ-DSFs. However, the disadvantage is that these fibers have a larger chromatic dispersion slope, about 0.11 ps/nm-km^2

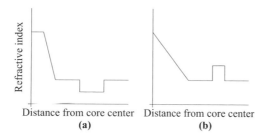

Figure 2.21   Refractive index profile of (a) normal NZ-DSF and (b) LEAF.

compared to about 0.07 ps/nm-km^2 for other NZ-DSFs, and about 0.05 ps/nm-km^2 for reduced slope fiber. Another trade-off is that a large effective area also reduces the efficiency of distributed Raman amplification (see Sections 2.4.3 and 5.8.3).

A typical refractive index profile of LEAF is shown in Figure 2.21. The core region consists of three parts. In the innermost part, the refractive index has a triangular variation. In the annular (middle) part, the refractive index is equal to that of the cladding. This is surrounded by the outermost part of the core, which is an annular region of higher refractive index. The middle part of the core, being a region of lower refractive index, does not confine the power, and thus the power gets distributed over a larger area. This reduces the peak power in the core and increases the effective area of the fiber. Figure 2.22 shows the distribution of power in the cores of DSF and LEAF.

Positive and Negative Dispersion Fibers

Fibers can be designed to have either positive chromatic dispersion or negative chromatic dispersion in the 1.55 μm band. Typical chromatic dispersion profiles of fibers having positive and negative chromatic dispersion in the 1.55 μm band are shown in Figure 2.23. Positive chromatic dispersion fiber is used for terrestrial systems, and negative chromatic dispersion fiber in submarine systems. (For chromatic dispersion compensation, the opposite is true: negative chromatic dispersion fiber is used for terrestrial systems, and positive chromatic dispersion fiber for submarine systems.) Both positive and negative chromatic dispersion cause pulse spreading, and the amount of pulse spreading depends only on the magnitude of the chromatic dispersion, and not on its sign (in the absence of chirping and nonlinearities). Then, why the need for fibers with different signs of chromatic dispersion, positive for terrestrial systems and negative for undersea links? To understand the motivation for this, we need to understand another nonlinear phenomenon: *modulation instability*.

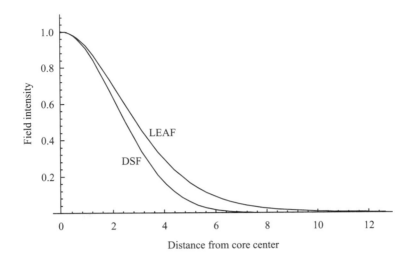

Figure 2.22   Distribution of power in the cores of DSF and LEAF. Note that the power in the case of LEAF is distributed over a larger area. (After [Liu98].)

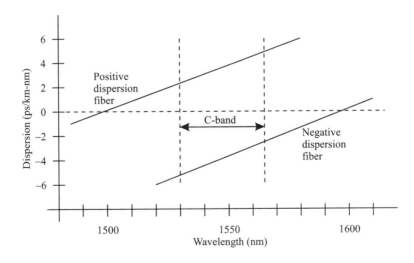

Figure 2.23   Typical chromatic dispersion profiles of fibers with positive and negative chromatic dispersion in the 1.55 μm band.

We have already seen in Section 2.3 (Figure 2.10) that pulse compression occurs for a positively chirped pulse when the chromatic dispersion is positive ($D > 0$ and $\beta_2 < 0$). We have also seen that SPM causes positive chirping of pulses (Figure 2.18). When the power levels are high, the interaction between these two phenomena—chromatic dispersion and SPM-induced chirp—leads to a breakup of a relatively broad pulse (of duration, say, 100 ps, which approximately corresponds to 10 Gb/s transmission) into a stream of short pulses (of duration a few picoseconds). This phenomenon is referred to as *modulation instability* and leads to a significantly increased bit error rate. Modulation instability occurs only in positive chromatic dispersion fiber and thus can be avoided by the use of negative chromatic dispersion fiber. Its effects in positive chromatic dispersion fiber can be minimized by using lower power levels. (In the next section, we will see that due to the same interaction between SPM and chromatic dispersion that causes modulation instability, a family of narrow, high-power pulses with specific shapes, called solitons, can propagate without pulse broadening.)

WDM systems cannot operate around the zero-dispersion wavelength of the fiber due to the severity of four-wave mixing. For positive chromatic dispersion fiber, the dispersion zero lies below the 1.55 μm band, and not in the L-band. Hence, systems using positive chromatic dispersion fiber can be upgraded to use the L-band (see Figure 2.7). This upgradability is an important feature for terrestrial systems. Thus, positive chromatic dispersion fiber is preferred for terrestrial systems, and the power levels are controlled so that modulation instability is not significant. For undersea links, however, the use of higher power levels is very important due to the very long link lengths. These links are not capable of being upgraded anyway—since they are buried on the ocean floor—so the use of the L-band in these fibers at a later date is not possible. Hence negative chromatic dispersion fiber is used for undersea links.

Since negative chromatic dispersion fiber is used for undersea links, the chromatic dispersion can be compensated using standard single-mode fiber (SMF), which has positive chromatic dispersion; that is, alternating lengths of negative chromatic dispersion fiber and (positive chromatic dispersion) SMF can be used to keep the total chromatic dispersion low. This is preferable to using dispersion compensating fibers since they are more susceptible to nonlinear effects because of their lower effective areas.

Note that all the fibers we have considered have positive chromatic dispersion slope; that is, the chromatic dispersion increases with increasing wavelength. This is mainly because the material dispersion slope of silica is positive and usually dominates the negative chromatic dispersion slope of waveguide dispersion (see Figure 2.12). Negative chromatic dispersion slope fiber is useful in chromatic dispersion slope compensation, a topic that we discuss in Section 5.7.3. While it is possible to build a negative chromatic dispersion fiber (in the 1.55 μm band) with negative

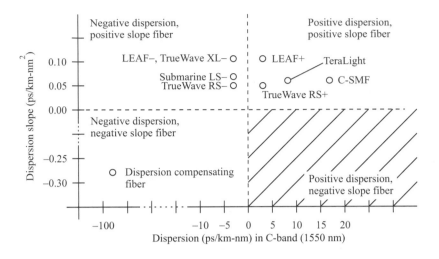

Figure 2.24   Chromatic dispersion in the C-band, and the chromatic dispersion slope, for various fiber types.

slope, it is considered difficult to design a positive chromatic dispersion fiber with negative slope.

In Figure 2.24, we summarize the chromatic dispersion in the C-band, and the chromatic dispersion slope, for all the fibers we have discussed.

2.5   Solitons

Solitons are narrow pulses with high peak powers and special shapes. The most commonly used soliton pulses are called *fundamental solitons*. The shape of these pulses is shown in Figure 2.25. As we have seen in Section 2.3, most pulses undergo broadening (spreading in time) due to group velocity dispersion when propagating through optical fiber. However, the soliton pulses take advantage of nonlinear effects in silica, specifically self-phase modulation discussed in Section 2.4.5, to overcome the pulse-broadening effects of group velocity dispersion. Thus these pulses can propagate for long distances with no change in shape.

We mentioned in Section 2.3, and discuss in greater detail in Appendix E, that a pulse propagates with the group velocity $1/\beta_1$ along the fiber and that, in general, because of the effects of group velocity dispersion, the pulse progressively broadens as it propagates. If $\beta_2 = 0$, all pulse shapes propagate without broadening, but if $\beta_2 \neq 0$, is there any pulse shape that propagates without broadening? The key to

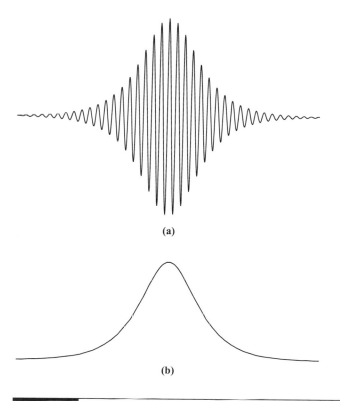

(a)

(b)

Figure 2.25   (a) A fundamental soliton pulse and (b) its envelope.

the answer lies in the one exception to this pulse-broadening effect that we already encountered in Section 2.3, namely, that if the chirp parameter of the pulse has the right sign (opposite to that of β_2), the pulse initially undergoes compression. But we have seen that even in this case (Problem 2.11), the pulse subsequently broadens. This happens in all cases where the chirp is *independent* of the pulse envelope. However, when the chirp is induced by SPM, the degree of chirp depends on the pulse envelope. If the relative effects of SPM and GVD are controlled just right, and the appropriate pulse shape is chosen, the pulse compression effect undergone by chirped pulses can exactly offset the pulse-broadening effect of dispersion. The pulse shapes for which this balance between pulse compression and broadening occurs so that the pulse either undergoes no change in shape or undergoes periodic changes in shape only are called *solitons*. The family of pulses that undergo no change in shape are called *fundamental solitons*, and those that undergo periodic changes in shape are called *higher-order solitons*. A brief quantitative discussion of soliton propagation in optical fiber appears in Section E.3.

The significance of solitons for optical communication is that they overcome the detrimental effects of chromatic dispersion completely. Optical amplifiers can be used at periodic intervals along the fiber so that the attenuation undergone by the pulses is not significant, and the higher powers and the consequent soliton properties of the pulses are maintained. Solitons and optical amplifiers, when used together, offer the promise of very high-bit-rate, repeaterless data transmission over very large distances. By the combined use of solitons and erbium-doped fiber amplifiers (Section 3.4.3), repeaterless data transmission at a bit rate of 80 Gb/s over a distance of 10,000 km has been demonstrated in the laboratory [NSK99].

The use of soliton pulses is key to the realization of the very high bit rates required in OTDM systems. These aspects of solitons will be explored in Chapter 12.

The main advantage of soliton systems is their relative immunity to fiber dispersion, which in turn allows transmission at high speeds of a few tens of gigabits per second. On the other hand, in conventional on-off–keyed systems, dispersion can be managed in a much simpler manner by alternating fibers with positive and negative dispersion. We encountered this in Section 2.4.9 and we will study this further in Chapter 5. Such systems, when using special pulses called chirped RZ pulses, can also be viewed as soliton systems, albeit of a different kind, and we discuss this in the next section.

2.5.1   Dispersion-Managed Solitons

Solitons can also be used in conjunction with WDM, but significant impairments arise when two pulses at different wavelengths overlap in time and position in the fiber. Such collisions, which occur frequently in the fiber, add timing jitter to the pulses. Although methods to overcome this timing jitter have been devised, commercial deployment of soliton-based systems has not been widespread for two main reasons. First, solitons require new disperson-shifted fiber with a small value of anomalous dispersion ($0 < D < 1$ ps/nm-km). Thus soliton-based systems cannot be used on existing fiber plants, whether based on SMF or on the popular NZ-DSF fibers. Second, solitons require amplification about every 20 km or so, which is an impracticably small spacing compared to today's WDM systems, which work with amplifier hut spacings of the order of 60–80 km. Larger values of dispersion lead to higher levels of timing jitter, higher peak pulse powers, and even closer amplifier spacings.

High-bit-rate transmission on widely deployed fiber plants, with reasonable amplifier spacings, has been achieved through a combination of (1) using pulses narrower than a bit period but much wider than solitons, and (2) dispersion compensation of the fiber plant at periodic intervals to keep the average dispersion low. The pulses used in such systems are called *chirped return-to-zero* (RZ) *pulses* and will be discussed in Section 4.1. When the characteristics of such a *dispersion-managed* system are mathematically analyzed, it can be shown that such a system is indeed "soliton-like" in the sense that a specific chirped Gaussian pulse shape will be transmitted through such a system with only periodic changes in shape, that is, with no net broadening due to dispersion, in the absence of loss. Such pulses are also called *dispersion-managed* (DM) *solitons*. We will discuss the performance of systems employing such pulses in Chapter 5. By the use of chirped RZ pulses, repeaterless data transmission in a 25-channel WDM system at a bit rate of 40 Gb/s per channel, over a distance of 1500 km, has been demonstrated in the laboratory [SKN01].

Summary

The understanding of light propagation in optical fiber is key to the appreciation of not only the significant advantages of using optical fiber as a propagation medium but also of the problems that we must tackle in designing high-bit-rate WDM systems. We started by understanding how light propagates in multimode fibers using a simple ray theory approach. This introduced the concept of pulse broadening due to multimode dispersion and motivated the use of single-mode fibers. After describing the elements of light propagation in single-mode fibers, we studied the limitations imposed on optical communication systems due to the pulse-broadening effects of chromatic dispersion.

Although dispersion is the most important phenomenon limiting the performance of systems at bit rates of 2.5 Gb/s and below, nonlinear effects become important at higher bit rates. The main nonlinear effects that impair high-speed WDM transmission are self-phase modulation and four-wave mixing. We studied the origin of these, as well as other nonlinear effects, and briefly outlined the constraints on optical communication systems imposed by them. We will return to the system limitations of both dispersion and nonlinearities when we discuss the design of optical transmission systems in Chapter 5.

We also studied the new types of fibers that have been introduced to mitigate the effects of dispersion and nonlinearities. Finally, we discussed solitons, which are special pulses designed to play off dispersion and nonlinearities against each other to achieve high-bit-rate, ultra-long-haul transmission.

Further Reading

The propagation of light in optical fiber is treated in several books at varying levels of detail. One of the earliest books on this subject is by Marcuse [Mar74]. The book by Green [Gre93] starts with the fundamentals of both geometrical optics and electromagnetics and describes the propagation of light using both the ray and wave theory approaches. The concepts of polarization and birefringence are also treated in some detail. However, the effects of dispersion and nonlinearities are described only qualitatively. The book on fiber optic communication by Agrawal [Agr97] focuses on the wave theory approach and treats the evolution of chirped Gaussian pulses in optical fiber and the pulse-broadening effects of chromatic dispersion in detail. Chromatic dispersion and intermodal dispersion are also treated at length in the books edited by Miller and Kaminow [MK88] and Lin [Lin89]. We recommend the book by Ramo, Whinnery, and van Duzer [RWv93] for an in-depth study of electromagnetic theory leading up to the description of light propagation in fiber. The books by Jeunhomme [Jeu90] and Neumann [Neu88] are devoted to the propagation of light in single-mode fibers. Jeunhomme treats fiber modes in detail and has a more mathematical treatment. We recommend Neumann's book for its physical explanations of the phenomena involved. The paper by Gloge [Glo71] on fiber modes is a classic.

In all these books, nonlinear effects are only briefly mentioned. The book by Agrawal [Agr95] is devoted to nonlinear fiber optics and contains a very detailed description of light propagation in optical fiber, including all the nonlinear effects we have discussed. Soliton propagation is also discussed. One of the earliest papers on four-wave mixing is [HJKM78]. Note that cgs units are used in this paper. The units used in the description of nonlinear effects are a source of confusion. The relationships between the various units and terminologies used in the description of nonlinear effects are described in the book by Butcher and Cotter [BC90]. This book also contains a particularly clear exposition of the fundamentals of nonlinear effects. The system impact of dispersion and nonlinearities and their interplay is discussed in detail in [KK97, Chapter 8].

Information on the new types of fibers that have been introduced to combat dispersion and nonlinearities can be found on the Web pages of the manufacturers: Corning and Lucent. Much of the data on the new fiber types for this chapter was gathered from these Web pages. The ITU has standardized three fiber types. ITU-T recommendation (standard) G.652 specifies the characteristics of standard single-mode fiber, G.653 that of DSF, and G.655 that of NZ-DSF.

A nice treatment of the basics of solitons appears in [KBW96]. Issues in the design of WDM soliton communication systems are discussed at length in [KK97, Chapter

12]. A summary of soliton field trials appears in [And00]. DM solitons are discussed in [Nak00].

Problems

Note that some of these problems require an understanding of the material in the appendices referred to in this chapter.

2.1 Derive (2.2).

2.2 A step-index multimode glass fiber has a core diameter of 50 μm and cladding refractive index of 1.45. If it is to have a limiting intermodal dispersion δT of 10 ns/km, find its acceptance angle. Also calculate the maximum bit rate for transmission over a distance of 20 km.

2.3 Derive equation (2.11) for the evolution of the magnetic field vector $\tilde{\mathbf{H}}$.

2.4 Derive an expression for the cutoff wavelength λ_{cutoff} of a step index fiber with core radius a, core refractive index n_1, and cladding refractive index n_2. Calculate the cutoff wavelength of a fiber with core radius $a = 4$ μm and $\Delta = 0.003$.

2.5 Consider a step-index fiber with a core radius of 4 μm and a cladding refractive index of 1.45.

    (a) For what range of values of the core refractive index will the fiber be single moded for all wavelengths in the 1.2–1.6 μm range?

    (b) What is the value of the core refractive index for which the V parameter is 2.0 at $\lambda = 1.55$ μm? What is the propagation constant of the single mode supported by the fiber for this value of the core refractive index?

2.6 Assume that, in the manufacture of single-mode fiber, the tolerance in the core radius a is $\pm 5\%$ and the tolerance in the normalized refractive index difference Δ is $\pm 10\%$, from their respective nominal values. If the nominal value of Δ is specified to be 0.005, what is the largest nominal value that you can specify for a while ensuring that the resulting fiber will be single moded for $\lambda > 1.2$ μm even in the presence of the worst-case (but within the specified tolerances) deviations of a and Δ from their nominal values? Assume that the refractive index of the core is 1.5.

2.7 In a reference frame moving with the pulse, the basic propagation equation that governs pulse evolution inside a dispersive fiber is

$$\frac{\partial A}{\partial z} + \frac{i}{2}\beta_2 \frac{\partial^2 A}{\partial t^2} = 0,$$

where $A(z, t)$ is the pulse envelope. If $A(0, t) = A_0 \exp(-t^2/2T_0^2)$ for some constants A_0 and T_0, solve this propagation equation to find an expression for $A(z, t)$.
Note: You may use the following result without proof:

$$\int_{-\infty}^{\infty} \exp(-(x - m)^2/2\alpha) \, dx = \sqrt{2\pi\alpha}$$

for all *complex m* and α provided $\Re(\alpha) > 0$.
Hint: Consider the Fourier transform $\tilde{A}(z, \omega)$ of $A(z, t)$.

2.8 Starting from (E.8), derive the expression (2.13) for the width T_z of a chirped Gaussian pulse with initial width T_0 after it has propagated a distance z.

2.9 Show that an unchirped Gaussian pulse launched at $z = 0$ remains Gaussian for all z but acquires a distance-dependent chirp factor

$$\kappa(z) = \frac{\mathrm{sgn}(\beta_2)z/L_D}{1 + (z/L_D)^2}.$$

2.10 Show that the rms width of a Gaussian pulse whose half-width at the $1/e$-intensity point is T_0 is given by $T_0/\sqrt{2}$.

2.11 Consider a chirped Gaussian pulse for which the product $\kappa\beta_2$ is negative that is launched at $z = 0$. Let $\kappa = 5$.
  (a) For what value of z (as a multiple of L_D) does the launched pulse attain its minimum width?
  (b) For what value of z is the width of the pulse equal to that of an unchirped pulse, for the same value of z? (Assume the chirped and unchirped pulses have the same initial pulse width.)

2.12 Show that in the case of four-wave mixing, the nonlinear polarization is given by terms (2.28) through (2.32).

2.13 You want to design a soliton communication system at 1.55 μm, at which wavelength the fiber has $\beta_2 = -2$ ps^2/km and $\gamma = 1$/W-km. The peak power of the pulses you can generate is limited to 50 mW. If you must use fundamental solitons and the bit period must be at least 10 times the full width at half-maximum (T_{FWHM}) of the soliton pulses, what is the largest bit rate you can use? (This problem requires familiarity with the material in Appendix E.)

▬ References

[Agr95] G. P. Agrawal. *Nonlinear Fiber Optics*, 2nd edition. Academic Press, San Diego, CA, 1995.

[Agr97]   G. P. Agrawal. *Fiber-Optic Communication Systems*. John Wiley, New York, 1997.

[And00]   P. A. Andrekson. High speed soliton transmission on installed fibers. In *OFC 2000 Technical Digest*, pages TuP2–1/229–231, 2000.

[BC90]   P. N. Butcher and D. Cotter. *The Elements of Nonlinear Optics*, volume 9 of *Cambridge Studies in Modern Optics*. Cambridge University Press, Cambridge, 1990.

[Buc95]   J. A. Buck. *Fundamentals of Optical Fibers*. John Wiley, New York, 1995.

[BW99]   M. Born and E. Wolf. *Principles of Optics: Electromagnetic Theory of Propagation, Diffraction and Interference of Light*. Cambridge University Press, 1999.

[Glo71]   D. Gloge. Weakly guiding fibers. *Applied Optics*, 10:2252–2258, 1971.

[Gre93]   P. E. Green. *Fiber-Optic Networks*. Prentice Hall, Englewood Cliffs, NJ, 1993.

[HJKM78]   K. O. Hill, D. C. Johnson, B. S. Kawasaki, and R. I. MacDonald. CW three-wave mixing in single-mode optical fibers. *Journal of Applied Physics*, 49(10):5098–5106, Oct. 1978.

[Jeu90]   L. B. Jeunhomme. *Single-Mode Fiber Optics*. Marcel Dekker, New York, 1990.

[Kan99]   J. Kani et al. Interwavelength-band nonlinear interactions and their suppression in multiwavelength-band WDM transmission systems. *IEEE/OSA Journal on Lightwave Technology*, 17:2249–2260, 1999.

[KBW96]   L. G. Kazovsky, S. Benedetto, and A. E. Willner. *Optical Fiber Communication Systems*. Artech House, Boston, 1996.

[KK97]   I. P. Kaminow and T. L. Koch, editors. *Optical Fiber Telecommunications IIIA*. Academic Press, San Diego, CA, 1997.

[Lin89]   C. Lin, editor. *Optoelectronic Technology and Lightwave Communications Systems*. Van Nostrand Reinhold, New York, 1989.

[Liu98]   Y. Liu et al. Advanced fiber designs for high capacity DWDM systems. In *Proceedings of National Fiber Optic Engineers Conference*, 1998.

[Mar74]   D. Marcuse. *Theory of Dielectric Optical Waveguides*. Academic Press, New York, 1974.

[MK88]   S. D. Miller and I. P. Kaminow, editors. *Optical Fiber Telecommunications II*. Academic Press, San Diego, CA, 1988.

[Nak00]   M. Nakazawa et al. Ultrahigh-speed long-distance TDM and WDM soliton transmission technologies. *IEEE Journal of Selected Topics in Quantum Electronics*, 6:363–396, 2000.

[Neu88]   E.-G. Neumann. *Single-Mode Fibers*. Springer-Verlag, Berlin, 1988.

[NSK99] M. Nakazawa, K. Suzuki, and H. Kubota. Single-channel 80 Gbit/s soliton transmission over 10000 km using in-line synchronous modulation. *Electronics Letters*, 35:1358–1359, 1999.

[RN76] H.-D. Rudolph and E.-G. Neumann. Approximations for the eigenvalues of the fundamental mode of a step-index glass fiber waveguide. *Nachrichtentechnische Zeitschrift*, 29(14):328–329, 1976.

[RWv93] S. Ramo, J. R. Whinnery, and T. van Duzer. *Fields and Waves in Communication Electronics*. John Wiley, New York, 1993.

[SKN01] K. Suzuki, H. Kubota, and M. Nakazawa. 1 Tb/s (40 Gb/s x 25 channel) DWDM quasi-DM soliton transmission over 1,500 km using dispersion-managed single-mode fiber and conventional C-band EDFAs. In *OFC 2001 Technical Digest*, pages TuN7/1–3, 2001.

3
chapter

Components

I N THIS CHAPTER, we will discuss the physical principles behind the operation of the most important components of optical communication systems. For each component, we will give a simple descriptive treatment followed by a more detailed mathematical treatment.

The components used in modern optical networks include couplers, lasers, photodetectors, optical amplifiers, optical switches, and filters and multiplexers. Couplers are simple components used to combine or split optical signals. After describing couplers, we will cover filters and multiplexers, which are used to multiplex and demultiplex signals at different wavelengths in WDM systems. We then describe various types of optical amplifiers, which are key elements used to overcome fiber and other component losses and, in many cases, can be used to amplify signals at multiple wavelengths. Understanding filters and optical amplifiers is essential to understanding the operation of lasers, which comes next. Semiconductor lasers are the main transmitters used in optical communication systems. Then we discuss photodetectors, which convert the optical signal back into the electrical domain. This is followed by optical switches, which play an important role as optical networks become more agile. Finally, we cover wavelength converters, which are used to convert signals from one wavelength to another, at the edges of the optical network, as well as inside the network.

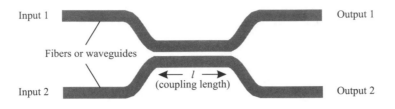

Figure 3.1   A directional coupler. The coupler is typically built by fusing two fibers together. It can also be built using waveguides in integrated optics.

3.1   Couplers

A *directional coupler* is used to combine and split signals in an optical network. A 2×2 coupler consists of two input ports and two output ports, as is shown in Figure 3.1. The most commonly used couplers are made by fusing two fibers together in the middle—these are called fused fiber couplers. Couplers can also be fabricated using waveguides in integrated optics. A 2×2 coupler, shown in Figure 3.1, takes a fraction α of the power from input 1 and places it on output 1 and the remaining fraction $1 - \alpha$ on output 2. Likewise, a fraction $1 - \alpha$ of the power from input 2 is distributed to output 1 and the remaining power to output 2. We call α the coupling ratio.

The coupler can be designed to be either wavelength selective or wavelength independent (sometimes called wavelength flat) over a usefully wide range. In a wavelength-independent device, α is independent of the wavelength; in a wavelength-selective device, α depends on the wavelength.

A coupler is a versatile device and has many applications in an optical network. The simplest application is to combine or split signals in the network. For example, a coupler can be used to distribute an input signal equally among two output ports if the coupling length, l in Figure 3.1, is adjusted such that half the power from each input appears at each output. Such a coupler is called a 3 dB coupler. An $n \times n$ star coupler is a natural generalization of the 3 dB 2×2 coupler. It is an n-input, n-output device with the property that the power from each input is divided equally among all the outputs. An $n \times n$ star coupler can be constructed by suitably interconnecting a number of 3 dB couplers, as shown in Figure 3.2. A star coupler is useful when multiple signals need to be combined and broadcast to many outputs. However, other constructions of an $n \times n$ coupler in integrated optics are also possible (see, for example, [Dra89]).

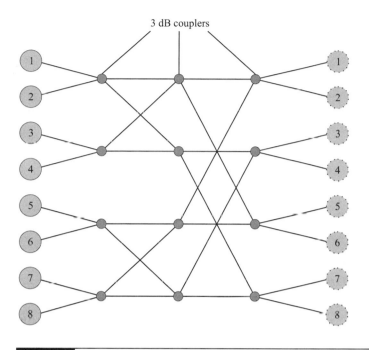

Figure 3.2   A star coupler with eight inputs and eight outputs made by combining 3 dB couplers. The power from each input is split equally among all the outputs.

Couplers are also used to tap off a small portion of the power from a light stream for monitoring purposes or other reasons. Such couplers are also called taps and are designed with values of α close to 1, typically 0.90–0.95.

Couplers are the building blocks for several other optical devices. We will explore the use of directional couplers in modulators and switches in Sections 3.5.4 and 3.7. Couplers are also the principal components used to construct *Mach-Zehnder interferometers,* which can be used as optical filters, multiplexers/demultiplexers, or as building blocks for optical modulators, switches, and wavelength converters. We will study these devices in Section 3.3.7.

So far, we have looked at wavelength-independent couplers. A coupler can be made wavelength selective, meaning that its coupling coefficient will then depend on the wavelength of the signal. Such couplers are widely used to combine signals at 1310 nm and 1550 nm into a single fiber without loss. In this case, the 1310 nm signal on input 1 is passed through to output 1, whereas the 1550 nm signal on input 2 is passed through also to output 1. The same coupler can also be used to separate the two signals coming in on a common fiber. Wavelength-dependent couplers are also used to combine 980 nm or 1480 nm pump signals along with a 1550 nm signal into an erbium-doped fiber amplifier; see Figures 3.34 and 3.37.

In addition to the coupling ratio α, we need to look at a few other parameters while selecting couplers for network applications. The *excess loss* is the loss of the device above the fundamental loss introduced by the coupling ratio α. For example, a 3 dB coupler has a nominal loss of 3 dB but may introduce additional losses of, say, 0.2 dB. The other parameter is the variation of the coupling ratio α compared to its nominal value, due to tolerances in manufacturing, as well as wavelength dependence. In addition, we also need to maintain low *polarization-dependent loss* (PDL) for most applications.

3.1.1   Principle of Operation

When two waveguides are placed in proximity to each other, as shown in Figure 3.1, light "couples" from one waveguide to the other. This is because the propagation modes of the combined waveguide are quite different from the propagation modes of a single waveguide due to the presence of the other waveguide. When the two waveguides are identical, which is the only case we consider in this book, light launched into one waveguide couples to the other waveguide completely and then back to the first waveguide in a periodic manner. A quantitative analysis of this coupling phenomenon must be made using *coupled mode theory* [Yar97] and is beyond the scope of this book. The net result of this analysis is that the electric fields, E_{o1} and E_{o2}, at the outputs of a directional coupler may be expressed in terms of the electric fields at the inputs E_{i1} and E_{i2} as follows:

$$\begin{pmatrix} E_{o1}(f) \\ E_{o2}(f) \end{pmatrix} = e^{-i\beta l} \begin{pmatrix} \cos(\kappa l) & i\sin(\kappa l) \\ i\sin(\kappa l) & \cos(\kappa l) \end{pmatrix} \begin{pmatrix} E_{i1}(f) \\ E_{i2}(f) \end{pmatrix}. \tag{3.1}$$

Here, l denotes the coupling length (see Figure 3.1), and β is the propagation constant in each of the two waveguides of the directional coupler. The quantity κ is called the *coupling coefficient* and is a function of the width of the waveguides, the refractive indices of the waveguiding region (core) and the substrate, and the proximity of the two waveguides. Equation (3.1) will prove useful in deriving the transfer functions of more complex devices built using directional couplers (see Problem 3.1).

Though the directional coupler is a two-input, two-output device, it is often used with only one active input, say, input 1. In this case, the power transfer function of the directional coupler is

$$\begin{pmatrix} T_{11}(f) \\ T_{12}(f) \end{pmatrix} = \begin{pmatrix} \cos^2(\kappa l) \\ \sin^2(\kappa l) \end{pmatrix}. \tag{3.2}$$

Here, $T_{ij}(f)$ represents the power transfer function from input i to output j and is defined by $T_{ij}(f) = |E_{oj}|^2/|E_{ii}|^2$. Equation (3.2) can be derived from (3.1) by setting $E_{i2} = 0$.

Note from (3.2) that for a 3 dB coupler the coupling length must be chosen to satisfy $\kappa l = (2k + 1)\pi/4$, where k is a nonnegative integer.

3.1.2 Conservation of Energy

The general form of (3.1) can be derived merely by assuming that the directional coupler is lossless. Assume that the input and output electric fields are related by a general equation of the form

$$\begin{pmatrix} E_{o1} \\ E_{o2} \end{pmatrix} = \begin{pmatrix} s_{11} & s_{12} \\ s_{21} & s_{22} \end{pmatrix} \begin{pmatrix} E_{i1} \\ E_{i2} \end{pmatrix}. \tag{3.3}$$

The matrix

$$S = \begin{pmatrix} s_{11} & s_{12} \\ s_{21} & s_{22} \end{pmatrix}$$

is the transfer function of the device relating the input and output electric fields and is called the *scattering matrix*. We use complex representations for the input and output electric fields, and thus the s_{ij} are also complex. It is understood that we must consider the real part of these complex fields in applications. This complex representation for the s_{ij} allows us to conveniently represent any induced phase shifts.

For convenience, we denote $\mathbf{E}_o = (E_{o1}, E_{o2})^T$ and $\mathbf{E}_i = (E_{i1}, E_{i2})^T$, where the superscript T denotes the transpose of the vector/matrix. In this notation, (3.3) can be written compactly as $\mathbf{E}_o = S\mathbf{E}_i$.

The sum of the powers of the input fields is proportional to $\mathbf{E}_i^T \mathbf{E}_i^* = |E_{i1}|^2 + |E_{i2}|^2$. Here, * represents the complex conjugate. Similarly, the sum of the powers of the output fields is proportional to $\mathbf{E}_o^T \mathbf{E}_o^* = |E_{o1}|^2 + |E_{o2}|^2$. If the directional coupler is lossless, the power in the output fields must equal the power in the input fields so that

$$\begin{aligned} \mathbf{E}_o^T \mathbf{E}_o &= (S\mathbf{E}_i)^T (S\mathbf{E}_i)^* \\ &= \mathbf{E}_i^T (S^T S^*) \mathbf{E}_i^* \\ &= \mathbf{E}_i^T \mathbf{E}_i^*. \end{aligned}$$

Since this relationship must hold for arbitrary $\mathbf{E}_i$, we must have

$$S^T S^* = I, \tag{3.4}$$

where **I** is the identity matrix. Note that this relation follows merely from conservation of energy and can be readily generalized to a device with an arbitrary number of inputs and outputs.

For a 2×2 directional coupler, by the symmetry of the device, we can set $s_{21} = s_{12} = a$ and $s_{22} = s_{11} = b$. Applying (3.4) to this simplified scattering matrix, we get

$$|a|^2 + |b|^2 = 1 \qquad (3.5)$$

and

$$ab^* + ba^* = 0. \qquad (3.6)$$

From (3.5), we can write

$$|a| = \cos(x) \text{ and } |b| = \sin(x). \qquad (3.7)$$

If we write $a = \cos(x)e^{i\phi_a}$ and $b = \sin(x)e^{i\phi_b}$, (3.6) yields

$$\cos(\phi_a - \phi_b) = 0. \qquad (3.8)$$

Thus ϕ_a and ϕ_b must differ by an odd multiple of $\pi/2$. The general form of (3.1) now follows from (3.7) and (3.8).

The conservation of energy has some important consequences for the kinds of optical components that we can build. First, note that for a 3 dB coupler, though the electric fields at the two outputs have the same magnitude, they have a relative phase shift of $\pi/2$. This relative phase shift, which follows from the conservation of energy as we just saw, plays a crucial role in the design of devices such as the Mach-Zehnder interferometer that we will study in Section 3.3.7.

Another consequence of the conservation of energy is that *lossless combining* is not possible. Thus we cannot design a device with three ports where the power input at two of the ports is completely delivered to the third port. This result is demonstrated in Problem 3.2.

3.2   Isolators and Circulators

Couplers and most other passive optical devices are *reciprocal* devices, in that the devices work exactly the same way if their inputs and outputs are reversed. However, in many systems there is a need for a passive *nonreciprocal* device. An *isolator* is an example of such a device. Its main function is to allow transmission in one direction through it but block all transmission in the other direction. Isolators are used in systems at the output of optical amplifiers and lasers primarily to prevent reflections from entering these devices, which would otherwise degrade their performance. The

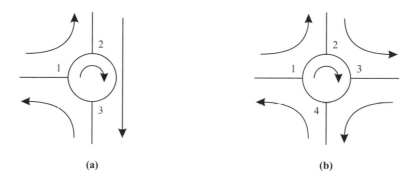

Figure 3.3   Functional representation of circulators: (a) three-port and (b) four-port. The arrows represent the direction of signal flow.

two key parameters of an isolator are its *insertion loss*, which is the loss in the forward direction, and which should be as small as possible, and its *isolation*, which is the loss in the reverse direction, and which should be as large as possible. The typical insertion loss is around 1 dB, and the isolation is around 40–50 dB.

A *circulator* is similar to an isolator, except that it has multiple ports, typically three or four, as shown in Figure 3.3. In a three-port circulator, an input signal on port 1 is sent out on port 2, an input signal on port 2 is sent out on port 3, and an input signal on port 3 is sent out on port 1. Circulators are useful to construct optical add/drop elements, as we will see in Section 3.3.4. Circulators operate on the same principles as isolators; therefore we only describe the details of how isolators work next.

3.2.1   Principle of Operation

In order to understand the operation of an isolator, we need to understand the notion of *polarization*. Recall from Section 2.1.2 that the *state of polarization* (SOP) of light propagating in a single-mode fiber refers to the orientation of its electric field vector on a plane that is orthogonal to its direction of propagation. At any time, the electric field vector can be expressed as a linear combination of the two orthogonal linear polarizations supported by the fiber. We will call these two polarization modes the horizontal and vertical modes.

The principle of operation of an isolator is shown in Figure 3.4. Assume that the input light signal has the vertical SOP shown in the figure. It is passed through a *polarizer*, which passes only light energy in the vertical SOP and blocks light energy in the horizontal SOP. Such polarizers can be realized using crystals, called *dichroics,*

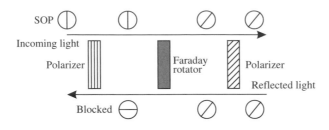

Figure 3.4 Principle of operation of an isolator that works only for a particular state of polarization of the input signal.

which have the property of selectively absorbing light with one SOP. The polarizer is followed by a *Faraday rotator*. A Faraday rotator is a nonreciprocal device, made of a crystal that rotates the SOP, say, clockwise, by 45°, regardless of the direction of propagation. The Faraday rotator is followed by another polarizer that passes only SOPs with this 45° orientation. Thus the light signal from left to right is passed through the device without any loss. On the other hand, light entering the device from the right due to a reflection, with the same 45° SOP orientation, is rotated another 45° by the Faraday rotator, and thus blocked by the first polarizer.

Note that the preceding explanation above assumes a particular SOP for the input light signal. In practice we cannot control the SOP of the input, and so the isolator must work regardless of the input SOP. This requires a more complicated design, and many different designs exist. One such design for a miniature polarization-independent isolator is shown in Figure 3.5. The input signal with an arbitrary SOP is first sent through a *spatial walk-off polarizer* (SWP). The SWP splits the signal into its two orthogonally polarized components. Such an SWP can be realized using *birefringent* crystals whose refractive index is different for the two components. When light with an arbitrary SOP is incident on such a crystal, the two orthogonally polarized components are refracted at different angles. Each component goes through a Faraday rotator, which rotates the SOPs by 45°. The Faraday rotator is followed by a *half-wave plate*. The half-wave plate (a reciprocal device) rotates the SOPs by 45° in the clockwise direction for signals propagating from left to right, and by 45° in the counterclockwise direction for signals propagating from right to left. Therefore, the combination of the Faraday rotator and the half-wave plate converts the horizontal polarization into a vertical polarization and vice versa, and the two signals are combined by another SWP at the output. For reflected signals in the reverse direction, the half-wave plate and Faraday rotator cancel each other's effects, and the SOPs remain unchanged as they pass through these two devices and are thus not recombined by the SWP at the input.

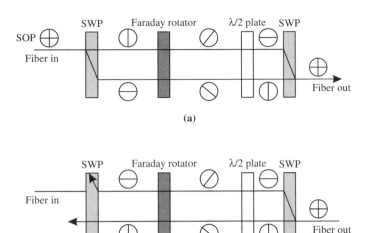

Figure 3.5 A polarization-independent isolator. The isolator is constructed along the same lines as a polarization dependent isolator but uses spatial walk-off polarizers at the inputs and outputs. (a) Propagation from left to right. (b) Propagation from right to left.

$\underline{3.3}$ Multiplexers and Filters

In this section, we will study the principles underlying the operation of a variety of wavelength selection technologies. Optical filters are essential components in transmission systems for at least two applications: to multiplex and demultiplex wavelengths in a WDM system—these devices are called multiplexers/demultiplexers—and to provide equalization of the gain and filtering of noise in optical amplifiers. Further, understanding optical filtering is essential to understanding the operation of lasers later in this chapter.

The different applications of optical filters are shown in Figure 3.6. A simple filter is a two-port device that selects one wavelength and rejects all others. It may have an additional third port on which the rejected wavelengths can be obtained. A multiplexer combines signals at different wavelengths on its input ports onto a common output port, and a demultiplexer performs the opposite function. Multiplexers and demultiplexers are used in WDM terminals as well as in larger *wavelength crossconnects* and *wavelength add/drop multiplexers*.

Demultiplexers and multiplexers can be cascaded to realize *static* wavelength crossconnects (WXCs). In a static WXC, the crossconnect pattern is fixed at the time

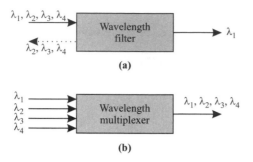

Figure 3.6   Different applications for optical filters in optical networks. (a) A simple filter, which selects one wavelength and either blocks the remaining wavelengths or makes them available on a third port. (b) A multiplexer, which combines multiple wavelengths into a single fiber. In the reverse direction, the same device acts as a demultiplexer to separate the different wavelengths.

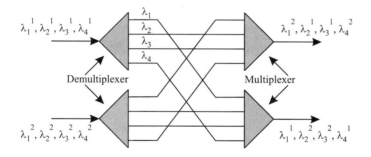

Figure 3.7   A static wavelength crossconnect. The device routes signals from an input port to an output port based on the wavelength.

the device is made and cannot be changed dynamically. Figure 3.7 shows an example of a static WXC. The device routes signals from an input port to an output port based on the wavelength. *Dynamic* WXCs can be constructed by combining using optical switches with multiplexers and demultiplexers. Static WXCs are highly limited in terms of their functionality. For this reason, the devices of interest are dynamic rather than static WXCs. We will study different dynamic WXC architectures in Chapter 7.

A variety of optical filtering technologies are available. Their key characteristics for use in systems are the following:

1. Good optical filters should have low *insertion losses*. The insertion loss is the input-to-output loss of the filter.

2. The loss should be independent of the state of polarization of the input signals. The state of polarization varies randomly with time in most systems, and if the filter has a polarization-dependent loss, the output power will vary with time as well—an undesirable feature.

3. The passband of a filter should be insensitive to variations in ambient temperature. The *temperature coefficient* is measured by the amount of wavelength shift per unit degree change in temperature. The system requirement is that over the entire operating temperature range (about 100°C typically), the wavelength shift should be much less than the wavelength spacing between adjacent channels in a WDM system.

4. As more and more filters are cascaded in a WDM system, the passband becomes progressively narrower. To ensure reasonably broad passbands at the end of the cascade, the individual filters should have very flat passbands, so as to accommodate small changes in operating wavelengths of the lasers over time. This is measured by the 1 dB bandwidth, as shown in Figure 3.8.

5. At the same time, the passband skirts should be sharp to reduce the amount of energy passed through from adjacent channels. This energy is seen as *crosstalk* and degrades the system performance. The crosstalk suppression, or *isolation* of the filter, which is defined as the relative power passed through from the adjacent channels, is an important parameter as well.

In addition to all the performance parameters described, perhaps the most important consideration is cost. Technologies that require careful hand assembly tend to be more expensive. There are two ways of reducing the cost of optical filters. The first is to fabricate them using integrated-optic waveguide technology. This is analogous to semiconductor chips, although the state of integration achieved with optics is significantly less. These waveguides can be made on many substrates, including silica, silicon, InGaAs, and polymers. Waveguide devices tend to be inherently polarization dependent due to the geometry of the waveguides, and care must be taken to reduce the PDL in these devices. The second method is to realize all-fiber devices. Such devices are amenable to mass production and are inherently polarization independent. It is also easy to couple light in and out of these devices from/into other fibers. Both of these approaches are being pursued today.

All the filters and multiplexers we study use the property of *interference* among optical waves. In addition, some filters, for example, gratings, use the *diffraction* property—light from a source tends to spread in all directions depending on the

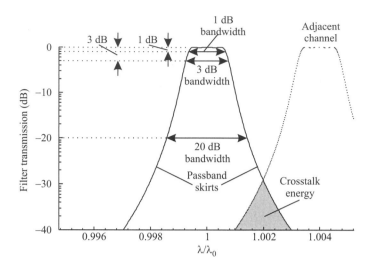

Figure 3.8 Characterization of some important spectral-shape parameters of optical filters. λ_0 is the center wavelength of the filter, and λ denotes the wavelength of the light signal.

incident wavelength. Table 3.1 compares the performance of different filtering technologies.

3.3.1 Gratings

The term *grating* is used to describe almost any device whose operation involves interference among multiple optical signals originating from the same source but with different relative *phase shifts*. An exception is a device where the multiple optical signals are generated by repeated traversals of a single cavity; such devices are called *etalons*. An electromagnetic wave (light) of angular frequency ω propagating, say, in the z direction has a dependence on z and t of the form $\cos(\omega t - \beta z)$. Here, β is the propagation constant and depends on the medium. The *phase* of the wave is $\omega t - \beta z$. Thus a relative phase shift between two waves from the same source can be achieved if they traverse two paths of different lengths.

Two examples of gratings are shown in Figure 3.9(a) and (b). Gratings have been widely used for centuries in optics to separate light into its constituent wavelengths. In WDM communication systems, gratings are used as demultiplexers to separate the individual wavelengths or as multiplexers to combine them. The Stimax grating of Table 3.1 is a grating of the type we describe in this section.

Table 3.1   Comparison of passive wavelength multiplexing/demultiplexing technologies. A 16-channel system with 100 GHz channel spacing is assumed. Other key considerations include center wavelength accuracy and manufacturability. All these approaches face problems in scaling with the number of wavelengths. TFMF is the dielectric thin-film multicavity filter, and AWG is the arrayed waveguide grating. For the fiber Bragg grating and the arrayed waveguide grating, the temperature coefficient can be reduced to 0.001 nm/°C by passive temperature compensation. The fiber Bragg grating is a single channel filter, and multiple filters need to be cascaded in series to demultiplex all 16 channels.

Filter Property	Fiber Bragg Grating	TFMF	AWG	Stimax Grating
1 dB BW (nm)	0.3	0.4	0.22	0.1
Isolation (dB)	25	25	25	30
Loss (dB)	0.2	7	5.5	6
PDL (dB)	0	0.2	0.5	0.1
Temp. coeff. (nm/°C)	0.01	0.0005	0.01	0.01

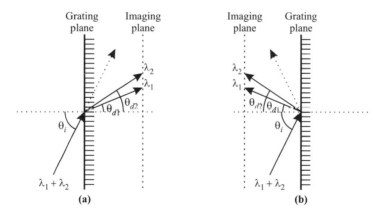

Figure 3.9   (a) A transmission grating and (b) a reflection grating. θ_i is the angle of incidence of the light signal. The angle at which the signal is diffracted depends on the wavelength ($\theta_d 1$ for wavelength λ_1 and $\theta_d 2$ for λ_2).

Consider the grating shown in Figure 3.9(a). Multiple narrow slits are spaced equally apart on a plane, called the *grating plane*. The spacing between two adjacent slits is called the *pitch* of the grating. Light incident from a source on one side of the grating is transmitted through these slits. Since each slit is narrow, by the phenomenon known as *diffraction*, the light transmitted through each slit spreads out in all directions. Thus each slit acts as a secondary source of light. Consider some other plane parallel to the grating plane at which the transmitted light from all the slits interferes. We will call this plane the *imaging plane*. Consider any point on this imaging plane. For wavelengths for which the individual interfering waves at this point are in phase, we have constructive interference and an enhancement of the light intensity at these wavelengths. For a large number of slits, which is the case usually encountered in practice, the interference is not constructive at other wavelengths, and there is little light intensity at this point from these wavelengths. Since different wavelengths interfere constructively at different points on the imaging plane, the grating effectively separates a WDM signal spatially into its constituent wavelengths. In a fiber optic system, optical fibers could be placed at different imaging points to collect light at the different wavelengths.

Note that if there were no diffraction, we would simply have light transmitted or reflected along the directed dotted lines in Figure 3.9(a) and (b). Thus the phenomenon of diffraction is key to the operation of these devices, and for this reason they are called *diffraction gratings*. Since multiple transmissions occur in the grating of Figure 3.9(a), this grating is called a *transmission grating*. If the transmission slits are replaced by narrow reflecting surfaces, with the rest of the grating surface being nonreflecting, we get the *reflection grating* of Figure 3.9(b). The principle of operation of this device is exactly analogous to that of the transmission grating. A majority of the gratings used in practice are reflection gratings since they are somewhat easier to fabricate. In addition to the plane geometry we have considered, gratings are fabricated in a concave geometry. In this case, the slits (for a transmission grating) are located on the arc of a circle. In many applications, a concave geometry leads to fewer auxiliary parts like lenses and mirrors needed to construct the overall device, say, a WDM demultiplexer, and is thus preferred.

The Stimax grating [LL84] is a reflection grating that is integrated with a concave mirror and the input and output fibers. Its characteristics are described in Table 3.1, and it has been used in commercially available WDM transmission systems. However, it is a bulk device that cannot be easily fabricated and is therefore relatively expensive. Attempts have been made to realize similar gratings in optical waveguide technology, but these devices are yet to achieve loss, PDL, and isolation comparable to the bulk version.

Principle of Operation

To understand quantitatively the principle of operation of a (transmission) grating, consider the light transmitted through adjacent slits as shown in Figure 3.10. The distance between adjacent slits—the *pitch* of the grating—is denoted by a. We assume that the light source is far enough away from the grating plane compared to a so that the light can be assumed to be incident at the same angle θ_i to the plane of the grating at each slit. We consider the light rays diffracted at an angle θ_d from the grating plane. The imaging plane, like the source, is assumed to be far away from the grating plane compared to the grating pitch. We also assume that the slits are small compared to the wavelength so that the phase change across a slit is negligible. Under these assumptions, it can be shown (Problem 3.4) that the path length difference between the rays traversing through adjacent slits is the difference in lengths between the line segments $\overline{AB}$ and $\overline{CD}$ and is given approximately by $a[\sin(\theta_i) - \sin(\theta_d)]$. Thus constructive interference at a wavelength λ occurs at the imaging plane among the rays diffracted at angle θ_d if the following *grating equation* is satisfied:

$$a[\sin(\theta_i) - \sin(\theta_d)] = m\lambda \tag{3.9}$$

for some integer m, called the *order* of the grating. The grating effects the separation of the individual wavelengths in a WDM signal since the grating equation is satisfied at different points in the imaging plane for different wavelengths. This is illustrated in Figure 3.9, where different wavelengths are shown being diffracted at the angles at which the grating equation is satisfied for that wavelength. For example, θ_{d1} is the angle at which the grating equation is satisfied for λ_1.

Note that the energy at a single wavelength is distributed over all the discrete angles that satisfy the grating equation (3.9) at this wavelength. When the grating is used as a demultiplexer in a WDM system, light is collected from only one of these angles, and the remaining energy in the other orders is lost. In fact, most of the energy will be concentrated in the zeroth-order ($m = 0$) interference maximum, which occurs at $\theta_i = \theta_d$ for all wavelengths. The light energy in this zeroth-order interference maximum is wasted since the wavelengths are not separated. Thus gratings must be designed so that the light energy is maximum at one of the other interference maxima. This is done using a technique called *blazing* [KF86, p. 386].

Figure 3.11 shows a blazed reflection grating with blaze angle α. In such a grating, the reflecting slits are inclined at an angle α to the grating plane. This has the effect of maximizing the light energy in the interference maximum whose order corresponds to the blazing angle. The grating equation for such a blazed grating can be derived as before; see Problem 3.5.

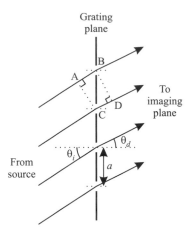

Figure 3.10   Principle of operation of a transmission grating. The reflection grating works in an analogous manner. The path length difference between rays diffracted at angle θ_d from adjacent slits is $\overline{AB} - \overline{CD} = a[\sin(\theta_i) - \sin(\theta_d)]$.

Figure 3.11   A blazed grating with blaze angle α. The energy in the interference maximum corresponding to the blaze angle is maximized.

3.3.2   Diffraction Pattern

So far, we have only considered the position of the diffraction maxima in the diffraction pattern. Often, we are also interested in the distribution of the intensity in the diffraction maxima. We can derive the distribution of the intensity by relaxing the assumption that the slits are much smaller than a wavelength, so that the phase change across a slit can no longer be neglected. Consider a slit of length w stretching from $y = -w/2$ to $y = w/2$. By reasoning along the same lines as we did in Figure 3.10, the light diffracted from position y at angle θ from this slit has a relative phase shift of $\phi(y) = (2\pi y \sin\theta)/\lambda$ compared to the light diffracted from $y = 0$. Thus, at the imaging plane, the amplitude $A(\theta)$ at angle θ is given by

$$
\begin{aligned}
\frac{A(\theta)}{A(0)} &= \frac{1}{w} \int_{-w/2}^{w/2} \exp\left(i\phi(y)\right) dy \\
&= \frac{1}{w} \int_{-w/2}^{w/2} \exp\left(i2\pi(\sin\theta)y/\lambda\right) dy \\
&= \frac{\sin\left(\pi w \sin\theta/\lambda\right)}{\pi w \sin\theta/\lambda}.
\end{aligned} \tag{3.10}
$$

Observe that the amplitude distribution at the imaging plane is the Fourier transform of the rectangular slit. This result holds for a general diffracting aperture, and not just a rectangular slit. For this more general case, if the diffracting aperture or slit is described by $f(y)$, the amplitude distribution of the diffraction pattern is given by

$$
A(\theta) = A(0) \int_{-\infty}^{\infty} f(y) \exp(2\pi i(\sin\theta)y/\lambda)\, dy. \tag{3.11}
$$

The intensity distribution is given by $|A(\theta)|^2$. Here, we assume $f(y)$ is normalized so that $\int_{-\infty}^{\infty} f(y)\, dy = 1$. For a rectangular slit, $f(y) = 1/w$ for $|y| < w/2$ and $f(y) = 0$, otherwise, and the diffraction pattern is given by (3.10). For a pair of narrow slits spaced distance d apart,

$$
f(y) = 0.5(\delta(y - d/2) + \delta(y + d/2))
$$

and

$$
A(\theta) = A(0) \cos\left(\pi(\sin\theta)\lambda/d\right).
$$

The more general problem of N narrow slits is discussed in Problem 3.6.

3.3.3   Bragg Gratings

Bragg gratings are widely used in fiber optic communication systems. In general, any periodic perturbation in the propagating medium serves as a Bragg grating. This perturbation is usually a periodic variation of the refractive index of the medium. We will see in Section 3.5.1 that lasers use Bragg gratings to achieve single frequency operation. In this case, the Bragg gratings are "written" in waveguides. Bragg gratings written in fiber can be used to make a variety of devices such as filters, add/drop multiplexers, and dispersion compensators. We will see later that the Bragg grating principle also underlies the operation of the acousto-optic tunable filter. In this case, the Bragg grating is formed by the propagation of an acoustic wave in the medium.

Principle of Operation

Consider two waves propagating in opposite directions with propagation constants β_0 and β_1. Energy is coupled from one wave to the other if they satisfy the Bragg *phase-matching* condition

$$|\beta_0 - \beta_1| = \frac{2\pi}{\Lambda},$$

where Λ is the period of the grating. In a Bragg grating, energy from the forward propagating mode of a wave at the right wavelength is coupled into a backward propagating mode. Consider a light wave with propagation constant β_1 propagating from left to right. The energy from this wave is coupled onto a scattered wave traveling in the opposite direction at the same wavelength provided

$$|\beta_0 - (-\beta_0)| = 2\beta_0 = \frac{2\pi}{\Lambda}.$$

Letting $\beta_0 = 2\pi n_{\text{eff}}/\lambda_0$, λ_0 being the wavelength of the incident wave and n_{eff} the effective refractive index of the waveguide or fiber, the wave is reflected provided

$$\lambda_0 = 2n_{\text{eff}}\Lambda.$$

This wavelength λ_0 is called the Bragg wavelength. In practice, the reflection efficiency decreases as the wavelength of the incident wave is detuned from the Bragg wavelength; this is plotted in Figure 3.12(a). Thus if several wavelengths are transmitted into a fiber Bragg grating, the Bragg wavelength is reflected while the other wavelengths are transmitted.

The operation of the Bragg grating can be understood by reference to Figure 3.13, which shows a periodic variation in refractive index. The incident wave is reflected from each period of the grating. These reflections add in phase when the path length in wavelength λ_0 each period is equal to half the incident wavelength λ_0. This is equivalent to $n_{\text{eff}}\Lambda = \lambda_0/2$, which is the Bragg condition.

The reflection spectrum shown in Figure 3.12(a) is for a grating with a uniform refractive index pattern change across its length. In order to eliminate the undesirable side lobes, it is possible to obtain an *apodized* grating, where the refractive index change is made smaller toward the edges of the grating. (The term *apodized* means "to cut off the feet.") The reflection spectrum of an apodized grating is shown in Figure 3.12(b). Note that, for the apodized grating, the side lobes have been drastically reduced but at the expense of increasing the main lobe width.

The index distribution across the length of a Bragg grating is analogous to the grating aperture discussed in Section 3.3.2, and the reflection spectrum is obtained as the Fourier transform of the index distribution. The side lobes in the case of a uniform refractive index profile arise due to the abrupt start and end of the grating,

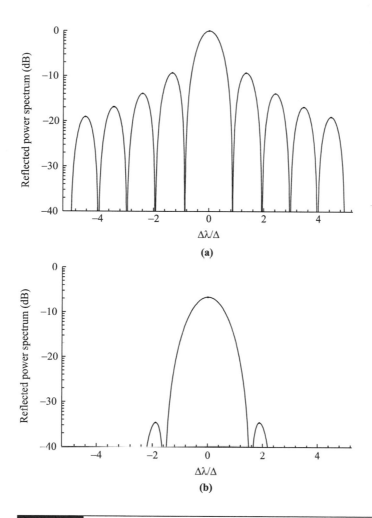

Figure 3.12 Reflection spectra of Bragg gratings with (a) uniform index profile and (b) apodized index profile. Δ is a measure of the bandwidth of the grating and is the wavelength separation between the peak wavelength and the first reflection minimum, in the uniform index profile case. Δ is inversely proportional to the length of the grating. $\Delta\lambda$ is the detuning from the phase-matching wavelength.

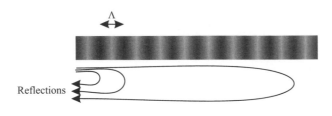

Figure 3.13   Principle of operation of a Bragg grating.

which result in a sinc(.) behavior for the side lobes. Apodization can be achieved by gradually starting and ending the grating. This technique is similar to pulse shaping used in digital communication systems to reduce the side lobes in the transmitted spectrum of the signal.

The bandwidth of the grating, which can be measured, for example, by the width of the main lobe, is inversely proportional to the length of the grating. Typically, the grating is a few millimeters long in order to achieve a bandwidth of 1 nm.

3.3.4   Fiber Gratings

Fiber gratings are attractive devices that can be used for a variety of applications, including filtering, add/drop functions, and compensating for accumulated dispersion in the system. Being all-fiber devices, their main advantages are their low loss, ease of coupling (with other fibers), polarization insensitivity, low temperature coefficient, and simple packaging. As a result, they can be extremely low-cost devices.

Gratings are written in fibers by making use of the *photosensitivity* of certain types of optical fibers. A conventional silica fiber doped with germanium becomes extremely photosensitive. Exposing this fiber to ultraviolet (UV) light causes changes in the refractive index within the fiber core. A grating can be written in such a fiber by exposing its core to two interfering UV beams. This causes the radiation intensity to vary periodically along the length of the fiber. Where the intensity is high, the refractive index is increased; where it is low, the refractive index is unchanged. The change in refractive index needed to obtain gratings is quite small—around 10^{-4}. Other techniques, such as *phase masks*, can also be used to produce gratings. A phase mask is a diffractive optical element. When it is illuminated by a light beam, it splits the beams into different diffractive orders, which then interfere with one another to write the grating into the fiber.

Fiber gratings are classified as either *short-period* or *long-period* gratings, based on the period of the grating. Short-period gratings are also called Bragg gratings and have periods that are comparable to the wavelength, typically around 0.5 μm. We discussed the behavior of Bragg gratings in Section 3.3.3. Long-period gratings, on the other hand, have periods that are much greater than the wavelength, ranging from a few hundred micrometers to a few millimeters.

Fiber Bragg Gratings

Fiber Bragg gratings can be fabricated with extremely low loss (0.1 dB), high wavelength accuracy ($\pm$ 0.05 nm is easily achieved), high adjacent channel crosstalk suppression (40 dB), as well as flat tops.

The temperature coefficient of a fiber Bragg grating is typically 1.25×10^{-2} nm/°C due to the variation in fiber length with temperature. However, it is possible to compensate for this change by packaging the grating with a material that has a negative thermal expansion coefficient. These passively temperature-compensated gratings have temperature coefficients of around 0.07×10^{-2} nm/°C. This implies a very small 0.07 nm center wavelength shift over an operating temperature range of 100°C, which means that they can be operated without any active temperature control.

These properties of fiber Bragg gratings make them very useful devices for system applications. Fiber Bragg gratings are finding a variety of uses in WDM systems, ranging from filters and optical add/drop elements to dispersion compensators. A simple optical drop element based on fiber Bragg gratings is shown in Figure 3.14(a). It consists of a three-port circulator with a fiber Bragg grating. The circulator transmits light coming in on port 1 out on port 2 and transmits light coming in on port 2 out on port 3. In this case, the grating reflects the desired wavelength λ_2, which is then dropped at port 3. The remaining three wavelengths are passed through. It is possible to implement an add/drop function along the same lines, by introducing a coupler to add the same wavelength that was dropped, as shown in Figure 3.14(b). Many variations of this simple add/drop element can be realized by using gratings in combination with couplers and circulators. A major concern in these designs is that the reflection of these gratings is not perfect, and as a result, some power at the selected wavelength leaks through the grating. This can cause undesirable crosstalk, and we will study this effect in Chapter 5.

Fiber Bragg gratings can also be used to compensate for dispersion accumulated along the link. We will study this application in Chapter 5 in the context of dispersion compensation.

Long-Period Fiber Gratings

Long-period fiber gratings are fabricated in the same manner as fiber Bragg gratings and are used today primarily as filters inside erbium-doped fiber amplifiers to compensate for their nonflat gain spectrum. As we will see, these devices serve as very

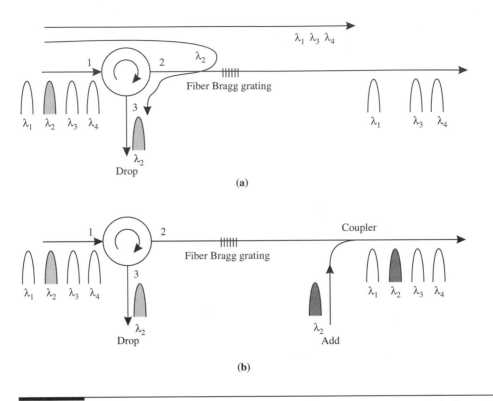

Figure 3.14   Optical add/drop elements based on fiber Bragg gratings. (a) A drop element. (b) A combined add/drop element.

efficient band rejection filters and can be tailored to provide almost exact equalization of the erbium gain spectrum. Figure 3.15 shows the transmission spectrum of such a grating. These gratings retain all the attractive properties of fiber gratings and are expected to become widely used for several filtering applications.

Principle of Operation

These gratings operate on somewhat different principles than Bragg gratings. In fiber Bragg gratings, energy from the forward propagating mode in the fiber core at the right wavelength is coupled into a backward propagating mode. In long-period gratings, energy is coupled from the forward propagating mode in the fiber core onto other forward propagating modes in the cladding. These cladding modes are extremely lossy, and their energy decays rapidly as they propagate along the fiber, due to losses at the cladding-air interface and due to microbends in the fiber. There are many cladding modes, and coupling occurs between a core mode at a given

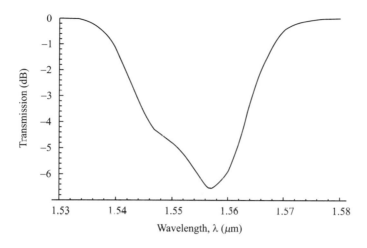

Figure 3.15 Transmission spectrum of a long-period fiber Bragg grating used as a gain equalizer for erbium-doped fiber amplifiers. (After [Ven96a].)

wavelength and a cladding mode depending on the pitch of the grating Λ, as follows: if β denotes the propagation constant of the mode in the core (assuming a single-mode fiber) and β_{cl}^p that of the pth-order cladding mode, then the phase-matching condition dictates that

$$\beta - \beta_{cl}^p = \frac{2\pi}{\Lambda}.$$

In general, the difference in propagation constants between the core mode and any one of the cladding modes is quite small, leading to a fairly large value of Λ in order for coupling to occur. This value is usually a few hundred micrometers. (Note that in Bragg gratings the difference in propagation constants between the forward and backward propagating modes is quite large, leading to a small value for Λ, typically around 0.5 μm.) If n_{eff} and n_{eff}^p denote the effective refractive indices of the core and pth-order cladding modes, then the wavelength at which energy is coupled from the core mode to the cladding mode can be obtained as

$$\lambda = \Lambda(n_{eff} - n_{eff}^p),$$

where we have used the relation $\beta = 2\pi n_{eff}/\lambda$.

Therefore, once we know the effective indices of the core and cladding modes, we can design the grating with a suitable value of Λ so as to cause coupling of energy out of a desired wavelength band. This causes the grating to act as a wavelength-dependent loss element. Methods for calculating the propagation

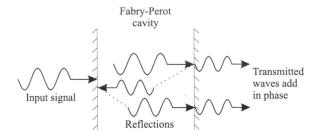

Figure 3.16   Principle of operation of a Fabry-Perot filter.

constants for the cladding modes are discussed in [Ven96b]. The amount of wavelength-dependent loss can be controlled during fabrication by controlling the UV exposure time. Complicated transmission spectra can be obtained by cascading multiple gratings with different center wavelengths and different exposures. The example shown in Figure 3.15 was obtained by cascading two such gratings [Ven96a]. These gratings are typically a few centimeters long.

3.3.5   Fabry-Perot Filters

A Fabry-Perot filter consists of the cavity formed by two highly reflective mirrors placed parallel to each other, as shown in Figure 3.16. This filter is also called a Fabry-Perot interferometer or etalon. The input light beam to the filter enters the first mirror at right angles to its surface. The output of the filter is the light beam leaving the second mirror.

This is a classical device that has been used widely in interferometric applications. Fabry-Perot filters have been used for WDM applications in several optical network testbeds. There are better filters today, such as the thin-film resonant multicavity filter that we will study in Section 3.3.6. These latter filters can be viewed as Fabry-Perot filters with wavelength-dependent mirror reflectivities. Thus the fundamental principle of operation of these filters is the same as that of the Fabry-Perot filter. The Fabry-Perot cavity is also used in lasers (see Section 3.5.1).

Compact Fabry-Perot filters are commercially available components. Their main advantage over some of the other devices is that they can be tuned to select different channels in a WDM system, as discussed later.

Principle of Operation

The principle of operation of the device is illustrated in Figure 3.16. The input signal is incident on the left surface of the cavity. After one pass through the cavity, as shown in Figure 3.16, a part of the light leaves the cavity through the right facet and a part is reflected. A part of the reflected wave is again reflected by the left facet to the right facet. For those wavelengths for which the cavity length is an integral multiple of half the wavelength in the cavity—so that a round trip through the cavity is an integral multiple of the wavelength—all the light waves transmitted through the right facet *add in phase*. Such wavelengths are called the *resonant wavelengths* of the cavity. The determination of the resonant wavelengths of the cavity is discussed in Problem 3.7.

The power transfer function of a filter is the fraction of input light power that is transmitted by the filter as a function of optical frequency f, or wavelength. For the Fabry-Perot filter, this function is given by

$$T_{FP}(f) = \frac{\left(1 - \frac{A}{1-R}\right)^2}{\left(1 + \left(\frac{2\sqrt{R}}{1-R}\sin(2\pi f\tau)\right)^2\right)}. \tag{3.12}$$

This can also be expressed in terms of the optical free-space wavelength λ as

$$T_{FP}(\lambda) = \frac{\left(1 - \frac{A}{1-R}\right)^2}{\left(1 + \left(\frac{2\sqrt{R}}{1-R}\sin(2\pi nl/\lambda)\right)^2\right)}.$$

(By a slight abuse of notation, we use the same symbol for the power transfer function in both cases.) Here A denotes the absorption loss of each mirror, which is the fraction of incident light that is absorbed by the mirror. The quantity R denotes the *reflectivity* of each mirror (assumed to be identical), which is the fraction of incident light that is reflected by the mirror. The one-way propagation delay across the cavity is denoted by τ. The refractive index of the cavity is denoted by n and its length by l. Thus $\tau = nl/c$, where c is the velocity of light in vacuum. This transfer function can be derived by considering the sum of the waves transmitted by the filter after an odd number of passes through the cavity. This is left as an exercise (Problem 3.8).

The power transfer function of the Fabry-Perot filter is plotted in Figure 3.17 for $A = 0$ and $R = 0.75$, 0.9, and 0.99. Note that very high mirror reflectivities are required to obtain good isolation of adjacent channels.

The power transfer function $T_{FP}(f)$ is periodic in f, and the peaks, or *passbands*, of the transfer function occur at frequencies f that satisfy $f\tau = k/2$ for some positive integer k. Thus in a WDM system, even if the wavelengths are spaced sufficiently far apart compared to the width of each passband of the filter transfer function, several frequencies (or wavelengths) may be transmitted by the filter if

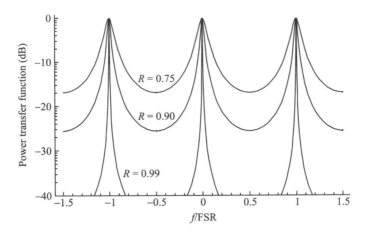

Figure 3.17   The transfer function of a Fabry-Perot filter. FSR denotes the free spectral range, f the frequency, and R the reflectivity.

they coincide with different passbands. The spectral range between two successive passbands of the filter is called the *free spectral range* (FSR). A measure of the width of each passband is its *full width* at the point where the transfer function is *half* of its *maximum* (FWHM). In WDM systems, the separation between two adjacent wavelengths must be at least a FWHM in order to minimize crosstalk. (More precisely, as the transfer function is periodic, adjacent wavelengths must be separated by a FWHM plus an integral multiple of the FSR.) Thus the ratio FSR/FWHM is an approximate (order-of-magnitude) measure of the number of wavelengths that can be accommodated by the system. This ratio is called the *finesse*, F, of the filter and is given by

$$F = \frac{\pi \sqrt{R}}{1 - R}. \tag{3.13}$$

This expression can be derived from (3.12) and is left as an exercise (Problem 3.9).

If the mirrors are highly reflective, won't virtually all the input light get reflected? Also, how does light get out of the cavity if the mirrors are highly reflective? To resolve this paradox, we must look at the light energy over all the frequencies. When we do this, we will see that only a small fraction of the input light is transmitted through the cavity because of the high reflectivities of the input and output facets, but at the right frequency, all the power is transmitted. This aspect is explored further in Problem 3.10.

Tunability

A Fabry-Perot filter can be tuned to select different wavelengths in one of several ways. The simplest approach is to change the cavity length. The same effect can be achieved by varying the refractive index within the cavity. Consider a WDM system, all of whose wavelengths lie within one FSR of the Fabry-Perot filter. The frequency f_0 that is selected by the filter satisfies $f_0\tau = k/2$ for some positive integer k. Thus f_0 can be changed by changing τ, which is the one-way propagation time for the light beam across the cavity. If we denote the length of the cavity by l and its refractive index by n, $\tau = ln/c$, where c is the speed of light in vacuum. Thus τ can be changed by changing either l or n.

Mechanical tuning of the filter can be effected by moving one of the mirrors so that the cavity length changes. This permits tunability only in times of the order of a few milliseconds. For a mechanically tuned Fabry-Perot filter, a precise mechanism is needed in order to keep the mirrors parallel to each other in spite of their relative movement. The reliability of mechanical tuning mechanisms is also relatively poor.

Another approach to tuning is to use a piezoelectric material within the cavity. A piezoelectric filter undergoes compression on the application of a voltage. Thus the length of the cavity filled with such a material can be changed by the application of a voltage, thereby effecting a change in the resonant frequency of the cavity. The piezo material, however, introduces undesirable effects such as thermal instability and hysteresis, making such a filter difficult to use in practical systems.

3.3.6 *Multilayer Dielectric Thin-Film Filters*

A thin-film resonant cavity filter (TFF) is a Fabry-Perot interferometer, or etalon (see Section 3.3.5), where the mirrors surrounding the cavity are realized by using multiple reflective dielectric thin-film layers (see Problem 3.13). This device acts as a bandpass filter, passing through a particular wavelength and reflecting all the other wavelengths. The wavelength that is passed through is determined by the cavity length.

A thin-film resonant multicavity filter (TFMF) consists of two or more cavities separated by reflective dielectric thin-film layers, as shown in Figure 3.18. The effect of having multiple cavities on the response of the filter is illustrated in Figure 3.19. As more cavities are added, the top of the passband becomes flatter and the skirts become sharper, both very desirable filter features.

In order to obtain a multiplexer or a demultiplexer, a number of these filters can be cascaded, as shown in Figure 3.20. Each filter passes a different wavelength and reflects all the others. When used as a demultiplexer, the first filter in the cascade

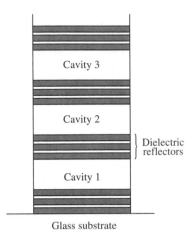

Figure 3.18   A three-cavity thin-film resonant dielectric thin-film filter. (After [SS96].)

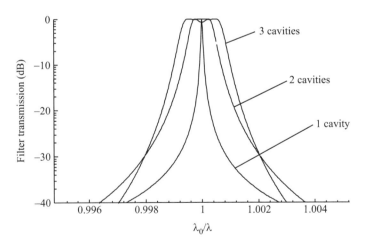

Figure 3.19   Transfer functions of single-cavity, two-cavity, and three-cavity dielectric thin-film filters. Note how the use of multiple cavities leads to a flatter passband and a sharper transition from the passband to the stop band.

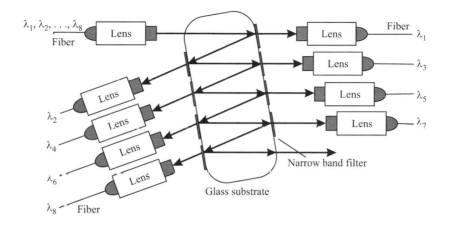

Figure 3.20   A wavelength multiplexer/demultiplexer using multilayer dielectric thin-film filters. (After [SS96].)

passes one wavelength and reflects all the others onto the second filter. The second filter passes another wavelength and reflects the remaining ones, and so on.

This device has many features that make it attractive for system applications. It is possible to have a very flat top on the passband and very sharp skirts. The device is extremely stable with regard to temperature variations, has low loss, and is insensitive to the polarization of the signal. Typical parameters for a 16-channel multiplexer are shown in Table 3.1. For these reasons, TFMFs are becoming widely used in commercial systems today. Understanding the principle of operation of these devices requires some knowledge of electromagnetic theory, and so we defer this to Appendix G.

3.3.7   Mach-Zehnder Interferometers

A Mach-Zehnder interferometer (MZI) is an interferometric device that makes use of two interfering paths of different lengths to resolve different wavelengths. Devices constructed on this principle have been around for some decades. Today, Mach-Zehnder interferometers are typically constructed in integrated optics and consist of two 3 dB directional couplers interconnected through two paths of differing lengths, as shown in Figure 3.21(a). The substrate is usually silicon, and the waveguide and cladding regions are silica (SiO_2).

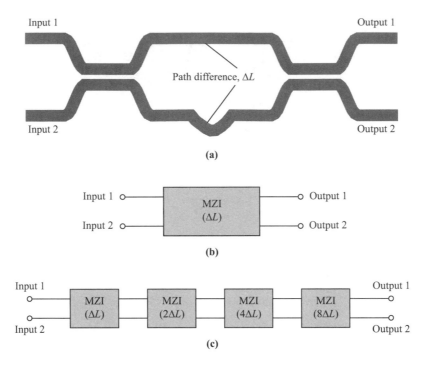

Figure 3.21 (a) An MZI constructed by interconnecting two 3 dB directional couplers. (b) A block diagram representation of the MZI in (a). ΔL denotes the path difference between the two arms. (c) A block diagram of a four-stage Mach-Zehnder interferometer, which uses different path length differences in each stage.

Mach-Zehnder interferometers are useful as both filters and (de)multiplexers. Even though there are better technologies for making narrow band filters, for example, dielectric multicavity thin-film filters, MZIs are still useful in realizing wide band filters. For example, MZIs can be used to separate the wavelengths in the 1.3 μm and 1.55 μm bands. Narrow band MZI filters are fabricated by cascading a number of stages, as we will see, and this leads to larger losses. In principle, very good crosstalk performance can be achieved using MZIs if the wavelengths are spaced such that the undesired wavelengths occur at, or close to, the nulls of the power transfer function. However, in practice, the wavelengths cannot be fixed precisely (for example, the wavelengths drift because of temperature variations or age). Moreover, the coupling ratio of the directional couplers is not 50:50 and could be wavelength dependent. As a result, the crosstalk performance is far from the ideal situation. Also the passband of narrow band MZIs is not flat. In contrast, the dielectric multicavity thin-film filters can have flat passbands and good stop bands.

MZIs are useful as two-input, two-output multiplexers and demultiplexers. They can also be used as tunable filters, where the tuning is achieved by varying the temperature of one of the arms of the device. This causes the refractive index of that arm to change, which in turn affects the phase relationship between the two arms and causes a different wavelength to be coupled out. The tuning time required is of the order of several milliseconds. For higher channel-count multiplexers and demultiplexers, better technologies are available today. One example is the *arrayed waveguide grating* (AWG) described in the next section. Since understanding the MZI is essential to understanding the AWG, we will now describe the principle of operation of MZIs.

Principle of Operation

Consider the operation of the MZI as a demultiplexer; so only one input, say, input 1, has a signal (see Figure 3.21(a)). After the first directional coupler, the input signal power is divided equally between the two arms of the MZI, but the signal in one arm has a phase shift of $\pi/2$ with respect to the other. Specifically, the signal in the lower arm lags the one in the upper arm in phase by $\pi/2$, as discussed in Section 3.1. This is best understood from (3.1). Since there is a length difference of ΔL between the two arms, there is a further phase lag of $\beta \Delta L$ introduced in the signal in the lower arm. In the second directional coupler, the signal from the lower arm undergoes another phase delay of $\pi/2$ in going to the first output relative to the signal from the upper arm. Thus the total relative phase difference at the first or upper output between the two signals is $\pi/2 + \beta \Delta L + \pi/2$. At the output directional coupler, in going to the second output, the signal from the upper arm lags the signal from the lower arm in phase by $\pi/2$. Thus the total relative phase difference at the second or lower output between the two signals is $\pi/2 + \beta \Delta L - \pi/2 = \beta \Delta L$.

If $\beta \Delta L = k\pi$ and k is odd, the signals at the first output add in phase, whereas the signals at the second output add with opposite phases and thus cancel each other. Thus the wavelengths passed from the first input to the first output are those wavelengths for which $\beta \Delta L = k\pi$ and k is odd. The wavelengths passed from the first input to the second output are those wavelengths for which $\beta \Delta L = k\pi$ and k is even. This could have been easily deduced from the transfer function of the MZI in the following equation (3.14), but this detailed explanation will help in the understanding of the arrayed waveguide grating (Section 3.3.8).

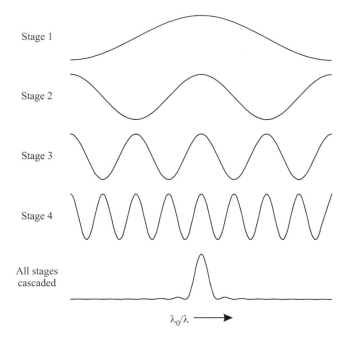

Figure 3.22   Transfer functions of each stage of a multistage MZI.

Assume that the difference between these path lengths is ΔL and that only one input, say, input 1, is active. Then it can be shown (see Problem 3.14) that the power transfer function of the Mach-Zehnder interferometer is given by

$$\begin{pmatrix} T_{11}(f) \\ T_{12}(f) \end{pmatrix} = \begin{pmatrix} \sin^2(\beta \Delta L/2) \\ \cos^2(\beta \Delta L/2) \end{pmatrix}. \tag{3.14}$$

Thus the path difference between the two arms, ΔL, is the key parameter characterizing the transfer function of the MZI. We will represent the MZI of Figure 3.21(a) using the block diagram of Figure 3.21(b).

Now consider k MZIs interconnected, as shown in Figure 3.21(c) for $k = 4$. Such a device is termed a *multistage Mach-Zehnder interferometer*. The path length difference for the kth MZI in the cascade is assumed to be $2^{k-1} \Delta L$. The transfer function of each MZI in this multistage MZI together with the power transfer function of the entire filter is shown in Figure 3.22. The power transfer function of the multistage MZI is also shown on a decibel scale in Figure 3.23.

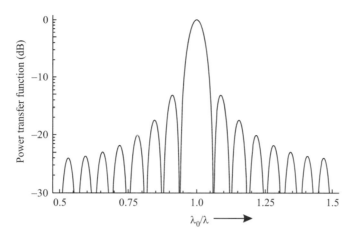

Figure 3.23 Transfer function of a multistage Mach-Zehnder interferometer.

We will now describe how an MZI can be used as a 1×2 demultiplexer. Since the device is reciprocal, it follows from the principles of electromagnetics that if the inputs and outputs are interchanged, it will act as a 2×1 multiplexer.

Consider a single MZI with a fixed value of the path difference ΔL. Let one of the inputs, say, input 1, be a wavelength division multiplexed signal with all the wavelengths chosen to coincide with the peaks or troughs of the transfer function. For concreteness, assume the propagation constant $\beta = 2\pi n_{\text{eff}}/\lambda$, where n_{eff} is the effective refractive index of the waveguide. The input wavelengths λ_i would have to be chosen such that $n_{\text{eff}}\Delta L/\lambda_i = m_i/2$ for some positive integer m_i. The wavelengths λ_i for which m is odd would then appear at the first output (since the transfer function is $\sin^2(m_i\pi/2)$), and the wavelengths for which m_i is even would appear at the second output (since the transfer function is $\cos^2(m_i\pi/2)$).

If there are only two wavelengths, one for which m_i is odd and the other for which m_i is even, we have a 1×2 demultiplexer. The construction of a $1 \times n$ demultiplexer when n is a power of two, using $n-1$ MZIs, is left as an exercise (Problem 3.15). But there is a better method of constructing higher channel count demultiplexers, which we describe next.

3.3.8 Arrayed Waveguide Grating

An *arrayed waveguide grating* (AWG) is a generalization of the Mach-Zehnder interferometer. This device is illustrated in Figure 3.24. It consists of two multiport couplers interconnected by an array of waveguides. The MZI can be viewed as a device where *two* copies of the same signal, but shifted in phase by different amounts, are added together. The AWG is a device where *several* copies of the same signal, but shifted in phase by different amounts, are added together.

The AWG has several uses. It can be used as an $n \times 1$ *wavelength multiplexer.* In this capacity, it is an n-input, 1-output device where the n inputs are signals at different wavelengths that are combined onto the single output. The inverse of this function, namely, $1 \times n$ *wavelength demultiplexing,* can also be performed using an AWG. Although these wavelength multiplexers and demultiplexers can also be built using MZIs interconnected in a suitable fashion, it is preferable to use an AWG. Relative to an MZI chain, an AWG has lower loss, flatter passband, and is easier to realize on an integrated-optic substrate. The input and output waveguides, the multiport couplers, and the arrayed waveguides are all fabricated on a single substrate. The substrate material is usually silicon, and the waveguides are silica, Ge-doped silica, or SiO_2-Ta_2O_5. Thirty-two–channel AWGs are commercially available, and smaller AWGs are being used in WDM transmission systems. Their temperature coefficient (0.01 nm/°C) is not as low as those of some other competing technologies such as fiber gratings and multilayer thin-film filters. So we will need to use active temperature control for these devices.

Another way to understand the working of the AWG as a demultiplexer is to think of the multiport couplers as lenses and the array of waveguides as a prism. The input coupler collimates the light from an input waveguide to the array of waveguides. The array of waveguides acts like a prism, providing a wavelength-dependent phase shift, and the output coupler focuses different wavelengths on different output waveguides.

The AWG can also be used as a static wavelength crossconnect. However, this wavelength crossconnect is not capable of achieving an arbitrary routing pattern. Although several interconnection patterns can be achieved by a suitable choice of

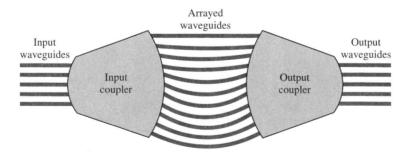

Figure 3.24   An arrayed waveguide grating.

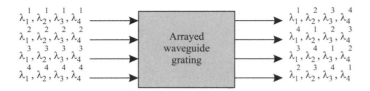

Figure 3.25   The crossconnect pattern of a static wavelength crossconnect constructed from an arrayed waveguide grating. The device routes signals from an input to an output based on their wavelength.

the wavelengths and the FSR of the device, the most useful one is illustrated in Figure 3.25. This figure shows a 4×4 static wavelength crossconnect using four wavelengths with one wavelength routed from each of the inputs to each of the outputs.

In order to achieve this interconnection pattern, the operating wavelengths and the FSR of the AWG must be chosen suitably. The FSR of the AWG is derived in Problem 3.17. Given the FSR, we leave the determination of the wavelengths to be used to achieve this interconnection pattern as another exercise (Problem 3.18).

Principle of Operation

Consider the AWG shown in Figure 3.24. Let the number of inputs and outputs of the AWG be denoted by n. Let the couplers at the input and output be $n \times m$ and $m \times n$ in size, respectively. Thus the couplers are interconnected by m waveguides. We will call these waveguides *arrayed waveguides* to distinguish them from the input and output waveguides. The lengths of these waveguides are chosen such that the difference in length between consecutive waveguides is a constant denoted by ΔL. The MZI is a special case of the AWG, where $n = m = 2$. We will now determine which wavelengths will be transmitted from a given input to a given output. The first coupler splits the signal into m parts. The relative phases of these parts are determined by the distances traveled in the coupler from the input waveguides to the arrayed waveguides. Denote the differences in the distances traveled (relative to any one of the input waveguides and any one of the arrayed waveguides) between input waveguide i and arrayed waveguide k by d_{ik}^{in}. Assume that arrayed waveguide k has a path length larger than arrayed waveguide $k-1$ by ΔL. Similarly, denote the differences in the distances traveled (relative to any one of the arrayed waveguides and any one of the output waveguides) between arrayed waveguide k and output waveguide j by d_{kj}^{out}. Then, the relative phases of the signals from input i to output j traversing the m different paths between them are given by

$$\phi_{ijk} = \frac{2\pi}{\lambda}(n_1 d_{ik}^{\text{in}} + n_2 k \Delta L + n_1 d_{kj}^{\text{out}}), \qquad k = 1, \ldots, m. \tag{3.15}$$

Here, n_1 is the refractive index in the input and output directional couplers, and n_2 is the refractive index in the arrayed waveguides. From input i, those wavelengths λ, for which ϕ_{ijk}, $k = 1, \ldots, m$, differ by a multiple of 2π will add in phase at output j. The question is, Are there any such wavelengths?

If the input and output couplers are designed such that $d_{ik}^{\text{in}} = d_i^{\text{in}} + k\delta_i^{\text{in}}$ and $d_{kj}^{\text{out}} = d_j^{\text{out}} + k\delta_j^{\text{out}}$, then (3.15) can be written as

$$\begin{aligned}
\phi_{ijk} &= \frac{2\pi}{\lambda}(n_1 d_i^{\text{in}} + n_1 d_j^{\text{out}}) \\
&\quad + \frac{2\pi k}{\lambda}(n_1 \delta_i^{\text{in}} + n_2 \Delta L + n_1 \delta_j^{\text{out}}), \qquad k = 1, \ldots, m.
\end{aligned} \tag{3.16}$$

Such a construction is possible and is called the *Rowland circle construction*. It is illustrated in Figure 3.26 and discussed further in Problem 3.16. Thus wavelengths λ that are present at input i and that satisfy $n_1 \delta_i^{\text{in}} + n_2 \Delta L + n_1 \delta_j^{\text{out}} = p\lambda$ for some integer p add in phase at output j.

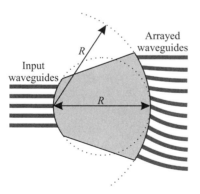

Figure 3.26   The Rowland circle construction for the couplers used in the AWG. The arrayed waveguides are located on the arc of a circle, called the *grating circle*, whose center is at the end of the central input (output) waveguide. Let the *radius* of this circle be denoted by R. The other input (output) waveguides are located on the arc of a circle whose *diameter* is equal to R; this circle is called the *Rowland circle*. The vertical spacing between the arrayed waveguides is chosen to be constant.

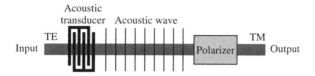

Figure 3.27 A simple AOTF. An acoustic wave introduces a grating whose pitch depends on the frequency of the acoustic wave. The grating couples energy from one polarization mode to another at a wavelength that satisfies the Bragg condition.

For use as a demultiplexer, all the wavelengths are present at the same input, say, input i. Therefore, if the wavelengths, $\lambda_1, \lambda_2, \ldots, \lambda_n$ in the WDM system satisfy $n_1\delta_i^{\text{in}} + n_2\Delta L + n_1\delta_j^{\text{out}} = p\lambda_j$ for some integer p, we infer from (3.16) that these wavelengths are demultiplexed by the AWG. Note that though δ_i^{in} and ΔL are necessary to define the precise set of wavelengths that are demultiplexed, the (minimum) spacing between them is independent of δ_i^{in} and ΔL, and determined primarily by δ_j^{out}.

Note in the preceding example that if wavelength λ_j' satisfies $n_1\delta_i^{\text{in}} + n_2\Delta L + n_1\delta_j^{\text{out}} = (p+1)\lambda_j'$, then both λ_j and λ_j' are "demultiplexed" to output j from input i. Thus like many of the other filter and multiplexer/demultiplexer structures we have studied, the AWG has a periodic response (in frequency), and all the wavelengths must lie within one FSR. The derivation of an expression for this FSR is left as an exercise (Problem 3.17).

3.3.9 Acousto-Optic Tunable Filter

The acousto-optic tunable filter is a versatile device. It is probably the only known *tunable* filter that is capable of selecting several wavelengths simultaneously. This capability can be used to construct a wavelength crossconnect, as we will explain later in this section.

The acousto-optic tunable filter (AOTF) is one example of several optical devices whose construction is based on the interaction of sound and light. Basically, an acoustic wave is used to create a Bragg grating in a waveguide, which is then used to perform the wavelength selection. Figure 3.27 shows a simple version of the AOTF. We will see that the operation of this AOTF is dependent on the state of polarization of the input signal. Figure 3.28 shows a more realistic polarization-independent implementation in integrated optics.

Principle of Operation

Consider the device shown in Figure 3.27. It consists of a waveguide constructed from a birefringent material and supporting only the lowest-order TE and TM modes (see Section 2.1.2). We assume that the input light energy is entirely in the TE mode. A *polarizer*, which selects only the light energy in the TM mode, is placed at the other end of the channel waveguide. If, somehow, the light energy in a narrow spectral range around the wavelength to be selected is converted to the TM mode, while the rest of the light energy remains in the TE mode, we have a wavelength-selective filter. This conversion is effected in an AOTF by launching an acoustic wave along, or opposite to, the direction of propagation of the light wave.

As a result of the propagation of the acoustic wave, the density of the medium varies in a periodic manner. The period of this density variation is equal to the wavelength of the acoustic wave. This periodic density variation acts as a Bragg grating. From the discussion of such gratings in Section 3.3.3, it follows that if the refractive indices n_{TE} and n_{TM} of the TE and TM modes satisfy the Bragg condition

$$\frac{n_{TM}}{\lambda} = \frac{n_{TE}}{\lambda} \pm \frac{1}{\Lambda}, \tag{3.17}$$

then light couples from one mode to the other. Thus light energy in a narrow spectral range around the wavelength λ that satisfies (3.17) undergoes TE to TM mode conversion. Thus the device acts as a narrow bandwidth filter when only light energy in the TE mode is input and only the light energy in the TM mode is selected at the output, as shown in Figure 3.27.

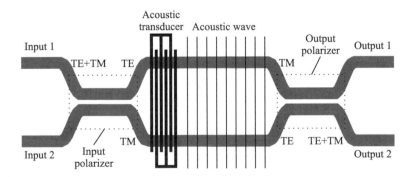

Figure 3.28   A polarization-independent integrated-optics AOTF. A polarizer splits the input signal into its constituent polarization modes, and each mode is converted in two separate arms, before being recombined at the output.

In LiNbO$_3$, the TE and TM modes have refractive indices n_{TE} and n_{TM} that differ by about 0.07. If we denote this refractive index difference by (Δn), the Bragg condition (3.17) can be written as

$$\lambda = \Lambda(\Delta n). \tag{3.18}$$

The wavelength that undergoes mode conversion and thus lies in the passband of the AOTF can be selected, or tuned, by suitably choosing the acoustic wavelength Λ. In order to select a wavelength of 1.55 μm, for (Δn) = 0.07, using (3.18), the acoustic wavelength is about 22 μm. Since the velocity of sound in LiNbO$_3$ is about 3.75 km/s, the corresponding RF frequency is about 170 MHz. Since the RF frequency is easily tuned, the wavelength selected by the filter can also be easily tuned. The typical insertion loss is about 4 dB.

The AOTF considered here is a polarization-dependent device since the input light energy is assumed to be entirely in the TE mode. A polarization-independent AOTF, shown in Figure 3.28, can be realized in exactly the same manner as a polarization-independent isolator by decomposing the input light signal into its TE and TM constituents and sending each constituent separately through the AOTF and recombining them at the output.

Transfer Function

Whereas the Bragg condition determines the wavelength that is selected, the width of the filter passband is determined by the length of the acousto-optic interaction. The longer this interaction, and hence the device, the narrower the passband. It can be shown that the wavelength dependence of the fraction of the power transmitted by the AOTF is given by

$$T(\lambda) = \frac{\sin^2\left((\pi/2)\sqrt{1 + (2\Delta\lambda/\Delta)^2}\right)}{1 + (2\Delta\lambda/\Delta)^2}.$$

This is plotted in Figure 3.29. Here $\Delta\lambda = \lambda - \lambda_0$, where λ_0 is the optical wavelength that satisfies the Bragg condition, and $\Delta = \lambda_0^2/l\Delta n$ is a measure of the filter passband width. Here, l is the length of the device (or, more correctly, the length of the acousto-optic interaction). It can be shown that the full width at half-maximum (FWHM) bandwidth of the filter is $\approx 0.8\Delta$ (Problem 3.20). This equation clearly shows that the longer the device, the narrower the passband. However, there is a trade-off here: the tuning speed is inversely proportional to l. This is because the tuning speed is essentially determined by the time it takes for a sound wave to travel the length of the filter.

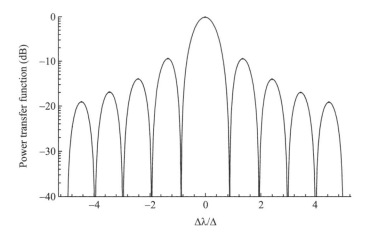

Figure 3.29   The power transfer function of the acousto-optic tunable filter.

AOTF as a Wavelength Crossconnect

The polarization-independent AOTF illustrated in Figure 3.28 can be used as a two-input, two-output dynamic wavelength crossconnect. We studied the operation of this device as a filter earlier; in this case, only one of the inputs was active. We leave it as an exercise (Problem 3.21) to show that when the second input is also active, the energy at the wavelength λ satisfying the Bragg phase-matching condition (3.18) is *exchanged* between the two ports. This is illustrated in Figure 3.30(a), where the wavelength λ_1 satisfies the Bragg condition and is exchanged between the ports.

Now the AOTF has one remarkable property that is not shared by any other tunable filter structure we know. By launching multiple acoustic waves *simultaneously*, the Bragg condition (3.18) can be satisfied for multiple optical wavelengths simultaneously. Thus multiple wavelength exchanges can be accomplished simultaneously between two ports with a single device of the form shown in Figure 3.28. This is illustrated in Figure 3.30(b), where the wavelengths λ_1 and λ_4 are exchanged between the ports. Thus this device performs the same routing function as the static crossconnect of Figure 3.7. However, the AOTF is a completely general two-input, two-output *dynamic* crossconnect since the routing pattern, or the set of wavelengths to be exchanged, can be changed easily by varying the frequencies of the acoustic waves launched in the device. In principle, larger dimensional dynamic crossconnects (with more input and output ports) can be built by suitably cascading 2×2 crossconnects. We will however, see in Section 3.7 that there are better ways of building large-scale crossconnects.

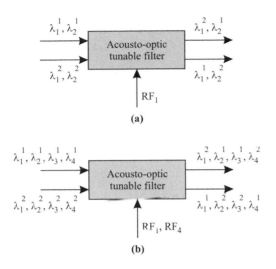

Figure 3.30 Wavelength crossconnects constructed from acousto-optic tunable filters. (a) The wavelength λ_1 is exchanged between the two ports. (b) The wavelengths λ_1 and λ_4 are simultaneously exchanged between the two ports by the simultaneous launching of two appropriate acoustic waves.

As of this writing, the AOTF has not yet lived up to its promise either as a versatile tunable filter or a wavelength crossconnect. One reason for this is the high level of crosstalk that is present in the device. As can be seen from Figure 3.29, the first side lobe in its power transfer function is not even 10 dB below the peak transmission. This problem can be alleviated to some extent by cascading two such filters. In fact, the cascade can even be built on a single substrate. But even then the first side lobe would be less than 20 dB below the peak transmission. It is harder to cascade more such devices without facing other problems such as an unacceptably high transmission loss. Another reason for the comparative failure of the AOTF today is that the passband width is fairly large (100 GHz or more) even when the acousto-optic interaction length is around 1 inch (Problem 3.22). This makes it unsuitable for use in dense WDM systems where channel spacings are now down to 50 GHz. Devices with larger interaction lengths are more difficult to fabricate. However, some recent theoretical work [Son95] indicates that some of these problems, particularly crosstalk, may be solvable. The crosstalk problems that arise in AOTFs when used as wavelength crossconnects are discussed in detail in [Jac96].

3.3.10 High Channel Count Multiplexer Architectures

With the number of wavelengths continuously increasing, designing multiplexers and demultiplexers to handle large numbers of wavelengths has become an important problem. The desired attributes of these devices are the same as what we saw at the beginning of Section 3.3. Our discussion will be based on demultiplexers, but these demultiplexers can all be used as multiplexers as well. In fact, in bidirectional applications, where some wavelengths are transmitted in one direction over a fiber and others in the opposite direction over the same fiber, the same device acts as a multiplexer for some wavelengths and a demultiplexer for others. We describe several architectural approaches to construct high channel count demultiplexers below.

Serial

In this approach, the demultiplexing is done one wavelength at a time. The demultiplexer consists of W filter stages in series, one for each of the W wavelengths. Each filter stage demultiplexes a wavelength and allows the other wavelengths to pass through. The architecture of the dielectric thin-film demultiplexer shown in Figure 3.20 is an example. One advantage of this architecture is that the filter stages can potentially be added one at a time, as more wavelengths are added. This allows a "pay as you grow" approach.

Serial approaches work for demultiplexing relatively small numbers of channels but do not scale to handle a large number of channels. This is because the insertion loss (in decibels) of the demultiplexer increases almost linearly with the number of channels to be demultiplexed. Moreover, different channels see different insertion losses based on the order in which the wavelengths are demultiplexed, which is not a desirable feature.

Single-Stage

Here, all the wavelengths are demultiplexed together in a single stage. The AWG shown in Figure 3.24 is an example of such an architecture. This approach provides relatively lower losses and better loss uniformity, compared to the serial approach. However, the number of channels that can be demultiplexed is limited by the maximum number of channels that can be handled by a single device, typically around 40 channels in commercially available AWGs today.

Multistage Banding

Going to larger channel counts requires the use of multiple demultiplexing stages, due to the limitations of the serial and single-stage approaches discussed above. A popular approach used today is to divide the wavelengths into *bands*. For example,

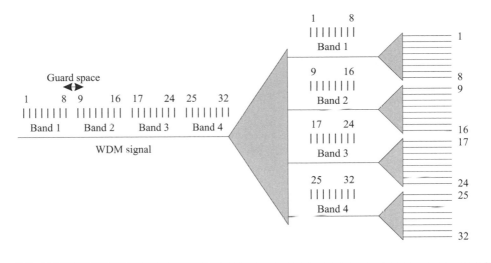

Figure 3.31 A two-stage demultiplexing approach using bands. A 32-channel demultiplexer is realized using four bands of 8 channels each.

a total of 32 wavelengths may be divided into four bands, each with 8 wavelengths. The demultiplexing is done in two stages, as shown in Figure 3.31. In the first the set of wavelengths is demultiplexed into bands. In the second stage, the bands are demultiplexed, and individual wavelengths are extracted. The scheme can be extended to more than two stages as well. It is also modular in that the demultiplexers in the second stage (or last stage in a multistage scheme) can be populated one band at a time.

One drawback with the banding approach is that we will usually need to leave a "guard" space between bands, as shown in Figure 3.31. This guard space allows the first-stage filters to be designed to provide adequate crosstalk suppression while retaining a low insertion loss.

Multistage Interleaving

Interleaving provides another approach to realizing large channel count demultiplexers. A two-stage *interleaver* is shown in Figure 3.32. In this approach the first stage separates the wavelengths into two groups. The first group consists of wavelengths 1, 3, 5, ... and the second group consists of wavelengths 2, 4, 6, ... The second stage extracts the individual wavelengths. This approach is also modular in the sense that the last stage of demultiplexers can be populated as needed. More than two stages can be used if needed as well.

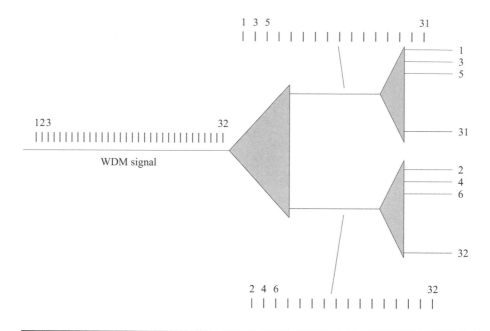

Figure 3.32   A two-stage multiplexing approach using interleaving. In this 32-channel demultiplexer, the first stage picks out every alternate wavelength, and the second stage extracts the individual wavelength.

A significant benefit of this approach is that the filters in the last stage can be much wider than the channel width. As an example, suppose we want to demultiplex a set of 32 channels spaced 50 GHz apart. After the first stage of demultiplexing, the channels are spaced 100 GHz apart, as shown in Figure 3.32. So demultiplexers with a broader passband suitable for demultiplexing 100 GHz spaced channels can be used in the second stage. In contrast, the single-stage or serial approach would require the use of demultiplexers capable of demultiplexing 50 GHz spaced channels, which are much more difficult to build. Carrying this example further, the second stage itself can in turn be made up of two stages. The first stage extracts every other 100 GHz channel, leading to a 200 GHz interchannel spacing after this stage. The final stage can then use even broader filters to extract the individual channels. Another advantage of this approach is that no guard bands are required in the channel plan.

The challenges with the interleaving approach lie in realizing the demultiplexers that perform the interleaving at all the levels except the last level. In principle, any periodic filter can be used as an interleaver by matching its period to the desired channel spacing. For example, a fiber-based Mach-Zehnder interferometer is a common choice. These devices are now commercially available, and interleaving is becoming a popular approach toward realizing high channel count multiplexers and demultiplexers.

$\underline{3.4}$ Optical Amplifiers

In an optical communication system, the optical signals from the transmitter are attenuated by the optical fiber as they propagate through it. Other optical components, such as multiplexers and couplers, also add loss. After some distance, the cumulative loss of signal strength causes the signal to become too weak to be detected. Before this happens, the signal strength has to be restored. Prior to the advent of optical amplifiers over the last decade, the only option was to regenerate the signal, that is, receive the signal and retransmit it. This process is accomplished by *regenerators*. A regenerator converts the optical signal to an electrical signal, cleans it up, and converts it back into an optical signal for onward transmission.

Optical amplifiers offer several advantages over regenerators. Regenerators are specific to the bit rate and modulation format used by the communication system. On the other hand, optical amplifiers are insensitive to the bit rate or signal formats. Thus a system using optical amplifiers can be more easily upgraded, for example, to a higher bit rate, without replacing the amplifiers. In contrast, in a system using regenerators, such an upgrade would require all the regenerators to be replaced. Furthermore, optical amplifiers have fairly large gain bandwidths, and as a consequence, a single amplifier can simultaneously amplify several WDM signals. In contrast, we would need a regenerator for each wavelength. Thus optical amplifiers have become essential components in high-performance optical communication systems.

Amplifiers, however, aren't perfect devices. They introduce additional noise, and this noise accumulates as the signal passes through multiple amplifiers along its path due to the analog nature of the amplifier. The spectral shape of the gain, the output power, and the transient behavior of the amplifier are also important considerations for system applications. Ideally we would like to have a sufficiently high output power to meet the needs of the network application. We would also like the gain to be flat over the operating wavelength range, and for the gain to be insensitive to variations in input power of the signal. We will study the impact of optical amplifiers on the physical layer design of the system in Chapters 4 and 5. Here we explore their principle of operation.

We will consider three different types of amplifiers: *erbium-doped fiber amplifiers*, *Raman amplifiers*, and *semiconductor optical amplifiers*.

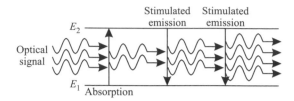

Figure 3.33   Stimulated emission and absorption in an atomic system with two energy levels.

3.4.1   Stimulated Emission

In all the amplifiers we consider, the key physical phenomenon behind signal amplification is *stimulated emission* of radiation by atoms in the presence of an electromagnetic field. (This is not true of fiber Raman or fiber Brillouin amplifiers, which make use of fiber nonlinearities, but we do not treat these here.) This field is an optical signal in the case of optical amplifiers. Stimulated emission is the principle underlying the operation of lasers as well; we will study lasers in Section 3.5.1.

According to the principles of quantum mechanics, any physical system (for example, an atom) is found in one of a discrete number of energy levels. Accordingly, consider an atom and two of its energy levels, E_1 and E_2, with $E_2 > E_1$. An electromagnetic field whose frequency f_c satisfies $hf_c = E_2 - E_1$ induces transitions of atoms between the energy levels E_1 and E_2. Here, h is Planck's constant (6.63×10^{-34} J s). This process is depicted in Figure 3.33. Both kinds of transitions, $E_1 \rightarrow E_2$ and $E_2 \rightarrow E_1$, occur. $E_1 \rightarrow E_2$ transitions are accompanied by *absorption* of photons from the incident electromagnetic field. $E_2 \rightarrow E_1$ transitions are accompanied by the *emission* of photons of energy hf_c, the same energy as that of the incident photons. This emission process is termed *stimulated emission* to distinguish it from another kind of emission called *spontaneous emission*, which we will discuss later. Thus if stimulated emission were to dominate over absorption—that is, the incident signal causes more $E_2 \rightarrow E_1$ transitions than $E_1 \rightarrow E_2$ transitions—we would have a net increase in the number of photons of energy hf_c and an amplification of the signal. Otherwise, the signal will be attenuated.

It follows from the theory of quantum mechanics that the rate of the $E_1 \rightarrow E_2$ transitions per atom *equals* the rate of the $E_2 \rightarrow E_1$ transitions *per atom*. Let this common rate be denoted by r. If the populations (number of atoms) in the energy levels E_1 and E_2 are N_1 and N_2, respectively, we have a net increase in power (energy per unit time) of $(N_2 - N_1)rhf_c$. Clearly, for amplification to occur, this must be positive, that is, $N_2 > N_1$. This condition is known as *population inversion*. The reason for this term is that, at thermal equilibrium, lower energy levels are more highly populated, that is, $N_2 < N_1$. Therefore, at thermal equilibrium, we have only absorption of the input signal. In order for amplification to occur, we must *invert* the relationship between the populations of levels E_1 and E_2 that prevails under thermal equilibrium.

Population inversion can be achieved by supplying additional energy in a suitable form to pump the electrons to the higher energy level. This additional energy can be in optical or electrical form.

3.4.2   Spontaneous Emission

Before describing the operation of the different types of amplifiers, it is important to understand the impact of spontaneous emission. Consider again the atomic system with the two energy levels discussed earlier. Independent of any external radiation that may be present, atoms in energy level E_2 transit to the lower energy level E_1, emitting a photon of energy hf_c. The spontaneous emission rate per atom from level E_2 to level E_1 is a characteristic of the system, and its reciprocal, denoted by τ_{21}, is called the *spontaneous emission lifetime*. Thus, if there are N_2 atoms in level E_2, the rate of spontaneous emission is N_2/τ_{21}, and the spontaneous emission power is $hf_c N_2/\tau_{21}$.

The spontaneous emission process does not contribute to the gain of the amplifier (to first order). Although the emitted photons have the same energy hf_c as the incident optical signal, they are emitted in random directions, polarizations, and phase. This is unlike the stimulated emission process, where the emitted photons not only have the same energy as the incident photons but also the same direction of propagation, phase, and polarization. This phenomenon is usually described by saying that the stimulated emission process is *coherent*, whereas the spontaneous emission process is *incoherent*.

Spontaneous emission has a deleterious effect on the system. The amplifier treats spontaneous emission radiation as another electromagnetic field at the frequency hf_c, and the spontaneous emission also gets amplified, in addition to the incident optical signal. This *amplified spontaneous emission* (ASE) appears as noise at the output of the amplifier. The implications of ASE on the design of optical communication systems are discussed in Chapters 4 and 5. In addition, in some amplifier designs, the ASE can be large enough so as to *saturate* the amplifier. Saturation effects are explored in Chapter 5.

3.4.3   Erbium-Doped Fiber Amplifiers

An erbium-doped fiber amplifier (EDFA) is shown in Figure 3.34. It consists of a length of silica fiber whose core is doped with ionized atoms (ions), Er^{3+}, of the rare earth element erbium. This fiber is pumped using a pump signal from a laser, typically at a wavelength of 980 nm or 1480 nm. In order to combine the output of the pump laser with the input signal, the doped fiber is preceded by a wavelength-selective coupler.

At the output, another wavelength-selective coupler may be used if needed to separate the amplified signal from any remaining pump signal power. Usually, an isolator is used at the input and/or output of any amplifier to prevent reflections into the amplifier—we will see in Section 3.5 that reflections can convert the amplifier into a laser, making it unusable as an amplifier.

A combination of several factors has made the EDFA the amplifier of choice in today's optical communication systems: (1) the availability of compact and reliable high-power semiconductor pump lasers, (2) the fact that it is an all-fiber device, making it polarization independent and easy to couple light in and out of it, (3) the simplicity of the device, and (4) the fact that it introduces no crosstalk when amplifying WDM signals. This last aspect is discussed later in the context of semiconductor optical amplifiers.

Principle of Operation

Three of the energy levels of erbium ions in silica glass are shown in Figure 3.35 and are labeled E_1, E_2, and E_3 in order of increasing energy. Several other levels in Er^{3+} are not shown. Each energy level that appears as a discrete line in an isolated ion of erbium is split into multiple energy levels when these ions are introduced into silica glass. This process is termed *Stark splitting*. Moreover, glass is not a crystal and thus does not have a regular structure. Thus the Stark splitting levels introduced are slightly different for individual erbium ions, depending on the local surroundings seen by those ions. Macroscopically, that is, when viewed as a collection of ions,

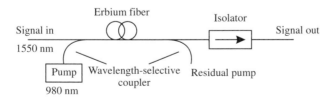

Figure 3.34   An erbium-doped fiber amplifier.

this has the effect of spreading each discrete energy level of an erbium ion into a continuous *energy band*. This spreading of energy levels is a useful characteristic for optical amplifiers since they increase the frequency or wavelength range of the signals that can be amplified. Within each energy band, the erbium ions are distributed in the various levels within that band in a nonuniform manner by a process known as *thermalization*. It is due to this thermalization process that an amplifier is capable of amplifying several wavelengths simultaneously. Note that Stark splitting denotes the phenomenon by which the energy levels of free erbium ions are split into a number of levels, or into an energy band, when the ion is introduced into silica glass. Thermalization refers to the process by which the erbium ions are distributed within the various (split) levels constituting an energy band.

Recall from our discussion of the two-energy-level atomic system that only an optical signal at the frequency f_c satisfying $hf_c = E_2 - E_1$ could be amplified in that case. If these levels are spread into bands, all frequencies that correspond to the energy difference between some energy in the E_2 band and some energy in the E_1 band can be amplified. In the case of erbium ions in silica glass, the set of frequencies that can be amplified by stimulated emission from the E_2 band to the E_1 band corresponds to the wavelength range 1525–1570 nm, a bandwidth of 50 nm, with a peak around

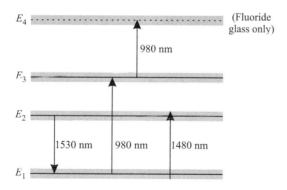

Figure 3.35 Three energy levels E_1, E_2, and E_3 of Er^{3+} ions in silica glass. The fourth energy level, E_4, is present in fluoride glass but not in silica glass. The energy levels are spread into bands by the Stark splitting process. The difference between the energy levels is labeled with the wavelength in nm of the photon corresponding to it. The upward arrows indicate wavelengths at which the amplifier can be pumped to excite the ions into the higher energy level. The 980 nm transition corresponds to the band gap between the E_1 and E_3 levels. The 1480 nm transition corresponds to the the gap between the bottom of the E_1 band to the top of the E_2 band. The downward transition represents the wavelength of photons emitted due to spontaneous and stimulated emission.

1532 nm. By a lucky coincidence, this is exactly one of the low-attenuation windows of standard optical fiber that optical communication systems use.

Denote ionic population in level E_i by N_i, $i = 1, 2, 3$. In thermal equilibrium, $N_1 > N_2 > N_3$. The population inversion condition for stimulated emission from E_2 to E_1 is $N_2 > N_1$ and can be achieved by a combination of absorption and spontaneous emission as follows. The energy difference between the E_1 and E_3 levels corresponds to a wavelength of 980 nm. So if optical power at 980 nm—called the *pump power*—is injected into the amplifier, it will cause transitions from E_1 to E_3 and vice versa. Since $N_1 > N_3$, there will be a net absorption of the 980 nm power. This process is called *pumping*.

The ions that have been raised to level E_3 by this process will quickly transit to level E_2 by the spontaneous emission process. The lifetime for this process, τ_{32}, is about 1 μs. Atoms from level E_2 will also transit to level E_1 by the spontaneous emission process, but the lifetime for this process, τ_{21}, is about 10 ms, which is much larger than the E_3 to E_2 lifetime. Moreover, if the pump power is sufficiently large, ions that transit to the E_1 level are rapidly raised again to the E_3 level only to transit to the E_2 level again. The net effect is that most of the ions are found in level E_2, and thus we have population inversion between the E_2 and E_1 levels. Therefore, if simultaneously a signal in the 1525–1570 nm band is injected into the fiber, it will be amplified by stimulated emission from the E_2 to the E_1 level.

Several levels other than E_3 are higher than E_2 and, in principle, can be used for pumping the amplifier. But the pumping process is more efficient, that is, uses less pump power for a given gain, at 980 nm than these other wavelengths. Another possible choice for the pump wavelength is 1480 nm. This choice corresponds to absorption from the bottom sublevel of the E_1 band to the top sublevel of the E_2 band itself. Pumping at 1480 nm is not as efficient as 980 nm pumping. Moreover, the degree of population inversion that can be achieved by 1480 nm pumping is lower. The higher the population inversion, the lower the noise figure of the amplifier. Thus 980 nm pumping is preferred to realize low-noise amplifiers. However, higher-power pump lasers are available at 1480 nm, compared to 980 nm, and thus 1480 nm pumps find applications in amplifiers designed to yield high output powers. Another advantage of the 1480 nm pump is that the pump power can also propagate with low loss in the silica fiber that is used to carry the signals. Therefore, the pump laser can be located remotely from the amplifier itself. This feature is used in some systems to avoid placing any active components in the middle of the link.

Gain Flatness

Since the population levels at the various levels within a band are different, the gain of an EDFA becomes a function of the wavelength. In Figure 3.36, we plot

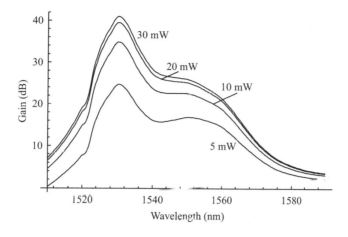

Figure 3.36 The gain of a typical EDFA as a function of the wavelength for four different values of the pump power, obtained through simulations. The length of the doped fiber is taken to be 15 m and 980 nm pumping is assumed.

the gain of a typical EDFA as a function of the wavelength for different values of the pump power. When such an EDFA is used in a WDM communication system, different WDM channels undergo different degrees of amplification. This is a critical issue, particularly in WDM systems with cascaded amplifiers, and is discussed in Section 5.5.2.

One way to improve the flatness of the amplifier gain profile is to use fluoride glass fiber instead of silica fiber, doped with erbium [Cle94]. Such amplifiers are called erbium-doped fluoride fiber amplifiers (EDFFAs). The fluoride glass produces a naturally flatter gain spectrum compared to silica glass. However, there are a few drawbacks to using fluoride glass. The noise performance of EDFFAs is poorer than EDFAs. One reason is that they must be pumped at 1480 nm and cannot be pumped at 980 nm. This is because fluoride glass has an additional higher energy level E_4 above the E_3 level, as shown in Figure 3.35, with the difference in energies between these two levels corresponding to 980 nm. This causes the 980 nm pump power to be absorbed for transitions from the E_3 to E_4 level, which does not produce useful gain. This phenomenon is called *excited state absorption*.

In addition to this drawback, fluoride fiber itself is difficult to handle. It is brittle, difficult to splice with conventional fiber, and susceptible to moisture. Nevertheless, EDFFAs are now commercially available devices.

Another approach to flatten the EDFA gain is to use a filter inside the amplifier. The EDFA has a relatively high gain at 1532 nm, which can be reduced by using a

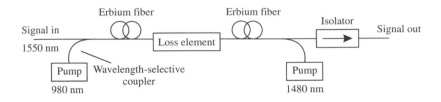

Figure 3.37   A two-stage erbium-doped fiber amplifier with a loss element inserted between the first and second stage.

notch filter in that wavelength region inside the amplifier. Some of the filters described in Section 3.3 can be used for this purpose. Long-period fiber gratings and dielectric thin-film filters are currently the leading candidates for this application.

Multistage Designs

In practice, most amplifiers deployed in real systems are more complicated than the simple structure shown in Figure 3.34. Figure 3.37 shows a more commonly used two-stage design. The two stages are optimized differently. The first stage is designed to provide high gain and low noise, and the second stage is designed to produce high output power. As we will see in Problem 4.5 in Chapter 4, the noise performance of the whole amplifier is determined primarily by the first stage. Thus this combination produces a high-performance amplifier with low noise and high output power. Another important consideration in the design is to provide redundancy in the event of the failure of a pump, the only active component of the amplifier. The amplifier shown in the figure uses two pumps and can be designed so that the failure of one pump has only a small impact on the system performance. Another feature of the two-stage design that we will address in Problem 4.5 is that a loss element can be placed between the two stages with negligible impact on the performance. This loss element may be a gain-flattening filter, a simple optical add/drop multiplexer, or a dispersion compensation module used to compensate for accumulated dispersion along the link.

L-Band EDFAs

So far, we have mostly focused on EDFAs operating in the C-band (1530–1565 nm). Erbium-doped fiber, however, has a relatively long tail to the gain shape extending well beyond this range to about 1605 nm. This has stimulated the development of systems in the so-called L-band from 1565 to 1625 nm. Note that current L-band EDFAs do not yet cover the top portion of this band from 1610 to 1625 nm.

L-band EDFAs operate on the same principle as C-band EDFAs. However, there are significant differences in the design of L- and C-band EDFAs. The gain spectrum of erbium is much flatter intrinsically in the L-band than in the C-band. This makes it easier to design gain-flattening filters for the L-band. However, the erbium gain coefficient in the L-band is about three times smaller than in the C-band. This necessitates the use of either much longer doped fiber lengths or fiber with higher erbium doping concentrations. In either case, the pump powers required for L-band EDFAs are much higher than their C-band counterparts. Due to the smaller absorption cross sections in the L-band, these amplifiers also have higher amplified spontaneous emission. Finally, many of the other components used inside the amplifier, such as isolators and couplers, exhibit wavelength-dependent losses and are therefore specified differently for the L-band than for the C-band. There are several other subtleties associated with L-band amplifiers; see [Flo00] for a summary.

As a result of the significant differences between C- and L-band amplifiers, these amplifiers are usually realized as separate devices, rather than as a single device. In a practical system application, the C- and L-band wavelengths on a fiber are first separated by a demultiplexer, then amplified by separate amplifiers, and recombined together afterwards.

3.4.4 Raman Amplifiers

In Section 2.4.3, we studied stimulated Raman scattering (SRS) as one of the nonlinear impairments that affect signals propagating through optical fiber. The same nonlinearity can be exploited to provide amplification as well. As we saw in Figure 2.17, the Raman gain spectrum is fairly broad and the peak of the gain is centered about 13 THz below the frequency of the pump signal used. In the near-infrared region of interest to us, this corresponds to a wavelength separation of about 100 nm. Therefore, by pumping a fiber using a high-power pump laser, we can provide gain to other signals, with a peak gain obtained 13 THz below the pump frequency. For instance, using pumps around 1460–1480 nm provides Raman gain in the 1550–1600 nm window.

A few key attributes distinguish Raman amplifiers from EDFAs. Unlike EDFAs, we can use the Raman effect to provide gain at any wavelength. An EDFA provides gain in the C- and L-bands (1528–1605 nm). Thus Raman amplification can potentially open up other bands for WDM, such as the 1310 nm window, or the so-called S-band lying just below 1528 nm. Also, we can use multiple pumps at different wavelengths and different powers simultaneously to tailor the overall Raman gain shape.

Second, Raman amplification relies on simply pumping the same silica fiber used for transmitting the data signals, so it can be used to produce a *lumped* or *discrete*

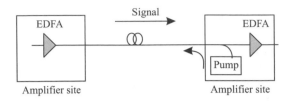

Figure 3.38   Distributed Raman amplifier using a backward propagating pump, shown operating along with discrete erbium-doped fiber amplifiers.

amplifier, as well as a *distributed* amplifier. In the lumped case, the Raman amplifier consists of a sufficiently long spool of fiber along with the appropriate pump lasers in a package. In the distributed case, the fiber can simply be the fiber span of interest, with the pump attached to one end of the span, as shown in Figure 3.38.

Today the most popular use of Raman amplifiers is to complement EDFAs by providing additional gain in a distributed manner in ultra-long-haul systems. The biggest challenge in realizing Raman amplifiers lies in the pump source itself. These amplifiers require high-power pump sources of the order of 1 W or more, at the right wavelength. We will study some techniques for realizing these pump sources in Section 3.5.5.

The noise sources in Raman amplifiers are somewhat different from EDFAs. The Raman gain responds instantaneously to the pump power. Therefore fluctuations in pump power will cause the gain to vary and will appear as crosstalk to the desired signals. This is not the case with EDFAs. We will see in Section 3.4.6 that the response time of the gain is much slower—on the order of milliseconds—in those devices. Therefore, for Raman amplifiers, it is important to keep the pump at a constant power. Having the pump propagate in the opposite direction to the signal helps dramatically because fluctuations in pump power are then averaged over the propagation time over the fiber. To understand this, first consider the case where the pump propagates along with the signal in the same direction. The two waves travel at approximately the same velocity. In this case, when the pump power is high at the input, the signal sees high gain, and when the power is low, the signal sees a lower gain. Now consider the case when the signal and pump travel in opposite directions. To keep things simple, suppose that the pump power varies between two states: high and low. As the signal propagates through the fiber, whenever it overlaps with the pump signal in the high power state, it sees a high gain. When it overlaps with the pump signal in the low power state, it sees a lower gain. If the pump fluctuations are relatively fast compared to the propagation time of the signal across the fiber, the gain variations average out, and by the time the signal exits the fiber, it has seen a constant gain.

Another major concern with Raman amplifiers is crosstalk between the WDM signals due to Raman amplification. A modulated signal at a particular wavelength depletes the pump power, effectively imposing the same modulation on the pump signal. This modulation on the pump then affects the gain seen by the next wavelength, effectively appearing as crosstalk on that wavelength. Again, having the pump propagate in the opposite direction to the signal dramatically reduces this effect. For these reasons, most Raman amplifiers use a counterpropagating pump geometry.

Another source of noise is due to the back-reflections of the pump signal caused by Rayleigh scattering in the fiber. Spontaneous emission noise is relatively low in Raman amplifiers. This is usually the dominant source of noise because, by careful design, we can eliminate most of the other noise sources.

3.4.5 Semiconductor Optical Amplifiers

Semiconductor optical amplifiers (SOAs) actually preceded EDFAs, although we will see that they are not as good as EDFAs for use as amplifiers. However, they are finding other applications in switches and wavelength converter devices. Moreover, the understanding of SOAs is key to the understanding of semiconductor lasers, the most widely used transmitters today.

Figure 3.39 shows the block diagram of a semiconductor optical amplifier. The SOA is essentially a *pn*-junction. As we will explain shortly, the depletion layer that is formed at the junction acts as the *active region*. Light is amplified through stimulated emission when it propagates through the active region. For an amplifier, the two ends of the active region are given an antireflection (AR) coating to eliminate ripples in the amplifier gain as a function of wavelength. Alternatively, the facets may also be angled slightly to reduce the reflection. In the case of a semiconductor laser, there would be no AR coating.

SOAs differ from EDFAs in the manner in which population inversion is achieved. First, the populations are not those of ions in various energy states but of *carriers—electrons* or *holes—*in a semiconductor material. Holes can be thought of also as charge carriers similar to electrons except that they have a positive charge. A semiconductor consists of two bands of electron energy levels: a band of low mobility levels called the *valence band* and a band of high mobility levels called the *conduction band*. These bands are separated by an energy difference called the *bandgap* and denoted by E_g. No energy levels exist in the bandgap. Consider a *p*-type semiconductor material. At thermal equilibrium, there is only a very small concentration of electrons in the conduction band of the material, as shown in Figure 3.40(a). With reference to the previous discussion of EDFAs, it is convenient to

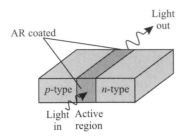

Figure 3.39   Block diagram of a semiconductor optical amplifier. Amplification occurs when light propagates through the active region. The facets are given an antireflective coating to prevent undesirable reflections, which cause ripple in the amplifier gain.

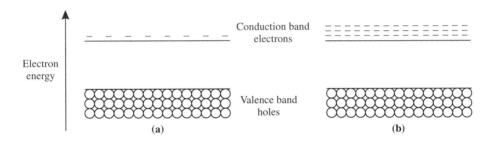

Figure 3.40   The energy bands in a p-type semiconductor and the electron concentration at (a) thermal equilibrium and (b) population inversion.

think of the conduction band as the higher energy band E_2, and the valence band as the lower energy band E_1. The terms *higher* and *lower* refer to the electron energy in these bands. (Note that if we were considering an n-type semiconductor, we would be considering hole energies rather than electron energies, the conduction band would be the lower energy band E_1, and the valence band, the higher energy band E_2.) In the population inversion condition, the electron concentration in the conduction band is much higher, as shown in Figure 3.40(b). This increased concentration is such that, in the presence of an optical signal, there are more electrons transiting from the conduction band to the valence band by the process of stimulated emission than there are electrons transiting from the valence band to the conduction band by the process of absorption. In fact, for SOAs, this condition must be used as the defining one for population inversion, or optical gain.

Population inversion in an SOA is achieved by forward-biasing a *pn*-junction. A *pn-junction* consists of two semiconductors: a *p*-type semiconductor that is doped with suitable impurity atoms so as to have an excess concentration of holes, and an *n*-type semiconductor that has an excess concentration of electrons. When the two semiconductors are in juxtaposition, as in Figure 3.41(a), holes diffuse from the *p*-type semiconductor to the *n*-type semiconductor, and electrons diffuse from the *n*-type semiconductor to the *p*-type semiconductor. This creates a region with net negative charge in the *p*-type semiconductor and a region with net positive charge in the *n*-type semiconductor, as shown in Figure 3.41(b). These regions are devoid of free charge carriers and are together termed the *depletion region*. When no voltage (bias) is applied to the *pn*-junction, the minority carrier concentrations (electrons in the *p*-type region and holes in the *n*-type region) remain at their thermal equilibrium values. When the junction is *forward biased*—positive bias is applied to the *p*-type and negative bias to the *n*-type—as shown in Figure 3.41(c), the width of the depletion region is reduced, and there is a drift of electrons from the *n*-type region to the *p*-type region. This drift increases the electron concentration in the conduction band of the *p*-type region. Similarly, there is a drift of holes from the *p*-type to the *n*-type region that increases the hole concentration in the valence band of the *n*-type region. When the forward bias voltage is sufficiently high, these increased minority carrier concentrations result in population inversion, and the *pn*-junction acts as an optical amplifier.

In practice, a simple *pn*-junction is not used, but a thin layer of a different semiconductor material is sandwiched between the *p*-type and *n*-type regions. Such a device is called a *heterostructure*. This semiconductor material then forms the *active region* or *layer*. The material used for the active layer has a slightly smaller bandgap and a higher refractive index than the surrounding *p*-type and *n*-type regions. The smaller bandgap helps to confine the carriers injected into the active region (electrons from the *n*-type region and holes from the *p*-type region). The larger refractive index helps to confine the light during amplification since the structure now forms a dielectric waveguide (see Section 2.1.2).

In semiconductor optical amplifiers, the population inversion condition (stimulated emission exceeds absorption) must be evaluated as a function of optical frequency or wavelength. Consider an optical frequency f_c such that $hf_c > E_g$, where E_g is the bandgap of the semiconductor material. The lowest optical frequency (or largest wavelength) that can be amplified corresponds to this bandgap. As the forward bias voltage is increased, the population inversion condition for this wavelength is reached first. As the forward bias voltage increases further, the electrons injected into the *p*-type region occupy progressively higher energy levels, and signals with smaller wavelengths can be amplified. In practice, bandwidths on the order of 100 nm can be achieved with SOAs. This is much larger than what is achievable

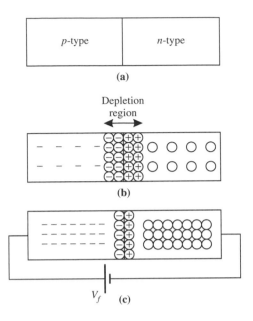

Figure 3.41   A forward-biased *pn*-junction used as an amplifier. (a) A *pn*-junction. (b) Minority carrier concentrations and depletion region with no bias voltage applied. (c) Minority carrier concentrations and depletion region with a forward bias voltage, V_f.

with EDFAs. Signals in the 1.3 and 1.55 μm bands can even be simultaneously amplified using SOAs. Nevertheless, EDFAs are widely preferred to SOAs for several reasons. The main reason is that SOAs introduce severe crosstalk when they are used in WDM systems. This is discussed next. The gains and output powers achievable with EDFAs are higher. The coupling losses and the polarization-dependent losses are also lower with EDFAs since the amplifier is also a fiber. Due to the higher input coupling loss, SOAs have higher *noise figures* relative to EDFAs. (We will discuss noise figure in Section 4.4.5—for our purposes here, we can think of it as a measure of the noise introduced by the amplifier.) Finally, the SOA requires very high-quality antireflective coatings on its facets (reflectivity of less than 10^{-4}), which is not easy to achieve. Higher values of reflectivity create ripples in the gain spectrum and cause gain variations due to temperature fluctuations. (Think of this device as a Fabry-Perot filter with very poor reflectivity, and the spectrum as similar to the one plotted in Figure 3.17 for the case of poor reflectivity.) Alternatively, the SOA facets can be angled to obtain the desired reflectivities, at the cost of an increased polarization dependence.

3.4.6   **Crosstalk in SOAs**

Consider an SOA to which is input the sum of two optical signals at different wavelengths. Assume that both wavelengths are within the bandwidth of the SOA. The presence of one signal will deplete the minority carrier concentration by the stimulated emission process so that the population inversion seen by the other signal is reduced. Thus the other signal will not be amplified to the same extent and, if the minority carrier concentrations are not very large, may even be absorbed! (Recall that if the population inversion condition is not achieved, there is net absorption of the signal.) Thus, for WDM networks, the gain seen by the signal in one channel varies with the presence or absence of signals in the other channels. This phenomenon is called *crosstalk*, and it has a detrimental effect on the system performance.

This crosstalk phenomenon depends on the spontaneous emission lifetime from the high-energy to the low-energy state. If the lifetime is large enough compared to the rate of fluctuations of power in the input signals, the electrons cannot make the transition from the high-energy state to the lower-energy state in response to these fluctuations. Thus there is no crosstalk whatsoever. In the case of SOAs, this lifetime is on the order of nanoseconds. Thus the electrons can easily respond to fluctuations in power of signals modulated at gigabit/second rates, resulting in a major system impairment due to crosstalk. In contrast, the spontaneous emission lifetime in an EDFA is about 10 ms. Thus crosstalk is introduced only if the modulation rates of the input signals are less than a few kilohertz, which is not usually the case. Thus EDFAs are better suited for use in WDM systems than SOAs.

There are several ways of reducing the crosstalk introduced by SOAs. One way is to operate the amplifier in the small signal region where the gain is relatively independent of the input power of the signal. Another is to *clamp* the gain of the amplifier using a variety of techniques, so that even at high signal powers, its gain remains relatively constant, independent of the input signal. Also, if a sufficiently large number of signals at different wavelengths are present, although each signal varies in power, the total signal power into the amplifier can remain fairly constant.

The crosstalk effect is not without its uses. We will see in Section 3.8.2 that it can be used to make a *wavelength converter*.

$\underline{3.5}$ **Transmitters**

We will study many different types of light sources in this section. The most important one is the laser, of which there are many different types. Lasers are used as transmitters and also to pump both erbium-doped and Raman amplifiers.

When using a laser as a light source for WDM systems, we need to consider the following important characteristics:

1. Lasers need to produce a reasonably high output power. For WDM systems, the typical laser output powers are in the 0–10 dBm range. Related parameters are the threshold current and slope efficiency. Both of these govern the efficiency of converting electrical power into optical power. The *threshold current* is the drive current at which the laser starts to emit optical power, and the *slope efficiency* is the ratio of output optical power to drive current.

2. The laser needs to have a narrow *spectral width* at a specified operating wavelength so that the signal can pass through intermediate filters and multiple channels can be placed close together. The side-mode suppression ratio is a related parameter, which we will discuss later. In the case of a tunable laser, the operating wavelength can be varied.

3. Wavelength stability is an important criterion. When maintained at constant temperature, the wavelength drift over the life of the laser needs to be small relative to the wavelength spacing between adjacent channels.

4. For lasers that are modulated, chromatic dispersion can be an important limiting factor that affects the link length. We will see in Chapter 5 that the dispersion limit can be stated in terms of a penalty as a function of the total accumulated dispersion along the link.

Pump lasers are required to produce much higher power levels than lasers used as WDM sources. Pump lasers used in erbium-doped fiber amplifiers put out 100–200 mW of power, and pump lasers for Raman amplifiers may go up to a few watts.

3.5.1   Lasers

A laser is essentially an optical amplifier enclosed within a reflective cavity that causes it to oscillate via positive feedback. *Semiconductor lasers* use semiconductors as the gain medium, whereas *fiber lasers* typically use erbium-doped fiber as the gain medium. Semiconductor lasers are by far the most popular light sources for optical communication systems. They are compact, usually only a few hundred micrometers in size. Since they are essentially *pn*-junctions, they can be fabricated in large volumes using highly advanced integrated semiconductor technology. The lack of any need for optical pumping, unlike fiber lasers, is another advantage. In fact, a fiber laser typically uses a semiconductor laser as a pump! Semiconductor lasers are also highly efficient in converting input electrical (pump) energy into output optical energy.

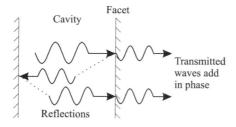

Figure 3.42   Reflection and transmission at the facets of a Fabry-Perot cavity.

Both semiconductor and erbium fiber lasers are capable of achieving high output powers, typically between 0 and 20 dBm, although semiconductor lasers used as WDM sources typically have output powers between 0 and 10 dBm. Fiber lasers are used mostly to generate periodic trains of very short pulses (by using a technique called mode locking, discussed later in this section).

Principle of Operation

Consider any of the optical amplifiers described, and assume that a part of the optical energy is reflected at the ends of the amplifying or *gain medium,* or *cavity,* as shown in Figure 3.42. Further assume that the two ends of the cavity are plane and parallel to each other. Thus the gain medium is placed in a *Fabry-Perot cavity* (see Section 3.3.5). Such an optical amplifier is called a *Fabry-Perot amplifier.* The two end faces of the cavity (which play the role of the mirrors) are called *facets.*

The result of placing the gain medium in a Fabry-Perot cavity is that the gain is high only for the resonant wavelengths of the cavity. The argument is the same as that used in the case of the Fabry-Perot filter (Section 3.3.5). After one pass through the cavity, as shown in Figure 3.42, a part of the light leaves the cavity through the right facet, and a part is reflected. A part of the reflected wave is again reflected by the left facet to the right facet. For the resonant wavelengths of the cavity, all the light waves transmitted through the right facet *add in phase.* As a result of in-phase addition, the amplitude of the transmitted wave is greatly increased for these resonant wavelengths compared to other wavelengths. Thus, when the facets are at least partially reflecting, the gain of the optical amplifier becomes a function of the wavelength.

If the combination of the amplifier gain and the facet reflectivity is sufficiently large, the amplifier will start to "oscillate," or produce light output, even in the absence of an input signal. For a given device, the point at which this happens is called its *lasing threshold*. Beyond the threshold, the device is no longer an amplifier but an oscillator or *laser*. This occurs because the stray spontaneous emission, which is always present at all wavelengths within the bandwidth of the amplifier, gets amplified even without an input signal and appears as the light output. This process is quite similar to what happens in an electronic oscillator, which can be viewed as an (electronic) amplifier with positive feedback. (In electronic oscillators, the thermal noise current due to the random motion of electrons serves the same purpose as spontaneous emission.) Since the amplification process is due to stimulated emission, the light output of a laser is *coherent*. The term *laser* is an acronym for *light amplification* by *stimulated emission* of *radiation*.

Longitudinal Modes

For laser oscillation to occur at a particular wavelength, two conditions must be satisfied. First, the wavelength must be within the bandwidth of the gain medium that is used. Thus, if a laser is made from erbium-doped fiber, the wavelength must lie in the range 1525–1560 nm. The second condition is that the length of the cavity must be an integral multiple of half the wavelength in the cavity. For a given laser, all the wavelengths that satisfy this second condition are called the *longitudinal modes* of that laser. The adjective "longitudinal" is used to distinguish these from the waveguide modes (which should strictly be called spatial modes) that we studied in Section 2.1.

The laser described earlier is called a *Fabry-Perot laser* (FP laser) and will usually oscillate simultaneously in several longitudinal modes. Such a laser is termed a *multiple-longitudinal mode* (MLM) laser. MLM lasers have large spectral widths, typically around 10 nm. A typical spectrum of the output of an MLM laser is shown in Figure 3.43(a). We saw in Section 2.3 that for high-speed optical communication systems, the spectral width of the source must be as narrow as possible to minimize the effects of chromatic dispersion. Likewise, a narrow spectral width is also needed to minimize crosstalk in WDM systems (see Section 3.3). Thus it is desirable to design a laser that oscillates in a single-longitudinal mode (SLM) only. The spectrum of the output of an SLM laser is shown in Figure 3.43(b). Single-longitudinal mode oscillation can be achieved by using a filtering mechanism in the laser that selects the desired wavelength and provides loss at the other wavelengths. An important attribute of such a laser is its *side-mode suppression ratio,* which determines the level to which the other longitudinal modes are suppressed, compared to the main mode. This ratio is typically more than 30 dB for practical SLM lasers. We will now consider some mechanisms that are commonly employed for realizing SLM lasers.

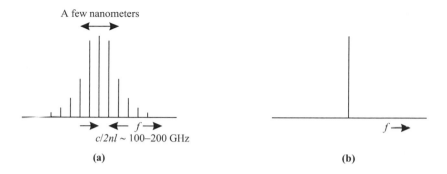

Figure 3.43 The spectrum of the output of (a) an MLM laser and (b) an SLM laser. The laser cavity length is denoted by l, and its refractive index by n. The frequency spacing between the modes of an MLM laser is then $c/2nl$.

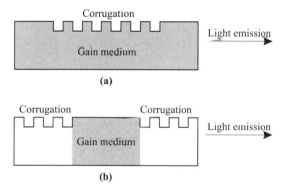

Figure 3.44 The structure of (a) a DFB laser and (b) a DBR laser. In a DFB laser, the gain and wavelength selection are obtained in the same region, whereas in a DBR laser, the wavelength selection region is outside the gain region.

Distributed-Feedback Lasers

In the Fabry-Perot laser described earlier, the feedback of the light occurs from the reflecting facets at the ends of the cavity. Thus the feedback can be said to be *localized* at the facets. Light feedback can also be provided in a *distributed* manner by a series of closely spaced reflectors. The most common means of achieving this is by providing a periodic variation in the width of the cavity, as shown in Figure 3.44(a) and (b).

In the corrugated section of the cavity, the incident wave undergoes a series of reflections. The contributions of each of these reflected waves to the resulting transmitted wave from the cavity add in phase if the period of the corrugation is an integral multiple of half the wavelength in the cavity. The reasoning for this condition is the same as that used for the Fabry-Perot cavity. This condition is called the Bragg condition and was discussed in Section 3.3.3. The Bragg condition will be satisfied for a number of wavelengths, but the strongest transmitted wave occurs for the wavelength for which the corrugation period is *equal* to half the wavelength, rather than some other integer multiple of it. Thus this wavelength gets preferentially amplified at the expense of the other wavelengths. By suitable design of the device, this effect can be used to suppress all other longitudinal modes so that the laser oscillates in a single-longitudinal mode whose wavelength is equal to twice the corrugation period. By varying the corrugation period at the time of fabrication, different operating wavelengths can be obtained.

Any laser that uses a corrugated waveguide to achieve single-longitudinal mode operation can be termed a distributed-feedback laser. However, the acronym *DFB laser* is used only when the corrugation occurs within the gain region of the cavity, as shown in Figure 3.44(a). When the corrugation is outside the gain region, as in Figure 3.44(b), the laser is called a *distributed Bragg reflector* (DBR) laser. The main advantage of DBR lasers is that the gain region is decoupled from the wavelength selection region. Thus it is possible to control both regions independently. For example, by changing the refractive index of the wavelength selection region, the laser can be tuned to a different wavelength without affecting its other operating parameters. Indeed, this is how many of the tunable lasers that we will study in Section 3.5.3 are realized.

DFB lasers are inherently more complex to fabricate than FP lasers and thus relatively more expensive. However, DFB lasers are required in almost all high-speed transmission systems today. FP lasers are used for shorter-distance data communication applications.

Reflections into a DFB laser cause its wavelength and power to fluctuate and are prevented by packaging the laser with an isolator in front of it. The laser is also usually packaged with a thermoelectric (TE) cooler and a photodetector attached to its rear facet. The TE cooler is necessary to maintain the laser at a constant operating temperature to prevent its wavelength from drifting. The temperature sensitivity of a semiconductor DFB laser operating in the 1.55 μm wavelength region is about 0.1 nm/°C. The photodetector monitors the optical power leaking out of the rear facet, which is proportional to the optical power coming out of the laser.

The packaging of a DFB laser contributes a significant fraction of the overall cost of the device. For WDM systems, it is very useful to package multiple DFB lasers at different wavelengths inside a single package. This device can then serve as a

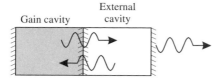

Figure 3.45 The structure of an external cavity laser.

multiwavelength light source or, alternatively, as a tunable laser (only one of the lasers in the array is turned on, depending on the desired wavelength). These lasers can all be grown on a single substrate in the form of an array. Four- and eight-wavelength laser arrays have been fabricated in research laboratories, but have not quite progressed to volume manufacturing. The primary reason for this is the relatively low yield of the array as a whole. If one of the lasers doesn't meet specifications, the entire array will have to be discarded.

External Cavity Lasers

Suppression of oscillation at more than one longitudinal mode can also be achieved by using another cavity—called an *external cavity*—following the primary cavity where gain occurs. This is illustrated in Figure 3.45. Just as the primary cavity has resonant wavelengths, so does the external cavity. This effect can be achieved, for example, by using reflecting facets for the external cavity as well. The net result of having an external cavity is that the laser is capable of oscillating only at those wavelengths that are resonant wavelengths of *both* the primary and external cavity. By suitable design of the two cavities, it can be ensured that only one wavelength in the gain bandwidth of the primary cavity satisfies this condition. Thus the laser oscillation can be confined to a single longitudinal mode.

Instead of another Fabry-Perot cavity, as shown in Figure 3.45, we can use a diffraction grating (see Section 3.3.1) in the external cavity, as shown in Figure 3.46. Such a laser is called a *grating external cavity* laser. In this case, the facet of the gain cavity facing the grating is given an antireflection coating. The wavelengths reflected by the diffraction grating back to the gain cavity are determined by the pitch of the grating (see Section 3.3.1) and its tilt angle (see Figure 3.46) with respect to the gain cavity. An external cavity laser, in general, uses a *wavelength-selective mirror* instead of a wavelength-flat mirror. (A highly polished and/or metal-coated facet used in conventional lasers acts as a wavelength-flat mirror.) The reflectivity of a wavelength-selective mirror is a function of the wavelength. Thus only certain wavelengths experience high reflectivities and are capable of lasing. If the

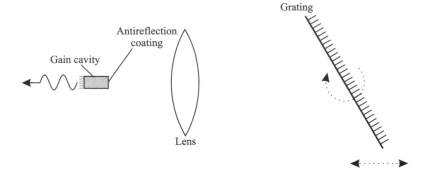

Figure 3.46   The structure of a grating external cavity laser. By rotating the grating, we can tune the wavelength of the laser.

wavelength-selective mirror is chosen suitably, only one such wavelength will occur within the gain bandwidth, and we will have a single-mode laser.

Several of the filters discussed in Section 3.3 can be used as wavelength-selective mirrors in external cavity lasers. We have already seen the use of the diffraction grating (Section 3.3.1) and Fabry-Perot filter (Section 3.3.5) in external cavity lasers. These laser structures are used today primarily in optical test instruments and are not amenable to low-cost volume production as SLM light sources for transmission systems. One version of the external cavity laser, though, appears to be particularly promising for this purpose. This device uses a fiber Bragg grating in front of a conventional FP laser with its front facet AR coated. This device then acts as an SLM DBR laser. It can be fabricated at relatively low cost compared to DFB lasers and is inherently more temperature stable in wavelength due to the low temperature coefficient of the fiber grating.

One disadvantage of external cavity lasers is that they cannot be modulated directly at high speeds. This is related to the fact that the cavity length is large.

Vertical Cavity Surface-Emitting Lasers

In this section, we will study another class of lasers that achieve single-longitudinal mode operation in a slightly different manner. As we saw in Figure 3.43, the frequency spacing between the modes of an MLM laser is $c/2nl$, where l is the length of the cavity and n is its refractive index. If we were to make the length of the cavity sufficiently small, the mode spacing increases such that only one longitudinal mode occurs within the gain bandwidth of the laser. It turns out that making a very thin active layer is much easier if the active layer is deposited on a semiconductor substrate, as illustrated in Figure 3.47. This leads to a vertical cavity with the mirrors

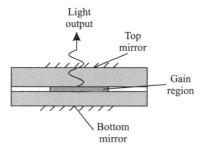

Figure 3.47 The structure of a VCSEL.

being formed on the top and bottom surfaces of the semiconductor wafer. The laser output is also taken from one of these (usually top) surfaces. For these reasons, such lasers are called *vertical cavity surface-emitting lasers* (VCSELs). The other lasers that we have been discussing hitherto can thus be referred to as *edge-emitting lasers*.

Since the gain region has a very short length, very high mirror reflectivities are required in order for laser oscillation to occur. Such high mirror reflectivities are difficult to obtain with metallic surfaces. A stack of alternating low- and high-index dielectrics serves as a highly reflective, though wavelength-selective, mirror. The reflectivity of such a mirror is discussed in Problem 3.13. Such dielectric mirrors can be deposited at the time of fabrication of the laser.

One problem with VCSELs is the large ohmic resistance encountered by the injected current. This leads to considerable heating of the device and the need for efficient thermal cooling. Many of the dielectric materials used to make the mirrors have low thermal conductivity. So the use of such dielectric mirrors makes room temperature operation of VCSELs difficult to achieve since the heat generated by the device cannot be dissipated easily. For this reason, for several years after they were first demonstrated in 1979, VCSELs were not capable of operating at room temperature. However, significant research effort has been expended on new materials and techniques, and as of this writing, VCSELs operating at 1.3 μm at room temperature have been demonstrated [Har00].

The advantages of VCSELs, compared to edge-emitting lasers, include simpler and more efficient fiber coupling, easier packaging and testing, and their ability to be integrated into multiwavelength arrays. As of this writing, VCSELs operating at 0.85 μm are commercially available and used for low-cost, short-distance multimode fiber interconnections. For single-mode fiber applications, 1.3 μm VCSELs are now becoming commercially available. There is work under way to produce 1.5 μm lasers as well. We will see one example in Section 3.5.3.

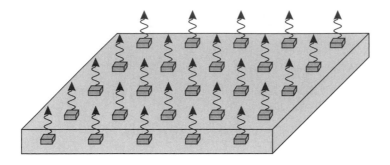

Figure 3.48   A two-dimensional array of vertical cavity surface-emitting lasers.

In a WDM system, many wavelengths are transmitted simultaneously over each link. Usually, this requires a separate laser for each wavelength. The cost of the transmitters can be significantly reduced if all the lasers can be integrated on a single substrate. This is the main motivation for the development of arrayed lasers such as the DFB laser arrays that we discussed earlier. Moreover, an arrayed laser can be used as a tunable laser simply by turning on only the one required laser in the array. The use of surface-emitting lasers enables us to fabricate a two-dimensional array of lasers, as shown in Figure 3.48. Much higher array packing densities can be achieved using surface-emitting lasers than edge-emitting ones because of this added dimension. However, it is harder to couple light from the lasers in this array onto optical fiber since multiplexers that work conveniently with this two-dimensional geometry are not readily available. These arrayed lasers have the same yield problem as other arrayed laser structures; if one of the lasers doesn't meet specifications, the entire array will have to be discarded.

Mode-Locked Lasers

Mode-locked lasers are used to generate narrow optical pulses that are needed for the high-speed TDM systems that we will study in Chapter 12. Consider a Fabry-Perot laser that oscillates in N longitudinal modes, which are adjacent to each other. This means that if the wavelengths of the modes are $\lambda_0, \lambda_1, \ldots, \lambda_{N-1}$, the cavity length l satisfies $l = (k+i)\lambda_i/2, i = 0, 1, \ldots, N-1$, for some integer k. From this condition, it can be shown (see Problem 3.7) that the corresponding frequencies $f_0, f_1, \ldots, f_{N-1}$ of these modes must satisfy $f_i = f_0 + i\Delta f, i = 0, 1, \ldots, N-1$. The oscillation at frequency f_i is of the form $a_i \cos(2\pi f_i t + \phi_i)$, where a_i is the amplitude and ϕ_i the phase of mode i. (Strictly speaking, this is the distribution in time of the electric field

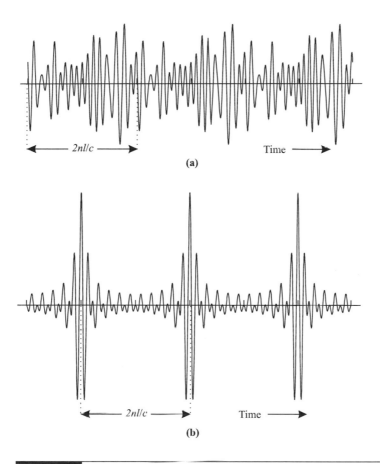

Figure 3.49 Output oscillation of a laser oscillating simultaneously in 10 longitudinal modes. (a) The phases of the modes are chosen at random. (b) All the phases are equal to each other; such a laser is said to be mode locked.

associated with the longitudinal mode.) Thus the total laser output oscillation takes the form

$$\sum_{i=0}^{N-1} a_i \cos(2\pi f_i t + \phi_i).$$

This expression is plotted in Figure 3.49 for $N = 10$, for different sets of values of the ϕ_i. In Figure 3.49(a), the ϕ_i are chosen at random, and in Figure 3.49(b), they are chosen to be equal to each other. All the a_i are chosen to be equal in both cases, and the frequency f_0 has been diminished from its typical value for the purpose of illustration.

From Figure 3.49(a), we observe that the output amplitude of an MLM laser varies rapidly with time when it is not mode locked. We have also seen in Figure 3.43(a) that the frequency spacing between adjacent longitudinal modes is $c/2nl$. If $n = 3$ and $l = 200$ μm, which are typical values for semiconductor lasers, this frequency spacing is 250 GHz. Thus these amplitude fluctuations occur extremely rapidly (at a time scale on the order of a few picoseconds) and pose no problems for on-off modulation even at bit rates of a few tens of gigabits per second.

We see from Figure 3.49(b) that when the ϕ_i are chosen to be equal to each other, the output oscillation of the laser takes the form of a periodic train of narrow pulses. A laser operating in this manner is called a *mode-locked laser* and is the most common means of generating narrow optical pulses.

The time interval between two pulses of a mode-locked laser is $2nl/c$, as indicated in Figure 3.49(b). For a typical semiconductor laser, as we have seen earlier, this corresponds to a few picoseconds. For modulation in the 1–10 GHz range, the interpulse interval should be in the 0.1–1 ns range. Cavity lengths, l, of the order of 1–10 cm (assuming $n = 1.5$) are required in order to realize mode-locked lasers with interpulse intervals in this range. These large cavity lengths are easily obtained using fiber lasers, which require the length anyway to obtain sufficient gain to induce lasing.

The most common means of achieving mode lock is by modulating the gain of the laser cavity. Either amplitude or frequency modulation can be used. Mode locking using amplitude modulation is illustrated in Figure 3.50. The gain of the cavity is modulated with a period equal to the interpulse interval, namely, $2nl/c$. The amplitude of this modulation is chosen such that the average gain is insufficient for any single mode to oscillate. However, if a large number of modes are in phase, there can be a sufficient buildup in the energy inside the cavity for laser oscillation to occur at the instants of high gain, as illustrated in Figure 3.50.

Gain modulation of the fiber laser can be achieved by introducing an external modulator inside the cavity.

3.5.2   Light-Emitting Diodes

Lasers are expensive devices and are not affordable for many applications where the data rates are low and distances are short. This is the case in many data communications applications (see Chapter 6) and in some access networks (Chapter 11). In such cases, *light-emitting diodes* (LEDs) provide a cheaper alternative.

An LED is a forward-biased *pn*-junction in which the recombination of the injected minority carriers (electrons in the *p*-type region and holes in the *n*-type

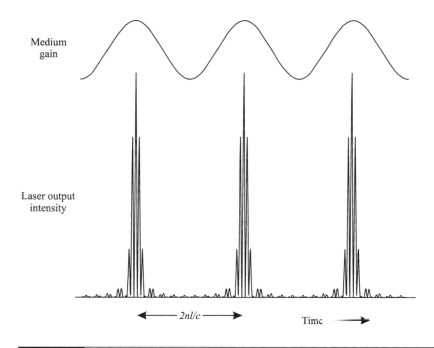

Figure 3.50   Illustration of mode locking by amplitude modulation of the cavity gain.

region) by the spontaneous emission process produces light. (Unwanted nonradiative recombination is also possible and is an important factor affecting the performance of LEDs.) Because spontaneous emission occurs within the entire bandwidth of the gain medium (corresponding to all energy differences between the valence and conduction bands for an LED), the light output of an LED has a broad spectrum, unlike that of a laser. We can crudely think of an LED as a laser with facets that are not very reflective. Increasing the pump current simply increases the spontaneous emission, and there is no chance to build up stimulated emission due to the poor reflectivity of the facets. For this reason, LEDs are also not capable of producing high-output powers like lasers, and typical output powers are on the order of −20 dBm. They cannot be directly modulated (see Section 3.5.4) at data rates higher than a few hundred megabits per second.

In some low-speed, low-budget applications, there is a requirement for a source with a narrow spectral width. DFB lasers provide narrow spectral widths but may be too expensive for these applications. In such cases, *LED slicing* provides a cheaper alternative. An LED slice is the output of a narrow passband optical filter placed in front of the LED. The optical filter selects a portion of the LED's output. Different filters can be used to select (almost) nonoverlapping spectral slices of the LED output. Thus one LED can be shared by a number of users. We will see an application for this technique in Chapter 11.

3.5.3   Tunable Lasers

Tunable lasers are highly desirable components for WDM networks for several reasons. Fixed-wavelength DFB lasers work very well for today's applications. However, each wavelength requires a different, unique laser. This implies that in order to supply a 100-channel WDM system, we need to stock 100 different laser types. The inventory and sparing issues associated with this are expensive and affect everybody from laser manufacturers to network operators. Laser manufacturers need to set up multiple production and test lines for each laser wavelength (or time-share the same production and test line but change the settings each time a different laser is made). Equipment suppliers need to stock these different lasers and keep inventories and spares for each wavelength. Finally, network operators need to stockpile spare wavelengths in the event transmitters fail in the field and need to be replaced. Having a tunable laser alleviates this problem dramatically.

Tunable lasers are also one of the key enablers of reconfigurable optical networks. They provide the flexibility to choose the transmit wavelength at the source of a lightpath. For instance, if we wanted to have a total of, say, four lightpaths starting at a node, we would equip that node with four tunable lasers. This would allow us to choose the four transmit wavelengths in an arbitrary manner. In contrast, if we were to use fixed-wavelength lasers, either we would have to preequip the node with a large number of lasers to cover all the possible wavelengths, or we would have to manually equip the appropriate wavelength as needed. We will see more of this application in Chapter 7. The tuning time required for such applications is on the order of milliseconds because the wavelength selection happens only at the times where the lightpath is set up, or when it needs to be rerouted in the event of a failure.

Another application for tunable lasers is in optical packet-switched networks, where data needs to be transmitted on different wavelengths on a packet-by-packet basis. These networks are primarily in their early stages of research today, but supporting such an application would require tuning times on the order of nanoseconds to microseconds, depending on the bit rate and packet size used.

Finally, tunable lasers are a staple in most WDM laboratories and test environments, where they are widely used for characterizing and testing various types of optical equipment. These lasers are typically tabletop-type devices and are not suitable for use in telecom applications, which call for compact, low-cost semiconductor lasers.

The InGaAsP/InP material used for most long-wavelength lasers is enhanced by the use of *quantum well* structures and has an overall gain bandwidth of about 250 nm at 1.55 μm, large enough for the needs of current WDM systems. However, the tuning mechanisms available potentially limit the tuning range to a small fraction of this number. The following tuning mechanisms are typically used:

- Injecting current into a semiconductor laser causes a change in the refractive index of the material, which in turn changes the lasing wavelength. This effect is fairly small—about a 0.5–2% change in the refractive index (and the wavelength) is possible. This effect can be used to effect a tuning range of approximately 10–15 nm in the 1.55 μm wavelength window.

- Temperature tuning is another possibility. The wavelength sensitivity of a semiconductor laser to temperature is approximately 0.1 nm/°C. In practice, the allowed range for temperature tuning is about 1 nm, corresponding to a 10°C temperature variation. Operating the laser at significantly higher temperatures than room temperature causes it to age rapidly, degrading its lifetime.

- Mechanical tuning can be used to provide a wide tunable range in lasers that use a separate external cavity mechanism. Many of these tend to be bulky. We will look at one laser structure of this type using a micro-electro-mechanical tuning mechanism, which is quite compact.

As we will see, the tuning mechanisms are complex, and in many cases, interact with the modulation mechanisms, making it difficult to directly modulate most of the tunable lasers that we will study here.

The ideal tunable laser is a device that can tune rapidly over a wide continuous tuning range of over 100 nm. It should be stable over its lifetime and easily controllable and manufacturable. Many of the tunable laser technologies described here have been around for many years, but we are only now beginning to see commercially available devices due to the complexity of manufacturing and controlling these devices and solving the reliability challenges. The strong market demand for these devices has stimulated a renewed effort toward solving these problems.

External Cavity Lasers

External cavity lasers can be tuned if the center wavelength of the grating or other wavelength-selective mirror used can be changed. Consider the grating external cavity laser shown in Figure 3.46. The wavelength selected by the grating for reflection to the gain cavity is determined by the pitch of the diffraction grating, its tilt angle with respect to the gain cavity, and its distance from the gain cavity (see Section 3.3.1, specifically, (3.9)). Thus by varying the tilt angle and the distance of the diffraction grating from the gain cavity (shown by the dotted arrows in Figure 3.46), the laser

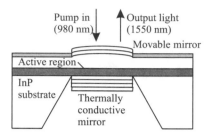

Figure 3.51   Structure of a tunable micro-electro-mechanical vertical cavity surface-emitting laser (MEM-VCSEL) (from [Vak99]).

wavelength can be changed. This is a slow method of tuning since the tilt and position of the diffraction grating have to be changed by mechanical means. However, a very wide tuning range of about 100 nm can be obtained for semiconductor lasers by this method. This method of tuning is appropriate for test instruments but not for a compact light source for communication systems.

Tunable VCSELs

We studied VCSELs in Section 3.5.1. There we saw that the main challenges in realizing long-wavelength 1.55 μm VCSELs were in obtaining sufficient cavity gain, obtaining highly reflective mirror surfaces, dealing with the heat dissipation, and making the laser operate in a single-longitudinal mode. Figure 3.51 shows a VCSEL design [Vak99] that attempts to solve these problems, while also making the laser itself tunable. The tunability is achieved by having the upper mirror be a movable micro-electro-mechanical (MEM) membrane. The cavity spacing can be adjusted by moving the upper mirror by applying a voltage across the upper and lower mirrors. The upper mirror is curved to prevent beam walk-off in the cavity, leading to better stability of the lasing mode.

To conduct the heat away from the bottom mirror, a hole is etched in the InP substrate. The design uses a 980 nm pump laser to pump the VCSEL cavity. Any pump wavelength lower than the desired lasing wavelength can be used to excite the semiconductor electrons to the conduction band. For example, the 980 nm semiconductor pumps used to pump erbium-doped fiber amplifiers can be used here as well. By designing the pump spot size to match the size of the fundamental lasing mode, the laser can be made single mode while suppressing the higher-order Fabry-Perot cavity modes. Using gain to perform this function is better than trying to design the cavity to provide higher loss at the higher-order modes. The high gain also allows the output coupling reflectivity to be reduced, while still maintaining sufficient inversion inside the cavity to prevent excessive recombination.

The laser described in [Vak99] was able to put out about 0 dBm of power in continuous-wave (CW) mode over a tuning range of 50 nm.

Two- and Three-Section DBR Lasers

We saw earlier that we can change the refractive index of a semiconductor laser by injecting current into it. This can result in an overall tuning range of about 10 nm. The DFB laser shown in Figure 3.44 can be tuned by varying the forward-bias current, which changes the refractive index, which in turn changes the effective pitch of the grating inside the laser cavity. However, changing the forward-bias current also changes the output power of the device, making this technique unsuitable for use in a DFB laser.

A conventional DBR laser also has a single gain region, which is controlled by injecting a forward-bias current I_g, as shown in Figure 3.44(b). Varying this current only changes the output power and doesn't affect the wavelength. This structure can be modified by adding another electrode to inject a separate current I_b into the Bragg region that is decoupled from the gain region, as shown in Figure 3.52(a). This allows the wavelength to be controlled independently of the output power.

As in a conventional DBR laser, the laser has multiple closely spaced cavity modes corresponding to the cavity length, of which the one that lases corresponds to the wavelength peak of the Bragg grating. As the wavelength peak of the grating is varied by varying I_b, the laser hops from one cavity mode to another. This effect is shown in Figure 3.52(a). As the current I_b is varied, the Bragg wavelength changes. At the same time, there is also a small change in the cavity mode spacing due to the change in refractive index in the grating portion of the overall cavity. The two changes don't track each other, however. As a result, as I_b is varied and the Bragg wavelength changes, the laser wavelength changes, with the laser remaining on the same cavity mode for some time. As the current is varied further, the laser hops to the next cavity mode. By careful control over the cavity length, we can make the wavelength spacing between the cavity modes equal to the WDM channel spacing.

In order to obtain continuous tuning over the entire wavelength range, an additional third *phase* section can be added to the DBR, as shown in Figure 3.52(b). Injecting a third current I_p into this section allows us to obtain control of the cavity mode spacing, independent of the other effects that are present in the laser. Recall from Section 3.3.5 that it is sufficient to vary the effective cavity length by half a wavelength (or equivalently, the phase by π) in order to obtain tuning across an entire free spectral range. This is a small fraction of the overall cavity length and is easily achieved by current injection into the phase section. By carefully controlling I_p

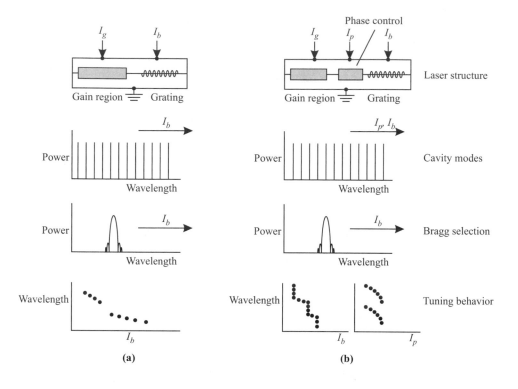

Figure 3.52   Two- and three-section DBR lasers and their principle of wavelength selection. (a) Two-section DBR showing separate control of the gain and Bragg sections. (c) Three-section DBR, which adds an additional control for the cavity phase.

to line up a cavity mode to correspond to the wavelength peak of the Bragg grating determined by I_b, the wavelength can be tuned continuously over the tunable range.

Two- and three-section DBRs capable of tuning over 32 channels in 50 GHz increments were demonstrated several years ago [KK90, Kam96] and are nearing commercial availability.

Clearly a major problem that needs to be solved is in the control of these lasers, which can be quite complicated. As the laser ages, or temperature changes, the control currents may need to be recaliberated; otherwise the laser could end up hopping to another wavelength. The hopping could happen back and forth rapidly, and could manifest itself as relative intensity noise (RIN) at the laser output. In a sense, we are eliminating the very fact that made DFB lasers so wavelength stable—a fixed grating.

These problems are only compounded further in the more complex laser structures that we will discuss next.

The DBRs that we have looked at so far are all limited to about a 10–15 nm tuning range by the 0.5–2% change in refractive index possible. Increasing the tuning range beyond this value requires a new bag of tricks. One trick makes the laser wavelength dependent on the *difference* between the refractive indices of two different regions. The overall variation possible is much higher than the variation of each of the individual regions. The so-called vertical grating-assisted coupler filter (VGF) lasers [AKB+92, AI93] make use of this principle. The second trick is to make use of the Vernier effect, where we have two combs of wavelengths, each with slightly different wavelength spacing. The combination of the two combs yields another periodic comb with a much higher wavelength spacing between its peaks. Problem 3.28 explains this effect in more detail. Even if each comb can be tuned only to a small extent, the combination of the two combs yields a much higher tuning range. The *sampled grating* (SG) DBRs and the *super-structure grating* (SSG) DBRs [JCC93, Toh93] use this approach. Finally, the *grating coupler sampled reflector* (GCSR) laser [WMB92, Rig95] is a combination of both approaches.

VGF Lasers

Figure 3.53 shows the schematic of a VGF laser. It consists of two waveguides, with a coupling region between them. Its operation is similar to that of the acousto-optic tunable filter of Section 3.3.9. Using (3.17), wavelength λ is coupled from one waveguide of refractive index n_1 to the other of refractive index n_2 if

$$\lambda = \Lambda_B(n_1 - n_2)$$

where Λ_B is the period of the Bragg grating. Changing the refractive index of one region, say, n_1 by Δn_1, therefore results in a wavelength tuning of $\Delta \lambda$ where

$$\frac{\Delta \lambda}{\lambda} \approx \frac{\Delta n_1}{n_1 - n_2}.$$

This is significantly larger than the $\Delta n_1 / n_1$ ratio that is achievable in the two- and three-section DBRs that we studied earlier.

In Figure 3.53, current I_c controls the index n_1, and current I_g provides the current to the gain region in the other waveguide. Just as with the two- and three-section DBRs, in order to obtain continuous tuning, the cavity mode spacing needs to be controlled by a third current I_p. Lasers with tuning ranges over 70 nm have been demonstrated using this approach.

One major problem with this approach is that the cavity length needs to be fairly long (typically 800–1000 μm) to get good coupling between the waveguides. This causes the cavity modes to be spaced very closely together. The laser therefore tends

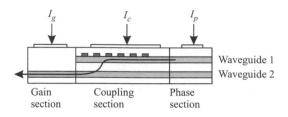

Figure 3.53   A vertical grating-assisted coupler filter tunable laser.

to hop fairly easily from one cavity mode to another even though all the control currents are held steady. This effectively results in a poor side-mode suppression, making the laser not as suitable for high-bit-rate long-distance transmission.

Sampled Grating and Super-Structure Grating DBR Lasers

A sampled grating DBR laser is shown in Figure 3.54. It has two gratings, one in the front and one in the back. The Bragg grating in front is interrupted periodically (or *sampled*) with a period Λ_1. This results in a periodic set of Bragg reflector peaks, spaced apart in wavelength by $\lambda^2/2n_{\text{eff}}\Lambda_1$, as shown in Figure 3.54, where λ is the nominal center wavelength. The peaks gradually taper off in reflectivity, with the highest reflection occurring at the Bragg wavelength $2n_{\text{eff}}\Lambda$, where Λ is the period of the grating. The grating in the back is sampled with a different period Λ_2, which results in another set of reflection peaks spaced apart in wavelength by $\lambda^2/2n_{\text{eff}}\Lambda_2$. In order for lasing to occur, we need to have an overlap between the two reflection peaks of the Bragg gratings and a cavity mode. Even though the tuning range of each reflection peak is limited to 10–15 nm, combining the two sets of reflection peaks results in a large tuning range. Just as with the two- and three-section DBR lasers, a separate phase section controls the cavity mode spacing to ensure continuous tuning. An additional complication with this approach is that because the reflection peaks taper off, the current in the gain region needs to be increased to compensate for the poorer reflectivity as the laser is tuned away from the primary Bragg reflection peak.

Another way of getting the same effect is to use periodically chirped gratings instead of the gratings shown in Figure 3.54. This structure is called a super-structure grating DBR laser. The advantage of this structure is that the chirped gratings provide a highly reflective set of peaks over a wider wavelength range than the sampled grating structure.

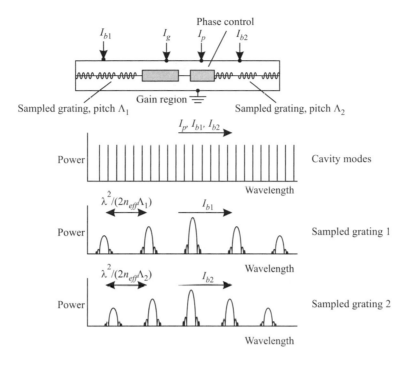

Figure 3.54   A sampled grating DBR laser and its principle of wavelength selection.

Grating Coupled Sampled Reflector Laser

The GCSR laser is a combination of a VGF and a sampled or super-structure grating, as shown in Figure 3.55. The VGF provides a wide tuning range, and the SSG grating provides high selectivity to eliminate side modes. In a sense, the VGF provides coarse tuning to select a wavelength band with multiple cavity modes in the band, and the SSG grating provides the wavelength selection within the band. Just as in the two- and three-section DBR lasers, an additional phase section provides the fine control over the cavity modes to provide continuous tuning within the band to suppress side modes.

Laser Arrays

Another way to obtain a tunable laser source is to use an array of wavelength-differentiated lasers and turn one of them on at any time. Arrays could also be used to replace individual light sources.

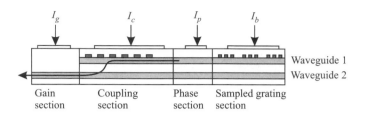

Figure 3.55   A grating coupled sampled reflector laser.

One approach is to fabricate an array of DFB lasers, each of them at a different wavelength. Combined with temperature tuning, we can use this method to obtain fairly continous tuning. A major problem with this approach is in the wavelength accuracy of the individual lasers in the array, making it difficult to obtain a comb of accurately spaced wavelengths out of the array. However, if only one laser is to be used at any given time, we can use temperature tuning to make up for this inaccuracy. Lasers using this approach have been demonstrated and used in system experiments [Zah92, You95].

Another approach is to use Fabry-Perot–type laser arrays and use an external mechanism for selecting the lasing wavelength. Several structures have been proposed [Soo92, ZJ94], one using an external waveguide grating and the other using an external arrayed waveguide grating. With these structures, the wavelength accuracy is determined by the external grating. The long cavity length results in potentially a large number of cavity modes within the grating wavelength selection window, which could cause the laser to hop between cavity modes during operation.

3.5.4   *Direct and External Modulation*

The process of imposing data on the light stream is called *modulation*. The simplest and most widely used modulation scheme is called *on-off keying* (OOK), where the light stream is turned on or off, depending on whether the data bit is a 1 or 0. We will study this in more detail in Chapter 4.

OOK modulated signals are usually realized in one of two ways: (1) by *direct modulation* of a semiconductor laser or an LED, or (2) by using an *external modulator*. The direct modulation scheme is illustrated in Figure 3.56. The drive current into the semiconductor laser is set well above threshold for a 1 bit and below (or slightly above) threshold for a 0 bit. The ratio of the output powers for the 1 and 0 bits is called the *extinction ratio*. Direct modulation is simple and inexpensive since no other components are required for modulation other than the light source

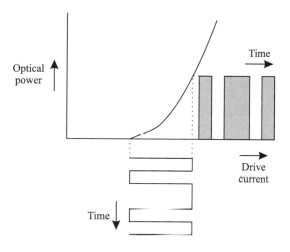

Figure 3.56 Direct modulation of a semiconductor laser.

(laser/LED) itself. In fact, a major advantage of semiconductor lasers is that they can be directly modulated. In contrast, many other lasers are continuous wave sources and cannot be modulated directly at all. These lasers require an external modulator. For example, because of the long lifetime of the erbium atoms at the E_2 level in Figure 3.35, erbium lasers cannot be directly modulated even at speeds of a few kilobits per second.

The disadvantage of direct modulation is that the resulting pulses are considerably *chirped*. Chirp is a phenomenon wherein the carrier frequency of the transmitted pulse varies with time, and it causes a broadening of the transmitted spectrum. As we saw in Section 2.3, chirped pulses have much poorer dispersion limits than unchirped pulses. The amount of chirping can be reduced by increasing the power of a 0 bit so that the laser is always kept well above its threshold; the disadvantage is that this reduces the extinction ratio, which in turn, degrades the system performance, as we will see in Section 5.3. In practice, we can realize an extinction ratio of around 7 dB while maintaining reasonable chirp performance. This enhanced pulse broadening of chirped pulses is significant enough to warrant the use of *external modulators* in high-speed, dispersion-limited communication systems.

An OOK external modulator is placed in front of a light source and turns the light signal on or off based on the data to be transmitted. The light source itself is continuously operated. This has the advantage of minimizing undesirable effects, particularly chirp. Several types of external modulators are commercially available and are increasingly being integrated with the laser itself inside a single package to reduce the packaging cost. In fact, transmitter packages that include a laser, external modulator, and wavelength stabilization circuits are becoming commercially available for use in WDM systems.

External modulators become essential in transmitters for communication systems using solitons or return-to-zero (RZ) modulation (see Section 2.5). As shown in Figure 3.57(a), to obtain a modulated train of RZ pulses, we can use a laser generating a train of periodic pulses, such as a mode-locked laser (see Section 3.5.1) followed by an external modulator. The modulator blocks the pulses corresponding to a 0 bit. (Usually we cannot directly modulate a pulsed laser emitting periodic pulses.) Unfortunately, cost-effective and compact solid-state lasers for generating periodic pulses are not yet commercially available. More commonly, as shown in Figure 3.57(b), practical RZ systems today use a continuous-wave DFB laser followed by a two-stage external modulator. The first stage creates a periodic train of short (RZ) pulses, and the second stage imposes the modulation by blocking out the 0 bits. Dispersion-managed soliton systems (see Section 2.5.1) require the generation of RZ pulses with a carefully controlled amount and sign of chirp. This can be accomplished by using another phase modulation stage.

Two types of external modulators are widely used today: lithium niobate modulators and semiconductor electro-absorption (EA) modulators. The lithium niobate modulator makes use of the electro-optic effect, where an applied voltage induces a change in refractive index of the material. The device itself is configured either as a directional coupler or as a Mach-Zehnder interferometer (MZI). Figure 3.58 shows the directional coupler configuration. Applying a voltage to the coupling region changes its refractive index, which in turn determines how much power is coupled from the input waveguide 1 to the output waveguide 1 in the figure.

Figure 3.59 shows the MZI configuration, which operates on the principles that we studied in Section 3.3.7. Compared to a directional coupler, the MZI offers a higher modulation speed for a given drive voltage and provides a higher extinction ratio. For these reasons, it is the more popular configuration. In one state, the signals in the two arms of the MZI are in phase and interfere constructively and appear at the output. In the other state, applying a voltage causes a π phase shift between the two arms of the MZI, leading to destructive interference and no output signal. These modulators have very good extinction ratios ranging from 15 to 20 dB, and we can control the chirp very precisely. Due to the high polarization dependence of the device, a polarization maintaining fiber is used between the laser and the modulator.

The EA modulator is an attractive alternative to lithium niobate modulators because it can be fabricated using the same material and techniques used to fabricate semiconductor lasers. This allows an EA modulator to be integrated along with a DFB laser in the same package and results in a very compact, lower-cost solution, compared to using an external lithium niobate modulator. In simple terms, the EA

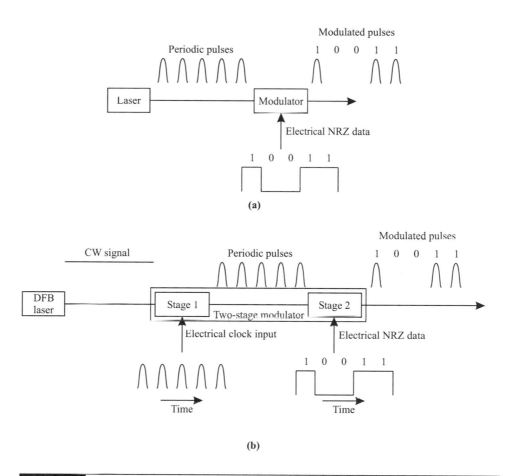

Figure 3.57 Using external modulators to realize transmitters for systems using RZ or soliton pulses. (a) A laser emitting a periodic pulse train, with the external modulator used to block the 0 bits and pass through the 1 bits. (b) A more common approach using a continuous-wave (CW) DFB laser followed by a two-stage modulator.

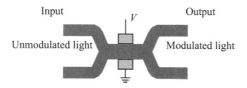

Figure 3.58 A lithium niobate external modulator using a directional coupler configuration.

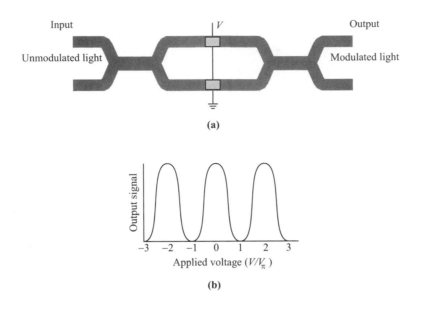

Figure 3.59 A lithium niobate external modulator using a Mach-Zehnder interferometer (MZI) configuration. (a) Device configuration. (b) Theoretical switching response as a function of applied voltage, V. V_π denotes the voltage required to achieve a π phase shift between the two arms. Note that the MZI has a periodic response.

modulator uses a material such that under normal conditions, its band gap is higher than the photon energy of the incident light signal. This allows the light signal to propagate through. Applying an electric field to the modulator results in shrinking the band gap of the material, causing the incident photons to be absorbed by the material. This effect is called the *Franz-Keldysh effect* or the *Stark effect*. The response time of this effect is sufficiently fast to enable us to realize 2.5 Gb/s and 10 Gb/s modulators. The chirp performance of EA modulators, while much better than directly modulated lasers, is not as good as that of lithium niobate MZI modulators. (While ideally there is no chirp in an external modulator, in practice, some chirp is induced in EA modulators because of residual phase modulation effects. This chirp can be controlled precisely in lithium niobate modulators.)

3.5.5 Pump Sources for Raman Amplifiers

One of the biggest challenges in realizing the Raman amplifiers that we discussed in Section 3.4.4 is a practical high-power pump source at the right wavelength. Since

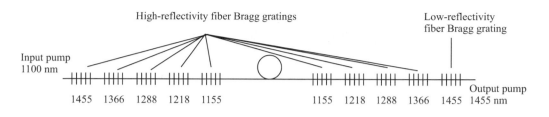

Figure 3.60 A high-power pump laser obtained by cascading resonators (after [Gru95]).

the Raman effect is only seen with very high powers in the fiber, pump powers on the order of several watts are required to provide effective amplification.

Several approaches have been proposed to realize high-power pump sources. One method is to combine a number of high-power semiconductor pump lasers. The power that can be extracted from a single semiconductor pump laser diode is limited to a few hundred milliwatts. Multiple semiconductor pump lasers can be combined using a combination of wavelength and/or polarization multiplexing to obtain a composite pump with sufficiently high power.

The other challenge lies in realizing the laser at the desired pump wavelength. One interesting approach is the cascaded Raman laser, shown in Figure 3.60.

Starting with a high-power pump laser at a conveniently available wavelength, we can generate pump sources at higher wavelengths using the Raman effect itself in fiber, by successively cascading a series of resonator structures. The individual resonators can be realized conveniently using fiber Bragg gratings or other filter structures. In Figure 3.60, a pump input at 1100 nm provides Raman gain into a fiber. A Fabry-Perot resonator is created in the fiber between by using a pair of matched fiber Bragg gratings that serve as wavelength-selective mirrors (see Section 3.3.5 for how the resonator works). The innermost resonator converts the initial pump signal into another pump signal at 1155 nm. It passes through signals at other wavelengths. The next resonator converts the 1155 nm pump into a 1218 nm pump. In principle, we can obtain any desired pump wavelength by cascading the appropriate series of resonators. The figure shows a series of resonators cascaded to obtain a 1455 nm pump output. The fiber Bragg grating at the end is designed to have lower reflectivity, allowing the 1455 nm pump signal to be output. This pump signal can then be used to provide Raman gain around 1550 nm. Due to the low fiber loss and high reflectivity of the fiber Bragg gratings, 80% of the input light is converted to the output.

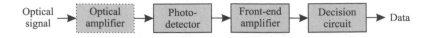

Figure 3.61   Block diagram of a receiver in a digital communication system.

3.6   Detectors

A receiver converts an optical signal into a usable electrical signal. Figure 3.61 shows the different components within a receiver. The *photodetector* generates an electrical current proportional to the incident optical power. The *front-end amplifier* increases the power of the generated electrical signal to a usable level. In digital communication systems, the front-end amplifier is followed by a *decision circuit* that estimates the data from the output of the front-end amplifier. The design of this decision circuit depends on the modulation scheme used to transmit the data and will be discussed in Section 4.4. An optical amplifier may be optionally placed before the photodetector to act as a *preamplifier*. The performance of optically preamplified receivers will be discussed in Chapter 4. This section covers photodetectors and front-end amplifiers.

3.6.1   Photodetectors

The basic principle of photodetection is illustrated in Figure 3.62. Photodetectors are made of semiconductor materials. Photons incident on a semiconductor are absorbed by electrons in the valence band. As a result, these electrons acquire higher energy and are excited into the conduction band, leaving behind a hole in the valence band. When an external voltage is applied to the semiconductor, these electron-hole pairs give rise to an electrical current, termed the *photocurrent*.

It is a principle of quantum mechanics that each electron can absorb only one photon to transit between energy levels. Thus the energy of the incident photon must be at least equal to the band gap energy in order for a photocurrent to be generated. This is also illustrated in Figure 3.62. This gives us the following constraint on the frequency f_c or the wavelength λ at which a semiconductor material with band gap E_g can be used as a photodetector:

$$hf_c = \frac{hc}{\lambda} \geq eE_g. \tag{3.19}$$

Here, c is the velocity of light, and e is the electronic charge.

The largest value of λ for which (3.19) is satisfied is called the *cutoff wavelength* and is denoted by λ_{cutoff}. Table 3.2 lists the band gap energies and the corresponding

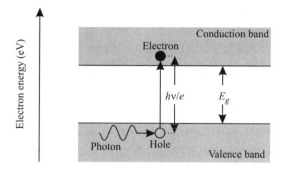

Figure 3.62   The basic principle of photodetection using a semiconductor. Incident photons are absorbed by electrons in the valence band, creating a free or mobile electron-hole pair. This electron-hole pair gives rise to a photocurrent when an external voltage is applied.

cutoff wavelengths for a number of semiconductor materials. We see from this table that the well-known semiconductors silicon (Si) and gallium arsenide (GaAs) cannot be used as photodetectors in the 1.3 and 1.55 μm bands. Although germanium (Ge) can be used to make photodetectors in both these bands, it has some disadvantages that reduce its effectiveness for this purpose. The new compounds indium gallium arsenide (InGaAs) and indium gallium arsenide phosphide (InGaAsP) are commonly used to make photodetectors in the 1.3 and 1.55 μm bands. Silicon photodetectors are widely used in the 0.8 μm band.

The fraction of the energy of the optical signal that is absorbed and gives rise to a photocurrent is called the *efficiency* η of the photodetector. For transmission at high bit rates over long distances, optical energy is scarce, and thus it is important to design the photodetector to achieve an efficiency η as close to 1 as possible. This can be achieved by using a semiconductor slab of sufficient thickness. The power absorbed by a semiconductor slab of thickness L μm can be written as

$$P_{\text{abs}} = (1 - e^{-\alpha L}) P_{\text{in}}, \tag{3.20}$$

where P_{in} is the incident optical signal power, and α is the absorption coefficient of the material; therefore,

$$\eta = \frac{P_{\text{abs}}}{P_{\text{in}}} = 1 - e^{-\alpha L}. \tag{3.21}$$

The absorption coefficient depends on the wavelength and is zero for wavelengths $\lambda > \lambda_{\text{cutoff}}$. Thus a semiconductor is transparent to wavelengths greater than its cutoff

Table 3.2 Band gap energies and cutoff wavelengths for a number of semiconductor materials. $In_{1-x}Ga_xAs$ is a ternary compound semiconductor material where a fraction $1-x$ of the Ga atoms in GaAs are replaced by In atoms. $In_{1-x}Ga_xAs_yP_{1-y}$ is a quaternary compound semiconductor material where, in addition, a fraction $1-y$ of the As atoms are replaced by P atoms. By varying x and y, the band gap energies and cutoff wavelengths can be varied.

Material	E_g (eV)	λ_{cutoff} (μm)
Si	1.17	1.06
Ge	0.775	1.6
GaAs	1.424	0.87
InP	1.35	0.92
$In_{0.55}Ga_{0.45}As$	0.75	1.65
$In_{1-0.45y}Ga_{0.45y}As_yP_{1-y}$	0.75–1.35	1.65–0.92

wavelength. Typical values of α are on the order of 10^4/cm, so to achieve an efficiency $\eta > 0.99$, a slab of thickness on the order of 10 μm is needed. The area of the photodetector is usually chosen to be sufficiently large so that all the incident optical power can be captured by it. Photodetectors have a very wide operating bandwidth since a photodetector at some wavelength can also serve as a photodetector at all smaller wavelengths. Thus a photodetector designed for the 1.55 μm band can also be used in the 1.3 μm band.

Photodetectors are commonly characterized by their *responsivity* $\mathcal{R}$. If a photodetector produces an average current of I_p amperes when the incident optical power is P_{in} watts, the responsivity

$$\mathcal{R} = \frac{I_p}{P_{\text{in}}} \text{ A/W}.$$

Since an incident optical power P_{in} corresponds to an incidence of P_{in}/hf_c photons/s on the average, and a fraction η of these incident photons are absorbed and generate an electron in the external circuit, we can write

$$\mathcal{R} = \frac{e\eta}{hf_c} \text{ A/W}.$$

The responsivity is commonly expressed in terms of λ; thus

$$\mathcal{R} = \frac{e\eta\lambda}{hc} = \frac{\eta\lambda}{1.24} \text{ A/W},$$

where λ in the last expression is expressed in μm. Since η can be made quite close to 1 in practice, the responsivities achieved are on the order of 1 A/W in the 1.3 μm band and 1.2 A/W in the 1.55 μm band.

In practice, the mere use of a slab of semiconductor as a photodetector does not realize high efficiencies. This is because many of the generated conduction band electrons recombine with holes in the valence band before they reach the external circuit. Thus it is necessary to sweep the generated conduction band electrons rapidly out of the semiconductor. This can be done by imposing an electric field of sufficient strength in the region where the electrons are generated. This is best achieved by using a semiconductor *pn*-junction (see Section 3.4.5) instead of a homogeneous slab and applying a *reverse bias* voltage (positive bias to the *n*-type and negative bias to the *p*-type) to it, as shown in Figure 3.63. Such a photodetector is called a *photodiode*.

The depletion region in a *pn*-junction creates a built-in electric field. Both the depletion region and the built-in electric field can be enhanced by the application of a reverse bias voltage. In this case, the electrons that are generated by the absorption of photons within or close to the depletion region will be swept into the *n*-type semiconductor before they recombine with the holes in the *p*-type semiconductor. This process is called *drift* and gives rise to a current in the external circuit. Similarly, the generated holes in or close to the depletion region drift into the *p*-type semiconductor because of the electric field.

Electron-hole pairs that are generated far away from the depletion region travel primarily under the effect of diffusion and may recombine without giving rise to a current in the external circuit. This reduces the efficiency η of the photodetector. More importantly, since diffusion is a much slower process than drift, the *diffusion current* that is generated by these electron-hole pairs will not respond quickly to changes in the intensity of the incident optical signal, thus reducing the frequency response of the photodiode.

pin Photodiodes

To improve the efficiency of the photodetector, a very lightly doped *intrinsic* semiconductor is introduced between the *p*-type and *n*-type semiconductors. Such photodiodes are called *pin* photodiodes, where the *i* in *pin* is for intrinsic. In these photodiodes, the depletion region extends completely across this intrinsic semiconductor (or region). The width of the *p*-type and *n*-type semiconductors is small compared to the intrinsic region so that much of the light absorption takes place in this region. This increases the efficiency and thus the responsivity of the photodiode.

A more efficient method of achieving this is to use a semiconductor material for these regions that is *transparent* at the wavelength of interest. Thus the wavelength

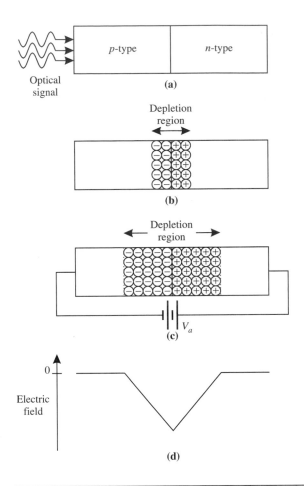

Figure 3.63   A reverse-biased *pn*-junction used as a photodetector. (a) A *pn*-junction photodiode. (b) Depletion region with no bias voltage applied. (c) Depletion region with a reverse bias voltage, V_a. (d) Built-in electric field on reverse bias.

of interest is larger than the cutoff wavelength of this semiconductor, and no absorption of light takes place in these regions. This is illustrated in Figure 3.64, where the material InP is used for the *p*-type and *n*-type regions, and InGaAs for the intrinsic region. Such a *pin* photodiode structure is termed a *double heterojunction* or a *heterostructure* since it consists of two junctions of completely different semiconductor materials. From Table 3.2, we see that the cutoff wavelength for InP is 0.92 μm, and that for InGaAs is 1.65 μm. Thus the *p*-type and *n*-type regions are transparent in the

Figure 3.64 A *pin* photodiode based on a heterostructure. The *p*-type and *n*-type regions are made of InP, which is transparent in the 1.3 and 1.55 μm wavelength bands. The intrinsic region is made of InGaAs, which strongly absorbs in both these bands.

1.3–1.6 μm range, and the diffusion component of the photocurrent is completely eliminated.

Avalanche Photodiodes

The responsivities of the photodetectors we have described thus far has been limited by the fact that one photon can generate only one electron when it is absorbed. However, if the generated electron is subjected to a very high electric field, it can acquire sufficient energy to knock off more electrons from the valence band to the conduction band. These secondary electron-hole pairs can generate even further electron-hole pairs when they are accelerated to sufficient levels. This process is called *avalanche multiplication.* Such a photodiode is called an *avalanche photodiode,* or simply an *APD*.

The number of secondary electron-hole pairs generated by the avalanche multiplication process by a single (primary) electron is random, and the mean value of this number is termed the *multiplicative gain* and denoted by G_m. The multiplicative gain of an APD can be made quite large and even infinite—a condition called *avalanche breakdown.* However, a large value of G_m is also accompanied by a larger variance in the generated photocurrent, which adversely affects the noise performance of the APD. Thus there is a trade-off between the multiplicative gain and the noise factor. APDs are usually designed to have a moderate value of G_m that optimizes their performance. We will study this issue further in Section 4.4.

3.6.2   Front-End Amplifiers

Two kinds of front-end amplifiers are used in optical communication systems: the *high-impedance* front end and the *transimpedance* front end. The equivalent circuits for these amplifiers are shown in Figure 3.65.

The capacitances C in this figure include the capacitance due to the photodiode, the amplifier input capacitance, and other parasitic capacitances. The main design issue is the choice of the load resistance R_L. We will see in Chapter 4 that the *thermal*

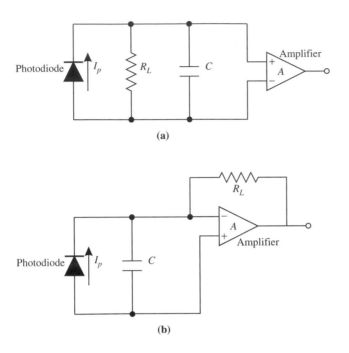

Figure 3.65   (a) Equivalent circuit for a high-impedance front-end amplifier. (b) Equivalent circuit for a transimpedance front-end amplifier.

noise current that arises due to the random motion of electrons and contaminates the photocurrent is inversely proportional to the load resistance. Thus, to minimize the thermal noise, we must make R_L large. However, the bandwidth of the photodiode, which sets the upper limit on the usable bit rate, is inversely proportional to the output load resistance seen by the photodiode, say, R_p. First consider the high-impedance front end. In this case, $R_p = R_L$, and we must choose R_L small enough to accommodate the bit rate of the system. Thus there is a trade-off between the bandwidth of the photodiode and its noise performance. Now consider the transimpedance front end for which $R_p = R_L/(A + 1)$, where A is the gain of the amplifier. The bandwidth is increased by a factor of $A + 1$ for the same load resistance. However, the thermal noise current is also higher than that of a high-impedance amplifier with the same R_L (due to considerations beyond the scope of this book), but this increase is quite moderate—a factor usually less than two. Thus the transimpedance front end is chosen over the high-impedance one for most optical communication systems.

There is another consideration in the choice of a front-end amplifier: *dynamic range*. This is the difference between the largest and smallest signal levels that the front-end amplifier can handle. This may not be an important consideration for many optical communication links since the power level seen by the receivers is usually more or less fixed. However, dynamic range of the receivers is a very important consideration in the case of networks where the received signal level can vary by a few orders of magnitude, depending on the location of the source in the network. The transimpedance amplifier has a significantly higher dynamic range than the high-impedance one, and this is another factor in favor of choosing the transimpedance amplifier. The higher dynamic range arises because large variations in the photocurrent I_p translate into much smaller variations at the amplifier input, particularly if the amplifier gain is large. This can be understood with reference to Figure 3.65(b). A change ΔI_p in the photocurrent causes a change in voltage $\Delta I_p R_L$ across the resistance R_L (ignoring the current through the capacitance C). This results in a voltage change across the inputs of the amplifier of only $\Delta I_p R_L/(A + 1)$. Thus if the gain, A, is large, this voltage change is small. In the case of the high-impedance amplifier, however, the voltage change across the amplifier inputs would be $\Delta I_p R_L$ (again ignoring the current through the capacitance C).

A *field-effect transistor* (FET) has a very high input impedance and for this reason is often used as the amplifier in the front end. A *pin* photodiode and an FET are often integrated on the same semiconductor substrate, and the combined device is called a *pinFET*.

3.7 Switches

Optical switches are used in optical networks for a variety of applications. The different applications require different switching times and number of switch ports, as summarized in Table 3.3. One application of optical switches is in the *provisioning* of lightpaths. In this application, the switches are used inside wavelength crossconnects to reconfigure them to support new lightpaths. In this application, the switches are replacements for manual fiber patch panels, but with significant added software for end-to-end network management, a subject that we will cover in detail in Chapters 9 and 10. Thus, for this application, switches with millisecond switching times are acceptable. The challenge here is to realize large switch sizes.

Another important application is that of *protection switching*, the subject of Chapter 10. Here the switches are used to switch the traffic stream from a primary fiber onto another fiber in case the primary fiber fails. The entire operation must typically be completed in several tens of milliseconds, which includes the time to detect the failure, communicate the failure to the appropriate network elements handling the switching, and the actual switch time. Thus the switching time required is on the order of a few milliseconds. Different types of protection switching are

Table 3.3 Applications for optical switches and their switching time and port count requirements.

Application	Switching Time Required	Number of Ports
Provisioning	1–10 ms	> 1000
Protection switching	1–10 ms	2–1000
Packet switching	1 ns	> 100
External modulation	10 ps	1

possible, and based on the scheme used, the number of switch ports needed may vary from two ports to several hundreds to thousands of ports when used in a wavelength crossconnect.

Switches are also important components in high-speed optical *packet-switched* networks. In these networks, switches are used to switch signals on a packet-by-packet basis. For this application, the switching time must be much smaller than a packet duration, and large switches will be needed. For example, a 53-byte packet (one *cell* in an ATM network) at 10 Gb/s is 42 ns long, so the switching time required for efficient operation is on the order of a few nanoseconds. Optical packet switching is still in its infancy and is the subject of Chapter 12.

Yet another use for switches is as external modulators to turn on and off the data in front of a laser source. In this case, the switching time must be a small fraction of the bit duration. So an external modulator for a 10 Gb/s signal (with a bit duration of 100 ps) must have a switching time (or, equivalently, a rise and fall time) of about 10 ps.

In addition to the switching time and the number of ports, the other important parameters used to characterize the suitability of a switch for optical networking applications are the following:

1. The *extinction ratio* of an on-off switch is the ratio of the output power in the on state to the output power in the off state. This ratio should be as large as possible and is particularly important in external modulators. Whereas simple mechanical switches have extinction ratios of 40–50 dB, high-speed external modulators tend to have extinction ratios of 10–25 dB.

2. The *insertion loss* of a switch is the fraction of power (usually expressed in decibels) that is lost because of the presence of the switch and must be as small as possible. Some switches have different losses for different input-output connections. This is an undesirable feature because it increases the dynamic range of the signals in the network. With such switches, we may need to include variable optical attenuators to equalize the loss across different paths. This *loss uniformity*

is determined primarily by the architecture used to build the switch, rather than the inherent technology itself, as we will see in several examples below.

3. Switches are not ideal. Even if input x is nominally connected to output y, some power from input x may appear at the other outputs. For a given switching state or interconnection pattern, and output, the *crosstalk* is the ratio of the power at that output from the desired input to the power from all other inputs. Usually, the *crosstalk of a switch* is defined as the worst-case crosstalk over all outputs and interconnection patterns.

4. As with other components, switches should have a low polarization-dependent loss (PDL). When used as external modulators, polarization dependence can be tolerated since the switch is used immediately following the laser, and the laser's output state of polarization can be controlled by using a special polarization-preserving fiber to couple the light from the laser into the external modulator.

5. A *latching* switch maintains its switch state even if power is turned off to the switch. This is a somewhat desirable feature because it enables traffic to be passed through the switch even in the event of power failures.

6. The switch needs to have a readout capability wherein its current state can be monitored. This is important to verify that the right connections are made through the switch.

7. The reliability of the switch is an important factor in telecommunications applications. The common way of establishing reliability is to cycle the switch through its various states a large number of times, perhaps a few million cycles. However, in the provisioning and protection-switching applications discussed above, the switch remains in one state for a long period, say, even a few years, and is then activated to change state. The reliability issue here is whether the switch will actually switch after it has remained untouched for a long period. This property is more difficult to establish without a long-term history of deployment.

3.7.1 Large Optical Switches

Switches with port counts ranging from a few hundred to a few thousand are being sought by carriers for their next-generation networks. Given that a single central office handles multiple fibers, with each fiber carrying several tens to hundreds of wavelengths, it is easy to imagine the need for large-scale switches to provision and protect these wavelengths. We will study the use of such switches as wavelength crossconnects in Chapter 7.

The main considerations in building large switches are the following:

Number of switch elements required. Large switches are made by using multiple switch elements in some form or the other, as we will see below. The cost and complexity of the switch to some extent depends on the number of switch elements required. However, this is only one of the factors that affects the cost. Other factors include packaging, splicing, and ease of fabrication and control.

Loss uniformity. As we mentioned in the context of switch characteristics earlier, switches may have different losses for different combinations of input and output ports. This situation is exacerbated for large switches. A measure of the loss uniformity can be obtained by considering the minimum and maximum number of switch elements in the optical path, for different input and output combinations.

Number of crossovers. Some of the optical switches that we will study next are fabricated by integrating multiple switch elements on a single substrate. Unlike integrated electronic circuits (ICs), where connections between the various components can be made at multiple layers, in integrated optics, all these connections must be made in a single layer by means of waveguides. If the paths of two waveguides cross, two undesirable effects are introduced: power loss and crosstalk. In order to have acceptable loss and crosstalk performance for the switch, it is thus desirable to minimize, or completely eliminate, such waveguide crossovers. Crossovers are not an issue with respect to free-space switches, such as the MEMS switches that we will describe later in this section.

Blocking characteristics. In terms of the switching function achievable, switches are of two types: *blocking* or *nonblocking*. A switch is said to be *nonblocking* if an unused input port can be connected to any unused output port. Thus a nonblocking switch is capable of realizing every interconnection pattern between the inputs and the outputs. If some interconnection pattern(s) cannot be realized, the switch is said to be *blocking*. Most applications require nonblocking switches. However, even nonblocking switches can be further distinguished in terms of the effort needed to achieve the nonblocking property. A switch is said to be *wide-sense nonblocking* if any unused input can be connected to any unused output, without requiring any existing connection to be rerouted. Wide-sense nonblocking switches usually make use of specific routing algorithms to route connections so that future connections will not be blocked. A *strict-sense non-blocking* switch allows any unused input to be connected to any unused output regardless of how previous connections were made through the switch.

A nonblocking switch that may require rerouting of connections to achieve the nonblocking property is said to be *rearrangeably nonblocking*. Rerouting of connections may or may not be acceptable depending on the application since the

Table 3.4   Comparison of different switch architectures. The switch count for the Spanke architecture is made in terms of $1 \times n$ switches, whereas 2×2 switches are used for the other architectures.

	Nonblocking Type	No. Switches	Max. Loss	Min. Loss
Crossbar	Wide sense	n^2	$2n - 1$	1
Clos	Strict sense	$4\sqrt{2}n^{1.5}$	$5\sqrt{2n} - 5$	3
Spanke	Strict sense	$2n$	2	2
Beneš	Rearrangeable	$\frac{n}{2}(2\log_2 n - 1)$	$2\log_2 n - 1$	$2\log_2 n - 1$
Spanke-Beneš	Rearrangeable	$\frac{n}{2}(n - 1)$	n	$\frac{n}{2}$

connection must be interrupted, at least briefly, in order to switch it to a different path. The advantage of rearrangeably nonblocking switch architectures is that they use fewer small switches to build a larger switch of a given size, compared to the wide-sense nonblocking switch architectures.

While rearrangeably nonblocking architectures use fewer switches, they require a more complex control algorithm to set up connections, but this control complexity is not a significant issue, given the power of today's microprocessors used in these switches that would execute such an algorithm. The main drawback of rearrangeably nonblocking switches is that many applications will not allow existing connections to be disrupted, even temporarily, to accommodate a new connection.

Usually, there is a trade-off between these different aspects. We will illustrate this when we study different architectures for building large switches next. Table 3.4 compares the characteristics of these architectures.

Crossbar

A 4×4 crossbar switch is shown in Figure 3.66. This switch uses 16 2×2 switches, and the interconnection between inputs and outputs is achieved by appropriately setting the states of these 2×2 switches. The settings of the 2×2 switches required to connect input 1 to output 3 are shown in Figure 3.66. This connection can be viewed as taking a path through the network of 2×2 switches making up the 4×4 switch. Note that there are other paths from input 1 to output 3; however, this is the preferred path as we will see next.

The crossbar architecture is wide-sense nonblocking. To connect input i to output j, the path taken traverses the 2×2 switches in row i till it reaches column j and then traverses the switches in column j till it reaches output j. Thus the 2×2 switches on this path in row i and column j must be set appropriately for this connection to be made. We leave it to you to be convinced that *if this connection rule is used,* this switch is nonblocking and doesn't require existing connections to be rerouted.

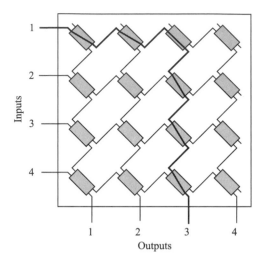

Figure 3.66   A 4×4 crossbar switch realized using 16 2×2 switches.

In general, an $n \times n$ crossbar requires n^2 2×2 switches. The shortest path length is 1 and the longest path length is $2n - 1$, and this is one of the main drawbacks of the crossbar architecture. The switch can be fabricated without any crossovers.

Clos

The Clos architecture provides a strict-sense nonblocking switch and is widely used in practice to build large port count switches. A three-stage 1024-port Clos switch is shown in Figure 3.67. An $n \times n$ switch is constructed as follows. We use three parameters, m, k, and p. Let $n = mk$. The first and third stage consist of k $(m \times p)$ switches. The middle stage consists of p $(k \times k)$ switches. Each of the k switches in the first stage is connected to all the switches in the middle stage. (Each switch in the first stage has p outputs. Each output is connected to the input of a different switch in the middle stage.) Likewise, each of the k switches in the third stage is connected to all the switches in the middle stage. We leave it to you to verify that if $p \geq 2m - 1$, the switch is strictly nonblocking (see Problem 3.29).

To minimize the cost of the switch, let us pick $p = 2m - 1$. Usually the individual switches in each stage are designed using crossbar switches. Thus each of the $m \times (2m - 1)$ switches requires $m(2m - 1)$ 2×2 switch elements, and each of the $k \times k$ switches in the middle stage requires k^2 2×2 switch elements. The total number of switch elements needed is therefore

$$2km(2m - 1) + (2m - 1)k^2.$$

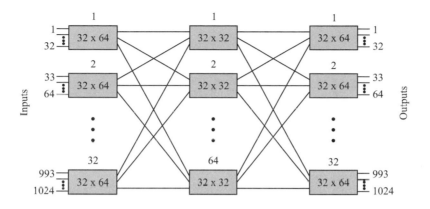

Figure 3.67   A strict-sense nonblocking 1024 × 1024 switch realized using 32 × 64 and 32 × 32 switches interconnected in a three-stage Clos architecture.

Using $k = n/m$, we leave it to you to verify that the number of switch elements is minimized when

$$m \approx \sqrt{\frac{n}{2}}.$$

Using this value for m, the number of switch elements required for the minimum cost configuration is approximately

$$4\sqrt{2}n^{3/2} - 4n,$$

which is significantly lower than the n^2 required for a crossbar.

The Clos architecture has several advantages that make it suitable for use in a multistage switch fabric. The loss uniformity between different input-output combinations is better than a crossbar, and the number of switch elements required is significantly smaller than a crossbar.

Spanke

The Spanke architecture shown in Figure 3.68 is turning out to be a popular architecture for building large switches. An $n \times n$ switch is made by combining n $1 \times n$ switches along with n $n \times 1$ switches, as shown in the figure.   The architecture is strict-sense nonblocking. So far we have been counting the number of 2×2 switch elements needed to build large switches as a measure of the switch cost. What makes the Spanke architecture attractive is that, in many cases, a $1 \times n$ optical switch can be built using a single switch element and does not need to be built out of 1×2 or 2×2 switch elements. This is the case with the MEMS analog beam steering

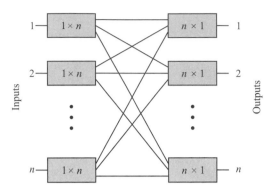

Figure 3.68   A strict-sense nonblocking $n \times n$ switch realized using $2n$ $1 \times n$ switches interconnected in the Spanke architecture.

mirror technology that we will discuss later in this section. Therefore only $2n$ such switch elements are needed to build an $n \times n$ switch. This implies that the switch cost scales linearly with n, which is significantly better than other switch architectures. In addition, each connection passes through two switch elements, which is significantly smaller than the number of switch elements in the path for other multistage designs. This approach provides a much lower insertion loss than the multistage designs. Moreover the optical path length for all the input-output combinations can be made essentially the same, so that the loss is the same regardless of the specific input-output combination.

Beneš

The Beneš architecture is a rearrangeably nonblocking switch architecture and is one of the most efficient switch architectures in terms of the number of 2×2 switches it uses to build larger switches. A rearrangeably nonblocking 8×8 switch that uses only 20 2×2 switches is shown in Figure 3.69. In comparison, an 8×8 crossbar switch requires 64 2×2 switches. In general, an $n \times n$ Beneš switch requires $(n/2)(2 \log_2 n - 1)$ 2×2 switches, n being a power of two. The loss is the same through every path in the switch—each path goes through $2 \log_2 n - 1$ 2×2 switches. Its two main drawbacks are that it is not wide-sense nonblocking, and that a number of waveguide crossovers are required, making it difficult to fabricate in integrated optics.

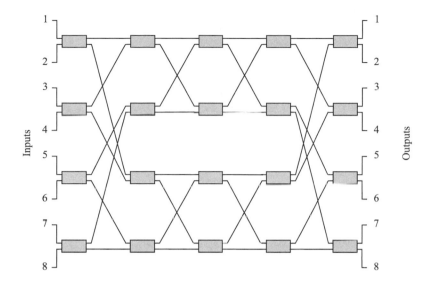

Figure 3.69 A rearrangeably nonblocking 8×8 switch realized using 20 2×2 switches interconnected in the Beneš architecture.

Spanke-Beneš

A good compromise between the crossbar and Beneš switch architectures is shown in Figure 3.70, which is a rearrangeably nonblocking 8×8 switch using 28 2×2 switches and *no* waveguide crossovers. This switch architecture was discovered by Spanke and Bencš [SB87] and is called the *n-stage planar architecture* since it requires n stages (columns) to realize an $n \times n$ switch. It requires $n(n - 1)/2$ switches, the shortest path length is $n/2$, and the longest path length is n. There are no crossovers. Its main drawbacks are that it is not wide-sense nonblocking and the loss is nonuniform.

3.7.2 Optical Switch Technologies

Many different technologies are available to realize optical switches. These are compared in Table 3.5. With the exception of the large-scale MEMS switch, the switch elements described below all use the crossbar architecture.

Bulk Mechanical Switches

In mechanical switches, the switching function is performed by some mechanical means. One such switch uses a mirror arrangement whereby the switching state

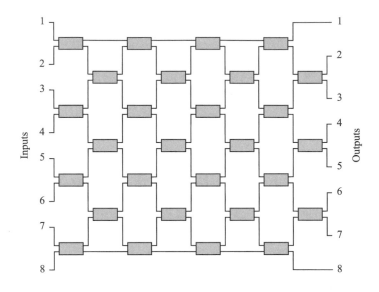

Figure 3.70   A rearrangeably nonblocking 8 × 8 switch realized using 28 2 × 2 switches and no waveguide crossovers interconnected in the *n*-stage planar architecture.

Table 3.5   Comparison of different optical switching technologies. The mechanical, MEMS, and polymer-based switches behave in the same manner for 1.3 and 1.55 μm wavelengths, but other switches are designed to operate at only one of these wavelength bands. The numbers represent parameters for commercially available switches in early 2001.

Type	Size	Loss (dB)	Crosstalk (dB)	PDL (dB)	Switching Time
Bulk mechanical	8 × 8	3	55	0.2	10 ms
2D MEMS	32 × 32	5	55	0.2	10 ms
3D MEMS	1000 × 1000	5	55	0.5	10 ms
Thermo-optic silica	8 × 8	8	40	Low	3 ms
Bubble-based	32 × 32	7.5	50	0.3	10 ms
Liquid crystal	2 × 2	1	35	0.1	4 ms
Polymer	8 × 8	10	30	Low	2 ms
Electro-optic LiNbO$_3$	4 × 4	8	35	1	10 ps
SOA	4 × 4	0	40	Low	1 ns

is controlled by moving a mirror in and out of the optical path. Another type of mechanical switch uses a directional coupler. Bending or stretching the fiber in the interaction region changes the coupling ratio of the coupler and can be used to switch light from an input port between different output ports.

Bulk mechanical switches have low insertion losses, low PDL, low crosstalk, and are relatively inexpensive devices. In most cases, they are available in a crossbar configuration, which implies somewhat poor loss uniformity. However, their switching speeds are on the order of a few milliseconds and the number of ports is fairly small, say, 8 to 16. For these reasons, they are particularly suited for use in small wavelength crossconnects for provisioning and protection-switching applications but not for the other applications discussed earlier. As with most mechanical components, long-term reliability for these switches is of some concern, but they are still more mature by far than all the other optical switching technologies available today. Larger switches can be realized by cascading small bulk mechanical switches, as we saw in Section 3.7.1, but there are better ways of realizing larger port count switches, as we will explore next.

Micro-Electro-Mechanical System (MEMS) Switches

Micro-electro-mechanical systems (MEMS) are miniature mechanical devices typically fabricated using silicon substrates. In the context of optical switches, MEMS usually refers to miniature movable mirrors fabricated in silicon, with dimensions ranging from a few hundred micrometers to a few millimeters. A single silicon wafer yields a large number of mirrors, which means that these mirrors can be manufactured and packaged as arrays. Moreover, the mirrors can be fabricated using fairly standard semiconductor manufacturing processes. These mirrors are deflected from one position to another using a variety of electronic actuation techniques, such as electromagnetic, electrostatic, or piezoelectric methods, hence the name MEMS. Of these methods, electrostatic deflection is particularly power efficient but is relatively hard to control over a wide deflection range.

The simplest mirror structure is a so-called two-state pop-up mirror, or 2D mirror, shown in Figure 3.71. In one state, the mirror is flat in line with the substrate. In this state, the light beam is not deflected. In the other state, the mirror pops up to a vertical position and the light beam if present is deflected. Such a mirror can be used in a crossbar arrangement discussed below to realize an $n \times n$ switch. Practical switch module sizes are limited by wafer sizes and processing constraints to be around 32×32. These switches are particularly easy to control through digital means, as only two mirror positions need to be supported.

Another type of mirror structure is shown in Figure 3.72. The mirror is connected through flexures to an inner frame, which in turn is connected through another set

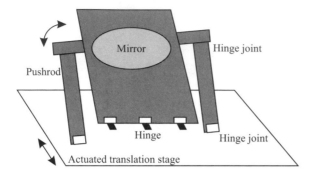

Figure 3.71   A two-state pop-up MEMS mirror, from [LGT98], shown in the popped-up position. The mirror can be moved to fold flat in its other position.

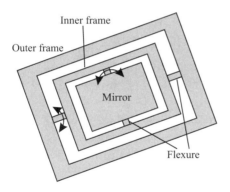

Figure 3.72   An analog beam steering mirror. The mirror can be freely rotated on two axes to deflect an incident light beam.

of flexures to an outer frame. The flexures allow the mirror to be rotated freely on two distinct axes. This mirror can be controlled in an analog fashion to realize a continuous range of angular deflections. This type of mirror is sometimes referred to as an analog beam steering mirror, a gimbel mirror, or a 3D mirror. A mirror of this type can be used to realize a 1 × n switch. The control of these mirrors is not a trivial matter, with fairly sophisticated servo control mechanisms required to deflect the mirrors to their correct position and hold them there.

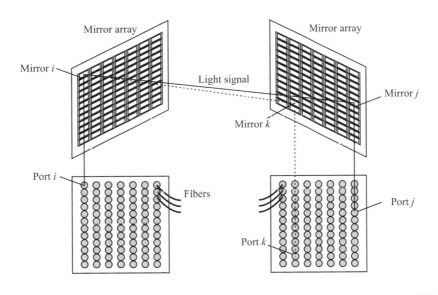

Figure 3.73 An $n \times n$ switch built using two arrays of analog beam steering MEMS mirrors.

Figure 3.73 shows a large $n \times n$ switch using two arrays of analog beam steering mirrors. This architecture corresponds to the Spanke architecture, which we discussed in Section 3.7.1. Each array has n mirrors, one associated with each switch port. An input signal is coupled to its associated mirror in the first array using a suitable arrangement of collimating lenses. The first mirror can be deflected to point the beam to any of the mirrors in the second array. To make a connection from port i to port j, the mirror i in the first array is pointed to mirror j in the second array and vice versa. Mirror j then allows the beam to be coupled out of port j. To make a connection from port i to another port, say, port k, mirror i in the first array and mirror k in the second array are pointed at each other. Note that in order to switch this connection from port i to port k, the beam is scanned from output mirror j to output mirror k, passing over other mirrors along the way. This does not lead to additional crosstalk because a connection is established only when the two mirrors are pointed at each other and not under any other circumstances. Note also that beams corresponding to multiple connections cross each other inside the switch but do not interfere.

There are two types of fabrication techniques used to make MEMS structures: *surface micromachining* and *bulk micromachining*. In surface micromachining, multiple layers are deposited on top of a silicon substrate. These layers are partially etched away and pieces are left anchored to the substrate to produce various structures. In bulk micromachining, the MEMS structures are crafted directly from the bulk of the silicon wafer. The type of micromachining used and the choice of the appropriate type of silicon substrate directly influence the properties of the resulting structure. For a more detailed discussion on some of the pros and cons of these approaches, see [NR01]. Today we are seeing the simple 2D MEMS mirrors realized using surface micromachining and the 3D MEMS mirrors realized using bulk micromachining.

Among the various technologies discussed in this section, the 3D MEMS analog beam steering mirror technology offers the best potential for building large-scale optical switches. These switches are compact, have very good optical properties (low loss, good loss uniformity, negligible dispersion), and can have extremely low power consumption. Most of the other technologies are limited to small switch sizes. Indeed, as of this writing, 3D MEMS switches ranging from 256 to over 1000 ports are becoming commercially available, as vendors are addressing the challenges of high-yield fabrication, control, and reliability and stability of these switches with respect to temperature, humidity, and vibration.

Bubble-Based Waveguide Switch

Another type of optical switch from Agilent Technologies uses an interesting planar waveguide approach where the switch actuation is based on a technology that is similar to what is used in inkjet printers. Figure 3.74 shows a picture of this switch. It consists of waveguides that cross each other. The switch also has lengthwise trenches as shown, and the crossover points of the waveguides align with the trenches. The trenches are filled with index matching fluid. Under normal conditions, light propagating in one waveguide continues along the same waveguide at the crossover points. However if the fluid at a crossover point is heated, an air bubble is formed. This air bubble breaks the index matching, and as a result, the light is now reflected at that crossover point. Therefore each crossover point behaves as a 2×2 crossbar switch. Using this approach, up to 32×32 switches can be fabricated on a single substrate. This technology offers the promise of realizing relatively low-cost, easily manufacturable small switch arrays with switching times on the order of tens of milliseconds.

Liquid Crystal Switches

Liquid crystal cells offer another way for realizing small optical switches. These switches typically make use of polarization effects to perform the switching function. By applying a voltage to a suitably designed liquid crystal cell, we can cause the polarization of the light passing through the cell either to be rotated or not. This

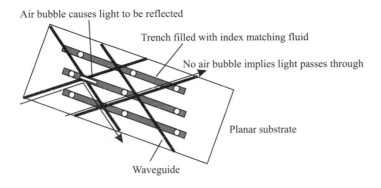

Figure 3.74 A planar waveguide switch using inkjet technology to actuate the switching.

can then be combined with passive polarization beam splitters and combiners to yield a polarization-independent switch, as shown in Figure 3.75. The principle of operation is similar to the polarization-independent isolator of Figure 3.5. Typically, the passive polarization beam splitter, combiner, and the active switch element can all be realized using an array of liquid crystal cells. The polarization rotation in the liquid crystal cell does not have to be digital in nature—it can be controlled in an analog fashion by controlling the voltage. Thus this technology can be used to realize a variable optical attenuator (VOA) as well. In fact the VOA can be incorporated in the switch itself to control the output power being coupled out. The switching time is on the order of a few milliseconds. Like the bubble-based waveguide switch, a liquid crystal switch is a solid-state device and can potentially be manufactured in volume at low cost.

Electro-Optic Switches

A 2×2 electro-optic switch can be realized using one of the external modulator configurations that we studied in Section 3.5.4. One commonly used material is lithium niobate (LiNbO$_3$). In the directional coupler configuration, the coupling ratio is varied by changing the voltage and thus the refractive index of the material in the coupling region. In the Mach-Zehnder configuration, the relative path length between the two arms of the Mach-Zehnder is varied. An electro-optic switch is capable of changing its state extremely rapidly; typically, in less than 1 ns. This switching time limit is determined by the capacitance of the electrode configuration.

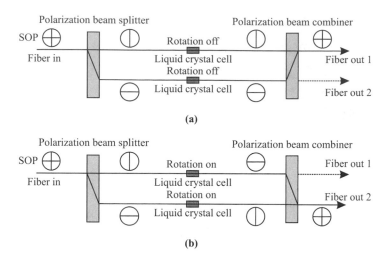

Figure 3.75   A 1 × 2 liquid crystal switch. (a) The rotation is turned off, causing the light beam to exit on output port 1. (b) The rotation is turned on by applying a voltage to the liquid crystal cell, causing the light beam to exit on output port 2.

Among the advantages of lithium niobate switches are that they allow modest levels of integration, compared to mechanical switches. Larger switches can be realized by integrating several 2 × 2 switches on a single substrate. However, they tend to have a relatively high loss and PDL, and are more expensive than mechanical switches.

Thermo-Optic Switches

These switches are essentially 2 × 2 integrated-optic Mach-Zehnder interferometers, constructed on waveguide material whose refractive index is a function of the temperature. By varying the refractive index in one arm of the interferometer, the relative phase difference between the two arms can be changed, resulting in switching an input signal from one output port to another. These devices have been made on silica as well as polymer substrates, but have relatively poor crosstalk. Also the thermo-optic effect is quite slow, and switching speeds are on the order of a few milliseconds.

Semiconductor Optical Amplifier Switches

The SOA described in Section 3.4.5 can be used as an on-off switch by varying the bias voltage to the device. If the bias voltage is reduced, no population inversion is achieved, and the device absorbs input signals. If the bias voltage is present, it amplifies the input signals. The combination of amplification in the on state and absorption in the off state makes this device capable of achieving very large extinction ratios. The switching speed is on the order of 1 ns. Larger switches can be fabricated by integrating SOAs with passive couplers. However, this is an expensive component, and it is difficult to make it polarization independent because of the highly directional orientation of the laser active region, whose width is almost always much greater than its height (except for VCSELs).

3.7.3   Large Electronic Switches

We have focused primarily on optical switch technologies in this section. However, many of the practical "optical" or wavelength crossconnects today actually use electronic switch fabrics. The main reason for this approach is that large-scale optical switch fabrics are only now beginning to be available.

Typically a large electronic switch uses a multistage design, and in many cases, the Clos approach is the preferred approach as it provides a strict-sense nonblocking architecture with a relatively small number of crosspoint switches. Two approaches are possible. In the first approach, the input signal at 2.5 Gb/s or 10 Gb/s is converted into a parallel bit stream at a manageable rate, say, 51 Mb/s, and all the switching is done at the latter bit rate. This approach makes sense if we need to switch the signal in units of 51 Mb/s for other reasons. Also in many cases, the overall cost of an electronic switch is dominated by the cost of the optical to electrical converters, rather than the switch fabric itself. This implies that once the signal is available in the electrical domain, it makes sense to switch signals at a fine granularity.

The other approach is to design the switch to operate at the line rate in a serial fashion without splitting the signal into lower-speed bit streams. The basic unit of this serial approach is a crossbar fabricated as a single IC. Today, 64 × 64 crossbar ICs operating at 2.5 Gb/s line rates are commercially available. The practical considerations related to building larger switches using these ICs have to do with managing the power dissipation and the interconnects between switch stages. A typical 64 × 64 switch IC may dissipate 25 W. About 100 such switches are required to build a 1024 × 1024 switch. The total power dissipated is therefore around 25 kW. (In contrast, a 1024 × 1024 optical switch using 3D MEMS may consume only about 3 kW and is significantly more compact overall, compared to an equivalent electrical switch.) Cooling such a switch is a significant problem. The other aspect has to do with the high-speed interconnect required between switch modules. As long as the switch modules are within a single printed circuit board, the interconnections are not difficult. However, practical considerations of power dissipation and board space dictate the necessity for having multiple printed circuit boards and perhaps multiple racks of equipment. The interconnects between these boards and racks need to operate at the line rate, which is typically 2.5 Gb/s or higher. High-quality electrical interconnects or optical interconnects can be used for this purpose. The drivers required for the electrical interconnects also dissipate a significant amount of power, and the distances possible are limited, typically to 5–6 m. Optical interconnects make use of arrayed lasers and receivers along with fiber optic ribbon cables. These offer lower power dissipation and significantly longer reach between boards, typically to about 100 m or greater.

3.8 Wavelength Converters

A wavelength converter is a device that converts data from one incoming wavelength to another outgoing wavelength. Wavelength converters are useful components in WDM networks for three major reasons. First, data may enter the network at a wavelength that is not suitable for use within the network. For example, the first-generation networks of Chapter 6 commonly transmit data in the 1310 nm wavelength window, using LEDs or Fabry-Perot lasers. Neither the wavelength nor the type of laser is compatible with WDM networks. So at the inputs and outputs of the network, data must be converted from these wavelengths to narrow band WDM signals in the 1550 nm wavelength range. A wavelength converter used to perform this function is sometimes called a *transponder.*

Second, wavelength converters may be needed within the network to improve the utilization of the available wavelengths on the network links. This topic is studied in detail in Chapter 8.

Finally, wavelength converters may be needed at boundaries between different networks if the different networks are managed by different entities and these entities do not coordinate the allocation of wavelengths in their networks.

Wavelength converters can be classified based on the range of wavelengths that they can handle at their inputs and outputs. A fixed-input, fixed-output device always takes in a fixed-input wavelength and converts it to a fixed-output wavelength. A variable-input, fixed-output device takes in a variety of wavelengths but always converts the input signal to a fixed-output wavelength. A fixed-input, variable-output device does the opposite function. Finally, a variable-input, variable-output device can convert any input wavelength to any output wavelength.

In addition to the range of wavelengths at the input and output, we also need to consider the range of input optical powers that the converter can handle, whether the converter is transparent to the bit rate and modulation format of the input signals, and whether it introduces additional noise or phase jitter to the signal. We will see that the latter two characteristics depend on the type of regeneration used in the converter. For all-optical wavelength converters, polarization-dependent loss should also be kept to a minimum.

There are four fundamental ways of achieving wavelength conversion: (1) optoelectronic, (2) optical gating, (3) interferometric, and (4) wave mixing. The latter three approaches are all-optical but not yet mature enough for commercial use. Optoelectronic converters today offer substantially better performance at lower cost than comparable all-optical wavelength converters.

3.8.1   Optoelectronic Approach

This is perhaps the simplest, most obvious, and most practical method today to realize wavelength conversion. As shown in Figure 3.76, the input signal is first converted to electronic form, regenerated, and then retransmitted using a laser at a different wavelength. This is usually a variable-input, fixed-output converter. The receiver does not usually care about the input wavelength, as long as it is in the 1310 or 1550 nm window. The laser is usually a fixed-wavelength laser. A variable output can be obtained by using a tunable laser.

The performance and transparency of the converter depend on the type of regeneration used. Figure 3.76 shows the different types of regeneration possible. In the simplest case, the receiver simply converts the incoming photons to electrons, which get amplified by an analog RF (radio-frequency) amplifier and drive the laser. This is called 1R regeneration. This form of conversion is truly transparent to the modulation format (provided the appropriate receiver is used to receive the signal) and can handle analog data as well. However, noise is added at the converter, and the effects of nonlinearities and dispersion (see Chapter 5) are not reset.

Another alternative is to use regeneration with reshaping but without retiming, also called 2R regeneration. This is applicable only to digital data. The signal is reshaped by sending it through a logic gate, but not retimed. Additional phase jitter is introduced because of this process and will eventually limit the number of stages that can be cascaded.

The final alternative is to use regeneration with reshaping and retiming (3R). This completely resets the effects of nonlinearities, fiber dispersion, and amplifier noise; moreover, it introduces no additional noise. However, retiming is a bit-rate-specific function, and we lose transparency. If transparency is not very important, this is a very attractive approach. (Note that we will discuss another way of maintaining some transparency with 3R using the so-called digital wrapper in Chapter 9). These types of regenerators often include circuitry to perform performance monitoring and process and modify associated management overheads associated with the signal. We will look at some of these overheads in Sections 6.1 and 9.5.7.

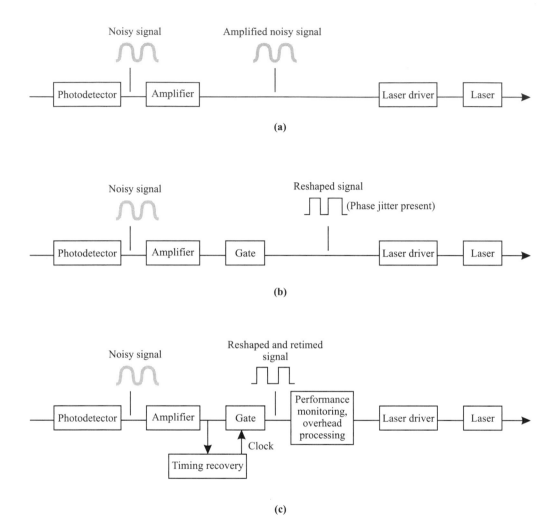

Figure 3.76   Different types of optoelectronic regeneration. (a) 1R (regeneration without reshaping or retiming). (b) 2R (regeneration with reshaping). (c) 3R (regeneration with reshaping and retiming).

3.8.2   Optical Gating

Optical gating makes use of an optical device whose characteristics change with the intensity of an input signal. This change can be transferred to another unmodulated *probe* signal at a different wavelength going through the device. At the output,

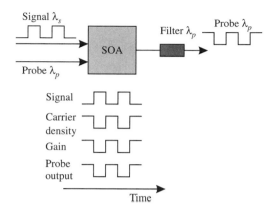

Figure 3.77   Wavelength conversion by cross-gain modulation in a semiconductor optical amplifier.

the probe signal contains the information that is on the input signal. Like the optoelectronic approach, these devices are variable-input and either fixed-output or variable-output devices, depending on whether the probe signal is fixed or tunable. The transparency offered by this approach is limited—only intensity-modulated signals can be converted.

The main technique using this principle is cross-gain modulation (CGM), using a nonlinear effect in a semiconductor optical amplifier (SOA). This approach works over a wide range of signal and probe wavelengths, as long as they are within the amplifier gain bandwidth, which is about 100 nm. Early SOAs were polarization sensitive, but by careful fabrication, it is possible to make them polarization insensitive. SOAs also add spontaneous emission noise to the signal.

CGM makes use of the dependence of the gain of an SOA on its input power, as shown in Figure 3.77. As the input power increases, the carriers in the gain region of the SOA get depleted, resulting in a reduction in the amplifier gain. What makes this interesting is that the carrier dynamics within the SOA are very fast, happening on a picosecond time scale. Thus the gain responds in tune with the fluctuations in input power on a bit-by-bit basis. The device can handle bit rates as high as 10 Gb/s. If a low-power probe signal at a different wavelength is sent into the SOA, it will experience a low gain when there is a 1 bit in the input signal and a higher gain when there is a 0 bit. This very same effect produces crosstalk when multiple signals at different wavelengths are amplified by a single SOA and makes the SOA relatively unsuitable for amplifying WDM signals.

The advantage of CGM is that it is conceptually simple. However, there are several drawbacks. The achievable extinction ratio is small (less than 10) since the gain does not really drop to zero when there is an input 1 bit. The input signal power must be high (around 0 dBm) so that the amplifier is saturated enough to produce a good variation in gain. This high-powered signal must be eliminated at the amplifier output by suitable filtering, unless the signal and probe are counterpropagating. Moreover, as the carrier density within the SOA varies, it changes the refractive index as well, which in turn affects the phase of the probe and creates a large amount of pulse distortion.

3.8.3   Interferometric Techniques

The same phase-change effect that creates pulse distortion in CGM can be used to effect wavelength conversion. As the carrier density in the amplifier varies with the input signal, it produces a change in the refractive index, which in turn modulates the phase of the probe—hence we use the term *cross-phase* modulation for this approach. This phase modulation can be converted into intensity modulation by using an interferometer such as a Mach-Zehnder interferometer (MZI) (see Section 3.3.7). Figure 3.78 shows one possible configuration of a wavelength converter using cross-phase modulation. Both arms of the MZI have exactly the same length, with each arm incorporating an SOA. The signal is sent in at one end (A) and the probe at the other end (B). If no signal is present, then the probe signal comes out unmodulated. The couplers in the MZI are designed with an asymmetric coupling ratio $\gamma \neq 0.5$. When the signal is present, it induces a phase change in each amplifier. The phase change induced by each amplifier on the probe is different because different amounts of signal power are present in the two amplifiers. The MZI translates this relative phase difference between its two arms on the probe into an intensity-modulated signal at the output.

This approach has a few interesting properties. The natural state of the MZI (when no input signal is present) can be arranged to produce either destructive or constructive interference on the probe signal. Therefore we can have a choice of whether the data coming out is the same as the input data or is complementary.

The advantage of this approach over CGM is that much less signal power is required to achieve a large phase shift compared to a large gain shift. In fact, a low signal power and a high probe power can be used, making this method more attractive than CGM. This method also produces a better extinction ratio because the phase change can be converted into a "digital" amplitude-modulated output signal by the interferometer. So this device provides regeneration with reshaping (2R) of

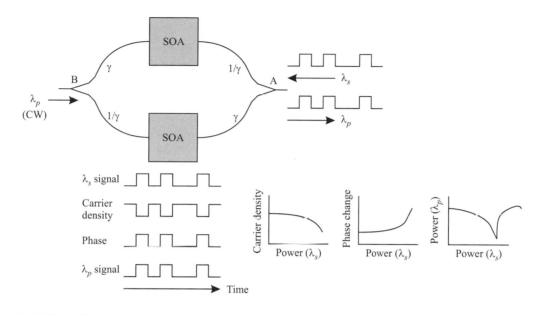

Figure 3.78 Wavelength conversion by cross-phase modulation using semiconductor optical amplifiers embedded inside a Mach-Zehnder interferometer.

the pulses. Depending on where the MZI is operated, the probe can be modulated with the same polarity as the input signal, or the opposite polarity. Referring to Figure 3.78, where we plot the power coupled out at the probe wavelength versus the power at the signal wavelength, depending on the slope of the demultiplexer, a signal power increase can either decrease or increase the power coupled out at the probe wavelength. Like CGM, the bit rate that can be handled is at most 10 Gb/s and is limited by the carrier lifetime. This approach, however, requires very tight control of the bias current of the SOA, as small changes in the bias current produce refractive index changes that significantly affect the phase of signals passing through the device.

We have seen above that the CPM interferometric approach provides regeneration with reshaping (2R) of the pulses. As we saw earlier, while 2R cleans up the signal shape, it does not eliminate phase (or equivalently timing) jitter in the signal, which would accumulate with each such 2R stage. In order to completely clean up the signal, including its temporal characteristics, we need regeneration with reshaping and retiming (3R). Figure 3.79 shows one proposal for accomplishing this in

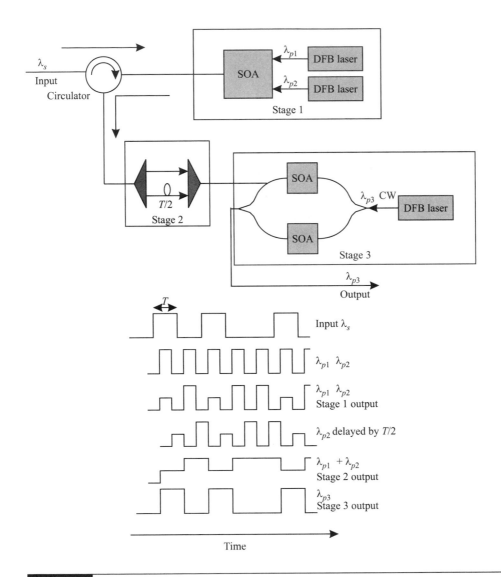

Figure 3.79 All-optical regeneration with reshaping and retiming (3R) using a combination of cross-gain modulation and cross-phase modulation in semiconductor optical amplifiers. (After [Chi97].)

the optical domain without resorting to electronic conversion [Chi97, Gui98]. The approach uses a combination of CGM and CPM. We assume that a local clock is available to sample the incoming data. This clock needs to be recovered from the data; we will study ways of doing this in Section 12.2. The regenerator consists of three stages. The first stage samples the signal. It makes use of CGM in an SOA. The incoming signal is probed using two separate signals at different wavelengths. The two probe signals are synchronized and modulated at twice the data rate of the incoming signal. Since the clock is available, the phase of the probe signals is adjusted to sample the input signal in the middle of the bit interval. At the output of the first stage, the two probe signals have reduced power levels when the input signal is present and higher power levels when the input signal is absent. In the second stage, one of the probe signals is delayed by half a bit period with respect to the other. At the output of this stage, the combined signal has a bit rate that matches the bit rate of the input signal and has been regenerated and retimed. This signal is then sent through a CPM-based interferometric converter stage, which then regenerates and reshapes the signal to create an output signal that has been regenerated, retimed, and reshaped.

3.8.4   Wave Mixing

The four-wave mixing phenomenon that occurs because of nonlinearities in the transmission medium (discussed in Section 2.4.8) can also be utilized to realize wavelength conversion. Recall that four-wave mixing causes three waves at frequencies f_1, f_2, and f_3 to produce a fourth wave at the frequency $f_1 + f_2 - f_3$; when $f_1 = f_2$, we get a wave at the frequency $2f_1 - f_3$. What is interesting about four-wave mixing is that the resulting waves can lie in the same band as the interacting waves. As we have seen in Section 2.4.8, in optical fibers, the generated four-wave mixing power is quite small but can lead to crosstalk if present (see Section 5.8.4).

For the purposes of wavelength conversion, the four-wave mixing power can be enhanced by using an SOA because of the higher intensities within the device. If we have a signal at frequency f_s and a probe at frequency f_p, then four-wave mixing will produce signals at frequencies $2f_p - f_s$ and $2f_s - f_p$, as long as all these frequencies lie within the amplifier bandwidth (Figure 3.80).

The main advantage of four-wave mixing is that it is truly transparent because the effect does not depend on the modulation format (since both amplitude and phase are preserved during the mixing process) and the bit rate. The disadvantages are that the other waves must be filtered out at the SOA output, and the conversion efficiency goes down significantly as the wavelength separation between the signal

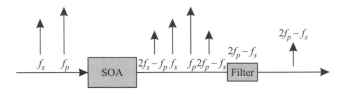

Figure 3.80 Wavelength conversion by four-wave mixing in a semiconductor optical amplifier.

and probe is increased. We will study the conversion efficiency of four-wave mixing in Section 5.8.4.

Summary

We have studied many different optical components in this chapter. Couplers, isolators, and circulators are all commodity components. Many of the optical filters that we studied are now commercially available, with fiber gratings, thin-film multicavity filters, and arrayed waveguide gratings all competing for use in commercial WDM systems.

Erbium-doped fiber amplifiers (EDFAs) are widely deployed and indeed served as a key enabler for WDM. EDFA designs today incorporate multiple stages and gain-flattening filters and provide midstage access between the multiple stages to insert other elements such as dispersion compensating modules and wavelength add/drop multiplexers. A new generation of EDFAs providing amplification in the L-band has recently emerged. We are also now seeing distributed Raman amplifiers used in conjunction with EDFAs in ultra-long-haul systems.

Semiconductor DFB lasers are used in most high-speed communication systems today although other single-longitudinal mode laser structures may eventually become viable commercially. Compact semiconductor tunable lasers are now emerging as viable commercial devices. High-speed APDs and pinFET receivers are both available today.

Large-scale MEMS-based optical switches for use in wavelength crossconnects are now emerging as commercial devices, and a variety of technologies are available to build smaller-scale switches. All-optical wavelength converters are still in the research laboratories, awaiting significant cost reductions and performance improvements before they can become practical.

Further Reading

The book by Green [Gre93] treats many of the optical components considered in this chapter in more detail, particularly tunable filters and lasers. See also [KK97] for more advanced coverage of a number of components.

Most of the filters we described are now commercially available. Gratings are described in detail in several textbooks on optics, for example, [KF86, BW99]. The Stimax grating is described in [LL84] and [Gre93]. See [CK94, Ben96, Kas99] for details on fiber grating fabrication and properties, and [Ven96b, Ven96a] for applications of long-period gratings. For a description of how dielectric thin-film multicavity filters work, see [SS96] and [Kni76]. The electromagnetics background necessary to understand their operation is provided, for example, by [RWv93]. Early papers on the arrayed waveguide grating are [DEK91] and [VS91]. The principle behind their operation is described in [McG98, TSN94, TOTI95, TOT96]. The integrated-optics AOTF is described in [SBJC90, KSHS01], and its systems applications are discussed in [Che90].

There is an extensive literature on optical amplifiers. See [BOS99, Des94] for EDFAs, [Flo00] for a summary of L-band EDFAs, and [O'M88] for a tutorial on SOAs. [Tie95, SMB00, FDW01] provide samples of some recent work on gain-clamped SOAs. See [NE01, NE00] and [KK97, Chapter 7] for an overview of Raman amplifiers.

There are several textbooks on the subject of lasers alone; see, for example, [AD93]. Laser oscillation and photodetection are covered in detail in [Yar97]. [JQE91] is a good reference for several laser-related topics. Other good tutorials on lasers appear in [BKLW00, LZ89, Lee91, SIA92]. A very readable and up-to-date survey of vertical cavity lasers can be found in [Har00]. See also [MZB97]. Most semiconductor lasers today make use of quantum well structures. See [AY86] for a good introduction to this subject. The mathematical theory behind mode locking is explained in [Yar89] and [Yar65]. There is an extensive discussion of various mode-locking methods for fiber lasers in [Agr95]. Lithium niobate external modulators are well described in [Woo00] and [KK97, Chapter 9], and electro-absorption modulators in [BKLW00] and [KK97, Chapter 4].

There is currently significant effort toward realizing commercially viable tunable lasers. We refer the reader to [Col00, Har00, AB98, Gre93, KK97] for more in-depth explorations of this subject. An early review of tunable laser approaches appeared in [KM88]. The VCSEL-based tunable laser is described in [Vak99]. Other types of tunable VCSELs have been demonstrated; see, for instance, [CH00, Har00]. The sampled grating laser structure is explained in [JCC93] and superstructure grating lasers in [Toh93]. See [WMB92, Rig95] for details on the GCSR laser. The arrayed external grating-based laser approaches were proposed in [Soo92, ZJ94, Zir96].

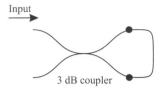

Input

3 dB coupler

Figure 3.81   A 3 dB coupler with the two outputs connected by a piece of fiber.

The tutorial article by Spanke [Spa87] is a good review of large switch architectures for optical switches. See also [MS88] for a good collection of papers on optical switching and [Clo53] for the original paper on the Clos switch architecture. The classic book by Beneš [Ben65] is the authoritative reference for the mathematical theory of large switch architectures developed for telephony applications.

A very accessible survey of mechanical switches can be found in [Kas95, Chapter 13]. Several papers [NR01, LGT98, Nei00, Ryf01, Lao99] describe MEMS-based switches. The inkjet-based waveguide switch is described in [Fou00]. See [WL96, PS95] for some early papers on liquid crystal switches.

Surveys and comparisons of different types of wavelength converters appear in [Stu00, EM00, NKM98, Yoo96, ISSV96, DMJ$^+$96, Chi97].

Problems

3.1   Consider the 3 dB 2 × 2 coupler shown in Figure 3.81. Suppose we connect the two outputs with a piece of fiber. Assume that the polarizations are preserved through the device. A light signal is sent in on the first input. What happens? Derive the field transfer function for the device. Assume the coupler used is a reciprocal device so that it works exactly the same way if its inputs and outputs are reversed. *Hint:* This device is called a loop mirror.

3.2   Consider a device with three ports where it is desired to send all the energy input at ports 1 and 2 to port 3. We assume, for generality, that all ports can be used as inputs and outputs. The scattering matrix of such a device can be written as

$$S = \begin{pmatrix} 0 & 0 & s_{13} \\ 0 & 0 & s_{23} \\ s_{31} & s_{32} & s_{33} \end{pmatrix}.$$

Show that a scattering matrix of this form cannot satisfy the conservation of energy condition, (3.4). Thus it is impossible to build a device that combines all the power from two input ports to a third port, without loss.

3.3 Consider an isolator that is a two-port device where power must be transferred from port 1 to port 2, but no power must be transferred from port 2 to port 1. The scattering matrix of such a device can be written as

$$S = \begin{pmatrix} s_{11} & s_{12} \\ 0 & s_{22} \end{pmatrix}.$$

Show that a scattering matrix of this form cannot satisfy the conservation of energy condition, (3.4). Thus the loss occurs in the isolator because the power input at port 2 must be absorbed by it. However, the power input at port 1 can be transferred to port 2 without loss.

3.4 In Figure 3.10, show that the path length difference between the rays diffracted at angle θ_d and traversing through adjacent slits is approximately $a[\sin(\theta_i) - \sin(\theta_d)]$ when the grating pitch a is small compared to the distance of the source and the imaging plane from the grating plane.

3.5 Derive the grating equation for a blazed reflection grating with blaze angle α, such as the one shown in Figure 3.11.

3.6 Derive the amplitude distribution of the diffraction pattern of a grating with N narrow slits spaced distance d apart. Show that we obtain diffraction maxima when $d \sin \theta = m\lambda$. Discuss what happens in the limit as $N \to \infty$.

3.7 Show that the resonant frequencies f_n of a Fabry-Perot cavity satisfy $f_n = f_0 + n\Delta f$, n integer, for some fixed f_0 and Δf. Thus these frequencies are spaced equally apart. Note that the corresponding wavelengths are *not* spaced equally apart.

3.8 Derive the power transfer function of the Fabry-Perot filter.

3.9 Derive the expression (3.13) for the finesse of the Fabry-Perot filter. Assume that the mirror reflectivity, R, is close to unity.

3.10 Show that the fraction of the input power that is transmitted through the Fabry-Perot filter, over all frequencies, is $(1 - R)/(1 + R)$. Note that this fraction is small for high values of R. Thus, when all frequencies are considered, only a small fraction of the input power is transmitted through a cavity with highly reflective facets.

3.11 Consider a cascade of two Fabry-Perot filters with cavity lengths l_1 and l_2, respectively. Assume the mirror reflectivities of both filters equal R, and the refractive index of their cavities is n. Neglect reflections from the second cavity to the first and vice versa. What is the power transfer function of the cascade? If $l_1/l_2 = k/m$, where k and m are relatively prime integers, find an expression for the FSR of the cascade. Express this FSR in terms of the FSRs of the individual filters.

3.12 Show that the transfer function of the dielectric slab filter shown in Figure G.1(b) is identical to that of a Fabry-Perot filter with facet reflectivity

$$\sqrt{R} = \frac{n_2 - n_1}{n_2 + n_1},$$

assuming $n_3 = n_1$.

3.13 Consider a stack of $2k$ alternating low-index (n_L) and high-index (n_H) dielectric films. Let each of these films have a quarter-wave thickness at λ_0. In the notation of Section 3.3.6, this stack can be denoted by $(HL)^k$. Find the reflectivity of this stack as a function of the optical wavelength λ. Thus a single-cavity dielectric thin-film filter can be viewed as a Fabry-Perot filter with wavelength-dependent mirror reflectivities.

3.14 Derive the power transfer function of the Mach-Zehnder interferometer, assuming only one of its two inputs is active.

3.15 Consider the Mach-Zehnder interferometer of Section 3.3.7.
   (a) With the help of a block diagram, show how a $1 \times n$ demultiplexer can be constructed using $n - 1$ MZIs. Assume n is a power of two. You must specify the path length differences ΔL that must be used in each of the MZIs.
   (b) Can you simplify your construction if only a specific one of the signals needs to be separated from the rest of the $n - 1$?

3.16 Consider the Rowland circle construction shown in Figure 3.26. Show that the differences in path lengths between a fixed-input waveguide and any two successive arrayed waveguides is a constant. Assume that the length of the arc on which the arrayed waveguides are located is much smaller than the diameter of the Rowland circle. *Hint:* Choose a Cartesian coordinate system whose origin is the point of tangency of the Rowland and grating circles. Now express the Euclidean distance between an arbitrary input (output) waveguide and an arbitrary arrayed waveguide in this coordinate system. Use the assumption stated earlier to simplify your expression. Finally, note that the vertical spacing between the arrayed waveguides is constant. In the notation of the book, this shows that $\delta_i = d \sin \theta_i$, where d is the vertical separation between successive arrayed waveguides, and θ_i is the angular separation of input waveguide i from the central input waveguide, as measured from the origin.

3.17 Derive an expression for the FSR of an AWG for a fixed-input waveguide i and a fixed-output waveguide j. The FSR depends on the input and output waveguides. But show that if the arc length of the Rowland circle on which the input and output waveguides are located (see Figure 3.26) is small, then the FSR is approximately constant. Use the result from Problem 3.16 that $\delta_i = d \sin \theta_i$.

3.18 Consider an AWG that satisfies the condition given in Problem 3.17 for its FSR to be approximately independent of the input and output waveguides. Given the FSR, determine the set of wavelengths that must be selected in order for the AWG to function as the wavelength router depicted in Figure 3.25. Assume that the angular spacing between the input (and output) waveguides is constant. Use the result from Problem 3.16 that $\delta_i = d \sin \theta_i$.

3.19 Design an AWG that can multiplex/demultiplex 16 WDM signals spaced 100 GHz apart in the 1.55 μm band. Your design must specify, among other things, the spacing between the input/output waveguides, the path length difference between successive arrayed waveguides, the radius R of the grating circle, and the FSR of the AWG. Assume the refractive index of the input/output waveguides and the arrayed waveguides is 1.5. Note that the design may not be unique, and you may have to make reasonable choices for some of the parameters, which will in turn determine the rest of the parameters.

3.20 Show that the FWHM bandwidth of the acousto-optic filter is $\approx 0.8 \lambda_0^2 / l \Delta n$.

3.21 Explain how the polarization-independent acousto-optic tunable filter illustrated in Figure 3.28 acts as a two-input, two-output wavelength router when both its inputs are active.

3.22 Calculate the acousto-optic interaction length that would be required for the AOTF to have a passband width (FWHM) of 1 nm at an operating wavelength of 1.55 μm. Assume $\Delta n = 0.07$.

3.23 Consider a 16-channel WDM system where the interchannel spacing is nominally 100 GHz. Assume that one of the channels is to be selected by a filter with a 1 dB bandwidth of 2 GHz. We consider three different filter structures for this purpose.

- Fabry-Perot filter: Assume the center wavelengths of the channels do not drift. What is the required finesse and the corresponding mirror reflectivity of a Fabry-Perot filter that achieves a crosstalk suppression of 30 dB from each adjacent channel? If the center wavelengths of the channels can drift up to ± 20 GHz from their nominal values, what is the required finesse and mirror reflectivity?

- Mach-Zehnder interferometer: Assume a cascade of MZIs, as shown in Figure 3.21(c), is used for this purpose and the same level of crosstalk suppression must be achieved. What is the path length difference ΔL and the number of stages required, when the channel center wavelengths are fixed and when they can drift by ± 20 GHz?

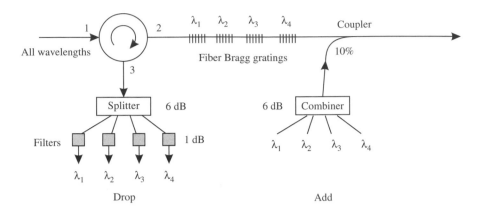

Figure 3.82   A four-channel add/drop multiplexer architecture.

- AOTF: Can an AOTF be used to achieve the same level of crosstalk suppression?

3.24   This problem compares different simple add/drop multiplexer architectures.

(a) First consider the fiber Bragg grating–based add/drop element shown in Figure 3.14(b). Suppose a 5% tap is used to couple the added signal into the output, and the grating induces a loss of 0.5 dB for the transmitted signals and no loss for the reflected signal. Assume that the circulator has a loss of 1 dB per pass. Carefully compute the loss seen by a channel that is dropped, a channel that is added, and a channel that is passed through the device. Suppose the input power per channel is −15 dBm. At what power should the add channel be transmitted so that the powers on all the channels at the output are the same?

(b) Suppose you had to realize an add/drop multiplexer that drops and adds four wavelengths. One possible way to do this is to cascade four add/drop elements of the type shown in Figure 3.14 in series. In this case, compute the best-case and worst-case loss seen by a channel that is dropped, a channel that is added, and a channel that is passed through the device.

(c) Another way to realize a four-channel add/drop multiplexer is shown in Figure 3.82. Repeat the preceding exercise for this architecture. Assume that the losses are as shown in the figure. Which of the two would you prefer from a loss perspective?

(d) Assume that fiber gratings cost $500 each, circulators $3000 each, filters $1000 each, and splitters, combiners, and couplers $100 each. Which of the two preceding architectures would you prefer from a cost point of view?

3.25 In a photodetector, why don't the conduction band electrons absorb the incident photons?

3.26 Consider an EDFA that is required to amplify wavelengths between 1532 nm and 1550 nm within the C-band (separated by 100 GHz).

(a) Draw a schematic of this basic EDFA, and assume the pump laser is selected to minimize ASE. Also, be sure to prevent backward reflections at the EDFA input/output.

(b) Draw the relevant energy bands and associated energy transitions between these bands.

(c) How many wavelengths could be amplified within this range (and spacing)?

(d) Compute the required range in energy transitions to support the entire range of wavelengths.

(e) Suppose we wanted to (1) add and drop a subset of these wavelengths at the EDFA and (2) add a second stage that would be best suited for maximum output powers. Please draw this new two-stage EDFA, with the add/drop multiplexing function drawn as a "black box" labeled "ADM."

(f) Now focusing on the "ADM," assume that two fiber Bragg gratings (along with associated circulator, splitters, and filters) are used to provide static drop capability of the lowest two contiguous wavelengths in the spectral range. In addition, a combiner is used to subsequently add these same wavelengths (of course, carrying different embedded signals). Sketch the architecture for this ADM (that is, the inside of the black box).

(g) If the effective refractive index of the ADM fiber segment is 1.5, calculate the associated fiber Bragg grating periods.

3.27 Consider the 4×4 switch shown in Figure 3.66 made up of 2×2 switches. Suppose each 2×2 switch has crosstalk suppression of 50 dB. What is the overall crosstalk suppression of the 4×4 switch? Assume for now that powers can be added and that we do not have to worry about individual electric fields adding in phase. If we wanted an overall crosstalk suppression of 40 dB, what should the crosstalk suppression of each switch be?

3.28 This problem looks at the Vernier effect, which is used to obtain a filter with a large periodicity given individual filters with smaller periodicities. Consider two periodic filters, one with period f_1 and the other with period f_2, both assumed to be integers. In other words, the first filter selects frequencies $f = mf_1$, where m is an integer, and the second filter selects wavelengths $f = mf_2$. If the two filters are cascaded, show that the resulting filtering function is periodic, with a period given by the least common multiple of f_1 and f_2. For example, if periods of the two filters are 500 GHz and 600 GHz, then the cascaded structure will be periodic with a period of 3000 GHz.

Now suppose the period of each filter can be tuned by 10%. For the numbers given above, the first filter's period can be tuned to 500 ± 25 GHz and the second filter's to 600 ± 30 GHz. Note that the two combs overlap at a frequency of 193,000 GHz. To get an idea of the tuning range of the cascaded structure, determine the nearest frequency to this initial frequency at which the two combs overlap when periods of the individual filters are tuned to (1) 525 GHz and 630 GHz, (2) 475 GHz and 630 GHz, (3) 475 GHz and 570 GHz, and (4) 525 GHz and 570 GHz.

To get an idea of how complex it is to tune this structure, also determine the periods of each filter to obtain an overlap at 193,100 GHz.

3.29   Consider the Clos switch architecture described in Section 3.7.1. Show that if $p \geq 2m - 1$, the switch is strictly nonblocking.

References

[AB98]   M.-C. Amann and J. Buus. *Tunable Laser Diodes*. Artech House, Boston, 1998.

[AD93]   G. P. Agrawal and N. K. Dutta. *Semiconductor Lasers*. Kluwer Academic Press, Boston, 1993.

[Agr95]   G. P. Agrawal. *Nonlinear Fiber Optics,* 2nd edition. Academic Press, San Diego, CA, 1995.

[AI93]   M.-C. Amann and S. Illek. Tunable laser diodes utilising transverse tuning scheme. *IEEE/OSA Journal on Lightwave Technology*, 11(7):1168–1182, July 1993.

[AKB⁺92] R. C. Alferness, U. Koren, L. L. Buhl, B. I. Miller, M. G. Young, T. L. Koch, G. Raybon, and C. A. Burrus. Widely tunable InGaAsP/InP laser based on a vertical coupler filter with 57-nm tuning range. *Applied Physics Letters*, 60:3209–3211, 1992.

[AY86]   Y. Arakawa and A. Yariv. Quantum well lasers—gain, spectra, dynamics. *IEEE Journal of Quantum Electronics*, 22(9):1887–1899, Sept. 1986.

[Ben65]   V. E. Beneš. *Mathematical Theory of Connecting Networks and Telephone Traffic*. Academic Press, New York, 1965.

[Ben96]   I. Bennion et al. UV-written in-fibre Bragg gratings. *Optical Quantum Electronics*, 28(2):93–135, Feb. 1996.

[BKLW00] W. F. Brinkman, T. L. Koch, D. V. Lang, and D. W. Wilt. The lasers behind the communications revolution. *Bell Labs Technical Journal*, 5(1):150–167, Jan.–March 2000.

[BOS99] P. C. Becker, N. A. Olsson, and J. R. Simpson. *Erbium-Doped Fiber Amplifiers: Fundamentals and Technology*. Academic Press, San Diego, CA, 1999.

[BW99] M. Born and E. Wolf. *Principles of Optics: Electromagnetic Theory of Propagation, Diffraction and Interference of Light*. Cambridge University Press, 1999.

[CH00] C. J. Chang-Hasnain. Tunable VCSEL. *IEEE Journal of Selected Topics in Quantum Electronics*, 6(6):978–987, Nov./Dec. 2000.

[Che90] K.-W. Cheung. Acoustooptic tunable filters in narrowband WDM networks: System issues and network applications. *IEEE Journal of Selected Areas in Communications*, 8(6):1015–1025, Aug. 1990.

[Chi97] D. Chiaroni et al. New 10 Gb/s 3R NRZ optical regenerative interface based on semiconductor optical amplifiers for all-optical networks. In *Proceedings of European Conference on Optical Communication*, pages 41–43, 1997. Postdeadline paper.

[CK94] R. J. Campbell and R. Kashyap. The properties and applications of photosensitive germanosilicate fibre. *International Journal of Optoelectronics*, 9(1):33–57, 1994.

[Cle94] B. Clesca et al. Gain flatness comparison between erbium-doped fluoride and silica fiber amplifiers with wavelength-multiplexed signals. *IEEE Photonics Technology Letters*, 6(4):509–512, April 1994.

[Clo53] C. Clos. A study of nonblocking switching networks. *Bell System Technical Journal*, 32:406–424, March 1953.

[Col00] L. A. Coldren. Monolithic tunable diode lasers. *IEEE Journal of Selected Topics in Quantum Electronics*, 6(6):988–999, Nov./Dec. 2000.

[DEK91] C. Dragone, C. A. Edwards, and R. C. Kistler. Integrated optics $N \times N$ multiplexer on silicon. *IEEE Photonics Technology Letters*, 3:896–899, Oct. 1991.

[Des94] E. Desurvire. *Erbium-Doped Fiber Amplifiers: Principles and Applications*. John Wiley, New York, 1994.

[DMJ+96] T. Durhuus, B. Mikkelsen, C. Joergensen, S. Lykke Danielsen, and K. E. Stubkjaer. All optical wavelength conversion by semiconductor optical amplifiers. *IEEE/OSA JLT/JSAC Special Issue on Multiwavelength Optical Technology and Networks*, 14(6):942–954, June 1996.

[Dra89] C. Dragone. Efficient $n \times n$ star couplers using Fourier optics. *IEEE/OSA Journal on Lightwave Technology*, 7(3):479–489, March 1989.

[EM00] J. M. H. Elmirghani and H. T. Mouftah. All-optical wavelength conversion technologies and applications in DWDM networks. *IEEE Communications Magazine*, 38(3):86–92, Mar. 2000.

[FDW01] D. A. Francis, S. P. Dijaili, and J. D. Walker. A single-chip linear optical amplifier. In *OFC 2001 Technical Digest*, pages PD13/1–3, 2001.

[Flo00] F. A. Flood. L-band erbium-doped fiber amplifiers. In *OFC 2000 Technical Digest*, pages WG1-1–WG1-4, 2000.

[Fou00] J. E. Fouquet. Compact optical cross-connect switch based on total internal reflection in a fluid-containing planar lightwave circuit. In *OFC 2000 Technical Digest*, pages TuM1-1–TuM1-4, 2000.

[Gre93] P. E. Green. *Fiber-Optic Networks*. Prentice Hall, Englewood Cliffs, NJ, 1993.

[Gru95] S. G. Grubb et al. High power 1.48 μm cascaded Raman laser in germanosilicate fibers. In *Optical Amplifiers and Applications*, page 197, 1995.

[Gui98] C. Guillemot et al. Transparent optical packet switching: The European ACTS KEOPS project approach. *IEEE/OSA Journal on Lightwave Technology*, 16(12):2117–2134, Dec. 1998.

[Har00] J. S. Harris. Tunable long-wavelength vertical-cavity lasers: The engine of next generation optical networks? *IEEE Journal of Selected Topics in Quantum Electronics*, 6(6):1145–1160, Nov./Dec. 2000.

[ISSV96] E. Iannone, R. Sabella, L. De Stefano, and F. Valeri. All optical wavelength conversion in optical multicarrier networks. *IEEE Transactions on Communications*, 44(6):716–724, June 1996.

[Jac96] J. L. Jackel et al. Acousto-optic tunable filters (AOTFs) for multiwavelength optical cross-connects: Crosstalk considerations. *IEEE/OSA JLT/JSAC Special Issue on Multiwavelength Optical Technology and Networks*, 14(6):1056–1066, June 1996.

[JCC93] V. Jayaraman, Z.-M. Chuang, and L. A. Coldren. Theory, design and performance of extended tuning range semiconductor lasers with sampled gratings. *IEEE Journal of Quantum Electronics*, 29:1824–1834, July 1993.

[JQE91] *IEEE Journal of Quantum Electronics*, June 1991.

[Kam96] I. P. Kaminow et al. A wideband all-optical WDM network. *IEEE JSAC/JLT Special Issue on Optical Networks*, 14(5):780–799, June 1996.

[Kas95] N. Kashima. *Passive Optical Components for Optical Fiber Transmission*. Artech House, Boston, 1995.

[Kas99] R. Kashyap. *Fibre Bragg Gratings*. Academic Press, San Diego, CA, 1999.

[KF86] M. V. Klein and T. E. Furtak. *Optics*, 2nd edition. John Wiley, New York, 1986.

[KK90]   T. L. Koch and U. Koren. Semiconductor lasers for coherent optical fiber communications. *IEEE/OSA Journal on Lightwave Technology*, 8(3):274–293, 1990.

[KK97]   I. P. Kaminow and T. L. Koch, editors. *Optical Fiber Telecommunications IIIB*. Academic Press, San Diego, CA, 1997.

[KM88]   K. Kobayashi and I. Mito. Single frequency and tunable laser diodes. *IEEE/OSA Journal on Lightwave Technology*, 6(11):1623–1633, November 1988.

[Kni76]   Z. Knittl. *Optics of Thin Films*. John Wiley, New York, 1976.

[KSHS01] A. M. J. Koonen, M. K. Smit, H. Herrmann, and W. Sohler. Wavelength selective devices. In H. Venghaus and N. Grote, editors, *Devices for Optical Communication Systems*. Springer-Verlag, Heidelberg, 2001.

[Lao99]   H. Laor. 576 × 576 optical cross connect for single-mode fiber. In *Proceedings of Annual Multiplexed Telephony Conference*, 1999.

[Lee91]   T. P. Lee. Recent advances in long-wavelength semiconductor lasers for optical fiber communication. *Proceedings of IEEE*, 79(3):253–276, March 1991.

[LGT98]   L. Y. Liu, E. L. Goldstein, and R. W. Tkach. Free-space micromachined optical switches with submillisecond switching time for large-scale optical crossconnects. *IEEE Photonics Technology Letters*, 10(4):525–528, Apr. 1998.

[LL84]   J. P. Laude and J. M. Lerner. Wavelength division multiplexing/demultiplexing (WDM) using diffraction gratings. *SPIE-Application, Theory and Fabrication of Periodic Structures*, 503:22–28, 1984.

[LZ89]   T. P. Lee and C-N. Zah. Wavelength-tunable and single-frequency lasers for photonic communication networks. *IEEE Communications Magazine*, 27(10):42–52, Oct. 1989.

[McG98]   K. A. McGreer. Arrayed waveguide gratings for wavelength routing. *IEEE Communications Magazine*, 36(12):62–68, Dec. 1998.

[MS88]   J. E. Midwinter and P. W. Smith, editors. *IEEE JSAC: Special Issue on Photonic Switching*, volume 6, Aug. 1988.

[MZB97]   N. M. Margalit, S. Z. Zhang, and J. E. Bowers. Vertical cavity lasers for telecom applications. *IEEE Communications Magazine*, 35(5):164–170, May 1997.

[NE00]   S. Namiki and Y. Emori. Recent advances in ultra-wideband Raman amplifiers. In *OFC 2000 Technical Digest*, pages FF-1–FF-2, 2000.

[NE01]   S. Namiki and Y. Emori. Ultra-broadband Raman amplifiers pumped and gain-equalized by wavelength-division-multiplexed high-power laser diodes. *IEEE Journal of Selected Topics in Quantum Electronics*, 7(1):3–16, Jan./Feb. 2001.

[Nei00] D. T. Neilson et al. Fully provisioned 112 × 112 micro-mechanical optical crossconnect with 35.8 Tb/s demonstrated capacity. In *OFC 2000 Technical Digest*, pages 204–206, 2000. Postdeadline paper PD-12.

[NKM98] D. Nesset, T. Kelly, and D. Marcenac. All-optical wavelength conversion using SOA nonlinearities. *IEEE Communications Magazine*, 36(12):56–61, Dec. 1998.

[NR01] A. Neukermans and R. Ramaswami. MEMS technology for optical networking applications. *IEEE Communications Magazine*, 39(1):62–69, Jan. 2001.

[O'M88] M. J. O'Mahony. Semiconductor laser amplifiers for future fiber systems. *IEEE/OSA Journal on Lightwave Technology*, 6(4):531–544, April 1988.

[PS95] J. S. Patel and Y. Silberberg. Liquid crystal and grating-based multiple-wavelength cross-connect switch. *IEEE Photonics Technology Letters*, 7(5):514–516, May 1995.

[Rig95] P.-J. Rigole et al. 114-nm wavelength tuning range of a vertical grating assisted codirectional coupler laser with a super structure grating distributed Bragg reflector. *IEEE Photonics Technology Letters*, 7(7):697–699, July 1995.

[RWv93] S. Ramo, J. R. Whinnery, and T. van Duzer. *Fields and Waves in Communication Electronics*. John Wiley, New York, 1993.

[Ryf01] R. Ryf et al. 1296-port MEMS transparent optical crossconnect with 2.07 Petabit/s switch capacity. In *OFC 2001 Technical Digest*, 2001. Postdeadline paper PD28.

[SB87] R. A. Spanke and V. E. Beneš. An *n*-stage planar optical permutation network. *Applied Optics*, 26, April 1987.

[SBJC90] D. A. Smith, J. E. Baran, J. J. Johnson, and K.-W. Cheung. Integrated-optic acoustically-tunable filters for WDM networks. *IEEE Journal of Selected Areas in Communications*, 8(6):1151–1159, Aug. 1990.

[SIA92] Y. Suematsu, K. Iga, and S. Arai. Advanced semiconductor lasers. *Proceedings of IEEE*, 80:383–397, 1992.

[SMB00] D. T. Schaafsma, E. Miles, and E. M. Bradley. Comparison of conventional and gain-clamped semiconductor optical amplifiers for wavelength-division-multiplexed transmission systems. *IEEE/OSA Journal on Lightwave Technology*, 18(7):922–925, July 2000.

[Son95] G. H. Song. Toward the ideal codirectional Bragg filter with an acousto-optic-filter design. *IEEE/OSA Journal on Lightwave Technology*, 13(3):470–480, March 1995.

[Soo92] J. B. D. Soole et al. Wavelength selectable laser emission from a multistripe array grating integrated cavity laser. *Applied Physics Letters*, 61:2750–2752, 1992.

[Spa87]   R. A. Spanke. Architectures for guided-wave optical space switching systems. *IEEE Communications Magazine*, 25(5):42–48, May 1987.

[SS96]   M. A. Scobey and D. E. Spock. Passive DWDM components using microplasma optical interference filters. In *OFC'96 Technical Digest*, pages 242–243, San Jose, Feb. 1996.

[Stu00]   K. E. Stubkjaer. Semiconductor optical amplifier-based all-optical gates for high-speed optical processing. *IEEE Journal of Selected Topics in Quantum Electronics*, 6(6):1428–1435, Nov./Dec. 2000.

[Tie95]   L. F. Tiemeijer et al. Reduced intermodulation distortion in 1300 nm gain-clamped MQW laser amplifiers. *IEEE Photonics Technology Letters*, 7(3):284–286, Mar. 1995.

[Toh93]   Y. Tohmori et al. Over 100 nm wavelength tuning in superstructure grating (SSG) DBR lasers. *Electronics Letters*, 29:352–354, 1993.

[TOT96]   H. Takahashi, K. Oda, and H. Toba. Impact of crosstalk in an arrayed-waveguide multiplexer on $n \times n$ optical interconnection. *IEEE/OSA JLT/JSAC Special Issue on Multiwavelength Optical Technology and Networks*, 14(6):1097–1105, June 1996.

[TOTI95] H. Takahashi, K. Oda, H. Toba, and Y. Inoue. Transmission characteristics of arrayed $n \times n$ wavelength multiplexer. *IEEE/OSA Journal on Lightwave Technology*, 13(3):447–455, March 1995.

[TSN94]   H. Takahashi, S. Suzuki, and I. Nishi. Wavelength multiplexer based on SiO_2–Ta_2O_5 arrayed-waveguide grating. *IEEE/OSA Journal on Lightwave Technology*, 12(6):989–995, June 1994.

[Vak99]   D. Vakhshoori et al. 2 mW CW singlemode operation of a tunable 1550 nm vertical cavity surface emitting laser. *Electronics Letters*, 35(11):900–901, May 1999.

[Ven96a] A. M. Vengsarkar et al. Long-period fiber-grating-based gain equalizers. *Optics Letters*, 21(5):336–338, 1996.

[Ven96b] A. M. Vengsarkar et al. Long-period gratings as band-rejection filters. *IEEE/OSA Journal on Lightwave Technology*, 14(1):58–64, Jan. 1996.

[VS91]   A. R. Vellekoop and M. K. Smit. Four-channel integrated-optic wavelength demultiplexer with weak polarization dependence. *IEEE/OSA Journal on Lightwave Technology*, 9:310–314, 1991.

[WL96]   K.-Y. Wu and J.-Y. Liu. Liquid-crystal space and wavelength routing switches. In *Proceedings of Lasers and Electro-Optics Society Annual Meeting*, pages 28–29, 1996.

[WMB92]   J. Willems, G. Morthier, and R. Baets. Novel widely tunable integrated optical filter with high spectral selectivity. In *Proceedings of European Conference on Optical Communication*, pages 413–416, 1992.

[Woo00] E. L. Wooten et al. A review of lithium niobate modulators for fiber-optic communication systems. *IEEE Journal of Selected Topics in Quantum Electronics*, 6(1):69–82, Jan./Feb. 2000.

[Yar65] A. Yariv. Internal modulation in multimode laser oscillators. *Journal of Applied Physics*, 36:388, 1965.

[Yar89] A. Yariv. *Quantum Electronics*, 3rd edition. John Wiley, New York, 1989.

[Yar97] A. Yariv. *Optical Electronics in Modern Communications*. Oxford University Press, 1997.

[Yoo96] S. J. B. Yoo. Wavelength conversion techniques for WDM network applications. *IEEE/OSA JLT/JSAC Special Issue on Multiwavelength Optical Technology and Networks*, 14(6):955–966, June 1996.

[You95] M. G. Young et al. Six-channel WDM transmitter module with ultra-low chirp and stable λ selection. In *Proceedings of European Conference on Optical Communication*, pages 1019–1022, 1995.

[Zah92] C. E. Zah et al. Monolithic integration of multiwavelength compressive strained multiquantum-well distributed-feedback laser array with star coupler and optical amplifiers. *Electronics Letters*, 28:2361–2362, 1992.

[Zir96] M. Zirngibl et al. An 18-channel multifrequency laser. *IEEE Photonics Technology Letters*, 8:870–872, 1996.

[ZJ94] M. Zirngibl and C. H. Joyner. A 12-frequency WDM laser source based on a transmissive waveguide grating router. *Electronics Letters*, 30:700–701, 1994.

chapter 4

Modulation and Demodulation

OUR GOAL IN THIS CHAPTER is to understand the processes of modulation and demodulation of digital signals. We start by discussing *modulation,* which is the process of converting digital data in electronic form to an optical signal that can be transmitted over the fiber. We then study the *demodulation* process, which is the process of converting the optical signal back into electronic form and extracting the data that was transmitted.

Mainly due to various kinds of noise that get added to the signal in the transmission process, decisions about the transmitted bit (0 or 1) based on the received signal are subject to error. We will derive expressions for the bit error rate introduced by the whole transmission process. Subsequently, we discuss how the bit error rate can be reduced, for the same level of noise (more precisely, signal-to-noise ratio) by the use of forward error-correcting codes. We also discuss clock recovery or synchronization, which is the process of recovering the exact transmission rate at the receiver.

With this background, in the next chapter, we will tackle transmission system engineering, which requires careful attention to a variety of impairments that affect system performance.

4.1 Modulation

The most commonly used modulation scheme in optical communication is *on-off keying* (OOK), which is illustrated in Figure 4.1. In this modulation scheme, a 1 bit is

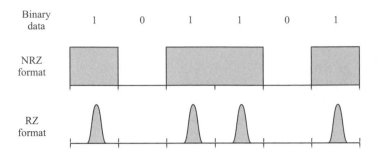

Figure 4.1   On-off keying modulation of binary digital data.

encoded by the presence of a light pulse in the bit interval or by turning a light source (laser or LED) "on." A 0 bit is encoded (ideally) by the absence of a light pulse in the bit interval or by turning a light source "off." The bit interval is the interval of time available for the transmission of a single bit. For example, at a bit rate of 1 Gb/s, the bit interval is 1 ns. As we saw in Section 3.5.4, we can either *directly* modulate the light source by turning it on or off, or use an external modulator in front of the source to perform the same function. Using an external modulator results in less chirp, and thus less of a penalty due to dispersion, and is the preferred approach for high-speed transmission over long distances.

4.1.1 Signal Formats

The OOK modulation scheme can use many different signal formats. The most common signal formats are non-return-to-zero (NRZ) and return-to-zero (RZ). These formats are illustrated in Figure 4.1. In the NRZ format, the pulse for a 1 bit occupies the entire bit interval, and no pulse is used for a 0 bit. If there are two successive 1s, the pulse occupies two successive bit intervals. In the RZ format, the pulse for a 1 bit occupies only a fraction of the bit interval, and no pulse is used for a 0 bit. In electronic (digital) communication, the RZ format has meant that the pulse occupies exactly half the bit period. However, in optical communication, the term RZ is used in a broader sense to describe the use of pulses of duration shorter than the bit period. Thus, there are several variations of the RZ format. In some of them, the pulse occupies a substantial fraction (say, 30%) of the bit interval. The term RZ, without any qualification, usually refers to such systems. If, in addition, the pulses are chirped, they are also sometimes termed dispersion-managed (DM) solitons. In other RZ systems, the pulse occupies only a small fraction of the bit interval. The primary example of such a system is a (conventional) soliton system.

The major advantage of the NRZ format over the other formats is that the signal occupies a much smaller bandwidth—about half that of the RZ format. The problem with the NRZ format is that long strings of 1s or 0s will result in a total absence of any transitions, making it difficult for the receiver to acquire the bit clock, a problem we discuss in Section 4.4.8. The RZ format ameliorates this problem somewhat since long strings of 1s (but not strings of 0s) will still produce transitions. However, the RZ format requires a higher peak transmit power in order to maintain the same energy per bit, and hence the same bit error rate as the NRZ format.

A problem with all these formats is the lack of *DC balance*. An OOK modulation scheme is said to have DC balance if, for all sequences of data bits that may have to be transmitted, the average transmitted power is constant. It is important for an OOK modulation scheme to achieve DC balance since this makes it easier to set the decision threshold at the receiver (see Section 5.2).

To ensure sufficient transitions in the signal and to provide DC balance, either *line coding* or *scrambling* is used in the system. There are many different types of line codes. One form of a *binary block line code* encodes a block of k data bits into $n > k$ bits that are then modulated and sent over the fiber. At the receiver, the n bits are mapped back into the original k data bits (assuming there were no errors). Line codes can be designed so that the encoded bit sequence is DC balanced and provides sufficient transitions irrespective of the input data bit sequence. An example of such a line code is the (8, 10) code that is used in the Fibre Channel standard [WF83, SV96]. This code has $k = 8$ and $n = 10$. The fiber distributed data interface (FDDI) [Ros86] uses a (4, 5) code that is significantly less complex than this (8, 10) code but does not quite achieve DC balance; the worst-case DC imbalance is 10% [Bur86].

An alternative to using line coding is to use *scrambling*. Scrambling is a one-to-one mapping of the data stream into another data stream before it is transmitted on the link. At the transmitter, a scrambler takes the incoming bits and does an EXOR operation with another carefully chosen sequence of bits. The latter sequence is chosen so as to minimize the likelihood of long sequences of 1s or 0s in the transmitted stream. The data is recovered back at the receiver by a descrambler that extracts the data from the scrambled stream. The advantage of scrambling over line coding is that it does not require any additional bandwidth. The disadvantages are that it does not guarantee DC balance, nor does it guarantee a maximum length for a sequence of 1s or 0s. However, the probability of having long run lengths or DC imbalance is made very small by choosing the mapping so that likely input sequences with long run lengths are mapped into sequences with a small run length. However, since the mapping is one to one, it is possible to choose an input sequence that results in a bad output sequence. The mapping is chosen so that only very rare input sequences produce bad output sequences. See Problem 4.2 for an example of how scrambling is implemented and its properties.

In practice, the NRZ format is used in most high-speed communication systems, ranging from speeds of 155 Mb/s to 10 Gb/s. Scrambling is widespread and used in most communication equipment ranging from PC modems to high-speed telecommunications links. High-speed computer data links (for example, Fibre Channel, which operates at 800 Mb/s, and Gigabit Ethernet, which operates at 1 Gb/s) use line codes. See Chapter 6 for a discussion of these protocols.

The RZ format is used in certain high-bit-rate communication systems, such as chirped RZ or DM soliton systems (see Section 2.5.1). In these systems, the pulse occupies about half the bit interval, though this is usually not precise as in digital/electronic communication. The use of RZ pulses also minimizes the effects of chromatic dispersion (see Section 5.7.2). RZ modulation with pulses substantially shorter than the bit interval is used in soliton communication systems (see Section 2.5). The pulses need to be very short in such systems because they must be widely separated (by about five times their width) in order to realize the dispersion-free propagation properties of solitons.

4.2   Subcarrier Modulation and Multiplexing

The optical signal emitted by a laser operating in the 1310 or 1550 nm wavelength band has a center frequency around 10^{14} Hz. This frequency is the *optical carrier* frequency. In what we have studied so far, the data modulates this optical carrier. In other words, with an OOK signal, the optical carrier is simply turned on or off, depending on the bit to be transmitted.

Instead of modulating the optical carrier directly, we can have the data first modulate an electrical carrier in the microwave frequency range, typically ranging from 10 MHz to 10 GHz, as shown in Figure 4.2. The upper limit on the carrier frequency is determined by the modulation bandwidth available from the transmitter. The modulated microwave carrier then modulates the optical transmitter. If the transmitter is directly modulated, then changes in the microwave carrier amplitude get reflected as changes in the transmitted optical power envelope, as shown in Figure 4.2. The microwave carrier can itself be modulated in many different ways, including amplitude, phase, and frequency modulation, and both digital and analog modulation techniques can be employed. The figure shows an example where the microwave carrier is amplitude modulated by a binary digital data signal. The microwave carrier is called the *subcarrier*, with the optical carrier being considered the main carrier. This form of modulation is called *subcarrier modulation*.

The main motivation for using subcarrier modulation is to multiplex multiple data streams onto a single optical signal. This can be done by combining multiple microwave carriers at different frequencies and modulating the optical transmitter

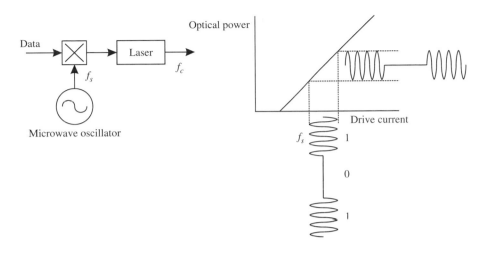

Figure 4.2   Subcarrier modulation. The data stream first modulates a microwave carrier, which, in turn, modulates the optical carrier.

with the combined signal. At the receiver, the signal is detected like any other signal, and the rest of the processing, to separate the subcarriers and extract the data from each subcarrier, is done electronically. This form of multiplexing is called *subcarrier multiplexing* (SCM).

4.2.1   Clipping and Intermodulation Products

The main issue in the design of SCM systems is the trade-off between power efficiency and signal fidelity. Consider Figure 4.2. The SCM system operates around a mean drive current that determines the average optical power. If the mean drive current is increased, for the same SCM waveform, the output optical power is increased. Thus, to keep the output optical power low, the mean drive current must be kept as low as possible. However, the fidelity of the signal depends critically on the linearity of the laser power as a function of the drive current. If f_i, f_j, and f_k denote microwave subcarrier frequencies that are being used, any nonlinearity in laser's power versus drive current characteristic leads to signals at the frequencies $f_i \pm f_j \pm f_k$, which leads to crosstalk, just as in the case of four-wave mixing (see Section 2.4.8). These spurious signals are called *intermodulation products*. Note that the frequencies in the case of SCM are microwave frequencies and those in the FWM case are optical frequencies. But the principle is the same in both cases. For a typical laser, the power–drive current relationship is more linear if the variation in the drive current

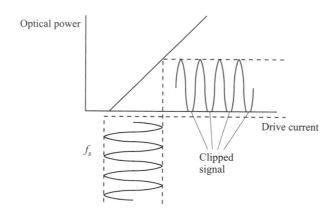

Figure 4.3 Clipping of a subcarrier modulated signal. When the drive current goes below a threshold, the laser output power goes to zero and the signal is said to be clipped.

is a smaller fraction of the average drive current. This means that we must operate at a higher output optical power in order to keep the intermodulation products low. SCM systems use lasers that are specially designed to be highly linear.

The microwave frequencies that are being multiplexed are usually chosen to lie within one octave; that is, if f_L is the lowest frequency and f_H is the highest frequency, these satisfy the condition, $f_H < 2f_L$. When this is the case, all sums and differences of two frequencies—which constitute the second-order intermodulation products—lie either below f_L or above f_H. Thus the second-order intermodulation products produce no crosstalk, and the dominant crosstalk is from third-order intermodulation products, which have much lower power.

A second source of signal distortion in SCM systems is *clipping*. To understand this phenomenon, assume k sinusoids with equal (drive current) amplitude a are being multiplexed. The maximum amplitude of the resulting signal will be ka, and this occurs when all the k signals are in phase. Ideally, the mean operating drive current must be chosen to be greater than ka so that the drive current is nonzero even if all the sinusoids line up in phase. If the operating current is less than ka and all the signals add in phase, there will be no output power for a brief period, when the total current exceeds ka. During this period, the signal is said to be *clipped*. Clipping is illustrated in Figure 4.3 for a single sinusoidal signal.

If k is large, the drive current ka may correspond to a very large optical power. Since the sinusoids are of different frequencies. the probability that they will all add in phase is quite small, particularly for large k. Thus SCM systems are designed to allow a small clipping probability (a few percent), which substantially reduces the power requirement while introducing only a small amount of signal distortion.

4.2.2   Applications of SCM

SCM is widely used by cable operators today for transmitting multiple analog video signals using a single optical transmitter. SCM is also being used in metropolitan-area networks to combine the signals from various users using electronic FDM followed by SCM. This reduces the cost of the network since each user does not require an optical transmitter/laser. We will study these applications further in Chapter 11.

SCM is also used to combine a control data stream along with the actual data stream. For example, most WDM systems that are deployed carry some control information about each WDM channel along with the data that is being sent. This control information has a low rate and modulates a microwave carrier that lies above the data signal bandwidth. This modulated microwave carrier is called a *pilot tone*. We will discuss the use of pilot tones in Chapter 9.

Often it is necessary to receive the pilot tones from all the WDM channels for monitoring purposes, but not the data. This can be easily done if the pilot tones use different microwave frequencies. If this is the case, and the combined WDM signal is photodetected, the detector output will contain an electronic FDM signal consisting of all the pilot tones from which the control information can be extracted. The information from all the data channels will overlap with one another and be lost.

4.3   Spectral Efficiency

We saw in Chapter 2 that the ultimate bandwidth available in silica optical fiber is about 400 nm from 1.2 μm to 1.6 μm, or about 50 THz. The natural question that arises is, therefore, what is the total capacity at which signals can be transmitted over optical fiber?

There are a few different ways to look at this. The *spectral efficiency* of a digital signal is defined as the ratio of the bit rate to the bandwidth used by the signal. The spectral efficiency depends on the type of modulation and coding scheme used. Today's systems primarily use on-off keying of digital data, and in theory can achieve a spectral efficiency of 1 b/s/Hz. In practice, the spectral efficiency of these systems is more like 0.4 b/s/Hz. Using this number, the maximum capacity of optical fiber is about 20 Tb/s. The spectral efficiency can be improved by using more sophisticated modulation and coding schemes, leading to higher channel capacities than the number above. As spectral efficiency becomes increasingly important, such new schemes are being invented, typically based on proven electrical counterparts.

One such scheme that we discuss in the next section is *optical duobinary modulation*. It can increase the spectral efficiency by a factor of about 1.5, typically, achieving a spectral efficiency of 0.6 b/s/Hz.

4.3.1   Optical Duobinary Modulation

The fundamental idea of duobinary modulation (electrical or optical) is to deliberately introduce intersymbol interference (ISI) by overlapping data from adjacent bits. This is accomplished by adding a data sequence to a 1-bit delayed version of itself. For example, if the (input) data sequence is $(0, 0, 1, 0, 1, 0, 0, 1, 1, 0)$, we would instead transmit the (output) data sequence $(0, 0, 1, 0, 1, 0, 0, 1, 1, 0) + (*, 0, 0, 1, 0, 1, 0, 0, 1, 1) = (0, 0, 1, 1, 1, 1, 0, 1, 2, 1)$. Here the $*$ denotes the initial value of the input sequence, which we assume to be zero.

Note that while the input sequence is binary and consists of 0s and 1s, the output sequence is a ternary sequence consisting of 0s, 1s, and 2s. Mathematically, if we denote the input sequence by $x(nT)$ and the output sequence by $y(nT)$, duobinary modulation results if

$$y(nT) = x(nT) + x(nT - T),$$

where T is the bit period. In the example above, $x(nT) = (0, 0, 1, 0, 1, 0, 0, 1, 1, 0)$, $1 \leq n \leq 10$, and $y(nT) = (0, 0, 1, 1, 1, 1, 0, 1, 2, 1)$, $1 \leq n \leq 10$.

Since the bits overlap with each other, how do we recover the input sequence $x(nT)$ at the receiver from $y(nT)$? This can be done by constructing the signal $z(nT) = y(nT) - z(nT - T)$ at the receiver. Note that here we subtract a delayed version of $z(nT)$ from $y(nT)$, and not a delayed version of $y(nT)$ itself. This operation recovers $x(nT)$ since $z(nT) = x(nT)$, assuming we also initialize the sequence $z(0) = 0$. (For readers familiar with digital filters, $y(nT)$ is obtained from $x(nT)$ by a digital filter, and $z(nT)$ from $y(nT)$ by using the inverse of the same digital filter.) The reader should verify this by calculating $z(nT)$ for the example sequence above. To see that this holds generally, just calculate as follows:

$$
\begin{aligned}
z(nT) &= y(nT) - z(nT - T) \\
&= y(nT) - y(nT - T) + z(nT - 2T) \\
&= y(nT) - y(nT - T) + y(nT - 2T) - z(nT - 3T) \\
&= y(nT) - y(nT - T) + y(nT - 2T) - \ldots + (-1)^{n-1} y(T) \\
&= [x(nT) + x(nT - T)] - [x(nT - T) - x(nT - 2T)] + \ldots \\
&= x(nT) \tag{4.1}
\end{aligned}
$$

There is one problem with this scheme, however; a single transmission error will cause all further bits to be in error, until another transmission error occurs to correct the first one! This phenomenon is known as *error propagation*. To visualize error propagation, assume a transmission error occurs in some ternary digit in the example sequence $y(nT)$ above, and calculate the decoded sequence $z(nT)$.

The solution to the error propagation problem is to encode the actual data to be transmitted, not by the absolute value of the input sequence $x(nT)$, but by changes in the sequence $x(nT)$. Thus the sequence $x(nT) = (0, 0, 1, 0, 1, 0, 0, 1, 1, 0)$ would correspond to the data sequence $d(nT) = (0, 0, 1, 1, 1, 1, 0, 1, 0, 1)$. A 1 in the sequence $d(nT)$ is encoded by changing the sequence $x(nT)$ from a 0 to a 1, or from a 1 to a 0. To see how differential encoding solves the problem, observe that if a sequence of consecutive bits are all in error, their differences will still be correct, modulo 2.

Transmission of a ternary sequence using optical intensity modulation (the generalization of OOK for nonbinary sequences) will involve transmitting three different optical powers, say, 0, P, and $2P$. Such a modulation scheme will also considerably complicate the demodulation process. We would like to retain the advantage of binary signaling while employing duobinary signaling to reduce the transmission bandwidth.

To see how this can be done, compare $y(nT) = (0, 0, 1, 1, 1, 1, 0, 1, 2, 1)$ and $d(nT) = (0, 0, 1, 1, 1, 1, 0, 1, 0, 1)$ in our example, and observe that $y(nT) \bmod 2 = d(nT)$! This result holds in general, and thus we may think that we could simply map the 2s in $y(nT)$ to 0s and transmit the resulting binary sequence, which could then be detected using the standard scheme. However, such an approach would eliminate the bandwidth advantage of duobinary signaling, as it should, because in such a scheme the differential encoding and the duobinary encoding have done nothing but cancel the effects of each other. The bandwidth advantage of duobinary signaling can only be exploited by using a ternary signaling scheme. A ternary signaling alternative to using three optical power levels is to use a combination of amplitude and phase modulation. Such a scheme is dubbed optical AM-PSK, and most studies of optical duobinary signaling today are based on AM-PSK.

Conceptually, the carrier is a continuous wave signal, a sinusoid, which we can denote by $a\cos(\omega t)$. The three levels of the ternary signal correspond to $-a\cos(\omega t) = a\cos(\omega t + \pi)$, $0 = 0\cos(\omega t)$, and $a\cos(\omega t)$, which we denote by -1, 0, and $+1$, respectively. The actual modulation is usually accomplished using an external modulator in the Mach-Zehnder arrangement (see Sections 3.3.7 and 3.5.4). These are the three signal levels corresponding to 0, 1, and 2, respectively, in $y(nT)$. This modulation scheme is clearly a combination of amplitude and phase modulation; hence the term AM-PSK. The AM-PSK signal retains the bandwidth advantage of duobinary signaling. However, for a direct detection receiver, the signals

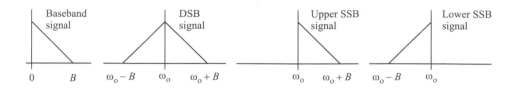

Figure 4.4 Spectrum of a baseband signal compared with the spectra of double sideband (DSB) and single sideband (SSB) modulated signals. The spectral width of the SSB signals is the same as that of the baseband signal, whereas the DSB signal has twice the spectral width of the baseband signal.

$\pm a \cos(\omega t)$ are indistinguishable so that the use of such a receiver merely identifies $2 = 0$ in $y(nT)$ naturally performing the mod 2 operation required to recover $d(nT)$ from $y(nT)$.

4.3.2   Optical Single Sideband Modulation

Another technique for increasing the spectral efficiency is *optical single sideband (SSB) modulation.* Such a scheme can improve the spectral efficiency by a factor of 2, if practical implementations capable of supporting transmission at 10 Gb/s and above can be found. Before we can define what optical SSB modulation is, we need to understand the concept of sidebands in a digital signal.

Consider a sinusoidal carrier signal $\cos(\omega_o t)$. Assume this is directly modulated by a data signal that is also a sinusoid, $\cos(\omega_d t)$, for simplicity. Typically $\omega_d \ll \omega_o$ since ω_o is an optical carrier frequency of the order of 200 THz and ω_d is of the order of 10 GHz. Direct modulation amounts to forming the product $\cos(\omega_o t) \cos(\omega_d t) = 0.5 \cos((\omega_o + \omega_d)t) + 0.5 \cos((\omega_o - \omega_d)t)$. Thus the transmitted signal contains two sinusoids at $\omega_o + \omega_d$ and $\omega_o + \omega_d$ for a data signal consisting of a single sinusoid at ω_d. In general, for a digital signal with a (baseband) frequency spectrum extending from 0 to B Hz, the modulated signal has a spectrum covering the frequency range from $\omega_o - B$ Hz to $\omega_o + B$ Hz, that is, a range of $2B$ Hz around the carrier frequency ω_o. Each of the spectral bands of width B Hz on either side of the carrier frequency ω_o is called a *sideband,* and such a signal is said to be a *double sideband* (DSB) signal. By appropriate filtering, we can eliminate one of these sidebands: either the lower or the upper one. The resulting signals are called *single sideband* (SSB) signals. DSB and SSB signals are illustrated in Figure 4.4.

The difficulty in implementing optical SSB modulation lies in designing the filters to eliminate one of the sidebands—they have to be very sharp. Instead of filtering it entirely, allowing a small part, or vestige, of one of the sidebands to remain makes implementation easier. Such a scheme is called *vestigial sideband* (VSB) modulation. This is the modulation scheme used in television systems, and its use is currently being explored for optical systems, mainly for analog signal transmission.

Optical SSB modulation is also being explored today either for analog signal transmission or, equivalently, for SCM systems, which are analog systems from the viewpoint of optical modulation.

4.3.3 Multilevel Modulation

The main technique used in digital communication to achieve spectral efficiencies greater than 1 b/s/Hz is *multilevel modulation*. The simplest multilevel modulation scheme uses $M > 2$ amplitude levels of a sinusoidal carrier to represent M possible signal values. In such a scheme, each signal represents $\log_2 M$ bits. However, the bandwidth occupied by a digital communication system transmitting R such symbols per second is nearly the same as that occupied by an R b/s digital system employing binary signals. Therefore, the bandwidth efficiency of such a multilevel scheme is $\log_2 M$ times higher, and about $\log_2 M$ b/s/Hz. However, such multilevel schemes have not been used in practical optical communication systems to date due to the complexities of detecting such signals at high bit rates. Another potential advantage of multilevel modulation is that the signaling rate on the channel is lower than the data rate. For example, a 16-level modulation scheme would be able to transmit at a date rate of 40 Gb/s but at a signaling rate of 10 Gbaud; that is, each signal occupies a period of 100 ps, and not 25 ps. This, in turn, helps mitigate the effects of dispersion and nonlinearities.

4.3.4 Capacity Limits of Optical Fiber

An upper limit on the spectral efficiency and the channel capacity is given by Shannon's theorem [Sha48]. Shannon's theorem says that the channel capacity C for a binary linear channel with additive noise is given by

$$C = B \log_2 \left(1 + \frac{S}{N} \right).$$

Here B is the available bandwidth and S/N is the signal-to-noise ratio. A typical value of S/N is 100. Using this number yields a channel capacity of 350 Tb/s or an equivalent spectral efficiency of 7 b/s/Hz. Clearly, such efficiencies can only be achieved through the use of multilevel modulation schemes.

In practice, today's long-haul systems operate at high power levels to overcome fiber losses and noise introduced by optical amplifiers. At these power levels, nonlinear effects come into play. These nonlinear effects can be thought of as adding

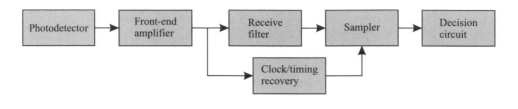

Figure 4.5   Block diagram showing the various functions involved in a receiver.

additional noise, which increases as the transmitted power is increased. Therefore they in turn impose additional limits on channel capacity. Recent work to quantify the spectral efficiency, taking into account mostly cross-phase modulation [Sta99, MS00], shows that the achievable efficiencies are of the order of 3–5 b/s/Hz. Other nonlinearities such as four-wave mixing and Raman scattering may place further limitations. At the same time, we are seeing techniques to reduce the effects of these nonlinearities.

Another way to increase the channel capacity is by reducing the noise level in the system. The noise figure in today's amplifiers is limited primarily by random spontaneous emission, and these are already close to theoretically achievable limits. Advances in quantum mechanics [Gla00] may ultimately succeed in reducing these noise limits.

4.4   Demodulation

The modulated signals are transmitted over the optical fiber where they undergo attenuation and dispersion, have noise added to them from optical amplifiers, and sustain a variety of other impairments that we will discuss in Chapter 5. At the receiver, the transmitted data must be recovered with an acceptable *bit error rate* (BER). The required BER for high-speed optical communication systems today is in the range of 10^{-9} to 10^{-15}, with a typical value of 10^{-12}. A BER of 10^{-12} corresponds to one allowed bit error for every terabit of data transmitted, on average.

Recovering the transmitted data involves a number of steps, which we will discuss in this section. Our focus will be on the demodulation of OOK signals. Figure 4.5 shows the block diagram of a receiver. The optical signal is first converted to an electrical current by a *photodetector*. This electrical current is quite weak and thus we use a *front-end amplifier* to amplify it. The photodetector and front-end amplifier were discussed in Sections 3.6.1 and 3.6.2, respectively.

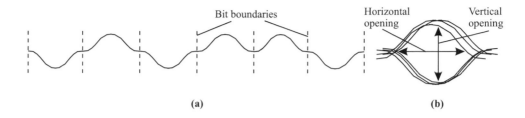

Figure 4.6   Eye diagram. (a) A typical received waveform along with the bit boundaries. (b) The received waveform of (a), wrapped around itself, on the bit boundaries to generate an eye diagram. For clarity, the waveform has been magnified by a factor of 2 relative to (a).

The amplified electrical current is then filtered to minimize the noise outside the bandwidth occupied by the signal. This filter is also designed to suitably shape the pulses so that the bit error rate is minimized. This filter may also incorporate additional functionality, such as minimizing the intersymbol interference due to pulse spreading. If the filter performs this function, it is termed an *equalizer*. The name denotes that the filter equalizes, or cancels, the distortion suffered by the signal. Equalization is discussed in Section 4.4.9.

The signal must then be sampled at the midpoints of the bit intervals to decide whether the transmitted bit in each bit interval was a 1 or a 0. This requires that the bit boundaries be recovered at the receiver. A waveform that is periodic with period equal to the bit interval is called a *clock*. This function is termed *clock recovery*, or *timing recovery*, and is discussed in Section 4.4.8.

A widely used experimental technique to determine the goodness of the received signal is the *eye diagram*. Consider the received waveform shown in Figure 4.6(a). This is a typical shape of the received signal for NRZ modulation, after it has been filtered by the receive filter and is about to be sampled (see Figure 4.5). The bit boundaries are also shown on the figure. If the waveform is cut along at the bit boundaries and the resulting pieces are superimposed on each other, we get the resulting diagram shown in Figure 4.6(b). Such a diagram is called an *eye diagram* because of its resemblance to the shape of the human eye. An eye diagram can be easily generated experimentally using an oscilloscope to display the received signal while it is being triggered by the (recovered) clock. The vertical opening of the eye indicates the margin for bit errors due to noise. The horizontal opening of the eye indicates the margin for timing errors due to an imperfectly recovered clock.

In Section 1.5, we saw that there could be different types of repeaters, specifically 2R (regeneration with reshaping) and 3R (regeneration with reshaping and retiming). The difference between these primarily lies in the type of receiver used. A 2R receiver does not have the timing recovery circuit shown in Figure 4.5, whereas a 3R does. Also a 3R receiver may use a multirate timing recovery circuit, which is capable of recovering the clock at a variety of data rates.

4.4.1 An Ideal Receiver

In principle, the demodulation process can be quite simple. Ideally, it can be viewed as "photon counting," which is the viewpoint we will take in this section. In practice, there are various impairments that are not accounted for by this model, and we discuss them in the next section.

The receiver looks for the presence or absence of light during a bit interval. If no light is seen, it infers that a 0 bit was transmitted, and if any light is seen, it infers that a 1 bit was transmitted. This is called *direct detection*. Unfortunately, even in the absence of other forms of noise, this will not lead to an ideal error-free system because of the random nature of photon arrivals at the receiver. A light signal arriving with power P can be thought of as a stream of photons arriving at average rate P/hf_c. Here, h is Planck's constant (6.63×10^{-34} J/Hz), f_c is the carrier frequency, and hf_c is the energy of a single photon. This stream can be thought of as a Poisson random process.

Note that our simple receiver does not make any errors when a 0 bit is transmitted. However, when a 1 bit is transmitted, the receiver may decide that a 0 bit was transmitted if no photons were received during that bit interval. If B denotes the bit rate, then the probability that n photons are received during a bit interval $1/B$ is given by

$$e^{-(P/hf_c B)} \frac{\left(\frac{P}{hf_c B}\right)^n}{n!}.$$

Thus the probability of not receiving any photons is $e^{-(P/hf_c B)}$. Assuming equally likely 1s and 0s, the bit error rate of this ideal receiver would be given as

$$\mathrm{BER} = \frac{1}{2} e^{-\frac{P}{hf_c B}}.$$

Let $M = P/hf_c B$. The parameter M represents the average number of photons received during a 1 bit. Then the bit error rate can be expressed as

$$\mathrm{BER} = \frac{1}{2} e^{-M}.$$

This expression represents the error rate of an ideal receiver and is called the *quantum limit*. To get a bit error rate of 10^{-12}, note that we would need an average of $M = 27$ photons per 1 bit.

In practice, most receivers are not ideal, and their performance is not as good as that of the ideal receiver because they must contend with various other forms of noise, as we shall soon see.

4.4.2   A Practical Direct Detection Receiver

As we have seen in Section 3.6 (see Figure 3.61), the optical signal at the receiver is first photodetected to convert it into an electrical current. The main complication in recovering the transmitted bit is that in addition to the photocurrent due to the signal there are usually three other additional noise currents. The first is the *thermal noise* current due to the random motion of electrons that is always present at any finite temperature. The second is the *shot noise* current due to the random distribution of the electrons generated by the photodetection process even when the input light intensity is constant. The shot noise current, unlike the thermal noise current, is not added to the generated photocurrent but is merely a convenient representation of the variability in the generated photocurrent as a separate component. The third source of noise is the spontaneous emission due to optical amplifiers that may be used between the source and the photodetector. The amplifier noise currents are treated in Section 4.4.5 and Appendix I. In this section, we will consider only the thermal noise and shot noise currents.

The thermal noise current in a resistor R at temperature T can be modeled as a Gaussian random process with zero mean and autocorrelation function $(4k_B T/R)\delta(\tau)$. Here k_B is Boltzmann's constant and has the value 1.38×10^{-23} J/°K, and $\delta(\tau)$ is the Dirac delta function, defined as $\delta(\tau) = 0$, $\tau \neq 0$ and $\int_{-\infty}^{\infty} \delta(\tau)d\tau = 1$. Thus the noise is white, and in a bandwidth or frequency range B_e, the thermal noise current has the variance

$$\sigma_{\text{thermal}}^2 = (4k_B T/R)B_e.$$

This value can be expressed as $I_t^2 B_e$, where I_t is the parameter used to specify the current standard deviation in units of pA/$\sqrt{\text{Hz}}$. Typical values are of the order of 1 pA/$\sqrt{\text{Hz}}$.

The electrical bandwidth of the receiver, B_e, is chosen based on the bit rate of the signal. In practice, B_e varies from $1/2T$ to $1/T$, where T is the bit period. We will also be using the parameter B_o to denote the optical bandwidth seen by the receiver. The optical bandwidth of the receiver itself is very large, but the value of B_o is usually determined by filters placed in the optical path between the transmitter and receiver.

By convention, we will measure B_e in baseband units and B_o in passband units. Therefore, the minimum possible value of $B_o = 2B_e$, to prevent signal distortion.

As we saw in the previous section, the photon arrivals are accurately modeled by a Poisson random process. The photocurrent can thus be modeled as a stream of electronic charge impulses, each generated whenever a photon arrives at the photodetector. For signal powers that are usually encountered in optical communication systems, the photocurrent can be modeled as

$$I = \bar{I} + i_s,$$

where $\bar{I}$ is a constant current, and i_s is a Gaussian random process with mean zero and autocorrelation $\sigma^2_{\text{shot}}\delta(\tau)$. For *pin* diodes, $\sigma^2_{\text{shot}} = 2e\bar{I}$. This is derived in Appendix I. The constant current $\bar{I} = \mathcal{R}P$, where $\mathcal{R}$ is the responsivity of the photodetector, which was discussed in Section 3.6. Here, we are assuming that the dark current, which is the photocurrent that is present in the absence of an input optical signal, is negligible. Thus the shot noise current is also white and in a bandwidth B_e has the variance

$$\sigma^2_{\text{shot}} = 2e\bar{I}B_e. \tag{4.2}$$

If we denote the load resistor of the photodetector by R_L, the total current in this resistor can be written as

$$I = \bar{I} + i_s + i_t,$$

where i_t has the variance $\sigma^2_{\text{thermal}} = (4k_BT/R_L)B_e$. The shot noise and thermal noise currents are assumed to be independent so that, if B_e is the bandwidth of the receiver, this current can be modeled as a Gaussian random process with mean $\bar{I}$ and variance

$$\sigma^2 = \sigma^2_{\text{shot}} + \sigma^2_{\text{thermal}}.$$

Note that both the shot noise and thermal noise variances are proportional to the bandwidth B_e of the receiver. Thus there is a trade-off between the bandwidth of a receiver and its noise performance. A receiver is usually designed so as to have just sufficient bandwidth to accommodate the desired bit rate so that its noise performance is optimized. In most practical direct detection receivers, the variance of the thermal noise component is much larger than the variance of the shot noise and determines the performance of the receiver.

4.4.3   Front-End Amplifier Noise

We saw in Chapter 3 (Figure 3.61) that the photodetector is followed by a front-end amplifier. Components within the front-end amplifier, such as the transistor, also contribute to the thermal noise. This noise contribution is usually stated by giving the *noise figure* of the front-end amplifier. The noise figure F_n is the ratio of the input signal-to-noise ratio (SNR_i) to the output signal-to-noise ratio (SNR_o). Equivalently, the noise figure F_n of a front-end amplifier specifies the factor by which the thermal noise present at the input of the amplifier is enhanced at its output. Thus the thermal noise contribution of the receiver has variance

$$\sigma_{thermal}^2 = \frac{4k_B T}{R_L} F_n B_e \tag{4.3}$$

when the front-end amplifier noise contribution is included. Typical values of F_n are 3–5 dB.

4.4.4 APD Noise

As we remarked in Section 3.6.1, the avalanche gain process in APDs has the effect of increasing the noise current at its output. This increased noise contribution arises from the random nature of the avalanche multiplicative gain, $G_m(t)$. This noise contribution is modeled as an increase in the shot noise component at the output of the photodetector. If we denote the responsivity of the APD by $\mathcal{R}_{APD}$, and the average avalanche multiplication gain by G_m, the average photocurrent is given by $\bar{I} = \mathcal{R}_{APD} P = G_m \mathcal{R} P$, and the shot noise current at the APD output has variance

$$\sigma_{shot}^2 = 2e G_m^2 F_A(G_m) \mathcal{R} P B_e. \tag{4.4}$$

The quantity $F_A(G_m)$ is called the *excess noise factor* of the APD and is an increasing function of the gain G_m. It is given by

$$F_A(G_m) = k_A G_m + (1 - k_A)(2 - 1/G_m).$$

The quantity k_A is called the ionization coefficient ratio and is a property of the semiconductor material used to make up the APD. It takes values in the range (0–1). The excess noise factor is an increasing function of k_A, and thus it is desirable to keep k_A small. The value of k_A for silicon (which is used at 0.8 μm wavelength) is $\ll 1$, and for InGaAs (which is used at 1.3 and 1.55 μm wavelength bands) is 0.7.

Note that $F_A(1) = 1$, and thus (4.4) also yields the shot noise variance for a *pin* receiver if we set $G_m = 1$.

4.4.5 Optical Preamplifiers

As we have seen in the previous sections, the performance of simple direct detection receivers is limited primarily by thermal noise generated inside the receiver. The performance can be improved significantly by using an optical (pre)amplifier after the receiver, as shown in Figure 4.7. The amplifier provides added gain to the input signal. Unfortunately, as we saw in Section 3.4.2, the spontaneous emission present in the amplifier appears as noise at its output. The amplified spontaneous (ASE) noise power at the output of the amplifier for each polarization mode is given by

$$P_N = n_{sp}hf_c(G-1)B_o, \tag{4.5}$$

where n_{sp} is a constant called the spontaneous emission factor, G is the amplifier gain, and B_o is the optical bandwidth. Two fundamental polarization modes are present in a single-mode fiber, as we saw in Chapter 2. Hence the total noise power at the output of the amplifier is $2P_N$.

The value of n_{sp} depends on the level of population inversion within the amplifier. With complete inversion $n_{sp} = 1$, but it is typically higher, around 2–5 for most amplifiers.

For convenience in the discussions to follow, we define

$$P_n = n_{sp}hf_c.$$

To understand the impact of amplifier noise on the detection of the received signal, consider the optical preamplifier system shown in Figure 4.7, used in front of a standard *pin* direct detection receiver. The photodetector produces a current that is proportional to the incident power. The signal current is given by

$$I = \mathcal{R}GP, \tag{4.6}$$

where P is the received optical power.

The photodetector produces a current that is proportional to the optical power. The optical power is proportional to the square of the electric field. Thus the noise field beats against the signal and against itself, giving rise to noise components referred to as the *signal-spontaneous* beat noise and *spontaneous-spontaneous* beat noise, respectively. In addition, shot noise and thermal noise components are also present.

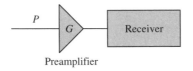

Preamplifier

Figure 4.7   A receiver with an optical preamplifier.

The variances of the thermal noise, shot noise, signal-spontaneous noise, and spontaneous-spontaneous noise currents at the receiver are, respectively,

$$\sigma^2_{\text{thermal}} = I_t^2 B_e, \tag{4.7}$$

$$\sigma^2_{\text{shot}} = 2e\mathcal{R}[GP + P_n(G-1)B_o]B_e, \tag{4.8}$$

$$\sigma^2_{\text{sig-spont}} = 4\mathcal{R}^2 GPP_n(G-1)B_e, \tag{4.9}$$

and

$$\sigma^2_{\text{spont-spont}} = 2\mathcal{R}^2[P_n(G-1)]^2(2B_o - B_e)B_e. \tag{4.10}$$

These variances are derived in Appendix I. Here I_t is the receiver thermal noise current. Provided the amplifier gain is reasonably large (> 10 dB), which is usually the case, the shot noise and thermal noise are negligible compared to the signal-spontaneous and spontaneous-spontaneous beat noise. In the bit error rate regime of interest to us (10^{-9} to 10^{-15}), these noise processes can be modeled adequately as Gaussian processes. The spontaneous-spontaneous beat noise can be made very small by reducing the optical bandwidth B_o. This can be done by filtering the amplifier noise before it reaches the receiver. In the limit, B_o can be made as small as $2B_e$. So the dominant noise component is usually signal-spontaneous beat noise.

The amplifier noise is commonly specified by the easily measurable parameter known as the noise figure. Recall from Section 4.4.3 that the noise figure F_n is the ratio of the input signal-to-noise ratio (SNR$_i$) to the output signal-to-noise ratio (SNR$_o$). At the amplifier input, assuming that only signal shot noise is present, using (4.2) and (4.6), the SNR is given by

$$\text{SNR}_i = \frac{(\mathcal{R}P)^2}{2\mathcal{R}ePB_e}.$$

At the amplifier output, assuming that the dominant noise term is the signal-spontaneous beat noise, using (4.6) and (4.9), the SNR is given by

$$\text{SNR}_o \approx \frac{(\mathcal{R}GP)^2}{4\mathcal{R}^2 PG(G-1)n_{\text{sp}}hf_cB_e}.$$

The noise figure of the amplifier is then

$$F_n = \frac{\text{SNR}_i}{\text{SNR}_o} \approx 2n_{\text{sp}} \tag{4.11}$$

In the best case, with full population inversion, $n_{\text{sp}} = 1$. Thus the best-case noise figure is 3 dB. Practical amplifiers have a somewhat higher noise figure, typically in the 4–7 dB range. This derivation assumed that there are no coupling losses between the amplifier and the input and output fibers. Having an input coupling loss degrades the noise figure of the amplifier (see Problem 4.5).

4.4.6   Bit Error Rates

Earlier, we calculated the bit error rate of an ideal direct detection receiver. Next, we will calculate the bit error rate of the practical receivers already considered, which must deal with a variety of different noise impairments.

The receiver makes decisions as to which bit (0 or 1) was transmitted in each bit interval by sampling the photocurrent. Because of the presence of noise currents, the receiver could make a wrong decision resulting in an erroneous bit. In order to compute this bit error rate, we must understand the process by which the receiver makes a decision regarding the transmitted bit.

First, consider a *pin* receiver without an optical preamplifier. For a transmitted 1 bit, let the received optical power $P = P_1$, and let the mean photocurrent $\bar{I} = I_1$. Then $I_1 = \mathcal{R}P_1$, and the variance of the photocurrent is

$$\sigma_1^2 = 2eI_1B_e + 4k_BTB_e/R_L.$$

If P_0 and I_0 are the corresponding quantities for a 0 bit, $I_0 = \mathcal{R}P_0$, and the variance of the photocurrent is

$$\sigma_0^2 = 2eI_0B_e + 4k_BTB_e/R_L.$$

For ideal OOK, P_0 and I_0 are zero, but we will see later (Section 5.3) that this is not always the case in practice.

Let I_1 and I_0 denote the photocurrent sampled by the receiver during a 1 bit and a 0 bit, respectively, and let σ_1^2 and σ_0^2 represent the corresponding noise variances. The noise signals are assumed to be Gaussian. The actual variances will depend on the type of receiver, as we saw earlier. So the bit decision problem faced by the receiver has the following mathematical formulation. The photocurrent for a 1 bit is a sample of a Gaussian random variable with mean I_1 and variance σ_1 (and similarly for the 0 bit as well). The receiver must look at this sample and decide whether the transmitted bit is a 0 or a 1. The possible probability density functions of the sampled photocurrent are sketched in Figure 4.8. There are many possible *decision rules* that the receiver can use; the receiver's objective is to choose the one that minimizes the bit error rate. This *optimum decision rule* can be shown to be the one that, given the observed photocurrent I, chooses the bit (0 or 1) that was *most likely* to have been transmitted. Furthermore, this optimum decision rule can be implemented as follows. Compare the observed photocurrent to a decision threshold I_{th}. If $I \geq I_{th}$, decide that a 1 bit was transmitted; otherwise, decide that a 0 bit was transmitted.

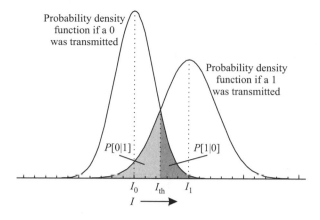

Figure 4.8   Probability density functions for the observed photocurrent.

For the case when 1 and 0 bits are equally likely (which is the only case we consider in this book), the threshold photocurrent is given approximately by

$$I_{\text{th}} = \frac{\sigma_0 I_1 + \sigma_1 I_0}{\sigma_0 + \sigma_1}. \tag{4.12}$$

This value is very close but not exactly equal to the optimal value of the threshold. The proof of this result is left as an exercise (Problem 4.7). Geometrically, I_{th} is the value of I for which the two densities sketched in Figure 4.8 cross. The probability of error when a 1 was transmitted is the probability that $I < I_{\text{th}}$ and is denoted by $P[0|1]$. Likewise, $P[1|0]$ is the probability of deciding that a 1 was transmitted when actually a 0 was transmitted and is the probability that $I \geq I_{\text{th}}$. Both probabilities are indicated in Figure 4.8.

Let $Q(x)$ denote the probability that a zero mean, unit variance Gaussian random variable exceeds the value x. Thus

$$Q(x) = \frac{1}{\sqrt{2\pi}} \int_x^\infty e^{-y^2/2}\, dy. \tag{4.13}$$

It now follows that

$$P[0|1] = Q\left(\frac{I_1 - I_{\text{th}}}{\sigma_1}\right)$$

and

$$P[1|0] = Q\left(\frac{I_{\text{th}} - I_0}{\sigma_0}\right).$$

Using (4.12), it can then be shown that the BER (see Problem 4.6) is given by

$$\text{BER} = Q\left(\frac{I_1 - I_0}{\sigma_0 + \sigma_1}\right). \tag{4.14}$$

The Q function can be numerically evaluated. Let $\gamma = Q^{-1}(\text{BER})$. For a BER rate of 10^{-12}, we need $\gamma \approx 7$. For a BER rate of 10^{-9}, $\gamma \approx 6$.

Note that it is particularly important to have a variable threshold setting in receivers if they must operate in systems with signal-dependent noise, such as optical amplifier noise. Many high-speed receivers do incorporate such a feature. However, many of the simpler receivers do not have a variable threshold adjustment and set their threshold corresponding to the average received current level, namely, $(I_1 + I_0)/2$. This threshold setting yields a higher bit error rate given by

$$\text{BER} = \frac{1}{2}\left[Q\left(\frac{(I_1 - I_0)}{2\sigma_1}\right) + Q\left(\frac{(I_1 - I_0)}{2\sigma_0}\right)\right].$$

We can use (4.14) to evaluate the BER when the received signal powers for a 0 bit and a 1 bit and the noise statistics are known. Often, we are interested in the inverse problem, namely, determining what it takes to achieve a specified BER. This leads us to the notion of *receiver sensitivity*. The receiver sensitivity $\bar{P}_{\text{sens}}$ is defined as the minimum average optical power necessary to achieve a specified BER, usually 10^{-12} or better. Sometimes the receiver sensitivity is also expressed as the number of photons required per 1 bit, M, which is given by

$$M = \frac{2\bar{P}_{\text{sens}}}{hf_c B},$$

where B is the bit rate.

In the notation introduced earlier, the receiver sensitivity is obtained by solving (4.14) for the average power per bit $(P_0 + P_1)/2$ for the specified BER, say, 10^{-12}. Assuming $P_0 = 0$, this can be obtained as

$$\bar{P}_{\text{sens}} = \frac{(\sigma_0 + \sigma_1)\gamma}{2G_m \mathcal{R}}. \tag{4.15}$$

Here, G_m is the multiplicative gain for APD receivers and is unity for *pin* photodiodes.

First consider an APD or a *pin* receiver, with no optical amplifier in the system. The thermal noise current is independent of the received optical power. However, the shot noise variance is a function of $\bar{P}_{\text{sens}}$. Assume that no power is transmitted for a 0 bit. Then $\sigma_0^2 = \sigma_{\text{thermal}}^2$ and $\sigma_1^2 = \sigma_{\text{thermal}}^2 + \sigma_{\text{shot}}^2$, where the shot noise variance

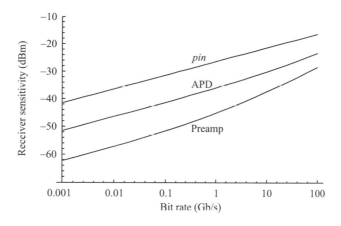

Figure 4.9 Sensitivity plotted as a function of bit rate for typical *pin*, APD, and optically preamplified receivers. The parameters used for the receivers are described in the text.

σ_{shot}^2 must be evaluated for the received optical power $P_1 = 2\bar{P}_{\text{sens}}$ that corresponds to a 1 bit. From (4.4),

$$\sigma_{\text{shot}}^2 = 4eG_m^2 F_A(G_m)\mathcal{R}\bar{P}_{\text{sens}}B_e.$$

Using this and solving (4.15) for the receiver sensitivity $\bar{P}_{\text{sens}}$, we get

$$\bar{P}_{\text{sens}} = \frac{\gamma}{\mathcal{R}}\left(eB_eF_A(G_m)\gamma + \frac{\sigma_{\text{thermal}}}{G_m}\right). \tag{4.16}$$

Assume that for a bit rate of B b/s, a receiver bandwidth $B_e = B/2$ Hz is required. Let the front-end amplifier noise figure $F_n = 3$ dB and the load resistor $R_L = 100\ \Omega$. Then, assuming the temperature $T = 300°$K, the thermal noise current variance, from (4.3), is

$$\sigma_{\text{thermal}}^2 = \frac{4k_BT}{R_L}F_nB_e = 1.656 \times 10^{-22}B\ \text{A}^2. \tag{4.17}$$

Assuming the receiver operates in the 1.55 μm band, the quantum efficiency $\eta = 1$, $\mathcal{R} = 1.55/1.24 = 1.25$ A/W. Using these values, we can compute the sensitivity of a *pin* receiver from (4.16), by setting $G_m = 1$. For BER $= 10^{-12}$ and thus $\gamma \approx 7$, the receiver sensitivity of a *pin* diode is plotted as a function of the bit rate in Figure 4.9. In the same figure, the sensitivity of an APD receiver with $k_A = 0.7$ and an avalanche multiplicative gain $G_m = 10$ is also plotted. It can be seen that the APD receiver has a sensitivity advantage of about 8–10 dB over a *pin* receiver.

We now derive the sensitivity of the optically preamplified receiver shown in Figure 4.7. In amplified systems, the signal-spontaneous beat noise component usually dominates over all the other noise components, unless the optical bandwidth B_o is large, in which case the spontaneous-spontaneous beat noise can also be significant. Making this assumption, the bit error rate can be calculated, using (4.6), (4.9), and (4.14), as

$$\text{BER} = Q\left(\frac{\sqrt{GP}}{2\sqrt{(G-1)P_nB_e}}\right). \tag{4.18}$$

Let us see what receiver sensitivity can be obtained for an ideal preamplified receiver. The receiver sensitivity is measured either in terms of the required power at a particular bit rate or in terms of the number of photons per bit required. As before, we can assume that $B_e = B/2$. Assuming that the amplifier gain G is large and that the spontaneous emission factor $n_{\text{sp}} = 1$, we get

$$\text{BER} = Q\left(\sqrt{\frac{M}{2}}\right).$$

To obtain a BER of 10^{-12}, the argument to the $Q(.)$ function γ must be 7. This yields a receiver sensitivity of $M = 98$ photons per 1 bit. In practice, an optical filter is used between the amplifier and the receiver to limit the optical bandwidth B_o and thus reduce the spontaneous-spontaneous and shot noise components in the receiver. For practical preamplified receivers, receiver sensitivities of a few hundred photons per 1 bit are achievable. In contrast, a direct detection pinFET receiver without a preamplifier has a sensitivity of the order of a few thousand photons per 1 bit.

Figure 4.9 also plots the receiver sensitivity for an optically preamplified receiver, assuming a noise figure of 6 dB for the amplifier and an optical bandwidth $B_o = 50$ GHz that is limited by a filter in front of the amplifier. From Figure 4.9, we see that the sensitivity of a *pin* receiver at a bit rate of 10 Gb/s is -21 dBm and that of an APD receiver is -30 dBm. For 10 Gb/s operation, commercial *pin* receivers with sensitivities of -18 dBm and APD receivers with sensitivities of -24 dBm are available today. From the same figure, at 2.5 Gb/s, the sensitivities of *pin* and APD receivers are -24 dBm and -34 dBm, respectively. Commercial *pin* and APD receivers with nearly these sensitivities at 2.5 Gb/s are available today.

In systems with cascades of optical amplifiers, the notion of sensitivity is not very useful because the signal reaching the receiver already has a lot of added amplifier noise. In this case, the two parameters that are measured are the average received signal power, $\bar{P}_{\text{rec}}$, and the received optical noise power, P_{ASE}. The *optical signal-to-noise ratio* (OSNR) is defined as $\bar{P}_{\text{rec}}/P_{\text{ASE}}$. In the case of an optically preamplified receiver, $P_{\text{ASE}} = 2P_n(G-1)B_o$. A system designer needs to relate the measured OSNR with the bit error rate. Neglecting the receiver thermal noise and shot noise, it can be shown using (4.6), (4.9), (4.10), and (4.14) that the argument to the $Q(.)$ function, γ, is related to the OSNR as follows:

$$\gamma = \frac{2\sqrt{\frac{B_o}{B_e}}\text{OSNR}}{1 + \sqrt{1 + 4\text{OSNR}}}. \tag{4.19}$$

Consider a typical 2.5 Gb/s system with $B_e = 2$ GHz, with an optical filter with bandwidth $B_o = 36$ GHz placed between the amplifier cascade and the receiver. For $\gamma = 7$, this system requires an OSNR = 4.37, or 6.4 dB. However, this is usually not sufficient because the system must deal with a variety of impairments, such as dispersion and nonlinearities. We will study these in Chapter 5. A rough rule of thumb used by system designers is to design the amplifier cascade to obtain an OSNR of at least 20 dB at the receiver, so as to allow sufficient margin to deal with the other impairments.

4.4.7   Coherent Detection

We saw earlier that simple direct detection receivers are limited by thermal noise and do not achieve the shot noise limited sensitivities of ideal receivers. We saw that the sensitivity could be improved significantly by using an optical preamplifier. Another way to improve the receiver sensitivity is to use a technique called *coherent detection*.

The key idea behind coherent detection is to provide gain to the signal by mixing it with another local light signal from a so-called local-oscillator laser. At the same time, the dominant noise in the receiver becomes the shot noise due to the local oscillator, allowing the receiver to achieve the shot noise limited sensitivity. (In fact, a radio receiver works very much in this fashion except that it operates at radio, rather than light, frequencies.)

A simple coherent receiver is shown in Figure 4.10. The incoming light signal is mixed with a local-oscillator signal via a 3 dB coupler and sent to the photodetector. (We will ignore the 3 dB splitting loss induced by the coupler since it can be eliminated by a slightly different receiver design—see Problem 4.15.) Assume that the phase and polarization of the two waves are perfectly matched. The power seen by the photodetector is then

$$\begin{aligned} P_r(t) &= \left[\sqrt{2aP}\cos(2\pi f_c t) + \sqrt{2P_{\text{LO}}}\cos(2\pi f_{\text{LO}}t)\right]^2 \\ &= aP + P_{\text{LO}} + 2\sqrt{aPP_{\text{LO}}}\cos[2\pi(f_c - f_{\text{LO}})t]. \end{aligned} \tag{4.20}$$

Here, P denotes the input signal power, P_{LO} the local-oscillator power, $a = 1$ or 0 depending on whether a 1 or 0 bit is transmitted (for an OOK signal), and f_c and

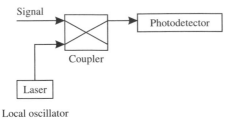

Local oscillator

Figure 4.10   A simple coherent receiver.

f_{LO} represent the carrier frequencies of the signal and local-oscillator waves. We have neglected the $2f_c$, $2f_{\mathrm{LO}}$, and $f_c + f_{\mathrm{LO}}$ components since they will be filtered out by the receiver. In a *homodyne* receiver, $f_c = f_{\mathrm{LO}}$, and in a *heterodyne* receiver, $f_c - f_{\mathrm{LO}} = f_{\mathrm{IF}} \neq 0$. Here, f_{IF} is called the intermediate frequency (IF), typically a few gigahertz.

To illustrate why coherent detection yields improved receiver sensitivities, consider the case of a homodyne receiver. For a 1 bit, we have

$$I_1 = \mathcal{R}(P + P_{\mathrm{LO}} + 2\sqrt{P P_{\mathrm{LO}}}),$$

and for a 0 bit,

$$I_0 = \mathcal{R} P_{\mathrm{LO}}.$$

The key thing to note here is that by making the local-oscillator power P_{LO} sufficiently large, we can make the shot noise dominate over all the other noise components in the receiver. Thus the noise variances are

$$\sigma_1^2 = 2e I_1 B_e$$

and

$$\sigma_0^2 = 2e I_0 B_e.$$

Usually, P_{LO} is around 0 dBm and P is less than -20 dBm. So we can also neglect P compared to P_{LO} when computing the signal power, and both P and $\sqrt{P P_{\mathrm{LO}}}$ compared to P_{LO} when computing the noise variance σ_1^2. With this assumption, using (4.14), the bit error rate is given by

$$\mathrm{BER} = Q\left(\sqrt{\frac{\mathcal{R} P}{2e B_e}}\right).$$

As before, assuming $B_e = B/2$, this expression can be rewritten as

$$\text{BER} = Q(\sqrt{M}),$$

where M is the number of photons per 1 bit as before. For a BER of 10^{-12}, we need the argument of the $Q(.)$ function γ to be 7. This yields a receiver sensitivity of 49 photons per 1 bit, which is significantly better than the sensitivity of a simple direct detection receiver.

However, coherent receivers are generally quite complex to implement and must deal with a variety of impairments. Note that in our derivation we assumed that the phase and polarization of the two waves match perfectly. In practice, this is not the case. If the polarizations are orthogonal, the mixing produces no output. Thus coherent receivers are highly sensitive to variations in the polarizations of the signal and local-oscillator waves as well as any phase noise present in the two signals. There are ways to get around these obstacles by designing more complicated receiver structures [KBW96, Grc93]. However, direct detection receivers with optical preamplifiers, which yield comparable receiver sensitivities, provide a simpler alternative and are widely used today.

There is another advantage to be gained by using coherent receivers in a multichannel WDM system. Instead of using a demultiplexer or filter to select the desired signal optically, with coherent receivers, this selection can be done in the IF domain using electronic filters, which can be designed to have very sharp skirts. This allows very tight channel spacings to be achieved. In addition, in a WDM system, the receiver can be tuned between channels in the IF domain, allowing for rapid tunability between channels, a desirable feature to support fast packet switching. However, we will require highly wavelength-stable and controllable lasers and components to make use of this benefit. Such improvements may result in the resurrection of coherent receivers when WDM systems with large numbers of channels are designed in the future.

4.4.8 Timing Recovery

The process of determining the bit boundaries is called *timing recovery*. The first step is to extract the clock from the received signal. Recall that the clock is a periodic waveform whose period is the bit interval (Section 4.4). This clock is sometimes sent separately by the transmitter, for example, in a different frequency band. Usually, however, the clock must be extracted from the received signal. Even if the extracted clock has a period equal to the bit interval, it may still be out of phase with the received signal; that is, the clock may be offset from the bit boundaries. Usually,

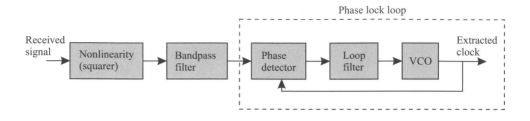

Figure 4.11   Block diagram illustrating timing, or clock, recovery at the receiver.

both the clock frequency (periodicity) and its phase are recovered simultaneously by a single circuit, as shown in Figure 4.11.

If we pass the received signal through a nonlinearity, typically some circuit that calculates the square of the received signal, it can be shown that the result contains a spectral component at $1/T$, where T is the bit period. Thus, we can filter the result using a bandpass filter as shown in Figure 4.11 to get a waveform that is approximately periodic with period T, and which we call a *timing signal*. However, this waveform will still have considerable *jitter;* that is, successive "periods" will have slightly different durations. A "clean" clock with low jitter can be obtained by using the *phase lock loop* (PLL) circuit shown in Figure 4.11.

A PLL consists of a voltage-controlled oscillator (VCO), a phase detector, and a loop filter. A VCO is an oscillator whose output frequency can be controlled by an input voltage. A *phase detector* produces an error signal that depends on the difference in phase between its two inputs. Thus, if the timing signal and the output of the VCO are input to the phase detector, it produces an error signal that is used to adjust the output of the VCO to match the (average) frequency and phase of the timing signal. When this adjustment is complete, the output of the VCO serves as the clock that is used to sample the filtered signal in order to decide upon the values of the transmitted bits. The *loop filter* shown in Figure 4.11 is a critical element of a PLL and determines the residual jitter in the output of the VCO, as well as the ability of the PLL to track changes in the frequency and phase of the timing signal.

4.4.9   Equalization

We remarked in Section 4.4 with reference to Figure 4.5 that the receive filter that is used just prior to sampling the signal can incorporate an *equalization filter* to cancel the effects of intersymbol interference due to pulse spreading. From the viewpoint of the electrical signal that has been received, the entire optical system (including the laser, the fiber, and the photodetector) constitutes the *channel* over which the signal

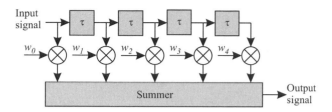

Figure 4.12   A transversal filter, a commonly used structure for equalization. The output (equalized) signal is obtained by adding together suitably delayed versions of the input signal, with appropriate weights.

has been transmitted. If nonlinearities are ignored, the main distortion caused by this channel is the dispersion-induced broadening of the (electrical) pulse. Dispersion is a linear effect, and hence the effect of the channel on the pulse, due to dispersion, can be modeled by the response of a filter with transfer function $H_D(f)$. Hence, in principle, by using the inverse of this filter, say, $H_D^{-1}(f)$, as the equalization filter, this effect can be canceled completely at the receiver. This is what an equalization filter attempts to accomplish.

The effect of an equalization filter is very similar to the effect of dispersion compensating fiber (DCF). The only difference is that in the case of DCF, the equalization is in the optical domain, whereas equalization is done electrically when using an equalization filter. As in the case of DCF, the equalization filter depends not only on the type of fiber used but also on the fiber length.

A commonly used filter structure for equalization is shown in Figure 4.12. This filter structure is called a *transversal filter*. It is essentially a tapped delay line: the signal is delayed by various amounts and added together with individual weights. The choice of the weights, together with the delays, determines the transfer function of the equalization filter. The weights of the tapped delay line have to be adjusted to provide the best possible cancellation of the dispersion-induced pulse broadening.

Electronic equalization involves a significant amount of processing that is difficult to do at higher bit rates, such as 10 Gb/s. Thus optical techniques for dispersion compensation, such as the use of DCF for chromatic dispersion compensation, are currently much more widely used compared to electronic equalization.

4.5   Error Detection and Correction

An error-correcting code is a technique for reducing the bit error rate on a communicaton channel. It involves transmitting additional bits, called *redundancy*, along with the data bits. These additional bits carry redundant information and are used by the receiver to correct most of the errors in the data bits. This method of reducing the error rate by having the transmitter send redundant bits (using an error-correcting code) is called *forward error correction* (FEC).

An alternative is for the transmitter to use a smaller amount of redundancy, which can be used by the receiver to detect the presence of an error, but there is not sufficient redundancy to identify/correct the errors. This approach is used in telecommunication systems based on SONET and SDH to monitor the bit error rate in the received signal. It is also widely used in data communication systems, where the receiver requests the transmitter to resend the data blocks that are detected to be in error. This technique is called *automatic repeat request* (ARQ).

A simple example of an error-detecting code is the *bit interleaved parity* (BIP) code. A BIP-N code adds N additional bits to the transmitted data. We can use either *even* or *odd* parity. With a BIP-N of even parity, the transmitter computes the code as follows: The first bit of the code provides even parity over the first bit of all N-bit sequences in the covered portion of the signal, the second bit provides even parity over the second bits of all N-bit sequences within the specified portion, and so on. Even parity is generated by setting the BIP-N bits so that there are an even number of 1s in each of all N-bit sequences including the BIP-N bit. Problem 4.16 provides more details on this code.

Optical communication systems are expected to operate at a very low residual BER: 10^{-12} or lower. When the demands on the communication system were lower due to relaxed channel spacing, negligible component crosstalk, negligible effect of nonlinearities, and so on, all that was required to achieve the specified BER was to increase the received power. No FEC techniques were necessary. However, this is not always possible in the very high capacity WDM systems that are being deployed today.

One reason for using FEC instead of higher power is that fiber nonlinearities prevent further increases in transmit power. A second reason for using an FEC code is simply the cost–performance trade-off. The use of an FEC enables a longer communication link before regeneration, since the link can now operate at a lower received power for the same BER. The price to be paid for this is the additional processing involved, mainly at the receiver.

Several communication systems suffer from a BER floor problem: the BER cannot be decreased further by increasing the received power. This is because the main impairment is not due to the various noises (thermal, shot, amplifier) but due to the crosstalk from adjacent WDM channels. Increasing the received power increases the crosstalk proportionately, and thus the BER cannot be decreased beyond a certain level, called the *BER floor.* However, FEC can be used to decrease the BER below this floor.

The use of an FEC code can sometimes provide an early warning for BER problems. Assume a link has a BER of 10^{-9} without the use of an FEC. Even though this may be adequate in some situations, it may be better to use an FEC to push the BER down much further, say, to 10^{-15} or lower. Suppose some component fails in such a way as to cause significantly more errors, but does not fail completely. For example, a switch may fail so as to cause significantly more crosstalk, or the output power of a laser may decrease considerably below the specified value. If the system were used without an FEC, the BER may immediately become unacceptable, but with the use of an FEC, the system may be able to continue operation at a much better BER, while alerting the network operator to the problem.

The simplest error-correcting code is a *repetition code.* In such a code, every bit is repeated some number of times, say, thrice. For example, a 1 is transmitted as 111, and a 0 as 000. Thus we have one data, or information bit, plus two redundant bits of the same value. The receiver can estimate the data bit based on the value of the majority of the three received bits. For example, the received bits 101 are interpreted to mean that the data bit is a 1, and the received bits 100 are interpreted to mean the data bit is a 0.

It is easy to see how the use of such a code improves the BER, if the same energy is transmitted per bit after coding, as in the uncoded system. This amounts to transmitting thrice the power in the above example, since three coded bits have to be transmitted for every data bit. In this case, the coded system has the same raw BER—the BER before error correction or decoding—as the uncoded system. However, after decoding, at least two bits in a block of three bits have to be in error for the coded system to make a wrong decision. This substantially decreases the BER of the coded system, as discussed in Problem 4.17. For example, the BER decreases from 10^{-6} for the uncoded system to 3×10^{-12} in the coded system.

However, this is not a fair assessment of the gains due to FEC, since the transmitted power has to be increased by a factor of 3. This may not be possible, for example, if nonlinearities pose a problem, or higher-power lasers are simply unavailable or too expensive. While such a code may have some application in the presence of BER floors, when there are no BER floors, using such a code may defeat the very purpose of using an FEC code. This is because the link length can be increased even further by simply increasing the transmit power and omitting the FEC code. Therefore, a better measure of the performance of an FEC code has to be devised, called the *coding gain.*

The coding gain of an FEC code is the decrease in the receiver sensitivity that it provides for the same BER compared to the uncoded system (for the same transmit power). In this sense, the repetition code is useless since it has a negative coding gain. However, codes with substantial coding gains, that is, which decrease the BER substantially for the same transmit power as in the uncoded system, have been designed by mathematicians and communication engineers over the last 50 years. In the next section, we discuss a popular and powerful family of such codes called Reed-Solomon codes.

4.5.1   Reed-Solomon Codes

A Reed-Solomon code, named after its inventors Irving Reed and Gus Solomon, does not operate on bits but on groups of bits, which we will call symbols. For example, a symbol could represent a group of 4 bits, or a group of 8 bits (a byte). A transmitter using a Reed-Solomon code considers k data symbols and calculates r additional symbols with redundant information, based on a mathematical formula: the code. The transmitter sends the $n = k + r$ symbols to the receiver.

If the transmitted power is kept constant, since $k + r$ symbols have to be transmitted in the same duration as k symbols, each symbol in the coded system has $k/k + r$ the duration, and hence $k/k + r$ the energy, of a symbol in the uncoded system.

The receiver considers a block of $n = k + r$ symbols, and knowing the code used by the transmitter, it can correctly decode the k data bits even if up to $r/2$ of the $k + r$ symbols are in error.

Reed-Solomon codes have the restriction that if a symbol consists of m bits, the length of the code $n = 2^m - 1$. Thus the code length $n = 255$ if (8-bit) bytes are used as symbols. The number of redundant bits r can take any even value. A popular Reed-Solomon code used in most recently deployed submarine systems has parameters $n = 255$ and $r = 16$, and hence $k = n - r = 239$. In this case, 16 redundant bytes are calculated for every block of 239 data bytes. The number of redundant bits added is less than 7% of the data bits, and the code is capable of correcting up to 8 errored bytes in a block of 239 bytes. This code provides a coding gain of about 6 dB. With this coding gain, the BER can be substantially reduced, for example, from 10^{-5} to 10^{-15}.

A discussion of the encoding and decoding processes involved in the use of Reed-Solomon codes is beyond the scope of this book. A number of references to this topic are listed at the end of this chapter. The principle of operation can be understood based on the following analogy with real numbers.

Assume two real numbers are to be transmitted. Consider a straight line (a polynomial of degree 1), say, $ax + b$, whose two coefficients a and b represent the real numbers to be transmitted. Instead of transmitting a and b, transmit five points on the straight line. The receiver knows that the transmitted points are on a straight line and can recover the straight line, and hence the transmitted data, even if two of the five points are in error: it just finds a straight line that fits at least three of the five points. Similarly, if the receiver is given n points but told that they all lie on a degree k polynomial ($k < n$) it can recover the polynomial, even if some of the received points are in error: it just fits the best possible degree-k polynomial to the set of received points.

A Reed-Solomon code works in a similar fashion except that the arithmetic is not over real numbers, but over the finite set of symbols (groups of bits) used in the code. For example, the finite set of symbols consists of the 256 possible 8-bit values for 8-bit symbols. All arithmetic operations are suitably defined over this finite set of symbols, which is called a *finite field*. (If we write $2 = 00000010$ and $3 = 00000011$, $3/2 \neq 1.5$ in finite field arithmetic: it is some other value in the set of symbols $[0, 255]$.) The $n = 2^m - 1$ transmitted symbols can be viewed as all the possible nonzero values of a degree-k polynomial whose coefficients lie in a finite field of size 2^m. For example, the 255 transmitted values in a Reed-Solomon code with $n = 255$ and $k = 239$ can be viewed as representing the 255 nonzero values of a degree-239 polynomial whose coefficients are 8-bit values that need to be transmitted. The receiver can recover the degree-239 polynomial, and hence the data bits, even if up to 8 of the 255 received values/symbols are in error. (In practice, the data bits are not encoded as the coefficients of such a polynomial, but as the first 239 of the 255 transmitted values/symbols as discussed above.)

FEC is currently used in 10 Gb/s systems and in undersea transmission systems. It is part of the digital wrapper defined by the ITU-T. The two codes standardized by the ITU-T are the (255, 239) and the (255, 223) Reed-Solomon codes. Both are popular codes used in many communication systems, and thus chipsets that implement the encoding and decoding functions for these codes are readily available. The (255, 239) Reed-Solomon code has less than 7% redundancy (16 bytes for 239 bytes) and can correct up to 8 errored bytes in a block of 239 bytes. The (255, 223) Reed-Solomon code has less than 15% redundancy and can correct up to 16 errored bytes in a block of 223 bytes. These codes, as well as much stronger ones, are used today in high-performance optical communication systems.

4.5.2   Interleaving

Frequently, when errors occur, they occur in bursts; that is, a large number of successive bits are in error. The Reed-Solomon codes we studied in the previous section are capable of correcting bursts of errors too. For example, since the (255, 223) code can correct up to 16 errored bytes, it can correct a burst of $16 \times 8 = 128$ bit errors. To correct larger bursts with a Reed-Solomon code, we would have to increase the redundancy. However, the technique of *interleaving* can be used along with the

Reed-Solomon codes to correct much larger bursts of errors, without increasing the redundancy.

Assume an (n, k) Reed-Solomon code is used and imagine the bytes are arranged in the following order:

1	2	3	...	k	($n - k$ redundant bits)
$k + 1$	$k + 2$	$k + 3$	...	$2k$	($n - k$ redundant bits)
$2k + 1$	$2k + 2$	$2k + 3$	...	$3k$	($n - k$ redundant bits)
...					

Without interleaving, the bytes would be transmitted in row order, that is, the bytes in row 1 are transmitted, followed by the bytes in row 2, and so on.

The idea of interleaving is to transmit the first d bytes in column 1, followed by the first d bytes in column 2, and so on. Thus byte 1 would be followed by byte $k + 1$. When d bytes have been transmitted from all n columns, we transmit the next d bytes in column 1 (from rows $(d + 1)$ to $2d$), followed by the next d bytes in column 2, and so on. The parameter d is called the *interleaving depth*.

Suppose there is a burst of b byte errors. Only $\lceil b/d \rceil$ of these bytes will occur in the same row due to interleaving. Thus, a $(255, 223)$ Reed-Solomon code will be able to correct any burst of b errors when interleaving to depth d is used, provided $\lceil b/d \rceil < 16$. For example, if interleaving to depth 4 is used ($d = 4$), a $(255, 223)$ Reed-Solomon code can correct a burst of 64 consecutive errored bytes in a block of 223 bytes, though if the errors occur at random byte positions, it can correct only 16 byte errors in the same block size of 223 bytes.

Summary

Modulation is the process of converting data in electronic form to optical form for transmission on the fiber. The simplest form of digital modulation is on-off keying, which most systems use today. Direct modulation of the laser or LED source can be used for transmission at low bit rates over short distances, whereas external modulation is needed for transmission at high bit rates over long distances. Some form of line coding or scrambling is needed to prevent long runs of 1s or 0s in the transmitted data stream to allow the clock to be recovered easily at the receiver and to maintain DC balance.

Subcarrier multiplexing is a technique where many signals are electronically multiplexed using FDM, and the combined signal is used to modulate an optical carrier. Multilevel modulation schemes are more spectrally efficient than on-off keying; optical duobinary signaling is an example of such a scheme.

A simple direct detection receiver looks at the energy received during a bit interval to decide whether it is a 1 or 0 bit. The receiver sensitivity is the average power required at the receiver to achieve a certain bit error rate. The sensitivity of a simple direct detection receiver is determined primarily by the thermal noise in the receiver. The sensitivity can be improved by using APDs instead of *pin* photodetectors or by using an optical preamplifier. Another technique to improve the sensitivity as well as the channel selectivity of the receiver is coherent detection. However, coherent detection is susceptible to a large number of impairments, and it requires a significantly more complicated receiver structure to overcome these impairments. For this reason, it is not practically implemented today.

Clock recovery is an important part of any receiver and is usually based on a phase lock loop.

Electronic equalization is another option to cancel the pulse spreading due to dispersion. This is accomplished by filtering the detected signal electrically to approximately invert the distortion undergone by it.

Error-correcting codes can be used to significantly lower the BER at the expense of additional processing. The most commonly used family of codes are Reed-Solomon codes.

Further Reading

Many books on optical communication cover modulation and detection in greater depth than we have. See, for example, [Gre93, MK88, Agr97]. See also [BL90] for a nice tutorial article on the subject. Subcarrier multiplexing and modulation are treated in depth in [WOS90, OLH89, Dar87, Gre93]. Line coding, scrambling, and bit clock recovery are covered extensively in [LM93]. Optical duobinary modulation is discussed in several recent papers [OY98, Ono98, Fra98]. Optical SSB modulation is discussed in [SNA97, Hui01]. For an excellent and current discussion of channel capacity and information theory in general, we recommend the textbook by Cover and Thomas [CT91]. These techniques have been applied to calculate the capacity limits of optical systems in [MS00].

The principles of signal detection are covered in the classic books by van Trees [vT68] and Wozencraft and Jacobs [WJ90]. For a derivation of shot noise statistics, see [Pap91]. The noise introduced by optical amplifiers has been studied extensively in the literature. Amplifier noise statistics have been derived using quantum mechanical approaches [Per73, Yam80, MYK82, Dan95] as well as semiclassical approaches [Ols89, RH90]. There was a great deal of effort devoted to realizing coherent receivers in the 1980s, but the advent of optical amplifiers in the late 1980s and early 1990s provided a simpler alternative. See [BL90, KBW96] for a detailed treatment of coherent receivers. Equalization is treated extensively in many books on digital communication; see, for example, [LM93, Pro00].

The field of error-correcting codes has developed rapidly since its founding by Hamming [Ham50] and Shannon [Sha48] more than a half-century ago. There are many textbooks on this topic; see, for example, [McE77, LC82]. A discussion of FEC techniques in submarine transmission systems appears in [Sab01].

Problems

4.1 A very simple line code used in early data networks is called *bit stuffing*. The objective of this code is to prevent long runs of 1s or 0s but not necessarily achieve DC balance. The encoding works as follows. Suppose the maximum number of consecutive 1s that we are allowed in the bit stream is k. Then the encoder inserts a 0 bit whenever it sees k consecutive 1 bits in the input sequence.

  (a) Suppose the incoming data to be transmitted is 11111111111001000000 (read left to right). What is the encoded bit stream, assuming $k = 5$?

  (b) What is the algorithm used by the decoder to recover the data? Suppose the received bit stream is 0111110101111100011 (read left to right). What is the decoded bit stream?

4.2 The SONET standard uses scrambling to prevent long runs of 1s and 0s from occurring in the transmitted bit stream. The scrambling is accomplished by a carefully designed feedback shift register shown in Figure 4.13. The shift register consists of flip-flops whose operation is controlled by a clock running at the bit rate and is reset at the beginning of each frame.

  (a) Suppose the incoming data to be transmitted is 11111111111001000000. Assume that the shift register contents are 1111111 at the beginning. What is the scrambled output?

  (b) Write a simulation program to compute the scrambled output as a function of the input. The input is a sequence of bits generated by a pseudo-random sequence with equal probabilities for a 1 and a 0. Plot the longest run length of 1s and the longest run length of 0s observed as a function of the sequence length for sequences up to 10 million bits long. Again assume that the shift register contents are 1111111 at the beginning of the sequence. What do you observe?

4.3 Consider the optical duobinary modulation scheme we discussed in Section 4.3.1. If the data sequence is $d(nT) = 10101011010111100001$, calculate (a) the differential encoding $x(nT)$ of $d(nT)$, and (b) the duobinary encoding $y(nT)$ of $x(nT)$. Recall that $y(nT) \bmod 2 = d(nT)$. How can you compute the sequence $y(nT)$ directly from $d(nT)$ without going through the two-stage differential and duobinary encoding processes?

4.4 Consider the SNR of an APD receiver when both shot noise and thermal noise are present. Assuming that the *excess noise factor* of the APD is given by $F_A(G_m) = G_m^x$ for some $x \in (0, 1)$, derive an expression for the optimum value G_m^{opt} of the APD gain G_m that maximizes the SNR.

4.5 This problem deals with the noise figure of a chain of optical amplifiers and placement of loss elements in the amplifier. The loss element may be an optical add/drop multiplexer, or a gain-flattening filter, or a dispersion compensation module used to compensate for accumulated dispersion along the link. The question is, where should this loss element be placed—in front of the amplifier, after the amplifier, or inside the amplifier?

(a) Consider an optical amplifier with noise figure F. Suppose we introduce a loss element in front of it, with loss $0 < \epsilon \le 1$ ($\epsilon = 0$ implies no loss, and $\epsilon = 1$ implies 100% loss). Show that the noise figure of the combination is $F/(1 - \epsilon)$. Note that this loss element may also simply reflect the coupling loss into the amplifier. Observe that this combination has a poor noise figure.

(b) Suppose the loss element is placed just after the amplifier. Show that the noise figure of the combination is still F; that is, placing a loss element after the amplifier doesn't affect the noise figure. However, the price we pay in this case is a reduction in optical output power, since the amplifier output is attenuated by the loss element placed after it.

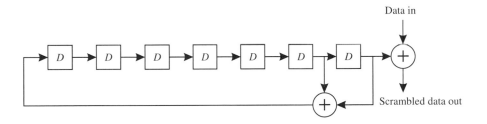

Figure 4.13 The feedback shift register used for scrambling in SONET.

(c) Consider an optical amplifier chain with two amplifiers, with gains G_1 and G_2, respectively, and noise figures F_1 and F_2, respectively, with no loss between the two amplifiers. Assuming $G_1 \gg 1$, show that the noise figure of the combined amplifier chain is

$$F = F_1 + \frac{F_2}{G_1}.$$

In other words, the noise figure of the chain is dominated by the noise figure of the first amplifier, provided its gain is reasonably large, which is usually the case.

(d) Now consider the case where a loss element with loss ϵ is introduced between the first and second amplifier. Assuming $G_1, G_2 \gg 1$, and $(1 - \epsilon)G_1G_2 \gg 1$, show that the resulting noise figure of the chain is given by

$$F = F_1 + \frac{F_2}{(1 - \epsilon)G_1}.$$

Observe that the loss element doesn't affect the noise figure of the cascade significantly as long as $(1 - \epsilon)G_1 \gg 1$, which is usually the case. This is an important fact that is made use of in designing systems. The amplifier is broken down into two stages, the first stage having high gain and a low noise figure, and the loss element is inserted between the two stages. This setup has the advantage that there is no reduction in the noise figure or the output power.

4.6 Show that the BER for an OOK direct detection receiver is given by

$$\mathrm{BER} = Q\left(\frac{I_1 - I_0}{\sigma_0 + \sigma_1}\right).$$

4.7 Consider a binary digital communication system with received signal levels m_1 and m_0 for a 1 bit and 0 bit, respectively. Let σ^2 and σ_0^2 denote the noise variances for a 1 and 0 bit, respectively. Assume that the noise is Gaussian and that a 1 and 0 bit are equally likely. In this case, the bit error rate BER is given by

$$\mathrm{BER} = \frac{1}{2}Q\left(\frac{m_1 - T_d}{\sigma_1}\right) + \frac{1}{2}Q\left(\frac{T_d - m_0}{\sigma_0}\right),$$

where T_d is the receiver's decision threshold. Show that the value of T_d that minimizes the bit error rate is given by

$$T_d = \frac{-m_1\sigma_0^2 + m_0\sigma_1^2 + \sqrt{\sigma_0^2\sigma_1^2(m_1 - m_0)^2 + 2(\sigma_1^2 - \sigma_0^2)\ln(\sigma_1/\sigma_0)}}{\sigma_1^2 - \sigma_0^2}. \tag{4.21}$$

For the case of high signal-to-noise ratios, it is reasonable to assume that

$$(m_1 - m_0)^2 \gg \frac{2(\sigma_1^2 - \sigma_0^2) \ln(\sigma_1/\sigma_0)}{\sigma_0^2 \sigma_1^2}.$$

In this case, (4.21) can be simplified to

$$T_d = \frac{m_0 \sigma_1 + m_1 \sigma_0}{\sigma_1 + \sigma_0}.$$

With $m_1 = RP_1$ and $m_0 = RP_0$, this is the same as (4.12).

4.8 Consider a *pin* direct detection receiver where the thermal noise is the main noise component, and its variance has the value given by (4.17). What is the receiver sensitivity expressed in photons per 1 bit at a bit rate of 100 Mb/s and 1 Gb/s for a bit error rate of 10^{-12}? Assume that the operating wavelength is 1.55 μm and the responsivity is 1.25 A/W.

4.9 Consider the receiver sensitivity, $\bar{P}_{rec}$ (for an arbitrary BER, not necessarily 10^{-9}), of an APD receiver when both shot noise and thermal noise are present but neglecting the dark current, for direct detection of on-off–keyed signals. Assume no power is transmitted for a 0 bit.
  (a) Derive an expression for $\bar{P}_{rec}$.
  (b) Find the optimum value G_m^{opt} of the APD gain G_m that minimizes $\bar{P}_{rec}$.
  (c) For $G_m = G_m^{opt}$, what is the (minimum) value of $\bar{P}_{rec}$?

4.10 Derive (4.18).

4.11 Plot the receiver sensitivity as a function of bit rate for an optically preamplified receiver for three different optical bandwidths: (a) the ideal case, $B_o = 2B_e$, (b) $B_o = 100$ GHz, and (c) $B_o = 30$ THz, that is, an unfiltered receiver. Assume an amplifier noise figure of 6 dB, and the electrical bandwidth B_e is half the bit rate, and use the thermal noise variance given by (4.17). What do you observe as the optical bandwidth is increased?

4.12 You are doing an experiment to measure the BER of an optically preamplified receiver. The setup consists of an optical amplifier followed by a variable attenuator to adjust the power going into the receiver, followed by a *pin* receiver. You plot the BER versus the power going into the receiver over a wide range of received powers. Calculate and plot this function. What do you observe regarding the slope of this curve? Assume that $B_o = 100$ GHz, $B_e = 2$ GHz, $B = 2.5$ Gb/s, a noise figure of 6 dB for the optical amplifier, and a noise figure of 3 dB for the front-end amplifier.

4.13 Derive (4.19).

4.14 Another form of digital modulation that can be used in conjunction with coherent reception is *phase-shift keying* (PSK). Here $\sqrt{2P}\cos(2\pi f_c t)$ is received for a 1 bit and $-\sqrt{2P}\cos(2\pi f_c t)$ is received for a 0 bit. Derive an expression for the bit error rate of a PSK homodyne coherent receiver. How many photons per bit are required to obtain a bit error rate of 10^{-9}?

4.15 A balanced coherent receiver is shown in Figure 4.14. The input signal and local oscillator are sent through a 3 dB coupler, and each output of the coupler is connected to a photodetector. This 3 dB coupler is different in that it introduces an additional phase shift of $\pi/2$ at its second input and second output. The detected current is the difference between the currents generated by the two photodetectors. Show that this receiver structure avoids the 3 dB penalty associated with the receiver we discussed in Section 4.4.7. Use the transfer function for a 3 dB coupler given by (3.1).

4.16 SONET and SDH systems use an 8-bit interleaved parity (BIP-8) check code with even parity to detect errors. The code works as follows. Let $b_0, b_1, b_2, \ldots$ denote the sequence of bits to be transmitted. The transmitter adds an 8-bit code sequence $c_0, c_1, \ldots, c_7$, to the end of this sequence where

$$c_i = b_i \oplus b_{i+8} \oplus b_{i+16} + \ldots.$$

Here $\oplus$ denotes an "exclusive OR" operation ($0 \oplus 0 = 0, 0 \oplus 1 = 1, 1 \oplus 1 = 0$).
  (a) Suppose the bits to be transmitted are 010111010111101111001110. What is the transmitted sequence with the additional parity check bits?
  (b) Suppose the received sequence (including the parity check bits at the end) is 010111010111101111001110. How many bits are in error? Assume that if a parity check indicates an error, it is caused by a single bit error in one of the bits over which the parity is computed.

4.17 If the BER of an uncoded system is p, show that the same system has a BER of $3p^2 + p^3$ when the repetition code (each bit is repeated thrice) is used. Note that the receiver makes its decision on the value of the transmitted bit by taking a majority vote on the corresponding three received coded bits. Assume that the energy per bit remains the same in both cases.

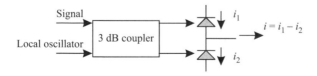

Figure 4.14   A balanced coherent receiver.

References

[Agr97]   G. P. Agrawal. *Fiber-Optic Communication Systems*. John Wiley, New York, 1997.

[BL90]   J. R. Barry and E. A. Lee. Performance of coherent optical receivers. *Proceedings of IEEE*, 78(8):1369–1394, Aug. 1990.

[Bur86]   W. E. Burr. The FDDI optical data link. *IEEE Communications Magazine*, 24(5):18–23, May 1986.

[CT91]   T. M. Cover and J. A. Thomas. *Elements of Information Theory*. Wiley, New York, 1991.

[Dan95]   S. L. Danielsen et al. Detailed noise statistics for an optically preamplified direct detection receiver. *IEEE/OSA Journal on Lightwave Technology*, 13(5):977–981, 1995.

[Dar87]   T. E. Darcie. Subcarrier multiplexing for multiple-access lightwave networks. *IEEE/OSA Journal on Lightwave Technology*, LT-5:1103–1110, 1987.

[Fra98]   T. Franck et al. Duobinary transmitter with low intersymbol interference. *IEEE Photonics Technology Letters*, 10:597–599, 1998.

[Gla00]   A. M. Glass et al. Advances in fiber optics. *Bell Labs Technical Journal*, 5(1):168–187, Jan.–March 2000.

[Gre93]   P. E. Green. *Fiber-Optic Networks*. Prentice Hall, Englewood Cliffs, NJ, 1993.

[Ham50]   R. W. Hamming. Error detecting and error correcting codes. *Bell System Technical Journal*, 29, 1950.

[Hui01]   R. Hui et al. 10 Gb/s SCM system using optical single side-band modulation. In *OFC 2001 Technical Digest*, pages MM4/1–4, 2001.

[KBW96]   L. G. Kazovsky, S. Benedetto, and A. E. Willner. *Optical Fiber Communication Systems*. Artech House, Boston, 1996.

[LC82]   S. Lin and D. J. Costello. *Error Correcting Codes*. Prentice Hall, Englewood Cliffs, NJ, 1982.

[LM93]   E. A. Lee and D. G. Messerschmitt. *Digital Communication*, 2nd edition. Kluwer, Boston, 1993.

[McE77]   R. J. McEliece. *The Theory of Information and Coding: A Mathematical Framework for Communication*. Addison-Wesley, Reading, MA, 1977.

[MK88]   S. D. Miller and I. P. Kaminow, editors. *Optical Fiber Telecommunications II*. Academic Press, San Diego, CA, 1988.

[MS00]   P. P. Mitra and J. B. Stark. Nonlinear limits to the information capacity of optical fibre communications. *Nature*, pages 1027–1030, 2000.

[MYK82]   T. Mukai, Y. Yamamoto, and T. Kimura. S/N and error-rate performance of AlGaAs semiconductor laser preamplifier and linear repeater systems. *IEEE Transactions on Microwave Theory and Techniques*, 30(10):1548–1554, 1982.

[OLH89]   R. Olshanksy, V. A. Lanzisera, and P. M. Hill. Subcarrier multiplexed lightwave systems for broadband distribution. *IEEE/OSA Journal on Lightwave Technology*, 7(9):1329–1342, Sept. 1989.

[Ols89]   N. A. Olsson. Lightwave systems with optical amplifiers. *IEEE/OSA Journal on Lightwave Technology*, 7(7):1071–1082, July 1989.

[Ono98]   T. Ono et al. Characteristics of optical duobinary signals in terabit/s capacity, high spectral efficiency WDM systems. *IEEE/OSA Journal on Lightwave Technology*, 16:788–797, 1998.

[OY98]   T. Ono and Y. Yano. Key technologies for terabit/second WDM systems high spectral efficiency of over 1 bit/s/hz. *IEEE Journal of Quantum Electronics*, 34:2080–2088, 1998.

[Pap91]   A. Papoulis. *Probability, Random Variables, and Stochastic Processes*, 3rd edition. McGraw-Hill, New York, 1991.

[Per73]   S. D. Personick. Applications for quantum amplifiers in simple digital optical communication systems. *Bell System Technical Journal*, 52(1):117–133, Jan. 1973.

[Pro00]   J. G. Proakis. *Digital Communications*, 4th edition. McGraw-Hill, New York, 2000.

[RH90]   R. Ramaswami and P. A. Humblet. Amplifier induced crosstalk in multi-channel optical networks. *IEEE/OSA Journal on Lightwave Technology*, 8(12):1882–1896, Dec. 1990.

[Ros86]   F. E. Ross. FDDI—a tutorial. *IEEE Communications Magazine*, 24(5):10–17, May 1986.

[Sab01]   O. A. Sab. FEC techniques in submarine transmission systems. In *OFC 2001 Technical Digest*, pages TuF1/1–3, 2001.

[Sha48]   C E. Shannon. A mathematical theory of communication. *Bell System Technical Journal*, 27(3):379–423, July 1948.

[SNA97]   G. H. Smith, D. Novak, and Z. Ahmed. Technique for optical SSB generation to overcome dispersion penalties in fibre-radio systems. *Electronics Letters*, 33:74–75, 1997.

[Sta99]   J. B. Stark. Fundamental limits of information capacity for optical communications channels. In *Proceedings of European Conference on Optical Communication*, pages 1–28, Nice, France, Sept. 1999.

[SV96]   M. W. Sachs and A. Varma. Fibre channel and related standards. *IEEE Communications Magazine*, 34(8):40–49, Aug. 1996.

[vT68]   H. L. van Trees. *Detection, Estimation, and Modulation Theory, Part I*. John Wiley, New York, 1968.

[WF83]   A. X. Widmer and P. A. Franaszek. A DC-balanced, partitioned-block, 8B-10B transmission code. *IBM Journal of Research and Development*, 27(5):440–451, Sept. 1983.

[WJ90]   J. M. Wozencraft and I. M. Jacobs. *Principles of Communication Engineering*. Waveland Press, Prospect Heights, IL, 1990. Reprint of the original 1965 edition.

[WOS90]   W. I. Way, R. Olshansky, and K. Sato, editors. Special issue on applications of RF and microwave subcarriers to optical fiber transmission in present and future broadband networks. *IEEE Journal of Selected Areas in Communications*, 8(7), Sept. 1990.

[Yam80]   Y. Yamamoto. Noise and error-rate performance of semiconductor laser amplifiers in PCM-IM transmission systems. *IEEE Journal of Quantum Electronics*, 16:1073–1081, 1980.

5 chapter

Transmission System Engineering

O UR GOAL IN THIS CHAPTER is to understand how to design the physical layer of an optical network. To this end, we will understand the various impairments that we must deal with, how to allocate margins for each of these impairments, how to reduce the effect of these impairments, and finally all the trade-offs that are involved between the different design parameters.

5.1 System Model

Figure 5.1 shows a block diagram of the various components of a unidirectional WDM link. The transmitter consists of a set of DFB lasers, with or without external modulators, one for each wavelength. The signals at the different wavelengths are combined into a single fiber by means of an optical multiplexer. An optical power amplifier may be used to increase the transmission power. After some distance along the fiber, the signal is amplified by an optical in-line amplifier. Depending on the distance, bit rate, and type of fiber used, the signal may also be passed through a dispersion-compensating module, usually at each amplifier stage. At the receiving end, the signal may be amplified by an optical preamplifier before it is passed through a demultiplexer. Each wavelength is then received by a separate photodetector.

Throughout this chapter, we will be focusing on digital systems, although it is possible to transmit analog signals over fiber as well. The physical layer of the system must ensure that bits are transmitted from the source to their destination reliably. The measures of quality are the bit error rate (BER) and the additional power budget

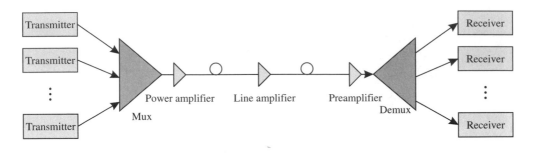

Figure 5.1   Components of a WDM link.

margin provided in the system. Usually the required bit error rates are of the order of 10^{-9} to 10^{-15}, typically 10^{-12}. The BER depends on the amount of noise as well as other impairments that are present in the system. Unless otherwise stated, we will assume that non-return-to-zero (NRZ) modulation is used. In some specific cases, such as chromatic dispersion, we consider both NRZ and return-to-zero (RZ) modulation.

The physical layer is also responsible for the link initialization and link take-down procedures, which are necessary to prevent exposure to potentially harmful laser radiation. This aspect is dealt with in Chapter 9.

We will look at the different components that are part of a system, including the transmitters, receivers, optical amplifiers, wavelength multiplexers, demultiplexers and switches, and the fiber itself, and we will discuss various forms of system impairments that arise from each of these components. Table B.1 in Appendix B summarizes the large number of parameters that are used in this chapter.

5.2   Power Penalty

The physical layer design must take into account the effect of a number of system impairments as previously discussed. Usually each impairment results in a *power penalty* to the system. In the presence of an impairment, a higher signal power will be required at the receiver in order to maintain a desired bit error rate. One way to define the power penalty is as the increase in signal power required (in dB) to maintain the same bit error rate in the presence of impairments. Another way to define the power penalty is as the reduction in signal-to-noise ratio as quantified by the value of γ (the argument to the $Q(.)$ function as defined in Section 4.4.6) due to a specific impairment. We will be using the latter definition since it is easier to calculate and consistent with popular usage.

Let P_1 denote the optical power received during a 1 bit, and P_0 the power received during a 0 bit without any system impairments. The corresponding electrical currents are given by $\mathcal{R}P_1$ and $\mathcal{R}P_0$, respectively, where $\mathcal{R}$ is the responsivity of the photodetector.

Let σ_1 and σ_0 denote the noise standard deviations during a 1 bit and a 0 bit, respectively. Assume that the noise is Gaussian. The bit error rate, assuming equally likely 1s and 0s, is obtained from (4.14) as

$$\text{BER} = Q\left(\frac{\mathcal{R}(P_1 - P_0)}{\sigma_1 + \sigma_0}\right). \tag{5.1}$$

This expression assumes that the receiver's decision threshold is set to the optimal value indicated by (4.12).

In the presence of impairments, let P_1', P_0', σ_1', σ_0' denote the received powers and noise standard deviations, respectively. Assuming an optimized threshold setting, the power penalty is given by

$$\text{PP} = -10\log\left(\frac{\frac{\mathcal{R}(P_1' - P_0')}{\sigma_1' + \sigma_0'}}{\frac{\mathcal{R}(P_1 - P_0)}{\sigma_1 + \sigma_0}}\right). \tag{5.2}$$

Calculating the power penalty in general for the simple AC-coupled receiver discussed in Section 4.4.6 is somewhat more complicated, but we will see that it is the same as the penalty for the optimized receiver for two important cases of interest.

The first case of interest is when the dominant noise component is receiver thermal noise, for which $\sigma_0 = \sigma_1 = \sigma_{\text{th}}$. This is usually the case in unamplified direct detection *pin* receivers. In this case, or in any situation where the noise is independent of the signal power, the power penalty is given by

$$\text{PP}_{\text{sig-indep}} = -10\log\left(\frac{P_1' - P_0'}{P_1 - P_0}\right) \tag{5.3}$$

and the best threshold setting corresponds to the setting of a simple AC-coupled receiver.

The other case of interest is amplified systems, or systems with APD receivers. In amplified systems, the dominant noise component is usually the amplifier signal-spontaneous beat noise (see Section 4.4.5). In APD receivers, the dominant noise component is the shot noise, which is enhanced because of the APD gain (see

Section 3.6.1). In amplified systems, and in systems with APD receivers, we can assume that $\sigma_1 \propto \sqrt{P_1}$; that is, the noise variance depends on the signal power. Assume also that $P_0 \ll P_1$. In this case, we can assume that $\sigma_1 \gg \sigma_0$. Here an optimized receiver would set its threshold close to the 0 level, whereas the simple receiver would still set its threshold at the average received power and would have a somewhat higher bit error rate. However, the power penalties turn out to be the same in both cases. This penalty is given by

$$\mathrm{PP}_{\text{sig-dep}} = -5 \log \left(\frac{P_1'}{P_1} \right). \tag{5.4}$$

Finally, it must be kept in mind that polarization plays an important role in many system impairments where signals interfere with each other. The worst case is usually when the interfering signals have the same state of polarization. However, the state of polarization of each signal varies slowly with time in a random manner, and thus we can expect the power penalties to vary with time as well. The system must be designed, however, to accommodate the worst case, usually identical polarizations.

System design requires careful budgeting of the power penalties for the different impairments. Here we sketch out one way of doing such a design for a transmission system with optical amplifiers. First we determine the ideal value of the parameter γ (see Section 4.4.6) that is needed. For a bit error rate of 10^{-12} typically assumed in high-speed transmission systems, we need $\gamma = 7$, or $20 \log \gamma = 17$ dB. This would be the case if there were no transmission impairments leading to power penalties. In practice, the various impairments result in power penalties that must be added onto this ideal value of γ, as shown in Table 5.1, to obtain the required value of γ that the system must be designed to yield. For instance, in the table, we allocate a 1 dB power penalty for an imperfect transmitter and a 2 dB power penalty for chromatic dispersion. (We will study these and several other impairments in the rest of this chapter.) The required value of γ after adding all these allocations is 31 dB. This is the value that we must obtain if we assume an ideal system to start with and compute γ based on only optical amplifier noise accumulation. The power penalty due to each impairment is then calculated one at a time assuming that the rest of the system is ideal. In practice, this is an approximate method because the different impairments may be related to each other, and we may not be able to isolate each one by itself. For example, the power penalties due to a nonideal transmitter and crosstalk may be related to each other, whereas chromatic dispersion may be treated as an independent penalty.

Table 5.1 An example system design that allocates power penalties for various transmission impairments.

Impairment	Allocation (dB)
Ideal γ	17
Transmitter	1
Crosstalk	1
Chromatic dispersion	2
Nonlinearities	1
Polarization-dependent loss	3
Component aging	3
Margin	3
Required γ	31

5.3 Transmitter

The key system design parameters related to the transmitter are its output power, rise/fall time, extinction ratio, modulation type, side-mode suppression ratio, relative intensity noise (RIN), and wavelength stability and accuracy.

The output power depends on the type of transmitter. DFB lasers put out about 1 mW (0 dBm) to 10 mW (10 dBm) of power. An optical power amplifier can be used to boost the power, typically to as much as 50 mW (17 dBm). The upper limits on power are dictated by nonlinearities (Section 5.8) and safety considerations (Section 9.7).

The extinction ratio is defined as the ratio of the power transmitted when sending a 1 bit, P_1, to the power transmitted when sending a 0 bit, P_0. Assuming that we are limited to an average transmitted power P, we would like to have $P_1 = 2P$ and $P_0 = 0$. This would correspond to an extinction ratio $r = \infty$. Practical transmitters, however, have extinction ratios between 10 and 20. With an extinction ratio r, we have

$$P_0 = \frac{2P}{r+1}$$

and

$$P_1 = \frac{2rP}{r+1}.$$

Reducing the extinction ratio reduces the difference between the 1 and 0 levels at the receiver and thus produces a penalty. The power penalty due to a nonideal extinction ratio in systems limited by signal-independent noise is obtained from (5.3) as

$$PP_{\text{sig-indep}} = -10 \log \frac{r-1}{r+1}.$$

Note that this penalty represents the decrease in signal-to-noise ratio performance of a system with a nonideal extinction ratio relative to a system with infinite extinction ratio, assuming the same *average* transmitted power for both systems. On the other hand, if we assume that the two systems have the same peak transmit power, that is, the same power for a 1 bit, then the penalty can be calculated to be

$$PP_{\text{sig-indep}} = -10 \log \frac{r-1}{r}.$$

Lasers tend to be physically limited by peak transmit power. Most nonlinear effects also set a limit on the peak transmit power. However, eye safety regulation limits (see Section 9.7.1), are stated in terms of average power. The formula to be used depends on which factor actually limits the power for a particular system.

The penalty is higher when the system is limited by signal-dependent noise, which is typically the case in amplified systems (Section 4.4.5)—see Problem 5.10. This is due to the increased amount of noise present at the 0 level. Other forms of signal-dependent noise may arise in the system, such as *laser relative intensity noise*, which refers to intensity fluctuations in the laser output caused by reflections from fiber splices and connectors in the link.

The laser at the transmitter may be modulated directly, or a separate external modulator can be used. Direct modulation is cheaper but results in a broader spectral width due to chirp (Section 2.3). This will result in an added power penalty due to chromatic dispersion (see Section 2.3). Broader spectral width may also result in penalties when the signal is passed through optical filters, such as WDM muxes and demuxes. This penalty can be reduced by reducing the extinction ratio, which, in turn, reduces the chirp and, hence, the spectral width.

Wavelength stability of the transmitter is an important issue and is addressed in Sections 5.9 and 5.12.8.

5.4   Receiver

The key system parameters associated with a receiver are its *sensitivity* and *overload parameter*. The sensitivity is the average optical power required to achieve a certain bit error rate at a particular bit rate. It is usually measured at a bit error rate of

Table 5.2 Typical sensitivities of different types of receivers in the 1.55 μm wavelength band. These receivers also operate in the 1.3 μm band, but the sensitivity may not be as good at 1.3 μm.

Bit Rate	Type	Sensitivity	Overload Parameter
155 Mb/s	pinFET	−36 dBm	−7 dBm
622 Mb/s	pinFET	−32 dBm	−7 dBm
2.5 Gb/s	pinFET	−23 dBm	−3 dBm
2.5 Gb/s	APD	−34 dBm	−8 dBm
10 Gb/s	pinFET	−18 dBm	−1 dBm
10 Gb/s	APD	−24 dBm	−6 dBm
40 Gb/s	pinFET	−7 dBm	3 dBm

10^{-12} using a pseudo-random $2^{23} - 1$ bit sequence. The overload parameter is the maximum input power that the receiver can accept. Typical sensitivities of different types of receivers for a set of bit rates are shown in Table 5.2; a more detailed evaluation can be found in Section 4.4.6. APD receivers have higher sensitivities than pinFET receivers and are typically used in high-bit-rate systems operating at and above 2.5 Gb/s. However, a pinFET receiver with an optical preamplifier has a sensitivity that is comparable to an APD receiver. The overload parameter defines the dynamic range of the receiver and can be as high as 0 dBm for 2.5 Gb/s receivers, regardless of the specific receiver type.

5.5 Optical Amplifiers

Optical amplifiers have become an essential component in transmission systems and networks to compensate for system losses. The most common optical amplifier today is the erbium-doped fiber amplifier (EDFA) operating in the C-band. In addition, L-band EDFAs and Raman amplifiers are also used. EDFAs are used in almost all amplified WDM systems, whereas Raman amplifiers are used in addition to EDFAs in many ultra-long-haul systems. These amplifiers are described in Section 3.4. In this section, we will focus mainly on EDFAs.

The EDFA has a gain bandwidth of about 35 nm in the 1.55 μm wavelength region. The great advantage of EDFAs is that they are capable of simultaneously amplifying many WDM channels. EDFAs spawned a new generation of transmission systems, and almost all optical fiber transmission systems installed over the last few years use EDFAs instead of repeaters. The newer L-band EDFAs are being installed today to increase the available bandwidth, and hence the number of wavelengths, in a single fiber.

Amplifiers are used in three different configurations, as shown in Figure 5.2. An optical *preamplifier* is used just in front of a receiver to improve its sensitivity. A *power amplifier* is used after a transmitter to increase the output power. A *line amplifier* is used typically in the middle of the link to compensate for link losses. The design of the amplifier depends on the configuration. A power amplifier is designed to provide the maximum possible output power. A preamplifier is designed to provide high gain and the highest possible sensitivity, that is, the least amount of additional noise. A line amplifier is designed to provide a combination of all of these.

Unfortunately, the amplifier is not a perfect device. There are several major imperfections that system designers need to worry about when using amplifiers in a system. First, an amplifier introduces noise, in addition to providing gain. Second, the gain of the amplifier depends on the total input power. For high input powers, the EDFA tends to saturate and the gain drops. This can cause undesirable power transients in networks. Finally, although EDFAs are a particularly attractive choice for WDM systems, their gain is not flat over the entire passband. Thus some channels see more gain than others. This problem gets worse when a number of amplifiers are cascaded.

We have studied optically preamplified receivers in Section 4.4.5. In this section, we will study the effect of gain saturation, gain nonflatness, noise, and power transients in systems with cascades of optical amplifiers.

5.5.1   Gain Saturation in EDFAs

An important consideration in designing amplified systems is the saturation of the EDFA. Depending on the pump power and the amplifier design itself, the output power of the amplifier is limited. As a result, when the input signal power is increased,

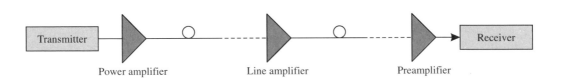

Figure 5.2   Power amplifiers, line amplifiers, and preamplifiers.

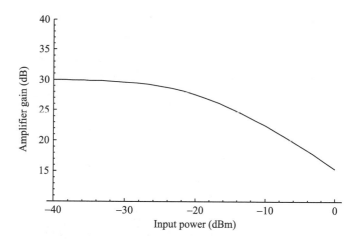

Figure 5.3   Gain saturation in an optical amplifier. Unsaturated gain $G_{max} = 30$ dB and saturation power $P^{sat} = 10$ dBm.

the amplifier gain drops. This behavior can be captured approximately by the following equation:

$$G = 1 + \frac{P^{sat}}{P_{in}} \ln \frac{G_{max}}{G}. \tag{5.5}$$

Here, G_{max} is the unsaturated gain, and G the saturated gain of the amplifier, P^{sat} is the amplifier's internal saturation power, and P_{in} is the input signal power. Figure 5.3 plots the amplifier gain as a function of the input signal power for a typical EDFA. For low input powers, the amplifier gain is at its unsaturated value, and at very high input powers, $G \rightarrow 1$ and the output power $P_{out} = P_{in}$. The output saturation power P_{out}^{sat} is defined to be the output power at which the amplifier gain has dropped by 3 dB. Using (5.5) and the fact that $P_{out} = GP_{in}$, and assuming that $G \gg 1$, the output saturation power is given by

$$P_{out}^{sat} \approx P^{sat} \ln 2.$$

The saturation power is a function of the pump power and other amplifier parameters. It is quite common to have output saturation powers on the order of 10 to 100 mW (10 to 20 dBm).

There is no fundamental problem in operating an EDFA in saturation, and power amplifiers usually do operate in saturation. The only thing to keep in mind is that the saturated gain will be less than the unsaturated gain.

5.5.2   Gain Equalization in EDFAs

The flatness of the EDFA passband becomes a critical issue in WDM systems with cascaded amplifiers. The amplifier gain is not exactly the same at each wavelength. Small variations in gain between channels in a stage can cause large variations in the power difference between channels at the output of the chain. For example, if the gain variation between the worst channel and the best channel is 1 dB at each stage, after 10 stages it will be 10 dB, and the worst channel will have a much poorer signal-to-noise ratio than the best channel. This effect is shown in Figure 5.4(a). Building amplifiers with flat gain spectra is therefore very important (see Section 3.4.3) and is the best way to solve this problem. In practice, it is possible to design EDFAs to be inherently flat in the 1545–1560 nm wavelength region, and this is where many early WDM systems operate. However, systems with a larger number of channels will need to use the 1530–1545 nm wavelength range, where the gain of the EDFA is not flat.

The gain spectrum of L-band EDFAs is relatively flat over the L-band from about 1565 nm to about 1625 nm so that gain flattening over this band is not a significant issue.

At the system level, a few approaches have been proposed to overcome this lack of gain flatness. The first approach is to use preequalization, or preemphasis, as shown in Figure 5.4(b). Based on the overall gain shape of the cascade, the transmitted power per channel can be set such that the channels that see low gain are launched with higher powers. The goal of preequalization is to ensure that all channels are received with approximately the same signal-to-noise ratios at the receiver and fall within the receiver's dynamic range. However, the amount of equalization that can be done is limited, and other techniques may be needed to provide further equalization. Also this technique is difficult to implement in a network, as opposed to a point-to-point link.

The second approach is to introduce equalization at each amplifier stage, as shown in Figure 5.4(c). After each stage, the channel powers are equalized. This equalization can be done in many ways. One way is to demultiplex the channels, attenuate each channel differently, and then multiplex them back together. This approach involves using a considerable amount of hardware. It adds wavelength tolerance penalties due to the added muxes and demuxes (see Section 5.6.6). For these reasons, such an approach is impractical. Another approach is to use a multichannel filter, such as an acousto-optic tunable filter (AOTF). In an AOTF, each channel can be attenuated differently by applying a set of RF signals with different frequencies. Each RF signal controls the attenuation of a particular center wavelength, and by controlling the RF powers of each signal, it is possible to equalize the channel powers. However, an AOTF requires a large amount of RF drive power (on the order of 1 W)

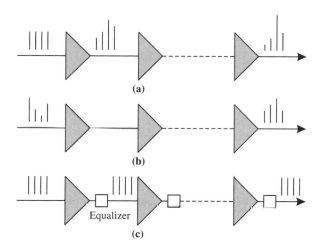

Figure 5.4   Effect of unequal amplifier gains at different wavelengths. (a) A set of channels with equal powers at the input to a cascaded system of amplifiers will have vastly different powers and signal-to-noise ratios at the output. (b) This effect can be reduced by preequalizing the channel powers. (c) Another way to reduce this effect is to introduce equalization at each amplifier stage. The equalization can be done using a filter inside the amplifier as well.

to equalize more than a few (2–4) channels. Both approaches introduce several decibels of additional loss and some power penalties due to crosstalk. The preferred solution today is to add an optical filter within the amplifier with a carefully designed passband to compensate for the gain spectrum of the amplifier so as to obtain a flat spectrum at its output. Both dielectric thin-film filters (Section 3.3.6) and long-period fiber gratings (Section 3.3.4) are good candidates for this purpose.

5.5.3   Amplifier Cascades

Consider a system of total length L with amplifiers spaced l km apart (see Figure 5.5). The loss between two stages is $e^{-\alpha l}$, where α is the fiber attenuation. Each amplifier adds some spontaneous emission noise. Thus the optical signal-to-noise ratio, OSNR (see Section 4.4.6 for the definition), gradually degrades along the chain.

    The amplifier gain must be at least large enough to compensate for the loss between amplifier stages; otherwise, the signal (and hence the OSNR) will degrade rapidly with the number of stages. Consider what happens when we choose the unsaturated amplifier gain to be larger than the loss between stages. For the first few stages, the total input power (signal plus noise from the previous stages) to a stage increases with the number of stages. Consequently, the amplifiers begin to saturate and their gains drop. Farther along the chain, a spatial steady-state condition is reached where the amplifier output power and gain remains the same from stage to stage. These values, $\overline{P}_{\text{out}}$ and $\overline{G}$, respectively, can be computed by observing that

$$(\overline{P}_{\text{out}}e^{-\alpha l})\overline{G} + 2P_n B_o(\overline{G} - 1) = \overline{P}_{\text{out}}. \tag{5.6}$$

Here $\overline{P}_{\text{out}}e^{-\alpha l}$ is the total input power to the amplifier stage, and the second term, from (4.5), is the spontaneous emission noise added at this stage. Also from (5.5) we must have

$$\overline{G} = 1 + \frac{P^{\text{sat}}}{\overline{P}_{\text{out}}e^{-\alpha l}} \ln \frac{G_{\text{max}}}{\overline{G}}. \tag{5.7}$$

Equations (5.6) and (5.7) can be solved simultaneously to compute the values of $\overline{P}_{\text{out}}$ and $\overline{G}$ (Problem 5.11). Observe from (5.6) that $\overline{G}e^{-\alpha l} < 1$; that is, the steady-state gain will be slightly smaller than the loss between stages, due to the added noise at each stage. Thus in designing a cascade, we must try to choose the saturated gain G to be as close to the loss between stages as possible.

Let us consider a simplified model of an amplifier cascade where we assume the saturated gain $G = e^{\alpha l}$. With L/l amplifiers in the system, the total noise power at the output, using (4.5), is

$$P_{\text{noise}}^{\text{tot}} = 2P_n B_o(G - 1)L/l = 2P_n B_o(e^{\alpha l} - 1)L/l. \tag{5.8}$$

Given a desired OSNR, the launched power P must satisfy

$$P \geq (\text{OSNR})P_{\text{noise}}^{\text{tot}} = (\text{OSNR})2P_n B_o(e^{\alpha l} - 1)L/l.$$

Figure 5.6 plots the required power P versus amplifier spacing l. If we don't worry about nonlinearities, we would try to maximize l subject to limitations on transmit power and amplifier output power. The story changes in the presence of nonlinearities, as we will see in Section 5.8.

5.5.4   Amplifier Spacing Penalty

In the preceding section, we saw that in an amplifier cascade the gain of each amplifier must approximately compensate for the span loss (the loss between two amplifier stages in the cascade). For a given span length, say, 80 km, this determines the gain of the amplifiers in the cascade. For example, for a span length of $l = 80$ km and a fiber loss of $\alpha(dB) = 0.25$ dB/km, we get an amplifier gain $G = 20$ dB. If the amplifier gain is smaller, we must choose a smaller span length. In this section, we will study

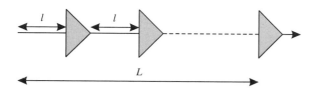

Figure 5.5 A system with cascaded optical amplifiers.

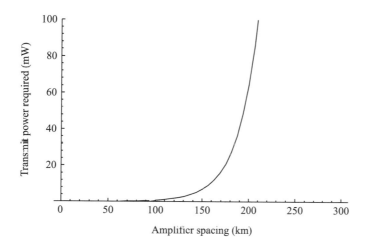

Figure 5.6 Power versus amplifier spacing. Required OSNR $= 50$, $n_{sp} = 2$, $B_o = 20$ GHz, $\alpha = 0.22$ dB/km, and the total link length $L = 1000$ km.

the effect of the span length, or, equivalently, the amplifier gain G, on the noise at the output of an amplifier cascade. This will enable us to then discuss quantitatively the penalty reduction we can obtain by the use of distributed amplifiers, in particular, distributed Raman amplifiers.

The ASE noise power at the output of a cascade of L/l amplifiers is given by (5.8). Rewriting this in terms of G, using $l = (\ln G)/\alpha$, we get

$$P_{\text{noise}}^{\text{tot}} = 2L P_n B_o \alpha (G - 1)/\ln G. \tag{5.9}$$

Ideally, the minimum noise power is achieved in an amplifier cascade with perfectly distributed gain, that is, $G = 1$ (and $N = \infty$ but $N \ln G = \alpha L$). The "power penalty" for using lumped amplifiers with gain $G > 1$, instead of an ideal distributed amplifier, is given by the factor

$$PP_{\text{lumped}} = \frac{G-1}{\ln G},$$

which is unity for $G = 1$. For $G = 20$ dB, $PP_{\text{lumped}} = 13.3$ dB, while for $G = 10$ dB, $PP_{\text{lumped}} = 5.9$ dB. Thus, assuming $\alpha = 0.25$ dB/km, the total ASE noise in an amplifier cascade can be reduced by more than 7 dB by reducing the amplifier spacing to 40 km from 80 km.

The reduction in ASE must be balanced against the increased system cost resulting from reducing the amplifier spacing, since twice the number of amplifier locations (huts) will be required when the amplifier spacing is halved from 80 km to 40 km. However, distributed amplification can reduce the ASE significantly without increasing the number of amplifier locations.

When a distributed amplifier is used, the amplification occurs continuously as the signal propagates in the fiber. The primary example of such an amplifier is the Raman amplifier we studied in Section 3.4.4.

Since system design engineers are accustomed to assuming lumped amplifiers, the increased ASE due to lumped amplification compared to distributed amplification is not viewed as a power penalty. Rather, the distributed amplifier is considered to have an equivalent (lower) noise figure, relative to a lumped amplifier, with the same total gain. For even moderate gains, this equivalent noise figure for the distributed amplifier can be negative! In our example above, we saw that the power penalty for using lumped amplifiers with gain $G = 20$ dB was 13.3 dB. A distributed amplifier with an actual noise figure ($2n_{\text{sp}}$) of 3.3 dB that provides the same total gain can also be viewed as having an effective noise figure of $3.3 - 13.3 = -10$ dB. This is because the accumulated ASE due to the use of such a distributed amplifier is the same as that of a lumped amplifier with a noise figure of -10 dB.

5.5.5   Power Transients and Automatic Gain Control

Power transients are an important effect to consider in WDM links and networks with a number of EDFAs in cascade. If some of the channels fail, the gain of each amplifier will increase because of the reduction in input power to the amplifier. In the worst case, $W - 1$ out of the W channels could fail, as shown in Figure 5.7. The surviving channels will then see more gain and will then arrive at their receivers with higher power. Likewise, the gain seen by existing channels will depend on what other channels are present. Thus setting up or taking down a new channel may affect the power levels in other channels. These factors drive the need for providing automatic gain control (AGC) in the system to keep the output power per channel at each amplifier constant, regardless of the input power.

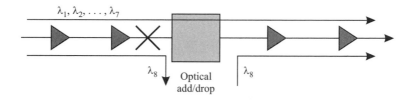

Figure 5.7 Illustrating the impact of failures in a network with optical amplifiers. In this example, λ_8, which is the only wavelength being added at the node, sees all the gain of the amplifier upon failure of the link preceding the node.

With only one EDFA in the cascade, the increase in power due to channel outages occurs rather slowly, in about 100 μs. However, with multiple amplifiers in the chain, the increase in power is much more rapid, with a rise-time of a few to tens of microseconds, and can result in temporary outages in the surviving channels. To prevent this, the AGC system must work very fast, within a few microseconds, to prevent these power transients from occurring.

Several types of AGC systems have been proposed. A simple AGC circuit monitors the signal power into the amplifier and adjusts the pump power to vary the gain if the input signal power changes. The response time of this method is limited ultimately by the lifetime of the electrons from the third energy level to the second energy level in erbium (see Section 3.4.3), which is around 1 μs.

Another interesting AGC circuit uses an optical feedback loop, as shown in Figure 5.8. A portion of the amplifier output is tapped off, filtered by a bandpass filter, and fed back into the amplifier. The gain of the loop is controlled carefully by using an attenuator in the loop. This feedback loop causes the amplifier to lase at the wavelength passed by the filter in the loop. This has the effect of clamping the amplifier gain seen by other wavelengths to a fixed value, irrespective of the input signal power. Moreover, it is usually sufficient to have this loop in the first amplifier in the cascade. This is because the output lasing power at the loop wavelength becomes higher as the input signal power decreases, and acts as a compensating signal to amplifiers farther down the cascade. Therefore, amplifiers farther down the cascade do not see a significant variation in the input power. Because of the additional couplers required for the AGC at the input and output, the amplifier noise figure is slightly increased and its output power is reduced.

Yet another approach is to introduce an additional wavelength on the link to act as a compensating wavelength. This wavelength is introduced at the beginning of the link and tapped off at the end of the link. The power on this wavelength is increased to compensate for any decrease in power seen at the input to the link. This

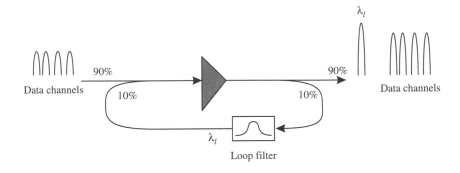

Figure 5.8   Optical automatic gain control circuit for an optical amplifier.

method requires an additional laser and is not as cost-effective as the other ones. It can compensate for only a few channels.

5.5.6   Lasing Loops

In systems with amplifiers, if we are not careful, we may end up with closed fiber loops that may lase. In our designs so far, we have tried to make the amplifier gain almost exactly compensate for the span losses encountered. If for some reason a closed fiber loop is encountered with amplifiers in the loop, and the total gain in the loop is comparable to the total loss in the loop, the loop may begin to lase. The effect here is similar to the optical automatic gain control circuit that we discussed in Section 5.5.5, but in this case lasing loops can cause power to be taken away from live channels and distributed to the channel that is lasing—a highly undesirable attribute. Note that this phenomenon may occur even if the loop is closed only for a single wavelength and not closed for the other wavelengths. Lasing loops are particularly significant problems in ring networks (which are inherently closed loops!) with optical add/drop multiplexers. In this case, even the amplified spontaneous emission traveling around the ring may be sufficient to cause the ring to lase.

We can deal with lasing loops in a few different ways. The preferred safe method is to ensure that the amplifier gain is always slightly lower than the loss being compensated for. The trade-off is that this would result in a small degradation of the signal-to-noise ratio. Another possibility is to ensure that closed loops never occur during operation of the system. For example, we could break a ring at a certain point and terminate all the wavelengths. Note, however, that it may not be sufficient to ensure loop freedom just under normal operation. We would not want a service person making a wrong fiber connection in the field to take down the entire network.

Therefore we need to make sure that loops aren't created even in the presence of human errors—not an easy problem to solve.

5.6 Crosstalk

Crosstalk is the general term given to the effect of other signals on the desired signal. Almost every component in a WDM system introduces crosstalk of some form or another. The components include filters, wavelength multiplexers/demultiplexers, switches, semiconductor optical amplifiers, and the fiber itself (by way of nonlinearities). Two forms of crosstalk arise in WDM systems: *interchannel crosstalk* and *intrachannel crosstalk*. The first case is when the crosstalk signal is at a wavelength sufficiently different from the desired signal's wavelength that the difference is larger than the receiver's electrical bandwidth. This form of crosstalk is called interchannel crosstalk. Interchannel crosstalk can also occur through more indirect interactions, for example, if one channel affects the gain seen by another channel, as with nonlinearities (Section 5.8). The second case is when the crosstalk signal is at the same wavelength as that of the desired signal or sufficiently close to it that the difference in wavelengths is within the receiver's electrical bandwidth. This form of crosstalk is called intrachannel crosstalk or, sometimes, *coherent crosstalk*. Intrachannel crosstalk effects can be much more severe than interchannel crosstalk, as we will see. In both cases, crosstalk results in a power penalty.

5.6.1 Intrachannel Crosstalk

Intrachannel crosstalk arises in transmission links due to reflections. This is usually not a major problem in such links since these reflections can be controlled. However, intrachannel crosstalk can be a major problem in networks. One source of this arises from cascading a wavelength demultiplexer (demux) with a wavelength multiplexer (mux), as shown in Figure 5.9(a). The demux ideally separates the incoming wavelengths to different output fibers. In reality, however, a portion of the signal at one wavelength, say, λ_i, leaks into the adjacent channel λ_{i+1} because of nonideal suppression within the demux. When the wavelengths are combined again into a single fiber by the mux, a small portion of the λ_i that leaked into the λ_{i+1} channel will also leak back into the common fiber at the output. Although both signals contain the same data, they are not in phase with each other, due to different delays encountered by them. This causes intrachannel crosstalk. Another source of this type of crosstalk arises from optical switches, as shown in Figure 5.9(b), due to the nonideal isolation of one switch port from the other. In this case, the signals contain different data.

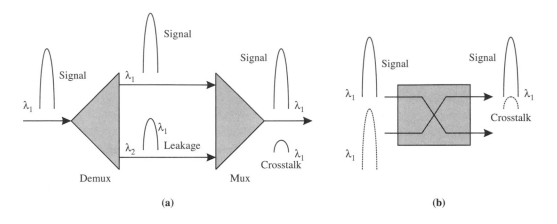

Figure 5.9   Sources of intrachannel crosstalk. (a) A cascaded wavelength demultiplexer and a multiplexer, and (b) an optical switch.

The crosstalk penalty is highest when the state of polarization (SOP) of the crosstalk signal is the same as the SOP of the desired signal. In practice, the SOPs vary slowly with time in a system using standard single-mode fiber (nonpolarization preserving). Similarly, the crosstalk penalty is highest when the crosstalk signal is exactly out of phase with the desired signal. The phase relationship between the two signals can vary over time due to several factors, including temperature variations. We must, however, design the system to work even if the two SOPs happen to match and the signals are exactly out of phase. Thus, for the calculations in this section, we will assume that the SOPs are the same and compute the penalty when the signals are out of phase, which is the worst-case scenario.

The power penalty due to intrachannel crosstalk can be determined as follows. Let P denote the average received signal power and ϵP the average received crosstalk power from a single other crosstalk channel. Assume that the signal and crosstalk are at the same optical wavelength. The electric field at the receiver can be written as

$$E(t) = \sqrt{2P}d_s(t)\cos[2\pi f_c t + \phi_s(t)] + \sqrt{2\epsilon P}d_x(t)\cos[2\pi f_c t + \phi_x(t)].$$

Here, $d_s(t) = \{0, 1\}$, depending on whether a 0 or 1 is being sent in the desired channel; $d_x(t) = \{0, 1\}$, depending on whether a 0 or 1 is being sent in the crosstalk channel; f_c is the frequency of the optical carrier; and $\phi_s(t)$ and $\phi_x(t)$ are the random phases of the signal and crosstalk channels, respectively. It is assumed that all channels have an ideal extinction ratio of ∞.

The photodetector produces a current that is proportional to the received power within its receiver bandwidth. This received power is given by

$$P_r = P d_s(t) + \epsilon P d_x(t) + 2\sqrt{\epsilon} P d_s(t) d_x(t) \cos[\phi_s(t) - \phi_x(t)]. \tag{5.10}$$

Assuming $\epsilon \ll 1$, we can neglect the ϵ term compared to the $\sqrt{\epsilon}$ term. Also the worst case above is when the $\cos(.) = -1$. Using this, we get the received power during a 1 bit as

$$P_r(1) = P(1 - 2\sqrt{\epsilon})$$

and the power during a 0 bit as

$$P_r(0) = 0.$$

First consider the case where the detection is limited by receiver thermal noise, which is independent of the received power. Using (5.3), the power penalty for this case is

$$\text{PP}_{\text{sig-indep}} = -10\log(1 - 2\sqrt{\epsilon}). \tag{5.11}$$

In amplified systems, or in systems with APD receivers, the dominant noise component is signal dependent (see Section 5.2). For this case, $\sigma_1 \propto \sqrt{P}$ and $\sigma_0 \ll \sigma_1$. Using (5.4), the power penalty in this case becomes

$$\text{PP}_{\text{sig-dep}} = -5\log(1 - 2\sqrt{\epsilon}). \tag{5.12}$$

If there are N interfering channels, each with average received power $\epsilon_i P$, then ϵ in (5.11) and (5.12) is given by $\sqrt{\epsilon} = \sum_{i=1}^{N} \sqrt{\epsilon_i}$ (see Problem 5.12).

Figure 5.10 shows the crosstalk penalties plotted against the crosstalk level for intrachannel and interchannel crosstalk, which we will consider next. If we allow a 1 dB penalty with signal-independent noise, then the intrachannel crosstalk level should be 20 dB below the desired signal.

5.6.2 Interchannel Crosstalk

Interchannel crosstalk can arise from a variety of sources. A simple example is an optical filter or demultiplexer that selects one channel and imperfectly rejects the others, as shown in Figure 5.11(a). Another example is in an optical switch, switching different wavelengths (shown in Figure 5.11(b)), where the crosstalk arises because of imperfect isolation between the switch ports.

Estimating the power penalty due to interchannel crosstalk is fairly straightforward. If the wavelength spacing between the desired signal and the crosstalk signal is large compared to the receiver bandwidth, (5.10) can be written as

$$P_r = P d_s(t) + \epsilon P d_x(t).$$

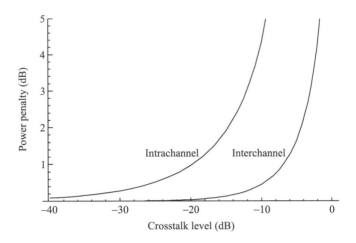

Figure 5.10 Thermal noise limited intrachannel and interchannel crosstalk power penalties as a function of crosstalk level, $-10 \log \epsilon$. Signal-spontaneous noise limited penalties would be reduced by half the values shown in the figure.

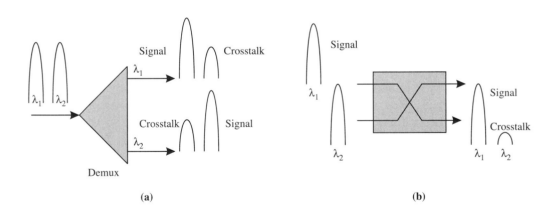

Figure 5.11 Sources of interchannel crosstalk. (a) An optical demultiplexer, and (b) an optical switch with inputs at different wavelengths.

Therefore, in the worst case, we have

$$P_r(1) = P,$$

and

$$P_r(0) = \epsilon P.$$

Using (5.3), the power penalty for the thermal noise limited case is given by

$$\text{PP}_{\text{sig-indep}} = -10 \log(1 - \epsilon). \tag{5.13}$$

For systems dominated by signal-dependent noise, the penalty is obtained from (5.4) as

$$\text{PP}_{\text{sig-dep}} = \quad 5 \log(1 - \epsilon). \tag{5.14}$$

If there are N interfering channels, each with average received power $c_i P$, then ϵ in (5.13) and (5.14) is given by $\epsilon = \sum_{i=1}^{N} \epsilon_i$ (see Problem 5.12).

Consider an unamplified WDM system with a filter receiving the desired channel and rejecting the others. The main crosstalk component usually comes from the two adjacent channels, and the crosstalk from the other channels is usually negligible. Assuming a 0.5 dB crosstalk penalty requirement, the adjacent channel suppression must be greater than 12.6 dB.

5.6.3   Crosstalk in Networks

Crosstalk suppression becomes particularly important in networks, where a signal propagates through many nodes and accumulates crosstalk from different elements at each node. Examples of such elements are muxes/demuxes and switches. In order to obtain an approximate idea of the crosstalk requirements, suppose that a signal accumulates crosstalk from N sources, each with crosstalk level ϵ_s. This neglects the fact that some interfering channels may have higher powers than the desired channel. Networks are very likely to contain amplifiers and to be limited by signal-spontaneous beat noise. Figure 5.12 plots the power penalties calculated from (5.12) and (5.14). For example, if we have 10 interfering equal-power crosstalk elements, each producing intrachannel crosstalk, then we must have a crosstalk suppression of below 35 dB in each element, in order to have an overall penalty of less than 1 dB.

5.6.4   Bidirectional Systems

In a bidirectional transmission system, data is transmitted in both directions over a single fiber, as shown in Figure 5.13. Additional crosstalk mechanisms arise in these systems. Although the laws of physics do not prevent the same wavelength from being used for both directions of transmissions, this is not a good idea in practice because of reflections. A back-reflection from a point close to the transmitter at one

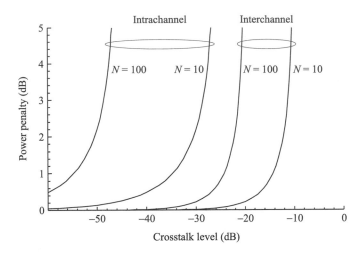

Figure 5.12   Signal-spontaneous noise limited intrachannel and interchannel crosstalk penalties as a function of crosstalk level $-10 \log \epsilon_s$ in a network. The parameter N denotes the number of crosstalk elements, all assumed to produce crosstalk at equal powers.

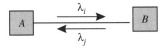

Figure 5.13   A bidirectional transmission system.

end, say, end A, will send a lot of power back into A's receiver, creating a large amount of crosstalk. In fact, the reflected power into A may be larger than the signal power received from the other end B. Reflections within the end equipment can be carefully controlled, but it is more difficult to restrict reflections from the fiber link itself. For this reason, bidirectional systems typically use different wavelengths in different directions. The two directions can be separated at the ends either by using an optical circulator or a WDM mux/demux, as in Figure 5.14. (If the same wavelength must be used in both directions, one alternative that is sometimes used in short-distance access networks is to use time division multiplexing where only one end transmits at a time.)

If a WDM mux/demux is used to handle both directions of transmission, crosstalk can also arise because a signal at a transmitted wavelength is reflected within the mux

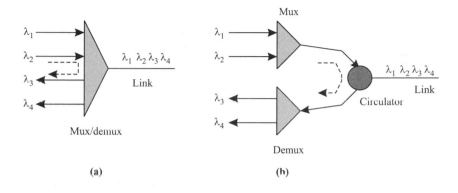

Figure 5.14 Separating the two directions in a bidirectional system: (a) using a wavelength multiplexer/demultiplexer, and (b) using an optical circulator. Both methods can introduce crosstalk, as shown by dashed lines in the figure.

into a port that is used to receive a signal from the other end, as in Figure 5.14(a). The mux/demux used should have adequate crosstalk suppression to ensure that this is not a problem. Likewise, if an optical circulator is used, crosstalk can arise because of imperfect isolation in the circulator, as shown in Figure 5.14(b). We have to be careful about these effects when designing bidirectional optical amplifiers as well.

5.6.5 Crosstalk Reduction

The simplest (and preferred) approach toward crosstalk reduction is to improve the crosstalk suppression at the device level; in other words, let the device designer worry about it. The network designer calculates and specifies the crosstalk suppression required for each device based on the number of such cascaded devices in the network and the allowable penalty due to crosstalk. However, there are a few architectural approaches toward reducing specific forms of crosstalk, particularly crosstalk arising in optical switches.

The first approach is to use *spatial dilation*, which is illustrated in Figure 5.15. Figure 5.15(a) shows a 2×2 optical switch with crosstalk ϵ. To improve the crosstalk suppression, we can *dilate* the switch, as shown in Figure 5.15(b), by adding some unused ports to it. Now the crosstalk is reduced to ϵ^2. The drawbacks of dilation are that it cannot be achieved without a significant increase in the number of switches. Usually, the number of switches is doubled.

Another approach to reduce switch crosstalk in a WDM network is to use *wavelength dilation* in the switches. This is particularly useful if a single switch is to handle

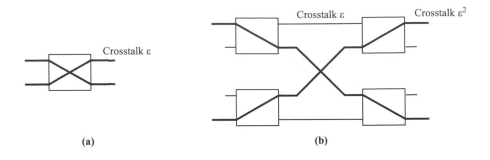

Figure 5.15   Using spatial dilation to reduce switch crosstalk. (a) A simple 2 × 2 switch. (b) A dilated version of a 2 × 2 switch.

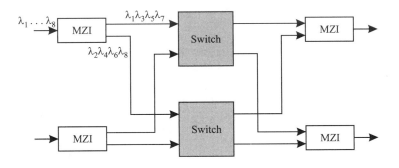

Figure 5.16   Using wavelength dilation to reduce switch crosstalk. MZI denotes a Mach-Zehnder interferometer that separates the channels into two groups or combines them.

multiple wavelengths, such as the acousto-optic tunable filter of Section 3.3.9. To reduce the interchannel crosstalk, you can use two switches instead of one, as shown in Figure 5.16. The first switch handles the odd-numbered channels, and the second the even-numbered channels. This effectively doubles the channel spacing as far as crosstalk is concerned. Again the cost is that twice as many switches are required. In the extreme case of wavelength dilation, we can have a separate switch for each wavelength.

The previous methods have dealt mainly with switch crosstalk. A simple method to reduce crosstalk in the mux/demux of Figure 5.9 is to add an additional filter

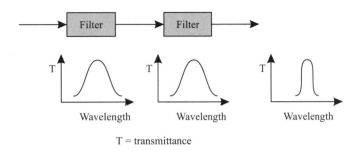

Figure 5.17 Bandwidth narrowing due to cascading of two filters.

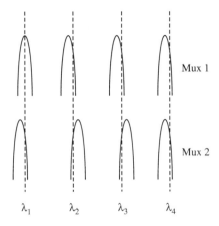

Figure 5.18 Wavelength misalignment between two mux/demuxes.

for each wavelength between the demux and mux stages. The extra filter stage produces an additional level of isolation and improves the overall crosstalk performance dramatically, but of course adds to the cost of the unit.

5.6.6 Cascaded Filters

Networks are likely to have several mux/demuxes or filters cascaded. When two mux/demuxes or filters are cascaded, the overall passband is much smaller than the passbands of the individual filters. Figure 5.17 shows this effect. The required wavelength stability and accuracy in these systems therefore goes up with the number of cascaded stages.

A related problem arises from the accuracy of wavelength registration in these mux/demuxes. If the center wavelengths of two units in a cascade are not identical (see Figure 5.18), the overall loss through the cascade for the desired signal will be higher, and the crosstalk from the adjacent channels could also be higher. If we are concerned only with one channel, we could align the center wavelengths exactly by temperature-tuning the individual mux/demuxes. However, other channels could become even more misaligned in the process (tuning one channel tunes the others as well). In addition, the lasers themselves will have a tolerance regarding their center wavelength. In a cascaded system, wavelength inaccuracies cause additional power penalties due to added signal loss and crosstalk (see Problems 5.18 and 5.19).

5.7   Dispersion

Dispersion is the name given to any effect wherein different components of the transmitted signal travel at different velocities in the fiber, arriving at different times at the receiver. A signal pulse launched into a fiber arrives smeared at the other end as a consequence of this effect. This smearing causes intersymbol interference, which in turn leads to power penalties. Dispersion is a cumulative effect: the longer the link, the greater the amount of dispersion.

Several forms of dispersion arise in optical communication systems. The important ones are *intermodal dispersion*, *polarization-mode dispersion*, and *chromatic dispersion*. Of these, we have already studied intermodal dispersion and chromatic dispersion in Chapter 2 and quantified the limitations that they impose on the link length and/or bit rate.

Intermodal dispersion arises only in multimode fiber, where the different modes travel with different velocities. Intermodal dispersion was discussed in Section 2.1. The link length in a multimode system is usually limited by intermodal dispersion and not by the loss. Clearly intermodal dispersion is not a problem with single-mode fiber.

Polarization-mode dispersion (PMD) arises because the fiber core is not perfectly circular, particularly in older installations. Thus different polarizations of the signal travel with different group velocities. PMD is proving to be a serious impediment in very high-speed systems operating at 10 Gb/s bit rates and beyond. We discuss PMD in Section 5.7.4.

The main form of dispersion that we are concerned with is *chromatic dispersion*, which has a profound impact in the design of single-mode transmission systems (so much so that we often use the term "dispersion" to mean "chromatic dispersion"). Chromatic dispersion arises because different frequency components of a pulse (and also signals at different wavelengths) travel with different group velocities in the fiber, and thus arrive at different times at the other end. We discussed the origin of chromatic dispersion in Section 2.3. Chromatic dispersion is a characteristic of the fiber, and different fibers have different chromatic dispersion profiles. We discussed the chromatic dispersion profiles of many different fibers in Section 2.4.9. As with other kinds of dispersion, the accumulated chromatic dispersion increases with the link length. Chromatic dispersion and the system limitations imposed by it are discussed in detail in the next two sections.

5.7.1 Chromatic Dispersion Limits: NRZ Modulation

In this section, we discuss the chromatic dispersion penalty for NRZ modulated signals. We will consider RZ modulated signals in Section 5.7.2.

The transmission limitations imposed by chromatic dispersion can be modeled by assuming that the pulse spreading due to chromatic dispersion should be less than a fraction ϵ of the bit period, for a given chromatic dispersion penalty. This fraction has been specified by both ITU (G.957) and Telcordia (GR-253). For a penalty of 1 dB, $\epsilon = 0.306$, and for a penalty of 2 dB, $\epsilon = 0.491$. If D is the fiber chromatic dispersion at the operating wavelength, B the bit rate, $\Delta\lambda$ the spectral width of the transmitted signal, and L the length of the link, this limitation can be expressed as

$$|D|LB(\Delta\lambda) < \epsilon. \tag{5.15}$$

D is usually specified in units of ps/nm-km. Here, the ps refers to the time spread of the pulse, the nm is the spectral width of the pulse, and km corresponds to the link length. For standard single-mode fiber, the typical value of D in the C-band is 17 ps/nm-km. For this value of D, $\lambda = 1.55$ μm, and $\epsilon = 0.491$ (2 dB penalty), (5.15) yields the condition $BL < 30$ (Gb/s)-km. This limit is plotted in Figure 5.19. Thus even at a bit rate of 1 Gb/s, the link length is limited to < 30 km, which is a severe limitation. This illustrates the importance of (1) using nearly monochromatic sources, for example, DFB lasers, for high-speed optical communication systems, and (2) devising methods of overcoming chromatic dispersion.

Narrow Source Spectral Width

We now consider the case of using sources with narrow spectral widths. Even for such a source, the spectral width of the transmitted signal depends on whether it is

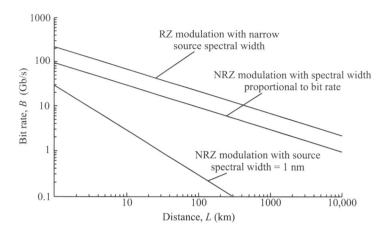

Figure 5.19   Chromatic dispersion limits on the distance and bit rate for transmission over standard single-mode fiber with a chromatic dispersion value of $D = 17$ ps/nm-km. A chromatic dispersion penalty of 2 dB has been assumed in the NRZ case; this implies that the rms width of the dispersion-broadened pulse must lie within a fraction 0.491 of the bit period. For sources with narrow spectral width, the spectral width of the modulated signal in GHz is assumed to be 2.5 times the bit rate in Gb/s. For RZ transmission, the rms output pulse width is assumed to be less than the bit interval.

directly modulated or whether an external modulator is used. SLM DFB lasers have unmodulated spectral widths of typically less than 50 MHz. Directly modulating a DFB laser would ideally cause its spectral width to correspond to the modulation bandwidth (for example, about 2.5 GHz for a 2.5 Gb/s on-off modulated signal). In practice, however, the spectral width can increase owing to chirp. As the modulation current (and thus optical power) varies, it is accompanied by changes in carrier density within the laser cavity, which, in turn, changes the refractive index of the cavity, causing frequency variations in its output. The magnitude of the effect depends on the variation in current (or power), but it is not uncommon to observe spectral widths over 10 GHz as a consequence of chirp. Chirp can be reduced by decreasing the extinction ratio. The spectral width can also be increased because of back-reflections from connectors, splices, and other elements in the optical path. To prevent this effect, high-speed lasers are typically packaged with built-in isolators.

For externally modulated sources, the spectral width is proportional to the bit rate. Assuming the spectral width is about 2.5 times the bit rate, a 10 Gb/s externally modulated signal has a spectral width of 25 GHz, which is a practical number today.

At 1.55 μm, this corresponds to a spectral width of 0.2 nm, using the relation $\Delta\lambda = (c/f^2)|\Delta f| = (\lambda^2/c)|\Delta f|$. Substituting $\Delta\lambda = (\lambda^2/c)2.5B$ in (5.15), we get

$$|D|LB^2\lambda^2/c < 0.4\epsilon,$$

or

$$B\lambda\sqrt{|D|L/c} < \sqrt{0.4\epsilon}. \tag{5.16}$$

For $D = 17$ ps/nm-km, $\lambda = 1.55$ μm, and $\epsilon = 0.491$ (2 dB penalty), (5.16) yields the condition $B^2L < 8327$ (Gb/s)^2-km. This limit is also plotted in Figure 5.19.

Note that the chromatic dispersion limitations are much more relaxed for narrow spectral width sources. This explains the widepsread use of narrow spectral width SLM lasers for high-bit-rate communication. In addition, external modulators are used for long-distance transmission (more than a few hundred kilometers) at 2.5 Gb/s, and in most 10 Gb/s systems.

5.7.2   Chromatic Dispersion Limits: RZ Modulation

In this section, we derive the system limitations imposed by chromatic dispersion for unchirped Gaussian pulses, which are used in RZ modulated systems. The results can be extended in a straightforward manner to chirped Gaussian pulses.

Consider a fiber of length L. From (2.13), the width of the output pulse is given by

$$T_L = \sqrt{T_0^2 + \left(\frac{\beta_2 L}{T_0}\right)^2}.$$

This is the half-width of the pulse at the $1/e$-intensity point. A different, and more commonly used, measure of the width of a pulse is its *root-mean square* (rms) *width* T^{rms}. For a pulse, $A(t)$, this is defined as

$$T^{\text{rms}} = \sqrt{\frac{\int_{-\infty}^{\infty} t^2 |A(t)|^2 \, dt}{\int_{-\infty}^{\infty} |A(t)|^2 \, dt}}. \tag{5.17}$$

We leave it as an exercise (Problem 2.10) to show that for Gaussian pulses whose half-width at the $1/e$-intensity point is T_0,

$$T^{\text{rms}} = T_0/\sqrt{2}.$$

If we are communicating at a bit rate of B bits/s, the bit period is $1/B$ s. We will assume that satisfactory communication is possible only if the width of the pulse as measured by its rms width T^{rms} is less than the bit period. (Satisfactory communication may be possible even if the output pulse width is larger than the bit period, with an associated power penalty, as in the case of NRZ systems.) Therefore, $T_L^{\mathrm{rms}} = T_L/\sqrt{2} < 1/B$ or

$$BT_L < \sqrt{2}.$$

Through this condition, chromatic dispersion sets a limit on the length of the communication link we can use at bit rate B without dispersion compensation. T_L is a function of T_0 and can be minimized by choosing T_0 suitably. We leave it as an exercise (Problem 5.21) to show that the optimum choice of T_0 is

$$T_0^{\mathrm{opt}} = \sqrt{\beta_2 L},$$

and for this choice of T_0, the optimum value of T_L is

$$T_L^{\mathrm{opt}} = \sqrt{2|\beta_2|L}.$$

The physical reason there is an optimum pulse width is as follows. If the pulse is made too narrow in time, it will have a wide spectral width and hence greater dispersion and more spreading. However, if the pulse occupies a large fraction of the bit interval, it has less room to spread. The optimum pulse width arises from a trade-off between these two factors. For this optimum choice of T_0, the condition $BT_L < \sqrt{2}$ becomes

$$B\sqrt{2|\beta_2|L} < \sqrt{2}. \tag{5.18}$$

Usually, the value of β_2 is specified indirectly through the *dispersion parameter D*, which is related to β_2 by the equation

$$D = -\frac{2\pi c}{\lambda^2}\beta_2. \tag{5.19}$$

Thus (5.18) can be written as

$$B\lambda\sqrt{\frac{|D|L}{2\pi c}} < 1. \tag{5.20}$$

For $D = 17$ ps/nm-km, (5.20) yields the condition $B^2L < 46152$ (Gb/s)^2-km. This limit is plotted in Figure 5.19. Note that this limit is higher than the limit for NRZ modulation when the spectral width is determined by the modulation bandwidth (for example, for external modulation of an SLM laser). However, for both RZ and NRZ transmission, the bit rate B scales as $1/\sqrt{L}$.

Note that we derived the dispersion limits for unchirped pulses. The situation is much less favorable in the presence of frequency chirp. A typical value of the chirp parameter κ of a directly modulated semiconductor laser at 1.55 μm is -6, and β_2 is also negative so that monotone pulse broadening occurs. We leave it as an exercise to the reader (Problem 5.30) to calculate the chromatic dispersion limit with this value of κ and compare it to the dispersion limit for an unchirped pulse at a bit rate of 2.5 Gb/s.

However, if the chirp has the right sign, it can interact with dispersion to cause pulse compression, as we saw in Section 2.3. Chirped RZ pulses can be used to take advantage of this effect.

Large Source Spectral Width

We derived (2.13) for the width of the output pulse by assuming a nearly monochromatic source, such as a DFB laser. In practice, this assumption is not satisfied for many sources such as MLM Fabry-Perot lasers. This formula must be modified to account for the finite spectral width of the optical source. Assume that the frequency spectrum of the source is given by

$$F(\omega) = B_0 W_0 e^{-(\omega - \omega_0)^2 / 2W_0^2}.$$

Thus the spectrum of the source has a Gaussian profile around the center frequency ω_0, and W_0 is a measure of the frequency spread or bandwidth of the pulse. The *rms spectral width* W^{rms}, which is defined in a fashion similar to that of the rms temporal width in (5.17), is given by $W^{rms} = W_0/\sqrt{2}$. As in the case of Gaussian pulses, the assumption of a Gaussian profile is chiefly for mathematical convenience; however, the results derived hold qualitatively for other source spectral profiles. From this spectrum, in the limit as $W_0 \to 0$, we obtain a monochromatic source at frequency ω_0. Equation (2.13) for the width of the output pulse is obtained under the assumption $W_0 << 1/T_0$. If this assumption does not hold, it must be modified to read

$$\frac{T_z}{T_0} = \sqrt{\left(1 + \frac{\kappa \beta_2 z}{T_0^2}\right)^2 + (1 + W_0^2 T_0^2)\left(\frac{\beta_2 z}{T_0^2}\right)^2}. \tag{5.21}$$

From this formula, we can derive the limitation imposed by chromatic dispersion on the bit rate B and the link length L. We have already examined this limitation for the case $W_0 \ll 1/T_0$. We now consider the case $W_0 \gg 1/T_0$ and again neglect chirp.

Consider a fiber of length L. With these assumptions, from (5.21), the width of the output pulse is given by

$$T_L = \sqrt{T_0^2 + (W_0 \beta_2 L)^2}.$$

In this case, since the spectral width of the pulse is dominated by the spectral width of the source and not by the temporal width of the pulse ($W_0 \gg 1/T_0$), we can make T_0 much smaller than the bit period $1/B$ provided the condition $W_0 \gg 1/T_0$ is still satisfied. For such short input pulses, we can approximate T_L by

$$T_L = W_0 |\beta_2| L.$$

Therefore, the condition $BT_L < \sqrt{2}$ translates to

$$BL\beta_2 W^{\text{rms}} < 1.$$

The key difference from the case of small source spectral width is that the bit rate B scales *linearly* with L. This is similar to the case of NRZ modulation using a source with a large spectral width, independent of the bit rate. As in the case of NRZ modulation, chromatic dispersion is much more of a problem when using sources with nonnegligible spectral widths.

In fact, the two conditions (for NRZ and RZ) are nearly the same. To see this, express the spectral width of the source in wavelength units rather than in angular frequency units. A spectral width of W in radial frequency units corresponds to a spectral width in wavelength units of $(\Delta\lambda) = -2\pi c W/\lambda^2$. Using this and the relation $D = -2\pi c\beta_2/\lambda^2$, the chromatic dispersion limit $BL\beta_2 W^{\text{rms}} < 1$ becomes

$$BL|D|(\Delta\lambda) < 1 \qquad (5.22)$$

which is the same as (5.15) with $\epsilon = 1$.

As we have seen, the parameter β_2 is the key to group velocity or chromatic dispersion. For a given pulse, the magnitude of β_2 governs the extent of pulse broadening due to chromatic dispersion and determines the system limitations. β_2 can be minimized by appropriate design of the fiber as discussed in Section 2.3.2.

5.7.3   Dispersion Compensation

Dispersion management is a very important part of designing WDM transmission systems, since dispersion affects the penalties due to various types of fiber nonlinearities, as we will see in Section 5.8. We can use several techniques to reduce the impact of chromatic dispersion: (1) using external modulation in conjunction with DFB lasers, (2) using fiber with small chromatic dispersion, and (3) by chromatic dispersion compensation. The first alternative is commonly used today in high-speed systems. New builds over the past few years have used nonzero-dispersion-shifted fibers (NZ-DSF) that have a small chromatic dispersion value in the C-band. Dispersion compensation can be employed when external modulation alone is not sufficient

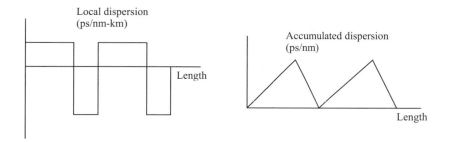

Figure 5.20   The chromatic dispersion map in a WDM link employing chromatic dispersion compensating fiber. (a) The (local) chromatic dispersion at each point along the fiber. (b) The accumulated chromatic dispersion from the beginning of the link up to each point along the fiber.

to reduce the chromatic dispersion penalty on the installed fiber type. We now discuss this option.

Along with the development of different fiber types, researchers have also developed various methods of compensating for chromatic dispersion. The two most popular methods use dispersion compensating fibers and chirped fiber Bragg gratings.

Dispersion Compensating Fibers

Special chromatic dispersion compensating fibers (DCFs) have been developed that provide negative chromatic dispersion in the 1550 nm wavelength range. For example, DCFs that can provide total chromatic dispersion of between −340 and −1360 ps/nm are commercially available. An 80 km length of standard single-mode fiber has an accumulated or total chromatic dispersion, at 17 ps/nm-km, of $17 \times 80 = 1360$ ps/nm. Thus a DCF with −1360 ps/nm can compensate for this accumulated chromatic dispersion, to yield a net zero chromatic dispersion. Between amplifier spans is standard single-mode fiber, but at each amplifier location, dispersion compensating fiber having a negative chromatic dispersion is introduced. The *chromatic dispersion map*—the variation of accumulated chromatic dispersion with distance—of such a system is shown in Figure 5.20. Even though the chromatic dispersion of the fibers used is high, because of the alternating signs of the chromatic dispersion, this approach leads to a small value of the accumulated chromatic dispersion so that we need not worry about penalties induced by chromatic dispersion.

One disadvantage of this approach is the added loss introduced in the system by the DCF. For instance the −1360 ps/nm DCF has a loss of 9 dB. Thus a commonly used measure for evaluating a DCF is the *figure of merit* (FOM), which is defined as the ratio of the absolute amount of chromatic dispersion per unit wavelength to the loss introduced by the DCF. The FOM is measured in ps/nm-dB, and the higher the FOM, the more efficient the fiber is at compensating for chromatic dispersion. The FOM for the DCF in the preceding example is thus 150 ps/nm-dB. DCF with a chromatic dispersion of −100 ps/nm-km and a loss of 0.5 dB/km is now available. The FOM of this fiber is 200. There is intensive research under way to develop DCFs with higher FOMs.

The FOM as defined here does not fully characterize the efficiency of the DCF since it does not take into account the added nonlinearities introduced by the DCF due to its smaller effective area. A modified FOM that does take this into account has been proposed in [FTCV96].

The preceding discussion has focused on standard single-mode fiber that has a large chromatic dispersion in the C-band, about 17 ps/nm-km. In systems that use NZ-DSF, the chromatic dispersion accumulates much more slowly, since this fiber has a chromatic dispersion in the C-band of only 2–4 ps/nm-km. Thus these systems need a much smaller amount of chromatic dispersion compensating fiber. In many newly designed submarine systems, NZ-DSF with a small but negative chromatic dispersion is used. The use of negative chromatic dispersion fibers permits higher transmit powers to be used since modulation instability is not an issue (see Section 2.4.9). In this case, the accumulated chromatic dispersion is negative and can be compensated with standard single-mode fiber. This avoids the use of special chromatic dispersion compensating fibers, with their higher losses and susceptibility to nonlinear effects. The use of standard single-mode fiber for chromatic dispersion compensation also reduces the cabling loss due to bending. Terrestrial systems do not adopt this approach since the use of negative chromatic dispersion fiber precludes the system from being upgraded to use the L-band since the chromatic dispersion zero for these fibers lies in the L-band. This is not an issue for submarine systems since these systems are not upgradable once they have been deployed.

Chirped Fiber Bragg Gratings

The fiber Bragg grating that we studied in Section 3.3.4 is a versatile device that can be used to compensate for chromatic dispersion. Such a device is shown in Figure 5.21. The grating itself is linearly chirped, in that the period of the grating varies linearly with position, as shown in Figure 5.21. This makes the grating reflect different wavelengths (or frequencies) at different points along its length. Effectively, a chirped Bragg grating introduces different delays at different frequencies.

In a regular fiber, chromatic dispersion introduces larger delays for the lower-frequency components in a pulse. To compensate for this effect, we can design

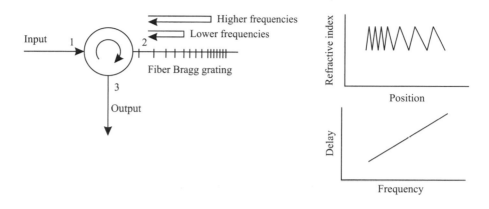

Figure 5.21 Chirped fiber Bragg grating for chromatic dispersion compensation.

chirped gratings that do exactly the opposite—namely, introduce larger delays for the higher-frequency components, in other words, compress the pulses. The delay as a function of frequency is plotted in Figure 5.21 for a sample grating.

Ideally, we want a grating that introduces a large amount of chromatic dispersion over a wide bandwidth so that it can compensate for the fiber chromatic dispersion over a large length as well as a wide range of wavelengths. In practice, the total length of the grating is limited by the size of the phase masks available. Until recently, this length used to be a few tens of centimeters. With a 10 cm long grating, the maximum delay that can be introduced is 1 ns. This delay corresponds to the product of the chromatic dispersion introduced by the grating and the bandwidth over which it is introduced. With such a grating, we introduce large chromatic dispersion over a small bandwidth, for example, 1000 ps/nm over a 1 nm bandwidth, or small chromatic dispersion over a wide bandwidth, for example, 100 ps/nm over a 10 nm bandwidth. Note that 100 km of standard single-mode fiber causes a total chromatic dispersion of 1700 ps/nm. When such chirped gratings are used to compensate for a few hundred kilometers of fiber chromatic dispersion, they must be very narrow band; in other words, we would need to use a different grating for each wavelength, as shown in Figure 5.22.

Chirped gratings are therefore ideally suited to compensate for individual wavelengths rather than multiple wavelengths. In contrast, DCF is better suited to compensate over a wide range of wavelengths. However, compared to chirped gratings, DCF introduces higher loss and additional penalties because of increased nonlinearities.

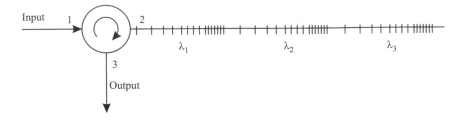

Figure 5.22   Chirped fiber Bragg gratings for compensating three wavelengths in a WDM system.

Recently, very long gratings, about 2 m in length, have been demonstrated [Bre01]. These gratings have been shown to compensate for the accumulated chromatic dispersion, over the entire C-band, after transmission over 40 km of standard single-mode fiber. Such a grating may prove to be a strong competitor to DCF.

Dispersion Slope Compensation

One problem with WDM systems is that since the chromatic dispersion varies for each channel (due to the nonzero slope of the chromatic dispersion profile), it may not be possible to compensate for the entire system using a common chromatic dispersion compensating fiber. A typical spread of the total chromatic dispersion, before and after compensation with DCF, across several WDM channels, is shown in Figure 5.23. This spread can be compensated by another stage of *chromatic dispersion slope compensation* where an appropriate length of fiber whose chromatic dispersion slope is opposite to that of the residual chromatic dispersion is used.

As we remarked in Section 2.4.9, it is difficult to fabricate positive chromatic dispersion fiber with negative slope (today), so that this technique can only be used for systems employing positive dispersion, positive slope fiber for transmission (and negative dispersion, negative slope fiber for dispersion, and dispersion slope, compensation). Thus, in submarine systems that use negative dispersion, positive slope fiber, dispersion slope compensation using dispersion compensating fiber is not possible. Moreover, if such systems employ large effective area fiber to mitigate nonlinear effects, the spread in chromatic dispersion slopes is enhanced, since large effective area fibers have larger dispersion slopes. One way to minimize the chromatic dispersion slope spread is to use a *hybrid fiber design*. In such a design, each span of, say, 50 km uses two kinds of fiber: large effective area fiber (with a consequent large dispersion slope) in the first half of the span and a reduced slope fiber in the second

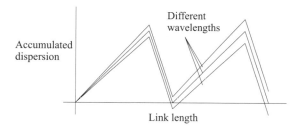

Figure 5.23 Variation of total chromatic dispersion in a WDM system across different channels, after chromatic dispersion compensation with a DCF.

half. Since nonlinear effects are significant only at the high power levels that occur in the first half of the span, the use of large effective area fiber in this half mitigates these effects, as effectively as using large effective area fiber for the whole span. The use of reduced slope fiber in the second half reduces (but does not eliminate) the overall spread in dispersion slope across channels (compared to using large effective area fiber in the whole span).

A second method of dispersion slope compensation is to provide the appropriate chromatic dispersion compensation for each channel separately at the receiver after the channels are demultiplexed. While individual channels can be compensated using appropriately different lengths of DCF, chirped fiber gratings (see Section 5.7.3) are commonly used to compensate individual channels since they are much more compact.

A third method of overcoming the dispersion slope problem is termed *mid-span spectral inversion* (MSSI). Roughly speaking, in this method, the spectrum of the pulse is inverted in the middle of the span, that is, the shorter and longer wavelengths of the pulse are interchanged. Recall that a pulse that is nominally at some frequency has a finite (nonzero) spectral width. Here we are referring to the different spectral components, or wavelengths, of a single pulse, and not the different wavelength channels in the system. This process is called *phase conjugation,* and it reverses the sign of the chromatic dispersion in the two halves of the span. Even if the chromatic dispersion values of different channels are equal, the chromatic dispersion in the two halves of the span cancels for each channel. Currently, the two other techniques, namely, chromatic dispersion compensating fiber and chirped fiber gratings, appear to be more suitable for commercial deployment.

5.7.4   Polarization-Mode Dispersion (PMD)

The origin of PMD lies in the fact that different polarizations travel with different group velocities because of the ellipticity of the fiber core; we discussed this in Section 2.1.2. Moreover, the distribution of signal energy over the different state of polarizations (SOPs) changes slowly with time, for example, because of changes in the ambient temperature. This causes the PMD penalty to vary with time as well. In addition to the fiber itself, PMD can arise from individual components used in the network.

The time-averaged differential time delay between the two orthogonal SOPs on a link is known to obey the relation [KK97a, Chapter 6]

$$\langle \Delta \tau \rangle = D_{PMD} \sqrt{L},$$

where $\langle \Delta \tau \rangle$ is called the differential group delay (DGD), L is the link length, and D_{PMD} is the fiber PMD parameter, measured in ps/$\sqrt{km}$. The PMD for typical fiber lies between 0.5 and 2 ps/$\sqrt{km}$. However, carefully constructed new links can have PMD as low as 0.1 ps/$\sqrt{km}$.

In reality, the SOPs vary slowly with time, and the actual DGD $\Delta \tau$ is a random variable. It is commonly assumed to have a Maxwellian probability density function (see Appendix H). This means that the square of the DGD is modeled by a more familiar distribution—the exponential distribution. The larger the DGD, the larger is the power penalty due to PMD. Thus, the power penalty due to PMD is also time varying, and it turns out that it is proportional to $\Delta \tau^2$ and thus obeys an exponential distribution (see Problem 5.22). If the power penalty due to PMD is large, it is termed a *PMD outage* and the link has effectively failed. For a DGD of $0.3T$, where T is the bit duration, the power penalty is approximately 0.5 dB for a receiver limited by thermal noise and 1 dB for a receiver with signal-dependent noise (ITU G.691).

Using the Maxwellian distribution, the probability that the actual delay will be greater than three times the average delay is about 4×10^{-5} (see Appendix H). Given our earlier reasoning, this means that in order to restrict the PMD outage probability (PMD $\geq$ 1 dB) to 4×10^{-5}, we must have the average DGD to be less than $0.1T$; that is,

$$\langle \Delta \tau \rangle = D_{PMD} \sqrt{L} < 0.1T. \tag{5.23}$$

This limit is plotted in Figure 5.24. Observe that for a bad fiber with PMD of 2 ps/$\sqrt{km}$, the limit is only 25 km. This is an extreme case, but it points out that PMD can impose a significant limitation.

Note that we have not said anything about the distribution of the length of time for which there is a PMD outage. In the above example, the DGD may exceed three times the average delay, and we may have one PMD outage with an average duration

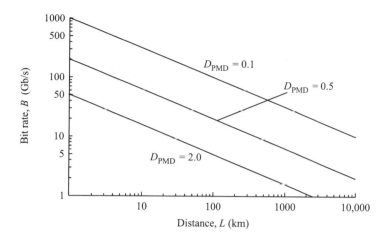

Figure 5.24   Limitations on the simultaneously achievable bit rates and distances imposed by PMD.

of one day once every 70 years, or one with an average duration of one minute every 17 days. This depends on the fiber cable in question, and typical outages last for a few minutes. Thus an outage probability of 4×10^{-5} can also be interpreted as a cumulative outage of about 20 minutes per year.

The limitations due to intermodal dispersion, chromatic dispersion, and PMD are compared in Figure 5.25.

PMD gives rise to intersymbol interference (ISI) due to pulse spreading, just as all other forms of dispersion. The traditional (electronic) technique for overcoming ISI in digital systems is *equalization,* discussed in Section 4.4.9. Equalization to compensate for PMD can be carried out in the electronic domain and is discussed in [WK92]. However, electronic equalization becomes more difficult as the bit rate increases and, to date, is not feasible for 40 Gb/s systems. At such high bit rates, optical PMD compensation must be used.

To understand how PMD can be compensated optically, recall that PMD arises due to the fiber birefringence and is illustrated in Figure 2.5. The transmitted pulse consists of a "fast" and a "slow" polarization component. The principle of PMD compensation is to split the received signal into its fast and slow polarization components, and to delay the fast component so that the DGD between the two components is compensated. Since the DGD varies in time, the delay that must be introduced in the fast component to compensate for PMD must be estimated in real time from the properties of the link.

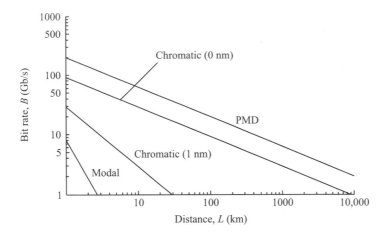

Figure 5.25   Limitations on the simultaneously achievable bit rates and distances imposed by intermodal dispersion, chromatic dispersion with a source spectral width of 1 nm, chromatic dispersion with spectral width proportional to the modulation bandwidth, and PMD with $D_{PMD} = 0.5$. NRZ modulation transmission over standard single-mode fiber with a chromatic dispersion value of 17 ps/nm-km is assumed.

The PMD effect we have discussed so far must strictly be called first-order polarization-mode dispersion. First-order PMD is a consequence of the fact that the two orthogonal polarization modes in optical fiber travel at slightly different speeds, which leads to a differential time delay between these two modes. However, this differential time delay itself is frequency dependent and varies over the bandwidth of the transmitted pulse. This effect is called second-order PMD. Second-order PMD is an effect that is similar to chromatic dispersion and thus can lead to pulse spreading.

PMD also depends on whether RZ or NRZ modulation is used; the discussion so far pertains to NRZ modulation. For RZ modulation, the use of short pulses enables more PMD to be tolerated since the output pulse has more room to spread—similar to the case of chromatic dispersion. However, second-order PMD depends on the spectral width of the pulse; narrower pulses have larger spectral widths. This is similar to the case of chromatic dispersion (Section 5.7.2). Again, as in the case of chromatic dispersion, there is an optimum input pulse width for RZ modulation that minimizes the output pulse width [SKA00, SKA01].

In addition to PMD, some other polarization-dependent effects influence system performance. One of these arises from the fact that many components have a polarization-dependent loss (PDL); that is, the loss through the component depends on the state of polarization. These losses accumulate in a system with many components in the transmission path. Again, since the state of polarization fluctuates with time, the signal-to-noise ratio at the end of the path will also fluctuate with time, and careful attention needs to be paid to maintain the total PDL on the path to within acceptable limits. An example of this is a simple angled-facet connector used in some systems to reduce reflections. This connector can have a PDL of about 0.1 dB, but hundreds of such connectors can be present in the transmission path.

5.8 Fiber Nonlinearities

As long as the optical power within an optical fiber is small, the fiber can be treated as a linear medium; that is, the loss and refractive index of the fiber are independent of the signal power. However, when power levels get fairly high in the system, we have to worry about the impact of nonlinear effects, which arise because, in reality, both the loss (gain) and refractive index depend on the optical power in the fiber. Nonlinearities can place significant limitations on high-speed systems as well as WDM systems.

As discussed in Chapter 2, nonlinearities can be classified into two categories. The first occurs because of scattering effects in the fiber medium due to the interaction of light waves with phonons (molecular vibrations) in the silica medium. The two main effects in this category are stimulated Brillouin scattering (SBS) and stimulated Raman scattering (SRS). The second set of effects occurs because of the dependence of refractive index on the optical power. This category includes four-wave mixing (FWM), self-phase modulation (SPM), and cross-phase modulation (CPM). In Chapter 2, we looked at the origins of all these effects. Here we will understand the limitations that all these nonlinearities place on system designers.

Except for SPM and CPM, all these effects provide gains to some channels at the expense of depleting power from other channels. SPM and CPM, on the other hand, affect only the phase of signals and can cause spectral broadening, which in turn, leads to increased chromatic dispersion penalties.

5.8.1 Effective Length in Amplified Systems

We discussed the notion of the effective length of a fiber span in Section 2.4.1. In systems with optical amplifiers, the signal gets amplified at each amplifier stage without resetting the effects due to nonlinearities from the previous span. Thus the effective length in such a system is the sum of the effective lengths of each span.

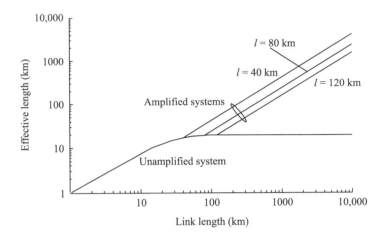

Figure 5.26   Effective transmission length as a function of link length, l.

In a link of length L with amplifiers spaced l km apart, the effective length is approximately given by

$$L_e = \frac{1 - e^{-\alpha l}}{\alpha} \frac{L}{l}. \tag{5.24}$$

Figure 5.26 shows the effective length plotted against the actual length of the transmission link for unamplified and amplified systems. The figure indicates that, in order to reduce the effective length, it is better to have fewer amplifiers spaced further apart. However, what matters in terms of the system effects of nonlinearities is not just the effective length; it is the product of the launched power P and the effective length L_e. Figure 5.6 showed how P varies with the amplifier spacing l. Now we are interested in finding out how PL_e grows with the amplifier spacing l. This is shown in Figure 5.27. The figure shows that the effect of nonlinearities can be reduced by reducing the amplifier spacing. Although this may make it easier to design the amplifiers (they need lower gain), we will also need more amplifiers, resulting in an increase in system cost.

The effect of a scattering nonlinearity depends on PL_e and thus increases with an increase in the input power and the link length. The longer the link, the greater is the amount of power that is coupled out from the signal (pump) into the Stokes wave. For a given link length, an approximate measure of the power level at which the effect of a nonlinearity starts becoming significant is the *threshold power*. For a given fiber length, the threshold power of a scattering nonlinearity is defined as the incident optical power per channel into the fiber at which the pump and Stokes

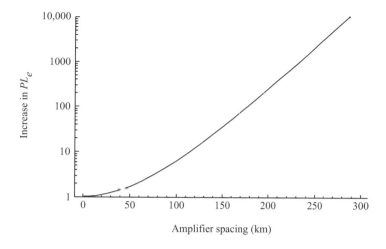

Figure 5.27 Relative value of PL_e versus amplifier spacing. The ordinate is the value relative to an amplifier spacing of 1 km. $\alpha = 0.22$ dB/km.

powers at the fiber output are equal. In amplified systems, the threshold power is reduced because of the increase in the effective length. This makes amplified systems more susceptible to impairments due to nonlinearities.

5.8.2 Stimulated Brillouin Scattering

The calculation of the threshold power for SBS P_{th} is quite involved, and we simply state the following approximation for it from [Smi72]:

$$P_{th} \approx \frac{21bA_e}{g_B L_e}.$$

Here, A_e and L_e are the effective area and length of the fiber, respectively (see Section 2.4.1), $g_B \approx 4 \times 10^{-11}$ m/W is called the Brillouin gain coefficient, and the value of b lies between 1 and 2 depending on the relative polarizations of the pump and Stokes waves. Assuming the worst-case value of $b = 1$, $A_e = 50 \ \mu\text{m}^2$, and $L_e = 20$ km, we get $P_{th} = 1.3$ mW. Since this is a low value, some care must be taken in the design of optical communication systems to reduce the SBS penalty.

The preceding expression assumes that the pump signal has a very narrow spectral width and lies within the narrow 20 MHz gain bandwidth of SBS. The threshold power is considerably increased if the signal has a broad spectral width, and thus much of the pump power lies outside the 20 MHz gain bandwidth of SBS. An approximate expression that incorporates this effect is given by

$$P_{\text{th}} \approx \frac{21bA_e}{g_B L_e}\left(1 + \frac{\Delta f_{\text{source}}}{\Delta f_B}\right),$$

where Δf_{source} is the spectral width of the source. With $\Delta f_{\text{source}} = 200$ MHz, and still assuming $b = 1$, the SBS threshold increases to $P_{\text{th}} = 14.4$ mW.

The SBS penalty can be reduced in several ways:

1. Keep the power per channel to much below the SBS threshold. The trade-off is that in a long-haul system, we may have to reduce the amplifier spacing.

2. Since the gain bandwidth of SBS is very small, its effect can be decreased by increasing the spectral width of the source. This can be done by directly modulating the laser, which causes the spectral width to increase because of chirp. This may cause a significant chromatic dispersion penalty. The chromatic dispersion penalty can, however, be reduced by suitable chromatic dispersion management, as we will see later. Another approach is to dither the laser slightly in frequency, say, at 200 MHz, which does not cause as high a penalty because of chromatic dispersion but increases the SBS threshold power by an order of magnitude, as we saw earlier. This approach is commonly employed in high-bit-rate systems transmitting at high powers. Irrespective of the bit rate, the use of an external modulator along with a narrow spectral width source increases the SBS threshold by only a small factor (between 2 and 4) for amplitude-modulated systems. This is because a good fraction of the power is still contained in the optical carrier for such systems.

3. Use phase modulation schemes rather than amplitude modulation schemes. This reduces the power present in the optical carrier, thus reducing the SBS penalty. In this case, the spectral width of the source can be taken to be proportional to the bit rate. However, this may not be a practical option in most systems.

5.8.3 Stimulated Raman Scattering

We saw in Section 2.4 that if two or more signals at different wavelengths are injected into a fiber, SRS causes power to be transferred from the shorter-wavelength channels to the longer-wavelength channels (see Figure 2.16). Channels up to 150 THz (125 nm) apart are coupled due to SRS, with the peak coupling occurring at a separation of about 13 THz. Coupling occurs for both copropagating and counterpropagating waves.

Coupling occurs between two channels only if both channels are sending 1 bits (that is, power is present in both channels). Thus the SRS penalty is reduced when chromatic dispersion is present because the signals in the different channels travel at different velocities, reducing the probability of overlap between pulses at different wavelengths at any point in the fiber. This is the same *pulse walk-off* phenomenon that we discussed in the case of CPM in Section 2.4.7. Typically, chromatic dispersion reduces the SRS effect by a factor of 2.

To calculate the effect of SRS in a multichannel system, following [Chr84], we approximate the Raman gain shape as a triangle, where the Raman gain coefficient as a function of wavelength spacing $\Delta\lambda$ is given by

$$g(\Delta\lambda) = \begin{cases} g_R \frac{\Delta\lambda}{\Delta\lambda_c}, & \text{if } 0 \le \Delta\lambda \le \Delta\lambda_c, \\ 0 & \text{otherwise.} \end{cases}$$

Here $\Delta\lambda_c = 125$ nm, and $g_R \approx 6 \times 10^{-14}$ m/W (at 1.55 μm) is the peak Raman gain coefficient.

Consider a system with W equally spaced channels $0, 1, \ldots, W - 1$, with $\Delta\lambda_s$ denoting the channel spacing. Assume that all the channels fall within the Raman gain bandwidth; that is, the system bandwidth $\Lambda = (W - 1)\Delta\lambda_s \le \Delta\lambda_c$. This is the case of practical interest given that the Raman gain bandwidth is 125 nm and the channels within a WDM system must usually be spaced within a 30 nm band dictated by the bandwidth of optical amplifiers. The worst affected channel is the channel corresponding to the lowest wavelength, channel 0, when there is a 1 bit in all the channels. Assume that the transmitted power is the same on all channels. Assume further that there is no interaction between the other channels, and the powers of the other channels remain the same (this approximation yields very small estimation errors). Assume also that the polarizations are scrambled. This is the case in practical systems. In systems that use polarization-maintaining fiber, the Raman interaction is enhanced, and the equation that follows does not have the factor of 2 in the denominator. The fraction of the power coupled from the worst affected channel, channel 0, to channel i is given approximately by [Buc95]

$$P_o(i) = g_R \frac{i\Delta\lambda_s}{\Delta\lambda_c} \frac{PL_e}{2A_e}.$$

This expression can be derived starting from the coupled wave equations for SRS that are similar in form to (2.14) and (2.15); see [Buc95] for details and [Zir98] for an alternate derivation with fewer assumptions. So the fraction of the power coupled out of channel 0 to all the other channels is

$$P_o = \sum_{i=1}^{W-1} P_o(i) = \frac{g_R \Delta\lambda_s PL_e}{2\Delta\lambda_c A_e} \frac{W(W-1)}{2}. \tag{5.25}$$

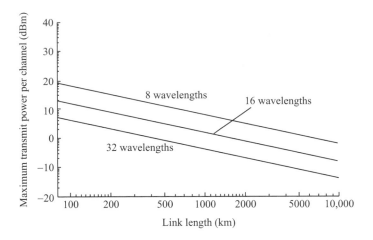

Figure 5.28   Limitation on the maximum transmit power per channel imposed by stimulated Raman scattering. The channel spacing is assumed to be 0.8 nm, and amplifiers are assumed to be spaced 80 km apart.

The power penalty for this channel is then

$$-10 \log(1 - P_o).$$

In order to keep the penalty below 0.5 dB, we must have $P_o < 0.1$, or, from (5.25),

$$WP(W - 1)\Delta\lambda_s L_e < 40{,}000 \text{ mW-nm-km}.$$

Observe that the total system bandwidth is $\Lambda = (W-1)\Delta\lambda_s$ and the total transmitted power is $P_{\text{tot}} = WP$. Thus the result can be restated as

$$P_{\text{tot}}\Lambda L_e < 40{,}000 \text{ mW-nm-km}.$$

The preceding formula was derived assuming that no chromatic dispersion is present in the system. With chromatic dispersion present, the right-hand side can be relaxed to approximately 80,000 mW-nm-km.

If the channel spacing is fixed, the power that can be launched decreases with W as $1/W^2$. For example, in a 32-wavelength system with channels spaced 0.8 nm (100 GHz) apart, and $L_e = 20$ km, $P \leq 2.5$ mW. Figure 5.28 plots the maximum allowed transmit power per channel as a function of the link length. The limit plotted here corresponds to $P_{\text{tot}}\Lambda L_e < 80{,}000$ mW-nm-km.

Although SRS is not a significant problem in systems with a small number of channels due to the relatively high threshold power, it can pose a serious problem

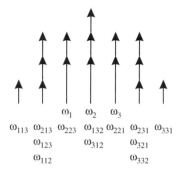

Figure 5.29   Four-wave mixing terms caused by the beating of three equally spaced channels at frequencies ω_1, ω_2, and ω_3.

in systems with a large number of wavelengths. To alleviate the effects of SRS, we can (1) keep the channels spaced as closely together as possible, and/or (2) keep the power levels below the threshold, which will require us to reduce the distance between amplifiers.

5.8.4   Four-Wave Mixing

We saw in Section 2.4 that the nonlinear polarization causes three signals at frequencies ω_i, ω_j, and ω_k to interact to produce signals at frequencies $\omega_i \pm \omega_j \pm \omega_k$. Among these signals, the most troublesome one is the signal corresponding to

$$\omega_{ijk} = \omega_i + \omega_j - \omega_k, \ i \neq k, j \neq k. \tag{5.26}$$

Depending on the individual frequencies, this beat signal may lie on or very close to one of the individual channels in frequency, resulting in significant crosstalk to that channel. In a multichannel system with W channels, this effect results in a large number $(W(W-1)^2)$ of interfering signals corresponding to i, j, k varying from 1 to W in (5.26). In a system with three channels, for example, 12 interfering terms are produced, as shown in Figure 5.29.

Interestingly, the effect of four-wave mixing depends on the phase relationship between the interacting signals. If all the interfering signals travel with the same group velocity, as would be the case if there were no chromatic dispersion, the effect is reinforced. On the other hand, with chromatic dispersion present, the different signals travel with different group velocities. Thus the different waves alternately overlap in and out of phase, and the net effect is to reduce the mixing efficiency. The velocity difference is greater when the channels are spaced farther apart (in systems with chromatic dispersion).

To quantify the power penalty due to four-wave mixing, we will use the results of the analysis from [SBW87, SNIA90, TCF+95, OSYZ95]. We start with (2.37) from Section 2.4.8:

$$P_{ijk} = \left(\frac{\omega_{ijk} \bar{n} d_{ijk}}{3cA_e} \right)^2 P_i P_j P_k L^2.$$

This equation assumes a link of length L without any loss and chromatic dispersion. Here P_i, P_j, and P_k denote the powers of the mixing waves and P_{ijk} the power of the resulting new wave, $\bar{n}$ is the nonlinear refractive index (3.0×10^{-8} $\mu m^2/W$), and d_{ijk} is the so-called degeneracy factor.

In a real system, both loss and chromatic dispersion are present. To take the loss into account, we replace L with the effective length L_e, which is given by (5.24) for a system of length L with amplifiers spaced l km apart. The presence of chromatic dispersion reduces the efficiency of the mixing, and we can model this by assuming a parameter η_{ijk}, which represents the efficiency of mixing of the three waves at frequencies ω_i, ω_j, and ω_k. Taking these two into account, the preceding equation can be modified to

$$P_{ijk} = \eta_{ijk} \left(\frac{\omega_{ijk} \bar{n} d_{ijk}}{3cA_e} \right)^2 P_i P_j P_k L_e^2.$$

For on-off keying (OOK) signals, this represents the worst-case power at frequency ω_{ijk}, assuming a 1 bit has been transmitted simultaneously on frequencies ω_i, ω_j, and ω_k.

The efficiency η_{ijk} goes down as the phase mismatch $\Delta\beta$ between the interfering signals increases. From [SBW87], we obtain the efficiency as

$$\eta_{ijk} = \frac{\alpha^2}{\alpha^2 + (\Delta\beta)^2} \left[1 + \frac{4e^{-\alpha l} \sin^2(\Delta\beta l/2)}{(1 - e^{-\alpha l})^2} \right].$$

Here, $\Delta\beta$ is the difference in propagation constants between the different waves, and D is the chromatic dispersion. Note that the efficiency has a component that varies periodically with the length as the interfering waves go in and out of phase. In our examples, we will assume the maximum value for this component. The phase mismatch can be calculated as

$$\Delta\beta = \beta_i + \beta_j - \beta_k - \beta_{ijk},$$

where β_r represents the propagation constant at wavelength λ_r.

Four-wave mixing manifests itself as intrachannel crosstalk. The total crosstalk power for a given channel ω_c is given as $\sum_{\omega_i + \omega_j - \omega_k = \omega_c} P_{ijk}$. Assume the amplifier gains are chosen to match the link loss so that the output power per channel is the same as the input power. The crosstalk penalty can therefore be calculated from (5.12).

Assume that the channels are equally spaced and transmitted with equal power, and the maximum allowable penalty due to FWM is 1 dB. Then if the transmitted power in each channel is P, the maximum FWM power in any channel must be $< \epsilon P$, where ϵ can be calculated to be 0.034 for a 1 dB penalty using (5.12). Since the generated FWM power increases with link length, this sets a limit on the transmit power per channel as a function of the link length. This limit is plotted in Figure 5.30 for both standard single-mode fiber (SMF) and dispersion-shifted fiber (DSF) for three cases: (1) 8 channels spaced 100 GHz apart, (2) 32 channels spaced 100 GHz apart, and (3) 32 channels spaced 50 GHz apart. For SMF the chromatic dispersion parameter is taken to be $D = 17$ ps/nm-km, and for DSF the chromatic dispersion zero is assumed to lie in the middle of the transmitted band of channels. The slope of the chromatic dispersion curve, $dD/d\lambda$, is taken to be 0.055 ps/nm-km^2. We leave it as an exercise (Problem 5.27) to compute the power limits in the case of NZ-DSF.

In Figure 5.30, first note that the limit is significantly worse in the case of dispersion-shifted fiber than it is for standard fiber. This is because the four-wave mixing efficiencies are much higher in dispersion-shifted fiber due to the low value of the chromatic dispersion. Second, the power limit gets worse with an increasing number of channels, as can be seen by comparing the limits for 8-channel and 32-channel systems for the same 100 GHz spacing. This effect is due to the much larger number of four-wave mixing terms that are generated when the number of channels is increased. In the case of dispersion-shifted fiber, this difference due to the number of four-wave mixing terms is imperceptible since, even though there are many more terms for the 32-channel case, the same 8 channels around the dispersion zero as in the 8-channel case contribute almost all the four-wave mixing power. The four-wave mixing power contribution from the other channels is small because there is much more chromatic dispersion at these wavelengths. Finally, the power limit decreases significantly if the channel spacing is reduced, as can be seen by comparing the curves for the two 32-channel systems with channel spacings of 100 GHz and 50 GHz. This decrease in the allowable transmit power arises because the four-wave mixing efficiency increases with a decrease in the channel spacing since the phase mismatch $\Delta\beta$ is reduced. (For SMF, though the efficiencies at both

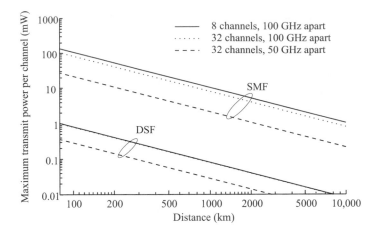

Figure 5.30 Limitation on the maximum transmit power per channel imposed by four-wave mixing for systems operating over standard single-mode fiber and dispersion-shifted fiber. For standard single-mode fiber, D is assumed to be 17 ps/nm-km, and for dispersion-shifted fiber, the chromatic dispersion zero is assumed to lie in the middle of the transmitted band of channels. The amplifiers are assumed to be spaced 80 km apart.

50 GHz and 100 GHz are small, the efficiency at 50 GHz is much higher than at 100 GHz.)

Four-wave mixing is a severe problem in WDM systems using dispersion-shifted fiber but does not usually pose a major problem in systems using standard fiber. In fact, it motivated the development of NZ-DSF fiber (see Section 5.7). In general, the following actions alleviate the penalty due to four-wave mixing:

1. Unequal channel spacing: The positions of the channels can be chosen carefully so that the beat terms do not overlap with the data channels inside the receiver bandwidth. This may be possible for a small number of channels in some cases, but needs careful computation of the exact channel positions.

2. Increased channel spacing: This increases the group velocity mismatch between channels. This has the drawback of increasing the overall system bandwidth, requiring the optical amplifiers to be flat over a wider bandwidth, and increases the penalty due to SRS.

3. Using higher wavelengths beyond 1560 nm with DSF: Even with DSF, a significant amount of chromatic dispersion is present in this range, which reduces the

effect of four-wave mixing. The newly developed L-band amplifiers can be used for long-distance transmission over DSF.

4. As with other nonlinearities, reducing transmitter power and the amplifier spacing will decrease the penalty.

5. If the wavelengths can be demultiplexed and multiplexed in the middle of the transmission path, we can introduce different delays for each wavelength. This randomizes the phase relationship between the different wavelengths. Effectively, the FWM powers introduced before and after this point are summed instead of the electric fields being added in phase, resulting in a smaller FWM penalty.

5.8.5   Self-/Cross-Phase Modulation

As we saw in Section 2.4, SPM and CPM also arise out of the intensity dependence of the refractive index. Fluctuations in optical power of the signal causes changes in the phase of the signal. This induces additional chirp, which in turn, leads to higher chromatic dispersion penalties. In practice, SPM can be a significant consideration in designing systems at 10 Gb/s and higher, and leads to a restriction that the maximum power per channel should not exceed a few milliwatts. CPM does not usually pose a problem in WDM systems unless the channel spacings are extremely tight (a few tens of gigahertz). In this section, we will study the system limitations imposed by SPM.

The combined effects of SPM-induced chirp and dispersion can be studied by numerically solving (E.15). For simplicity, we consider the following approximate expression for the width T_L of an initially unchirped Gaussian pulse after it has propagated a distance L:

$$\frac{T_L}{T_0} = \sqrt{1 + \sqrt{2}\frac{L_e}{L_{NL}}\frac{L}{L_D} + \left(1 + \frac{4}{3\sqrt{3}}\frac{L_e^2}{L_{NL}^2}\right)\frac{L^2}{L_D^2}}. \tag{5.27}$$

This expression is derived in [PAP86] starting from (E.15) and is also discussed in [Agr95]. Note the similarity of this expression to the broadening factor for chirped Gaussian pulses in (2.13); L_e/L_{NL} in (5.27) serves the role of the chirp factor in (2.13).

Consider a 10 Gb/s system operating over standard single-mode fiber at 1.55 μm. Since $\beta_2 < 0$ and the SPM-induced chirp is positive, from Figure 2.11 we expect that pulses will initially undergo compression and subsequently broaden. Since the SPM-induced chirp increases with the transmitted power, we expect both the extent of initial compression and the rate of subsequent broadening to increase with the

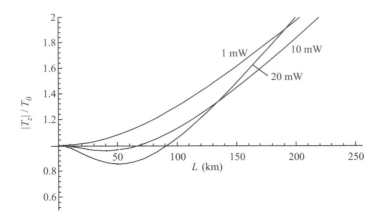

Figure 5.31 Evolution of pulse width as a function of the link length L for transmitted powers of 1 mW, 10 mW, and 20 mW, taking into account the chirp induced by SPM. A 10 Gb/s system operating over standard single-mode fiber at 1.55 μm with an initial pulse width of 50 ps is considered.

transmitted power. This is indeed the case, as can be seen from Figure 5.31, where we use (5.27) to plot the evolution of the pulse width as a function of the link length, taking into account the chirp induced by SPM. We consider an initially unchirped Gaussian pulse of width (half-width at $1/e$-intensity point) 50 ps, which is half the bit period. Three different transmitted powers, 1 mW, 10 mW, and 20 mW, are considered. As expected, for a transmit power of 20 mW, the pulse compresses more initially but subsequently broadens more rapidly so that the pulse width exceeds that of a system operating at 10 mW or even 1 mW. The optimal transmit power therefore depends on the link length and the amount of dispersion present. For standard single-mode fiber in the 1.55 μm band, the optimal power is limited to the 2–10 mW range for link lengths on the order of 100 km and is a real limit today for 10 Gb/s systems. We can use higher transmit powers to optimize other system parameters such as the signal-to-noise ratio (SNR) but at the cost of increasing the pulse broadening due to the combined effects of SPM and dispersion.

The system limits imposed by SPM can be calculated from (5.27) just as we did in Figure 5.31. We can derive an expression for the power penalty due to SPM, following the same approach as we did for chromatic dispersion. This is detailed in Problem 5.25. Since SPM can be beneficial due to the initial pulse compression it can cause, the SPM penalty can be negative. This occurs when the pulse at the end of the link is narrower due to the chirping caused by SPM than it would be in the presence of chromatic dispersion alone.

In amplified systems, as we saw in Section 5.5, two things happen: the effective length L_e is multiplied by the number of amplifier spans as the amplifier resets the power after each span, and in general, higher output powers are possible. Both of these serve to exacerbate the effects of nonlinearities.

In WDM systems, CPM aids the SPM-induced intensity dependence of the refractive index. Thus in WDM systems, these effects may become important even at lower power levels, particularly when dispersion-shifted fiber is used so that the dispersion-induced walk-off effects on CPM are minimized.

5.8.6 Role of Chromatic Dispersion Management

As we have seen, chromatic dispersion plays a key role in reducing the effects of nonlinearities, particularly four-wave mixing. However, chromatic dispersion by itself produces penalties due to pulse smearing, which leads to intersymbol interference. The important thing to note is that we can engineer systems with zero total chromatic dispersion but with chromatic dispersion present at all points along the link, as shown in Figure 5.20. This approach leads to reduced penalties due to nonlinearities, but the total chromatic dispersion is small so that we need not worry about dispersion-induced penalties.

5.9 Wavelength Stabilization

Luckily for us, it turns out that the wavelength drift due to temperature variations of some of the key components used in WDM systems is quite small. Typical multiplexers and demultiplexers made of silica/silicon have temperature coefficients of 0.01 nm/°C, whereas DFB lasers have a temperature coefficient of 0.1 nm/°C. Some of the other devices that we studied in Chapter 3 have even lower temperature coefficients.

The DFB laser source used in most systems is a key element that must be kept wavelength stabilized. In practice, it may be sufficient to maintain the temperature of the laser fairly constant to within ±0.1°C, which would stabilize the laser to within ±0.01 nm/°C. The laser comes packaged with a thermistor and a thermo-electric (TE) cooler. The temperature can be sensed by monitoring the resistance of the thermistor and can be kept constant by adjusting the drive current of the TE cooler. However, the laser wavelength can also change because of aging effects over a long period. Laser manufacturers usually specify this parameter, typically around ±0.1 nm. If this presents a problem, an external feedback loop may be required to stabilize the laser. A small portion of the laser output can be tapped off and sent to a wavelength discriminating element, such as an optical filter, called a *wavelength locker*. The output of the wavelength locker can be monitored to establish the laser wavelength, which can then be controlled by adjusting the laser temperature.

Depending on the temperature range needed (typically −10 to 60°C for equipment in telco central offices), it may be necessary to temperature-control the multiplexer/demultiplexer as well. For example, even if the multiplexer and demultiplexer are exactly aligned at, say, 25°C, the ambient temperature at the two ends of the link could be different by 70°C, assuming the given numbers. Assuming a temperature coefficient of 0.01 nm/°C, we would get a 0.7 nm difference between the center wavelengths of the multiplexer and demultiplexer, which is clearly intolerable if the interchannel spacing is only 0.8 nm (100 GHz). One problem with temperature control is that it reduces the reliability of the overall component because the TE cooler is often the least reliable component.

An additional factor to be considered is the dependence of laser wavelength on its drive current, typically between 100 MHz/mA and 1 GHz/mA. A laser is typically operated in one of two modes, constant output power or constant drive current, and the drive circuitry incorporates feedback to maintain these parameters at constant values. Keeping the drive current constant ensures that the laser wavelength does not shift because of current changes. However, as the laser ages, it will require more drive current to produce the same output power, so the output power may decrease with time. On the other hand, keeping the power constant may require the drive current to be increased as the laser ages, inducing a small wavelength shift. With typical channel spacings of 100 GHz or thereabouts, this is not a problem, but with tighter channel spacings, it may be desirable to operate the laser in constant current mode and tolerate the penalty (if any) due to the reduced output power.

5.10 Design of Soliton Systems

While much of our discussion in this chapter applies to the design of soliton systems as well, there are a few special considerations in the design of these systems, which we now briefly discuss.

We discussed the fundamentals of soliton propagation in Section 2.5. Soliton pulses balance the effects of chromatic dispersion and the nonlinear refractive index of the fiber, to preserve their shapes during propagation. In order for this balance to occur, the soliton pulses must not only have a specific shape but also a specific energy. Due to the inevitable fiber attenuation, the pulse energies are reduced, and thus the ideal soliton energy cannot be preserved. A theoretical solution to this problem is the use of dispersion-tapered fibers, where the chromatic dispersion of the fiber is varied suitably so that the balance between chromatic dispersion and nonlinearity is preserved in the face of fiber loss.

In practice, soliton propagation occurs reasonably well even in the case of systems with periodic amplification. However, the ASE added by these amplifiers causes a few detrimental effects. The first effect is that the ASE changes the energies of the pulses and causes bit errors. This effect is similar to the effect in NRZ systems although the quantitative details are somewhat different.

While solitons have a specific shape, they are resilient to changes in shape. For example, if a pulse with a slightly different energy is launched, it reshapes itself into a soliton component with the right shape and a nonsoliton component. When ASE is added, the effect is to change the pulse shape, but the solitons reshape themselves to the right shape.

A second effect of the ASE noise that is specific to soliton systems is that the ASE noise causes random changes to the center frequencies of the soliton pulses. For soliton propagation, per se, this would not be a problem because solitons can alter their frequency without affecting their shape and energy. (This is the key to their ability to propagate long distances without pulse spreading.) To see why this is the case, consider the soliton pulse shape given by

$$U(\xi, \tau) = e^{i\xi/2}\mathrm{sech}\tau. \tag{5.28}$$

Here, the distance ξ and time τ are measured in terms of the chromatic dispersion length of the fiber and the pulse width, respectively. The pulse

$$U(\xi, \tau + \Omega\xi)e^{i(\Omega t + \Omega^2\xi/2} \tag{5.29}$$

is also a soliton for any frequency shift Ω, and thus solitons can alter their frequency without affecting their shape and energy.

However, due to the chromatic dispersion of the fiber, changes in pulse frequencies are converted into changes in the pulse arrival times, that is, timing jitter. This jitter is called Gordon-Haus jitter, in honor of its discoverers, and is a significant problem for soliton communication systems.

A potential solution to this timing jitter problem is the addition of a bandpass filter whose center frequency is close to that of the launched soliton pulse. In the presence of these filters, the solitons change their center frequencies to match the passband of the filters. For this reason, these filters are called *guiding filters*. This has the effect of keeping the soliton pulse frequencies stable, and hence minimizing the timing jitter. This phenomenon is similar to the solitons reshaping themselves when their shape is perturbed by the added ASE.

The problem with the above solution is that the ASE noise accumulates within the passband of the chain of filters. As a result, the transmission length of the system, before the timing jitter becomes unacceptable, is only moderately improved compared to a system that does not use these filters. The solution to this problem is to change the center frequencies of the filters progressively along the link length. For example, if the filters are used every 20 km, each filter can be designed to have a center frequency that is 0.2 GHz higher than the previous one. Over a distance of 1000 km, this corresponds to a change of 10 GHz. The soliton pulses track the center frequencies of the filters, but the accumulation of ASE noise is lessened. This technique of using *sliding-frequency* guiding filters significantly minimizes timing jitter and makes transoceanic soliton transmission practical.

5.11 Design of Dispersion-Managed Soliton Systems

There are a few drawbacks associated with conventional soliton systems. First, soliton systems require fiber with a very low value of anomalous chromatic dispersion, typically $D < 0.2$ ps/nm-km. This rules out the possibility of using solitons over the existing fiber infrastructure, which primarily uses SMF or NZ-DSF, since these fibers have much higher values of dispersion. Second, solitons require amplifier spacings on the order of 20–25 km—much closer than what is typically used in practical WDM systems. Finally, cross-phase modulation (CPM) in WDM systems using conventional solitons causes soliton-soliton collisions, resulting in timing jitter. For these reasons, soliton systems have not been widely deployed.

The use of chirped RZ pulses (see Section 2.5.1), also called dispersion-managed (DM) solitons, overcomes all three problems associated with soliton transmission. First, these pulses can be used over a dispersion-managed fiber plant consisting of fiber spans with large local chromatic dispersion, but with opposite signs such that the total, or average, chromatic dispersion is small. This is typical of most fiber plants used today for 10 Gb/s transmission since they consist of SMF or NZ-DSF spans with dispersion compensation. Thus, no special fiber is required. Second, DM solitons require amplification only every 60–80 km, which is compatible with the amplifier spacings in today's WDM systems. Finally, the effect of CPM is vastly reduced because of the large local chromatic dispersion and thus there is no timing jitter problem. For the same reason, the Gordon-Haus jitter is also reduced, and the sliding-frequency guiding filters used in conventional soliton systems are not required.

In a dispersion-managed system, the spans between amplifiers consist of fibers with alternating chromatic dispersions, as shown in Figure 5.32. Each fiber could have a fairly high chromatic dispersion, but the total chromatic dispersion is small. For example, each span in a dispersion-managed system could consist of a 50 km anomalous chromatic dispersion segment with a chromatic dispersion of 17 ps/nm-km, followed by a 30 km normal chromatic dispersion segment with a chromatic dispersion of -25 ps/nm-km. The total chromatic dispersion over the span is $50 \times 17 - 30 \times 25 = 100$ ps/km. The average chromatic dispersion is

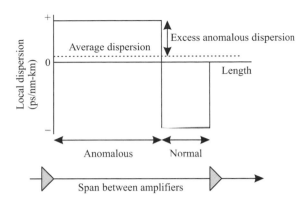

Figure 5.32 A typical dispersion-managed span consisting of a segment of fiber with anomalous chromatic dispersion followed by a segment with normal chromatic dispersion.

100/80 = 1.25 ps/nm-km, which is anomalous. A dispersion-managed system could have an average span dispersion that is normal or anomalous. In the same example, if the normal fiber had a chromatic dispersion of −30 ps/nm-km, the average span dispersion would have been −50/80 = −0.625 ps/nm-km, which is normal.

When NRZ pulses are used, the average chromatic dispersion can be anomalous or normal, without having a significant impact on system performance. However, in a DM soliton system, the average chromatic dispersion must be designed to be anomalous in order to maintain the shape of the DM solitons. This is similar to the case of conventional solitons, but with the crucial difference that the chromatic dispersion need not be uniformly low and anomalous.

An important aspect of the design of DM soliton systems is the choice of the peak transmit power and the average chromatic dispersion. Both should lie within a certain range in order to achieve low BER operation. This range can be plotted as a contour in a plot of peak transmit power versus average chromatic dispersion, as shown in Figure 5.33. In this figure, we show a typical contour for achieving a BER of 10^{-12} (or $\gamma = 7$) in a 5160 km system with 80 km spans. For values of the transmit power and average chromatic dispersion lying within this contour, the desired BER is achieved or exceeded. In the same plot, the contour for a 2580 km NRZ system with 80 km spans is also shown. In both NRZ and DM soliton systems, the allowed transmit power has both a lower bound, determined by OSNR requirements, and an upper bound determined by fiber nonlinear effects. From Figure 5.33, note that not only is the DM soliton system capable of achieving regeneration-free transmission for twice the distance as the NRZ system, it is also able to tolerate a much wider range of variation in the transmit power and the average chromatic dispersion.

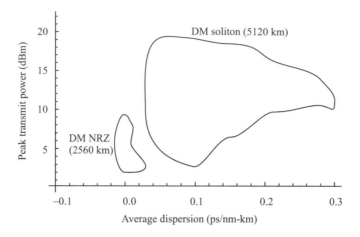

Figure 5.33   Typical contours of constant BER for a DM soliton and an NRZ modulated 10 Gb/s system. (After [Nak00].)

Another important factor influencing the performance of DM soliton systems is the peak-to-peak variation of the chromatic dispersion from the average over the span. In Figure 5.33, the peak-to-peak variation was chosen to be small (1.6 ps/nm-km), and thus both the anomalous and normal segments had very low chromatic dispersion. However, the achievable regeneration-free transmission distance is quite sensitive to the excess chromatic dispersion, relative to the average chromatic dispersion on the span, because of the delicate balancing of the chromatic dispersion against the nonlinearities in the fiber that occurs for soliton-like pulses. Figure 5.34 plots the maximum distance between regenerators as a function of the excess anomalous chromatic dispersion on the span, while maintaining a fixed value of the average chromatic dispersion, for DM solitons as well as NRZ and (unchirped) RZ systems. The excess anomalous chromatic dispersion is the excess of the chromatic dispersion in the anomalous segment over and above the average chromatic dispersion on the link, as indicated in Figure 5.32. Here we assume that the 80 km spans consist of a 50 km anomalous segment and a 30 km normal segment. The NRZ and RZ systems are assumed to be fully dispersion compensated so that the average chromatic dispersion on these spans is zero. For the DM soliton system, the average chromatic dispersion is 0.1 ps/nm-km, which is slightly anomalous. Since the average chromatic dispersion is zero for the NRZ and RZ systems, and quite small in the DM soliton case, the abscissa in Figure 5.34 is effectively the chromatic dispersion of the anomalous segment.

Note from Figure 5.34 that the NRZ system is not sensitive to the excess local chromatic dispersion. This is because the NRZ system essentially operates in the

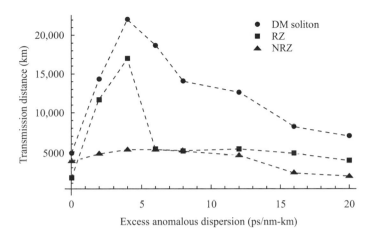

Figure 5.34   Performance of 10 Gb/s DM soliton systems compared with NRZ and (unchirped) RZ modulated systems. (After [Nak00].)

linear regime. Note also that the DM soliton system can achieve considerably higher transmission distances than NRZ and RZ systems for all values of the excess anomalous chromatic dispersion. Thus, DM soliton systems are superior to these systems over virtually all kinds of dispersion-managed fiber spans.

We saw in Section 5.7.4 that (unchirped) RZ systems have a smaller PMD penalty than NRZ systems. Chirped RZ, or DM soliton systems, have an even smaller PMD penalty and thus are more suitable for transmission rates of 40 Gb/s and above, from the PMD perspective as well.

5.12 Overall Design Considerations

We have seen that there is an interplay of many different effects that influence the system design parameters. We will summarize some of these effects in this section. In addition, two key issues in this regard, (1) the trade-off between higher bit rates per channel versus more channels, and (2) whether to use bidirectional or unidirectional systems, will be discussed in Chapter 13.

5.12.1   Fiber Type

Among the many issues facing system designers is what type of fiber should be deployed in new installations. This very much depends on the type of system that is going to be deployed. For single-channel systems operating at very high bit rates (10 Gb/s and above) over long distances, DSF is the best choice. However, DSF makes it much harder to use WDM for upgrading the link capacity in the future, primarily due to four-wave mixing, and thus is not a practical choice for most links. For WDM systems, the choice of fiber type depends on the distance and bit rate per channel. DSF is clearly a bad choice. If the system is not chromatic dispersion limited, then standard single-mode fiber is the best choice because such a system is least susceptible to degradation from nonlinearities. As the distance and bit rate increase in future upgrades, the system will eventually become chromatic dispersion limited (for example, over 600 km at 2.5 Gb/s), and chromatic dispersion compensation must be incorporated into the system. For WDM systems operating at high bit rates over long distances, NZ-DSF provides a good alternative to using standard single-mode fiber with dispersion compensation.

If the residual dispersion slope after chromatic dispersion compensation is the main problem, you can use reduced slope fiber, such as Lucent's TrueWave RS fiber. On the other hand, if nonlinearities are the significant problem, large effective area fiber, such as Corning's LEAF, can be used. For terrestrial systems, NZ-DSF fiber with positive dispersion in the 1.55 μm band can be used in order to be able to upgrade the system to use the L-band wavelengths. For submarine systems, NZ-DSF with negative dispersion fiber can be used in order to avoid modulation instability.

Some sample transmission numbers that have been reported to date are as follows. Using carefully dispersion-managed fiber spans, transmission of 120 channels each running at 20 Gb/s over a distance of 6200 km has been demonstrated [VPM01]. This experiment used only C-band EDFAs. Using both the C-band and the L-band, and combining distributed Raman amplification with EDFAs, transmission of 77 42.7 Gb/s channels over 1200 km has been demonstrated [Zhu01]. Over short distances, about 100 km, and using all three bands (S-band, C-band, and L-band), transmission of over 250 40 Gb/s channels has been demonstrated [Fuk01, Big01].

5.12.2 Transmit Power and Amplifier Spacing

The upper limit on the transmitted power per channel P is determined by the saturation power of the optical amplifiers, the effect of nonlinearities, and safety considerations. From a cost point of view, we would like to maximize the distance l between amplifier stages, so as to minimize the number of amplifiers. The transmitted power per channel, P, and the total link length L, along with the amplifier noise figure and receiver sensitivity, determines the maximum value of l possible. In addition, as l

increases, the penalty due to nonlinearities also increases, which by itself may play a role in limiting the value of l.

The amplifier spacing in existing systems must also conform to the repeater hut spacing, typically about 80 km, though this is not an issue for new installations.

5.12.3 Chromatic Dispersion Compensation

In systems that have to operate over standard single-mode fiber, chromatic dispersion must be compensated frequently along the link, since the total chromatic dispersion usually cannot be allowed to accumulate beyond a few thousand ps/nm. Systems employing NZ-DSF can span longer lengths before chromatic dispersion compensation is required. In addition to chromatic dispersion compensation, chromatic dispersion slope also needs to be compensated. The ultimate limits of link lengths before the wavelengths need to be demultiplexed and compensated individually is set by the variation in dispersion slope since dispersion slope cannot usually be compensated exactly for all the channels. The use of reduced slope fiber increases this length. By careful span engineering using a large effective area fiber followed by a carefully tailored dispersion compensating fiber, to minimize the dispersion slope, transmission of 120 WDM channels at 20 Gb/s each over 6200 km has been demonstrated [Cai01]. Using similar techniques, transmission of 101 WDM channels at 10 Gb/s each over 9000 km has also been demonstrated [Bak01].

5.12.4 Modulation

Most systems in use today employ NRZ modulation. However, chirped RZ modulation is being considered for ultra-long-haul systems, operating at 10 Gb/s and above. The main motivation for chirped RZ systems is that by the appropriate combination of chirping and chromatic dispersion compensation, such systems achieve very long, regeneration-free transmission. The penalties due to PMD are also lower for RZ modulation than they are for NRZ modulation.

Within NRZ systems, direct modulation is less expensive but leads to chirping, which in turn increases the chromatic dispersion penalties. External modulation is required in chromatic dispersion–limited systems, particularly 10 Gb/s systems. Today, most long-haul systems use external modulation. Metro WDM systems usually employ direct modulation up to bit rates of 2.5 Gb/s to keep costs low, and try to achieve distances of 100–200 km before reaching the chromatic dispersion limit.

Prechirping can be used to increase the link lengths by taking advantage of the pulse compression effects that occur when positively (negatively) chirped pulses are used in positive (negative) dispersion fiber.

5.12.5 Nonlinearities

Nonlinear effects can be minimized by using lower transmit powers. The use of a large effective area fiber allows the use of higher transmit powers, and hence longer links, in the presence of nonlinearities. The trade-off is the higher dispersion slope of these fibers.

Some nonlinear effects can actually be beneficial. For example, SPM can sometimes lead to longer link lengths since the positive chirping due to SPM over positive dispersion fiber leads to pulse compression.

5.12.6 Interchannel Spacing and Number of Wavelengths

Another design choice is the interchannel spacing. On the one hand, we would like to make the spacing as large as possible, since it makes it easier to multiplex and demultiplex the channels and relaxes the requirements on component wavelength stability. Larger interchannel spacing also reduces the four-wave mixing penalty if that is an issue (for example, in systems with dispersion-shifted fiber). It also allows future upgrades to higher bit rates per channel, which may not be feasible with very tight channel spacings. For example, today's systems operate with 100 GHz channel spacing with bit rates per channel up to 10 Gb/s. Such a system can be upgraded by introducing additional wavelengths between two successive wavelengths leading to 50 GHz channel spacing. Alternatively, the channel spacing can be maintained at 100 GHz and the bit rate per channel increased to 40 Gb/s. If the initial channel spacing is reduced to 50 GHz, it becomes much harder to upgrade the system to operate the channels at 40 Gb/s.

On the other hand, we would like to have as many channels as possible within the limited amplifier gain bandwidth, which argues for having a channel spacing as tight as possible. For a given number of channels, it is easier to flatten the amplifier gain profile over a smaller total bandwidth. Moreover, the smaller the total system bandwidth, the lesser the penalty due to stimulated Raman scattering (although this is not a limiting factor unless the number of channels is fairly large).

Other factors also limit the number of wavelengths that can be supported in the system. The total amplifier output power that can be obtained is limited typically to 20–25 dBm, and this power must be shared among all the channels in the system. So as the number of wavelengths increases, the power per channel decreases, and this limits the total system span. Another limiting factor is the stability and wavelength selectivity of the multiplexers and demultiplexers.

Two other techniques are worthy of mention in the context of designing high channel count systems. The first is the interleaving of wavelengths transmitted in the two directions. Thus, if λ_i^E and λ_i^W denote the wavelengths to be transmitted in the east and west directions, we transmit $\lambda_1^E, \lambda_2^W, \lambda_3^E, \ldots$ on one fiber, and $\lambda_1^W, \lambda_2^E, \lambda_3^W, \ldots$ on the other fiber. This technique effectively doubles the spacing between the wavelengths as far as the nonlinear interactions are concerned.

The second technique is similar but is applicable when both the C-band and L-band are used. In this case, the nonlinear interactions between the signals in the two bands can be avoided by transmitting the signals in one band in one direction over the fiber, and the signals in the other band in the other direction. If this is done, the nonlinear interactions effectively "see" only one of the bands.

Taking all this into consideration, 160-channel systems operating at 10 Gb/s per channel, with 50 GHz spacings, have been designed and are commercially available today. Even larger numbers of channels can be obtained by reducing the channel spacing and improving the stability and selectivity of the wavelength multiplexers and demultiplexers.

5.12.7   *All-Optical Networks*

All-optical networks consist of optical fiber links between nodes with all-optical switching and routing of signals at the nodes, without electronic regeneration. The various aspects of system design that we studied in this chapter apply to point-to-point links as well as all-optical networks, and we have attempted to consider several factors that affect networks more than point-to-point links. Designing networks is significantly harder than designing point-to-point links for the following reasons:

- The reach required for all-optical networks is considerably more than the reach required for point-to-point links, since lightpaths must traverse multiple links. In addition, loss, chromatic dispersion, and nonlinearities do not get reset at each node.

- The network is more susceptible to crosstalk, which is accumulated at each node along the path.

- Misalignment of multiplexers and demultiplexers along the path is more of a problem in networks than in links.

- Because of bandwidth narrowing of cascaded multiplexers and demultiplexers, the requirements on laser wavelength stability and accuracy are much higher than in point-to-point links.

- The system designer must deal with the variation of signal powers and signal-to-noise ratios among different lightpaths traveling through different numbers of nodes and having different path lengths. This can make system design particularly difficult. A common approach used to solve this problem is to equalize the powers of each channel at each node individually. Thus, at each node the powers in all the channels are set to a common value. This ensures that all lightpaths reach their receivers with the same power, regardless of their origin or their path through the network.

- Rapid dynamic equalization of the amplifier gains will be needed to compensate for fluctuations in optical power as lightpaths are taken down or set up, or in the event of failures.

5.12.8 Wavelength Planning

The International Telecommunications Union (ITU) has been active in trying to standardize a set of wavelengths for use in WDM networks. This is necessary to ensure eventual interoperability between systems from different vendors (although this is very far away). An important reason for setting these standards is to allow component vendors to manufacture to a fixed standard, which allows volume cost reductions, as opposed to producing custom designs for different system vendors.

The first decision to be made is whether to standardize channels at equal wavelength spacing or at equal frequency spacing. At $\lambda = 1550$ nm, $c = 3 \times 10^8$ m/s, a 1 nm wavelength spacing corresponds to approximately 120 GHz of frequency spacing. Equal frequency spacing results in somewhat unequal wavelength spacing. Certain components used in the network, such as AWGs and Mach-Zehnder filters, naturally accept channels at equal frequency spacings, whereas other components, including other forms of gratings, accept channels more naturally at equal wavelength spacings. There is no major technical reason to favor one or the other. The ITU has picked equal frequency spacing for their standard, and this is specified in ITU G.692. The channels are to be placed in a 50 GHz grid (0.4 nm wavelength spacing) with a nominal center frequency of 193.1 THz (1552.52 nm) in the middle of the 1.55 μm fiber and EDFA passband, as shown in Figure 5.35. For systems with channel spacings of 100 GHz or more, the frequencies are to be placed on a 100 GHz grid, with the same reference frequency of 193.1 THz. This latter grid was the first standard, before the 50 GHz grid was introduced.

The choice of the 50 GHz frequency spacing is based on what is feasible with today's technology in terms of mux/demux resolutions, frequency stability of lasers and mux/demuxes, and so on. As the technology improves, and systems with more channels become practical, the grid spacing may have to be reduced. Moreover, in systems that must operate over dispersion-shifted fiber, it may be desirable to have unequal channel spacings to alleviate the effects of four-wave mixing. This will also require a finer grid spacing since all these unequal spacings must be accommodated

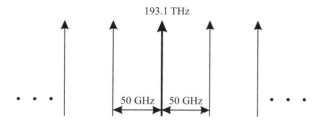

Figure 5.35 Wavelength grid selected by the ITU.

within the same total bandwidth, which in turn necessitates a finer grid. For example, a system using the channels 193.1, 193.2, 193.3, and 193.4 THz is spaced on a 100 GHz grid and the channel spacings are all equal to 100 GHz. If the channel spacings are made unequal and are, say, 50, 100, and 150 GHz, we can use the channels 193.1, 193.15, 193.25, and 193.4 THz. This system occupies the same bandwidth from 193.1 to 193.4 THz as the equally spaced system, but the channels are on a 50 GHz grid instead of a 100 GHz grid. (If we do not place the channels on this finer 50 GHz grid but still use a 100 GHz grid, we will end up using more total bandwidth to achieve the unequal channel spacing; see Problem 5.26.) In fact, to tackle the unequal spacing requirement due to four-wave mixing on dispersion-shifted fibers, ITU allows such systems to have some wavelengths that are on a 25 GHz grid; see ITU G.692 for details.

That being said, a much more difficult decision is to pick a standard set of wavelengths for use in 4-, 8-, 16-, and 32-wavelength systems to ensure interoperability. This is because different manufacturers have different optimized channel configurations and different upgrade plans to go from a system with a small number of channels to a system with a larger number of channels. As of this writing, ITU is standardizing (ITU G.959) the set of 16 wavelengths starting with 192.1 THz, and spaced 200 GHz apart, for multichannel interfaces between WDM equipment.

It is not enough to specify the nominal center frequencies of the channels alone. A maximum deviation must also be specified because of manufacturing tolerances and aging over the system's lifetime. The deviation should not be too large; otherwise, we would get significant penalties due to crosstalk, additional loss, chirp, and the like. The deviation is a function of the interchannel spacing, Δf. For $\Delta f \geq 200$ GHz, the ITU has specified that the deviation should be no more than $\pm \Delta f/5$ GHz. For $\Delta f = 50$ GHz and $\Delta f = 100$ GHz, the frequency deviation values have not been standardized by the ITU at the time of writing.

5.12.9 Transparency

Among the advantages touted for WDM systems is the fact that they are transparent to bit rate, protocol, and modulation formats. It is true to a large extent that a wavelength can carry arbitrary data protocols. Providing transparency to bit rate and modulation formats is much more difficult. For instance, analog transmission requires much higher signal-to-noise ratios and linearity in the system than digital transmission and is much more susceptible to impairments. A WDM system can be designed to operate at a maximum bit rate per channel and can support all bit rates below that maximum. We cannot assume that the system is transparent to increases in the maximum bit rate. The maximum bit rate affects the choice of amplifier spacings, filter bandwidths, and dispersion management, among other parameters. Thus the system must be designed up front to support the maximum possible bit rate.

Summary

This chapter was devoted to studying the effects of various impairments on the design of the new generation of WDM and high-speed TDM transmission systems and networks. Although impairments due to amplifier cascades, dispersion, nonlinearities, and crosstalk may not be significant in lower-capacity systems, they play significant roles in the new generation of systems, particularly in networks, as opposed to point-to-point links. We learned how to compute the penalty due to each impairment and budget for the penalty in the overall system design. We also studied how to reduce the penalty due to each impairment. Transmission system design requires careful attention to each impairment because requirements on penalties usually translate into specifications on the components that the system is built out of, which in turn translate to system cost. Design considerations for transmission systems are summarized in the last section of this chapter.

Further Reading

We recommend the recent books by Kaminow and Koch [KK97a, KK97b] for an in-depth coverage of the advanced aspects of lightwave system design. For authoritative treatments of EDFAs, see [BOS99, Des94]. Gain equalization of amplifiers is an important problem, and several approaches have been proposed [Des94]. Amplifier cascades are discussed in several papers; see, for example, [Ols89, RL93, MM98].

Amplifier power transients are discussed in [Zys96, LZNA98]. The optical feedback loop for automatic gain control (AGC) illustrated in Figure 5.8 was first described in [Zir91].

Crosstalk is analyzed extensively in several papers. Intrachannel crosstalk is considered in [ZCC+96, GEE94, TOT96]. Interchannel crosstalk is analyzed in [ZCC+96, HH90]. Dilation in switches is discussed in [Jac96, PN87].

Chromatic dispersion and intermodal dispersion are treated at length in the aforementioned books. The different types of single-mode fiber have been standardized; see ITU G.652, ITU G.653, and ITU G.655. Polarization-mode dispersion is studied in [PTCF91, CDdM90, BA94, ZO94]; see also [KK97a, Chapter 6]. For recent work on PMD compensation, see [Kar01, PL01]. PMD compensation is analyzed in [SKA00] and the effects of PMD on NRZ and RZ pulses are compared in [SKA01].

Good surveys of fiber nonlinearities appear in [Chr90, Agr95, Buc95, SNIA90]. See also [TCF+95, FTC95, SBW87, Chr84, OSYZ95].

The standards bodies have given a lot of thought in defining the system parameters for WDM systems. The 50 GHz wavelength grid is specified in ITU G.692. It is instructive to read this and other related standards: ITU G.691, ITU G.681, ITU G.692, Telcordia GR-253, Telcordia GR-192, and Telcordia GR 2918, which provide values for most of the system parameters used in this chapter.

For a discussion of the design issues in achieving 40 Gb/s WDM transmission, see [Nel01]. The design of transoceanic WDM systems is discussed in [Gol00]. Our treatment of the design of DM soliton systems is based on [Nak00].

Problems

5.1 In an experiment designed to measure the attenuation coefficient α of optical fiber, the output power from an optical source is coupled onto a length of the fiber and measured at the other end. If a 10 km–long spool of fiber is used, the received optical power is −20 dBm. Under identical conditions but with a 20 km–long spool of fiber (instead of the 10 km–long spool), the received optical power is −23 dBm. What is the value of α (in dB/km)? If the source-fiber coupling loss is 3 dB, the fiber-detector coupling loss is 1 dB, and there are no other losses, what is the output power of the source (expressed in mW)?

5.2 The following problems relate to simple link designs. Assume that the bit rate on the link is 1 Gb/s, the dispersion at 1.55 μm is 17 ps/nm-km, and the attenuation is 0.25 dB/km, and at 1.3 μm, the dispersion is 0 and the attenuation is 0.5 dB/km. (Neglect all losses except the attenuation loss in the fiber.) Assume that NRZ modulation is used.

(a) You have a transmitter that operates at a wavelength of 1.55 μm, has a spectral width of 1 nm, and an output power of 0.5 mW. The receiver requires -30 dBm of input power in order to achieve the desired bit error rate. What is the length of the longest link that you can build?

(b) You have another transmitter that operates at a wavelength of 1.3 μm, has a spectral width of 2 nm, and an output power of 1 mW. Assume the same receiver as before. What is the length of the longest link that you can build?

(c) You have the same 1.3 μm transmitter as before, and you must achieve an SNR of 30 dB using an APD receiver with a responsivity of 8 A/W, a gain of 10, an excess noise factor of 5 dB, negligible dark current, a load resistance of 50 Ω, and an amplifier noise figure of 3 dB. Assume that a receiver bandwidth of $B/2$ Hz is sufficient to support a bit rate of B b/s. What is the length of the longest link you can build?

(d) Using the same 1.3 μm transmitter as before, you must achieve an SNR of 20 dB using a *pin* receiver with a responsivity of 0.8 A/W, a load resistance of 300 Ω, and an amplifier noise figure of 5 dB. Assume that a receiver bandwidth of $B/2$ Hz is sufficient to support a bit rate of B b/s. What is the length of the longest link you can build?

5.3 Compute the dispersion-limited transmission distance for links with standard single-mode fiber at 1550 nm as a function of the bit rate (100 Mb/s, 1 Gb/s, and 10 Gb/s) for the following transmitters: (a) a Fabry-Perot laser with a spectral width of 10 nm, (b) a directly modulated DFB laser with a spectral width of 0.1 nm, and (c) an externally modulated DFB laser with a spectral width of 0.01 nm. Assume that the modulation bandwidth equals the bit rate and the dispersion penalty is 2 dB. Assume that NRZ modulation is used.

5.4 Repeat Problem 5.3 for NZ-DSF assuming a dispersion parameter of 5 ps/nm-km.

5.5 Consider a length L of step-index multimode fiber having a core diameter of 50 μm and a cladding diameter of 200 μm. The refractive indices of the core and cladding are 1.50 and 1.49, respectively. A fixed-wavelength, 1310 nm DFB laser (operating at 0 dBm) is used at one end of the fiber to serve as a 155.52 Mb/s transmitter source. At the far end, a photodetector is used as a receiver. Assume that NRZ modulation is used.

(a) Draw and label a diagram that illustrates the above configuration.

(b) What would be the corrugation period of the DFB laser at this wavelength?

(c) Compute the numerical aperture for this fiber.

(d) What would be the maximum acceptable fiber length when operating at this bit rate?

(e) Assuming an attenuation of 0.40 dB/km, what would be the output power (in dBm) at the receive end of the fiber?

(f) Assuming a perfectly efficient photodetector, what would be the resulting photocurrent?

(g) If we instead used single-mode fiber for this application, what would be the new requirement on its core diameter?

Note that this problem requires you to understand the material in Chapters 2, 3, and 4 as well.

5.6 Consider a passive WDM link of length L, consisting of single-mode fiber into which five wavelengths are launched through an optical combiner such that the aggregate launch power at its output is 5 mW. These five wavelengths are *centered* on the 193.1 THz ITU grid, with uniform 100 GHz interchannel spacing. The transmitters all use directly modulated DFB lasers with a spectral width of 0.1 nm. Each channel is transporting a SONET OC-48 (2.5 Gb/s) signal. At the end of this link, the channels are optically demultiplexed and are each received by a direct detection *pin* receiver. For this problem, neglect all losses and crosstalk associated with the WDM mux and demux. Assume that NRZ modulation is used.

(a) Draw and label a diagram illustrating this configuration.

(b) Calculate the wavelengths (in nm, to two decimal places) associated with these five channels.

(c) Calculate the average launch power per channel at the output of the WDM combiner.

(d) Assuming $\alpha_{dB} = 0.25$ dB/km, $D = 17$ ps/nm-km, and $D_{PMD} = 0.5$ ps/$\sqrt{km}$, calculate the worst-case dispersion, PMD, and loss limits for this link.

(e) What is the maximum value of L meeting all of these requirements?

5.7 Consider a point-to-point link connecting two nodes separated by 60 km. This link was constructed with standard single-mode fiber, and a 2.5 Gb/s system is deployed over the link. The transmitter uses a directly modulated 1310 nm DFB laser. The receiver uses perfectly efficient *pin* photodiodes, and we will assume, for this problem, that they can be modeled as ideal receivers. The bit error rate requirement for this system is 10^{-12}. Assume $\alpha_{dB} = 0.4$ dB/km and that NRZ modulation is used.

(a) Draw and label a diagram illustrating this configuration.

(b) Is this system loss limited or dispersion limited? Briefly explain your reasoning.

(c) What is the required receiver sensitivity (in mW and dBm)?

(d) What would be the resulting photocurrent?

(e) What would be the required launch power (in dBm)?

5.8 The link of Problem 5.7 is at full capacity, and we must design a solution that will enable capacity expansion and accommodate further growth. After considering the options, we determine that the most cost-effective solution is to add a 1550 nm point-to-point system over the existing set of fibers, thereby realizing a two-wavelength (1310 nm/1550 nm) passive WDM configuration. Assume that 3 dB couplers are used to combine the two signals just after the transmitters and separate the two signals just before the receivers. The next step is to determine what bit rate can be supported by this 1550 nm system. Assume that the 1550 nm transmitter uses a directly modulated DFB laser (with spectral width of 0.1 nm). At 1550 nm, assume $\alpha_{\text{dB}} = 0.25$ dB/km, $D = 17$ ps/nm-km, and $D_{\text{PMD}} = 1$ ps/$\sqrt{\text{km}}$.

(a) Draw and label a diagram illustrating this new configuration.

(b) What is the launch power now required for the original 2.5 Gb/s 1310 nm system to maintain the same level of receiver performance?

(c) If we assume an ideal receiver with the same 10^{-12} bit error rate performance for the 1550 nm system, determine the associated receiver sensitivities for both 2.5 Gb/s and 10 Gb/s signals.

(d) Calculate bit rate limits based on loss, dispersion, and PMD for the new system.

(e) Can 10 Gb/s be suitably transported by this new system? Briefly explain your reasoning.

(f) For the 2.5 Gb/s and 10 Gb/s (if it is possible) line rates, calculate the required launch power to successfully transport the signal.

5.9 Derive equation (5.4).

5.10 Show that the extinction ratio penalty in amplified systems limited by signal-spontaneous beat noise and spontaneous-spontaneous beat noise is

$$\text{PP} = -10 \log \left(\frac{r-1}{r+1} \frac{\sqrt{r+1}}{\sqrt{r}+1} \right).$$

Assume that other noise terms are negligible.

5.11 Consider the amplifier chain discussed in Section 5.5.3. Using equations (5.6) and (5.7), compute the steady-state values of $\overline{P}_{\text{out}}$ and $\overline{G}$ in a long chain of amplifiers. Assume $G_{\text{max}} = 35$ dB, $l = 120$ km, $\alpha = 0.25$ dB/km, $n_{\text{sp}} = 2$, $P^{\text{sat}} = 10$ mW, and $B_o = 50$ GHz. How do these values compare against the unsaturated gain G_{max} and the output saturation power of the amplifier $P_{\text{out}}^{\text{sat}}$? Plot the evolution of the signal power and optical signal-to-noise ratio as a function of distance along the link.

5.12 Derive equations (5.11), (5.12), (5.13), and (5.14) when there are N interfering signals rather than just one.

5.13 Why is equation (5.24) an approximation? Derive a precise form of this equation.

5.14 Consider the WDM link shown in Figure 5.1. Each multiplexer and demultiplexer introduces crosstalk from adjacent channels that is C dB below the desired channel.

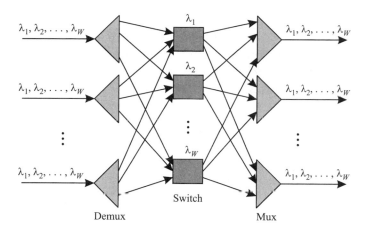

Figure 5.36   A node in a WDM network for Problems 5.16–5.19.

(a) Compute the crosstalk at the output when N such stages are cascaded.

(b) What must C be so that the overall crosstalk penalty after five stages is less than 1 dB?

5.15   Consider a WDM system with W channels, each with average power P and extinction ratio $P_1/P_0 = r$. Derive the interchannel crosstalk power penalty in (5.13) for this system compared to a system with ideal extinction and no crosstalk. What should the crosstalk level be for a maximum 1 dB penalty if the extinction ratio is 10 dB?

5.16   Consider the WDM network node shown in Figure 5.36. Assume the node has two inputs and two outputs. The multiplexers/demultiplexers are ideal (no crosstalk), but each switch has a crosstalk level C dB below the desired channel. Assume that in the worst case, crosstalk in each stage adds coherently to the signal.

(a) Compute the crosstalk level after N nodes.

(b) What must C be so that the overall crosstalk penalty after five nodes is less than 1 dB?

5.17   Consider the WDM network node shown in Figure 5.36. Assume the node has two inputs and two outputs. The mux/demuxes have adjacent channel crosstalk suppressions of -25 dB, and crosstalk from other channels is negligible. The switches have a crosstalk specification of -40 dB. How many nodes can be cascaded in a network without incurring more than a 1 dB penalty due to crosstalk? Consider only intrachannel crosstalk from the switches and the multiplexers/demultiplexers.

5.18   Consider a WDM system with N nodes, each node being the one shown in Figure 5.36. The center wavelength λ'_c for each channel in a mux/demux has an accuracy of $\pm\Delta\lambda$ nm around the nominal center wavelength λ_c. Assume a Gaussian passband shape for each channel in a mux; that is, the ratio of output power to input power, called the transmittance, is given by

$$T_R(\lambda) = e^{-\frac{(\lambda-\lambda'_c)^2}{2\sigma^2}},$$

where σ is a measure of the channel bandwidth and λ'_c is the center wavelength. This passband shape is typical for an arrayed waveguide grating.

   (a) Plot the worst-case and best-case peak transmittance in decibels as a function of the number of nodes N for $\sigma = 0.2$ nm, $\Delta\lambda = 0.05$ nm. Assume that the laser is centered exactly at λ_c.

   (b) What should $\Delta\lambda$ be if we must have a worst-case transmittance of 3 dB after 10 nodes?

5.19   Consider a system with the same parameters as in Problem 5.18. Suppose the WDM channels are spaced 0.8 nm apart. Consider only crosstalk from the two adjacent channels. Compute the interchannel crosstalk power relative to the signal power in decibels, as a function of N, assuming all channels are at equal power and exactly centered. Compute the crosstalk also for the case where the desired channel is exactly centered at λ_i, but the adjacent channels are centered at $\lambda_{i-1} + \Delta\lambda$ and $\lambda_{i+1} - \Delta\lambda$.

5.20   Consider the simple add/drop element shown in Figure 3.14(b). Suppose we use another circulator instead of the coupler shown in the figure to add the wavelength. This eliminates the loss due to the coupler. Let the input power on the wavelength to be dropped be -30 dBm and the transmitted power on the added wavelength be 0 dBm. Suppose the grating has a reflectivity of 99%. Compute the intrachannel crosstalk power arising from (a) leakage of the added wavelength into the dropped wavelength, and (b) leakage of the dropped wavelength into the added wavelength. Assume that each circulator has a loss of 1 dB. Will the element work?

5.21   Show that the optimum choice of the pulse width of an unchirped Gaussian pulse (with narrow spectral width) that minimizes the pulse-broadening effects of chromatic dispersion over a fiber of length L is

$$T_0^{\text{opt}} = \sqrt{\beta_2 L}.$$

5.22   If $0 \leq \epsilon \leq 1$ is the power-splitting ratio between the two polarization components, the random power penalty in decibels due to PMD is related to the random differential time delay as

$$PP(dB) = \alpha \frac{\Delta \tau^2}{T^2} \epsilon (1 - \epsilon),$$

where T is the bit period and α is a constant depending on the pulse shape and takes values in the range 12–25 for commonly studied pulse shapes [KK97a, Chapter 6]. Note that we have already taken logarithms in the above equation. Thus the random variable PP is a function of the random variables $\Delta \tau$ and ϵ. Assuming a Maxwellian distribution for $\Delta \tau$ with mean $\langle \Delta \tau \rangle$ and a uniform distribution for ϵ, show that PP has an exponential distribution. What is the mean value of PP? What is the probability that PP $\geq$ 1 dB?

5.23 Neglecting the depletion of the pump wave, solve (2.14) and (2.15) to obtain the evolution of the SBS pump and Stokes waves.

5.24 Compute the SBS threshold power for the following systems: (a) a single-channel system using a Fabry-Perot laser with 10 lines, each line having a modulated line width of 1 GHz, (b) a multichannel system with a DFB laser having a modulated line width of 1 GHz, and (c) same as (b) except that the line width is 10 GHz.

5.25 Consider (5.27) as expressing T_L, the pulse width after a distance L, in terms of the initial pulse width T_0.
      (a) As in the case of chromatic dispersion, there is an optimum initial pulse width (for a given link length L). Find an expression for this optimum initial pulse width.
      (b) Assuming a pulse with this optimum width is used, find the maximum link length for a power penalty of 1 dB. Note that this power penalty is due to both SPM and chromatic dispersion.
      (c) Assume that a pulse of the same initial width is used but that the link has no SPM but only chromatic dispersion. Using (2.13), calculate the pulse width at the end of the link and hence the penalty due to chromatic dispersion.
      The remainder of the 1 dB penalty is due to SPM. Note that the SPM penalty can be negative for some combinations of link, dispersion, and nonlinear lengths. This occurs when the initial pulse compression due to the chirping caused by SPM results in a narrower pulse at the end of the link, compared to the case when SPM is absent and only chromatic dispersion is present.

5.26 You are required to design a four-wavelength transmission system operating over dispersion-shifted fiber. The four wavelengths are to be placed in a band from 193.1 THz to 194.1 THz. The possible slots are spaced 100 GHz apart in this band. Pick the four wavelengths carefully so that no four-wave mixing component falls on any of the chosen wavelengths.

5.27 Compute and plot the four-wave mixing limit on the transmit power per channel for a WDM system operating over NZ-DSF. Assume that the channels are equally spaced and transmitted with equal power, and the maximum allowable penalty due to FWM is 1 dB. For the fiber, assume the dispersion parameter $D = 3$ ps/nm-km in the middle of the transmitted band of channels, and the slope of the dispersion curve is $dD/d\lambda = 0.055$ ps/nm-km^2. Consider the same three cases as in Figure 5.30: (a) 8 channels spaced 100 GHz apart, (b) 32 channels spaced 100 GHz apart, and (c) 32 channels spaced 50 GHz apart.

5.28 Why do second-order nonlinearities typically not affect a lightwave system?

5.29 In discussing the chromatic dispersion penalty, the Telcordia standard for SONET systems [Tel99] specifies the spectral width of a pulse, for *single-longitudinal mode* (SLM) lasers, as its 20 dB spectral width divided by 6.07. We studied these lasers in Section 3.5.1. Show that for SLM lasers whose spectra have a Gaussian profile, this is equivalent to the rms spectral width.

5.30 For a narrow but chirped Gaussian pulse with chirp factor $\kappa = -6$, calculate the chromatic dispersion limit at a bit rate of 1 Gb/s, in the 1.55 μm band, for a penalty of 2 dB. Compare this with the chromatic dispersion limit for unchirped pulses plotted in Figure 5.19.

▬ References

[Agr95] G. P. Agrawal. *Nonlinear Fiber Optics*, 2nd edition. Academic Press, San Diego, CA, 1995.

[BA94] F. Bruyère and O. Audouin. Assessment of system penalties induced by polarization mode dispersion in a 5 Gb/s optically amplified transoceanic link. *IEEE Photonics Technology Letters*, 6(3):443–445, March 1994.

[Bak01] B. Bakhshi et al. 1 Tb/s (101 × 10 Gb/s) transmission over transpacific distance using 28 nm C-band EDFAs. In *OFC 2001 Technical Digest*, pages PD21/1–3, 2001.

[Big01] S. Bigo et al. 10.2 Tb/s (256 × 42.7 Gbit/s PDM/WDM) transmission over 100 km TeraLight fiber with 1.28bit/s/Hz spectral efficiency. In *OFC 2001 Technical Digest*, pages PD25/1–3, 2001.

[BOS99] P. C. Becker, N. A. Olsson, and J. R. Simpson. *Erbium-Doped Fiber Amplifiers: Fundamentals and Technology*. Academic Press, San Diego, CA, 1999.

[Bre01] J. F. Brennan III et al. Dispersion and dispersion-slope correction with a fiber Bragg grating over the full C-band. In *OFC 2001 Technical Digest*, pages PD12/1–3, 2001.

[Buc95]   J. A. Buck. *Fundamentals of Optical Fibers.* John Wiley, New York, 1995.

[Cai01]   J.-X. Cai et al. 2.4 Tb/s (120 × 20 Gb/s) transmission over transoceanic distance with optimum FEC overhead and 48% spectral efficiency. In *OFC 2001 Technical Digest,* pages PD20/1–3, 2001.

[CDdM90]   F. Curti, B. Daino, G. de Marchis, and F. Matera. Statistical treatment of the evolution of the principal states of polarization in single-mode fibers. *IEEE/OSA Journal on Lightwave Technology,* 8(8):1162–1166, Aug. 1990.

[Chr84]   A. R. Chraplyvy. Optical power limits in multichannel wavelength-division-multiplexed systems due to stimulated Raman scattering. *Electronics Letters,* 20:58, 1984.

[Chr90]   A. R. Chraplyvy. Limitations on lightwave communications imposed by optical-fiber nonlinearities. *IEEE/OSA Journal on Lightwave Technology,* 8(10):1548–1557, Oct. 1990.

[Des94]   E. Desurvire. *Erbium-Doped Fiber Amplifiers: Principles and Applications.* John Wiley, New York, 1994.

[FTC95]   F. Forghieri, R. W. Tkach, and A. R. Chraplyvy. WDM systems with unequally spaced channels. *IEEE/OSA Journal on Lightwave Technology,* 13(5):889–897, May 1995.

[FTCV96]   F. Forghieri, R. W. Tkach, A. R. Chraplyvy, and A. M. Vengsarkar. Dispersion compensating fiber: Is there merit in the figure of merit? In *OFC'96 Technical Digest,* pages 255–257, 1996.

[Fuk01]   K. Fukuchi et al. 10.92 Tb/s (273 × 40 Gb/s) triple-band/ultra-dense WDM optical-repeatered transmission experiment. In *OFC 2001 Technical Digest,* pages PD24/1–3, 2001.

[GEE94]   E. L. Goldstein, L. Eskildsen, and A. F. Elrefaie. Performance implications of component crosstalk in transparent lightwave networks. *IEEE Photonics Technology Letters,* 6(5):657–670, May 1994.

[Gol00]   E. A. Golovchenko et al. Modeling of transoceanic fiber-optic WDM communication systems. *IEEE Journal of Selected Topics in Quantum Electronics,* 6:337–347, 2000.

[HH90]   P. A. Humblet and W. M. Hamdy. Crosstalk analysis and filter optimization of single- and double-cavity Fabry-Perot filters. *IEEE Journal of Selected Areas in Communications,* 8(6):1095–1107, Aug. 1990.

[Jac96]   J. L. Jackel et al. Acousto-optic tunable filters (AOTFs) for multiwavelength optical cross-connects: Crosstalk considerations. *IEEE/OSA JLT/JSAC Special Issue on Multiwavelength Optical Technology and Networks,* 14(6):1056–1066, June 1996.

[Kar01]   M. Karlsson et al. Higher order polarization mode dispersion compensator with three degrees of freedom. In *OFC 2001 Technical Digest*, pages MO1/1–3, 2001.

[KK97a]   I. P. Kaminow and T. L. Koch, editors. *Optical Fiber Telecommunications IIIA*. Academic Press, San Diego, CA, 1997.

[KK97b]   I. P. Kaminow and T. L. Koch, editors. *Optical Fiber Telecommunications IIIB*. Academic Press, San Diego, CA, 1997.

[LZNA98]   G. Luo, J. L. Zyskind, J. A. Nagel, and M. A. Ali. Experimental and theoretical analysis of relaxation-oscillations and spectral hole burning effects in all-optical gain-clamped EDFA's for WDM networks. *IEEE/OSA Journal on Lightwave Technology*, 16:527–533, 1998.

[MM98]   A. Mecozzi and D. Marcenac. Theory of optical amplifier chains. *IEEE/OSA Journal on Lightwave Technology*, 16:745–756, 1998.

[Nak00]   M. Nakazawa et al. Ultrahigh-speed long-distance TDM and WDM soliton transmission technologies. *IEEE Journal of Selected Topics in Quantum Electronics*, 6:363–396, 2000.

[Nel01]   Lynn E. Nelson. Challenges of 40 Gb/s WDM transmission. In *OFC 2001 Technical Digest*, pages ThF1/1–3, 2001.

[Ols89]   N. A. Olsson. Lightwave systems with optical amplifiers. *IEEE/OSA Journal on Lightwave Technology*, 7(7):1071–1082, July 1989.

[OSYZ95]   M. J. O'Mahony, D. Simeonidou, A. Yu, and J. Zhou. The design of a European optical network. *IEEE/OSA Journal on Lightwave Technology*, 13(5):817–828, May 1995.

[PAP86]   M. J. Potasek, G. P. Agrawal, and S. C. Pinault. Analytic and numerical study of pulse broadening in nonlinear dispersive optical fibers. *Journal of Optical Society of America B*, 3(2):205–211, Feb. 1986.

[PL01]   D. Penninckx and S. Lanne. Reducing PMD impairments. In *OFC 2001 Technical Digest*, pages TuP1/1–3, 2001.

[PN87]   K. Padmanabhan and A. N. Netravali. Dilated networks for photonic switching. *IEEE Transactions on Communications*, 35:1357–1365, 1987.

[PTCF91]   C. D. Poole, R. W. Tkach, A. R. Chraplyvy, and D. A. Fishman. Fading in lightwave systems due to polarization-mode dispersion. *IEEE Photonics Technology Letters*, 3(1):68–70, Jan. 1991.

[RL93]   R. Ramaswami and K. Liu. Analysis of effective power budget in optical bus and star networks using erbium-doped fiber amplifiers. *IEEE/OSA Journal on Lightwave Technology*, 11(11):1863–1871, Nov. 1993.

[SBW87] N. Shibata, R. P. Braun, and R. G. Waarts. Phase-mismatch dependence of efficiency of wave generation through four-wave mixing in a single-mode optical fiber. *IEEE Journal of Quantum Electronics*, 23:1205–1210, 1987.

[SKA00] H. Sunnerud, M. Karlsson, and P. A. Andrekson. Analytical theory for PMD-compensation. *IEEE Photonics Technology Letters*, 12:50–52, 2000.

[SKA01] H. Sunnerud, M. Karlsson, and P. A. Andrekson. A comparison between NRZ and RZ data formats with respect to PMD-induced system degradation. *IEEE Photonics Technology Letters*, 13:448–450, 2001.

[Smi72] R. G. Smith. Optical power handling capacity of low loss optical fibers as determined by stimulated Raman and Brillouin scattering. *Applied Optics*, 11(11):2489–2160, Nov. 1972.

[SNIA90] N. Shibata, K. Nosu, K. Iwashita, and Y. Azuma. Transmission limitations due to fiber nonlinearities in optical FDM systems. *IEEE Journal of Selected Areas in Communications*, 8(6):1068–1077, Aug. 1990.

[TCF+95] R. W. Tkach, A. R. Chraplyvy, F. Forghieri, A. H. Gnauck, and R. M. Derosier. Four-photon mixing and high-speed WDM systems. *IEEE/OSA Journal on Lightwave Technology*, 13(5):841–849, May 1995.

[Tel99] Telcordia Technologies. *SONET Transport Systems: Common Generic Criteria*, 1999. GR-253-CORE Issue 2, Revision 2.

[TOT96] H. Takahashi, K. Oda, and H. Toba. Impact of crosstalk in an arrayed-waveguide multiplexer on $n \times n$ optical interconnection. *IEEE/OSA JLT/JSAC Special Issue on Multiwavelength Optical Technology and Networks*, 14(6):1097–1105, June 1996.

[VPM01] G. Vareille, F. Pitel, and J. F. Marcerou. 3 Tb/s (300 × 11.6 Gbit/s) transmission over 7380 km using 28 nm C+L-band with 25 GHz channel spacing and NRZ format. In *OFC 2001 Technical Digest*, pages PD22/1–3, 2001.

[WK92] J. H. Winters and S. Kasturia. Adaptive nonlinear cancellation for high-speed fiber-optic systems. *IEEE/OSA Journal on Lightwave Technology*, 10:971–977, 1992.

[ZCC+96] J. Zhou, R. Cadeddu, E. Casaccia, C. Cavazzoni, and M. J. O'Mahony. Crosstalk in multiwavelength optical cross-connect networks. *IEEE/OSA JLT/JSAC Special Issue on Multiwavelength Optical Technology and Networks*, 14(6):1423–1435, June 1996.

[Zhu01] B. Zhu et al. 3.08 Tb/s (77 × 42.7 Gb/s) transmission over 1200 km of non-zero dispersion-shifted fiber with 100-km spans using C- L-band distributed Raman amplification. In *OFC 2001 Technical Digest*, pages PD23/1–3, 2001.

[Zir91] M. Zirngibl. Gain control in erbium-doped fiber amplifiers by an all-optical feedback loop. *Electronics Letters*, 27:560, 1991.

[Zir98] M. Zirngibl. Analytical model of Raman gain effects in massive wavelength division multiplexed transmission systems. *Electronics Letters*, 34:789, 1998.

[ZO94] J. Zhou and M. J. O'Mahony. Optical transmission system penalties due to fiber polarization mode dispersion. *IEEE Photonics Technology Letters*, 6(10):1265–1267, Oct. 1994.

[Zys96] J. L. Zyskind et al. Fast power transients in optically amplified multiwavelength optical networks. In *OFC'96 Technical Digest*, 1996. Postdeadline paper PD31.

part

II

Networks

6
chapter

Client Layers of the Optical Layer

THIS CHAPTER DESCRIBES several networks that use optical fiber as their underlying transmission mechanism. These networks can be thought of as client layers of the optical layer. As we saw in Chapter 1, the optical layer provides lightpaths to the client layers. To the client layer, these lightpaths look like physical links between client layer network elements. All the client layers that we will study process the data in the electrical domain, performing functions such as fixed time division multiplexing or statistical time division multiplexing (packet switching). These client layers aggregate and bring a variety of lower-speed voice, data, and private line services into the network. Each of these client networks is important in its own right and can operate over point-to-point fiber links as well as over a more sophisticated optical layer, using the lightpaths provided by the optical layer.

The predominant client layers in backbone networks today are SONET/SDH, IP, and ATM. SONET/SDH is particularly adept at dealing with lower-speed time division multiplexed streams, whereas IP and ATM are adept at dealing with statistically multiplexed packet streams. In many cases, IP and ATM use SONET/SDH as the underlying transport mechanism. With the emergence of high-speed interfaces on IP and ATM equipment, we are also seeing IP and ATM mapped directly into the optical layer, without requiring separate SONET/SDH equipment. In the metro network, we are seeing a proliferation of several types of client layers, such as Gigabit Ethernet, ESCON, and Fibre Channel. Many of the latter networks are used to interconnect computers and their peripherals in so-called storage-area networks.

In this chapter, we provide descriptions of these various networks, focusing primarily on a qualitative understanding, as well as characteristics that are important in the context of the optical layer. For instance, we describe SONET/SDH in some detail, including the SONET sublayers, frame structure, and the various overhead bytes. We will see in Chapters 9 and 10 that many functions in the optical layer are somewhat analogous to those in the SONET layer. In particular, the control, management, and survivability built into SONET/SDH networks is the basis for how these functions are being implemented in the optical layer. Similarly, in the context of IP, we discuss the IP routing and signaling protocols. There is a great deal of interest in reusing these protocols to control the optical layer.

6.1 SONET/SDH

SONET (Synchronous Optical Network) is the current transmission and multiplexing standard for high-speed signals within the carrier infrastructure in North America. A closely related standard, SDH (Synchronous Digital Hierarchy), has been adopted in Europe and Japan and for most submarine links.

In order to understand the factors underlying the evolution and standardization of SONET and SDH, we need to look back in time and understand how multiplexing was done in the public network. Prior to SONET and SDH, the existing infrastructure was based on the *plesiochronous digital hierarchy* (PDH), dating back to the mid-1960s. (North American operators refer to PDH as the *asynchronous* digital hierarchy.) At that time the primary focus was on multiplexing digital voice circuits. An analog voice circuit with a bandwidth of 4 kHz could be sampled at 8 kHz and quantized at 8 bits per sample, leading to a bit rate of 64 kb/s for a digital voice circuit. This became the widely accepted standard. Higher-speed streams were defined as multiples of this basic 64 kb/s stream. Different sets of standards emerged in different parts of the world for these higher-speed streams, as shown in Table 6.1. In North America, the 64 kb/s signal is called DS0 (digital signal-0), the 1.544 Mb/s signal is DS1, the 44.736 Mb/s is DS3, and so on. In Europe, the hierarchy is labeled E0, E1, E2, E3, and so on, with the E0 rate being the same as the DS0 rate. These rates are widely prevalent today in carrier networks and are offered as leased line services by carriers to customers, more often than not to carry data rather than voice traffic.

PDH suffered from several problems, which led carriers and vendors alike to seek a new transmission and multiplexing standard in the late 1980s. This resulted in the the SONET/SDH standards, which solved many problems associated with PDH. We explain some of the benefits of SONET/SDH below and contrast it with PDH.

1. **Multiplexing simplification:** In asynchronous multiplexing, each terminal in the network runs its own clock, and while we can specify a nominal clock rate for the

Table 6.1 Transmission rates for asynchronous and plesiochronous signals, adapted from [SS96].

Level	North America	Europe	Japan
0	0.064 Mb/s	0.064 Mb/s	0.064 Mb/s
1	1.544 Mb/s	2.048 Mb/s	1.544 Mb/s
2	6.312 Mb/s	8.448 Mb/s	6.312 Mb/s
3	44.736 Mb/s	34.368 Mb/s	32.064 Mb/s
4	139.264 Mb/s	139.264 Mb/s	97.728 Mb/s

signal, there can be significant differences in the actual rates between different clocks. For example, in a DS3 signal, a 20 ppm (parts per million) variation in clock rate between different clocks, which is not uncommon, can produce a difference in bit rate of 1.8 kb/s between two signals. So when lower-speed streams are multiplexed by interleaving their bits, extra bits may need to be stuffed in the multiplexed stream to account for differences between the clock rates of the individual streams. As a result, the bit rates in the asynchronous hierarchy are not integral multiples of the basic 64 kb/s rate, but rather slightly higher to account for this bit stuffing. For instance, a DS1 signal is designed to carry 24 64 kb/s signals, but its bit rate (1.544 Mb/s) is slightly higher than 24×64 kb/s.

With asynchronous multiplexing, it is very difficult to pick out a low-bit-rate stream, say, at 64 kb/s, from a higher-speed stream passing through, say, a DS3 stream, without completely demultiplexing the higher-speed stream down to its individual component streams. This results in the need for "multiplexer mountains," or stacked-up multiplexers, each time a low-bit-rate stream needs to be extracted, as shown in Figure 6.1. This is a relatively expensive proposition and also compromises network reliability because of the large amount of electronics needed overall.

The synchronous multiplexing structure of SONET/SDH provides significant reduction in the cost of multiplexing and demultiplexing. All the clocks in the network are perfectly synchronized to a single master clock, and as a consequence, the rates defined in SONET/SDH are integral multiples of the basic rate and no bit stuffing is needed when multiplexing streams together. As a result, a lower-speed signal can be extracted from a multiplexed SONET/SDH stream in a single step by locating the appropriate positions of the corresponding bits in the multiplexed signal. This makes the design of SONET multiplexers and demultiplexers much easier than their asynchronous equivalents. We will explore this in more detail in Section 6.1.1.

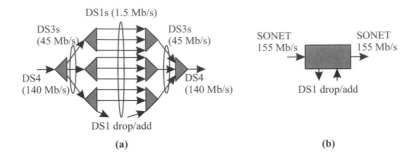

Figure 6.1   Comparison of asynchronous and synchronous multiplexing. (a) In the asynchronous case, demultiplexers must be stacked up to extract a lower-speed stream from a multiplexed stream. (b) In the synchronous case, this can be done in a single step using relatively simple circuitry.

2. **Management:** The SONET and SDH standards incorporate extensive management information for managing the network, including extensive performance monitoring, identification of connectivity and traffic type, identification and reporting of failures, and a data communication channel for transporting management information between the nodes. This is mostly lacking in the PDH standards.

3. **Interoperability:** Although PDH defined multiplexing methods, it did not define a standard format on the transmission link. Thus different vendors used different line coding, optical interfaces, and so forth to optimize their products, which made it very difficult to connect one vendor's equipment to another's via a transmission link. SONET and SDH avoid this problem by defining standard optical interfaces that enable interoperability between equipment from different vendors on the link. Unfortunately, certain aspects of SONET and SDH were only recently standardized, such as the data communication channel mentioned above. As a result, even today, it is not trivial to interconnect SONET equipment from different vendors.

4. **Network availability:** The SONET and SDH standards have evolved to incorporate specific network topologies and specific protection techniques and associated protocols to provide high-availability services. As a consequence, the service restoration time after a failure with SONET and SDH is much smaller—less than 60 ms—than the restoration time in PDH networks, which typically took several seconds to minutes.

6.1.1 Multiplexing

SONET and SDH employ a sophisticated multiplexing scheme, which can, however, be easily implemented in today's very large-scale integrated (VLSI) circuits. Although SONET and SDH are basically similar, the terms used in SONET and SDH are different, and we will use the SONET version in what follows and introduce the SDH version wherever appropriate.

For SONET, the basic signal rate is 51.84 Mb/s, called the synchronous transport signal level-1 (STS-1). Higher-rate signals (STS-N) are obtained by interleaving the bytes from N frame-aligned STS-1s. Because the clocks of the individual signals are synchronized, no bit stuffing is required. For the same reason, a lower-speed stream can be extracted easily from a multiplexed stream without having to demultiplex the entire signal.

The currently defined SONET and SDH rates are shown in Table 6.2. Note that an STS signal is an electrical signal and in many cases (particularly at the higher speeds) may exist only inside the SONET equipment. The interface to other equipment is usually optical and is essentially a scrambled version of the STS signal in optical form. Scrambling is used to prevent long runs of 0s or 1s in the data stream. (See Section 4.1.1 for a more detailed explanation of scrambling.) Each SONET transmitter scrambles the signal before it is transmitted over the fiber, and the next SONET receiver descrambles the signal. The optical interface corresponding to the STS-3 rate is called OC-3 (optical carrier-3), and similar optical interfaces have been defined for OC-12, OC-48, OC-192, and OC-768 corresponding to the STS-12, STS-48, STS-192, and STS-768 signals.

For SDH, the basic rate is 155 Mb/s and is called STM-1 (synchronous transport module-1). Note that this is higher than the basic SONET bit rate. The SONET bit rate was chosen to accommodate the commonly used asynchronous signals, which are DS1 and DS3 signals. The SDH bit rate was chosen to accommodate the commonly used PDH signals, which are E1, E3, and E4 signals. Higher-bit-rate signals are defined analogous to SONET, as shown in Table 6.2.

A SONET frame consists of some overhead bytes called the *transport* overhead and the payload bytes. The payload data is carried in the so-called synchronous payload envelope (SPE). The SPE includes a set of additional *path* overhead bytes that are inserted at the source node and remain with the data until it reaches its destination node. For instance, one of these bytes is the *path trace*, which identifies the SPE and can be used to verify connectivity in the network. We will study the frame structure in more detail in Section 6.1.3.

SONET and SDH make extensive use of pointers to indicate the location of multiplexed payload data within a frame. The SPE doesn't have a fixed starting point within a frame. Instead, its starting point is indicated by a pointer in the line overhead.

Table 6.2 Transmission rates for SONET/SDH, adapted from [SS96].

SONET Signal	SDH Signal	Bit Rate (Mb/s)
STS-1		51.84
STS-3	STM-1	155.52
STS-12	STM-4	622.08
STS-24		1244.16
STS-48	STM-16	2488.32
STS-192	STM-64	9953.28
STS-768	STM-256	39, 814.32

Even though the clocks in SONET are all derived from a single source, there can be small transient variations in frequency between different signals. Such a difference between the incoming signal and the local clock used to generate an outgoing signal translates into accumulated phase differences between the two signals. This problem is easily solved by allowing the payload to be shifted earlier or later in a frame and indicating this by modifying the associated pointer. This avoids the need for bit stuffing or additional buffering. However, it does require a fair amount of pointer processing, which can be performed easily in today's integrated circuits.

Lower-speed non-SONET streams below the STS-1 rate are mapped into *virtual tributaries* (VTs). Each VT is designed to have sufficient bandwidth to carry its payload. In SONET, VTs have been defined in four sizes: VT1.5, VT2, VT3, and VT6. These VTs are designed to carry 1.5, 2, 3, and 6 Mb/s asynchronous/plesiochronous streams, as shown in Figure 6.2. Of these, the VT1.5 signal is the most common, as it holds the popular DS1 asynchronous signal. At the next level in the hierarchy, a VT group consists of either four VT1.5s, three VT2s, two VT3s, or a single VT6. Seven such VT groups are *byte* interleaved along with a set of path overheads to create a *basic* SONET SPE. Just as an SPE floats within a SONET frame, the VT payload (called VT SPE) can also float within the STS-1 SPE, and a VT pointer is used to point to the VT SPE. The pointer is located in two designated bytes within each VT group. Figure 6.3 illustrates this pointer structure.

In many cases, it is necessary to map higher-speed non-SONET signals into an SPE for transport over SONET. The most common examples today are probably high-speed packet streams from IP routers or ATM switches. For this purpose, an STS-*N*c signal with a *locked* payload is also defined in the standards. The "c" stands for *concatenated*. The concatenated or locked payload implies that this signal cannot be demultiplexed into lower-speed streams. For example, a 150 Mb/s ATM signal is

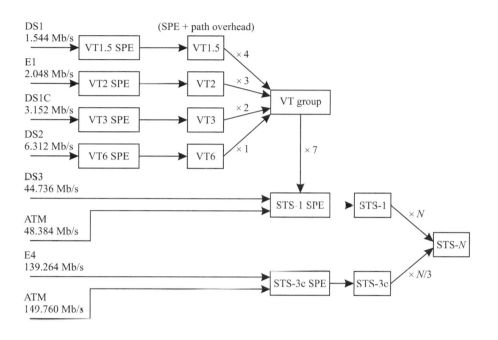

Figure 6.2 The mapping of lower-speed asynchronous streams into virtual tributaries in SONET.

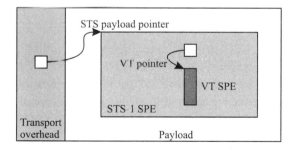

Figure 6.3 The use of pointers in a SONET STS-1 signal carrying virtual tributaries (VTs). The STS payload pointer in the transport overhead points to the STS-1 synchronous payload envelope (SPE) and the VT pointer inside the STS-1 SPE points to the VT SPE.

mapped into an STS-3c signal. Mappings have been defined in the standards for a variety of signals, including IP, ATM, and FDDI (fiber distributed data interface).

While SDH employs the same philosophy as SONET, there are some differences in terminology and in the multiplexing structure for sub-STM-1 signals. Analogous to SONET virtual tributaries, SDH uses *virtual containers* (VCs) to accommodate lower-speed non-SDH signals. VCs have been defined in five sizes: VC-11, VC-12, VC-2, VC-3, and VC-4. These VCs are designed to carry 1.5 Mb/s (DS1), 2 Mb/s (E1), 6 Mb/s (E2), 45 Mb/s (E3 and DS3), and 140 Mb/s (E4) asynchronous/plesiochronous streams, respectively. However, a two-stage hierarchy is defined here, where VC-11s, VC-12s, and VC-2s can be multiplexed into VC-3s or VC-4s, and VC-3s and VC-4s are then multiplexed into an STM-1 signal.

6.1.2   SONET/SDH Layers

The SONET layer consists of four sublayers—the *path, line, section,* and *physical* layers. Figure 6.4 shows the top three layers. Each layer, except for the physical layer, has a set of associated overhead bytes that are used for several purposes. These overhead bytes are added whenever the layer is introduced and removed whenever the layer is terminated in a network element. The functions of these layers will become clearer when we discuss the frame structure and overheads associated with each layer in the next section.

The path layer in SONET (and SDH) is responsible for end-to-end connections between nodes and is terminated only at the ends of a SONET connection. It is possible that intermediate nodes may do performance monitoring of the path layer signals, but the path overhead itself is inserted at the source node of the connection and terminated at the destination node.

Each connection traverses a set of links and intermediate nodes in the network. The line layer (*multiplex section* layer in SDH) multiplexes a number of path-layer connections onto a single link between two nodes. Thus the line layer is terminated at each intermediate line terminal multiplexer (TM) or add/drop multiplexer (ADM) along the route of a SONET connection. The line layer is also responsible for performing certain types of protection switching to restore service in the event of a line failure.

Each link consists of a number of sections, corresponding to link segments between regenerators. The section layer (*regenerator-section* layer in SDH) is terminated at each regenerator in the network.

Finally, the physical layer is responsible for actual transmission of bits across the fiber.

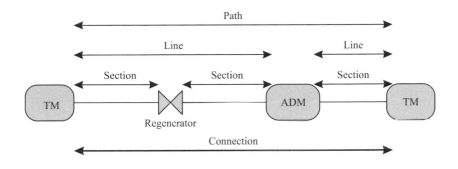

Figure 6.4 SONET/SDH layers showing terminations of the path, line, and section layers for a sample connection passing through terminal multiplexers (TMs) and add/drop multiplexers (ADMs). The physical layer is not shown.

6.1.3 SONET Frame Structure

Figure 6.5 shows the structure of an STS-1 frame. A frame is 125 μs in duration (which corresponds to a rate of 8000 frames/s), regardless of the bit rate of the SONET signal. This time is set by the 8 kHz sampling rate of a voice circuit. The frame is a specific sequence of 810 bytes, including specific bytes allocated to carry overhead information and other bytes carrying the payload. We can visualize this frame as consisting of 9 rows and 90 columns, with each cell holding an 8-bit byte. The bytes are transmitted row by row, from left to right, with the most significant bit in each byte being transmitted first.

The first three columns are reserved for section and line overhead bytes. The remaining bytes carry the STS-1 SPE. The STS-1 SPE itself includes one column of overhead bytes for carrying the path overhead.

An STS-N frame is obtained by byte-interleaving N STS-1 frames, as shown in Figure 6.6. The transport overheads are in the first $3N$ columns and the remaining $87N$ columns contain the payload. The transport overheads need to be frame aligned before they are interleaved. However, because each STS-1 has an associated payload pointer to indicate the location of its SPE, the payloads do not have to be frame aligned. An STS-Nc frame looks like an STS-N frame, except that the payload cannot be broken up into lower-speed signals in the SONET layer. The same $87N$ columns contain the payload, and special values in the STS-payload pointers are used to indicate that the payload is concatenated.

Figure 6.7 shows the overhead bytes in an STS-1 frame or an STS-Nc frame. In an STS-N frame, there are N sets of overhead bytes, one for each STS-1. Each STS-1

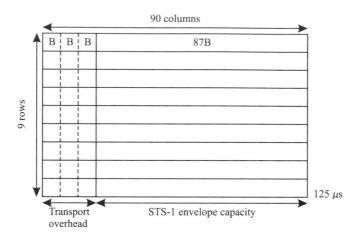

Figure 6.5   Structure of an STS-1 frame. B denotes an 8-bit byte.

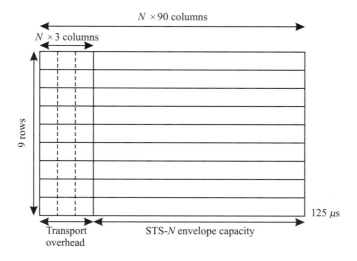

Figure 6.6   Structure of an STS-*N* frame, which is obtained by byte-interleaving *N* STS-1 frames.

			Path overhead
Framing A1	Framing A2	Trace/Growth J0/Z0	Trace J1
BIP-8 B1/undefined	Orderwire E1/undefined	User F1/undefined	BIP-8 B3
Datacom D1/undefined	Datacom D2/undefined	Datacom D3/undefined	Signal label C2
Pointer H1	Pointer H2	Pointer H3	Path status G1
BIP-8 B2	Datacom D5/undefined	Datacom D6/undefined	User channel F2
Datacom D4/undefined	APS K1/undefined	APS K2/undefined	Indicator H4
Datacom D7/undefined	Datacom D8/undefined	Datacom D9/undefined	Growth Z3
Datacom D10/undefined	Datacom D11/undefined	Datacom D12/undefined	Growth Z4
Sync status/Growth S1/Z1	REI-L/Growth M0 or M1/Z2	Orderwire E2/undefined	Tandem connection Z5

Section overhead covers the first three rows; Line overhead covers the lower six rows.

Figure 6.7 SONET overhead bytes. Entries of the form X/Y indicate that the first label X applies to the first STS-1 within an STS-N signal and the second label Y applies to the remaining STS-1's in the STS-N.

has its own set of section and line overheads. An STS-Nc, on the other hand, has only a single set of overhead bytes, due to the fact that its payload has to be carried intact from its source to its destination with the SONET network.

We cover the overhead bytes here because they provide some key management functions that make SONET so attractive for network operators. In the following discussion, the actual locations and formatting of the bytes is not as important as understanding the functions they perform. We will look at these functions in more detail in the context of the optical layer in Chapter 9. The section and line overheads in particular are of great interest to the optical layer. Some if not all these bytes are monitored by optical layer equipment. In addition, some of the overhead bytes are currently undefined, and these bytes are now being considered as possible candidates to carry optical layer overhead information. We will discuss this aspect in more detail in Chapter 9. For a more detailed description of the overhead bytes, see [Tel99].

Section Overhead

Framing (A1/A2). These two bytes are used for delineating the frame and are set to prespecified values in each STS-1 within an STS-N. Network elements use these bytes to determine the start of a new frame.

Section Trace(J0)/Section Growth(Z0). The J0 byte is present in the first STS-1 in an STS-N and is used to carry an identifier, which can be monitored to verify connectivity between adjacent section-terminating nodes in the network. The Z0 byte is present in the remaining STS-1s, and its use is still to be determined.

Section BIP-8 (B1). This byte is located in the first STS-1 in an STS-N and is used to monitor the bit error rate performance of each section. The byte locations in the remaining frames within an STS-N are currently undefined. The transmitter computes a bit interleaved parity (BIP) computed over all bytes in the previous STS-N frame after scrambling and places it in the B1 byte of the current frame before it is scrambled. An odd parity value indicates an error. We studied how this code works in Section 4.5 and Problem 4.16 in Chapter 4.

Orderwire (E1). This byte (located in the first STS-1 in a frame) is used to carry a voice channel between nodes, for use by maintainence personnel in the field.

Section User Channel (F1). This byte (located in the first STS-1 in a frame) is made available to the user for inserting additional user-specific information.

Section Data Communication Channel (D1, D2, D3). These bytes (located in the first STS-1 in a frame) are used to carry a data communication channel (DCC) for maintenance purposes such as alarms, monitoring, and control.

Line Overhead

We give below a brief outline of the functions of some of the line overhead bytes.

STS Payload Pointer (H1 and H2). The H1 and H2 bytes in the line overhead carry a two-byte pointer that specifies the location of the STS SPE. More precisely, these bytes carry a value corresponding to the offset in bytes between the pointer and the first byte of the STS SPE.

Line BIP-8 (B2). The B2 byte carries a bit interleaved parity check value for each STS-1 within the STS-N. It is computed by taking the parity over all bits of the line overhead and the envelope capacity of the previous STS-1 frame before it is scrambled. This byte is checked by line terminating equipment. The intermediate section terminating equipment checks and resets the B1 byte in the section overhead but does not alter the B2 byte.

APS channel (K1, K2). The K1 and K2 bytes are used to provide a channel for carrying signaling information during automatic protection switching (APS). We will study the different types of SONET APS schemes in Chapter 10. The K2 byte is also used to detect a specific kind of a signal called a *forward defect indicator* and to carry a *return defect indicator* signal. These defect indicator signals are used for maintenance purposes in the network; we will study their use in detail in Section 9.5.4.

Line Data Communication Channel. Bytes D4 through D12 (located in the first STS-1 in a frame) are used to carry a line data communication channel for maintenance purposes such as alarms, monitoring, and control.

Path Overhead

STS Path trace (J1). Just as in the section overhead, the path overhead includes a byte (J1) to carry a path identifier that can be monitored to verify connectivity in the network.

STS Path BIP-8 (B3). The B3 byte provides bit error rate monitoring at the path layer. It carries a bit interleaved parity check value calculated over all bits of the previous STS SPE before scrambling.

STS Path Signal Label (C2). The C2 byte is used to indicate the content of the STS SPE. Specific labels are assigned to denote each type of signal mapped into a SONET STS-1.

Path Status (G1). The G1 byte is used to convey the performance of the path from the destination back to the source node. The destination inserts the current error count in the received signal into this byte, which is then monitored by the source node. Part of this byte is also used to carry a defect indicator signal back to the source. We will study the use of defect indicator signals in Section 9.5.4.

6.1.4   SONET/SDH Physical Layer

A variety of physical layer interfaces are defined for SONET/SDH, depending on the bit rates and distances involved, as shown in Table 6.3. We have used the SDH version standardized by the ITU, as it is more current. The interfaces defined for SONET systems generally align with the SDH versions. Generally, we can classify the different applications based on the target distance and loss on the link between the transmitter and receiver. With this in mind, the applications defined fit into one of the following categories:

- *Intraoffice* connections (I) corresponding to distances of less than approximately 2 km (the SONET term for this is *short reach*)

Table 6.3 Different physical interfaces for SDH. Adapted from ITU recommendations G.957 and G.691. No optical amplifiers are used in the spans. The first letter in the application code specifies the target reach and the following number indicates the bit rate. The number after the period indicates the fiber type and operating wavelength: a blank or 1 indicates 1310 nm transmission over standard single-mode fiber (G.652), 2 indicates 1550 nm transmission over for standard single-mode fiber (G.652), 3 indicates 1550 nm transmission over dispersion-shifted fiber (G.653), and 5 indicates 1550 nm transmission over nonzero-dispersion-shifted fiber (G.655). The transmitters include multilongitudinal mode (MLM) Fabry-Perot lasers and single-longitudinal mode (SLM) DFB lasers, as well as light-emitting diodes (LEDs). The two values of the dispersion limit correspond, respectively, to the two choices of the transmitter. ffs indicates that the specification is for further study. This is the case for dispersion-limited links using directly modulated SLM lasers where no agreement has been reached on how to specify the chirp limits. Some of the applications are loss limited, and therefore the dispersion limit is not applicable (NA).

Bit Rate	Code	Wavelength (nm)	Fiber	Loss (dB)	Transmitter	Dispersion (ps/nm)
STM-1	I-1	1310	G.652	0-7	LED/MLM	18/25
	S-1.1	1310	G.652	0-12	MLM	96
	S-1.2	1550	G.652	0-12	MLM/SLM	296/NA
	L-1.1	1310	G.652	10-28	MLM/SLM	246/NA
	L-1.2	1550	G.652	10-28	SLM	NA
	L-1.3	1550	G.653	10-28	MLM/SLM	296/NA
STM-4	I-4	1310	G.652	0-7	LED/MLM	14/13
	S-4.1	1310	G.652	0-12	MLM	74
	S-4.2	1310	G.652	0-12	SLM	NA
	L-4.1	1310	G.652	10-24	MLM/SLM	109/NA
	L-4.2	1550	G.652	10-24	SLM	ffs
	L-4.3	1550	G.653	10-24	SLM	NA
	V-4.1	1310	G.652	22-33	SLM	200
	V-4.2	1550	G.652	22-33	SLM	2400
	V-4.3	1550	G.653	22-33	SLM	400
	U-4.2	1550	G.652	33-44	SLM	3200
	U-4.3	1550	G.653	33-44	SLM	530
STM-16	I-16	1310	G.652	0-7	MLM	12
	S-16.1	1310	G.652	0-12	SLM	NA
	S-16.2	1550	G.652	0-12	SLM	ffs
	L-16.1	1310	G.652	10-24	SLM	NA
	L-16.2	1550	G.652	10-24	SLM	1600
	L-16.3	1550	G.653	10-24	SLM	ffs
	V-16.2	1550	G.652	22-33	SLM	2400
	V-16.3	1550	G.653	22-33	SLM	400
	U-4.2	1550	G.652	33-44	SLM	3200
	U-4.3	1550	G.653	33-44	SLM	530

Table 6.3 Different physical interfaces for SDH *(continued)*.

Bit Rate	Code	Wavelength (nm)	Fiber	Loss (dB)	Transmitter	Dispersion (ps/nm)
STM-64	I-64.1r	1310	G.652	0-4	MLM	3.8
	I-64.1	1310	G.652	0-4	SLM	6.6
	I-64.2r	1550	G.652	0-7	SLM	40
	I-64.2	1550	G.652	0-7	SLM	500
	I-64.3	1550	G.653	0-7	SLM	80
	I-64.5	1550	G.655	0-7	SLM	ffs
	S-64.1	1550	G.652	6-11	SLM	70
	S-64.2	1550	G.652	3/7-11	SLM	800
	S-64.3	1550	G.653	3/7-11	SLM	130
	S-64.5	1550	G.655	3/7-11	SLM	130
	L-64.1	1310	G.652	17-22	SLM	130
	L-64.2	1550	G.652	11/16-22	SLM	1600
	L-64.3	1550	G.653	16-22	SLM	260
	L-64.3	1550	G.653	0-7	SLM	ffs
	V-64.2	1550	G.652	22-33	SLM	2400
	V-64.3	1550	G.653	22-33	SLM	400

- *Short-haul* interoffice connections (S) corresponding to distances of approximately 15 km at 1310 nm operating wavelength and 40 km at 1550 nm operating wavelength (the SONET term for this is *intermediate reach*)

- *Long-haul* interoffice connections (L) corresponding to distances of approximately 40 km at 1310 nm operating wavelength and 80 km at 1550 nm operating wavelength (the SONET term for this is *long reach*)

- *Very-long-haul* interoffice connections (V) corresponding to distances of approximately 60 km at 1310 nm operating wavelength and 120 km at 1550 nm operating wavelength

- *Ultra-long-haul* interoffice connections (U) corresponding to distances of approximately 160 km

The other variables include the type of fiber and the type of transmitter used. The fiber types are the ones we covered in Section 2.4.9 and include standard single-mode fiber (G.652), dispersion-shifted fiber (G.653), and nonzero dispersion-shifted fiber (G.655). The transmitter types include LEDs or multilongitudinal mode (MLM) Fabry-Perot lasers at 1310 nm for short distances at the lower bit rates to 1550 nm single-longitudinal mode (SLM) DFB lasers for the higher bit rates and longer distances. The physical layer uses scrambling to prevent long runs of 1s or 0s in the data (see Section 4.1.1).

The applications specify many transmission-related parameters, of which the main ones are the allowed loss range and the maximum chromatic dispersion on the link. The loss includes connectors and splices along the path. The relative contribution of the latter to the overall loss is particularly high in intraoffice connections, where a number of patch panels and connectors can be present in the interconnect. We can translate the loss numbers into target distances by assuming a loss of approximately 3.5 dB/km for intraoffice connections, 0.8 dB/km for short-haul, and 0.5 dB/km at 1310 nm and 0.3 dB/km at 1550 nm for the other longer-distance applications. Likewise the chromatic dispersion numbers can be translated into target distances based on the dispersion parameter of the fiber used in the relevant operating range.

These standards allow the use of optical power amplifiers and preamplifiers but do not include optical line amplifiers. With optical line amplifiers, we are now seeing spans without regeneration well in excess of the distance limits specified here. Today's long-haul WDM systems with line amplifiers have regenerator spacings of about 400 to 600 km, with some ultra-long-haul systems extending this distance to a few thousand kilometers. The spans for such systems are vendor dependent and have not yet been standardized. (Note that the use of "long-haul" and "ultra-long-haul" in the context of WDM systems is different from their use in SDH terminology.)

6.1.5   Elements of a SONET/SDH Infrastructure

Figure 6.8 shows different types of SONET equipment deployed in a network. SONET is deployed in three types of network configurations: rings, linear configurations, and point-to-point links. The early deployments were in the form of point-to-point links, and this topology is still used today for many applications. In this case, the nodes at the ends of the link are called *terminal multiplexers* (TMs). TMs are also sometimes called *line terminating equipment* (LTE). In many cases, it is necessary to pick out one or more low-speed streams from a high-speed stream and, likewise, add one or more low-speed streams to a high-speed stream. This function is performed by an *add/drop multiplexer* (ADM). For example, an OC-48 ADM can drop and add OC-12 or OC-3 streams from/to an OC-48 stream. Similarly, an OC-3 ADM can drop/add DS3 streams from/to an OC-3 stream. ADMs are now widely used in the SONET infrastructure. ADMs can be inserted in the middle of a point-to-point link between TMs to yield a linear configuration.

Maintaining service availability in the presence of failures has become a key driver for SONET deployment. The most common topology used for this purpose is a ring. Rings provide an alternate path to reroute traffic in the event of link or node failures, while being topologically simple. The rings are made up of ADMs, which in addition to performing the multiplexing and demultiplexing operations, incorporate

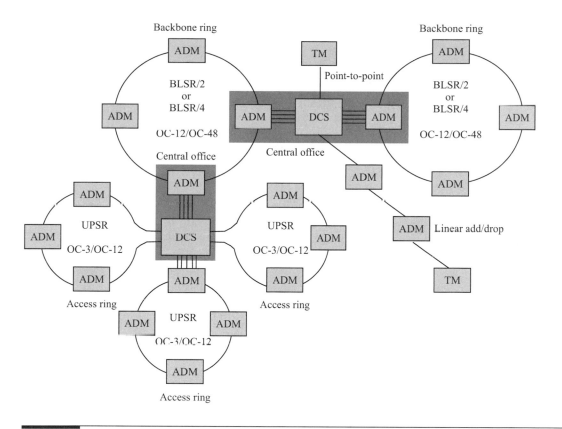

Figure 6.8 Elements of a SONET infrastructure. Several different SONET configurations are shown, including point-to-point, linear add/drop, and ring configurations. Both access and interoffice (backbone) rings are shown. The figure also explains the role of a DCS in the SONET infrastructure, to crossconnect lower-speed streams, to interconnect multiple rings, and to serve as a node on rings by itself.

the protection mechanisms needed to handle failures. Usually, SONET equipment can be configured to work in any of these three configurations: ring ADM, linear ADM, or as a terminal multiplexer.

Rings are used both in the access part of the network and in the backbone (interoffice) part of the network to interconnect central offices. Today, most access rings run at OC-3/OC-12 speeds, and interoffice rings at OC-12/OC-48/OC-192 speeds. Clearly these ring speeds will increase in the future, and 40 Gb/s rings should become available soon. Given the capacity requirements in today's networks, it is quite common to use multiple overlaid rings, particularly in backbone networks, each operating over a different wavelength provided by an underlying optical layer.

Two types of ring architectures are used: *unidirectional path-switched rings* (UPSRs) and *bidirectional line-switched rings* (BLSRs). The BLSRs can use either two fibers (BLSR/2) or four fibers (BLSR/4). We will discuss these architectures and the protection mechanisms that they incorporate in detail in Chapter 10. In general, UPSRs are used in the access part of the network to connect multiple nodes to a hub node residing in a central office, and BLSRs are used in the interoffice part of the network to interconnect multiple central offices.

Another major component in the SONET infrastructure is a *digital crossconnect* (DCS). A DCS is used to manage all the transmission facilities in the central office. Before DCSs arrived, the individual DS1s and DS3s in a central office were manually patched together using a patch panel. Although this worked fine for a small number of traffic streams, it is quite impossible to manage today's central offices, which handle thousands of such streams, using this approach. A DCS automates this process and replaces a patch panel by crossconnecting these individual streams under software control. It also does performance monitoring and has grown to incorporate multiplexing as well. DCSs started out handling only PDH streams but have evolved to handle SONET streams as well. Although the overall network topology including the DCSs is a mesh, note that only rings have been standardized so far.

A variety of DCSs are available today, as shown in Figure 6.9. Typically, these DCSs have hundreds to thousands of ports. The term *grooming* refers to the grouping together of traffic with similar destinations, quality of service, or traffic type. It includes multiplexing of lower-speed streams into high-speed streams, as well as extracting lower-speed streams from different higher-speed streams and combining them based on specific attributes. In this context, the type of grooming that a DCS performs is directly related to the granularity at which it switches traffic. If a DCS is switching traffic at granularities of DS1 rates, then we say that it grooms the traffic at the DS1 level. At the bottom of the hierarchy is a *narrowband* DCS, which grooms traffic at the DS0 level. Next up is a *wideband* DCS, which grooms traffic at DS1 rates, and then a *broadband* DCS, which grooms traffic at DS3/STS-1 rates. These DCSs typically have interfaces ranging from the grooming rate to much higher-speed interfaces. For instance, a wideband DCS will have interfaces ranging from DS1 to OC-12, while a broadband DCS will have interfaces ranging from DS3 to OC-48 or OC-192. Today we are also seeing a new generation of DCSs that groom at DS3 rates and above, with primarily high-speed optical interfaces. While such a box could be called broadband DCS, it is more commonly called an *optical crossconnect*. However, we also have other types of optical crossconnects that groom traffic at STS-48 rates, and yet others that use purely optical switch fabrics and groom traffic in units of wavelengths or more.

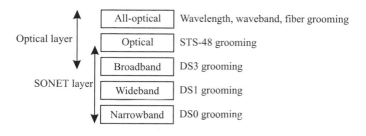

Figure 6.9 Different types of crossconnect systems.

Instead of having this hierarchy of crossconnect systems, why not have a single DCS with high-speed interfaces, which grooms at the lowest desired rate, say, DS0? This is not possible due to practical considerations of scalability, cost, and footprint. For instance, it is difficult to imagine building a crossconnect with hundreds to thousands of 10 Gb/s OC-192 ports that grooms down to the DS1 level. In general, the higher the speed of the desired interfaces on the crossconnect, the higher up it will reside in the grooming hierarchy of Figure 6.9.

DCSs can also incorporate ADM functions and perform other network functions such as restoration against failures, the topic of Chapter 10.

6.2 ATM

Voice and data networks have traditionally been separate even though almost the entire telephone network is digital. ATM (asynchronous transfer mode) is a networking standard that was developed with many goals, one of which was the integration of voice and data networks. An ATM network uses packets or *cells* with a fixed size of 53 bytes; this packet size is a compromise between the conflicting requirements of voice and data applications. A small packet size is preferable for voice since the packets must be delivered with only a short delay. A large packet size is preferable for data since the overheads involved in large packets are smaller. Of the 53 bytes in an ATM packet, 5 bytes constitute the header, which is the overhead required to carry information such as the destination of the packet. ATM networks span the whole gamut from local-area networks (LANs) to metropolitan-area networks (MANs) to wide-area networks (WANs).

One of the key advantages of ATM is its ability to provide quality-of-service guarantees, such as bandwidth and delay, to applications even while using statistical multiplexing of packets to make efficient use of the link bandwidth (see Chapter 1).

ATM achieves this by using a priori information about the characteristics of a connection (say, a virtual circuit), for example, the peak and average bandwidth required by it. ATM uses *admission control* to block new connections when necessary to satisfy the guaranteed quality-of-service requirements.

Another advantage of ATM is that it employs switching even in a local-area environment, unlike other LAN technologies like Ethernets, token rings, and FDDI, which use a shared medium such as a bus or a ring. This enables it to provide quality-of-service guarantees more easily than these other technologies. The fixed size of the packets used in an ATM network is particularly advantageous for the development of low-cost, high-speed switches.

Various lower or physical layer standards are specified for ATM. These range from 25.6 Mb/s over twisted-pair copper cable to 622.08 Mb/s over single-mode optical fiber. Among the optical interfaces is a 100 Mb/s interface whose specifications, such as transmit power, maximum allowed attenuation, and line coding, are identical to that of FDDI, which we have described. A 155.52 Mb/s optical interface that operates over distances up to 2 km using LEDs over multimode fiber in the 1300 nm band is also defined. Using the specified minimum transmit and receive powers, the loss budget for this interface is 9 dB. The line code used in this case is the (8, 10) line code specified by the Fibre Channel standard.

These two interfaces are called *private user–network interfaces* in ATM terminology, since they are meant for interconnecting ATM users and switches in networks that are owned and managed by private enterprises. A number of *public user–network interfaces,* which are meant for connecting ATM users and switches to the public or carrier network, are also defined. In these latter interfaces, ATM uses either PDH or SONET/SDH as the immediately lower layer. These interfaces are defined at many of the standard PDH and SONET/SDH rates shown in Tables 6.1 and 6.2, respectively. Among these are DS3, STS-3c, STS-12c, and STS-48c interfaces. In the terminology of the ATM standards, since the layer below ATM is called the physical layer, these interfaces to PDH and SONET/SDH are called *physical layer interfaces*. On the other hand, in the classical layered view of networks, which we discussed in Section 1.4, PDH and SONET/SDH must be viewed as *data link layers* when ATM is viewed as a network layer.

6.2.1   Functions of ATM

ATM data can either be transmitted from an ATM user to an ATM network across a user-to-network interface (UNI) or the data can be transmitted across a network-to-network interface (NNI) between two ATM switches. Of the 53 bytes in an ATM cell, 48 bytes form the payload, that is, carry information sent from the higher layers, and 5 bytes constitute the header inserted by the ATM layer. The

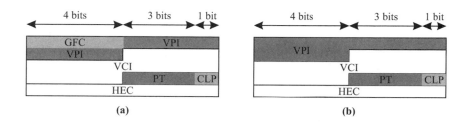

Figure 6.10 The header structure of ATM cells across (a) the UNI and (b) the NNI. The GFC field is used for flow control across the UNI. The VPI and VCI fields are used for forwarding the cells within the network. PT indicates the payload type and CLP is the cell loss priority bit. The HEC field provides error checking for the ATM header.

structure of the 5-byte ATM header is slightly different for the UNI and NNI. The two headers are shown in Figure 6.10.

The fields in the ATM header are as follows.

- GFC or Generic Flow Control: 4 bits on UNI, not present on NNI.
- VPI or Virual Path Identifier: 8 bits on UNI, 12 bits on NNI.
- VCI or Virtual Circuit Identifier: 16 bits.
- PT or Payload Type: 3 bits.
- CLP or Cell Loss Priority: 1 bit.
- HEC or Header Error Control: 8 bits. The HEC constitutes a CRC on the 5 ATM header bytes and is used to detect corrupted ATM cells.

The functions of each of these fields are described in the following sections.

Connections and Cell Forwarding

ATM establishes a *connection* between two end points for the purpose of transferring data between them. This is unlike IP (which we study in the next section), which transfers data in a connectionless manner. ATM connections are termed *virtual channels* and are assigned a virtual channel identifier (VCI). The VCI for a connection is unique for each link that the ATM connection traverses between its end points but can vary from link to link on the path, as illustrated in Figure 6.11(a). For example, the top connection has a VCI of a1, a2, and b on the three links it traverses. The VCIs for each connection on every link of the path are determined at the time of connection setup and released when the connection is torn down.

Each node (switch) maintains a VCI table as illustrated in Figure 6.11(b). The table specifies, for each incoming VCI, the outgoing link and the outgoing VCI. For

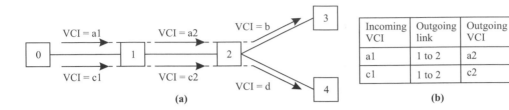

Figure 6.11   The use of ATM VCIs for cell forwarding across a path. The ATM switches use the VCI to determine the outgoing link for a cell. The switches also rewrite the VCI field with the value assigned to the virtual channel on the outgoing link. (a) Illustration of the cell forwarding and VCI swapping. (b) The VCI table maintained at node 1 of (a).

example, at node 1, incoming cells with a VCI of a1 are sent on the link 1–2 with a VCI of a2.

Virtual Paths

There could be millions of virtual channels sharing a link. Looking up a VCI table larger than $2^{16} = 65,536$ entries for forwarding every single cell is expensive. Thus we need to have some mechanism for bundling or aggregating virtual channels for the purpose of forwarding. It is quite likely that thousands of virtual channels will have the same path, if not end to end, at least over significant parts of the network. This property of virtual channels can be used for aggregation and is accomplished by the use of VPIs. The use of VPIs can be understood through the following example.

Consider Figure 6.12. Here we have four links, connecting the nodes 0, 1, 2, and 3, as shown. The two virtual circuits shown share the links 0–1 and 1–2. These virtual channels can be assigned a common VPI on each of these links (which can be, and generally is, different on the two individual links). For example, a VPI of x can be assigned on link 0–1, and a VPI of y on link 1–2. The set of two links constitutes a virtual path in the network, with node 0 constituting the beginning of the virtual path, and node 2 constituting the end of the virtual path. All cells belonging to any virtual circuit assigned to this path are routed on these links based on the smaller VPI value. When the cells reach the end of the virtual path, node 2 in this example, they are again forwarded based on the VCI values. Simply put, the virtual channels treat each virtual path as a segment in their route between the source and destination: the switches within a virtual path forward cells based only on the VPI field.

The use of the two level labels, VPI and VCI, simplifies the cell forwarding process and enables the development of cost-effective ATM switches. If a single field were

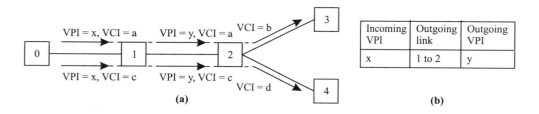

Figure 6.12 The use of ATM VPIs for simplifying cell forwarding across a shared route segment. Virtual channels sharing a common route segment are assigned the same VPI values on the links of this segment and routing within this segment is based on the smaller VPI field rather than on the VCI field. (a) The two virtual channels are assigned the same VPIs x and y, on the links 0–1 and 1–2, respectively. (b) The switching at node 1 is now based on the VPI field and thus results in a smaller table, enabling more efficient switching.

used, it would be 24 bits long across the UNI and 28 bits long across the NNI. Such a large field would make the cell forwarding process expensive.

Another advantage of the use of virtual paths is that it enables the creation of logical links between nodes: the virtual path between two nodes is treated like a logical link by the virtual channels. In the example of Figure 6.12, the virtual path from node 0 to node 2 is treated as a logical link by the virtual channels.

6.2.2 Adaptation Layers

ATM uses fixed-size cells for transport, but applications using ATM either are continuous media such as voice or video, or use variable (and large) packets like IP. In this case, it is necessary to map the user data (voice, video, IP packets) into ATM cells. This is accomplished by an ATM adaptation layer (AAL). The main function of an AAL is segmentation and reassembly (SAR): an AAL *segments* the user data at the source into ATM cells and *reassembles* the ATM cells into user data at the destination.

Four ATM adaptation layers, AAL-1, AAL-2, AAL-3/4 and AAL-5, are described in ITU recommendation I.363. (AAL-3 and AAL-4 started life separately but have since been merged into a single AAL.) We briefly describe AAL-1 and AAL-5.

AAL-1

AAL-1 is meant for transport of constant bit rate data such as circuits, voice, and video. Here, the source can be considered to send a continuous stream of data. This data is segmented by AAL-1 into 47-byte AAL payloads. AAL-1 adds a 1-byte header, containing a sequence number field, and sends the resulting 48-byte packet, which constitutes the ATM payload, to the ATM layer for transport to the peer AAL-1 process at the destination node in the network. While the sequence number field is protected by a CRC (4 bits of SN are protected by a 3-bit CRC and a 1-bit parity check), the 47-byte payload is unprotected. This is considered adequate for the circuit emulation and voice applications that AAL-1 is designed to support.

AAL-5

AAL-5 is designed to transport variable-sized packets, up to $2^{16} = 65,536$ bytes in length, over an ATM network. Its most significant use is for the transport of IP packets over an ATM network. AAL-5 segments the user packets into cells but does not add any overhead (AAL header or trailer) in every cell. Instead, it uses the Payload Type field in the ATM header to indicate whether a cell is the last cell of a segmented IP packet or not. If a cell is the last cell of a segmented IP packet, the last 2 bytes of the cell constitute the AAL-5 trailer and contain the length of the IP packet and a CRC covering the entire IP packet. Thus, in all but one cell, the AAL-5 payload is equal to the 48-byte ATM payload, and AAL-5 has lower overhead compared to AAL-1. Also note that AAL-5 provides error detection for its payload through the use of a CRC, whereas AAL-1 does not.

6.2.3   Quality of Service

The primary motivation for use of ATM is that it is capable of providing quality-of-service (QoS) guarantees for connections. These guarantees take the form of bounds on cell loss, cell delay, and jitter. ATM is able to provide such guarantees through a combination of traffic shaping and admission control. Roughly speaking, this works as follows:

1. **Traffic Shaping:** ATM requires that all user traffic adhere to a contract that has been established between the user and the network. This contract usually specifies the peak cell rate, the average cell rate, and the burst size (number of consecutive cells at the peak cell rate) that the user can transfer across the UNI. The ATM network may monitor these contracted parameters for each connection across the UNI and can drop those cells that violate this contract. Alternatively, it can admit the violating cells but mark the CLP bit for these cells so that they are preferentially dropped in the event of congestion. As a result of this, ATM can carefully control the traffic from each connection that enters the network. The network's half of this bargain is the QoS guarantees that it provides to the user in terms of cell loss, delay, and jitter.

2. **Admission Control:** Based on the knowledge of the user traffic characteristics that are enforced through traffic shaping, the ATM network can determine the set of connections it can admit without violating the guranteed QoS for the connections when the cells from these connections are transferred across the network. A new connection will not be admitted if it would potentially result in the violation of QoS guarantees provided to connections that have already been established.

Based on the QoS parameters that the network can guarantee (cell loss, delay, jitter) and the traffic parameters that the user can specify (peak cell rate, average cell rate, burst size), ATM identifies a number of service classes to which a connection can belong. Among these are the constant bit rate (CBR) and the unspecified bit rate (UBR) service classes. A CBR connection specifies only the peak cell rate and is guaranteed a specified cell loss, delay, and jitter. A UBR connection also specifies only the peak cell rate but has no QoS guarantees. AAL-1 has been designed specifically to support CBR connections, whereas AAL-5 is used for UBR connections.

Another aspect of guaranteeing QoS, in addition to traffic shaping and admission control, is the use of queueing policies. ATM uses sophisticated queueing techniques to ensure that the QoS guarantees for each service class are met in the face of misbehaving traffic from other service classes. ATM also uses sophisticated mathematical techniques to determine the admission control policy so that QoS guarantees are met.

6.2.4   Flow Control

ATM also provides a mechanism to control the traffic from a user, not based on a prespecified contract, but based on feedback about congestion levels in the network. Such a mechanism is applicable to some service classes designed primarily for data traffic, such as file transfers, which are capable of being flow controlled (but not for CBR). The flow control is implemented across the UNI using the GFC bytes in the ATM UNI header. Using messages encoded by these bytes, the ATM network can instruct the user across the UNI whether data can be transmitted, or if data transmission should be halted.

6.2.5   Signaling and Routing

While the VCI and VPI fields are used for forwarding ATM cells on a given route, the determination of this route is the responsibility of a routing protocol. The routing protocols used in ATM networks are the PNNI (private network-to-network interface) and B-ICI (broadband intercarrier interface) protocols standardized by the ATM forum. Here we provide a brief overview of PNNI routing.

The goal of PNNI routing is to determine a path through the network from the source to the destination. This path should be capable of meeting the QoS requirements of the user. Each link in the network is characterized by a set of parameters, which describes the state of the link. Examples of link state parameters include cell loss, maximum cell delay, and available link bandwidth. Another parameter for each link is its administrative cost or weight. This is meant to reflect the cost to the network for using this link. These parameters are advertised by each ATM switch for all the links outgoing from it. The link state advertisements are flooded to all other ATM switches in the network. As a result of these link state advertisements, each ATM switch has the current topology of the network with the states of all the links. Using this topology and link state information, the ingress switch in the network that receives an ATM connection request can calculate a path through the network that is capable of satisfying the QoS requested by the connection and that also minimizes some administrative cost in the network.

Once a route has been computed, each switch on the route should be informed of the new connection and its QoS requirement. The VCI/VPI labels also need to be set up at each switch. This is accomplished by the PNNI signaling protocol. Once the signaling protocol terminates successfully, the connection setup is complete and data traffic can begin to flow. The signaling protocol is invoked again to tear down the connection.

6.3 IP

IP (Internet Protocol) is by far the most widely used wide-area networking technology today. IP is the underlying network protocol used in the all-pervasive Internet and is equally important in most private intranets to link up computers. IP is a networking technology, or protocol, that is designed to work above a wide variety of lower layers, which are termed *data link layers* in the classical layered view of networks (Section 1.4). This is one of the important reasons for its widespread success.

Figure 6.13 shows IP within the layered architecture framework. Some traditional data link layers over which IP operates are those associated with popular local-area networks such as Ethernet and token ring. IP also operates over low-speed serial lines as well as high-speed optical fiber lines using well-known data link layer protocols—for example, high-level data link control (HDLC) or point-to-point protocol (PPP).

Several layering structures are possible to map IP into the optical layer. The term "IP over WDM" is commonly used to refer to a variety of possible mappings shown in Figure 6.14. Figure 6.14(a) shows an implementation where IP packets are mapped into ATM cells, which are then encoded using SONET framing. The

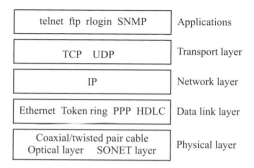

Figure 6.13   IP in the layered hierarchy, working along with a variety of data link layers and transport layers.

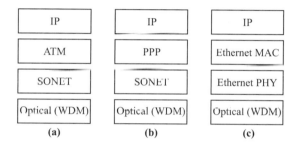

Figure 6.14   Various implementations of IP over WDM. (a) A traditional implementation, which maps IP packets into ATM cells, which are then encoded using SONET framing, before being transmitted over a wavelength. (b) The packet-over-SONET (POS) variant, where IP packets are mapped into PPP frames and then encoded using SONET framing. (c) Using Gigabit or 10-Gigabit Ethernet media access control (MAC) as the link layer and Gigabit or 10-Gigabit Ethernet physical layer (PHY) for encoding the frames for transmission over a wavelength.

SONET-framed signal is then transmitted over a wavelength. Figure 6.14(b) shows the packet-over-SONET (POS) implementation. Here, IP packets are mapped into PPP frames, and then encoded into SONET frames for transmission over a wavelength. Figure 6.14(c) shows an implementation using Gigabit or 10-Gigabit Ethernet as the underlying link (media access control) layer and Gigabit/10-Gigabit Ethernet physical layer (PHY) for encoding the frames for transmission over a wavelength. We will study the implications of these different approaches in Chapter 13.

IP, being a network layer protocol, does not guarantee reliable, in-sequence delivery of data from source to destination. This job is performed by a transport protocol, typically the *transmission control protocol* (TCP). Another commonly used transport protocol for simple message transfers over IP is the *user datagram protocol* (UDP). Commonly used applications, such as telnet, file transfer protocol (FTP), and rlogin, use TCP as their transport protocol, whereas certain other applications, such as the network file system (NFS) used to share files across a network and the simple network management protocol (SNMP) used for management, use UDP for transport. (We will talk about SNMP in Chapter 9.) UDP is also the transport protocol of choice for streaming media.

6.3.1   Routing and Forwarding

IP was one of the earliest packet-switching protocols. IP transports information in the form of packets, which are of variable length. An IP router is the key network element in an IP network. A router forwards packets from an incoming link onto an outgoing link. Figure 6.15 illustrates how packets are forwarded in an IP network. The nature of this routing is fundamental to IP. Here we describe the classical routing mechanism used by IP. Each router maintains a routing table. The routing table has one or more entries for each destination router in the network. The entry indicates the next node adjacent to this router to which packets need to be forwarded. The forwarding process works as follows. The router looks at the header in a packet arriving on an incoming link. The header contains the identity of the destination router for that packet. The router then does a lookup of its routing table to determine the next adjacent node for that packet, and forwards the packet on the link leading to that node. In the example shown in Figure 6.15, consider a packet from node 1 destined for node 4. Node 1 looks at its table and forwards this packet to node 5. Node 5 forwards the packet to node 3, which in turn forwards the packet to node 4, its ultimate destination.

Clearly, maintaining these routing tables at the routers is central to the operation of the network. It is likely that links and nodes in the network may fail, or reappear, and new links and nodes may be added over the course of time. The routers detect these changes automatically and update their routing tables using a distributed *routing protocol*. The protocol works as follows. Each router is assumed to be capable of determining whether its links to its neighbors are up or down. Whenever a router detects a change in the status of these links, it generates a *link state packet* and *floods* it to all the routers in the network. Flooding is a technique used to disseminate information across the network. Each node, upon receiving a flood packet, forwards the packet on all its adjacent links except the link it came from. Thus these packets eventually reach all the nodes in the network. A node receiving a link state packet

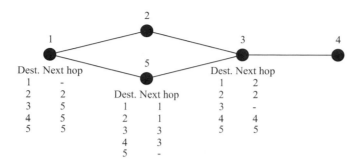

Figure 6.15   Routing in an IP network. The routing tables at some of the nodes are also shown. The tables contain the identity of the next hop node for each destination.

updates its routing table based on the new information. Over time, all nodes in the network have updated routing tables that reflect the current network topology.

There are a number of subtle enhancements needed to make the flooding process work reliably. For example, link state packets could take different paths through the network and undergo different delays. As a result, an older link state packet might arrive after a more recent up-to-date version. If left unchecked, this could cause damage. Consider what happens when a link goes down and comes back up. The first link state packet (packet X) says that the link is down and the subsequent one (packet Y) indicates that the link is up. A node receiving packet X after packet Y will think that the link is down, even after it has come up! To prevent this phenomenon, the link state packets have a sequence number. If a router receives a link state packet whose sequence number is lower than a previously received link state packet, it simply discards the packet. Packets could also be lost in the network, so link state updates are generated periodically and not just after a link up/down event occurs.

Using these link state packets, each router can construct its view of the entire network topology. On this topology, each router then computes the shortest path from itself to all the other routers and stores the identity of the next router in the path for each destination node in its routing table. A typical shortest-path algorithm used for this purpose is the Dijkstra algorithm [Dij59].

The routing protocol that we have described above is an example of an *intradomain routing protocol*. One of the most commonly used intradomain routing protocols in the Internet—*Open Shortest Path First* (OSPF)—works just as we have described above.

The Internet is a very large network, and it is impractical to expect each router to maintain a topology of the entire Internet. For this purpose, the network is divided into multiple interconnected domains. Each domain is called an *autonomous system* (AS). A separate *interdomain routing protocol* is used to route between domains in a large network. One example of such a protocol is the *border gateway protocol* (BGP), details of which the reader can find in the references at the end of this chapter.

6.3.2   Quality of Service

IP networks traditionally offer "best-effort" services. IP tries its best to get a packet from its source to its destination. However, different packets may take different routes through the network and experience random delays, and some packets will be dropped if there is congestion in the network. There has been a great deal of effort to improve this state of affairs so as to offer some quality-of-service (QoS) assurance to the users of the network. Within IP, a mechanism called Diff-Serv (differentiated services) has been proposed. In Diff-Serv, packets are grouped into different classes, with the class type indicated in the IP header. The class type specifies how packets are treated within each router. Packets marked as *expedited forwarding* (EF) are handled in a separate queue and routed through as quickly as possible. Several additional priority levels of *assured forwarding* (AF) are also specified. An AF has two attributes: xy. The attribute x typically indicates the queue to which the packet is held in the router prior to switching. The attribute y indicates the drop preference for the packets. Packets with $y = 3$ have a higher likelihood of being dropped, compared to packets with $y = 1$.

While Diff-Serv attempts to tackle the QoS issue, it does not provide any end-to-end method to guarantee QoS. For example, we cannot determine a priori if sufficient bandwidth is available in the network to handle a new traffic stream with real-time delay requirements. This is one of the benefits of multiprotocol label switching, which we will study next.

6.3.3   Multiprotocol Label Switching (MPLS)

MPLS is a new technology in the IP world and has a wide variety of applications. MPLS can be thought of as a layer sandwiched between the IP layer and the data link layer. MPLS provides a label-switched path (LSP) between nodes in the network. A router implementing MPLS is called a label-switched router (LSR). Each packet now carries a label that is associated with a label-switched path. Each LSR maintains a label-forwarding table, which specifies the outgoing link and outgoing label for each incoming label. When an LSR receives a packet, it extracts the label, uses it to index into the forwarding table, replaces the incoming label with the outgoing label, and forwards the packet on to the link specified in the forwarding table. Note that the processing of actually setting up label-switched paths is a control function that is completely decoupled from the forwarding action taking place within each LSR.

This very simple MPLS paradigm has several applications in an IP network. One of the fundamental design philosophies in MPLS is that the label-switching and packet-forwarding process at each router is completely decoupled from how LSPs are set up and taken down in the network. We can think of the latter as a network control function, which involves first deciding what LSPs to set up or take down and then actually setting them up and taking them down. This simple separation allows us to build optimized hardware for packet forwarding, independent of the network control mechanisms, and allows for LSPs to be set up and taken down based on different criteria and using different protocols.

An LSR doing label forwarding can potentially process a much larger number of packets per second compared to a regular router because the label switching and forwarding process is much simpler than classical IP routing and can be implemented almost entirely in hardware. While many of the functions of classical IP routing discussed in the previous section can also be implemented in hardware, there is a close coupling between the routing function and the control function in IP. Any changes to the control framework get reflected in the routing behavior. As a result, existing hardware will not continue to remain optimized for routing if the control framework changes. In contrast, in MPLS, we can optimize the forwarding hardware in the LSRs, independent of how label-switched paths are set up or taken down.

Another major benefit of MPLS is that it introduces the notion of a path in an IP network. IP traditionally switches packets, or datagrams, and has no notion of end-to-end paths. Different packets between the same pair of routers could take different routes through the network, based on the current state of the routing tables at the routers. The ability to specify paths along which packets can be routed has several implications. First, a service provider owning a network can now plan end-to-end routes for packets based on a variety of criteria. For example, it could plan routes so as to optimize the use of bandwidth in its network. It could plan routes to prevent some links from getting congested while other links are idle.

The ability to have explicit routed paths also allows a service provider to offer certain QoS assurances for selected traffic in the network. IP itself has traditionally offered "best-effort" service. As we said earlier, different packets could take different routes and could therefore arrive at their destinations with random delays. Moreover, it is quite possible and likely that this type of routing can cause congestion, or hot spots in parts of the network, causing a large number of packets to be dropped in the network. With MPLS, we could potentially reserve bandwidth along the links at the time an LSP is set up, to enable QoS guarantees.

Packets belonging to an LSP can be rerouted rapidly onto another LSP if there is a failure in the network. For example, we could set up two LSPs between a pair of nodes along diverse paths. If an LSP fails, we can reroute packets from that LSP to the other LSP and ensure rapid restoration of service. We will see in Chapter 10 that the IP routing mechanism itself cannot be relied upon to provide rapid rerouting of packets in case of a failure. Thus, MPLS can be used to provide rapid restoration times in an IP network in the case of failures.

Finally, MPLS can also be used to support multiple *virtual private networks* (VPNs) over a single IP network. Each VPN is carried over a separate set of LSPs, allowing the service provider to provide QoS, security, and other policy measures on a VPN-specific basis.

Deciding which LSPs to set up in a network can be a complicated process, depending on the objectives and the application. Luckily, as we indicated earlier, this function is completely decoupled from the label-switching mechanism in the LSRs. For example, if the objective is simply to reduce packet delay, we might set up LSPs between pairs of nodes with a lot of traffic between them. If the objective is to provide QoS guarantees, we would set up LSPs based on the bandwidth availability in the network.

Two signaling protocols are now available to set up LSPs across a network—the *resource reservation protocol* (RSVP) and the *label distribution protocol with constrained routing* (CR-LDP). Both protocols operate by sending a setup message from the source of the LSP to the destination of the LSP along the desired path on a hop-by-hop basis. Each LSR along the path determines if resources to support the LSP are available before passing on the setup message to the next LSR in the path. An acknowledgment message then flows back from the destination to the source along the path to complete the process.

6.3.4   Whither ATM?

Readers will note that almost all the capabilities of MPLS are offered by ATM. In fact, prior to the development of MPLS, ATM was viewed as the layer below IP, with which all the MPLS functions described above could be provided. Indeed, as of this writing, the various ATM protocols are much better developed and standardized, compared to MPLS. ATM was initially viewed as a replacement for IP—its fixed cell size allows high-speed switches to be designed, and its connection-oriented nature and superior QoS capabilities allow better transport of voice, video, and other real-time traffic over packet networks. However, IP appears to have won the day primarily because of its ubiquity—it is widely deployed in everything ranging from desktop computers to core routers and is hard to displace. As a result, the ATM standards have defined interfaces so that IP can operate using ATM as its immediately lower layer. The development of MPLS appears to threaten this use of ATM as well. You could argue that MPLS is better optimized for use in data networks because it allows larger packet sizes

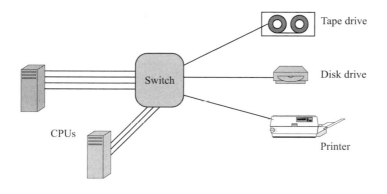

Figure 6.16 Architecture of a storage-area network.

(1500 bytes) compared to ATM's 53-byte cells. Interestingly, many MPLS routers use ATM switch fabrics internally to perform high-speed packet forwarding, so at least this aspect of ATM may continue to exist.

6.4 Storage-Area Networks

Storage-area networks (SANs) are networks used to interconnect computer systems with other computer systems and peripheral equipment, such as disk drives, printers, and tape drives. These networks are built by enterprises having medium to large data centers. Figure 6.16 shows a typical SAN interconnecting multiple CPUs and various types of peripheral devices. A key part of a SAN is a switch, which provides reconfigurable connectivity between the various attached devices. The SANs that we consider below all use a circuit-switched approach, where connections are rapidly established and taken down between the attached devices as needed.

In early installations, the entire SAN was located within a building or campus, but today the network is distributed over a wider metropolitan area, with some links extending into the long-haul network. One reason to do so is to be able to provide resilience against disasters. A common technique is to maintain two data centers, with data from one center backed up onto the other. Another reason to distribute the network is to locate peripherals and other equipment away from major downtown areas into cheaper suburban areas where real estate is less expensive.

SANs today typically operate at bit rates ranging from 200 Mb/s to 1 Gb/s and operate over fiber optic links in most cases. While the bit rate itself is relatively modest, what makes SANs important from the perspective of the optical layer is that

Table 6.4 Different storage-area networks. The fiber interfaces use either light emitting diodes (LEDs) and multimode fiber (MMF) or multilongitudinal mode laser (MLM) transmitters and standard single-mode fiber (SMF).

Network	Data Rate (MBytes/s)	Transmission Rate (Mbaud)	Physical Interface
ESCON	17	200	LED/MMF
			MLM/SMF
HIPPI	100		parallel copper
HIPPI (serialized)	100	1200	MLM/SMF
Fibre Channel	100	1063	MLM/SMF or copper
	50	531	MLM/SMF or copper
	25	266	MLM/SMF or copper
	12.5	133	MLM/SMF or copper
	200	2126	MLM/SMF
	400	4252	MLM/SMF

there can be a huge number of such connections between two data centers. Large mainframes have hundreds of I/O channels to connect them to other devices. It is not uncommon to see networks with hundreds to thousands of these links between two data centers.

The two main SAN technologies today are ESCON (enterprise serial connection) and Fibre Channel, with Fibre Channel dominating new installations. In addition, an older standard, called HIPPI (high performance parallel interface) is widely used in supercomputer and high-end computing installations.

Table 6.4 summarizes the salient attributes of ESCON, HIPPI, and Fibre Channel. These protocols typically add overhead to the data and then use line coding to encode the signal for transmission over the fiber. In each case, we have indicated the data rate as well as the actual transmission rate over the fiber, which is obtained after adding overheads and line coding. The latter rate is usually called the *baud rate;* thus we say the transmission rate is 1062.5 Mbaud rather than 1062.5 Mb/s.

6.4.1 ESCON

ESCON was developed by IBM in the late 1980s to replace the cumbersome, low-speed, and limited number of copper-based I/O interfaces on mainframe computers. It is widely deployed in large mainframe installations.

The data rate per ESCON channel is 17 MBytes/s. The transmission rate over the fiber after line coding and overheads is 200 Mbaud. LEDs at 1.3 μm are used over multimode fiber if the link length is less than 3 km. Longer distances, up to 20 km, are supported by using 1.3 μm MLM lasers over single-mode fiber.

One of the limiting factors of ESCON is that it uses a stop-and-wait link layer protocol. After sending a block of data, the sender waits for an acknowledgment from the receiver before sending the next block. As a result, the throughput on the links drops as the length of the link increases. To some extent, this can be offset by using larger block sizes. For this reason, many ESCON devices specify the maximum interconnection distance with other devices separately from the allowed link loss on the fiber link.

ESCON uses an (8, 10) line code (see Section 4.1.1) to avoid long runs of 0s or 1s and to achieve DC balance, that is, equal numbers of transmitted 0 and 1 bits.

6.4.2 Fibre Channel

Fibre Channel is a standard developed in the early 1990s, used for the same set of applications as ESCON. Like ESCON, the Fibre Channel architecture includes I/O ports on computers and peripherals, as well as an electronic switch. Fibre Channel is now widely deployed. The standard allows a variety of data rates. The most popular rate in use today is the "full speed" 100 MBytes/s rate; higher rates have been defined as shown in Table 6.4. Quarter-speed (25 MBytes/s) interfaces have also been deployed. Fibre Channel uses the same (8, 10) line code (see Section 4.1.1) as ESCON.

Both copper and fiber interfaces have been defined, with the fiber interface widely used in practice. Shielded twisted-pair copper interfaces are also deployed up to the 100 MBytes/s rate.

6.4.3 HIPPI

HIPPI is a 100 MBytes/s parallel electrical I/O interface standard. Owing to clock skew, the maximum distance is limited to 25 m. For longer distances, HIPPI is serialized and transmitted over single-mode fiber. A modified standard called Serial HIPPI, which includes an optical interface at 1.2 Gbaud, has been defined recently for this purpose. The standard also supports a 200 MBytes/s serial interface using two 100 MBytes/s serial interfaces in parallel. Work is also under way toward defining a 12-fiber version of the protocol, each fiber supporting a 100 MBytes/s data rate.

HIPPI predates Fibre Channel and is widely deployed in supercomputer installations. Like ESCON and Fibre Channel, a HIPPI network consists of hosts and peripherals connected via HIPPI switches and, in many cases, serial fiber optic links.

6.5   Gigabit and 10-Gigabit Ethernet

Ethernet is the most popular local-area packet-switched network today. The original Ethernet operated at 10 Mb/s and was then upgraded to 100 Mb/s. Ethernet is based on a bus architecture where all the nodes are connected to a single bus. The nodes use a simple media access control protocol called carrier-sense multiple access with collision detect (CSMA/CD). A node wanting to send a packet senses the bus to see if it is idle. Upon detecting that it is idle, it transmits the packet. If another node happens to sense the bus at the same time and transmits a packet, the two packets collide and get corrupted. In this case, both nodes back off and attempt to transmit again after waiting for a randomized delay interval. At higher speeds and longer bus lengths, the efficiency of the protocol drops. For this reason, Ethernet is also deployed in point-to-point configurations with only two nodes on the bus. An Ethernet switch is used to interconnect multiple such busses.

Gigabit Ethernet is an extension of the same standard to 1 Gb/s. It operates over both copper and fiber interfaces. Gigabit Ethernet over fiber is becoming a popular choice in metro networks to interconnect multiple enterprise networks. It is also extending its tentacles into the long-haul network.

Currently, there is work under way to extend the Ethernet standard to 10 Gb/s. This standard is being developed with the intent of enabling long-haul interconnections, with the data rate being aligned to the OC-192/STM-64 SONET/SDH rates for better compatibility with wide-area transport.

Summary

In this chapter, we studied several important client layers of the optical layer. These have been deployed widely in public telecommunications networks as well as private enterprise networks. The public transmission infrastructure in North America is dominated by SONET; SDH is used in most other parts of the world. SONET/SDH provides efficient time division multiplexing for low-speed streams and allows these streams to be transported across the network in a reliable, well-managed way. The predominant network layer protocol today is IP. Most of the data traffic entering the network is IP traffic, spurred by the growth of the Internet and corporate intranets. IP provides primarily best-effort routing of packets from their source to destination and has no notion of connections. A new link layer, MPLS, is emerging below the IP layer, to expand the scope of IP to allow explicit routing of packets along defined paths through the network. ATM is another protocol that provides similar capabilities.

Storage-area networks area constitute another important class of networks using optical fiber for transmission. These are used to link up computers to other computers and their peripherals. ESCON, HIPPI, and Fibre Channel are all widely deployed, with Fibre Channel being more popular for new deployments.

Further Reading

A general reference that covers SONET, IP, and ATM is the book by Walrand and Varaiya [WV00]. There is an extensive body of literature dealing with SONET/SDH. A comprehensive set of papers that cover the multiplexing standards, network topologies, and performance and management is collected in [SS96]. See also the book by Sexton and Reid [SR97] for an advanced treatment of the subject and [Gor00] as well. SONET/SDH has been extensively standardized by the American National Standards Institute (ANSI) and the International Telecommunications Union (ITU). In addition, Telcordia publishes generic criteria for equipment vendors. A list of the standards documents may be obtained on the World Wide Web at *www.itu.ch*, *www.ansi.org*, and *www.telcordia.com*; some of them are listed in Appendix C. Telcordia's GR-253 [Tel99] contains an extensive description of SONET, which we have made liberal use of in this chapter.

Readers wanting to learn about ATM and IP will be deluged with information. There are several books on ATM. See, for instance, [dP95, MS98]. The ATM forum (*www.atmforum.com*) maintains and makes available the ATM standards.

For an introductory overview of IP, see [PD99, Per99]. See [Com00, Ste94] for a more detailed treatment of TCP/IP, and [DR00] for MPLS. The Internet Engineering Task Force (*www.ietf.org*) develops and maintains standards, with all standards documents (RFCs—request for comments) being readily available.

ESCON was invented at IBM in the late 1980s [CdLS92, ES92]. It was subsequently standardized by ANSI as SBCON [Ame97]. ANSI standards have been established for HIPPI and Fibre Channel as well. [Cla99, TS00] provide primers on storage-area networks in general, focusing on Fibre Channel solutions. See *www.hippi.org* for HIPPI, including pointers to the HIPPI standards, and [Ame98, Ben96, SV96] as well as *www.fibrechannel.org* for Fibre Channel. Finally, the Ethernet standards are available from ANSI. See *www.gigabit-ethernet.org* for details on Gigabit Ethernet.

Problems

6.1   Which sublayer within the SONET or optical layer would be responsible for handling the following functions?

(a) A SONET path fails and the traffic must be switched over to another path.

Table 6.5 Specifications for STM-16 intraoffice and short-haul interfaces (from ITU G.707).

Parameter	I-16	S-16.1
Transmitter	MLM	SLM
Wavelength range	1.3 μm	1.3 μm
Transmit power (max)	−3 dBm	0 dBm
Transmit power (min)	−10 dBm	−5 dBm
Receive sensitivity (min)	−18 dBm	−27 dBm
Receive overload (min)	−18 dBm	−27 dBm

    (b) Many SONET streams are to be multiplexed onto a higher-speed stream and transmitted over a SONET link.

    (c) A fiber fails and SONET line terminals at the end of the link reroute all the traffic on the failed fiber onto another fiber.

    (d) The error rate on a SONET link between regenerators is to be monitored.

    (e) The connectivity of an STS-1 stream through a network needs to be verified.

6.2 In Table 6.3, calculate the equivalent distance limitations of the different types of SONET systems. Assume a loss of 0.25 dB/km at 1550 nm and 0.5 dB/km at 1310 nm.

6.3 You have to connect two SDH boxes operating at STM-16 line rate over a link that can have a loss of anywhere from 0 to 7 dB. Unfortunately they do not support the same interfaces. One of them supports an I-16 interface and the other has an S-16.1 interface. The detailed specifications for these interfaces, extracted from ITU Recommendation G.707, are given in Table 6.5. Can you find a way to interconnect these boxes and make the link budget work? You are allowed to use variable optical attenuators in the link.

6.4 Consider an ESCON link operating at a data rate of 17 MBytes/s. The sender transmits a block of data and waits for an acknowledgment before sending the next block of data. Compute the throughput on the link for the following sets of parameters:

    (a) Block size of 1 KByte, link length of 1 km

    (b) Block size of 1 KByte, link length of 10 km

    (c) Block size of 1 KByte, link length of 100 km

    (d) Block size of 4 KBytes, link length of 10 km

    (e) Block size of 4 KBytes, link length of 100 km

Assume that the speed of light in fiber is 2×10^5 km/s.

References

[Ame97]   American National Standards Institute. X3.296. *Single-Byte Command Code Sets CONnection (SBCON) Architecture*, 1997.

[Ame98]   American National Standards Institute. X3.303. *Fibre Channel Physical and Signalling Interface-3 (FC-3)*, 1998.

[Ben96]   A. F. Benner. *Fibre Channel*. McGraw-Hill, New York, 1996.

[CdLS92]   S. A. Calta, S. A. deVeer, E. Loizides, and R. N. Strangewayes. Enterprise systems connection (ESCON) architecture—system overview. *IBM Journal of Research and Development*, 36(4):535–551, July 1992.

[Cla99]   T. Clark. *Designing Storage-Area Networks*. Addison-Wesley, Reading, MA, 1999.

[Com00]   D. E. Comer. *Internetworking with TCP/IP: Vol. I: Principles, Protocols and Architecture*. Prentice Hall, Englewood Cliffs, NJ, 2000.

[Dij59]   E. W. Dijkstra. A note on two problems in connexion with graphs. *Numerical Mathematics*, pages 269–271, 1959.

[dP95]   M. de Prycker. *Asynchronous Transfer Mode: Solution for Broadband ISDN* Prentice Hall, London, 1995.

[DR00]   B. S. Davie and Y. Rekhter. *MPLS Technology and Applications*. Morgan Kaufmann, San Francisco, 2000.

[ES92]   J. C. Elliott and M. W. Sachs. The IBM enterprise systems connection architecture. *IBM Journal of Research and Development*, 36(4):577–591, July 1992.

[Gor00]   W. J. Goralski. *SONET*. McGraw-Hill, New York, 2000.

[MS98]   D. E. McDysan and D. L. Spohn. *ATM: Theory and Application*. McGraw-Hill, New York, 1998.

[PD99]   L. L. Peterson and B. S. Davie. *Computer Networks: A Systems Approach*. Morgan Kaufmann, San Francisco, 1999.

[Per99]   R. Perlman. *Interconnections: Bridges, Routers, Switches, and Internetworking Protocols*. Addison-Wesley, Reading, MA, 1999.

[SR97]   M. Sexton and A. Reid. *Broadband Networking: ATM, SDH and SONET*. Artech House, Boston, 1997.

[SS96]   C. A. Siller and M. Shafi, editors. *SONET/SDH: A Sourcebook of Synchronous Networking*. IEEE Press, Los Alamitos, CA, 1996.

[Ste94]   W. R. Stevens. *TCP/IP Illustrated, Volume 1*. Addison-Wesley, Reading, MA, 1994.

[SV96]   M. W. Sachs and A. Varma. Fibre channel and related standards. *IEEE Communications Magazine*, 34(8):40–49, Aug. 1996.

[Tel99]   Telcordia Technologies. *SONET Transport Systems: Common Generic Criteria*, 1999. GR-253-CORE Issue 2, Revision 2.

[TS00]   R. H. Thornburg and B. J. Schoenborn. *Storage Area Networks: Designing and Implementing a Mass Storage System*. Prentice Hall, Englewood Cliffs, NJ, 2000.

[WV00]   J. Walrand and P. Varaiya. *High-Performance Communication Networks*. Morgan Kaufmann, San Francisco, 2000.

7
chapter

■ WDM Network Elements

W E HAVE ALREADY EXPLORED some of the motivations for deploying WDM networks in Chapter 1 and will go back to this issue in Chapter 13. These networks provide circuit-switched end-to-end *optical channels*, or *lightpaths*, between network nodes to their users, or *clients*. A lightpath consists of an optical channel, or wavelength, between two network nodes that is routed through multiple intermediate nodes. Intermediate nodes may switch and convert wavelengths. These networks may thus be thought of as *wavelength-routing* networks. Lightpaths are set up and taken down as dictated by the users of the network.

In this chapter we will explore the architectural aspects of the network elements that are part of this network. The architecture of such a network is shown in Figure 7.1. The network consists of *optical line terminals* (OLTs), *optical add/drop multiplexers* (OADMs), and *optical crossconnects* (OXCs) interconnected via fiber links. Not shown in the figure are optical line amplifiers, which are deployed along the fiber link at periodic locations to amplify the light signal. In addition, the OLTs, OADMs, and OXCs may themselves incorporate optical amplifiers to make up for losses. As of this writing, OLTs are widely deployed, and OADMs are deployed to a lesser extent. OXCs are just beginning to be deployed.

The architecture supports a variety of topologies, including ring and mesh topologies. OLTs multiplex multiple wavelengths into a single fiber and also demultiplex a composite WDM signal into individual wavelengths. OLTs are used at either end of a point-to-point link. OADMs are used at locations where some fraction of the wavelengths need to be terminated locally and others need to be routed to other

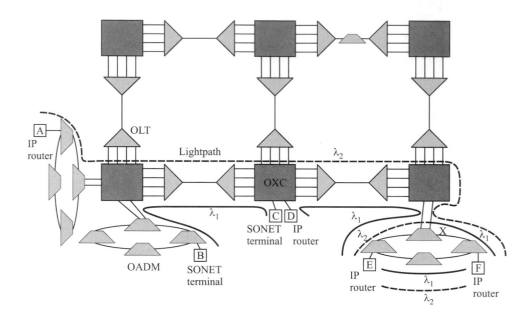

Figure 7.1   A wavelength-routing mesh network showing optical line terminals (OLTs), optical add/drop multiplexers (OADMs), and optical crossconnects (OXCs). The network provides lightpaths to its users, such as SONET boxes and IP routers. A lightpath is carried on a wavelength between its source and destination but may get converted from one wavelength to another along the way.

destinations. They are typically deployed in linear or ring topologies. OXCs perform a similar function but on a much larger scale in terms of number of ports and wavelengths involved, and are deployed in mesh topologies or in order to interconnect multiple rings. We will study these network elements in detail later in this chapter. The users (or clients) of this network are connected to the OLTs, OADMs, or OXCs. The network supports a variety of client types, such as IP routers, ATM switches, and SONET terminals and ADMs.

Each link can support a certain number of wavelengths. The number of wavelengths that can be supported depends on the component- and transmission-imposed limitations that we studied in Chapters 2, 3, and 5.

We next describe several noteworthy features of this architecture:

Wavelength reuse. Observe from Figure 7.1 that multiple lightpaths in the network can use the same wavelength, as long as they do not overlap on any link. This spatial reuse capability allows the network to support a large number of lightpaths using a limited number of wavelengths.

Wavelength conversion. Lightpaths may undergo *wavelength conversion* along their route. Figure 7.1 shows one such lightpath that uses wavelength λ_2 on link EX, gets converted to λ_1 at node X, and uses that wavelength on link XF. Wavelength conversion can improve the utilization of wavelengths inside the network. We will study this aspect in Section 7.4.1 and in Chapter 8. Wavelength conversion is also needed at the boundaries of the network to adapt signals from outside the network into a suitable wavelength for use inside the network.

Transparency. Transparency refers to the fact that the lightpaths can carry data at a variety of bit rates, protocols, and so forth and can, in effect, be made protocol insensitive. This enables the optical layer to support a variety of higher layers *concurrently*. For example, Figure 7.1 shows lightpaths between pairs of SONET terminals, as well as between pairs of IP routers. These lightpaths could carry data at different bit rates and protocols.

Circuit switching. The lightpaths provided by the optical layer can be set up and taken down upon demand. These are analogous to setting up and taking down circuits in circuit-switched networks, except that the rate at which the setup and take-down actions occur is likely to be much slower than, say, the rate for telephone networks with voice circuits. In fact, today these lightpaths, once set up, remain in the network for months to years. With the advent of new services and capabilities offered by today's network equipment, we are likely to see a situation where this process is more dynamic, both in terms of arrivals of lightpath requests and durations of lightpaths.

Note that packet switching is *not* provided within the optical layer. The technology for optical packet switching is still fairly immature; see Chapter 12 for details. It is left to the higher layer, for example, IP or ATM, to perform any packet-switching functions needed.

Survivability. The network can be configured such that, in the event of failures, lightpaths can be rerouted over alternative paths automatically. This provides a high degree of resilience in the network. We will study this aspect further in Chapter 10.

Lightpath topology. The *lightpath topology* is the graph consisting of the network nodes, with an edge between two nodes if there is a lightpath between them. The lightpath topology thus refers to the topology seen by the higher layers using the

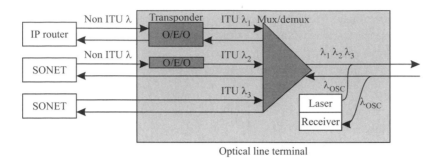

Figure 7.2   Block diagram of an optical line terminal. The OLT has wavelength multi-plexers and demultiplexers and adaptation devices called transponders. The transponders convert the incoming signal from the client to a signal suitable for transmission over the WDM link and an incoming signal from the WDM link to a suitable signal toward the client. Transponders are not needed if the client equipment can directly send and receive signals compatible with the WDM link. The OLT also terminates a separate optical supervisory channel (OSC) used on the fiber link.

optical layer. To an IP network residing above the optical layer, the lightpaths look like links between IP routers. The set of lightpaths can be tailored to meet the traffic requirements of the higher layers. This topic will be explored further in Chapter 8.

7.1   Optical Line Terminals

OLTs are relatively simple network elements from an architectural perspective. They are used at either end of a point-to-point link to multiplex and demultiplex wave-lengths. Figure 7.2 shows the three functional elements inside an OLT: *transponders, wavelength multiplexers,* and optionally, *optical amplifiers* (not shown in the figure). A transponder adapts the signal coming in from a client of the optical network into a signal suitable for use inside the optical network. Likewise, in the reverse direction, it adapts the signal from the optical network into a signal suitable for the client. The interface between the client and the transponder may vary depending on the client, bit rate, and the distance and/or loss between the client and the transponder. The most common interface is the SONET/SDH short-reach (SR) interface described in Section 6.1.4. We are also seeing the emergence of cheaper very short reach (VSR) interfaces at bit rates of 10 Gb/s and higher.

The adaptation includes several functions, which we will explore in detail in Section 9.6.3. The signal may need to be converted into a wavelength that is suited for use inside the optical network. The wavelengths generated by the transponder typically conform to standards set by the International Telecommunications Union (ITU) in the 1.55 μm wavelength window, as indicated in the figure, while the incoming signal may be a 1.3 μm signal. The transponder may add additional overhead for purposes of network management. It may also add forward error correction (FEC), particularly for signals at 10 Gb/s and higher rates. The transponder typically also monitors the bit error rate of the signal at the ingress and egress points in the network. For these reasons, the adaptation is typically done through an optical-to-electrical-to-optical (O/E/O) conversion. Down the road, we may see some of the all-optical wavelength-converting technologies of Section 3.8 being used in transponders—these are still in research laboratories.

In some situations, it is possible to have the adaptation enabled only in the incoming direction and have the ITU wavelength in the other direction directly sent to the client equipment. This is shown in the middle of Figure 7.2. In some other situations, we can avoid the use of transponders by having the adaptation function performed inside the client equipment that is using the optical network, such as a SONET network element. This is shown at the bottom of Figure 7.2. This reduces the cost and results in a more compact and power-efficient solution. However, this WDM interface specification is proprietary to each WDM vendor, and there are no standards. (More on this in Section 9.4.) Transponders typically constitute the bulk of the cost, footprint, and power consumption in an OLT. Therefore reducing the number of transponders helps minimize both the cost and the size of the equipment deployed.

The signal coming out of a transponder is multiplexed with other signals at different wavelengths using a wavelength multiplexer onto a fiber. Any of the multiplexing technologies described in Chapter 3, such as arrayed waveguide gratings, dielectric thin-film filters, or fiber Bragg gratings, can be used for this purpose. In addition, an optical amplifier may be used to boost the signal power if needed. In the other direction, the WDM signal is amplified again, if needed, before it is sent through a demultiplexer that extracts the individual wavelengths. These wavelengths are again terminated in a transponder (if present) or directly in the client equipment.

Finally, the OLT also terminates an *optical supervisory channel* (OSC). The OSC is carried on a separate wavelength, different from the wavelengths carrying the actual traffic. It is used to monitor the performance of amplifiers along the link as well as for a variety of other management functions that we will study in Chapter 9.

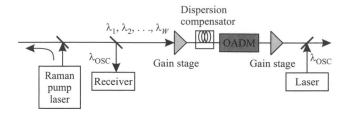

Figure 7.3   Block diagram of a typical optical line amplifier. Only one direction is shown. The amplifier uses multiple erbium gain stages and optionally includes dispersion compensators and OADMs between the gain stages. A Raman pump may be used to provide additional Raman gain over the fiber span. The OSC is filtered at the input and terminated, and added back at the output.

7.2   Optical Line Amplifiers

Optical line amplifiers are deployed in the middle of the optical fiber link at periodic intervals, typically 80–120 km. Figure 7.3 shows a block diagram of a fairly standard optical line amplifier. The basic element is an erbium-doped fiber gain block, which we studied in Chapter 3. Typical amplifiers use two or more gain blocks in cascade, with so-called midstage access. This feature allows some lossy elements to be placed between the two amplifier stages without significantly impacting the overall noise figure of the amplifier (see Problem 4.5 in Chapter 4). These elements include dispersion compensators to compensate for the chromatic dispersion accumulated along the link, and also the OADMs that we will discuss next. The amplifiers also include automatic gain control (see Chapter 5) and built-in performance monitoring of the signal, a topic we will discuss in Chapter 9.

We are also seeing the use of Raman amplifiers, where a high-power pump laser is used at each amplifier site to pump the fiber in the direction opposite to the signal. The optical supervisory channel is filtered at the input and terminated, and added back at the output. In a system using C- and L-bands, the bands are separated at the input to the amplifier and separate EDFAs are used for each band.

7.3   Optical Add/Drop Multiplexers

Optical add/drop multiplexers (OADMs) provide a cost-effective means for handling passthrough traffic in both metro and long-haul networks. OADMs may be used at

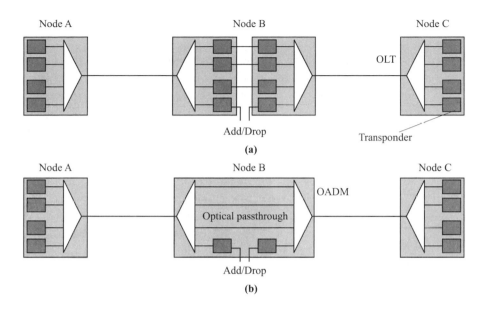

Figure 7.4 A three-node linear network example to illustrate the role of optical add/drop multiplexers. Three wavelengths are needed between nodes A and C, and one wavelength each between nodes A and B and between nodes B and C. (a) A solution using point-to-point WDM systems. (b) A solution using an optical add/drop multiplexer at node B.

amplifier sites in long-haul networks but can also be used as stand-alone network elements, particularly in metro networks. To understand the benefits of OADMs, consider a network between three nodes, say, A, B, and C, shown in Figure 7.4, with IP routers located at nodes A, B, and C. This network supports traffic between A and B, B and C, and A and C. Based on the network topology, traffic between A and C passes through node B. For simplicity, we will assume full-duplex links and full-duplex connections. This is the case for most networks today. Thus the network in Figure 7.4 actually consists of a pair of fibers carrying traffic in opposite directions.

Suppose the traffic requirement is as follows: one wavelength between A and B, one wavelength between B and C, and three wavelengths between A and C. Now suppose we deploy point-to-point WDM systems to support this traffic demand. The resulting solution is shown in Figure 7.4(a). Two point-to-point systems are deployed, one between A and B and the other between B and C. As we saw earlier in Section 7.1, each point-to-point system uses an OLT at each end of the link. The

OLT includes multiplexers, demultiplexers, and transponders. These transponders constitute a significant portion of the system cost.

Consider what is needed at node B. Node B has two OLTs. Each OLT terminates four wavelengths and therefore requires four transponders. However, only one out of those four wavelengths is destined for node B. The remaining transponders are used to support the passthrough traffic between A and C. These transponders are hooked back to back to provide this function. Therefore, six out of the eight transponders at node B are used to handle passthrough traffic—a very expensive proposition.

Consider the OADM solution shown in Figure 7.4(b). Instead of deploying point-to-point WDM systems, we now deploy a wavelength-routing network. The network uses an OLT at nodes A and C and an OADM at node B. The OADM drops one of the four wavelengths, which is then terminated in transponders. The remaining three wavelengths are passed through in the optical domain using relatively simple filtering techniques, without being terminated in transponders. The net effect is that only two transponders are needed at node B, instead of the eight transponders required for the solution shown in Figure 7.4(a). This represents a significant cost reduction. We will explore this subject of cost savings in detail in Section 8.1.

In typical carrier networks, the fraction of traffic that is to be passed through a node without requiring termination can be quite large at many of the network nodes. Thus OADMs perform a crucial function of passing through this traffic in a cost-effective manner.

Going back to our example, the reader may ask why transponders are needed in the solution of Figure 7.4(a) to handle the passthrough traffic. In other words, why can't we simply eliminate the transponders and connect the WDM multiplexers and demultiplexers between the two OLTs at node B directly, as shown in Figure 7.4(b), rather than designing a separate OADM? Indeed, this is possible, provided those OLTs are engineered to support such a capability. The physical layer engineering for networks is considerably more complex than that for point-to-point systems, as we saw in Chapter 5. For example, in a simple point-to-point system design, the power level of a signal coming into node B from node A might be so low that it cannot be passed through for another hop to node C. Also, in a network, the power of the signals added at a node must ideally be equal to the power of the signals passing through. However, there are also simpler and less expensive methods for building OADMs, as we will see in Section 7.3.1.

We will see in the next section that today's OADMs are rather inflexible. They are, for the most part, static elements and do not allow in-service selection under software control of what channels are dropped and passed through. We will see how *reconfigurable* OADMs can be built in Section 7.3.2, using tunable filters and lasers.

7.3.1 **OADM Architectures**

Several architectures have been proposed for building OADMs. These architectures typically use one or more of the multiplexers/filters that we studied in Chapter 3. Most practical OADMs use either fiber Bragg gratings, dielectric thin-film filters, or arrayed waveguide gratings. Here, we view an OADM as a black box with two line ports carrying the aggregate set of wavelengths and a number of local ports, each dropping and adding a specific wavelength. The key attributes to look for in an OADM are the following:

- What is the total number of wavelengths that can be supported?

- What is the maximum number of wavelengths that can be dropped/added at the OADM? Some architectures allow only a subset of the total number of wavelengths to be dropped/added.

- Are there constraints on whether specific wavelengths can be dropped/added? Some architectures only allow a certain set of wavelengths to be dropped/added and not any arbitrary wavelength. This capability ranges from being able to add/drop a single wavelength, to groups of wavelengths, to any arbitrary wavelength. This has a significant impact on how traffic can be routed in the network, as we will see below.

- How easy is it to add and drop additional channels? Is it necessary to take a service hit (i.e., disrupt existing channels) in order to add/drop an additional channel? This is the case with some architectures but not with others.

- Is the architecture modular, in the sense that the cost is proportional to the number of channels dropped? This is important to service providers because they prefer to "pay as they grow" as opposed to incurring a high front-end cost. In other words, service providers usually start with a small number of channels in the network and add additional channels as traffic demands increase.

- What is the complexity of the physical layer (transmission) path design with the OADM and how does adding new channels or nodes affect this design? Fundamentally, if the overall passthrough loss seen by the channels is independent of the number of channels dropped/added, then adding/dropping additional channels can be done with minimal impact to existing channels. (Other impairments like crosstalk would still have to be factored in, however.) This is an important aspect of the design that we will pay close attention to.

- Is the OADM reconfigurable, in the sense that selected channels can be dropped/added or passed through under remote software control? This is a desirable feature to minimize manual intervention. For instance, if we need to drop an additional channel at a node due to traffic growth at that node, it would be simpler to do so under remote software control rather than sending a craftsperson to that location. We will study this issue in Section 7.3.2.

Figure 7.5 shows three different OADM architectures, and Table 7.1 compares their salient attributes. Several other variants are possible, and some will be explored in Problem 7.1.

In the parallel architecture (Figure 7.5(a)), all incoming channels are demultiplexed. Some of the demultiplexed channels can be dropped locally and others are passed through. An arbitrary subset of channels can be dropped and the remaining passed through. So there are no constraints on what channels can be dropped and added. As a consequence this architecture imposes minimal constraints on planning lightpaths in the network. In addition, the loss through the OADM is fixed, independent of how many channels are dropped and added. So if the other transmission impairments discussed in Chapter 5 are taken care of by proper design, then adding and dropping additional channels does not affect existing channels. Unfortunately, this architecture is not very cost-effective for handling a small number of dropped channels because, regardless of how many channels are dropped, all channels need to be demultiplexed and multiplexed back together. Therefore we need to pay for all the demultiplexing and multiplexing needed for all channels, even if we need to drop only a single channel. This also results in incurring a higher loss through the OADM. However, the architecture becomes cost-effective if a large fraction of the total number of channels is to be dropped, or if complete flexibility is desired with respect to adding and dropping any channel. The other impact of this architecture is that since all channels are demultiplexed and multiplexed at all the OADMs, each lightpath passes through many filters before reaching its destination. As a result, wavelength tolerances on the multiplexers and lasers (see Section 5.6.6) can be fairly stringent.

Some cost improvements can be made by making the design modular as shown in Figure 7.5(b). Here, the multiplexing and demultiplexing is done in two stages. The first stage of demultiplexing separates the wavelengths into bands, and the second stage separates the bands into individual channels. For example, a 16-channel system might be implemented using four bands, each having 4 channels. If only 4 channels are to be dropped at a location, the remaining 12 channels can be expressed through at the band level, instead of being demultiplexed down to the individual channel level. In addition to the cost savings in the multiplexers and demultiplexers realized, the use of bands allows signals to be passed through with lower optical loss and better loss uniformity. Several commercially available OADMs use this approach. Moreover, as the number of channels becomes large, a modular multistage multiplexing approach (see Section 3.3.10) becomes essential. Parallel OADMs are

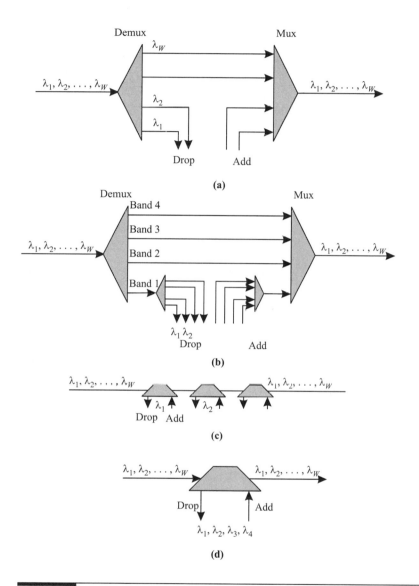

Figure 7.5   Different OADM architectures. (a) Parallel, where all the wavelengths are separated and multiplexed back; (b) modular version of the parallel architecture; (c) serial, where wavelengths are dropped and added one at a time; and (d) band drop, where a band of wavelengths are dropped and added together. W denotes the total number of wavelengths.

Table 7.1 Comparison of different OADM architectures. W is the total number of channels and D represents the maximum number of channels that can be dropped by a single OADM.

Attribute	Parallel	Serial	Band Drop
D	$= W$	1	$\ll W$
Channel constraints	None	Decide on channels at planning stage	Fixed set of channels
Traffic changes	Hitless	Requires hit	Partially hitless
Wavelength planning	Minimal	Required	Highly constrained
Loss	Fixed	Varies	Fixed up to D
Cost (small drops)	High	Low	Medium
Cost (large drops)	Low	High	Medium

typically realized using dielectric thin-film filters and arrayed waveguide gratings, and may use interleaver-type filters for large channel counts.

In the serial architecture (Figure 7.5(c)), a single channel is dropped and added from an incoming set of channels. We call this device a single-channel OADM (SC-OADM). These can be realized using fiber Bragg gratings or dielectric thin-film filters. In order to drop and add multiple channels, several SC-OADMs are cascaded. This architecture in many ways complements the parallel architecture described above. Adding and dropping additional channels disrupts existing channels. Therefore it is desirable to plan what set of wavelengths need to get dropped at each location ahead of time to minimize such disruptions. The architecture is highly modular in that the cost is proportional to the number of channels dropped. Therefore the cost is low if only a small number of channels are to be dropped. However, if a large number of channels are to be dropped, the cost can be quite significant since a number of individual devices must be cascaded. There is also an indirect impact on the cost because the loss increases as more channels are dropped, requiring the use of additional amplification.

The increase of loss with number of channels dropped plays a major role in increasing the complexity of deploying networks using serial OADMs. This is illustrated by the simple example shown in Figure 7.6. Suppose the allowed link budget for a lightpath between a transmitter and a receiver is 25 dB. Consider a situation where a lightpath from node B to node D is deployed with a loss of close to 25 dB between its transmitter and receiver. Now consider the situation when a new lightpath is to be supported at a different wavelength from node A to node C. In order to support this lightpath, an additional SC-OADM must be deployed at node C (and at node A) to drop the new lightpath. This OADM introduces an additional loss, say,

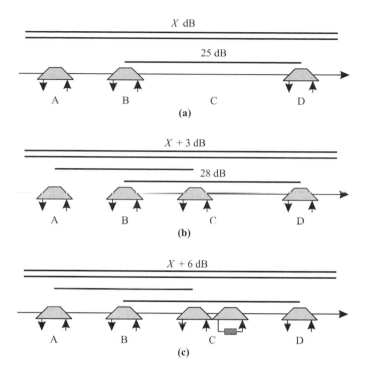

Figure 7.6   Impact of traffic changes on a network using serial OADMs. (a) Initial situation. (b) A new lightpath is added between node A and node C, causing lightpath BD to fail. (c) Lightpath BD is regenerated by adding a regenerator at node C. However, this causes other lightpaths flowing through C to be impacted.

of 3 dB, to the channels passing through node C. Introducing this OADM suddenly increases the loss on the lightpath from B to D to 28 dB, making it inoperative. The story doesn't end there, however! Suppose that in order to fix this problem we decide to regenerate this lightpath at node C. In order to regenerate this lightpath, we need to drop it at node C, send it through a regenerator, and add it back. This requires an additional SC-OADM at node C, which introduces 3 dB of additional loss for channels passing through node C. This in turn could disrupt other lightpaths passing through node C. Therefore adding or dropping additional channels can have a ripple effect on all the other lightpaths in the network. The use of optical amplifiers in conjunction with careful link engineering can alleviate some of these problems. For instance, a certain amount of loss can be allocated up front, after an optical amplifier is introduced. SC-OADMs can be added until the loss budget is met, after which another amplifier can be added.

Note that passthrough channels do not undergo any filtering. As a result, each lightpath only passes through two filters, one at the source node and one at the destination node. Thus wavelength tolerances on the multiplexers and lasers are less stringent, compared to the parallel architecture.

In the band drop architecture (Figure 7.5(d)), a fixed group of channels is dropped and added from the aggregate set of channels. The dropped channels then typically go through a further level of demultiplexing where they are separated out. The added channels are usually combined together with simple couplers and added to the passthrough channels. A typical implementation could drop, say, 4 adjacent channels out of 32 channels using a band filter.

This architecture tries to make a compromise between the parallel architecture and the serial architecture. The maximum number of channels that can be dropped is determined by the type of band filter used. Within this group of channels, adding/dropping additional channels doesn't affect other lightpaths in the network as the passthrough loss for all the other channels not in this group is fixed.

However, this architecture does complicate wavelength planning in the network and places several constraints on wavelength assignment because the same set of wavelengths are dropped at each location. For example, if wavelength λ_1 is added at a node and dropped at the next node, all other wavelengths, say, $\lambda_2, \lambda_3, \lambda_4$, in the same band as λ_1 will also be added at the same node and dropped at the next node. What makes this not so ideal is that once a wavelength is dropped as part of a band, it will likely need to be regenerated before it can be added back into the network. So in this example, wavelengths $\lambda_2, \lambda_3, \lambda_4$ will need to be regenerated at both nodes even if they are passing through. It is difficult to engineer the link budget to allow optical passthrough of these wavelengths without regeneration. This problem can be fixed by using different varieties of OADMs, each of which drops a different set of channels. As the reader can readily imagine, this makes network planning complicated. If wavelength drops can be planned in advance and the network remains static, then this may be a viable option. However, in networks where the traffic changes over time, this may not be easy to plan.

The architectures that we discussed above are the ones that are feasible based on today's technology, and commercial implementations of all of these exist today. It is clear that none of them offers a perfect solution that meets a full range of applications. Serial and band drop architectures have a low entry cost, but their deployment has been hindered due to the lack of flexibility in dealing with traffic changes in the network. There is certainly a trend toward building parallel architectures, while trying to retain a reasonable initial cost.

7.3.2   **Reconfigurable OADMs**

Reconfigurability is a very desirable attribute in an OADM. Reconfigurability refers to the ability to select the desired wavelengths to be dropped and added on the fly, as opposed to having to plan ahead and deploy appropriate equipment. This allows carriers to be flexible when planning their network and allows lightpaths to be set up and taken down dynamically as needed in the network. The architectures that we considered in Figure 7.5 were not reconfigurable in this sense.

Figure 7.7 shows a few different reconfigurable OADM architectures. Figure 7.7(a) shows a variation of the parallel architecture. It uses optical switches to add/drop specific wavelengths as and when needed. Figure 7.7(b) shows a variation of the serial architecture where each SC-OADM is now a tunable device that is capable of either dropping and adding a specific wavelength, or passing it through.

Both of these architectures only partially address the reconfigurability problem because transponders are still needed to provide the adaptation into the optical layer. We distinguish between two types of transponders: a fixed-wavelength transponder and a tunable transponder. A fixed-wavelength transponder is capable of transmitting and receiving at a particular fixed wavelength. This is the case with most of the transponders today. A tunable transponder, on the other hand, can be set to transmit at any desired wavelength and receive at any desired wavelength. A tunable transponder uses a tunable WDM laser and a broadband receiver capable of receiving any wavelength.

With fixed-wavelength transponders, in order to make use of the reconfigurable OADMs shown in Figure 7.7(a) and (b), we need to deploy the transponders ahead of time so that they are available when needed. This leads to two problems: First, it is expensive to have a transponder deployed and not used while the associated OADM is passing that wavelength through. But let us suppose that this cost is offset by the added value of being able to set up and take down lightpaths rapidly. The second problem is that although the OADMs are reconfigurable, the transponders are not. So we still need to decide ahead of time as to which set of wavelengths we will need to deploy transponders for, making the network planning problem more constrained.

Avoiding these problems requires the use of tunable transponders, and even more flexible architectures than the ones shown in Figure 7.7(a) and (b). For example, Figure 7.7(c) shows a serial architecture where we have full reconfigurability. Each tunable SC-OADM is capable of adding/dropping *any* single wavelength and passing the others through, as opposed to a fixed wavelength. The adaptation is performed using a tunable transponder. This provides a fully reconfigurable OADM. Likewise, Figure 7.7(d) shows a parallel architecture with full reconfigurability. Note that this

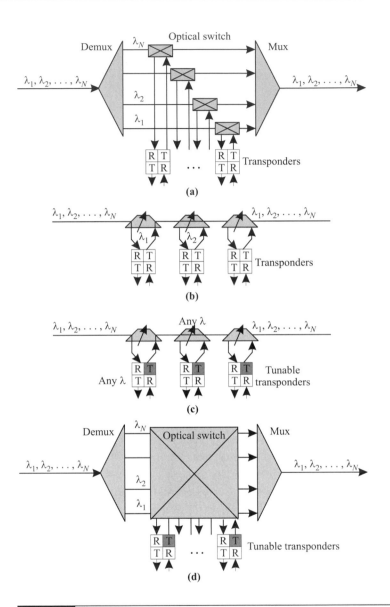

Figure 7.7 Reconfigurable OADM architectures. (a) A partially tunable OADM using a parallel architecture with optical add/drop switches and fixed-wavelength transponders. T indicates a transmitter and R indicates a receiver. (b) A partially tunable OADM using a serial architecture with fixed-wavelength transponders. (c) A fully tunable OADM using a serial architecture with tunable transponders. This transponder uses a tunable laser (marked T in the shaded box) and a broadband receiver. (d) A fully tunable OADM using a parallel architecture with tunable transponders.

architecture requires the use of a large optical switch. This is exactly the optical crossconnect that we will study next.

So what would an ideal OADM look like? Such an OADM (1) would be capable of being configured to drop a certain maximum number of channels, (2) would allow the user to select what specific channels are dropped/added and what are passed through under remote software control, including the transponders, without affecting the operation of existing channels, (3) would not require the user to plan ahead as to what channels may need to be dropped at a particular node, and (4) would maintain a low fixed loss regardless of how many channels are dropped/added versus passed through. The architecture of Figure 7.7(d) meets these criteria but may not be suitable for small-sized nodes where only a few channels need be dropped, due to its relatively high up-front cost. Other architectures will emerge as new component technologies such as tunable add/drops and tunable lasers become mature.

7.4   Optical Crossconnects

OADMs are useful network elements to handle simple network topologies, such as the linear topology shown in Figure 7.4 or ring topologies, and a relatively modest number of wavelengths. An additional network element is required to handle more complex mesh topologies and large numbers of wavelengths, particularly at hub locations handling a large amount of traffic. This element is the optical crossconnect (OXC). We will see that though the term "optical" is used, an OXC could internally use either a pure optical or an electrical switch fabric. An OXC is also the key network element enabling reconfigurable optical networks, where lightpaths can be set up and taken down as needed, without having to be statically provisioned.

Consider a large carrier central office hub. This might be an office in a large city for local service providers or a large node in a long-haul service provider's network. Such an office might terminate several fiber links, each carrying a large number of wavelengths. A number of these wavelengths might not need to be terminated in that location but rather passed through to another node. The OXC shown in Figure 7.8 performs this function. OXCs work alongside SONET/SDH network elements as well as IP routers and ATM switches, and WDM terminals and add/drop multiplexers as shown in Figure 7.8. Typically some OXC ports are connected to WDM equipment and other OXC ports to terminating devices such as SONET/SDH ADMs, IP routers, or ATM switches. Thus, the OXC provides cost-effective passthrough for express traffic not terminating at the hub as well as collects traffic from attached equipment into the network. Some people think of an OXC as a crossconnect switch together with the surrounding OLTs. However, our definition of OXC doesn't include

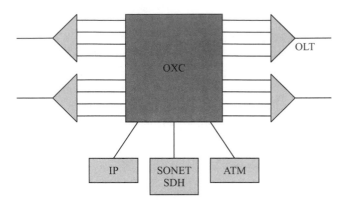

Figure 7.8   Using an OXC in the network. The OXC sits between the client equipment of the optical layer and the optical layer OLTs.

the surrounding OLTs, because carriers view crossconnects and OLTs as separate products and often buy OXCs and OLTs from different vendors.

An OXC provides several key functions in a large network:

Service provisioning.   An OXC can be used to provision lightpaths in a large network in an automated manner, without having to resort to performing manual patch panel connections. This capability becomes important when we deal with large numbers of wavelengths in a node or with a large number of nodes in the network. It also becomes important when the lightpaths in the network need to be reconfigured to respond to traffic changes. The manual operation of sending a person to each office to implement a patch panel connection is expensive and error prone. Remotely configurable OXCs take care of this function.

Protection.   Protecting lightpaths against fiber cuts and equipment failures in the network is emerging as one of the most important functions expected from a crossconnect. The crossconnect is an intelligent network element that can detect failures in the network and rapidly reroute lightpaths around the failure. Cross-connects enable true mesh networks to be deployed. These networks can provide particularly efficient use of network bandwidth, compared to the SONET/SDH rings we discussed in Chapter 6. We discuss this topic in detail in Chapter 10.

Bit rate transparency.   The ability to switch signals with arbitrary bit rates and frame formats is a desirable attribute of OXCs.

Performance monitoring, test access, and fault localization.   OXCs provide visibility to the performance parameters of a signal at intermediate nodes. They usually allow test equipment to be hooked up to a dedicated test port where the signals passing through the OXC can be monitored in a non-intrusive manner. Non-intrusive test access requires *bridging* of the input signal. In bridging, the input signal is split into two parts. One part is sent to the core, and the other part is made available at the test access port. OXCs also provide loopback capabilities. This allows a lightpath to be looped back at intermediate nodes for diagnostic purposes.

Wavelength conversion. In addition to switching a signal from one port to another port, OXCs may also incorporate wavelength conversion capabilities.

Multiplexing and grooming. OXCs typically handle input and output signals at optical line rates. However, they can incorporate multiplexing and grooming capabilities to switch traffic internally at much finer granularities, such as STS-1 (51 Mb/s). Note that this time division multiplexing has to be done in the electrical domain and is really SONET/SDH multiplexing, but incorporated into the OXC, rather than in a separate SONET/SDH box.

An OXC can be functionally divided into a switch core and a port complex. The switch core houses the switch that performs the actual crossconnect function. The port complex houses port cards that are used as interfaces to communicate with other equipment. The port interfaces may or may not include optical-to-electrical (O/E) or optical-to-electrical-to-optical (O/E/O) converters.

Figure 7.9 shows different types of OXCs and different configurations for interconnecting OXCs with OLTs or OADMs in a node. The scenarios differ in terms of whether the actual switching is done electrically or optically, in the use of O/E and O/E/O converters, and how the OXC is interconnected to the surrounding equipment. Table 7.2 summarizes the main differences between these architectures.

The first three configurations shown in Figure 7.9 are *opaque* configurations—the optical signal is converted into the electrical domain as it passes through the node. The last configuration (Figure 7.9(d)) is an *all-optical* configuration—the signal remains in the optical domain as it passes through the node.

Looking at Figure 7.9, observe that in the opaque configurations the switch core can be electrical or optical; that is, signals may be switched either in the electrical domain or in the optical domain. An electrical switch core can groom traffic at fine granularities and typically includes time division multiplexing of lower-speed circuits into the line rate at the input and output ports. Today, we have electrical core OXCs switching signals at granularities of STS-1 (51 Mb/s) or STS-48 (2.5 Gb/s). In contrast, a true optical switch core does not offer any grooming. It simply switches signals from one port to another.

An electrical switch core is designed to have a total switch capacity, for instance, 1.28 Tb/s. This capacity can be utilized to switch, say, up to 512 OC-48 (2.5 Gb/s)

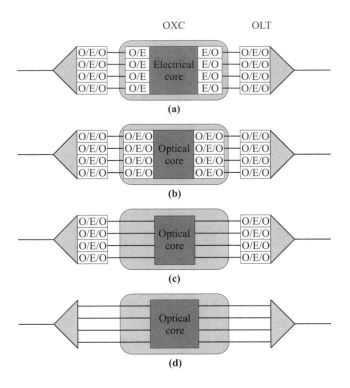

Figure 7.9 Different scenarios for OXC deployment. (a) Electrical switch core; (b) optical switch core surrounded by O/E/O converters; (c) optical switch core directly connected to transponders in WDM equipment; and (d) optical switch core directly connected to the multiplexer/demultiplexer in the OLT. Only one OLT is shown on either side in the figure, although in reality an OXC will be connected to several OLTs.

signals or 128 OC-192 (10 Gb/s) signals. The optical core is typically bit rate independent. Therefore a 1000-port optical switch core can switch 1000 OC-48 streams, 1000 OC-192 streams, or even 1000 OC-768 (40 Gb/s) streams, all at the same cost per port. The optical core is thus more scalable in capacity, compared to an electrical core, making it more future proof as bit rates increase in the future. In particular, the configuration of Figure 7.9(d) allows us to switch groups of wavelengths or all the wavelengths on a fiber together on a single OXC port, making that configuration capable of handling enormous overall capacities, and reducing the number of OXC ports required in a node.

As bit rates increase, the cost of a port on an electrical switch increases. For instance, an OC-192 port might cost twice as much as an OC-48 port. The cost of

Table 7.2 Comparison of different OXC configurations. Some configurations use optical to electrical converters as part of the crossconnect, in which case, they are able to measure electrical layer parameters such as the bit error rate (BER) and invoke network restoration based on this measurement. For the first two configurations, the interface on the OLTs is typically a SONET short-reach (SR), or very-short-reach (VSR) interface. For the opaque photonic configuration, it is an intermediate-reach (IR) or a special VSR interface. The cost, power, and footprint comparisons are made based on characteristics of commercially available equipment at OC-192 line rates.

Attribute	Opaque Electrical	Opaque Optical with O/E/Os	Opaque Optical	All-Optical
	Figure 7.9(a)	Figure 7.9(b)	Figure 7.9(c)	Figure 7.9(d)
Low-speed grooming	Yes	No	No	No
Switch capacity	Low	High	High	Highest
Wavelength conversion	Yes	Yes	Yes	No
Switching triggers	BER	BER	Optical power	Optical power
Interface on OLT	SR/VSR	SR/VSR	IR/serial VSR	Proprietary
Cost per port	Medium	High	Medium	Low
Power consumption	High	High	Medium	Low
Footprint	High	High	Medium	Low

a port on an optical core switch, on the other hand, is the same regardless of the bit rate. Therefore at higher bit rates, it will be more cost-effective to switch signals through an optical core OXC than an electrical core OXC.

An optical switch core is also transparent; it does not care whether it is switching a 10 Gb/s Ethernet signal or a 10 Gb/s SONET signal. In contrast, electrical switch cores require separate port cards for each interface type, which convert the input signal into a format suitable for the switch fabric.

Figure 7.9(a) shows an OXC consisting of an electrical switch core surrounded by O/E converters. The OXC interoperates with OLTs through standard non-WDM short-reach (SR) optical interfaces, typically at 1310 nm. We are also seeing the deployment of cheaper very-short-reach (VSR) interfaces. The OLT has transponders to convert this signal into the appropriate WDM wavelength. Alternatively the OXC itself may have wavelength-specific lasers that operate with the OLTs without requiring transponders between them.

Figure 7.9(b)–(d) show OXCs with an optical switch core. The differences between the figures lie in how they interoperate with the WDM equipment. In Figure 7.9(b), the interworking is done in a somewhat similar fashion as in Figure 7.9(a)—through the use of O/E/O converters with short-reach or very-short-reach optical interfaces between the OXC and the OLT. In Figure 7.9(c), there are no O/E/O converters and the optical switch core directly interfaces with the transponders in the OLT. Figure 7.9(d) shows a different scenario where there are no transponders in the OLT and the wavelengths in the fiber are directly switched by the optical switch core in the OXC after they are multiplexed/demultiplexed. The cost, power, and overall node footprint all improve as we go from Figure 7.9(b) to Figure 7.9(d). The electrical core option typically uses higher power and takes up more footprint, compared to the optical option, but the relative cost depends on how the different products are priced, as well as the operating bit rate on each port.

The OXCs in Figure 7.9(a) and (b) both have access to the signals in the electrical domain and can therefore perform extensive performance monitoring (signal identification and bit error rate measurements). The bit error rate measurement can also be used to trigger protection switching. Moreover, they can signal to other network elements by using inband overhead channels embedded in the data stream. (We will study signaling in more detail in Chapter 9.)

The OXCs in Figure 7.9(c) and (d) do not have the capability to look at the signal, and therefore they cannot do extensive signal performance monitoring. Thus, they cannot, for instance, invoke protection switching based on bit error rate monitoring, but instead could use optical power measurement as a trigger. These crossconnects need an out-of-band signaling channel to exchange control information with other network elements. With the configuration of Figure 7.9(c), the attached equipment needs to have optical interfaces that can deal with the loss introduced by the optical switch. These interfaces will also need to be single-mode fiber interfaces since that is what most optical switches are designed to handle. In addition, serial interfaces (single fiber pair) are preferred rather than parallel interfaces (multiple fiber pairs), as each fiber pair consumes a port on the optical switch.

The all-optical configuration of Figure 7.9(d) provides a truly all-optical network. However, it mandates a more complex physical layer design (see Chapter 5) as signals are now kept in the optical domain all the way from their source to their destination, being switched optically at intermediate nodes. Given that link engineering is complex and usually vendor proprietary, it is not easy to have one vendor's OXC interoperate with another vendor's OLT in this configuration.

Note also that the configurations of Figure 7.9(b), (c), and (d) can all be combined in a single OXC. We could have some ports having O/E/Os, others connected to OLTs with O/E/Os, and still others connected to OLTs without any O/E/Os.

It is possible to integrate the OXC and OLT systems together into one piece of equipment. Doing so provides some significant benefits. It eliminates the need for redundant O/E/Os in multiple network elements, allows tight coupling between the two to support efficient protection, and makes it easier to signal between multiple OXCs in a network using the optical supervisory channel available in the OLTs. For example, in Figure 7.9(a), we could have WDM interfaces directly on the crossconnect and eliminate the intraoffice short-reach interface. We would migrate from the configuration in Figure 7.9(b) to the configuration in Figure 7.9(c).

However, this integration also has the drawback of making it a single-vendor solution. Service providers must then buy all their WDM equipment, including OLTs and OXCs, from the same vendor in order to realize this simplification. Some service providers prefer to build their network by mixing and matching "best-in-class" equipment from multiple vendors. Moreover, this solution doesn't address the problem of dealing with legacy situations where the OLTs are already deployed and OXCs must be added later.

7.4.1 All-Optical OXC Configurations

We now focus the discussion on understanding some of the issues associated with the all-optical configuration of Figure 7.9(d). As shown, the configuration can be highly cost-effective relative to the other configurations, but lacks three key functions: low-speed grooming, wavelength conversion, and signal regeneration. Low-speed grooming is needed to aggregate the lower-speed traffic streams properly for transmission over the fiber. Optical signals need to be regenerated once they have propagated through a number of fiber spans and/or other lossy elements.

Wavelength conversion is needed to improve the utilization of the network. We illustrate this with the simple example shown in Figure 7.10. Each link in the three-node network can carry three wavelengths. We have two lightpaths currently set up on each link in the network as shown and need to set up a new lightpath from node A to node C. Figure 7.10(a) shows the case where node B cannot perform wavelength conversion. Even though there are free wavelengths available in the network, the same wavelength is not available on both links in the network. As a result, we cannot set up the desired lightpath. On the other hand, if node B can convert wavelengths, then we can set up the lightpath as shown in Figure 7.10(b).

Note the configurations of Figure 7.9(a), (b), and (c) all provide wavelength conversion and signal regeneration either in the OXC itself or by making use of the transponders in the attached OLTs. Figure 7.9(a) also provides low-speed grooming, assuming that the electrical core has been designed to support that capability. In order to provide grooming, signal regeneration, and wavelength conversion, the configuration of Figure 7.9(d) is modified to include an electrical core crossconnect as shown in Figure 7.11. This configuration allows most of the signals to be switched in the optical domain, minimizing the cost and maximizing the capacity of the network, while allowing us to route the signals down to the electrical layer whenever necessary. As we discussed earlier, we could save optical switch ports by switching signals through in wavelength bands or even entire fibers at a time.

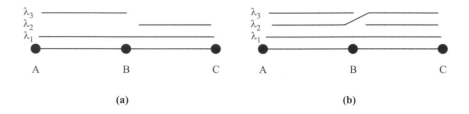

Figure 7.10 Illustrating the need for wavelength conversion. (a) Node B does not convert wavelengths. (b) Node B can convert wavelengths.

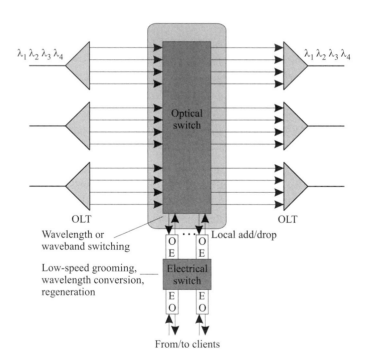

Figure 7.11 A realistic "all-optical" network node combining optical core crossconnects with electrical core crossconnects. Signals are switched in the optical domain whenever possible but routed down to the electrical domain whenever they need to be groomed, regenerated, or converted from one wavelength to another.

Looking at Figure 7.11, note that the optical switch does not have to switch signals from any input port to any output port. For example, it does not need to switch a signal entering at wavelength λ_1 to an output port that is connected to a multiplexer that takes in wavelength λ_2. This allows some potential simplification by making use of *wavelength planes*.

Figure 7.12 shows a wavelength plane OXC. The signals coming in over different fiber pairs are first demultiplexed by the OLTs. All the signals at a given wavelength are sent to a switch dedicated to that wavelength, and the signals from the outputs of the switches are multiplexed back together by the OLTs. In a node with F WDM fiber pairs and W wavelengths on each fiber pair, this arrangement uses F OLTs and W $2F \times 2F$ switches. This allows any or all signals on any input wavelength to be dropped locally. In contrast, the configuration of Figure 7.11 uses F OLTs and a $2WF \times 2WF$ switch to provide the same capabilities. Consider, for example, $F = 4$, $W = 32$, which are realistic numbers today. In this case, the configuration of Figure 7.12 uses four OLTs and 32 8×8 switches. In contrast, Figure 7.9(b) requires four OLTs and a 256×256 switch. As we saw in Section 3.7, larger optical switches are significantly harder to build than small ones and will need to use technologies like analog beam-steering micromirrors, whereas small optical switches can be realized using a variety of different technologies.

Based on the discussion above, it would appear that the wavelength plane approach offers a cheaper alternative to large-scale nonblocking optical switches. However, we did not consider how to optimize the number of add/drop terminations (which would be transponders or O/E interfaces on electrical switch cores). Both Figure 7.11 and Figure 7.12 assume that there are sufficient ports to terminate all WF signals. This is almost never the case, as only a fraction of traffic will need to be dropped, and the terminations are expensive. Moreover, observe that if we indeed do need WF terminations on an electrical switch, the best solution is to use the electrical core configuration of Figure 7.9(a), without having the wavelength plane switches!

If we have a total of T terminations, with all of them having tunable lasers, and we would like to drop any of the WF signals, this requires an additional $T \times WF$ optical switch between the wavelength plane switches and the terminations, as shown in Figure 7.13. In contrast, with a large nonblocking switch, we would simply connect the T terminations to T ports of this switch, resulting in a $(WF + T) \times (WF + T)$ switch overall. This situation somewhat reduces the appeal of a wavelength plane approach.

To summarize, the wavelength plane approach needs to take into account the number of fibers, fraction of add/drop traffic, number of terminations, and their tuning capabilities as separate parameters in the design. With a large-scale switch, we can partition the ports in a flexible way to account for variations in all these

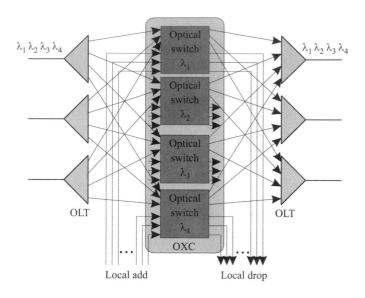

Figure 7.12   An optical core wavelength plane OXC, consisting of a plane of optical switches, one for each wavelength. With F fibers and W wavelengths on each fiber, each switch is a $2F \times 2F$ switch, if we want the flexibility to drop and add any wavelength at the node.

parameters—the only constraint is in the total number of ports available. See Problem 7.7 for another example of these types of trade-offs.

As of this writing, both electrical core and optical core OXCs are becoming available. Electrical core OXCs with total capacities up to a few Tb/s, capable of grooming down to STS-1 (51 Mb/s), are becoming available. Optical core OXCs with over 1000 ports are also emerging as commerical products, and wavelength plane OXCs are being offered by some vendors as well.

Summary

We studied the basic network elements constituting WDM networks in this chapter. We refer the reader back to Chapter 3 to get an understanding of the various technologies that are used to build these elements.

The WDM network provides circuit-switched lightpaths that can have varying degrees of transparency associated with them. Wavelengths can be reused in the network to support multiple lightpaths as long as no two lightpaths are assigned the same wavelength on a given link. Lightpaths may be protected by the network in

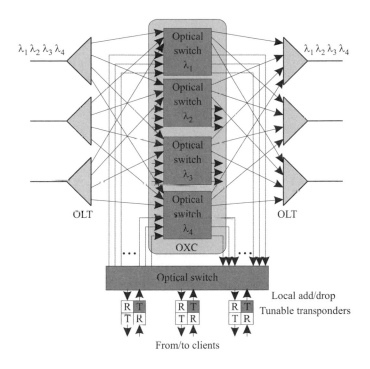

Figure 7.13 Dealing with add/drop terminations in a wavelength plane approach. An additional optical switch is required between the tunable transponders and the wavelength plane switches. Here, T denotes a transmitter, assumed to be a tunable transmitter on the WDM side, and R denotes a receiver.

the event of failures. Lightpaths can be used to provide flexible interconnections between users of the optical network, such as IP routers, allowing the router topology to be tailored to the needs of the router network.

An optical line terminal (OLT) multiplexes and demultiplexes wavelengths and is used for point-to-point applications. It typically includes transponders, multiplexers, and optical amplifiers. Transponders provide the adaptation of user signals into the optical layer. They also consitute a significant portion of the cost and footprint in an OLT. In some cases, transponders can be eliminated by deploying interfaces that provide already-adapted signals at the appropriate wavelengths in other equipment.

An optical add/drop multiplexer (OADM) drops and adds a selective number of wavelengths from a WDM signal, while allowing the remaining wavelengths to pass through. OADMs provide a cost-effective way of performing this function, compared to using OLTs interconnected back to back, or relying on other equipment to handle the passthrough traffic. OADMs are typically deployed in linear or ring topologies.

Several types of OADMs are possible with a range of capabilities based on the number of wavelengths they can add and drop, the ease of dropping and adding additional wavelengths, static or reconfigurable, and so on. We studied the basic architectural flavors of OADMs: parallel, serial, and band drop. Each of these has its pros and cons. We also looked at reconfigurable OADM architectures, which use tunable filters and/or multiplexers, as well as tunable lasers, in order to provide the maximum possible flexibility in the network.

An optical crossconnect (OXC) is the other key network element in the optical layer. OXCs are large switches used to provision services dynamically as well as provide network restoration. OXCs are typically deployed in a mesh network configuration. As with OADMs, several variants of OXCs exist, ranging from OXCs with electrical switch cores capable of grooming traffic at STS-1 rates to all-optical OXCs that can switch wavelengths, bands of wavelengths, and entire fibers. Optical core crossconnects can also be surrounded by optical-to-electrical-to-optical converters to provide some of the grooming and wavelength conversion capabilities offered by electrical core crossconnects, but are not suited for grooming traffic at fine granularities such as STS-1 rates. Each has its role in the network.

Further Reading

Information regarding the various types of OLTs, OADMs, and OXCs is not easy to come by because many of the commercial implementations are proprietary in nature. Browsing through network equipment vendors' Web pages will provide some illustration of the capabilities of the different products in this space. However, there are several early testbeds that explored various forms of these network elements. For instance, [Ale93, Kam96] used a static all-optical OXC that provided a fixed interconnection pattern without any switching or wavelength conversion. [Cha94, CEG+96] explored a parallel WADM architecture as well as an OXC with an electrical switch core. [Hil93] developed an all-optical OXC without wavelength conversion. [WASG96, Gar98] developed a parallel WADM as well as a small all-optical OXC without wavelength conversion. See also [HH96, OWS96, Der95, MS96, Ber96a, Ber96b, Bac96, RS95, Chb98, KWK+98] for other relevant testbeds and architectures. The use of wavelength bands has been discussed in various contexts in [Ste90, GRW00, SS99]. For a discussion of optical crossconnects and a comparison of them to electrical crossconnects, see [GR00, GRL00].

Problems

7.1   Consider a ring network with two intermediate adjacent nodes A and B, each with an OADM.

(a)   Consider the case where the OADM at node A adds wavelength λ_1 and the OADM at node B drops the adjacent wavelength λ_2. Suppose the minimum received power is set at -30 dBm and the transmit power is set at 0 dBm. Adjacent channel crosstalk at the receiver must be less than 15 dB. Assume that signals are added and dropped by the OADMs with no loss. What is the crosstalk suppression required at the OADM for the adjacent channel? How does this change with the link loss between the two nodes?

(b)   Next consider the case where both OADMs drop and add wavelength λ_1. We are worried about the case where some of the λ_1 power, instead of being dropped at the node, "leaks" through. The intrachannel crosstalk at the receiver must be at least -30 dB below the desired signal. For the same assumptions as above, what is the intrachannel crosstalk suppression required at the OADM? How does this change with the link loss between the two nodes?

7.2   This problem illustrates some of the difficulties facing network planners when they have to use OADMs that are constrained in what channels they can add and drop. Consider a four-node linear network with nodes A, B, C, and D in that order. We have three wavelengths $\lambda_1, \lambda_2, \lambda_3$ available and are given OADMs that drop two fixed channels. That is, we can put in OADMs that drop either λ_1, λ_2, or λ_2, λ_3, or λ_1, λ_3. Now consider the situation where we need to set up the following lightpaths: AB, BC, CD, AC, BD. What OADMs would you deploy at each of the nodes? Suppose at a later point the lightpath traffic changes and now we need to replace lightpaths AC and BD by AD and BC. What changes would you have to make to support this new traffic?

7.3   Consider a linear network with serial OADMs. Assume that the transmit power is 0 dBm; the minimum received power is -30 dBm; and each OADM has a passthrough loss of 2 dB, a loss of 1 dB for the drop path, and a loss of 1 dB for the add path. Assume that the adjacent channel suppression offered by each OADM is 20 dB and that at the receiver the adjacent channel power must be at least 15 dB less than the desired signal power.

(a)   Write a computer program that takes as its input the set of lightpaths in the network and their wavelengths, the loss between each pair of adjacent nodes, and determines whether each lightpath is feasible or not. The program should also determine any wavelength conflicts, that is, if two lightpaths overlap and are assigned the same wavelength.

(b) What is the maximum number of OADMs that a lightpath can pass through before it needs to be regenerated? Plot this number as a function of the total link loss in the network.

(c) Plot the output of your program for a network with five nodes numbered sequentially from 1 to 5, link loss of 5 dB between nodes, and the following sets of lightpaths and wavelength assignments: $(1, 5, \lambda_1)$, $(2, 4, \lambda_2)$, $(3, 5, \lambda_3)$, $(1, 4, \lambda_4)$.

7.4   This problem explores architectures for constructing fully reconfigurable OADMs. Consider the parallel architecture shown in Figure 7.7(d). How would you build a fully reconfigurable parallel OADM like this one, wihtout using a large optical switch? You are allowed to use tunable filters, passive splitters and combiners, and small (2×2) optical switches. These solutions need to meet properties (1) and (2) specified for the ideal OADM described in Section 7.3. With respect to property (3), you still need to keep the loss fixed, regardless of how many channels are dropped or added, but are allowed to have a reasonably high value for this loss. Compare the pros and cons of your solution versus the one in Figure 7.7(d).

7.5   You have to design a five-node ring network with a hub node and four remote nodes. Each remote node needs two wavelengths of traffic to/from the hub node on both sides of the ring; that is, you will need to dedicate two wavelengths to each remote node and terminate all the wavelengths at the hub node. You have to pick between two systems.

The first system uses eight channels in two bands, each with four channels. It provides band OADMs, which can drop one out of the two bands. Once a band is dropped, all four wavelengths in the band have to be regenerated. A band OADM costs $20,000 and a single-channel regenerator costs $10,000. No optical amplifiers are required with this system.

The second system also uses eight channels but has no bands. It provides SC-OADMs, which can drop any single wavelength. Each SC-OADM costs $10,000. For this system, two optical line amplifiers are required, each costing $30,000. Whose system would you select based on just equipment cost?

7.6   This problem illustrates the need for large OXCs and also illustrates the value of using wavelength bands.

Consider an all-optical OXC with 256 ports deployed in the configuration shown in Figure 7.9(d). Each WDM line system carries 32 wavelengths; 75% of the lightpaths pass through the node while the remaining 25% are dropped and added onto routers attached to the OXC. Each lightpath added and dropped onto a router takes up two OXC ports.

(a) How many WDM line systems can the OXC support?

(b) Next suppose that 25% of the lightpaths passing through need to be converted from one wavelength to another. This is done by sending the lightpath to one of a pool of regenerators/wavelength converters attached to the OXC. Each such regenerator uses two ports in the OXC. Thus a lightpath needing to be converted uses four OXC ports. Now, how many WDM line systems can the OXC support?

(c) Now suppose the WDM line systems are designed with eight bands, each with four wavelengths. Assume that all the lightpaths passing through can be passed through at the band level without having to be demultiplexed down to the individual channel. For lightpaths that are dropped and added, the entire band is dropped and demultiplexed after the bands are passed through the OXC. No wavelength conversion is needed. How many WDM line systems can the OXC support?

7.7 Consider the wavelength plane switch architecture of Figure 7.12. Consider the situation where we have a total of four fibers and 40 wavelengths on each fiber. We must design the node such that any four signals can be dropped. (Note that this implies we could potentially drop all the wavelengths on a particular fiber while passing through all the wavelengths on the other fibers.) The wavelengths are dropped onto transponders, which have tunable lasers. The 40 wavelengths are split into five bands of 8 wavelengths each, and a tunable laser can tune over a single band.

(a) Draw a block diagram of this node and indicate the minimum number of transponders needed. Compare this against an approach using large nonblocking switches.

(b) Now suppose we have tunable lasers that can tune over two bands instead of one. How does the situation change?

References

[Ale93] S. B. Alexander et al. A precompetitive consortium on wide-band all-optical networks. *IEEE/OSA Journal on Lightwave Technology*, 11:714–735, May/June 1993.

[Bac96] E.-J. Bachus et al. Coherent optical systems implemented for business traffic routing and access: The RACE COBRA project. *IEEE/OSA JLT/JSAC Special Issue on Multiwavelength Optical Technology and Networks*, 14(6):1309–1319, June 1996.

[Ber96a] L. Berthelon et al. Experimental assessment of node cascadability in a reconfigurable survivable WDM ring network. In *Proceedings of Topical Meeting on Broadband Optical Networks*, 1996.

[Ber96b] L. Berthelon et al. Over 40,000 km across a layered network by recirculation through an experimental WDM ring network. In *Proceedings of European Conference on Optical Communication*, 1996.

[CEG⁺96] G. K. Chang, G. Ellinas, J. K. Gamelin, M. Z. Iqbal, and C. A. Brackett. Multiwavelength reconfigurable WDM/ATM/SONET network testbed. *IEEE/OSA JLT/JSAC Special Issue on Multiwavelength Optical Technology and Networks*, 14(6):1320–1340, June 1996.

[Cha94] G. K. Chang et al. Experimental demonstration of a reconfigurable WDM/ATM/SONET multiwavelength network testbed. In *OFC'94 Technical Digest*, 1994. Postdeadline paper PD9.

[Chb98] M. W. Chbat et al. Towards wide-scale all-optical networking: The ACTS optical pan-European network (OPEN) project. *IEEE JSAC: Special Issue on High-Capacity Optical Transport Networks*, 16(7):1226–1244, Sept. 1998.

[Der95] F. Derr. Design of an 8 × 8 optical cross-connect switch: Results on subsystems and first measurements. In *ECOC'95 Optical Networking Workshop*, 1995. Paper S2.2.

[Gar98] L. D. Garrett et al. The MONET New Jersey demonstration network. *IEEE JSAC: Special Issue on High-Capacity Optical Transport Networks*, 16(7):1199–1219, Sept. 1998.

[GR00] J. Gruber and R. Ramaswami. Moving towards all-optical networks. *Lightwave*, 34(8):40–49, Dec. 2000.

[GRL00] J. Gruber, P. Roorda, and F. Lalonde. The photonic switch crossconnect (PSX)—its role in evolving optical networks. In *Proceedings of National Fiber Optic Engineers Conference*, pages 678–689, 2000.

[GRW00] O. Gerstel, R. Ramaswami, and W-K. Wang. Making use of a two stage multiplexing scheme in a WDM network. In *OFC 2000 Technical Digest*, pages ThD1-1–ThD1-3, 2000.

[HH96] A. M. Hill and A. J. N. Houghton. Optical networking in the European ACTS programme. In *OFC'96 Technical Digest*, pages 238–239, San Jose, CA, Feb. 1996.

[Hil93] G. R. Hill et al. A transport network layer based on optical network elements. *IEEE/OSA Journal on Lightwave Technology*, 11:667–679, May/June 1993.

[Kam96] I. P. Kaminow et al. A wideband all-optical WDM network. *IEEE JSAC/JLT Special Issue on Optical Networks*, 14(5):780–799, June 1996.

[KWK⁺98] M. Koga, A. Watanabe, T. Kawai, K. Sato, and Y. Ohmori. Large-capacity optical path cross-connect system for WDM photonic transport network. *IEEE JSAC: Special Issue on High-Capacity Optical Transport Networks*, 16(7):1260–1269, Sept. 1998.

[MS96]   W. C. Marra and J. Schesser. Africa ONE: The Africa optical network. *IEEE Communications Magazine*, 34(2):50–57, Feb. 1996.

[OWS96]   S. Okamoto, A. Watanabe, and K.-I. Sato. Optical path cross-connect node architectures for photonic transport network. *IEEE/OSA JLT/JSAC Special Issue on Multiwavelength Optical Technology and Networks*, 14(6):1410–1422, June 1996.

[RS95]   R. Ramaswami and K. N. Sivarajan. Routing and wavelength assignment in all-optical networks. *IEEE/ACM Transactions on Networking*, pages 489–500, Oct. 1995. An earlier version appeared in *Proceedings of IEEE Infocom'94*.

[SS99]   A. A. M. Saleh and J. M. Simmons. Architectural principles for optical regional and metropolitan access networks. *IEEE/OSA Journal on Lightwave Technology*, 17(12), Dec. 1999.

[Ste90]   T. E. Stern. Linear lightwave networks: How far can they go? In *Proceedings of IEEE Globecom*, pages 1866–1872, 1990.

[WASG96]   R. E. Wagner, R. C. Alferness, A. A. M. Saleh, and M. S. Goodman. MONET: Multiwavelength optical networking. *IEEE/OSA JLT/JSAC Special Issue on Multiwavelength Optical Technology and Networks*, 14(6):1349–1355, June 1996.

8
chapter

WDM Network Design

I N THE PREVIOUS CHAPTER, WE LEARNED that the optical layer provides high-speed circuit-switched connections, or lightpaths, between pairs of higher-layer equipment such as SONET/SDH muxes, IP routers, and ATM switches. The optical layer realizes these lightpaths over the physical fiber using elements such as optical line terminals (OLTs), optical add/drop multiplexers (OADMs), and optical crossconnects (OXCs). We called a network using such lightpaths a wavelength-routing network. In this chapter, our goal is to study how to design a wavelength-routing network. This involves studying not only how to design the optical layer but also how the higher-layer SONET or IP network is to be designed because the design of the two layers is closely coupled. We illustrate with an example.

> **Example 8.1** In Figure 8.1(a), there are three nodes labeled A, B, and C, connected by WDM fiber links. For simplicity, assume the traffic generated is in the form of IP packets from routers located at these nodes. Similar examples hold if the higher layer consists of SONET/SDH muxes or ATM switches. For concreteness, also assume that all router interfaces operate at 10 Gb/s, which is also the transmission capacity on each wavelength on the WDM links. Now suppose, based on estimates of the IP packet traffic, 50 Gb/s of capacity is required between all three pairs of routers: A–B, B–C, and A–C. The network can be designed to handle this traffic in two ways.
>
> 1. **No optical add/drop:** In the first method, we set up 10 wavelengths on each of the links A–B and B–C connecting the routers at the ends of these links. We observe that the traffic flowing on link A–B is 50 Gb/s (traffic from A–B)

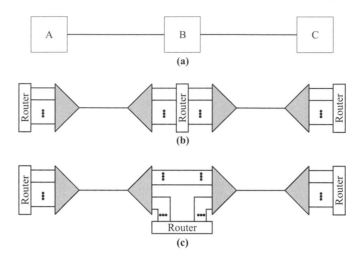

Figure 8.1   (a) A three-node network. (b) Nodes A–B and B–C are interconnected by WDM links. All wavelengths are dropped and added at node B. (c) Half the wavelengths pass through optically at node B, reducing the number of router ports at node B.

+ 50 Gb/s (traffic from A–C that must use link A–B) = 100 Gb/s. Similarly, the traffic flowing on link B–C is also 100 Gb/s. Thus the 10 wavelengths on each of the links A–B and B–C are sufficient to carry this traffic. In this case, we use 10 router ports at node A, 20 router ports at node B, and 10 router ports at node C, for a total of 40 router ports. At the optical layer, nodes A and C have OLTs, whereas node B has a pair of OLTs that terminate all the wavelengths passing through node B. This is illustrated in Figure 8.1(b).

2. **With optical add/drop:** In the second design, we set up only five lightpaths each on the routes A–B, B–C, and A–C. The five lightpaths on the route A–C pass through the node B within the optical layer, without being converted to an electrical signal. This design requires only 10 router ports at each of the three nodes, A, B, and C, for a total of 30 router ports, compared to 40 router ports in the design without optical add/drop. However, this design requires node B to have an OADM node that is capable of adding and dropping 10 of the 20 lightpaths that terminate at the router at node B, while passing the other 10 lightpaths through. This is illustrated in Figure 8.1(c).

Thus, in the design with optical add/drop capability, we can trade off the number of IP router ports at node B (10 versus 20) for optical add/drop capability at the same node. In general, as we will see later, the trade-off is between the cost of the higher-layer equipment (IP router ports) and the cost of the optical layer equipment (OADMs, or increased number of wavelengths as we will see in other examples later). Both designs are perfectly valid and will do the job as far as the user is concerned. The choice between them will be made based on the cost trade-off between the optical and higher-layer equipment. In this example, providing optical add/drop capability requires an OADM at node B instead of two OLTs. The cost of doing this is cheaper in many scenarios today than providing additional 10 Gb/s IP router ports. This situation is likely to prevail over the long run as well fundamentally because passing a wavelength through is a much simpler operation than routing all the packets that have been transmitted on a wavelength at the IP layer.

Note that in this example the transponder costs are bundled with the IP router port costs. Increasingly, routers are being equipped with optical interfaces with wavelengths on the ITU grid, so that no additional transponders are required. Thus, strictly speaking, in this example, we are comparing the cost of IP layer terminations versus keeping things optical.

In the same example, if the amount of passthrough traffic at node B was a small fraction of a wavelength, an entire wavelength with a capacity of 10 Gb/s would have to be used for the passthrough traffic if we used a design with optical passthrough. At the same time, a design without optical passthrough may be able to handle the passthrough traffic without an increase in the number of IP router ports. This would lead us to prefer to handle the passthrough packets using the IP router at node B. We will study this effect further in the next section in the context of rings.

From the point of view of the IP routers, the topology of the network when all the wavelengths are terminated at node B is shown in Figure 8.2(a). This is the topology seen by the IP layer packet-routing algorithm, such as OSPF. This is a linear topology with 10 parallel links between nodes A and B, and 10 parallel links between nodes B and C. In the optical add/drop case, the topology of the network seen by IP routers is a completely connected mesh with 5 parallel links between each of the three pairs of nodes, as shown in Figure 8.2(b). Note that both topologies are capable of meeting the traffic needs at the IP layer, which calls for 50 Gb/s of capacity between each pair of routers.

The topology seen by the IP routers, or the SONET/SDH muxes, is the topology of the lightpaths provided by the optical layer, and hence we will call it the *lightpath topology*. It is often called the *logical* or *virtual* topology, but we will not employ this terminology. In the same vein, the fiber topology upon which the lightpaths are created is called the *physical topology*, but we will not use this terminology either.

We can view the general problem of designing wavelength-routing networks as follows. The fiber topology and the traffic requirements (traffic matrix) are specified.

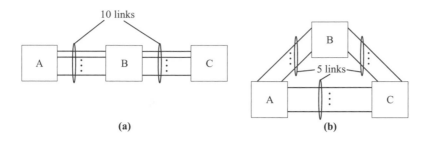

Figure 8.2   (a) The lightpath topology of the three-node network corresponding to Figure 8.1(a) that is seen by the routers. Routers A–B and B–C are connected by 10 parallel links. (b) The lightpath topology of the three-node network corresponding to Figure 8.1(b) that is seen by the routers. All pairs of routers, A–B, B–C, and C–A, are connected by 5 parallel links.

In our example the fiber topology is a linear one with three nodes, and the traffic requirement is 50 Gb/s between every pair of these nodes. The task is to design a lightpath topology that interconnects the IP routers and to realize this topology within the optical layer. In our example, two lightpath topologies that meet the traffic requirements are shown in Figure 8.2. We call the first problem the *lightpath topology design* (LTD) *problem.* We call the problem of realizing the lightpath topology within the optical layer the *routing and wavelength assignment* (RWA) *problem,* for reasons that will become clear shortly. The RWA problem is simple to solve in this example because there is only one route in the fiber topology between every pair of nodes. In a general topology, the RWA problem can be quite difficult. The realization of the two lightpath topologies of Figure 8.2 are shown in Figures 8.1(b) and (c).

Another problem we face in the design of wavelength-routing networks is that of grooming the higher-layer traffic. The term *grooming* is commonly used to refer to the packing of low-speed SONET/SDH circuits (for example, STS-1) into higher-speed circuits (for example, STS-48 or STS-192). This is the function provided by digital crossconnects. While the term is usually not applied to IP routers, conceptually IP routers can be considered to provide the grooming function at the packet level. In order to reap the benefits of optical passthrough, the higher-layer traffic must be groomed appropriately. For example, in Figure 8.1(c), all the traffic destined for node B must be groomed onto a few wavelengths, so that only these wavelengths need to be dropped at node B. Otherwise, node B will have to drop many wavelengths, and this will increase the network cost.

In the rest of this chapter we will discuss several aspects of the design of wavelength-routing networks in some detail. In Section 8.1, we will analyze the cost trade-offs between the higher-layer and optical-layer equipment in a ring network. We will then discuss the LTD and RWA problems, which we introduced in the discussion of the three-node network above, in Section 8.2. We then discuss the problem of dimensioning the WDM links, that is, determining the number of wavelengths to be provided on each link, in Section 8.3. We discuss statistical dimensioning methods in Section 8.4. In Section 8.5, we discuss a number of research results that have been obtained regarding the trade-offs between optical crossconnects with and without wavelength conversion capability. (We will discuss a practical long-haul network design example in Section 13.2.6.)

8.1 Cost Trade-Offs: A Detailed Ring Network Example

In this section, we will study the cost trade-offs in designing networks in different ways to meet the same traffic demand by varying the lightpath topology. We will consider the trade-offs between the cost of the higher-layer equipment and the optical layer equipment. We measure the higher-layer equipment cost by the number of IP router ports (or SONET line terminals). The number of IP router ports required is equal to twice the number of lightpaths that need to be established since each lightpath connects a pair of IP router ports. An important component of the optical layer cost is the number of transponders required in the OLTs and OADMs. Since every lightpath requires a pair of transponders, we club the cost of the transponders with that of the higher-layer equipment. This also covers the case where the transponders are present within the higher-layer equipment (see Figure 7.2). We measure the remainder of the cost of the optical layer equipment by the number of wavelengths used on a link.

Network topologies are usually designed to be *2-connected*, that is, to have two node-wise disjoint routes between every pair of nodes in the network. While fiber mesh topologies that are arbitrary, but 2-connected, are more cost-efficient for large networks than fiber ring topologies, the latter have been widely deployed and are good for a network that does not have a wide geographic spread. For this reason we will consider fiber ring topologies in this section. One reason for the wide deployment of rings is because a ring connecting N nodes has the minimum possible number of links (only N) for a network that is 2-connected, and thus tends to have a low fiber deployment cost.

We will consider a traffic matrix where t units of traffic are to be routed from one IP router to all other IP routers in the network. We denote the number of nodes in the network by N and assume the traffic is uniform; that is, $t/(N-1)$ units of traffic are to be routed between every pair of IP routers. For normalization purposes, the capacity of a wavelength is assumed to be 1 unit. As in the three-node linear

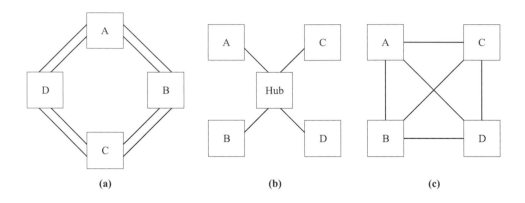

Figure 8.3   Three different lightpath topologies that can be deployed over a fiber ring topology. (a) A point-to-point WDM ring where adjacent routers on the ring are connected by one or more lightpaths. (b) A hub topology where all routers are connected to one central router (hub) by lightpaths. (c) A full mesh where each router is connected to every other router by lightpaths.

topology above, we divide the network design problem into two: the LTD and RWA problems. We will consider three different lightpath topologies, all of which are capable of meeting the traffic requirements. The general form of these topologies is shown in Figure 8.3.

The first lightpath topology, shown in Figure 8.3(a), is a ring, which we call a *point-to-point WDM* (PWDM) ring. In this case, the lightpath topology is also a ring, just like the fiber topology, except that we can have multiple lightpaths between adjacent nodes in the ring, in order to provide the required capacity between the IP routers.

The second lightpath topology, shown in Figure 8.3(b), is a *hub design.* All routers are connected to a central (hub) router by one or more lightpaths. Thus all packets traverse two lightpaths: from the source router to the hub, and from the hub to the destination router.

The third, and final, lightpath topology, shown in Figure 8.3(c), is an *all-optical design.* In this case, we establish direct lightpaths between all pairs of routers. Thus, packets traverse only one lightpath to get from the source router to the destination router.

We next consider how to realize these lightpath topologies on the fiber network; that is, we solve the RWA problem for these three designs. The RWA problem is to find a route for each lightpath and to assign it a wavelength on every link of

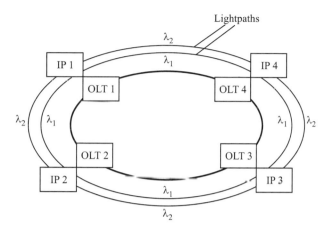

Figure 8.4   A PWDM ring architecture. The lightpaths and their wavelength assignment are shown in the figure for the case $t = 3$.

the route. We assume that a lightpath must be assigned the same wavelength on all the links it traverses; that is, the optical layer provides no wavelength conversion capability. In addition, no two lightpaths traversing the same link can be assigned the same wavelength.

Example 8.2   We first consider the *PWDM ring*. The network shown in Figure 8.4 is a PWDM ring. At each node, all the wavelengths are received and sent to the IP routers. For this network, all lightpaths are "single-hop" lightpaths between adjacent nodes in the ring. If W denotes the number of wavelengths on each link, then we can set up W lightpaths between each pair of adjacent nodes.

The number of IP router ports needed will depend on the algorithm used to route the traffic. Suppose we route each traffic stream along the shortest path between its source and destination, and N is the number of nodes in the network. Assuming N is even, we can calculate the traffic load (in units of lightpaths) on each link to be

$$L = \frac{N + 1 + \frac{1}{N-1}}{8}t. \tag{8.1}$$

In this case, since all lightpaths are single-hop lightpaths, the number of wavelengths needed to support this traffic is simply

$$W = \lceil L \rceil = \left\lceil \frac{N + 1 + \frac{1}{N-1}}{8}t \right\rceil. \tag{8.2}$$

Since all the wavelengths are received and retransmitted at each node, the number of router ports required per node, Q, is

$$Q = 2W. \tag{8.3}$$

This example has illustrated the following set of design parameters that need to be considered in determining the cost of the network:

Router ports. Clearly, we would like to use the minimum possible number of IP router ports to support the given traffic. Note that since a lightpath is established between two router ports, minimizing the number of ports is the same as minimizing the number of lightpaths that must be set up to support the traffic.

Wavelengths. At the same time, we would also like to use the minimum possible number of wavelengths since using more wavelengths incurs additional equipment cost in the optical layer.

Hops. This parameter refers to the maximum number of hops taken up by a lightpath. For the PWDM ring, each lightpath takes up exactly one hop. The reason this parameter becomes important is that it becomes more difficult to design the transmission system as the number of hops increases (see Chapter 5), which again increases the cost of optical layer equipment.

In general, we will see that there is a trade-off between these different parameters. For example, we will see that the PWDM ring uses a large number of router ports, but the smallest possible number of wavelengths. In the hub and all-optical design examples that follow we will use fewer router ports at the cost of requiring more wavelengths.

> **Example 8.3** Here, we will consider the hubbed network architecture shown in Figure 8.5. An additional hub router is added to the ring. At the hub router, the packets on all the wavelengths are received and routed appropriately. This node is identical to a PWDM ring node. The other N nodes are simpler nodes that contain just enough router ports to source and sink the traffic at that node. (To keep the example simple, we will assume that the hub router itself does not source or sink any traffic. This is, of course, not true in practice. In fact, the hub node could serve as a gateway node to the rest of the network.) Lightpaths are established between each node and the hub node h. Traffic from a nonhub node i to another nonhub node j is routed on two lightpaths—one from i to h and another from h to j. To support this traffic, we will set up $\lceil t \rceil$ lightpaths from

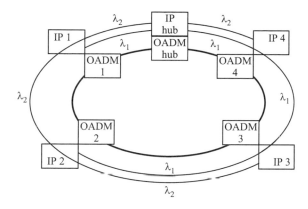

Figure 8.5   A hubbed WDM ring architecture. The lightpaths and their wavelength assignment are shown in the figure for the case $\lceil t \rceil = 1$.

each node to the hub node. Thus the number of router ports needed per node for this configuration is

$$Q = 2 \lceil t \rceil .\tag{8.4}$$

We assume that the lightpaths are routed and assigned wavelengths as follows: Two adjacent nodes use different paths along the ring and reuse the same set of wavelengths, as shown in Figure 8.5. For this RWA algorithm, the number of wavelengths required can be calculated to be

$$W = \frac{N}{2} \lceil t \rceil .\tag{8.5}$$

The worst-case hop length is

$$H = N - 1.\tag{8.6}$$

Example 8.4   The final example is the all-optical design shown in Figure 8.6, where data is transmitted on a single lightpath between its source and destination and never sent through an intermediate router enroute. In this case, we must set up $\lceil t/(N-1) \rceil$ lightpaths between each pair of nodes to handle the $t/(N-1)$ units of traffic between each node pair. The number of router ports per node is therefore

$$Q = (N-1) \left\lceil \frac{t}{N-1} \right\rceil .\tag{8.7}$$

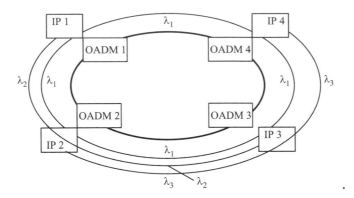

Figure 8.6 An all-optical four-node network configuration. The lightpaths and their wavelength assignment are shown in the figure for the case $t = 3$.

The number of wavelengths will depend on how the lightpaths are routed and assigned wavelengths (see Problem 8.5). It is possible to obtain a suitable routing and wavelength assignment such that (for N even)

$$W = \left\lceil \frac{t}{N-1} \right\rceil \left(\frac{N^2}{8} + \frac{N}{4} \right).$$ (8.8)

To understand the quality of the designs produced by the three preceding examples, we can compare them to some simple lower bounds on the number of router ports and wavelengths required for any design. Clearly, any design requires $Q \geq \lceil t \rceil$. We next derive a lower bound on the number of wavelengths required as follows. Let h_{ij} denote the minimum distance between nodes i and j in the network measured in number of hops. Define the minimum average number of hops between nodes as

$$H_{\min} = \frac{\sum_{i=1}^{N} \sum_{j=1}^{N} h_{ij}}{N(N-1)}.$$

For a ring network, we can derive the following equation on $H_{\min}$ (N even):

$$H_{\min} = \frac{N+1}{4} + \frac{1}{4(N-1)}.$$ (8.9)

Note that the maximum traffic load on any link is greater than the average traffic load, which is given by the equation

$$L \geq L_{\text{avg}} = \frac{H_{\min} \times \text{Total traffic}}{\text{Number of links}} = \frac{H_{\min} \times \frac{1}{2} N t}{N}$$

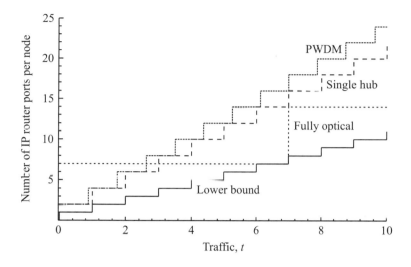

Figure 8.7   Number of IP router ports required for the different designs of Examples 8.2–8.4, for a ring with $N = 8$ nodes. The lower bound of $\lceil t \rceil$ is also shown.

$$= \left(\frac{N+1}{8} + \frac{1}{8(N-1)} \right) t. \tag{8.10}$$

Clearly, we need to have the number of wavelengths $W \geq L$.

Figure 8.7 plots the number of router ports required for the three different designs, as well as the lower bound, for a network with eight nodes. Observe that for small amounts of traffic, the hubbed network requires the smallest number of router ports. The PWDM design requires the largest number of router ports. This clearly demonstrates the value of routing signals within the optical layer, as opposed to having just point-to-point WDM links.

Unfortunately, the reduction in router ports is achieved at the expense of requiring a larger number of wavelengths to support the same traffic load. Figure 8.8 plots the number of wavelengths required for the three different designs, along with the lower bound derived earlier. The PWDM ring uses the smallest number of wavelengths—it achieves the lower bound and is the best possible design from this point of view. The hubbed architecture uses a relatively large number of wavelengths to support the same traffic load.

The all-optical design is a good design provided t is slightly less than or equal to $N - 1$ (or some multiple of $N - 1$). This is because, in these cases, an integral number of lightpaths is needed between each pair of nodes, which is best realized by having dedicated lightpaths between the node pairs without terminating any traffic

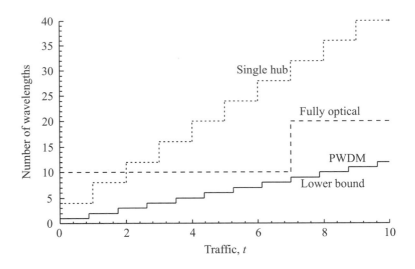

Figure 8.8   Number of wavelengths required for the different designs of Examples 8.2–8.4, for a ring with $N = 8$ nodes. The lower bound from (8.10) is also shown.

in intermediate nodes. This brings out an important point: denote the traffic between a pair of nodes by $m + t'$, where m is a nonnegative integer and $0 \leq t' < 1$. Then the best solution is to set up m lightpaths between that node pair to route m units of traffic, and to handle the residual t' units by some other methods such as the hubbed or PWDM architectures. If t' is close to one unit, then the best solution may be to have another direct lightpath between them.

Overall, we have learned that it is possible to save significantly in higher-layer (IP or SONET) equipment costs by providing networking functions (routing and switching of wavelengths) within the optical layer.

8.2   LTD and RWA Problems

The general approach of dividing the wavelength-routing network design problem into that of an LTD problem and an RWA problem, which we employed above in the three-node linear network and the ring network, is a good heuristic for practical problems because solving the two problems in a combined fashion is quite hard. In both the examples, we considered a few different lightpath topologies and examined the RWA problem for each of them. This clarified the cost trade-offs among the different designs. In practice, each lightpath topology together with its realization in the optical layer (the solution of the RWA problem) would result in a net, real (monetary) cost. We can then pick the design that results in the lowest cost. We will consider one such example in Chapter 13. We will now examine the two component problems, LTD and RWA, in greater detail.

8.2.1 Lightpath Topology Design

We now consider a specific, though rather simplified, lightpath topology design problem and examine how it can be solved. We will assume that no constraints are imposed by the underlying fiber topology or the optical layer. (Examples of such constraints are a limit on the length of a lightpath and a limit on the number of lightpaths traversing a link.)

We assume that all lightpaths are bidirectional (see Section 8.2.2); that is, if we use a lightpath from node i to node j, then we also use a lightpath from node j to node i. This is the case that most frequently occurs in practice since almost all higher-layer protocols, including IP and SONET, assume bidirectional physical layer links.

One constraint is that at each node we use an IP router with at most Λ ports connecting it to other IP routers. (In addition, each router would have local interfaces to Ethernet switches and the like.) This constrains the maximum number of ports per router to Δ and thus indirectly constrains the cost of the IP routers. This also constrains the number of lightpaths in the network to $n\Delta$, where n is the number of nodes in the network, since each lightpath starts and ends at an IP router port. This constraint is equivalent to a constraint on the lightpath costs if we assume that the tariff for a lightpath is the same regardless of its end points. This is an assumption that would clearly not hold in a wide-area environment where we expect longer lightpaths to be more expensive than shorter ones. However, it may hold in a regional network. (Many phone companies offer a single rate for all calls made within their region. So it is not inconceivable that we could have a single tariff for all lightpaths within a region.) The main reason for the assumption, of course, is that it simplifies the problem.

When we design the lightpath topology, we also have to solve the problem of routing packets (or connections) over the lightpath topology. This is because whether or not a given (lightpath) topology supports the traffic requirements depends on both the topology itself and the routing algorithm that is used.

To formulate the problem in mathematical terms, we need to introduce a number of definitions. We assume a statistical model for the IP packet traffic: the arrival rate for packets for source-destination (s-d) pair (s, d) is λ^{sd} (in packets/second), $s, d = 1, \ldots, n$. $b_{ij}, i, j = 1, \ldots, n, i \neq j$, are n^2 binary valued (0 or 1) variables, one for each possible lightpath. $b_{ij} = 1$ if the designed lightpath topology has a link from node i to node j; otherwise, $b_{ij} = 0$. The solution to the lightpath topology design problem will specify which of the b_{ij} are 1 and which are 0. We assume that we can arbitrarily split the traffic between the same pair of nodes over different paths through the network. This is not a problem if the traffic is IP packets, but if we were instead considering SONET circuits, this is tantamount to assuming that the traffic between nodes consists of a large number of such circuits. This assumption is satisfied when we are designing a backbone network to support a large number of private leased lines such as T1s or T3s.

Let the fraction of the traffic between s-d pair (s, d) that is routed over link (i, j) (if it exists) be a_{ij}^{sd}. Then $\lambda_{ij}^{sd} = a_{ij}^{sd} \lambda^{sd}$ is the traffic (in packets/second) between s-d pair (s, d) that is routed over link (i, j). The total traffic from all s-d pairs that is routed over link (i, j) is thus $\lambda_{ij} = \sum_{sd} \lambda_{ij}^{sd}$. We define a parameter called the *congestion* as $\lambda_{max} = \max_{ij} \lambda_{ij}$. Note that the λ_{ij}^{sd} (and thus the λ_{ij} and λ_{max}) are variables that we have to determine. Determining their values amounts to finding a routing algorithm.

To understand why the congestion is an important parameter, let us consider the case where the packet arrivals follow a Poisson process and the packet transmission times are exponentially distributed with mean time given by $1/\mu$ seconds. Making the standard assumption that the traffic offered to a link (lightpath) in the network is independent of the traffic offered to other links, each link can be modeled as an M/M/1 queue. The average queuing delay on link (i, j) is then given by [BG92, Section 3.6.1]

$$d_{ij} = \frac{1}{\mu - \lambda_{ij}}. \tag{8.11}$$

The *throughput* can be defined as the minimum value of the offered load for which the delay on any link becomes infinite. This happens when $\lambda_{max} = \max_{i,j} \lambda_{ij} = \mu$. Thus our performance objective will be to minimize the congestion λ_{max}.

We are now ready to state the problem formally as a mathematical program:

Objective function:

$$\min \lambda_{max}$$

subject to

Flow conservation at each node:

$$\sum_{j} \lambda_{ij}^{sd} - \sum_{j} \lambda_{ji}^{sd} = \begin{cases} \lambda^{sd} & \text{if } s = i, \\ -\lambda^{sd} & \text{if } d = i, \\ 0 & \text{otherwise,} \end{cases} \qquad \text{for all } s, d, i,$$

Total flow on a logical link:

$$\begin{aligned}
\lambda_{ij} &= \sum_{s,d} \lambda_{ij}^{sd}, &&\text{for all } i, j, \\
\lambda_{ij} &\leq \lambda_{\max}, &&\text{for all } i, j, \\
\lambda_{ij}^{sd} &\leq b_{ij} \lambda^{sd}, &&\text{for all } i, j, s, d,
\end{aligned}$$

Degree constraints:

$$\sum_i b_{ij} \leq \Delta, \qquad \text{for all } j,$$

$$\sum_j b_{ij} \leq \Delta, \qquad \text{for all } i,$$

Bidirectional lightpath constraint:

$$b_{ij} = b_{ji}, \qquad \text{for all } i, j,$$

Nonnegativity and integer constraints:

$$\lambda_{ij}^{sd}, \lambda_{ij}, \lambda_{\max} \geq 0, \qquad \text{for all } i, j, s, d,$$

$$b_{ij} \in \{0, 1\}, \qquad \text{for all } i, j.$$

We identify the packets to be routed between each s-d pair with the *flow* of a commodity. The left-hand side of the flow conservation constraint at node i in the network computes the *net* flow out of a node i for one commodity (sd). The net flow is the difference between the outgoing flow and the incoming flow. The right-hand side is 0 if that node is neither the source nor the destination for that commodity $(i \neq s, d)$. If node i is the source of the flow $(i = s)$, the net flow equals λ^{sd}, the arrival rate for those packets, and if node i is the destination, $i = d$, the net flow equals $-\lambda^{sd}$.

The constraint $\lambda_{ij} = \sum_{s,d} \lambda_{ij}^{sd}$ is just the definition of λ_{ij}. The constraint $\lambda_{ij} \leq \lambda_{\max}$, together with the fact that we are minimizing $\lambda_{\max}$, ensures that the minimum value of $\lambda_{\max}$ is the congestion. The constraint $\lambda_{ij}^{sd} \leq b_{ij} \lambda^{sd}$ ensures that if $b_{ij} = 0$, $\lambda_{ij}^{sd} = 0$ for all values of s and d. So if the link (i, j) doesn't exist in the topology, no packets can be routed on that link. If the link (i, j) exists in the topology $(b_{ij} = 1)$, this constraint simply states that $\lambda_{ij}^{sd} \leq \lambda^{sd}$, which is always true; thus it imposes no constraint on the values of λ_{ij}^{sd} in this case.

The degree constraints ensure that the designed topology has no more than Δ links into and out of each node. The bidirectional lightpath constraint $b_{ij} = b_{ji}$ ensures that the resulting topology has only bidirectional lightpaths; that is, if there is a lightpath from node i to node j, there is also a lightpath from node j to node

i. The constraints $b_{ij} \in \{0, 1\}$ restrict the b_{ij} to take on only the values 0 or 1. As we will see shortly, but for these constraints, the problem would have been easy to solve! Note that the objective function and the constraints are linear functions of the variables (λ_{ij}^{sd}, λ_{ij}, λ_{max}, b_{ij}). A mathematical program with this property is called a *linear program* (LP) if, in addition, all the variables are real. It is called an *integer linear program* (ILP) if all the variables are restricted to take integer values. In our case, some of the variables, for instance, the b_{ij}, are restricted to integer values. So our program is an example of a *mixed integer linear program* (MILP). We call it the LTD-MILP. Although many efficient algorithms are known for solving even very large LPs, no efficient algorithms are known for the solution of arbitrary ILPs and MILPs. In fact, a general ILP or MILP is an example of an NP-hard problem [GJ79]. Commercial packages are readily available to solve LPs, ILPs, and MILPs. In many cases, these are part of a larger package of mathematical and/or optimization routines.

Even with the use of such packages, ILPs and MILPs are too time consuming to solve, except for small-sized problems. Therefore, many heuristics have been developed for finding approximate solutions to these problems. These approximations are often based on specific features of the problem at hand. In the following, we describe one such heuristic for our problem. Our heuristic uses the fact the LPs are easy to solve and obtains an approximate solution to the LTD-MILP using the techniques of *LP-relaxation* and *rounding*. Before we can describe our method, we need to define a few terms used in mathematical programming.

A *feasible solution of a mathematical program* is any set of values of the variables that satisfy all the constraints. An *optimal solution*, or simply *solution*, of a mathematical program is a feasible solution that optimizes (minimizes or maximizes, as the case may be) the objective function. The *value* of a mathematical program is the value of the objective function achieved by any optimal solution.

Note that if we replace the constraints $b_{ij} \in \{0, 1\}$ by the constraints $0 \leq b_{ij} \leq 1$, the LTD-MILP reduces to an LP, which we will call the LTD-LP. Moreover, any feasible solution of the LTD-MILP is also a feasible solution of the LTD-LP, but the LTD-LP may (and usually will) have other feasible solutions. If some optimal solution of the LTD-LP happens to be a feasible solution of the LTD-MILP (that is, the b_{ij}s are 0 or 1), the values of the LTD-MILP and LTD-LP will be equal. Otherwise, the value of the LTD-LP will be a lower bound on the value of the LTD-MILP. (This is the case for minimization problems.) We call this lower bound the *LP-relaxation bound*.

Note that if the values of the b_{ij} are fixed at 0 or 1 such that the degree constraints are satisfied, the LTD-MILP again reduces to an LP. Fixing the values of the b_{ij} fixes the lightpath topology; the remaining problem is to route the packets over this lightpath topology to minimize the congestion. So we call the LP obtained in this manner the routing-LP. The value of any routing-LP is an upper bound on the value of the LTD-MILP. If we are clever (or lucky) in fixing the values of the b_{ij} so that the degree constraints are satisfied, this will be a good upper bound. For clues on how to fix the values of the b_{ij}, we turn again to the LTD-LP.

Consider any optimal solution of the LTD-LP. Intuitively, we expect that b_{ij}s that are close to 1 (respectively, close to 0) must be equal to 1 (respectively, 0) in the LTD-MILP. So we could try a heuristic approach to determine the values of b_{ij} in the LTD-MILP from the values of b_{ij} in the LTD-LP: *round* the b_{ij} in the LTD-LP to the closest integer. However, we have to be careful not to violate the degree constraints on the b_{ij}. So we modify the rounding approach to incorporate this in the following *rounding algorithm*.

Algorithm 8.1

1. Arrange the values of the b_{ij} obtained in an optimal solution of the LTD-LP in decreasing order.

2. Starting at the top of the list, set each $b_{ij} = 1$ if the degree constraints would not be violated. Otherwise, set the $b_{ij} = 0$.

3. Stop when all the degree constraints are satisfied or the b_{ij}s are exhausted.

If the LP-relaxation lower bound and the upper bound obtained by using the rounding algorithm and solving the routing-LP are close to each other, then we have a good approximation to the value of the MILP. We can then use the topology and routing algorithm obtained by the rounding algorithm and routing-LP as approximations to the optimal topology and routing algorithm. A modified version of this approach has been used in [RS96, Jai96] to solve the LTD-MILP approximately in a few examples. Table 8.1 shows the congestion as a function of the degree for one such example, which is a 14-node network with a sample traffic matrix given in [RS96]. In contrast to the work in [RS96, Jai96], which considered directed lightpaths, here we have considered bidirectional lightpaths. This imposes an additional constraint on the lightpath topology (the bidirectional lightpath constraint) and results in slightly higher values of the congestion. The three columns in Table 8.1 correspond to the LP-relaxation lower bound, an exact value obtained by solving the MILP, and the value obtained by the rounding algorithm. Note that the rounding algorithm yields a value that is quite close to the optimum value and in fact achieves the optimum value as the degree increases.

We have discussed the problem of designing a lightpath topology to minimize the maximum packet traffic on any lightpath, subject to a number of constraints. This problem is a very special case of the general problem, and even it is hard! Not unexpectedly, network design problems have been studied for many years [Ker93,

Table 8.1   Congestion versus node degree for a lightpath topology designed over a 14-node sample network with a given traffic pattern from [RS96], but with bidirectional lightpaths. Observe that the LP rounding algorithm yields very good results in this example.

Degree	LP-Relaxation	MILP	LP Rounding
2	284.67	388.59	440.20
3	189.78	189.78	194.56
4	142.33	142.33	142.33
5	113.87	113.87	113.87
6	94.89	94.89	94.89

Cah98] and are known to be hard. In many cases, even formulating the problem becomes hard because of a large number of parameters to be optimized and a large number of constraints to be dealt with. We illustrated one heuristic method for solving such an ILP, but several other techniques exist that can also be used—see, for example, [CMLF00, KS98, MBRM96, BG95, ZA95, JBM95, GW94, CGK93, LA91].

A practically important example of a lightpath topology that must be realized from a WDM network is a SONET ring. SONET rings come in two flavors: unidirectional path-switched rings (UPSR) and bidirectional line-switched rings (BLSR). We discuss these rings in Chapter 10. The problem of the combined design of the SONET rings (lightpaths) and the WDM layer, to minimize the cost of the SONET ADMs, is discussed in [GLS99].

In the traffic model considered above, we had only one traffic matrix whose values were denoted by λ^{sd}. In practice, the traffic can change over time, and thus it may be better to change the lightpath topology also to reduce the cost of the network. Lightpath topology changes can be disruptive and thus must be undertaken only occasionally, adding and dropping only a few lightpaths at a time. An iterative reconfiguration algorithm to change the lightpath topology gradually, in step with traffic changes, is discussed in [NTM00].

8.2.2   Routing and Wavelength Assignment

In Section 8.1, we saw that the overall design problem involves a trade-off between optical layer equipment (essentially, number of wavelengths) and higher-layer equipment (for example, IP router ports or SONET line terminals). In the previous section, we studied the LTD problem. Here we study the RWA problem, which is defined as follows. Given a network topology and a set of end-to-end lightpath requests (which could be obtained, for example, by solving the LTD problem), determine a route and wavelength(s) for the requests, using the minimum possible number of wavelengths.

Sometimes we are already given the routing, in which case, we are concerned only with the wavelength assignment (WA) problem. The wavelength assignment must obey the following constraints:

1. Two lightpaths must not be assigned the same wavelength on a given link.

2. If no wavelength conversion is available, then a lightpath must be assigned the same wavelength on all the links in its route.

In the rest of this chapter, we assume that the network as well as the lightpaths are *undirected*. Different combinations of types of lightpaths and types of networks are possible, as shown in Table 8.2. The combination that we explore here corresponds to a network that has a pair of unidirectional fiber links in opposite directions between nodes and assumes that all lightpaths are bidirectional, with the same route and wavelength chosen for both directions of the lightpath. From an operational viewpoint, most lightpaths will be full duplex, as the higher-level traffic streams that they carry (for example, SONET streams) are full duplex. Moreover, network operators would prefer to assign the same route and wavelength to both directions for operational simplicity. Note, however, that it is possible to reduce the number of wavelengths needed in some cases by assigning different wavelengths to different directions of the lightpath. This is treated in Problem 8.22.

An undirected edge can also represent a bidirectional fiber link with transmission in both directions over a single fiber. We consider this case in the example below.

The routing and wavelength problem can be also be studied in the context of the other combinations shown in Table 8.2. There has been a fair amount of theoretical work devoted to solving the routing and wavelength assignment problem

Table 8.2 Different combinations of types of lightpaths and network edges. We study the case of networks with undirected edges that support undirected lightpaths.

Undirected lightpaths, undirected edges	Directed lightpaths, undirected edges
Undirected lightpaths, directed edges	Directed lightpaths, directed edges

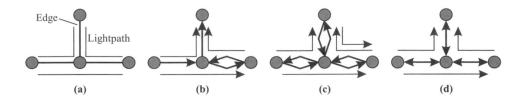

Figure 8.9   Different network models corresponding to undirected/directed edges and undirected/directed lightpaths. (a) Undirected edges, undirected lightpaths. (b) Directed edges, directed lightpaths. (c) and (d) show two different cases of undirected edges and directed lightpaths.

on networks with directed edges and directed lightpaths, which the reader can find in the References at the end of this chapter. This is the most general case, and the other three cases are special subsets of this general problem.

Example 8.5   In this example, we will illustrate the differences between the different network models considered in Table 8.2, using an example. (Assume that no wavelength conversion is available.) Consider the simple four-node star network shown in Figure 8.9. In Figure 8.9(a), we have a network with undirected edges that must support three lightpaths. Note that although only at most two lightpaths use each edge, we need three wavelengths to support this traffic pattern.

Next consider a network with directed edges and directed lightpaths shown in Figure 8.9(b). Note that the number of lightpaths on each edge is again no more than two, but only two wavelengths are required in this instance.

Figure 8.9(c) and (d) show two cases, both with undirected edges and directed lightpaths. In Figure 8.9(c), we represent the undirected edge by two unidirectional edges. This corresponds to having a fiber in each direction in the real network and having W wavelengths on each fiber. In this case, note that only two wavelengths are required to support this traffic pattern.

The final case is Figure 8.9(d), where we represent the undirected edge by a bidirectional edge. This corresponds to having a single fiber over which transmission takes place in both directions. There is a fixed total number of wavelengths; some wavelengths are transmitted in one direction, and the remaining ones in the opposite direction. We may assume that this assignment can be done in a flexible manner as required for the traffic pattern. In this case the wavelength assignment constraint is somewhat different. If a wavelength is used in one direction of the link, it cannot be used in the other direction. Note that in this case the direction

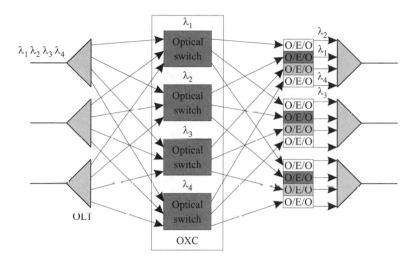

Figure 8.10   A node with fixed-wavelength conversion capability. Signals entering at wavelength λ_1 are converted to λ_2 and vice versa. Signals entering at wavelength λ_3 are converted to λ_4 and vice versa.

of a lightpath does not affect the wavelength assignment. In practice, bidirectional systems have a fixed set of wavelengths in each direction, which further constrains this case, as we will see in Problem 8.10.

8.2.3   Wavelength Conversion

We discussed wavelength conversion in Chapter 7, specifically with reference to optical crossconnects. This kind of wavelength conversion is called *full wavelength conversion*, and a node capable of full wavelength conversion can change the wavelength of an incoming lightpath to any of the outgoing wavelengths. The crossconnects shown in Figure 7.9(a)–(c) are capable of full wavelength conversion, while the crossconnects of Figures 7.9(d) and 7.12 have no wavelength conversion capability.

Two other kinds of wavelength conversion are *fixed conversion* and *limited conversion*. In fixed-wavelength conversion, a lightpath entering a node at a particular wavelength λ_i always exits the node at a given wavelength λ_j. The mapping between the input and output wavelength is fixed at the time the network is designed and cannot be varied. An implementation of this approach is shown in Figure 8.10. In limited wavelength conversion, a signal is allowed to be converted from one wavelength to a limited subset of other wavelengths. For instance, we may allow a signal to be converted from one wavelength to two other predetermined wavelengths.

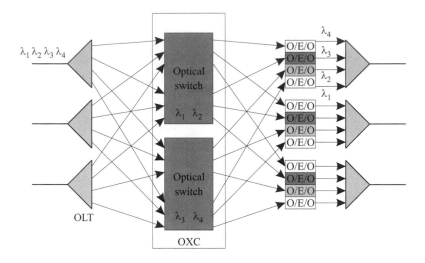

Figure 8.11   A node with limited wavelength conversion capability. Each input wavelength can be converted to one of two possible output wavelengths. Signals entering at wavelength λ_1 or λ_2 can be converted to λ_3 or λ_4. Signals entering at wavelength λ_3 or λ_4 can be converted to λ_1 or λ_2.

Figure 8.11 shows an implementation of this approach, where each input wavelength can be converted to one of two other wavelengths. In this case, we say that the node provides limited conversion of degree 2. We will see later in Section 8.5 that having a small amount of wavelength conversion in the network provides almost the same benefits as having full wavelength conversion at every node in the network.

The fixed and limited conversion models described above allow us to save on switch cost but still require an O/E/O for each signal. Since the O/E/Os dominate the cost, these models are mainly of theoretical interest today. However, two other factors make these models useful. The first is when we have practical all-optical wavelength converters. It is quite possible that these devices will inherently not allow converting a signal to an arbitrary output wavelength but only to one or a subset of other wavelengths (see Section 3.8). Thus limited conversion becomes very important in this case. The second is that networks with multiple fibers and no wavelength conversion can be modeled using this approach, as we will see next.

In many situations, networks may use multiple fiber pairs between nodes to provide higher capacities. We will now see that having multiple fiber pairs is equivalent to having a single fiber pair but with some limited wavelength conversion capabilities at the nodes. Figure 8.12(a) shows a network with two fiber pairs between nodes

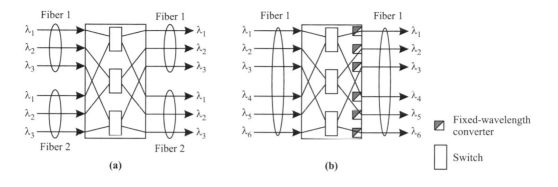

Figure 8.12   The equivalence between multiple fiber networks and single fiber networks.

and no wavelength conversion at the nodes. Each fiber pair carries W wavelengths. At each node, signals from one fiber pair can be switched to the other fiber pair. Figure 8.12(b) shows a network with one fiber pair between nodes, with that pair carrying $2W$ wavelengths. The nodes have limited conversion of degree 2. These two networks are equivalent in terms of their traffic-carrying capacity. Any set of lightpaths supported by one network can be supported by the other network as well. The proof of this is left as an exercise (Problem 8.11). Therefore, we can characterize multiple fiber networks with no conversion by equivalent single fiber networks with limited-degree wavelength conversion at the nodes. For this reason, we will not consider multiple fiber networks separately in this chapter.

We will use the suffixes NC, FC, C, and LC to denote no wavelength conversion, fixed conversion, full conversion, and limited conversion, respectively.

In the full, limited, and fixed conversion cases, the WA problem must be suitably modified. In the case of full conversion, the constraint on a lightpath being assigned the same wavelength on every link it traverses can be dispensed with entirely. In the case of limited wavelength conversion, the wavelength assigned to a lightpath can be changed but only to a limited set of other wavelengths. In the case of fixed-wavelength conversion, the wavelength assigned to a lightpath *must* be changed at each node.

Given a set of lightpath requests and a routing, let l_i denote the number of lightpaths on link i. Then we define the *load* of a request to be $L = \max_i l_i$. Clearly, from the first constraint, we need at least L wavelengths to accommodate this set of lightpath requests. If we have full wavelength conversion in the network, then the problem of wavelength assignment becomes trivial because it no longer matters what wavelength we assign to a lightpath on a given link. As long as no more than

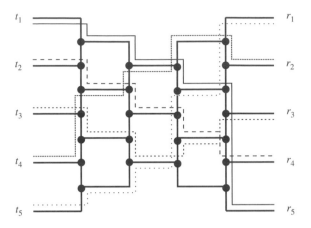

Figure 8.13 An example to illustrate the difference between having and not having wavelength conversion.

L lightpaths use this link, L wavelengths will clearly be sufficient to accommodate this request. However, without wavelength conversion, the number of wavelengths required could be much larger. The important question is, How much larger? We will study this problem in detail in Section 8.5, under various conditions, but we consider one (somewhat extreme) example now.

Example 8.6   Consider the network shown in Figure 8.13. The set of lightpath requests is shown in the figure to be the following. Transmitter t_i must be connected to receiver r_{N-i+1}, where N is the number of transmitters or receivers. Clearly, there are many routes for each lightpath. Interestingly, however, regardless of how we route each lightpath, any two lightpaths belonging to this set of requests must share a common link. Thus each lightpath must be assigned a different wavelength, requiring a total of N wavelengths to satisfy this set of requests.

If we are clever about how we route these lightpaths, we can arrange matters so that at most two lightpaths use a given link, as shown in the figure. This means that the load is 2. Thus two wavelengths are sufficient to satisfy this set of requests if full wavelength conversion is available at each node in the network.

Does this mean that full wavelength conversion is absolutely needed? Luckily for us, the example shown here is a worst-case scenario. We will quantify the benefit due to wavelength conversion in Section 8.5.

8.2.4 *Relationship to Graph Coloring*

It turns out that the WA problem described earlier is closely related to the problem of coloring the nodes in a graph. To understand this better, consider a graph representation of the network G, where the vertices of the graph represent nodes in the network, with an undirected edge between two vertices corresponding to an optical fiber link between the corresponding nodes. Figure 8.14(a) is an example. The route for each lightpath corresponds to a path in G, and thus the set of routes that have been specified for the lightpaths corresponds to a set of paths, say, P. Now consider another graph, the *path graph* of G, denoted by $P(G)$, which is constructed as follows. Each path in P corresponds to a node in $P(G)$, and two nodes in $P(G)$ are connected by an (undirected) edge if the corresponding paths in P share a common edge in G. Figure 8.14(b) shows the path graph of the graph in Figure 8.14(a).

Solving the WA problem is then equivalent to solving the classical graph coloring problem on $P(G)$; that is, we have to assign a color to each node of $P(G)$ such that adjacent nodes are assigned distinct colors and the total number of colors is minimized. These colors correspond to wavelengths used on the paths in G. The minimum number of colors needed to color the nodes of a graph in this manner is called the *chromatic number* of the graph. Thus the minimum number of wavelengths required to solve the WA problem is the chromatic number of $P(G)$.

> **Example 8.7** Let the graph G depicting the network be as shown in Figure 8.14. Since there is only one path between any pair of nodes in G, given the set of node pairs to be connected by lightpaths, the routes are uniquely determined. So we have only to solve the WA problem. Suppose we need to set up lightpaths between nodes 1 and 2, 2 and 3, and 1 and 3. The resulting path graph, $P(G)$, is also shown in the same figure. The chromatic number of $P(G)$ is 3, and a coloring of $P(G)$ in three colors is also shown. Thus we need three wavelengths to solve the WA problem in this example.

Coloring an arbitrary graph is a hard problem that has been intensively studied for several decades. In fact, it is an example of a class of problems, called *NP-complete* problems [GJ79], which are considered to be hard to solve. However, there are several special classes of graphs for which fast coloring algorithms have been found. If the $P(G)$ we are interested in belongs to one of these special classes, then we can find an exact solution to the WA problem. Otherwise, unless $P(G)$ has only a few nodes, we have to content ourselves with finding an approximate solution to the WA problem. Many fast but approximate (heuristic) algorithms have been devised for the general graph coloring problem (see, for example, [Big90, dW90]), and these algorithms can be used to find good but approximate solutions to the WA problem.

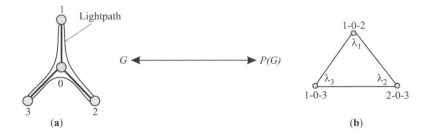

Figure 8.14   Illustrating the relationship to graph coloring.

Although this transformation illustrates the relationship to graph coloring, it does not prove that the WA problem is in itself hard or, specifically, NP-complete. To show this, you need to perform the transformation in the opposite direction, namely, take an instance of a graph coloring problem and convert it into an instance of the WA problem. This has been done in [CGK92], which proves that WA is indeed NP-complete. However, it is still possible to obtain useful bounds for this problem, as well as develop algorithms for several specific and important topologies such as ring networks, as we will see next.

8.3   Dimensioning Wavelength-Routing Networks

The key aspect of designing a wavelength-routing network is determining the number and, more generally, the set of wavelengths that must be provided on each WDM link. We call this the *wavelength dimensioning problem.*

In most practical situations today, the network is designed to support a certain, fixed traffic matrix. The traffic matrix may be in terms of lightpaths or in terms of higher-layer (IP, SONET) traffic. In the former case, only the RWA needs to be solved, while in the latter case, both the LTD and RWA problems must be solved (in conjunction or separately). By and large, this is the approach used in practice today to design wavelength-routing networks. The solution of the RWA problem determines the specific set of wavelengths that must be provided on each link to realize the required lightpath topology, and thus solves the dimensioning problem. This is the *offline* RWA problem since we are given all the lightpaths at once, and is useful in the network planning stage. Once a network is operational, the RWA problem has to be solved for one lightpath at a time, when the lightpath is required to be set up. This is the *online* RWA problem. With the reduction in lightpath service provisioning

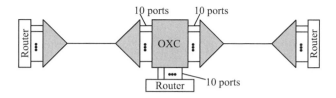

Figure 8.15 The three-node network of Figure 8.1(c) with the static OADM at the central node replaced by a reconfigurable OADM, or OXC. The OXC allows the set of lightpaths added/dropped at the node to be decided dynamically based on the lightpath/traffic requirements.

times that is being faced by carriers, it is becoming increasingly important to find good, rapid solutions to the online RWA problem.

While the specific sets of wavelengths obtained by solving the offline RWA problem can be provisioned in a network without OXCs, OXCs are used where flexibility in handling different traffic matrices is needed. Without OXCs, the lightpaths must be established by a static, or a priori, mapping of incoming wavelengths to outgoing wavelengths at each node. Since most of the wavelength-routing networks that have been deployed today do not use OXCs, the lightpaths have been established in such a static fashion in these networks. When OXCs are deployed, by appropriate configuration of the OXCs, the optical layer can change the lightpath topology and hence adapt to different traffic requirements. Thus this approach can support any one of several different lightpath topologies, and consequently, traffic requirements at the higher layer, on the same fiber topology with the same optical layer equipment. Since the higher-layer traffic requirements are usually unknown, this flexibility is quite important in building a future-proof optical network.

> **Example 8.8** To illustrate the flexibility obtained by using OXCs in the network, consider the three-node linear network example again. By replacing the static OADM in Figure 8.1(c) by a reconfigurable OADM, or OXC, with 30 ports, we obtain the node design shown in Figure 8.15. This design can handle any combination of traffic that does not require termination of more than 100 Gb/s of traffic at each node, in contrast to the design of Figure 8.1(c), which was designed for a specific traffic matrix: 50 Gb/s of traffic between each pair of nodes.

Solving the dimensioning problem determines not only the number of wavelengths that need to be supported on each link, but also determines the sizes of the OLTs and the OXCs. The size of the OXC also depends on the maximum number of lightpaths to be terminated at each node, which corresponds to the number of router interface cards provided at that node.

As discussed above, in contemporary practice, the design of wavelength-routing networks today is accomplished by forecasting a certain fixed traffic matrix between the nodes. This forecast is revised every six months or so, and based on this forecast, the network is upgraded with the addition of more capacities on the WDM links, or more links, or additional nodes, or a combination of these approaches. Solving the network upgrade problem is similar to solving the original problem, except that the lightpaths that have already been established are usually not disturbed.

We can view the above approach of forecasting a fixed traffic matrix and dimensioning the network to support the forecasted traffic as using a "deterministic" traffic model since the variations in traffic are not explicitly accounted for during the design phase, though the use of crossconnects in the network enables some of these variations to be handled at the time of actually setting up the lightpaths. Another approach to capacity planning is through the use of statistical traffic models, which we will discuss in Section 8.4.

In a wavelength-routing network, if the nodes have full conversion capability, the situation is the same as in classical circuit-switched telephone networks: a lightpath is equivalent to a phone call and must be assigned one circuit on each of the links it traverses. Another approach studied extensively by researchers is to dimension optical networks with no or limited conversion capabilities, to support the same traffic that would be supported using full conversion within the optical layer. We discuss these methods in Section 8.5. In this case, as well as in the case of statistical models, we consider only the RWA problem and not the LTD problem. Thus, grooming issues that are part of the LTD problem are not discussed. The problem of determining the location of regenerators is also outside the scope of our discussion.

8.4   Statistical Dimensioning Models

There are two classes of statistical traffic models that can be used in solving the dimensioning problem. These models differ in their assumptions regarding what is known about the set or sets of lightpaths that must be supported. In some cases, these models also assume that each link supports the same number (and set) of wavelengths, but this may not always be appropriate.

1. **First-passage model:** In this model, the network is assumed to start with no lightpaths at all. Lightpaths arrive randomly according to a statistical model and

have to be set up on the optical layer. Some lightpaths may depart as well, but it is assumed that, on average, the number of lightpaths will keep increasing and eventually we will have to reject a lightpath request. (Thus the rate of arrival of lightpath requests exceeds the rate of termination of lightpaths, and the network is not in equilibrium.) We are interested in dimensioning the WDM links so that the first lightpath request rejection will occur, with high probability, after a specified period of time, T. This is a reasonable model today since lightpaths are long lived. This longevity, combined with the cost of a high-bandwidth lightpath today, means that network operators are unlikely to reject a lightpath request. Rather, they would like to upgrade their network by the addition of more capacity on existing links, or by the addition of more links, in order to accommodate the lightpath request. The time period T corresponds approximately to the time by which the operators must institute such upgrades in order to avoid rejecting lightpath requests.

2. **Blocking model:** In this model, the lightpath requests are treated in the same way that a telephone network treats phone calls. Requests are assumed to arrive and depart at random instants according to a statistical model. (However, the network is assumed to be in equilibrium, that is, the rate of arrival and the rate of termination of lightpaths are equal.) The assumption is that most requests must be honored but occasionally requests may be blocked. The goal again is to dimension the WDM links so that the blocking events are relatively rare (say, a fraction of 1%). This is a futuristic model since lightpaths today are relatively long lived, but it is quite possible that lightpaths will be provided on demand by some operators in the future. In such a scenario, this would be a reasonable model to use in order to dimension the WDM links.

For these statistical models, the *analysis problem* is easier to solve than the *design problem*. For example, in the blocking model, it is easier to calculate the blocking probabilities on each of the links given the link capacities (and the traffic model) than it is to design the link capacities to achieve prespecified blocking probabilities. Similarly, in the first-passage model, it is easier to calculate the (statistics of the) first time at which the network operator will have to block a lightpath request for given link capacities than it is to design the link capacities to achieve a prespecified first-passage time. However, the capacity design or dimensioning problem can be solved by iterating on the analysis problem. For example, we can calculate the blocking probabilities for a given set of capacities, and if the blocking is not acceptable on some links, increase the capacities of those links and recalculate the blocking probabilities. In the rest of this section, we will address the analysis problems.

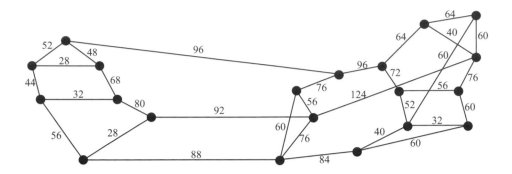

Figure 8.16 A 20-node, 32-link network representing a skeleton of the ARPANET. An average of one lightpath request is assumed to arrive every month, between every pair of nodes, and this lightpath is assumed to be in place for an average of one year. The link capacities shown are calculated such that no link will need a capacity upgrade within two years, with high (85%) probability.

8.4.1   First-Passage Model

In this model, the network is assumed to start with no lightpaths, but the link capacities are given. The model is analytically tractable only if we assume that lightpath requests follow a Poisson process and their durations are exponentially distributed. (This is the standard assumption in telephone networks for the statistics of phone calls. Thus, this is tantamount to assuming that lightpath requests are like phone calls.) The network can be modeled by a Markov chain where the state of the Markov chain represents the set of calls in progress. You can consider both fully wavelength-converting crossconnects and OXCs with no conversion capability. The Markov chain approach is somewhat tractable only in the case of full wavelength conversion. An approximate analysis of this model appears in [NS02].

We do not describe the mathematical details of the analytical model that can be found in [NS02], but we present the outcome of such an analysis for a moderate-sized network. The network considered is shown in Figure 8.16. It has 20 nodes and 32 links and represents a skeleton of the original ARPANET. The request for lightpaths on each of the possible 190 routes is assumed to arrive at a rate of one request per month (but with a Poisson distribution). The average lightpath lease time is assumed to be one year (with an exponential distribution). It is assumed that the capacity on each link can be a multiple of four wavelengths. The capacities of the links shown in Figure 8.16 are determined such that the probability that any of these links needs a capacity upgrade within two years is less than 15%.

8.4.2   Blocking Model

In this model, we assume that lightpath arrival and termination requests follow a statistical pattern. We may allow some lightpath requests to be blocked and are interested therefore in minimizing the blocking probability. In this case, a measure of the lightpath traffic is the *offered load*, which is defined as the arrival rate of lightpath requests multiplied by the average lightpath duration.

In practice, the maximum blocking probability is specified, say, 1%. We are then interested in determining the maximum offered load that the network can support. A more convenient metric is the wavelength *reuse factor*, R, which we define as the offered load per wavelength in the network that can be supported with the specified blocking probability. Clearly, R could depend on (1) the network topology, (2) the traffic distribution in the network, (3) the actual RWA algorithm used, and (4) the number of wavelengths available.

In principle, if we are given (1)–(4), we can determine the reuse factor R. However, this problem is difficult to solve analytically for specific RWA algorithms. When the routes between the source-destination nodes in the network are fixed (fixed routing) and an available wavelength is chosen randomly, the blocking probabilities (and hence the reuse factor) can be analytically estimated for a reasonable number of wavelengths (say, up to 64). A discussion of these analytical techniques is beyond the scope of this book but can be found in [SS00]. The results of such an analysis can be used to dimension the links for a given blocking probability just as in the case of the first-passage model discussed above.

When the routing is not fixed, estimating the blocking probabilities or reuse factors is analytically intractable, and in practice, the best way to estimate R even for small networks is by simulation. It is possible to analytically calculate the maximum value of R when the number of wavelengths is very large for small networks; this has been done in [RS95] and serves as an upper bound on the reuse factor for practical values of the number of wavelengths. When the number of wavelengths is small, simulation techniques can be used to compute the reuse factor. To this end, we summarize some of the simulation results from [RS95]. We will also compare the simulation results with the analytically calculated upper bound on the reuse factor. We will use randomly chosen graphs to model the network, assume a Poisson arrival process with exponential holding times, assume a uniform traffic distribution, and use the following RWA algorithm.

Algorithm 8.2

1. Number the W available wavelengths from 1 to W.

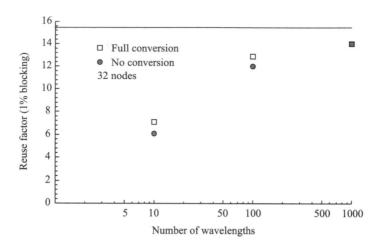

Figure 8.17   Reuse factor plotted against the number of wavelengths for a 32-node random graph with average degree 4, with full wavelength conversion and no wavelength conversion, from [RS95]. The horizontal line indicates the value of the reuse factor that can be achieved with an infinite number of wavelengths with full wavelength conversion, which can be calculated analytically.

2. For a lightpath request between two nodes, assign to it the first available wavelength on a fixed shortest path between the two nodes.

Figure 8.17 shows the reuse factor plotted against the number of wavelengths for a 32-node random graph with average node degree 4. The figure also shows the value of the blocking probability that can be achieved with an infinite number of wavelengths, which can be calculated analytically as mentioned before [RS95]. The reuse factor is slightly higher with full conversion. The interesting point to be noted is that the reuse factor improves as the number of wavelengths increases. This is due to a phenomenon known as *trunking efficiency*, which is familiar to designers of telephone networks. Essentially, the blocking probability is reduced if you scale up both traffic and link capacities by the same factor. To illustrate this phenomenon, consider a single link with Poisson arrivals with offered load ρ with W wavelengths. The blocking probability on this link is given by the famous Erlang-B formula:

$$P_b(\rho, W) = \frac{\frac{\rho^W}{W!}}{\sum_{i=0}^{W} \frac{\rho^i}{i!}}.$$

The reader can verify that if both the offered traffic and the number of wavelengths are scaled by a factor $\alpha > 1$, then

$$P_b(\alpha\rho, \alpha W) < P_b(\rho, W)$$

and

$$P_b(\alpha\rho, \alpha W) \to 0 \text{ as } \alpha \to \infty \text{ if } \rho \le W.$$

Figure 8.18 shows the reuse factor plotted against the number of nodes N. The value of R for each N is obtained by averaging the simulated results over three different random graphs, each of average degree 4. The figure shows that (1) R increases with N, and (2) the difference between not having conversion and having it also increases with N. Note that observation (1) is to be expected because the average lightpath length (in number of hops) in the network grows as $\log N$, whereas the number of links in the network grows as N. Thus we would expect the reuse factor to increase roughly as $N/\log N$. The reason for observation (2) is that the average path length (or hops) of a lightpath in the network increases with N. We will see next that wavelength converters are more effective when the network has longer paths.

A similar simulation has been performed in [KA96] for ring networks. In general, the increase in reuse factor obtained after using wavelength conversion was found to be very small. This may seem counterintuitive initially because hop lengths in rings are quite large compared to mesh networks. We will see next that hop length alone is not the sole criterion for determining the gain due to wavelength conversion. In rings, lightpaths that overlap tend to do so over a relatively large number of links, compared to mesh networks. We will see that the larger this overlap, the less the gain due to wavelength conversion.

Factors Governing Wavelength Reuse

We will next quantify the impact of the number of hops and the "overlap" between lightpaths on the wavelength conversion gain. We assume a statistical model for the lightpath requests and make a highly simplified comparison of the probability that a lightpath request will be denied (blocked) when the network uses wavelength converters and when it does not, based on [BH96]. We assume that the route through the network for each lightpath is specified. When the network does not use wavelength converters, the wavelength assignment algorithm assigns an arbitrary but identical wavelength on every link of the route when one such wavelength is free (not assigned to any other lightpath) on every link of the path. When the network uses wavelength

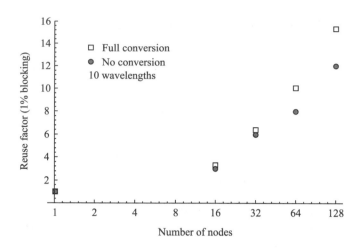

Figure 8.18 Reuse factor plotted against the number of nodes for random graphs with average degree 4, with full wavelength conversion and no wavelength conversion (from [RS95]).

converters, the wavelength assignment algorithm assigns an arbitrary free wavelength on every link in the route to the lightpath; thus we assume full wavelength conversion. In both cases, if the wavelength assignment algorithm is unable to find a suitable wavelength, the lightpath request is blocked.

In order to compute the blocking probability for lightpath requests, we make the simplifying assumption that the probability that a wavelength is used on a link is π and that this event is independent of the use of other wavelengths on the same link and the use of (the same and other) wavelengths on other links. If the network has W wavelengths on every link and a lightpath request chooses a route with H links, the probability that it is blocked is given by

$$P_{b,nc} = \left(1 - (1 - \pi)^H\right)^W \tag{8.12}$$

when the network does not use wavelength converters. To see this, note that the probability that a given wavelength is free on any given link is $(1 - \pi)$, and thus the probability that it is *free on all the H links* in the route is $(1 - \pi)^H$ by the assumed independence of the use of a wavelength on each link. Therefore, $(1 - (1 - \pi)^H)$ is the probability that a given wavelength is *not free on some link* of the route and, since the use of each wavelength is assumed to be independent of the use of other wavelengths, $(1 - (1 - \pi)^H)^W$ is the probability that all W wavelengths are not free on some link of the route, that is, $P_{b,nc}$.

When the network uses full wavelength conversion, the probability that a light-path request is blocked is given by

$$P_{b,fc} = 1 - \left(1 - \pi^W\right)^H. \tag{8.13}$$

The derivation of this equation using reasoning similar to that used in the derivation of (8.12) is left as an exercise (Problem 8.23).

Given the blocking probability, we denote the solution of (8.12) and (8.13) for π by π_{nc} and π_{fc}, respectively. Thus π_{nc} (respectively, π_{fc}) represents the achievable link utilization for a given blocking probability when wavelength converters are not used (respectively, used). It is easily seen that

$$\pi_{nc} = 1 - \left(1 - P_{b,nc}^{1/W}\right)^{1/H} \tag{8.14}$$

and

$$\pi_{fc} = \left(1 - (1 - P_{b,fc})^{1/H}\right)^{1/W}. \tag{8.15}$$

For small values of $P_{b,.}$ (which is the case of practical interest) and sufficiently small values of W such that $P_{b,.}^{1/W}$ is not too close to 1, π_{nc} and π_{fc} can be approximated by

$$\pi_{nc} = P_{b,nc}^{1/W} / H \tag{8.16}$$

and

$$\pi_{fc} = \left(P_{b,fc}/H\right)^{1/W}. \tag{8.17}$$

Thus for the same blocking probability, the ratio π_{fc}/π_{nc} can be approximated by $H^{1-1/W}$. Therefore, this simplified analysis predicts that even for moderately large values of W the achievable link utilization is lower by approximately a factor of H when wavelength converters are not used in the network.

Although the preceding analysis is highly simplified, ignores several important effects, and overestimates the efficacy of wavelength converters in improving the link utilization, it does predict correctly that the achievable link utilization is more sensitive to the path length (H) when wavelength converters are not used than otherwise.

We now remove the assumption that the probability of a wavelength being used on a link is independent of the use of the same wavelength on other links. However, we will continue to assume that the events on one wavelength are independent of the events on all other wavelengths. We first consider networks with no wavelength conversion and calculate the probability that a lightpath request that chooses a route with H links is blocked. Any lightpath that has already been established and uses one of these H links is termed an *interfering* lightpath. We assume that an interfering lightpath that uses one of these H links, say, link i, will *not use the next link $i + 1$* with probability π_l. (So with probability π_l a lightpath that interferes on link i of the route chosen by the lightpath request leaves after that link.) For any wavelength λ, we also assume that a *new* lightpath request (one that does not interfere on link $i - 1$) would interfere on link i of the route chosen by the lightpath request with probability π_n. This gives us the following conditional probabilities for the use of wavelength λ on link i:

Prob(λ used on link i | λ not used on link $i - 1$) $= \pi_n$,

and

Prob(λ used on link i | λ used on link $i - 1$) $= (1 - \pi_l) + \pi_l \pi_n$.

Note that under the assumption of independent use of the same wavelength on the links, both these conditional probabilities must equal π; thus this assumption corresponds to setting $\pi_l = 1$ and $\pi = \pi_n$.

Using the same reasoning as that used to derive (8.12), we can show that now

$$P_{b,nc} = \left(1 - (1 - \pi_n)^H\right)^W. \tag{8.18}$$

For networks with full wavelength conversion, the following expression for blocking probability can be derived under a set of assumptions that are similar to that used to derive (8.18):

$$P_{b,fc} = 1 - \prod_{i=1}^{H} \left(1 - \frac{\pi_i^W - (1 - \pi_l + \pi_l \pi_n)^W \pi_i^W}{1 - \pi_{i-1}^W}\right), \tag{8.19}$$

where

$$\pi_i = \frac{\pi_n}{\pi_n + \pi_l - \pi_n \pi_l}\left(1 - (1 - (\pi_l + \pi_n - \pi_l \pi_n))^i\right).$$

For a given blocking probability, we can solve (8.18) and (8.19) for π_{nc} and π_{fc}, respectively. Then we can approximate the *conversion gain* π_{fc}/π_{nc} for small blocking probabilities and $H \gg 1/\pi_l$ by

$$\frac{\pi_{fc}}{\pi_{nc}} \approx H^{1-1/W}(\pi_n + \pi_l - \pi_l \pi_n). \tag{8.20}$$

Define the *interference length* $L_i = 1/\pi_l$. L_i is an approximation to the expected number of links that an interfering lightpath uses on the route chosen by a lightpath request. The assumption $H \gg 1/\pi_l = L_i$ is thus equivalent to assuming that the number of hops in the path chosen by a lightpath request is much larger than the average number of hops that it shares with an interfering lightpath. This assumption is a good one when the network is well connected, but it is a poorer approximation to the behavior in, say, rings.

The conversion gain under the assumption of independent use of a wavelength on each link ($\pi_l = 1$) is approximately $H^{1-1/W}$. Thus the conversion gain given by (8.20) is lower than this by the factor ($\pi_n + \pi_l - \pi_l\pi_n$). This factor is the *mixing probability*: the probability that at a node along the route chosen by a lightpath request, an interfering lightpath leaves or a new interfering lightpath joins. Thus the conversion gain is more in networks where there is more mixing, for example, in dense mesh networks where the node degrees (switch sizes) are large, as opposed to ring networks where the mixing is small and the interference length is large.

In summary, path length is only one of the factors governing the amount of reuse we get by using wavelength conversion; interference length and switch sizes are other important factors.

An analysis of WDM ring networks, based on the techniques described above, can be found in [SM00].

Wavelength Assignment and Alternate Routes

So far, while studying the RWA problem using a statistical model for the traffic, we have assumed a fixed route between each source-destination pair. We will now present some simulation results to show the effect of using alternate routes. We will also consider two different ways of assigning wavelengths once the route has been selected. Thus we consider the following four RWA algorithms.

Random-1. For a lightpath request between two nodes, choose at random one of the available wavelengths on a fixed shortest path between the two nodes.

Random-2. Fix two shortest paths between every pair of nodes. For a lightpath request between two nodes, choose at random one of the available wavelengths on the first shortest path between the two nodes. If no such wavelength is available, choose a random one of the available wavelengths on the second shortest path.

Max-used-1. For a lightpath request between two nodes, among the available wavelengths on a fixed shortest path between the two nodes, choose the one that is used the most number of times in the network at that point of time.

Max-used-2. Fix two shortest paths between every pair of nodes. For a lightpath request between two nodes, among the available wavelengths on the first shortest

Table 8.3   Reuse factor for 1% blocking for different RWA algorithms for the
20-node network considered in [RS95].

RWA Algorithm	Reuse Factor
Random-1	6.9
Random-2	7.8
Max-used-1	7.5
Max-used-2	8.3

path between the two nodes, choose the one that is used the most number of times in the network at that point of time. If no such wavelength is available, among the available wavelengths on the second shortest path between the two nodes, choose the one that is used the most number of times in the network at that point of time.

The topology we consider is the 20-node, 39-link network from [RS95]. We assume 32 wavelengths are available on each link and that the traffic is uniform (same for every pair of nodes). The reuse factor obtained by using each of the above four RWA algorithms for a blocking probability of 1% is shown in Table 8.3. Observe that the reuse factor improves substantially when an alternate path is considered. Ideally we would like to have more alternate routes for longer routes and less for shorter routes. This will help reduce the blocking probability on longer routes and ensure better fairness overall. Otherwise, short routes tend to have much less blocking than long routes. Having more routes to consider usually increases the control traffic in the network and leads to an additional compuational burden on the network nodes, but this is not significant in networks with a moderate number of nodes where lightpaths are set up and taken down slowly.

In addition to the choice of routes, the wavelength assignment algorithm also plays an important role in determining the reuse factor. Note that for the same number of available paths, the max-used algorithms have a distinct advantage over the random algorithms. The intuitive reason for this phenomenon is that the max-used strategy provides a higher likelihood of finding the same free wavelength on all the links along a particular route. A drawback of the max-used algorithm is that it requires a knowledge of the wavelengths in use by all other connections in the network. When the routing and wavelength assignment is performed in a distributed manner, such information typically has to be obtained by means of periodic updates broadcast by each node. This again increases the control traffic load on the network.

8.5 Maximum Load Dimensioning Models

As discussed above, from a dimensioning perspective, the fundamental property that distinguishes wavelength-routing networks from traditional electronic circuit-switched networks is the absence of full wavelength conversion. A number of studies have been undertaken to determine how networks using no, or limited, wavelength conversion should be dimensioned in order to support the *same set, or sets, of lightpaths* as an optical layer with full conversion. In this section, we will present some of the results obtained in this direction. We assume that both the lightpaths and the network edges are undirected (see Table 8.2).

The results can be broadly classified into two categories: offline requests and online requests. The offline problem corresponds to a "static" network design problem, where only a single set of lightpaths is to be supported. This set is constrained to be such that it can be supported in a network with nodes capable of full wavelength conversion, with at most L wavelengths per link, since there is a routing that places no more than L routes on any link. Thus, the *maximum load* of this set of lightpaths is said to be L. In a network with nodes incapable of wavelength conversion, more than L wavelengths per link would be needed, in general, to support the same set of lightpaths. We are interested in determining the additional number of wavelengths that would be required to support every set of such lightpaths, with nodes that do not have any wavelength conversion capability.

Online RWA corresponds to the "dynamic" network design case where lightpaths arise one at a time and have to be assigned routes and wavelengths when the request arrives, without waiting for future requests to be known. However, the requests and routing are such that no more than L lightpaths use any link at any given time. Thus a network with fully wavelength-converting crossconnects that provides L wavelengths on each link would be able to support all the requests. In this case, the task is to compare the number of additional wavelengths that would be required to support the same sets of lightpaths with non-wavelength-converting crossconnects.

One shortcoming of this maximum load model is that the number of wavelengths required may be excessively large in order to support *all* sets of lightpaths with maximum load L. If we are permitted not to support a small fraction of these sets of lightpaths, it may be possible to considerably reduce the number of wavelengths required. In this sense, the maximum load model is a worst-case dimensioning method.

8.5.1 Offline Lightpath Requests

In this section, we will survey the results for offline lightpath requests.

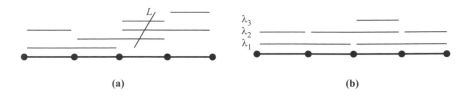

Figure 8.19   (a) A line network with a set of lightpaths, also called an interval graph. (b) Wavelength assignment done by Algorithm 8.3.

Theorem 8.1   [ABC+94] *Given a routing of a set of lightpaths with load L in a network G with M edges, with the maximum number of hops in a lightpath being D, the number of wavelengths sufficient to satisfy this request is $W \leq \min[(L-1)D + 1, (2L-1)\sqrt{M} - L + 2]$.*

Proof.   Observe that each lightpath can intersect with at most $(L-1)D$ other lightpaths. Thus the maximum degree of the path graph $P(G)$ is $(L-1)D$. Any graph with maximum degree Δ can be colored using $\Delta + 1$ colors by a simple greedy coloring algorithm, and hence the path graph can be colored using $(L-1)D + 1$ colors. So $W \leq (L-1)D + 1$.

To prove the remainder of the theorem, suppose there are K lightpaths of length $\geq \sqrt{M}$ hops. The average load due to these lightpaths on an edge is

$$\frac{K\sqrt{M}}{M} \leq L$$

so that $K \leq L\sqrt{M}$. Assign $L\sqrt{M}$ separate wavelengths to these lightpaths. Next consider the lightpaths of length $\leq \sqrt{M} - 1$ hops. Each of these intersects with at most $(L-1)(\sqrt{M} - 1)$ other such lightpaths, and so will need at most $(L-1)(\sqrt{M} - 1) + 1$ additional wavelengths. So we have

$$W \leq L\sqrt{M} + (L-1)(\sqrt{M} - 1) + 1 = (2L-1)\sqrt{M} - L + 2,$$

which proves the theorem.   ■

A line network, shown in Figure 8.19, is simply a network of nodes interconnected in a line. A sample set of lightpath requests is also shown in the figure. In this case, there is no routing aspect; only the wavelength assignment problem remains. We study this topology because the results will be useful in analyzing ring networks, which are practically important.

Our WA-NC problem (see Section 8.2.2) is equivalent to the problem of coloring intervals on a line. The following greedy algorithm accomplishes the coloring using

L wavelengths. The algorithm is greedy in the sense that it never backtracks and changes a color that it has already assigned when assigning a color to a new interval.

Algorithm 8.3   [Ber76, Section 16.5]

1. Number the wavelengths from 1 to L. Start with the first lightpath from the left and assign to it wavelength 1.

2. Go to the next lightpath starting from the left and assign to it the least numbered wavelength possible, until all lightpaths are colored.

Rings are perhaps the most important specific topology to consider. A ring is the simplest 2-connected topology and has been adopted by numerous standards (FDDI, SONET) as the topology of choice. We expect WDM networks to be first deployed as rings.

In a ring, we have two possible routes for each lightpath. Given a set of lightpath requests, there is an algorithm [FNS$^+$92] that does the routing with the minimum possible load L_{min}. This algorithm may involve some lightpaths taking the longest route around the ring. A simpler alternative is to use shortest-path routing for lightpaths, which, however, yields a higher load, as shown next.

Lemma 8.2   [RS97] *Suppose we are given a request of source-destination pairs and the minimum possible load for satisfying this request is L_{min}. Then shortest-path routing yields a load of at most $2L_{min}$.*

Proof.   Suppose shortest-path routing yields a load L_{sp}. Consider a link i with load L_{sp}. Rerouting k connections using link i on their longer routes on the ring reduces the load on link i by $L_{sp} - k$. Note that since all these connections are routed on paths on length $\leq \lfloor N/2 \rfloor$ initially, their longer routes on the ring will all use the link $\lfloor N/2 \rfloor + i$, increasing its load by k. Therefore, the load L_{min} of the optimal routing algorithm must satisfy $L_{min} \geq \min_k \max(L_{sp} - k, k)$, or $L_{min} \geq \lceil L_{sp}/2 \rceil$.   ∎

It turns out that the joint RWA-NC problem is hard, even in rings. However, we can get good bounds on how many wavelengths are needed.

Theorem 8.3   [Tuc75] *Given a set of lightpath requests and a routing on a ring with load L, WA-NC can be done with $2L - 1$ wavelengths.*

Proof.   Determine the node in the ring with a minimum number l of lightpaths passing through it (do not count lightpaths starting or terminating at the node). Cut the ring at this point (see Figure 8.20). Now we have an interval

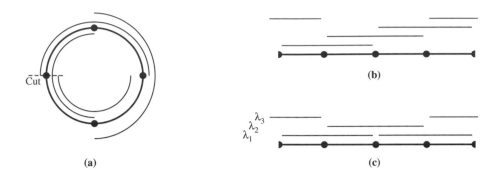

Figure 8.20   Wavelength assignment in a ring network. (a) A ring network and a set of lightpaths. (b) The ring is cut at a node that has a minimum number of lightpaths passing through it to yield a line network. (c) The lightpaths in the line network are assigned wavelengths according to Algorithm 8.3. The lightpaths going across the cut node are assigned separate additional wavelengths.

graph with a maximum load of L, which we can color with L wavelengths, using Algorithm 8.3. However, we still have to deal with the l lightpaths that may wrap around the edge of the line. In the worst case, we can always assign wavelengths to these lightpaths using l additional wavelengths, requiring a total of $L + l$ wavelengths.

Now with any routing, there is a node in the ring where $l \leq L - 1$. To see this, suppose all nodes have at least L paths flowing through them. There exists a node, say, node x, where a path terminates. Let y be the node adjacent to x on this path. Then link xy must have a load of at least $L + 1$, a contradiction.   ∎

It is possible to construct an example of a traffic pattern consisting of $2L - 1$ lightpaths, with each pair of lightpaths sharing at least one common link. This implies that all of them have to be assigned different wavelengths regardless of the algorithm used, showing that there are examples for which $2L - 1$ wavelengths will be required. However, this is not a scenario that occurs very often. In fact, it has been shown in [Tuc75] that if no three lightpaths in a given traffic pattern cover the entire ring, then $\frac{3}{2}L$ wavelengths are sufficient to perform the wavelength assignment. This is an example where the worst-case nonblocking model results in overdesigning the network. In order to support a few pathological patterns, we end up using approximately $\frac{L}{2}$ additional wavelengths.

Let us see what can be gained by having wavelength conversion capabilities in a ring network. Clearly, if we have full conversion capabilities at all the nodes, then we

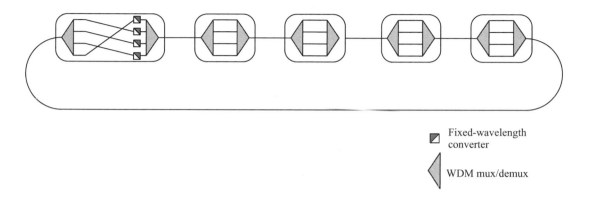

Figure 8.21   A ring network with fixed-wavelength conversion at one node and no conversion at the others that is able to support lightpath requests with load $L \leq W - 1$. One of the nodes is configured to convert wavelength i to wavelength $(i + 1) \bmod W$, and the other nodes provide no wavelength conversion.

can support all lightpath requests with load $L \leq W$. However, the same result can be achieved by providing much less conversion capabilities, as shown by the following results.

> **Theorem 8.4**   [RS97] *Consider a ring network that has full wavelength conversion at one node and no wavelength conversion at the other nodes. This network can support all lightpath requests with load $L \leq W$.*

The proof of this result is left as an exercise (Problem 8.19).

Limited-wavelength conversion can help significantly in improving the load that can be supported in many network configurations. The detailed derivations of the results for this case are beyond the scope of this book. We summarize the key results here.

> **Theorem 8.5**   [RS97] *Consider the ring network shown in Figure 8.21, which has fixed-wavelength conversion at one node where wavelength i is converted to wavelength $(i + 1) \bmod W$, and no wavelength conversion at the other nodes. This network can support all lightpath requests with load $L \leq W - 1$.*

By having $d = 2$ limited conversion at two nodes and no conversion at the others, it is possible to improve this result to $L \leq W$ [RS97], making such a network as good as a network with full wavelength conversion at each node.

Table 8.4   Number of wavelengths required to perform offline wavelength assignment as a function of the load L with and without wavelength converters. The fixed conversion result for arbitrary topologies applies only to one- and two-hop lightpaths.

Network	Conversion Type			
	None	Fixed	Full	Limited
Arbitrary	$\min[(L-1)D+1,$ $(2L-1)\sqrt{M} - L + 2]$	L	L	
Ring	$2L - 1$	$L + 1$	L	L
Star	$\frac{3}{2}L$	L	L	
Tree	$\frac{3}{2}L$		L	L

Other topologies such as star networks and tree networks have also been considered in the literature. In star and tree networks, $\frac{3}{2}L$ wavelengths are sufficient to do WA-NC [RU94]. In star networks, L wavelengths are sufficient for WA-FC [RS97]. The same result can be extended to arbitrary networks where lightpaths are at most two hops long. Table 8.4 summarizes the results to date on this problem. It is still a topic of intense research.

Multifiber Rings

The wavelength assignment problem in multifiber rings is considered in [LS00]. In a multifiber ring, each pair of adjacent nodes is connected by $k > 1$ fiber pairs: $k > 1$ fibers are used for each direction of transmission instead of 1 fiber. Recall that we are considering undirected edges and lightpaths, and each edge represents a pair of fibers, one for each direction of transmission. Thus, such a multifiber ring is represented by k edges between pairs of adjacent nodes. There is no wavelength conversion, but it is assumed that the same wavelength can be switched from an incoming fiber to any of the k outgoing fibers at each node. The following results on multifiber rings are proved in [LS00].

Theorem 8.6   [LS00] *Given a set of lightpath requests and a routing on a k-fiber-pair ring with load L on each multifiber link, the number of wavelengths, summed over all the fibers, required to solve the wavelength assignment problem is no more than* $\left\lceil \frac{k+1}{k}L - 1 \right\rceil$.

Thus, for a dual-fiber-pair ring ($k = 2$), the number of wavelengths required is no more than $\left\lceil \frac{3}{2}L - 1 \right\rceil$, which is a significant improvement over the bound of $2L - 1$ for a single-fiber-pair ring.

As in the case of the single-fiber-pair ring, you can come up with a set of lightpath requests with load L for which this upper bound on the number of wavelengths is tight, for all values of the fiber multiplicity, k.

8.5.2 Online RWA in Rings

We next consider the online wavelength assignment problem in rings. Assume that the routing of the lightpaths is already given and that lightpaths are set up as well as taken down, that is, the lightpaths are nonpermanent. Here, it becomes much more difficult to come up with smart algorithms that maximize the load that can be supported for networks without full wavelength conversion. (With full wavelength conversion at all the nodes, an algorithm that assigns an arbitrary free wavelength can support all lightpath requests with load up to W.) We describe an algorithm that provides efficient wavelength assignment for line and ring networks without wavelength conversion.

Lemma 8.7 [GSKR99] *Let $W(N, L)$ denote the number of wavelengths required to support all online lightpath requests with load L in a network with N nodes without wavelength conversion. In a line network, $W(N, L) \leq L + W(N/2, L)$, when N is a power of 2.*

Proof. Break the line network in the middle to realize two disjoint subline networks, each with $N/2$ nodes. Break the set of lightpath requests into two groups: one group consisting of lightpaths that lie entirely within the subline networks, and the other group consisting of lightpaths that go across between the two subline networks. The former group of lightpaths can be supported with at most $W(N/2, L)$ wavelengths (the same set of wavelengths can be used in both subline networks). The latter group of lightpaths can have a load of at most L. Dedicate L additional wavelengths to serving this group. This proves the lemma. ∎

The following theorem follows immediately from Lemma 8.7, with the added condition that $W(1, L) = 0$ (or $W(2, L) = L$).

Theorem 8.8 [GSKR99] *In a line network with N nodes, all online lightpath requests with load L can be supported using at most $L \lceil \log_2 N \rceil$ wavelengths without requiring wavelength conversion.*

The algorithm implied by this theorem is quite efficient since it is possible to come up with lightpath traffic patterns for which any algorithm will require at least $0.5L \log_2 N$ wavelengths [GSKR99].

Theorem 8.9 [GSKR99] *In a ring network with N nodes, all online lightpath requests with load L can be supported using at most $L \lceil \log_2 N \rceil + L$ wavelengths, without requiring wavelength conversion.*

The proof of this theorem is left as an exercise (Problem 8.21).

When we have permanent lightpaths being set up, it is possible to obtain somewhat better wavelength assignments, as given by the following theorem, the proof of which is beyond the scope of this book.

Theorem 8.10 [GSKR99] *In a ring network with N nodes, all online permanent lightpath requests with load L can be supported using (a) at most 2L wavelengths without wavelength conversion, and (b) with at most $\max(0, L - d) + L$ wavelengths with degree-d ($d \geq 2$) limited wavelength conversion.*

Table 8.5 summarizes the results to date on the offline and online RWA problem for ring networks, with the traffic model characterized by the maximum link load. For this model, observe that significant increases in the traffic load can be achieved by having wavelength converters in the network. For the offline case, very limited conversion provides almost as much benefit as full wavelength conversion. For the online cases, the loads that can be supported are much less than the offline case. The caveat is that, as illustrated in Figure 8.13, this model represents worst-case scenarios, and a majority of traffic patterns could perhaps be supported efficiently without requiring as many wavelengths or as many wavelength converters.

Summary

We studied the design of wavelength-routing networks in this chapter. We saw that there is a clear benefit to building wavelength-routing networks, as opposed to simple point-to-point WDM links. The main benefit is that traffic that is not to be terminated within a node can be passed through by the node, resulting in significant savings in higher-layer terminating equipment.

The design of these networks is more complicated than the design of traditional networks. It includes the design of the higher-layer topology (IP or SONET), which is the lightpath topology design problem, and its realization in the optical layer, which is the routing and wavelength assignment problem. These problems may need to be

Table 8.5 Bounds on the number of wavelengths required in rings to support all traffic patterns with maximum load L for different models, offline and online, from [GRS97, GSKR99]. d denotes the degree of wavelength conversion. The upper bound indicates the number of wavelengths that are sufficient to accommodate all traffic patterns with maximum load L, using some RWA algorithm. The lower bound indicates that there is some traffic pattern with maximum load L that requires this many wavelengths regardless of the RWA algorithm that is employed. For the online traffic model, we consider two cases, one where lightpaths are set up over time but never taken down, and another where lightpaths are both set up and taken down over time.

Conversion Degree	Lower Bound on W	Upper Bound on W
Offline traffic model		
No conversion	$2L - 1$	$2L - 1$
Fixed conversion	$L + 1$	$L + 1$
≥ 2	L	L
Online model without lightpath terminations		
No conversion	$3L$	$3L$
Fixed conversion	L	$3L$
Full conversion	L	L
Online model with lightpath terminations		
No conversion	$0.5L \lfloor \log_2 N \rfloor$	$L \lceil \log_2 N \rceil + L$
Full conversion	L	L

solved in conjunction if the carrier provides IP or SONET VTs over its own optical infrastructure. However, this is difficult to do, and a practical approach may be to iteratively solve these problems.

We then discussed the wavelength dimensioning problem. The problem here is to provide sufficient capacity on the links of the wavelength-routing network to handle the expected demand for lightpaths. This problem is solved today by periodically forecasting a traffic matrix and (re)designing the network to support the forecasted matrix. Alternatively, you can employ statistical traffic demand models to estimate the required capacities, and we discussed two such models.

The absence of wavelength conversion in the network can be overcome by providing more wavelengths on the links. In the last section, we studied this trade-off under various models.

Further Reading

The issue of how much cost savings is afforded by providing networking functions within the optical layer is only beginning to be understood. For some more insights into this issue, see [RLB95, Bal96, GRS98, SGS99, CM00, BM00]. The material in this chapter is based on [GRS98]. See [Wil96, WW98, Ber96] for a discussion of the problem of setting up connections between all pairs of nodes in a WDM ring network.

The lightpath topology design problem is discussed in [RS96, KS98, CMLF00, MBRM96, BG95, ZA95, JBM95, GW94, CGK93, LA91]. Our discussion is based on [RS96]. This is an example of a network flow problem; these problems are dealt with in detail in [AMO93].

Several papers [ABC+94, RU94, RS95, CGK92, RS97, MKR95, KS97, KPEJ97, ACKP97] study the offline routing and wavelength assignment problem. There is also a vast body of literature describing routing and wavelength assignment heuristics. See, for example, [CGK92, SBJS93, RS95, Bir96, WD96, SOW95].

The statistical blocking model for dimensioning is analyzed in [SS00, BK95, RS95, KA96, SAS96, YLES96, BH96].

The worst-case analysis of the maximum load model with online traffic is considered in [GK97].

Problems

8.1 In general there are several valid design options even for a three-node network. Consider the designs shown in Figure 8.1(c), but now assume that the number of dropped lightpaths is six instead of five as discussed in the text. The advantage of this design is that it provides more flexibility in handling surges in A–B and B–C traffic. For example, this design not only can handle the traffic requirement of 50 Gb/s between every pair of nodes, it can also handle a traffic requirement of 60 Gb/s between nodes A–B and B–C, and 40 Gb/s between nodes A–C. This latter traffic pattern cannot be handled if only five lightpaths/wavelengths are dropped.

Consider the design of Figure 8.1(c), and assume that x wavelengths are dropped at node B and y wavelengths pass through. Determine the range of traffic matrices that this design is capable of handling as a function of x and y.

8.2 Consider the network design approach using fixed-wavelength routing in a four-node ring network with consecutive nodes A, B, C, and D. Suppose the traffic requirements are as follows:

	A	B	C	D
A	–	3	–	3
B	3	–	2	3
C	–	2	–	2
D	3	3	2	–

(a) Do a careful routing of traffic onto each wavelength so as to minimize the number of wavelengths needed.

(b) How do you know that your solution uses the minimum possible number of wavelengths required to do this routing for any algorithm?

(c) How many ADMs are required at each node to support this traffic?

(d) How many ADMs are required at each node if instead of fixed-wavelength routing, you decided to use point-to-point WDM links and receive and re-transmit all the wavelengths at each node? How many ADMs does wavelength routing eliminate?

8.3 Derive (8.1). What is the value when N is odd?

8.4 Derive (8.5). What is the value when N is odd?

8.5 Derive (8.8) for the case where there is one full-duplex lightpath between each pair of nodes. *Hint:* Use induction. Start with two nodes on the ring, and determine the number of wavelengths required. Add two more nodes so that they are diametrically opposite to each other on the ring and continue.

8.6 Show that when N is odd, (8.8) is modified to

$$W = \left\lceil \frac{t}{N-1} \right\rceil \frac{N^2 - 1}{8}.$$

8.7 Derive (8.9). What is the value when N is odd?

8.8 Develop other network designs besides the ones shown in Examples 8.2, 8.3, and 8.4, and compare the number of LTs and wavelengths required for these designs against these three examples.

8.9 Consider the network shown in Figure 8.9(a). Assume that each undirected edge can be represented by a pair of directed edges as in Figure 8.9(c). Represent each undirected lightpath in Figure 8.9(a) by a pair of directed lightpaths with opposite directions. Consider the RWA problem in the resulting network and show that two wavelengths are sufficient to support these directed lightpaths. Note that three wavelengths were required to support the corresponding undirected lightpaths.

8.10 This problem illustrates the complexity of wavelength assignment in networks where the transmission is bidirectional over each fiber. Consider the two networks shown

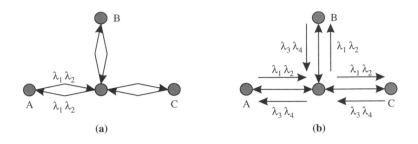

Figure 8.22 Two different scenarios of wavelength assignment in networks with bidirectional links.

in Figure 8.22. In Figure 8.22(a), the network uses two fibers on each link, with two wavelengths on each fiber, with unidirectional transmission on each fiber. In Figure 8.22(b), the network uses one fiber on each link, with four wavelengths. Transmission is bidirectional on each fiber, with two wavelengths in one direction and two in the other. No wavelength conversion is allowed in either network. Both networks have the same nominal capacity (four wavelengths/link). Which network utilizes the capacity more efficiently?

8.11 Show that a network having P fiber pairs between nodes and W wavelengths on each fiber with no wavelength conversion is equivalent to a network with one fiber pair between nodes with PW wavelengths, and degree P wavelength conversion capability at the nodes.

8.12 Generalize the example of Figure 8.13 to the case when the number of nodes is arbitrary, say, N. Compare the number of wavelengths required in this general case to the upper bound given by Theorem 8.1.

8.13 In order to prove that $W \le (2L - 1)\sqrt{M} - L + 2$ in Theorem 8.1, we supposed that there were K lightpaths of length $\ge \sqrt{M}$ hops. Instead suppose there are $K(x)$ lightpaths of length $\ge x$ hops, and derive an upper bound for W that holds for every x. Now, optimize x to get the least upper bound for W. Compare this bound with the bound obtained in Theorem 8.1.

8.14 Show that Algorithm 8.3 always does the wavelength assignment using L wavelengths. *Hint:* Use induction on the number of nodes.

8.15 Consider the following modified version of Algorithm 8.3. In step 2, the algorithm is permitted to assign any free wavelength from a fixed set of L wavelengths, instead of the least numbered wavelength. Show that this algorithm always succeeds in performing the wavelength assignment.

8.16 Prove that Theorem 8.3 can be tight in some cases. In other words, give an example of a ring network and a set of lightpath requests and routing with load L that requires $2L - 1$ wavelengths. *Hint:* First, give an example that requires $2L - 2$ wavelengths and then modify it by adding an additional lightpath without increasing the load. Note that the example in Figure 8.20 shows such an example for the case $L = 2$. Obtain an example for the case $L > 2$.

8.17 Consider a ring network with a lightpath request set of one lightpath between each source-destination pair. Compute the number of wavelengths sufficient to support this set with full wavelength conversion and without wavelength conversion. What do you conclude from this?

8.18 Give an example of a star network without wavelength conversion where $\frac{3}{2}L$ wavelengths are necessary to perform the wavelength assignment.

8.19 Prove Theorem 8.4.

8.20 Prove Theorem 8.8. Based on this proof, write pseudo-code for an algorithm to perform wavelength assignment.

8.21 Prove Theorem 8.9.

8.22 This problem relates to the wavelength assignment problem in networks without wavelength conversion. Let us assume that the links in the network are duplex, that is, consist of two unidirectional links in opposite directions. A set of duplex lightpath requests and their routing is given. In practice, each request between two nodes A and B is for a lightpath l from A to B and another lightpath l' from B to A, which we will assume are both routed along the same path in the network.

One wavelength assignment scheme (scheme 1) is to assign the same wavelength to both l and l'. Give an example to show that it is possible to do a better wavelength assignment (using fewer wavelengths) by assigning different wavelengths to l and l' (scheme 2). Show using this example that scheme 1 can need up to $\frac{3}{2}W$ wavelengths, where W is the number of wavelengths required for scheme 2. *Hint:* Consider a representation of the path graph corresponding to directed lightpaths.

8.23 Derive the expression (8.13) for the probability that a lightpath request is blocked when the network uses full wavelength conversion.

8.24 Derive the approximate expressions for π_{nc} and π_{fc} given by (8.16) and (8.17). Plot these approximations and the exact values given by (8.14) versus W for $P_b = 10^{-3}$, 10^{-4}, and 10^{-5}, and $H = 5$, 10, and 20 hops to study the behavior of π_{nc} and π_{fc}, and to verify the range of accuracy of these approximations.

8.25 Derive (8.18).

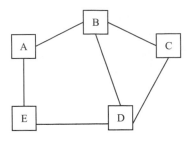

Figure 8.23   Network topology for Problem 8.26.

8.26   Consider the five-node fiber topology shown in Figure 8.23 on which IP bandwidth is to be routed between IP router node pairs over a WDM network. The bandwidth demands are given for each node pair in the following table. Assume that all demands are bidirectional, and both directions are routed along the same path using the same wavelengths in opposite directions.

Gb/s	B	C	D	E
A	15	25	5	15
B		5	35	15
C			15	25
D				5

(a) Assuming OC-192c (10 Gb/s) trunks are used, complete an equivalent table for the required number of lightpaths (that is, wavelengths) between each pair of nodes.

(b) Using the given physical topology, and assuming that there are no wavelength conversion capabilities contained within the optical crossconnects at the nodes, specify a reasonable wavelength-routing design for each lightpath. Clearly label each wavelength along its end-to-end path through the network.

(c) What is the maximum load on any link in the network, and how does it compare with the number of wavelengths you are using in total?

References

[ABC+94]   A. Aggarwal, A. Bar-Noy, D. Coppersmith, R. Ramaswami, B. Schieber, and M. Sudan. Efficient routing and scheduling algorithms for optical networks. In

Proceedings of 5th Annual ACM-SIAM Symposium on Discrete Algorithms, pages 412–423, Jan. 1994.

[ACKP97]   V. Auletta, I. Caragiannis, C. Kaklamanis, and P. Persiano. Bandwidth allocation algorithms on tree-shaped all-optical networks with wavelength converters. In *Proceedings of the 4th International Colloquium on Structural Information and Communication Complexity*, 1997.

[AMO93]   R. K. Ahuja, T. L. Magnanti, and J. B. Orlin. *Network Flows: Theory, Algorithms, and Applications*. Prentice Hall, Englewood Cliffs, NJ, 1993.

[Bal96]   K. Bala et al. WDM network economics. In *Proceedings of National Fiber Optic Engineers Conference*, pages 163–174, 1996.

[Ber76]   C. Berge. *Graphs and Hypergraphs*. North Holland, Amsterdam, 1976.

[Ber96]   J.-C. Bermond et al. Efficient collective communication in optical networks. In *23rd International Colloquium on Automata, Languages and Programming—ICALP '96, Paderborn, Germany*, pages 574–585, 1996.

[BG92]   D. Bertsekas and R. G. Gallager. *Data Networks*. Prentice Hall, Englewood Cliffs, NJ, 1992.

[BG95]   D. Bienstock and O. Gunluk. Computational experience with a difficult mixed-integer multicommodity flow problem. *Mathematical Programming*, 68:213–237, 1995.

[BH96]   R. A. Barry and P. A. Humblet. Models of blocking probability in all-optical networks with and without wavelength changers. *IEEE JSAC/JLT Special Issue on Optical Networks*, 14(5):858–867, June 1996.

[Big90]   N. Biggs. Some heuristics for graph colouring. In R. Nelson and R. J. Wilson, editors, *Graph Colourings*, Pitman Research Notes in Mathematics Series, pages 87–96. Longman Scientific & Technical, Burnt Mill, Harlow, Essex, UK, 1990.

[Bir96]   A. Birman. Computing approximate blocking probabilities for a class of optical networks. *IEEE JSAC/JLT Special Issue on Optical Networks*, 14(5):852–857, June 1996.

[BK95]   A. Birman and A. Kershenbaum. Routing and wavelength assignment methods in single-hop all-optical networks with blocking. In *Proceedings of IEEE Infocom*, pages 431–438, 1995.

[BM00]   R. Berry and E. Modiano. Reducing electronic multiplexing costs in SONET/WDM rings with dynamically changing traffic. *IEEE Journal of Selected Areas in Communications*, 18:1961–1971, 2000.

[Cah98]   R. Cahn. *Wide Area Network Design: Concepts and Tools for Optimization*. Morgan Kaufmann, San Francisco, 1998.

[CGK92] I. Chlamtac, A. Ganz, and G. Karmi. Lightpath communications: An approach to high-bandwidth optical WAN's. *IEEE Transactions on Communications*, 40(7):1171–1182, July 1992.

[CGK93] I. Chlamtac, A. Ganz, and G. Karmi. Lightnets: Topologies for high-speed optical networks. *IEEE/OSA Journal on Lightwave Technology*, 11(5/6):951–961, May/June 1993.

[CM00] A. L. Chiu and E. H. Modiano. Traffic grooming algorithms for reducing electronic multiplexing costs in WDM ring networks. *IEEE/OSA Journal on Lightwave Technology*, 18:2–12, 2000.

[CMLF00] T. Cinkler, D. Marx, C. P. Larsen, and D. Fogaras. Heuristic algorithms for joint configuration of the optical and electrical layer in multi-hop wavelength routing networks. In *Proceedings of IEEE Infocom*, 2000.

[dW90] D. de Werra. Heuristics for graph coloring. In G. Tinhofer, E. Mayr, and H. Noltemeier, editors, *Computational Graph Theory*, volume 7 of *Computing, Supplement*, pages 191–208. Springer-Verlag, Berlin, 1990.

[FNS⁺92] A. Frank, T. Nishizeki, N. Saito, H. Suzuki, and E. Tardos. Algorithms for routing around a rectangle. *Discrete Applied Mathematics*, 40:363–378, 1992.

[GJ79] M. R. Garey and D. S. Johnson. *Computers and Intractability—A Guide to the Theory of NP Completeness*. W. H. Freeman, San Francisco, 1979.

[GK97] O. Gerstel and S. Kutten. Dynamic wavelengh allocation in all-optical ring networks. In *Proceedings of IEEE International Conference on Communication*, 1997.

[GLS99] O. Gerstel, P. Lin, and G. Sasaki. Combined WDM and SONET network design. In *Proceedings of IEEE Infocom*, 1999.

[GRS97] O. Gerstel, R. Ramaswami, and G. H. Sasaki. Benefits of limited wavelength conversion in WDM ring networks. In *OFC'97 Technical Digest*, pages 119–120, 1997.

[GRS98] O. Gerstel, R. Ramaswami, and G. H. Sasaki. Cost effective traffic grooming in WDM rings. In *Proceedings of IEEE Infocom*, 1998.

[GSKR99] O. Gerstel, G. H. Sasaki, S. Kutten, and R. Ramaswami. Worst-case analysis of dynamic wavelength allocation in optical networks. *IEEE/ACM Transactions on Networking*, 7(6):833–846, Dec. 1999.

[GW94] A. Ganz and X. Wang. Efficient algorithm for virtual topology design in multihop lightwave networks. *IEEE/ACM Transactions on Networking*, 2(3):217–225, June 1994.

[Jai96] M. Jain. Topology designs for wavelength routed optical networks. Technical report, Indian Institute of Science, Bangalore, Jan. 1996.

[JBM95]   S. V. Jagannath, K. Bala, and M. Mihail. Hierarchical design of WDM optical networks for ATM transport. In *Proceedings of IEEE Globecom*, pages 2188–2194, 1995.

[KA96]   M. Kovacevic and A. S. Acampora. On the benefits of wavelength translation in all optical clear-channel networks. *IEEE JSAC/JLT Special Issue on Optical Networks*, 14(6):868–880, June 1996.

[Ker93]   A. Kershenbaum. *Telecommunications Network Design Algorithms*. McGraw-Hill, New York, 1993.

[KPEJ97] C. Kaklamanis, P. Persiano, T. Erlebach, and K. Jansen. Constrained bipartite edge coloring with applications to wavelength routing in all-optical networks. In *International Colloquium on Automata, Languages, and Programming*, 1997.

[KS97]   V. Kumar and E. Schwabe. Improved access to optical bandwidth in trees. In *Proceedings of the ACM Symposium on Distributed Algorithms*, 1997.

[KS98]   R. M. Krishnaswamy and K. N. Sivarajan. Design of logical topologies: A linear formulation for wavelength routed optical networks with no wavelength changers. In *Proceedings of IEEE Infocom*, 1998.

[LA91]   J.-F. P. Labourdette and A. S. Acampora. Logically rearrangeable multihop lightwave networks. *IEEE Transactions on Communications*, 39(8):1223–1230, Aug. 1991.

[LS00]   G. Li and R. Simha. On the wavelength assignment problem in multifiber WDM star and ring networks. In *Proceedings of IEEE Infocom*, 2000.

[MBRM96] B. Mukherjee, D. Banerjee, S. Ramamurthy, and A. Mukherjee. Some principles for designing a wide-area optical network. *IEEE/ACM Transactions on Networking*, 4(5):684–696, 1996.

[MKR95]   M. Mihail, C. Kaklamanis, and S. Rao. Efficient access to optical bandwidth. In *IEEE Symposium on Foundations of Computer Science*, pages 548–557, 1995.

[NS02]   T. K. Nayak and K. N. Sivarajan. A new approach to dimensioning optical networks. *IEEE Journal of Selected Areas in Communications*, to appear, 2002.

[NTM00]   A. Narula-Tam and E. Modiano. Dynamic load balancing for WDM-based packet networks. In *Proceedings of IEEE Infocom*, 2000.

[RLB95]   P. Roorda, C.-Y. Lu, and T. Boutlier. Benefits of all-optical routing in transport networks. In *OFC'95 Technical Digest*, pages 164–165, 1995.

[RS95]   R. Ramaswami and K. N. Sivarajan. Routing and wavelength assignment in all-optical networks. *IEEE/ACM Transactions on Networking*, pages 489–500, Oct. 1995. An earlier version appeared in *Proceedings of IEEE Infocom'94*.

[RS96]   R. Ramaswami and K. N. Sivarajan. Design of logical topologies for wavelength-routed optical networks. *IEEE JSAC/JLT Special Issue on Optical Networks*, 14(5):840–851, June 1996.

[RS97]   R. Ramaswami and G. H. Sasaki. Multiwavelength optical networks with limited wavelength conversion. In *Proceedings of IEEE Infocom*, pages 490–499, 1997.

[RU94]   P. Raghavan and E. Upfal. Efficient routing in all-optical networks. In *Proceedings of 26th ACM Symposium on Theory of Computing*, pages 134–143, May 1994.

[SAS96]   S. Subramaniam, M. Azizoglu, and A. K. Somani. Connectivity and sparse wavelength conversion in wavelength-routing networks. In *Proceedings of IEEE Infocom*, pages 148–155, 1996.

[SBJS93]   T. E. Stern, K. Bala, S. Jiang, and J. Sharony. Linear lightwave networks: Performance issues. *IEEE/OSA Journal on Lightwave Technology*, 11:937–950, May/June 1993.

[SGS99]   J. M. Simmons, E. L. Goldstein, and A. A. M. Saleh. Quantifying the benefit of wavelength add-drop in WDM rings with distance-independent and dependent traffic. *IEEE/OSA Journal on Lightwave Technology*, 17:48–57, 1999.

[SM00]   B. Schein and E. Modiano. Quantifying the benefit of configurability in circuit-switched WDM ring networks. In *Proceedings of IEEE Infocom*, 2000.

[SOW95]   K. I. Sato, S. Okamoto, and A. Watanabe. Photonic transport networks based on optical paths. *International Journal of Communication Systems (UK)*, 8(6):377–389, Nov./Dec. 1995.

[SS00]   A. Sridharan and K. N. Sivarajan. Blocking in all-optical networks. In *Proceedings of IEEE Infocom*, 2000.

[Tuc75]   A. Tucker. Coloring a family of circular arcs. *SIAM Journal on Applied Mathematics*, 29(3):493–502, 1975.

[WD96]   N. Wauters and P. Demeester. Design of the optical path layer in multiwavelength cross-connected networks. *IEEE JSAC/JLT Special Issue on Optical Networks*, 14(6):881–892, June 1996.

[Wil96]   G. Wilfong. Minimizing wavelengths in an all-optical ring network. In *7th International Symposium on Algorithms and Computation*, pages 346–355, 1996.

[WW98]   G. Wilfong and P. Winkler. Ring routing and wavelength translation. In *Proceedings of the Symposium on Discrete Algorithms (SODA)*, pages 334–341, 1998.

[YLES96]   J. Yates, J. Lacey, D. Everitt, and M. Summerfield. Limited-range wavelength translation in all-optical networks. In *Proceedings of IEEE Infocom*, pages 954–961, 1996.

[ZA95] Z. Zhang and A. S. Acampora. A heuristic wavelength assignment algorithm for multihop WDM networks with wavelength routing and wavelength reuse. *IEEE/ACM Transactions on Networking*, 3(3):281–288, June 1995.

9 chapter

Control and Management

NETWORK MANAGEMENT is an important part of any network. However attractive a specific technology might be, it can be deployed in a network only if it can be managed and interoperates with existing management systems. The cost of operating and managing a large network is a recurring cost and in many cases dominates the cost of the equipment deployed in the network. As a result, carriers are now paying a lot of attention to minimizing *life cycle* costs, as opposed to worrying just about up-front equipment costs. We start with a brief introduction to network management concepts in general and how they apply to managing optical networks. We follow this with a discussion of optical layer services and how the different aspects of the optical network are managed.

9.1 Network Management Functions

Classically, network management consists of several functions, all of which are important to the operation of the network:

1. *Performance management* deals with monitoring and managing the various parameters that measure the performance of the network. Performance management is an essential function that enables a service provider to provide quality-of-service guarantees to their clients and to ensure that clients comply

with the requirements imposed by the service provider. It is also needed to provide input to other network management functions, in particular, fault management, when anomalous conditions are detected in the network. This function is discussed further in Section 9.5.

2. *Fault management* is the function responsible for detecting failures when they happen and isolating the failed component. The network also needs to restore traffic that may be disrupted due to the failure, but this is usually considered a separate function and is the subject of Chapter 10. We will study fault management in Section 9.5.

3. *Configuration management* deals with the set of functions associated with managing orderly changes in a network. The basic function of managing the equipment in the network belongs to this category. This includes tracking the equipment in the network and managing the addition/removal of equipment, including any rerouting of traffic this may involve and the management of software versions on the equipment.

   Another aspect of configuration management is *connection management,* which deals with setting up, taking down, and keeping track of connections in a network. This function can be performed by a centralized management system. Alternatively, it can also be performed by a distributed *network control* entity. Distributed network control becomes necessary when connection setup/take-down events occur very frequently or when the network is very large and complex.

   Finally, the network needs to convert external client signals entering the optical layer into appropriate signals inside the optical layer. This function is *adaptation management.* We will study this and the other configuration management functions in Section 9.6.

4. *Security management* includes administrative functions such as authentication of users and setting attributes such as read and write permissions on a per-user basis. From a security perspective, the network is usually partitioned into domains, both horizontally and vertically. Vertical partitioning implies that some users may be allowed to access only certain network elements and not other network elements. For example, a local craftsperson may be allowed to access only the network elements he is responsible for and not other network elements. Horizontal partitioning implies that some users may be allowed to access some parameters associated with all the network elements across the network. For example, a user leasing a lightpath may be provided access to all the performance parameters associated with that lightpath across all the nodes that the lightpath traverses.

Security also involves protecting data belonging to network users from being tapped or corrupted by unauthorized entities. This part of the problem needs to be handled by encrypting the data before transmission and providing the decrypting capability to legitimate users.

5. *Accounting management* is the function responsible for billing and for developing lifetime histories of the network components. This function doesn't appear to be much different for optical networks, compared to other networks, and we will not be discussing this topic further.

For optical networks, an additional consideration is *safety management*, which is needed to ensure that optical radiation conforms to limits imposed for ensuring eye safety. This subject is treated in Section 9.7.

9.1.1 Management Framework

Most functions of network management are implemented in a centralized manner by a hierarchy of management systems. However, this method of implementation is rather slow, and it can take several hundreds of milliseconds to seconds to communicate between the management system and the different parts of the network because of the large software path overheads usually involved in this process. Decentralized methods are usually much faster than centralized methods, even in small networks with only a few nodes. Therefore, certain management functions that require rapid action may have to be decentralized, such as responding to failures and setting up and taking down connections if these must be done rapidly. For example, a SONET ring can restore failures within 60 ms, and this is possible only because this process is completely decentralized. For this reason, restoration is viewed as more of an autonomous control function rather than an integrated part of network management.

Another reason for decentralizing some of the functions arises when the network becomes very large. In this case, it becomes difficult for a single central manager to manage the entire network. Further, networks could include multiple domains administered by different managers. The managers of each domain will need to communicate with managers of other domains to perform certain functions in a coordinated manner.

Figure 9.1 provides an overview of how network management functions are implemented on a typical network. Management is performed in a hierarchical manner, involving multiple management systems in many cases. The individual components to be managed are called *network elements*. Network elements include optical line terminals (OLTs), optical add/drop multiplexers (OADMs), optical amplifiers, and optical crossconnects (OXCs). Each element is managed by its *element management system* (EMS). The element itself has a built-in *agent*, which communicates with

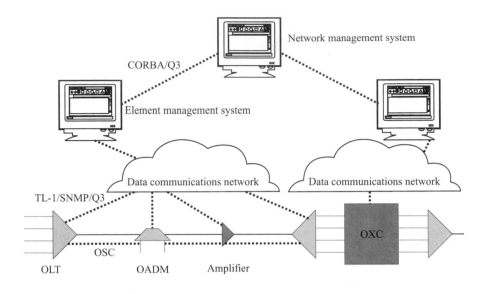

Figure 9.1   Overview of network management in a typical optical network, showing the network elements (OLTs, OADMs, OXCs, amplifiers), the management systems, and the associated interfaces.

its EMS. The agent is implemented in software, usually in a microprocessor in the network element.

The EMS is usually connected to one or more of the network elements and communicates with the other network elements in the network using a *data communication network* (DCN). In addition to the DCN, a fast *signaling channel* is also required between network elements to exchange real-time control information to manage protection switching and other functions. The DCN and signaling channel can be realized in many different ways, as will be discussed in Section 9.5.5. One example is the *optical supervisory channel* (OSC), shown in Figure 9.1, a separate wavelength dedicated to performing control and management functions, particularly for line systems with optical amplifiers.

Multiple EMSs may be used to manage the overall network. Typically each EMS manages a single vendor's network elements. For example, a carrier using WDM line systems from vendor A and crossconnects from vendor B will likely use two EMSs, one for managing the line systems and the other for managing the crossconnects, as shown in Figure 9.1.

The EMS itself typically has a view of one network element at a time and may not have a comprehensive view of the entire network, and also of other types of network elements that it cannot manage. Therefore the EMSs in turn communicate with a *network management system* (NMS) or an operations support system (OSS) through a management network. The NMS has a networkwide view and is capable of managing different types of network elements from possibly different vendors. In some cases, it is possible to have a multitiered hierarchy of management systems. Multiple OSSs may be used to perform different functions. For example, the regional Bell operating companies (RBOCs) in the United States—Verizon, Southwestern Bell, Bellsouth, and U.S. West (now part of Qwest)—use a set of OSSs from Telcordia Technologies: network monitoring and analysis (NMA) for fault management, trunk inventory and record keeping system (TIRKS) for inventorying the equipment in the network, and transport element management system (TEMS) for provisioning circuits. These systems date back a few decades, and introducing new network elements into these networks is often gated by the time taken to modify these systems to support the new elements.

In addition to the EMSs, a simplified local management system is usually provided to enable craftspeople and other service personnel to configure and manage individual network elements. This system is usually made available on a laptop or on a simple text-based terminal that can be plugged into individual elements to configure and provision them.

9.1.2   Information Model

The information to be managed for each network element is represented in the form of an *information model* (IM). The information model is typically an object-oriented representation that specifies the attributes of the system and the external behavior of the network element with respect to how it is managed. It is implemented in software inside the network element as well as in the element and network management systems used to manage the network element, usually in an object-oriented programming language.

An object provides an abstract way to model the parts of a system. It has certain attributes and functions associated with it. The functions describe the behavior of the object or describe operations that can be performed on the object. For example, the simplest function is to create a new object of a particular type. There may be many types, or *classes*, of objects representing different parts of a system. An important concept in object-oriented modeling is *inheritance*. One object class can be inherited from another parent object class if it has all the attributes and behaviors of the parent class but adds additional attributes and behaviors. To provide a concrete example in our context, an OLT typically consists of one or more racks of equipment. Each rack consists of multiple shelves and multiple types of shelves. Each shelf has several slots into which line cards can be plugged. Many different types of line cards exist, such as transponders, amplifiers, multiplexers, and so on. With respect to this, there may be an object class called *rack,* which has as one of its attributes another object class called *shelf.* Multiple types of shelves may be represented in the form of inherited object classes from the parent object *shelf.* For example, there may be a common equipment shelf and a transponder shelf, which are inherited from the generic shelf object.

A shelf object has as one of its attributes another object called *slot.* Each line card object is associated with a slot. Multiple types of line cards may be represented in the form of inherited object classes from the parent object *line card.* For example, the transponder shelf may house multiple transponder types (say, one to handle SONET signals and another to handle Gigabit Ethernet signals). The common equipment shelf may house multiple types of cards, such as amplifier cards, processor cards, and power supply cards.

Each object has a variety of attributes associated with it, including the set of parameters that can be set by the management system and the set of parameters that can be monitored by the management system. As an example, each line card object normally has a state attribute associated with it, which is one of *in service, out of service,* or *fault,* and there are detailed behaviors governing transitions between these states.

Another example that is part of a typical information model is the concept of *connection trails,* which are used to model lightpaths. Again multiple types of trails may be defined, and each trail has a variety of associated attributes, including ones that can be configured as well as others that can be used to monitor the trail's performance.

9.1.3   Management Protocols

Most network management systems use a master-slave sort of relationship between a manager and the agents managed by the manager. The manager queries the agent to obtain the status of parameters in the network element (called the *get* operation). For example, the manager may query the agent periodically for performance monitoring information. The manager can also change the values of variables in the network element (called the *set* operation) and uses this method to effect changes within the network element. For example, the manager may use this method to change the configuration of the switches inside a network element such as an OXC. In addition to these methods, it is necessary for the agent sometimes to initiate a message to its manager. This is essential if the agent detects problems in the network element and wants to alert its manager. The agent then sends a *notification* message to its manager. Notifications also take the form of *alarms* if the condition is serious and are sometimes called *traps*.

There are multiple standards relating to network management and perhaps thousands of acronyms describing them. Here is a brief summary. In most cases, the physical management interface to the network element is usually through an Ethernet or RS-232 serial interface.

The Internet world uses a management framework based on the *simple network management protocol* (SNMP). SNMP is an application protocol that runs over a standard Internet Protocol stack. The manager communicates with the agents using SNMP. The information model in SNMP is called a *management information base* (MIB).

In North America, the carrier world has been using for a few decades a simple textual (or ASCII) command and control language called *Transaction Language-1* (TL-1). TL-1 was invented in the days when the primary means of managing network elements was through a simple terminal interface using textual command sets. However, it is still widely used today and will probably remain for a while, as many of the existing legacy management systems still mainly support only TL-1.

Over the past decade, there has been a huge effort to standardize a management framework for the carrier world called the *telecommunications management network* (TMN). TMN defined a hierarchy of management systems and object-oriented ways to model the information to be managed, and also specified protocols for communicating between managers and their agents. The protocol is called the *common management information protocol* (CMIP), which usually runs over an *open systems interconnection* (OSI) protocol stack; the associated management interface is called a *Q3* interface. Adaptations have also been defined for running CMIP over the more commonly used TCP/IP protocol stack. The specific object model is based on a standard called *guidelines for description of managed objects* (GDMO). The first two concepts of TMN, namely, the hierarchical management view and the object-oriented way of modeling information, are widely used today, but the specific protocols, interfaces, and object models defined in TMN have not yet been widely adopted, mostly because of the perceived complexity of the entire system.

There is currently a significant effort under way to migrate toward a model where network elements from different vendors come with their own element management systems, and a common interface is specified between these element management systems and a centralized network management system. This interface is based on the *common object request broker* (CORBA) model. CORBA is a software industry standard developed to allow diverse systems to exchange and jointly process information and communicate with each other.

9.2   Optical Layer Services and Interfacing

The optical layer provides lightpaths to other layers such as the SONET, IP, or ATM layers. In this context, the optical layer can be viewed as a *server* layer, and the higher layer that makes use of the services provided by the optical layer as the *client* layer. From this perspective, we need to specify clearly the service interface between the optical layer and its client layers. The key attributes of such a managed lightpath service are the following:

- Lightpaths need to be set up and taken down as required by the client layer and as required for network maintenance.

- Lightpath bandwidths need to be negotiated between the client layer and the optical layer. Typically the client layer specifies the amount of bandwidth needed on the lightpath.

- An adaptation function may be required at the input and output of the optical network to convert client signals to signals that are compatible with the optical layer. This function is typically provided by transponders, as we discussed in Section 7.1. The specific range of signal types, including bit rates and protocols supported, need to be established between the client and the optical layer.

- Lightpaths need to provide a guaranteed level of performance, typically specified by the bit error rate (typical requirements are 10^{-12} or less). Adequate performance management needs to be in place inside the network to ensure this.

- Multiple levels of protection may need to be supported, as we will see in Chapter 10, for example, protected, unprotected, and protect on a best-effort basis, in addition to being able to carry low-priority data on the protection bandwidth in the network. In addition, restoration time requirements may also vary by application.

- Lightpaths may be unidirectional or bidirectional. Almost all lightpaths today are bidirectional. However, if more bandwidth is desired in one direction compared to the other, it may be desirable to support unidirectional lightpaths.

- A multicasting, or a *drop-and-continue*, function may need to be supported. Multicasting is useful to support distribution of video or conferencing information. In a drop-and-continue situation, a signal passing through a node is dropped locally, but a copy of it is also transmitted downstream to the next node. We will see in Chapter 10 that the drop-and-continue function is particularly useful for network survivability when multiple rings are interconnected.

- Jitter requirements exist, particularly for SONET/SDH connections. In order to meet these requirements, 3R regeneration may be needed in the network. Using

2R regeneration in the network increases the jitter, which may not be acceptable for some signals. We discussed 3R and 2R in the context of transparency in Section 1.5.

- There may be requirements on the maximum delay for some types of traffic, notably ESCON. In ESCON, the throughput of the protocol goes down as the propagation delay increases. This causes ESCON devices to place restrictions on the maximum allowed propagation delay (or equivalent link length) between them. This will need to be accounted for while designing the lightpaths.

- Extensive fault management needs to be supported so that root-cause alarms can be reported and adequate isolation of faults can be performed in the network. This is important because a single failure can trigger multiple alarms. The root-cause alarm reports the actual failure, and we need to suppress the remaining alarms. Not only are they undesirable from a management perspective, but they may also result in multiple entities in the network reacting to a single failure, which cannot be allowed. We will look at examples of this later.

Enabling the delivery of these services requires a control and management interface between the optical layer and the client layer. This interface allows the client to specify the set of lightpaths that are to be set up or taken down and set the service parameters associated with those lightpaths, and enables the optical layer to provide performance and fault management information to the client layer. This interface can take on one of two facets. The simple interface used today is through the management system. A separate management system communicates with the optical layer EMS, and the EMS in turn then manages the optical layer.

The present method of operation works fine as long as lightpaths are set up fairly infrequently and remain nailed down for long periods of time. It is quite possible that, in the future, lightpaths are provisioned and taken down more dynamically in large networks. In such a scenario, it would make sense to specify a *signaling* interface between the optical layer and the client layer. For instance, an IP router could signal to an associated optical crossconnect to set up and take down lightpaths and specify their levels of protection through such an interface. Different philosophies exist as to whether such an interface is desirable or not. Some carriers are of the opinion that they should decouple optical layer management from its client layers and plan and operate the optical network separately. This approach makes sense if the optical layer is to serve multiple types of client layers and allows them to decouple its management from a specific client layer. Others would like tight coupling between the client and optical layers. This makes sense if the optical layer primarily serves a single client layer, and also if there is a need to set up and take down connections rapidly as we discussed above. We will discuss this issue further in Section 9.6.

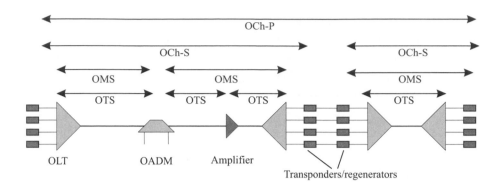

Figure 9.2   Layers within the optical layer, showing the optical channel-path (OCh-P) layer, optical channel-section layer (OCh-S), optical multiplex section (OMS) layer, and the optical transmission section (OTS) layer.

9.3   Layers within the Optical Layer

The optical layer is a complicated entity performing several functions, such as multiplexing wavelengths, switching and routing wavelengths, and monitoring network performance at various levels in the network. In order to help delineate management functions and in order to provide suitable boundaries between different equipment types, it is useful to further subdivide the optical layer into several sublayers. The International Telecommunications Union (ITU) has identified three such layers within the optical layer, as shown in Figure 9.2.   At the top is the *optical channel* (OCh) layer. This layer takes care of end-to-end routing of the lightpaths. We have been using the term *lightpath* to denote an optical connection. More precisely, a lightpath is an optical channel trail between two nodes that carries an entire wavelength's worth of traffic. A lightpath traverses many links in the network, wherein it is multiplexed with many other wavelengths carrying other lightpaths. It may also get regenerated along the way. Note that we do not include any electronic time division multiplexing functions in the optical layer. This is a higher-layer (for example SONET/SDH) function. So a 10 Gb/s connection between two nodes that is carried through without any electronic multiplexing/demultiplexing would be considered a lightpath.

Each link between OLTs or OADMs represents an *optical multiplex section* (OMS) carrying multiple wavelengths. Each OMS in turn consists of several link segments, each segment being the portion of the link between two optical amplifier stages. Each of these portions is an *optical transmission section* (OTS). The OTS consists of the OMS along with an additional optical supervisory channel (OSC), which we will study in Section 9.5.7.

The optical channel layer itself is further subdivided into multiple sublayers. ITU G.709 describes these sublayers. To keep the discussion simple, we will use some terms that differ slightly from the ITU definitions. An *optical channel-transparent section* (OCh-TS) represents the section of a lightpath within an all-optical subnetwork. Within this section, a lightpath is carried optically without any conversion into the electrical domain. At the boundary of an OCh-TS, a lightpath is regenerated. Just above the OCh-TS is the *optical channel-section* (OCh-S). This layer adds some overheads to the lightpath, such as forward error correction (FEC), to condition the signal for transport over an all-optical subnet. Finally, the *optical channel-path* (OCh-P) represents the end-to-end transport of a lightpath across multiple regenerators in the path.

In principle, once the interfaces between the different layers are defined, it is possible for vendors to provide standardized equipment ranging from just optical amplifiers to WDM links to entire WDM networks. Equally importantly, the layers help us break down the management functions necessary in the network, as we will see in this chapter and in Chapter 10. For example, dropping and adding wavelengths is a function performed at the optical channel layer. Monitoring optical power on each wavelength also belongs to this layer, but monitoring total power belongs either to the OTS layer or the OMS layer, depending on whether the optical supervisory channel is included or not.

The preceding definition of an optical layer does not include optical networks that may be able to provide more sophisticated packet-switched services, such as virtual circuits or datagrams. We will study photonic packet-switched networks in Chapter 12 that can potentially provide such services; however, these types of networks are several years away from commercial realization.

9.4 Multivendor Interoperability

Service providers like to deploy equipment from multiple vendors that operate together in a single network. This is desirable to reduce the dependence on any single vendor as well as to drive down costs and is one of the driving factors behind network standards. For instance, without standards, we would have to have special interoperability between every pair of vendors, rather than having to deal with a single standardized interface to which all vendors conform. Another important effect of standards is that they allow operations personnel to get trained on a single type of equipment and then become capable of managing that type of equipment from a

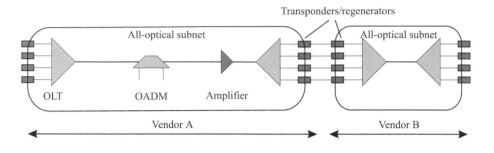

Figure 9.3 Interoperability between WDM systems from different vendors, showing all-optical subnets from different vendors interconnected through transponder/regenerators.

variety of vendors, in contrast to being trained separately to deal with each vendor's equipment.

However, interoperability between WDM equipment from different vendors is easier said than done. The SONET standards were established in the late 1980s, and only recently have we been able to achieve interoperability between equipment from different vendors. In the case of WDM, achieving interoperability at the optical level is made particularly difficult by the fact that the interface is a fairly complex analog interface, rather than a simple digital interface. The set of parameters that we would need to standardize to achieve interoperability include optical wavelength; optical power; signal-to-noise ratio; bit rate; and the supervisory channel wavelength, bit rate, and its contents. Different vendors use significantly different parameters in their link design and make different compromises among the various impairments that we studied in Chapter 5. For example, vendor A might choose to use directly modulated lasers and dispersion compensation inside the network to eliminate dispersion. Vendor B instead might choose to use externally modulated lasers and avoid dispersion compensation inside the network. This would make it difficult to have vendor A's equipment and vendor B's equipment on opposite sides of the same WDM link. Even if some interoperability can be achieved, it is quite difficult to locate and isolate faults in such an environment.

Rather than trying to solve this complex problem, the practical solution toward interoperability is to use regenerators or transponders to interconnect disparate all-optical subnetworks, as shown in Figure 9.3. While this approach may result in higher equipment costs, it provides clear-cut boundaries between all-optical subnets, making it easier to locate and identify faults. Each all-optical subnet would include equipment from a single vendor. For example, a subnet could simply be a WDM link with some intermediate add/drops. So a service provider could deploy vendor A's equipment on one link and vendor B's equipment on another link and have them interoperate through transponders. The interface between the transponders would be either SONET/SDH or the digital wrapper, which we will study in Section 9.5.7. Using the digital wrapper allows the service provider to manage the entire network effectively.

The standards bodies initially started with the goal of establishing optical interoperability and are still pursuing this (ITU G.959, Telcordia GR-2918), although it will be a while before this comes to fruition in a practical network. Meanwhile there is a consensus building around the digital wrapper standard (ITU G.709).

In addition to accomplishing interoperability at the data level, we also need to have interoperability as far as the control and signaling protocols are concerned, particularly if we are using distributed methods discussed in Section 9.6.2. This is a goal that appears to be accomplishable, given that similar functions have been standardized for other networks in the past.

9.5 Performance and Fault Management

As we stated earlier, the goal of performance management is to enable service providers to provide guaranteed quality of service to the users of their network. This usually requires monitoring of the performance parameters for all the connections supported in the network and taking any actions necessary to ensure that the desired performance goals are met. Performance management is closely tied in to fault management. Fault management involves detecting problems in the network and alerting the management systems appropriately through alarms. If a certain parameter is being monitored and its value falls outside its preset range, the network equipment generates an alarm. For example, we may monitor the power levels of an incoming signal and declare a loss-of-signal (LOS) alarm if we see the power level drop below a certain threshold. In other cases, alarms could be triggered by outright failures, such as the failure of a line card or other components in the system.

Fault management also includes restoring service in the event of failures, a subject that we will cover in detail in Chapter 10. This function is considered an autonomous network control function because it is typically a distributed application without network managment intervention (except for configuring various protection parameters up front, reporting events, and performing maintenance operations).

9.5.1   The Impact of Transparency

The lightpaths provided by the optical layer need to be managed just like SONET and SDH connections are managed. To a large extent how much management can be provided depends on the level of transparency provided by the optical layer. As we have seen in Chapter 1, different levels of transparency are possible, based on the range of signals, bit rates, and protocols that can be carried on a lightpath.

In a purely transparent network, a lightpath will be capable of carrying analog and digital signals with arbitrary bit rates and protocol formats. This is the utopian vision of optical networking and would allow service providers to offer a range of services without any constraints and provide future-proofing in case the service mix changes over time or when new services are added. However, such a network is very difficult to engineer and manage. It is difficult to engineer because the various physical layer impairments that must be taken into account in the network design are critically dependent on the type of signal (analog versus digital) and the bit rate. It is difficult to manage because the management system may have no prior knowledge of the protocols or bit rates being used in the network. Therefore, it is not possible to access overhead bits in the transmitted data to obtain performance-related measures. This makes it difficult to monitor the bit error rate. Other parameters such as optical power levels and optical signal-to-noise ratios can be measured. Most systems today only measure optical power levels. However, small, portable optical spectrum analyzers are now becoming available to measure the signal-to-noise ratio, making it practical to incorporate this measurement in newer systems. However, the acceptable values for these parameters depend on the type of signal. Unless the management system is told what type of signal is being carried on a lightpath, it will not be able to determine whether the measured power levels and signal-to-noise ratios fall within acceptable limits.

At the other exteme, we could design a network that carries data at a fixed bit rate (say, 2.5 Gb/s or 10 Gb/s) and of a particular format (say, SONET/SDH only). Such a network would be very cost-effective to build and manage. However, it does not offer service providers the flexibility they need to deliver a wide variety of services using a single network infrastructure and is not future-proof at all.

Most optical networks deployed today fall somewhere in between these two extremes. The network is designed to handle digital data at arbitrary bit rates up to a certain specified maximum (say, 10 Gb/s) and a variety of protocol formats such as SONET/SDH, IP, ATM, Gigabit Ethernet, and ESCON. These networks make use of a number of unique techniques to provide management functions, as we will see next.

9.5.2   BER Measurement

The bit error rate (BER) is the key performance attribute associated with a lightpath. The BER can be detected only when the signal is available in the electrical domain, typically at regenerator or transponder locations. As we saw in Chapter 6, framing protocols used in SONET and SDH include overhead bytes. Part of this overhead consists of parity check bytes by which the BER can be computed. This provides a direct measure of the BER. Similarly, the digital wrapper overhead developed specifically for the optical layer also allows the BER to be measured. We will study the digital wrapper in Section 9.5.7. As long as the client signal data is encapsulated using the SONET/SDH or digital wrapper overhead, we can measure the BER and guarantee the performance within the optical layer.

Given the complexity of optical physical layer designs, it is difficult to estimate the BER accurately based on indirect measurements of parameters such as the optical signal power or the optical signal-to-noise ratio. These parameters may be used to provide some measure of signal quality and may be used as triggers for events such as maintenance or possibly protection switching (which could be based, for example, on loss of power and signal detection) but not to measure BER.

9.5.3   Optical Trace

Lightpaths pass through multiple nodes and through multiple cards within the equipment deployed at each node. It is desirable to have a unique identifier associated with each lightpath. For example, this identifier may include the IP address of the originating network element along with the actual identity of the transponder card within that network element where the lightpath terminates. This identifier is called an *optical path trace*. The trace enables the management system to identify, verify, and manage the connectivity of a lightpath. In addition it provides the ability to perform fault isolation in the event that incorrect connections are made.

A trace can be used in different layers within the optical layer. For instance, a lightpath passes through multiple nodes and potentially gets regenerated along the way. We can verify the end-to-end connectivity of a lightpath using an *optical channel-path trace*. This trace is inserted at the beginning of the lightpath and monitored at various locations along the path of the lightpath. In order to localize and verify connectivity between regenerator locations, we make use of an additional identifier called the *optical channel-section trace*, which is associated between each adjacent pair of regeneration points of the lightpath. Within an all-optical subnet, we can use a *optical channel-transparent section trace*. The latter two traces are inserted and removed at regenerator locations in the network. We will look at different ways of carrying the trace information in Section 9.5.7.

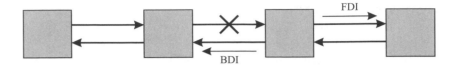

Figure 9.4   Forward and backward defect indicator signals and their use in a network.

9.5.4   Alarm Management

In a network, a single failure event may cause multiple alarms to be generated all over the network and incorrect actions to be taken in response to the failed condition. Consider, in particular, a simple example. When a link fails, all lightpaths on that link fail. This could be detected at the nodes at the end of the failed link, which would then issue alarms for each individual lightpath as well as report an entire link failure. In addition, all the nodes through which these lightpaths traverse could detect the failure of these lightpaths and issue alarms. For example, in a network with 32 lightpaths on a given link, each traversing through two intermediate nodes, the failure of a single link could trigger a total of 129 alarms (1 for the link failure and 4 for each lightpath at each of the nodes associated with the lightpath). It is clearly the management system's job to report the single root-cause alarm in this case, namely, the failure of the link, and suppress the remaining 128 alarms.

Alarm suppression is accomplished by using a set of special signals, called the *forward defect indicator* (FDI) and the *backward defect indicator* (BDI). Figure 9.4 shows the operation of the FDI and BDI signals. When a link fails, the node downstream of the failed link detects it and generates a *defect condition*. For instance, a defect condition could be generated because of a high bit error rate on the incoming signal or an outright loss of light on the incoming signal. If the defect persists for a certain time period (typically a few seconds), the node generates an alarm.

Immediately upon detecting a defect, the node inserts an FDI signal downstream to the next node. The FDI signal propagates rapidly and nodes further downstream receive the FDI and suppress their alarms. The FDI signal is also sometimes referred to as the *alarm indication signal* (AIS). A node detecting a defect also sends a BDI signal upstream to the previous node, to notify that node of the failure. If this previous node didn't send out an FDI, it then knows that the link to the next node downstream has failed.

Note further that separate FDI and BDI signals are needed for different sublayers within the optical layer, for example, to distinguish between link failures and failures of individual lightpaths, or to distinguish between the failure of a section of the link between amplifier locations and that of the entire link. The exact types

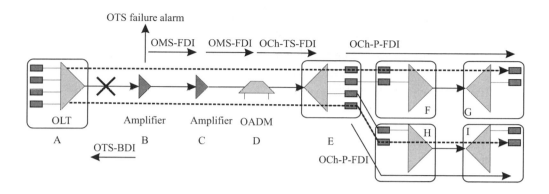

Figure 9.5 Using hierarchical defect indicator signals in a network. Defect indicators are used at the OTS, OMS, and the various OCh sublayers.

and behavior of defect indicators for the optical layer are being standardized currently (ITU G.709). Figure 9.5 illustrates one possible use of these different indicator signals in a network. Suppose there is a link cut between OLT A and amplifier B as shown. Amplifier B detects the cut. It immediately inserts an OMS-FDI signal downstream indicating that all channels in the multiplexed group have failed and also an OTS-BDI signal upstream to OLT A. The OMS-FDI is transmitted as part of the overhead associated with the OMS layer, and the OTS-BDI is transmitted as part of the overhead associated with the OTS layer.

Note that an OMS-FDI is transmitted downstream and not an OTS-FDI. This is because the defect information needs to be propagated all the way downstream to the network element where the OMS layer is terminated, which, in this case, is OADM D. Amplifier C downstream receives the OMS-FDI and passes it on. OADM D, which is the next node downstream, receives the OMS-FDI and determines that all the lightpaths on the incoming link have failed. Some of these lightpaths are dropped locally and others are passed through. For each lightpath passed through, the OADM generates OCh-TS-FDIs and sends them downstream. The OCh-TS-FDIs are transmitted as part of the OCh-TS overhead. At the end of the all-optical subnet, at OLT E, the wavelengths are demultiplexed and terminated in transponders/regenerators. Therefore the OCh-TS layer is terminated here. OLT E receives the OCh-TS-FDIs. It then generates OCh-P-FDI indicators for each failed lightpath and sends that downstream to the ultimate destination of each lightpath as part of the OCh-P overhead. Finally, the only node that issues an alarm is node B.

Another major reason for using the defect indicator signals is that defects are used to trigger protection switching. For example, nodes adjacent to a failure detect the failure and may trigger a protection-switching event to reroute traffic around the failure. At the same time, nodes further downstream and upstream of the failure may think that other links have failed and decide to reroute traffic as well. A node receiving an FDI knows whether it should or shouldn't initiate protection switching. For example, if the protection-switching method requires the nodes immediately adjacent to the failure to reroute traffic, other nodes receiving the FDI signal will not invoke protection switching. On the other hand, if protection switching is done by the nodes at the end of a lightpath, then a node receiving an FDI initiates protection switching if it is the end point of the associated lightpath.

9.5.5   Data Communication Network (DCN) and Signaling

The element management system (EMS) communicates with the different network elements through the DCN. This DCN is usually a standard TCP/IP or OSI network (see Chapter 6). If the DCN is sufficiently well connected (2-connected, to be more precise), then the DCN can stay up even if there is a failure in the network. The DCN can be transported in several ways:

1. Through a separate out-of-band network outside the optical layer. Carriers can make use of their existing TCP/IP or OSI networks for this purpose. If such a network is not available, dedicated leased lines could be used for this purpose. This option is viable for network elements that are located in big central offices where such connectivity is easily available, but not viable for network elements such as optical amplifiers that are located in remote huts in the field.

2. Through the OSC on a separate wavelength (see Section 9.5.7). This option is available for WDM line equipment that processes the optical transmission section and multiplex section layers, where the optical supervisory channel is made available. For example, optical amplifiers are managed using this approach. However, this option is not available to equipment that only looks at the optical channel layer, such as optical crossconnects.

3. Through the rate-preserving or digital wrapper inband optical channel layer overhead techniques to be described in Section 9.5.7. This option is useful for equipment that only looks at the optical channel layer and does not process the multiplex and transmission section layers, such as optical crossconnects. Also, it is available only at locations where the lightpath is processed in the electrical domain, that is, at regenerator or transponder locations.

Table 9.1 summarizes the applicability of different DCN options available for each type of network element. We assume that OADMs are part of the line system that

Table 9.1   Different ways of realizing the DCN for different network elements. The OADM is assumed to have transponders for channels that are dropped and added, but not for channels that are passed through.

Network Element	Out-of-Band	OSC	Rate-Preserving Overhead or Digital Wrapper
OLT with transponders	Yes	Yes	Yes
OADM	Yes	Yes	Yes (for dropped channels)
Amplifier	No	Yes	No
OXC with regenerators	Yes	No	Yes
All-optical OXC (no regenerators)	Yes	No	No

includes OLTs and amplifiers. Access to the optical supervisory channel is typically restricted to elements within a line system due to the proprietary nature of the OSC.

In addition to the DCN, in many cases, a fast signaling network is needed between network elements. This allows the network elements to exchange critical information between them in real time. For instance, the FDI and BDI signals need to be propagated quickly to the nodes along a lightpath. Other such signals include information needed to implement fast protection switching in the network, the topic of Chapter 10. Just as with the DCN, the signaling network can be implemented using dedicated out-of-band connections, the optical supervisory channel, or through one of the overhead techniques.

9.5.6   Policing

One function of the management system is to monitor the wavelength and power levels of signals being input to the network to ensure that they meet the requirements imposed by the network. As we discussed above, the acceptable power levels will depend on the signal types and bit rates. The types and bit rates are specified by the user, and the network can then set thresholds for the parameters as appropriate for each signal type and monitor them accordingly. This includes threshold values for the parameters at which alarms must be set off. The thresholds depend on the data rate, wavelength, and specific location along the path of the lightpath, and degradations may be measured relative to their original values.

Another more important function is to monitor the actual service being utilized by the user. For example, the service provider may choose to provide two services, say, an ESCON service and an OC-3 service, by leasing a transparent lightpath to the user. The two services may be tariffed differently. With a purely transparent network, it is difficult to prevent a user who opts for the ESCON service from sending OC-3

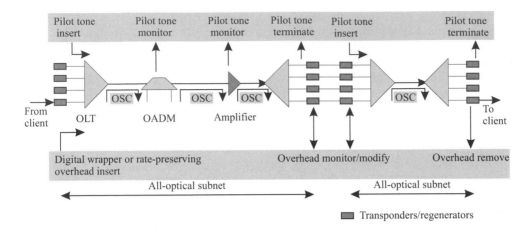

Figure 9.6 Different types of optical layer overhead techniques. The OSC is used hop by hop. The pilot tone is inserted by a transmitter and can be monitored at elements in an all-optical subnet until it is terminated at a receiver. The digital wrapper or rate-preserving overhead is used end to end across multiple subnets through intermediate regenerators.

traffic. What this implies is that services based on leasing wavelengths will likely be tariffed based on a specified maximum bit rate, with the user being allowed to send any signal up to the specified maximum bit rate.

9.5.7 Optical Layer Overhead

Supporting the optical path trace, defect indicators, and BER measurement requires the use of some sort of overhead in the optical layer. We have alluded indirectly to some of these overheads earlier, for example, the use of the SONET/SDH overhead to measure the BER and the use of the optical supervisory channel to carry some of the defect indicator signals. In this section, we describe four different methods for carrying the optical layer overhead. These methods are illustrated in Figure 9.6 and compared in Table 9.2. The pilot tone approach and the optical supervisory channel are useful to carry overhead information within an all-optical subnetwork. At the boundaries of each subnetwork, the signal is regenerated (3R) by converting into the electrical domain and back. The rate-preserving overhead and the digital wrapper can be used to carry overhead information across an entire optical network through multiple all-optical subnetworks.

Table 9.2   Applications of different optical layer overhead techniques. The different techniques apply to different sublayers within the optical layer—namely, the optical transmission section (OTS), optical multiplex section (OMS), or optical channel-section (OCh-S) or optical channel (OCh) layers. The trace and defect indicator (DI) signals are defined at multiple sublayers.

Application	All-Optical Subnet		End-to-End	
	OSC	Pilot Tone	Rate-Preserving	Digital Wrapper
Trace	OTS	OCh-TS	OCh-P	OCh-P
			OCh-S	OCh-S
DIs	OTS	None		
	OMS		OCh-P	OCh-P
	OCh-TS			
Performance monitoring	None	Optical power	BER	BER
Client signal compatibility	Any	Any	SONET/SDH	Any

Pilot Tone or Subcarrier Modulated Overhead

Here, the overhead is realized by modulating the optical carrier (wavelength) of a lightpath with an additional subcarrier signal, as described in Section 4.2. This signal is also sometimes called a *pilot tone*. As long as the modulation depth of this signal is kept small compared to the data, typically between 5–10%, and the subcarrier frequency is chosen carefully, the data is relatively unaffected as a result. The pilot tone itself may be amplitude or frequency modulated at a low rate, say, a few kilobits per second, to carry additional overhead information.

At intermediate locations, a small fraction of the optical power can be tapped off and the pilot tones extracted without receiving and retransmitting the entire signal. Note that the pilot tones on each wavelength can be extracted from the composite WDM signal carrying all the wavelengths without requiring each wavelength to be demultiplexed.

The pilot tone frequency needs to be chosen carefully. First, it should have minimal overlap with the data bandwidth. For instance, a lightpath carrying SONET data at 2.5 Gb/s has relatively little spectral content below 2 MHz, and a pilot tone in the 1–2 MHz range can be added with minimal impact to the data. The pilot tone frequency also needs to lie above the gain modulation cutoff of the erbium-doped optical amplifiers, which is typically around 100 kHz (see Section 3.4.3). Tones below this frequency will cause the amplifier gain to vary with the pilot tone amplitude, causing this modulation to be imposed on other channels as undesirable "ghost"

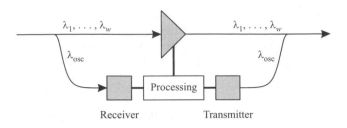

Figure 9.7   The optical supervisory channel, which is terminated at each amplifier location.

tones or crosstalk. The pilot tone frequency can also be chosen to lie above the data band, in this example, say, above 2.5 GHz, but it is relatively more expensive to process signals at higher frequencies than at lower frequencies.

The advantages of the pilot tone approach are that it is relatively inexpensive and that it allows monitoring of the overhead in transparent networks without requiring knowledge of the actual protocol or bit rate of the signal. The disadvantages are that it cannot be used to monitor the BER, and the pilot tone can be modified only at the transmitter or at a regenerator and not at the intermediate nodes. Thus it can be used for the OCh-TS trace function inside a transparent subnetwork between regenerator points, but cannot be used to insert FDI and BDI signals at intermediate nodes without a regenerator. The trace function can be accomplished using pilot tones in several possible ways. For example, each lightpath could have a unique pilot tone frequency, which by itself serves as the trace. Alternatively, we could have a unique pilot tone frequency for each wavelength, and the pilot tone can be modulated with a digital signal containing a unique lightpath identifier.

Optical Supervisory Channel

In systems with line amplifiers, a separate OSC is used to convey information associated with monitoring the state of the amplifiers along the link, particularly if these amplifiers are in remote locations where other direct access is not possible. The OSC is also used to control the line amplifiers, for example, turning them on or turning them off for test purposes. It can also be used to carry the DCN, as well as some of the overhead information.

The OSC is carried on a wavelength different from the wavelengths used for carrying traffic. It is separated from the other wavelengths at each amplifier stage and received, processed, and retransmitted, as shown in Figure 9.7.

The choice of the exact wavelength for the OSC involves a number of trade-offs. Figure 9.8 shows the usage of various wavelength bands in the network for carrying

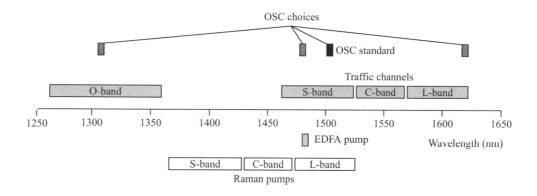

Figure 9.8   Usage of wavelengths in the network. Traffic is carried on the O (original), S (short), C (conventional), or L (long) wavelength bands. Raman pumps, if used, are located about 80–100 nm below the signal.

traffic, for pumping the erbium or Raman amplifiers, and for the OSC. The OSC could be located within the same band as the traffic-bearing channels, or in a separate band located away from the traffic-bearing channels. In the latter situation, it is easier to filter out and reinsert the OSC at each amplifier location. However, we need to locate the OSC away from the Raman pumps if they are used in the system.

Perhaps the only advantage of locating the OSC in the same band as the traffic-bearing channels is a slight reduction in amplifier noise. For instance, if a two-stage amplifier design is used, the in-band OSC can be filtered out after the first stage along with the amplifier noise that is present at this wavelength.

For WDM systems operating in the C-band, the popular choices for the OSC wavelength include 1310 nm, 1480 nm, 1510 nm, or 1620 nm. Using the 1310 nm band for the OSC precludes the use of this band for carrying traffic. The 1480 nm wavelength was considered only because of the easy availability of lasers at that wavelength—it happens to be one of the wavelengths used to pump an erbium-doped fiber amplifier (EDFA). For the same reason, however, there can be some undesirable interactions between the OSC laser and the EDFA pump, so this is not a popular choice.

After going through some of these trade-offs, the ITU has adopted the 1510 nm wavelength as the preferred choice. This wavelength is outside the EDFA passband, does not coincide with an EDFA pump wavelength, and lies outside the C- and L-bands. Note, however, that this wavelength falls in the S-band and may also overlap with Raman pumps for the L-band.

Yet another choice used by some vendors is the 1620 nm wavelength, on the outer edge of the L-band. This choice avoids most of the problems above, except that we have to be careful about separating this channel from a traffic-bearing channel toward the edge of the L-band.

The OSC can be used to carry OTS traces and defect indicators, as well as OMS and OCh-TS defect indicators.

Rate-Preserving Overhead

The idea here is to make use of the existing SONET/SDH overhead that is used with most of the signals entering the optical layer. This overhead includes several bytes that are currently unused. Some of these bytes can be used by the optical layer. These bytes can also be used to add forward error correction (FEC), which improves the optical layer link budget. This technique can be used only at locations where the signal is available in electrical form, that is, at regenerator locations or at the edges of the network. Unlike the pilot tone method, it cannot be used inside a transparent optical subnetwork.

The advantages of this method are the following: First, it can be used with the existing equipment in the network. For example, a new network element with this capability can communicate with other network elements of the same type through intermediate WDM and SONET equipment that is already present in the network. Second, it retains the existing hierarchy of bit rates in the SONET/SDH standards, without the need for creating a new hierarchy of rates that would be needed with the digital wrapper technique to be discussed next. This allows existing SONET/SDH chipsets, such as clock recovery circuits, receivers, modulators, and overhead processing chips, to be used without requiring the development of a new set of components to support the new rates.

The disadvantages of this method are the following: First, the number of unused bytes available is limited and may not offer sufficient bandwidth to carry all the optical layer overhead and FEC. Second, while the SONET/SDH standards specify the set of unused bytes, several vendors have already made use of some of these bytes for their own proprietary reasons, which makes it difficult to determine which set of bytes are truly unused! Third, it does not work with signals that don't use SONET/SDH framing, such as Fibre Channel or Gigabit Ethernet (see Chapter 6).

Digital Wrapper Overhead

Here, a new set of overhead bytes is added to the signal as it enters the optical layer and removed when the signal is handed back to the client layer. This scheme offers essentially the same capabilities as the rate-preserving overhead discussed above. The digital wrapper defines a new set of overheads associated with the optical layer and can be used instead of the SONET/SDH overhead. It is being standardized in the ITU.

The advantages of this method are the following: First, sufficient overhead bytes can be added so as to provide adequate FEC and support the DCN as well as to allow for future needs. Second, a new standard based on this technique would allow better interoperability among multiple vendors through regenerators. Third, the technique is not limited to SONET/SDH signals. The wrapper can be used to encapsulate a variety of different signals, such as Fibre Channel and Gigabit Ethernet.

The main disadvantages of the digital wrapper approach are that it is not suitable for use with legacy equipment, and that it requires the development of a new set of components to support the new hierarchy of bit rates. However, new components have already been developed to support the wrapper, and it is now available on many WDM products.

The digital wrapper is ideally suited to carrying OCh-section and path layer traces and defect indicators, as well as providing other overheads for management, such as those used by an automatic protection-switching (APS) protocol for signaling between network elements during failures.

9.6   Configuration Management

We can break down configuration management functions into three parts: managing the equipment in the network, managing the connections in the network, and managing the adaptation of client signals into the optical layer.

9.6.1   Equipment Management

In general, the principles of managing optical networking equipment are no different from those of managing other high-speed networking equipment. We must be able to keep track of the actual equipment in the system (for example, number and location of optical line amplifiers) as well as the equipment in each network element and its capabilities. For example, in a terminal of a point-to-point WDM system, we may want to keep track of the maximum number of wavelengths and the number of wavelengths currently equipped, whether there are optical pre- and power amplifiers or not, and so forth.

Among the considerations in designing network equipment is that we should be able to add to existing equipment in a modular fashion. For instance, we should be able to add additional wavelengths (up to a designed maximum number) without disrupting the operation of the existing wavelengths. Also, ideally the failure of one channel shouldn't affect other channels, and the failed channel should be capable of being serviced without affecting the other channels. An issue that comes up in this regard is the use of arrayed multiwavelength components versus separate components for individual wavelengths, such as multiwavelength laser arrays instead of individual lasers for each wavelength. Using arrayed components can reduce the cost and footprint of the equipment. However, if one element in the array fails, the entire array will have to be replaced. This reduces the system availability, as replacing the array will involve disrupting the operation of multiple channels, and not just a single channel. Using arrays also increases the replacement cost of the module. Therefore there is always a trade-off between obtaining reduced cost and footprint on one front against system availability and replacement cost on the other front.

We may also want to start out by deploying the equipment in the form of a point-to-point link and later upgrade it to handle ring or other network configurations. We may also desire flexibility in associating specific port cards in the equipment with specific wavelengths. For example, it is better to have a system where we can choose the wavelength transmitted out of a port card independently of what slot it is located in.

Another problem in WDM systems is the need to maintain an inventory of wavelength-specific spare cards. For example, each channel may be realized by using a card with a wavelength-specific laser in it. Thus you would need to stock spare cards for each wavelength. This can be avoided by using a wavelength-selectable (or tunable) laser on each card instead of a wavelength-specific laser; such devices are only now becoming commercially available at reasonable cost.

9.6.2   Connection Management

The optical network provides lightpaths, or more generally, circuit-switched connections, to its user. Connection management deals with setting up connections, keeping track of them, and taking them down when they are not needed anymore.

The traditional telecommunications way of providing this function is through a centralized management system, or rather a set of systems. However, this process has been extremely cumbersome and slow. The process usually involves configuring equipment from a variety of vendors, each with its own management system, and usually one network element at a time. Moreover, interoperability between management systems, while clearly feasible, has been difficult to achieve in practice. Finally, service providers in many cases deploy equipment only when needed. The net result of this process is that it can take months for a service provider to turn up a new connection in response to a user request. Given this fact, it is not surprising that once a connection is set up, it remains in effect for a fairly significant period of time, ranging from several months to years!

As optical networks evolve, connections are getting more dynamic and networks are becoming bigger and more complex. Service providers would like to provide connections to their customers rapidly, ideally in seconds to minutes, and not impose long-term holding time commitments on these connections. In other words, users would *dial up* bandwidth as needed.

Supporting all this requires carriers to predeploy equipment (and bandwidth) ahead of time in the network and having methods in place to be able to turn on the service rapidly when needed. This is becoming a significant competitive issue in differentiating one carrier from another. This method of operation also stimulates what is called *bandwidth trading*, where carriers trade their unused bandwidth with other carriers for increasingly shorter durations to improve the utilization of their networks and maximize their revenue.

Due to the reasons above, we are seeing a trend toward a more distributed form of control for connection management. Distributed control protocols have been used in IP and ATM networks. They have also had a fair degree of success with respect to standardization and accomplishing interoperability across vendor boundaries. We can make use of similar protocols for performing these functions in the optical layer.

Distributed connection control has several components to it:

Topology management. Each node in the network maintains a database of the network topology and the current set of resources available as well as the resources used to support traffic. In the event of any changes in the network, for example, a link capacity change, the updated topology information needs to be propagated to all the network nodes. We can use the same techniques used in IP networks for this purpose. Nodes periodically, or in the event of changes, *flood* the updated information to all the network nodes. We can use an Internet routing and topology management protocol such as OSPF or IS-IS (see Section 6.3), with suitable modifications to represent optical layer topology information, and update it automatically.

At the time the network is brought up, or whenever there is a topology change (link/node addition, removal), nodes will need to automatically discover the network topology. This is done typically by having adjacent nodes exchange information to determine their local connectivity (to their neighbors) and then broadcasting this information to all the network nodes using the same procedure used to convey topology changes.

Route computation. When a connection is requested from the network, the network needs to find a route and obtain resources along the route to support this connection. This can be done by applying a routing algorithm on the topology database of the network. The routing algorithm needs to take into account the various constraints imposed by the network, such as wavelength conversion ability, and the capacity available on each link of the network. We studied this aspect in Section 8.2.2. In addition to computing routes for carrying the *working* traffic, the algorithm may also have to compute *protection* routes for the connection, which are used in the event of failures.

Signaling protocol. Once routes are computed, the connection needs to be set up. This process involves reserving the resources required for the connection and setting the actual switches inside the network to set up the connection. The process requires nodes to exchange messages with other nodes. Typically, the destination or source of the connection signals to each of the nodes along the connection path to perform this function. Protocols based on MPLS Internet signaling protocols such as RSVP or CR-LDP (see Section 6.3) can be used for this purpose. The same protocols can also be used to take down connections when they are no longer needed.

The process of setting up or taking down a connection must be executed carefully. For example, if the connection is simply taken down by the source and destination, then the intermediate nodes may sense the loss of light on the connection as a failure condition and trigger unwanted alarms and protection switching. This can be avoided by suitable coordination among the nodes along the route of the lightpath.

Signaling network. Nodes need a signaling channel to exchange control information with other nodes. We described the many options available to realize this in Section 9.5.5.

Interaction with Other Layers

One important aspect of the connection management protocols is in how they interact with the client layers of the optical layer. With IP routers emerging as the dominant clients of the optical layer, and because the optical layer control protocols are based on Internet protocols, the issue of how these protocols interact in particular with the IP layer becomes a crucial issue.

Different types of interactions are likely needed for different scenarios, such as metro versus long-haul networks, incumbent versus new service providers, multiservice versus IP service-centric providers, and facility ownership versus leasing providers.

There are many schools of thought with respect to this interaction, ranging from the so-called overlay model to a peer model. Figure 9.9 shows a variety of models being considered today.

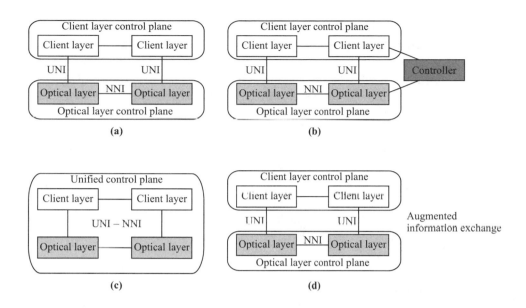

Figure 9.9   Different control plane models for interconnecting client layers with the optical layer. (a) Overlay model, (b) overlay+ model, (c) peer model, and (d) augmented model.

Figure 9.9(a) shows the *overlay model*. In this model, the optical layer has its own control plane, and the higher layers have their own independent control planes. The optical layer provides a *user network interface* (UNI), through which higher (client) layers can request connections from the optical layer. Within the optical layer, different subnetworks can interoperate through a standardized *network-to-network interface* (NNI). This approach allows the connection control software for the optical layer to be tailored specifically to the optical layer without having to worry about developing a single unified piece of control software. It also allows the optical layer and client layers to scale and evolve independently. Details of the optical network topology can be hidden from the client layer through the UNI. We can use this model to interconnect a variety of clients, including IP, ATM, Ethernet, and SONET/SDH clients, with the optical layer. The model is also appropriate for supporting private line lightpath service, transport bandwidth brokering, carrier's carrier trunking, and optical virtual private networks. Finally, this model can be applied to incumbent or new multiservice carriers who either own or lease their transport facilities.

An enhanced version of the overlay model is the *overlay plus model*, shown in Figure 9.9(b), which allows closer interaction between the layers. In this case, there is a trusted intermediate intelligent controller between the two layers that has available to it a suitably abstracted version of specific client and optical layer topology and status information. The controller can use this information to request and release lightpaths based on specific policies, such as specific service level agreements made between the client and optical layers. These requests can be rapidly invoked to avoid network abnormalities such as congestion and failures, increase infrastructure utilization, coordinate protection and restoration options, and automate engineering by rebalancing the network and forecasting needed resource (such as node and link capacity) upgrades for both the IP and optical layers.

Figure 9.9(c) shows the *peer model*, where IP routers and optical layer elements, such as OXCs and OADMs, run the same control plane software. This would allow routers to look at OXCs as if they were routers, effectively treating the IP layer and optical layer as peers. An OXC would simply be a special type of router, analogous to a label-switched router (LSR). Routers would have full topology awareness of the optical layer and could therefore control optical layer connections directly. While this is an elegant approach, it is made complicated by the fact that optical layer elements impose significantly different constraints with respect to routing and protection of connections, compared to the IP layer. In this case, we need to find a way to suitably abstract optical layer routing constraints into a form that can be used by route computation engines residing on IP routers.

Figure 9.9(d) shows another enhanced version of the overlay model, called an *augmented model,* where the IP layer has access to summarized routing, addressing, and topology information of the optical layer, but still operates as a separate control plane from the optical layer.

The models in Figure 9.9(c) and (d) tend to apply mainly to new IP-centric providers or IP-centric business units within established carriers who own their transport facilities. These models allow (or require) significantly more trust and closer coupling between the IP and optical layers, compared to the overlay models of Figure 9.9(a) and (b). All these models are being pursued today, but the overlay approach is likely to be the first one implemented. It has also been adopted for standardization by the ITU.

9.6.3   Adaptation Management

Adaptation management is the function of taking the client signals and converting them to a form that can be used inside the optical layer. This function includes the following:

- Converting the signal to the appropriate wavelength, optical power level, and other optical parameters associated with the optical layer. This is done through the use of transponders, which convert the signal to electrical form and retransmit the signal using a WDM-specific laser. In the other direction, the WDM signal is received and converted into a standardized signal, such as a short-reach SONET signal.

- Adding and removing appropriate overheads to enable the signal to be managed inside the optical layer. This could include one or more of the overhead techniques that we studied in Section 9.5.7.

- Policing the client signal to make sure that the client signal stays within boundaries that have been agreed upon as part of the service agreement. We discussed this in Section 9.5.

The WDM network must support different types of interfaces to accommodate a variety of different users requiring different functions. Figure 9.10 shows the different possible adaptation interfaces.

1. **Compliant wavelength interface:** One interface might be to allow the client to send in light at a wavelength that is supported in the network. In this case, the user would be expected to comply with a variety of criteria set by the network, such as the signal wavelength, power, modulation type, and so on. These wavelengths may be regarded as *compliant* wavelengths. In this case, the interface might be a purely optical interface, with no optoelectronic conversions required (a significant cost savings). For example, you might envision that SONET or IP equipment must incorporate WDM-capable lasers at wavelengths suitable for the WDM network. Likewise, it would be possible to directly send a wavelength from the WDM network into SONET equipment. Here the user complies to the requirements imposed by the network.

2. **Noncompliant wavelength interface:** This is the most common interface and encompasses a variety of different types of attached client equipment that use optical transmitters and/or receivers not compatible with the signals used inside the WDM network. For example, this would include SONET equipment using 1.3 μm lasers. Here until all-optical wavelength conversion (and perhaps all-optical regeneration) becomes feasible, optoelectronic conversion must be used, along with possibly regeneration, to convert the signal to a form suitable for the WDM network. This is likely to be the interface as well when we need to interconnect WDM equipment from different vendors adhering to different specifications, as we discussed in Section 9.4.

3. **Subrate multiplexing:** Additional adaptation functions include time division multiplexing of lower-speed streams into a higher-speed stream within the WDM

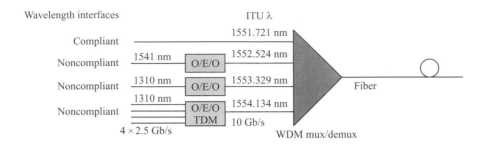

Figure 9.10   Different types of interfaces between a WDM optical network and its clients.

equipment prior to transmission. For example, the WDM equipment could include multiplexing of SONET OC-48 streams into OC-192 streams. This could reduce costs by eliminating the separate equipment that would normally be needed to perform this function.

The level of transparency offered by the network also affects the type of adaptation performed at the edges of the network. The network needs to be capable of transporting multiple bit rates. In general the optical path can be engineered to support signals up to a specified maximum bit rate. The adaptation devices and regenerators used within the network need to be capable of supporting a variety of bit rates as well. An important enabler for this purpose is a programmable clock data recovery chip that can be set to work at a variety of bit rates. The chips available today are capable of handling integral multiples of bit rates (for example, 155 Mb/s, 622 Mb/s, 1.25 Gb/s, and 2.5 Gb/s). They are also capable of handling a narrow range of bit rates around a mean value. For example, a single chip could deal with SONET OC-24 signals or with Gigabit Ethernet signals, which are both around 1.25 Gb/s but not exactly at the same rate. Finally, using a digital wrapper to encapsulate the client signal allows the network to transport multiple data rates and protocol formats in a supervised way.

9.7   Optical Safety

The semiconductor lasers used in optical communication systems are relatively low-power devices; nevertheless, their emissions can cause serious damage to the human eye, including permanent blindness and burns. The closer the laser wavelength is to the visible range, the more damage it can do, since the cornea is more transparent to these wavelengths. For this reason, systems with lasers must obey certain safety standards. Systems with lasers are classified according to their emission levels, and the relevant classes for communication systems are described next. These safety issues in some cases can limit the allowable optical power used in the system.

A *Class I* system cannot emit damaging radiation. The laser itself may be a high-power laser, but it is prevented from causing damage by enclosing it in a suitably interlocking enclosure. The maximum power limit in a fiber for a Class I system is about 10 mW (10 dBm) at 1.55 μm and 1 mW (0 dBm) at 1.3 μm. Moreover, the power must not exceed this level even under a single failure condition within the equipment. A typical home CD player, for example, is a Class I system.

A *Class IIIa* system allows higher emission powers—up to 17 dBm in the 1.55 μm wavelength range—but access must be restricted to trained service personnel. Class IIIa laser emissions are generally safe unless the laser beam is collected or focused onto the human eye. A *Class IIIb* system permits even higher emission powers, and the radiation can cause eye damage even if not focused or collected.

Under normal operation, optical communication systems are completely "enclosed" systems—laser radiation is confined to within the system and not seen outside. The problem arises during servicing or installation, or when there is a fiber cut, in which case the system is no longer completely enclosed and emission powers must be kept below the levels recommended for that particular system class. Communication systems deployed in the enterprise world must generally conform to Class I standards since untrained users are likely to be using them. Systems deployed within carrier networks, on the other hand, may likely be Class IIIa systems, since access to these systems is typically restricted to trained service personnel.

The safety issue thus limits the maximum power that can be launched into a fiber. For single-channel systems without optical power amplifiers using semiconductor lasers, the emission levels are small enough (−3 to 0 dBm typically) that we do not have to worry much about laser safety. However, with WDM systems, or with systems using optical power amplifiers, we must be careful to regulate the total power into the fiber at all times.

Simple safety mechanisms use shuttered optical connectors on the network equipment. This takes care of regulating emissions if a connector is removed from the equipment, but cannot prevent emissions on a cut fiber further away from the equipment. This is taken care of by a variety of automatic shutdown mechanisms that are designed into the network equipment. These mechanisms detect open connections and turn off lasers and/or optical amplifiers (the spontaneous emission from amplifiers may itself be large enough to cause damage). Several techniques are used to perform this function. If an amplifier senses a loss of signal at its input, it turns off its pump lasers to prevent any output downstream. There is some handshaking needed between the two ends of a failed link to handle unidirectional cuts. If one end

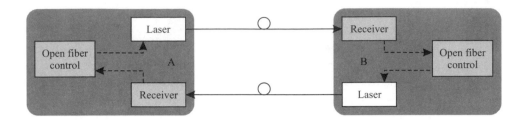

Figure 9.11   Open fiber control protocol in the Fibre Channel standard.

senses a loss of signal, it turns off its transmitter or amplifier in the other direction. This in turn allows the other end to detect a loss of signal and turn off its transmitter or amplifier. Another technique is to look at the back-reflected light. In the event of a fiber cut, the back-reflection increases and can be used to trigger a shutdown mechanism.

After the failure is repaired, the system can be brought up manually. More sophisticated *open fiber control* mechanisms allow the link to be brought back up automatically once the failure is repaired. These mechanisms typically pulse the link periodically to determine if the link has been repaired. The pulse power is maintained below the levels specified for the safety class. Here, we describe a particular protocol that has been chosen for the Fibre Channel standard.

9.7.1   Open Fiber Control Protocol

Figure 9.11 shows a block diagram of a system with two nodes A and B using the OFC protocol. Figure 9.12 shows the finite-state machine of the protocol.

The protocol works as follows:

1. Under normal operating conditions, A and B are in the ACTIVE state. If the link from A to B fails, receiver B detects a loss of light and turns off laser B, and B enters the DISCONNECT state. Receiver A subsequently detects a loss of light and turns off its laser and also enters the DISCONNECT state. Similarly, if the link from B to A fails, or if both links fail simultaneously, A and B both enter the DISCONNECT state.

2. In the DISCONNECT state, A transmits a pulse of duration τ every T seconds. B does the same. If A detects light while it is transmitting a pulse, it enters the STOP state and is called the *master*. If A detects light while it is not transmitting

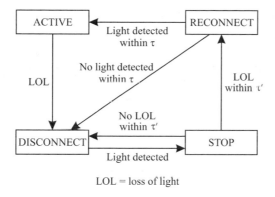

Figure 9.12 State machine run by each node for the open fiber control protocol in the Fibre Channel standard.

a pulse, it transmits a pulse for τ seconds and then enters the STOP state and is called the *slave*; likewise for B.

3. Upon entering the STOP state, the node turns off its laser for a period of τ' seconds. It remains in this state until a loss of light condition is detected on the incoming link. If this happens within the τ' seconds, it moves into the RECONNECT state. Otherwise, it moves back into the DISCONNECT state.

4. Upon entering the RECONNECT state, if the node is the master, it sends out a pulse of duration τ. If light is detected on the incoming link within this time period, the node enters the ACTIVE state. Otherwise, it shuts off its transmitter and enters the DISCONNECT state. If the node is the slave, it monitors the link for a period of τ seconds, and if light is detected on the incoming link within this period, it turns on its laser and enters the ACTIVE state. Otherwise, it goes back to the DISCONNECT state.

This is a fairly complex protocol. A simpler version of this protocol would not have the STOP and RECONNECT states. Instead, the nodes would directly enter the ACTIVE state from the DISCONNECT state upon detecting light. The reason for having the other states is to try to ensure that both nodes have functioning safety circuitry. If one of the nodes does not turn off its laser during the STOP period, it is assumed that the safety circuitry is not working and the other node goes back to the DISCONNECT state.

In order for the protocol to work, τ, τ', and T must be chosen carefully. In the DISCONNECT state, the average power transmitted is $\tau P/T$, where P is the transmitted power when the laser is turned on. This must be less than the allowed emission limits for the safety class. The values chosen for τ and τ' depend on the link propagation delay (see Problem 9.5).

Since the Class I safety standard also specifies that emission limits must be maintained during single fault conditions, the open fiber control circuitry at each node is duplicated for redundancy.

Summary

Network management is essential to operate and maintain any network. Operating costs dominate equipment costs for most telecom networks, making good network management imperative in ensuring the smooth operation of the network. The main functions of network management include configuration (of equipment and connections in the network), performance monitoring, and fault management. In addition, security and accounting are also management functions. Most functions of management are performed through a hierarchy of centralized management systems, but certain functions, such as restoration against failures, or the use of defect indicators to suppress alarms, are done in a distributed fashion. Several management protocols exist, the main ones being TL-1, SNMP, and CMIP.

It is useful to break down the optical layer into three sublayers: the optical channel layer, which deals with individual connections or lightpaths and is end to end across the network; the optical multiplex section layer, which deals with multiplexed wavelengths on a point-to-point link basis; and the optical transmission section layer, which deals with multiplexed wavelengths and the optical supervisory channel between adjacent amplifiers.

Interoperability between equipment from different vendors is a major issue facing the industry today. Initially the focus was on trying to get interoperability between vendors at the WDM level, but that has been recognized now as being very complex. Today the focus is on establishing interoperability by defining standard port-side single-wavelength interfaces at regenerator (or transponder) locations. There is also significant work under way to define optical layer overheads and their functions as well as to establish signaling and control protocol standards for controlling connections in the optical layer.

The level of transparency offered by the optical network affects the amount of management that can be performed. Key performance parameters such as the bit error rate can only be monitored in the electrical domain. Fast signaling methods need to be in place between network elements to perform some key management functions. These include the use of defect indicator signals to prevent the generation of unwanted alarms and protection-switching action, and other signaling bytes to control rapid protection switching. Optical path trace is another indicator that can be used to verify and manage connectivity in the network. Several methods exist for exchanging management information between nodes, including the optical supervisory channel, pilot tones, the use of certain overhead bytes in the SONET/SDH overhead, and the new digital wrapper overhead defined specifically for the optical layer.

Connection management in the optical network is slowly migrating from a centralized management-plane-based approach to a more distributed connection control plane approach using protocols similar to those used in IP and ATM networks.

Eye safety considerations are a unique feature of optical fiber communication systems. These considerations set an upper limit on the power that can be emitted from an open fiber, and these limits make it harder to design WDM systems, since they apply to the total power and not to the power per channel. Safety is maintained by using automated shutdown mechanisms in the network that detect failures and turn off lasers and amplifiers to prevent any laser radiation from exiting the system.

Further Reading

Network management is a vast subject, and several books have been written on the subject—see, for instance, [Sub00, Udu99, Bla95, AP94] for good introductions to the field, including descriptions of the various standards. [McG99, Wil00, Mae98] provide overviews of issues in optical network management.

There is currently a lot of interest in the standards bodies in standardizing many of the items we discussed in this chapter. The standards groups currently engaged in this are the International Telecommunications Union (ITU) study groups 13 and 15 (*www.itu.ch*), the American National Standards Institute (ANSI) T1X1.5 subcommittee (*www.ansi.org*), the Optical Internetworking Forum (OIF) (*www.oiforum.com*), the Internet Engineering Task Force (IETF) (*www.ietf.org*), Telcordia Technologies (*www.telcordia.com*), and the Network and Services Interoperability Forum (NSIF) (*www.atis.org/atis/sif/sifhom.htm*). The ITU defines the standards, including both SDH and the optical layer. ANSI provides the North American input to the ITU. IETF is the standards body for the Internet and is actively involved in defining optical layer control protocols. The OIF serves as a discussion forum for data communications equipment vendors, optical networking vendors, and service providers. Telcordia defined many of the SONET standards. NSIF has defined many of the management interfaces for facilitating interoperability in SONET. We have provided a list of relevant standards documents in Appendix C.

Pilot tones have been used in optical networks for several years now. See [Hil93, HFKV96, HK97] for a sampling of papers describing implementations of pilot tones for signal tracing and monitoring. [Epw95] uses pilot tones to control the gain of optical amplifiers.

ITU G.709 defines the digital wrapper including the associated maintenance signals such as the path trace and the defect indicators. Telcordia's GR-253 defines an equivalent set of signals for SONET.

Distributed protocols for connection management are commonly used in many types of networks; examples include PNNI in ATM networks [ATM96] and RSVP/CR-LDP [BZB+97, Abo01] in IP/MPLS networks. See [CGS93] for some early work and [RS97, Wei98] for related work on optical networks. Significant activity is under way currently toward defining extensions to IP control protocols to provide optical layer connection management. Many of these are contributions to the ITU, ANSI, IETF, and OIF and may be accessed from their Web sites. See also [GR00, AR01] for a discussion of the various types of control plane models.

Laser safety is covered by several standards, including ANSI, the International Electrotechnical Commission (IEC), the U.S. Food and Drug Administration (FDA), and the ITU [Ame88, Int93, Int00, US86, ITU99, ITU96].

Problems

9.1 Which sublayer within the optical layer would be responsible for handling the following functions?

(a) Setting up and taking down lightpaths in the network

(b) Monitoring and changing the digital wrapper overhead in a lightpath

(c) Rerouting all wavelengths (except the optical supervisory channel) from a failed fiber link onto another fiber link

(d) Detecting a fiber cable cut in a WDM line system

(e) Detecting failure of an individual lightpath

(f) Detecting bit errors in a lightpath

9.2 Consider the SONET network operating over the optical layer shown in Figure 9.13. Trace the path of the connection through the network, and show the termination of different layers at each network element.

9.3 Consider the network shown in Figure 9.14. Suppose the link segment between OLT A and amplifier B fails.

(a) Assume that each node detects loss of light in 2 ms and waits 5 ms before it sends an FDI signal downstream. Also, each node waits for 2 s after the loss of light is detected before it triggers an alarm. Assume that the propagation delay on each link segment (segment defined as the part of the link between adjacent amplifiers or between an OLT and adjacent amplifier) is 3 ms.

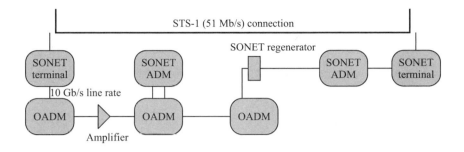

Figure 9.13 A combined SONET/WDM optical network for Problem 9.2.

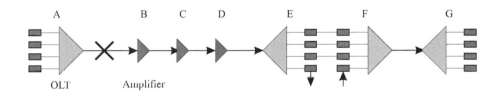

Figure 9.14 Example for Problem 9.3.

Draw a time line indicating the behavior of each node in the network after
the failure, including the transmission of OCh-FDI and OMS-FDI signals.

(b) Now assume that each node detects loss of light in 2 ms, immediately sends
an FDI signal downstream, and waits an additional 2 s after the loss of light
is detected before it triggers an alarm. Assume the same propagation delay
values as before. Redraw the time line indicating the behavior of each node
in the network after the failure, including the transmission of OCh-FDI and
OMS-FDI signals.

What do you observe as the difference between the two methods proposed
above?

9.4 Consider an OXC connected to multiple OLTs.

(a) If the OXC has an electronic switch core with optical-to-electrical conversions at its ports, what overhead techniques can it use? How would it communicate with other such OXCs in the network? What performance parameters
could it monitor?

(b) If the OXC is all optical, with no optical-to-electrical conversions, what overhead techniques can it use? How would it communicate with other such OXCs in the network? What performance parameters could it monitor?

9.5 Consider the open fiber control protocol in the Fibre Channel standard.

(a) How would you choose the parameters τ and τ' as a function of the maximum link propagation delay d_{prop}?

(b) What is the time taken for a node to go from the DISCONNECT state to the ACTIVE state, assuming a successful reconnection attempt, that is, it never has to go back to the DISCONNECT state?

▬ References

[Abo01]   O. Aboul-Magd et al. *Constraint-Based LSP Setup Using LDP*. Internet Engineering Task Force, 2001. *draft-ietf-mpls-cr-ldp-05.txt*.

[Ame88]   American National Standards Institute. Z136.2. *Safe Use of Optical Fiber Communication Systems Utilizing Laser Diodes and LED Sources*, 1988.

[AP94]   S. Aidarus and T. Plevyak, editors. *Telecommunications Network Management into the 21st Century*. IEEE Press, Los Alamitos, CA, 1994.

[AR01]   D. Awduche and Y. Rekhter. Multiprotocol lambda switching: Combining MPLS traffic engineering control with optical crossconnects. *IEEE Communications Magazine*, 39(4):111–116, Mar. 2001.

[ATM96]   ATM Forum. *Private Network-Network Interface Specification: Version 1.0*, 1996.

[Bla95]   U. Black. *Network Management Standards*. McGraw-Hill, New York, 1995.

[BZB+97]   R. Bradon, L. Zhang, S. Berson, S. Herzog, and S. Jamin. *Resource Reservation Protocol—Version 1 Functional Specification*. Internet Engineering Task Force, Sept. 1997.

[CGS93]   I. Cidon, I. S. Gopal, and A. Segall. Connection establishment in high-speed networks. *IEEE/ACM Transactions on Networking*, 1(4):469–482, Aug. 1993.

[Epw95]   R. E. Epworth. Optical transmission system. U.S. Patent 5463487, 1995.

[GR00]   J. Gruber and R. Ramaswami. Towards agile all-optical networks. *Lightwave*, Dec. 2000.

[HFKV96]   F. Heismann, M. T. Fatehi, S. K. Korotky, and J. J. Veselka. Signal tracking and performance monitoring in multi-wavelength optical networks. In *Proceedings of European Conference on Optical Communication*, pages 3.47–3.50, 1996.

[Hil93]   G. R. Hill et al. A transport network layer based on optical network elements. *IEEE/OSA Journal on Lightwave Technology*, 11:667–679, May/June 1993.

[HK97]   Y. Hamazumi and M. Koga. Transmission capacity of optical path overhead transfer scheme using pilot tone for optical path networks. *IEEE/OSA Journal on Lightwave Technology*, 15(12):2197–2205, Dec. 1997.

[Int93]   International Electrotechnical Commission. *60825-1: Safety of Laser Products—Part 1: Equipment Classification, Requirements and User's Guide*, 1993.

[Int00]   International Electrotechnical Commission. *60825-2: Safety of Laser Products—Part 2: Safety of Optical Fiber Communication Systems*, 2000.

[ITU96]   ITU-T SG15/WP 4. *Rec. G.681: Functional Characteristics of Interoffice and Long-Haul Line Systems Using Optical Amplifiers, Including Optical Multiplexing*, 1996.

[ITU99]   ITU-T. *Rec. G.664: Optical Safety Procedures and Requirements for Optical Transport Systems*, 1999.

[Mae98]   M. Maeda. Management and control of optical networks. *IEEE Journal of Selected Areas in Communications*, 16(6):1008–1023, Sept. 1998.

[McG99]   A. McGuire. Management of optical transport networks. *IEE Electronics and Communication Engineering Journal*, 11(3):155–163, June 1999.

[RS97]   R. Ramaswami and A. Segall. Distributed network control for optical networks. *IEEE/ACM Transactions on Networking*, Dec. 1997.

[Sub00]   M. Subramanian. *Network Management: Principles and Practice*. Addison-Wesley, Reading, MA, 2000.

[Udu99]   D. K. Udupa. *TMN Telecommunications Management Network*. McGraw-Hill, New York, 1999.

[US86]   U. S. Food and Drug Administration, Department of Radiological Health. *Requirements of 21 CFR Chapter J for Class 1 Laser Products*, Jan. 1986.

[Wei98]   Y. Wei et al. Connection management for multiwavelength optical networking. *IEEE Journal of Selected Areas in Communications*, 16(6):1097–1108, Sept. 1998.

[Wil00]   B. J. Wilson et al. Multiwavelength optical networking management and control. *IEEE/OSA Journal on Lightwave Technology*, 18(12):2038–2057, 2000.

chapter

Network Survivability

PROVIDING RESILIENCE AGAINST FAILURES is an important requirement for many high-speed networks. As these networks carry more and more data, the amount of disruption caused by a network-related outage becomes more and more significant. A single outage can disrupt millions of users and result in millions of dollars of lost revenue to users and operators of the network.

As part of the service-level agreement between a carrier and its customer leasing a connection, the carrier commits to providing a certain *availability* for the connection. A common requirement is that the connection be available 99.999% (five 9s) of the time. This requirement corresponds to a connection downtime of less than 5 minutes per year.

A connection is routed through many nodes in the network between its source and its destination, and there are many elements along its path that can fail. The only practical way of obtaining 99.999% availability is to make the network *survivable*, that is, able to continue providing service in the presence of failures. *Protection switching* is the key technique used to ensure survivability. These protection techniques involve providing some redundant capacity within the network and automatically rerouting traffic around the failure using this redundant capacity. A related term is *restoration*. Some people apply the term *protection* when the traffic is restored in the tens to hundreds of milliseconds, and use the term *restoration* to schemes where traffic is restored on a slower time scale. However, we do not distinguish between protection and restoration in this chapter.

Protection is usually implemented in a distributed manner without requiring centralized control in the network. This is necessary to ensure fast restoration of service after a failure.

We will be concerned with failures of network links, nodes, and individual channels (in the case of a WDM network). In addition, the software residing in today's network elements is immensely complex, and reliability problems arising from software bugs has become a serious issue. This is something that is usually dealt with by using proper software design and is hard to protect against in the network.

In most cases failures are triggered by human error, such as a backhoe cutting through a fiber cable, or an operator pulling out the wrong connection or turning off the wrong switch. Links fail mostly because of fiber cuts. This is the most likely failure event. There were 136 such failures reported by U.S. carriers to the Federal Communications Commission in 1997. Fiber that is deployed inside of oil and gas pipelines is less likely to be cut than fiber that is buried directly in the ground or strung on poles. For instance, Williams Communications, which runs fiber beside oil pipelines, has experienced only a single fiber cut since 1986.

The next most likely failure event is the failure of active components inside network equipment, such as transmitters, receivers, or controllers. In general, network equipment is designed with redundant controllers. Moreover, failure of controllers doesn't affect traffic but only impacts management visibility into the network.

Node failures are another possibility to be reckoned with. Entire central offices can fail, usually because of catastrophic events such as fires or flooding. These events are rare, but they cause widespread disruption when they occur. Examples include the fire at the Hinsdale central office of Illinois Bell in 1988 and the flooding of several central offices due to Hurricane Floyd in 1999.

Protection schemes are also used extensively to allow maintenance actions in the network. For example, in order to service a link, typically the traffic on the link is switched over to an alternate route using the protection scheme before it is serviced. The same technique is used when nodes or links are upgraded in the network.

In most cases, the protection schemes are engineered to protect against a single failure event or maintenance action. If the network is large, we may need to provide the capability to deal with more than one concurrent failure or maintenance action. One way to handle this is to break up the network into smaller subnetworks and restrict the operation of the protection scheme to within a subnetwork. This allows one failure per subnetwork at any given time. Another way to deal with this issue is to ensure that the mean time to repair a failure is much smaller than the mean time between failures. This ensures that, in most cases, the failed link will be repaired before another failure happens. Some of the protection schemes that we will study do, however, protect the network against some types of simultaneous multiple failures.

The restoration times required depend on the application/type of data being carried. For SONET/SDH networks, the maximum allowed restoration time is 60 ms. This restoration time requirement came from the fact that some equipment in the network drops voice calls if the connection is disrupted for a period significantly longer than 60 ms. Over time, operators have gotten used to being able to achieve restoration on these time scales. However, in a world dominated by data, rather than voice traffic, the 60 ms number may not be a hard requirement, and operators may be willing to tolerate somewhat larger restoration times, particularly if they see other benefits as a result, such as higher bandwidth efficiency, which in turn would lead to lower operating costs. On the other hand, another point of view is that the restoration time requirements could get more stringent as data rates in the network increase. A downtime of 1 second at 10 Gb/s corresponds to losing over a gigabyte of data. Most IP networks today provide services on a best-effort basis and do not guarantee availability; that is, they try to route traffic in the network as best as they can, but packets can have random delays through the network and can be dropped if there is congestion.

Survivability can be addressed within many layers in the network. Protection can be performed at the physical layer, or layer 1, which includes the SONET/SDH and the optical layers. Protection can also be performed at the link layer, or layer 2, which includes the ATM layer and the MPLS layer that is part of IP networks. Finally, protection can also be performed at the network layer, or layer 3, such as the IP layer. There are several reasons why this is the case. For instance, each layer can protect against certain types of failures but probably not protect against all types of failures effectively. We will focus primarily on layer 1 restoration in this chapter, but also briefly discuss the protection techniques applicable to layers 2 and 3.

The rest of this chapter is organized as follows. We start by outlining the basic concepts behind protection schemes. Many of the protection techniques used in today's telecommunication networks were developed for use in SONET and SDH networks, and we will explore these techniques in detail. We will also look at how protection is implemented in today's IP networks. Following this, we will look at protection functions in the optical layer in detail, and then discuss how protection functions in the different layers of the network can work together.

10.1 Basic Concepts

A great variety of protection schemes are used in today's networks. We will talk about *working* paths and *protect* paths. Working paths carry traffic under normal operation; protect paths provide an alternate path to carry the traffic in case of failures. Working and protection paths are usually diversely routed so that both paths aren't lost in case of a single failure.

Protection schemes are designed to operate over a range of network topologies. Some work on point-to-point links. Ring topologies are particularly popular in

SONET/SDH. A ring is the simplest topology offering an alternate route around a failure. In the optical layer, many protection schemes have been designed to operate over true mesh topologies.

Protection may be *dedicated* or *shared*. In dedicated protection, each working connection is assigned its own dedicated bandwidth in the network over which it can be rerouted in case of a failure. In shared protection, we make use of the fact that not all working connections in the network fail simultaneously (for example, if they are in different parts of the network). Therefore, by careful design, we can make multiple working connections share protection bandwidth among themselves. This helps reduce the amount of bandwidth needed in the network for protection. Another advantage of shared protection is that the protection bandwidth is available to carry low-priority traffic under normal conditions. This low-priority traffic is discarded in the event of a failure when the bandwidth is needed to protect a connection.

Protection schemes can either be *revertive* or *nonrevertive*. In both schemes, if a failure occurs, traffic is switched from the working path to the protect path. In a nonrevertive scheme, the traffic remains on the protect path until it is manually switched back onto the original working path, usually by a user through the network management system. In a revertive scheme, once the working path is repaired, the traffic is automatically switched back from the protect path onto the working path. Reversion allows the network to return to its original state once the failure is restored. Dedicated protection schemes may be revertive or nonrevertive; however, shared protection schemes are usually revertive. Since multiple working connections share a common protection bandwidth, the protection bandwidth must be freed up as soon as possible after the original failure has been repaired, so that it can be used to protect other connections in the event of another failure occurring.

To confuse terminology further, the protection switching can be *unidirectional* or *bidirectional*. This is not to be confused with unidirectional transmission or bidirectional transmission over a fiber. Figure 10.1 illustrates the two schemes for the case where two fiber pairs are used on the point-to-point link, with each fiber carrying traffic in one direction (unidirectional transmission). In unidirectional protection switching, each direction of traffic is handled independent of the other. Thus in the event of a single fiber cut, only one direction of traffic is switched over to the protection fiber and the other direction remains on the original working fiber. In bidirectional switching, both directions are switched over to the protection fibers. For the case where bidirectional transmission is used, the switching mostly becomes bidirectional by default because both directions of traffic are lost when a fiber is cut (both directions may not be lost if there is an equipment failure, rather than a fiber cut).

Unidirectional protection switching is used in conjuction with dedicated protection schemes since it can be implemented very easily by switching the traffic at the

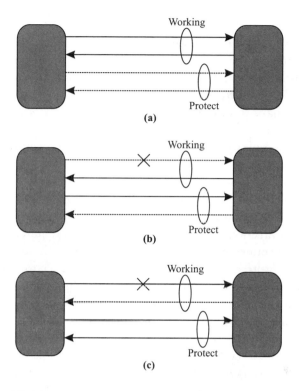

Figure 10.1   Unidirectional and bidirectional protection switching. (a) The link is shown under normal operation. (b) Unidirectional protection switching. After a unidirectional fiber cut, only the affected direction of traffic is switched over to the protection fiber. (c) Bidirectional protection switching. After a unidirectional fiber cut, both directions of traffic are switched over to the protection fibers.

receiving end from the working to the protect path, without requiring a signaling protocol between the receiver and the transmitter. For example, in Figure 10.1, if a fiber carrying traffic from left to right is cut, without affecting the fiber carrying traffic from right to left, the transmitter on the left is not aware that there has been a failure. In the case of unidirectional dedicated protection, if traffic is transmitted simultaneously on the working and protect paths, the receiver at the end of the paths simply selects the better of the two arriving signals. However, if bidirectional switching is required, the receiver needs to inform the transmitter that there has been a cut. This requires a signaling protocol, called an *automatic protection-switching* (APS) protocol.

A simple APS protocol works as follows: if a receiver in a node detects a fiber cut, it turns off its transmitter on the working fiber and then switches over to the protection fiber to transmit traffic. The receiver at the other node then also detects the loss of signal on the working fiber and then switches its traffic over to the protection fiber. Actual APS protocols used in SONET and optical networks are quite a bit more complicated because they have to deal with many different possible scenarios than the one described here.

In a bidirectional communication system, where traffic is transmitted in both directions over a single fiber, a fiber cut will be detected by both the source and the destination. While no APS protocol is required to deal with fiber cuts, an APS protocol will still be needed to deal with unidirectional equipment failures and to support other maintenance functions.

In the case of shared protection schemes, an APS protocol is required to coordinate access to the shared protection bandwidth. Therefore most shared protection schemes use bidirectional protection switching because it is easier to control and manage in a more complex network than unidirectional switching.

There is also the question of how and where the traffic is rerouted in the event of a failure. Here we distinguish between *path* switching, *span* switching, and *ring* switching. Figure 10.2 illustrates these concepts. In path switching (Figure 10.2(b)), the connection is rerouted end to end from its source to its destination along an alternate path. In span switching (Figure 10.2(c)), the connection is rerouted on a spare link between the nodes adjacent to the failure. In ring switching (Figure 10.2(d)), the connection is rerouted on a ring between the nodes adjacent to the failure.

Finally, different protection schemes operate at different layers in the network (for example, SONET/SDH, ATM, MPLS, IP) and at different sublayers within a layer. For example, there are schemes that protect one connection at a time, as well as schemes that protect all connections on a failed fiber together. In SONET/SDH networks, the former schemes operate at the path layer, and the latter schemes operate at the line (multiplex section in SDH) layer. In many cases, path layer schemes operate end to end, rerouting traffic along an alternate path all the way from the source to the destination. In contrast, line layer schemes are almost all localized—that is, they reroute traffic around the failed link. Similarly, in the optical layer, we have schemes operating either at the optical channel layer or the optical multiplex section layer.

10.2 Protection in SONET/SDH

A major accomplishment of SONET and SDH network deployment was to provide a significant improvement in the availability and reliability of the overall network. This was done through the use of an extensive set of protection techniques. Similar

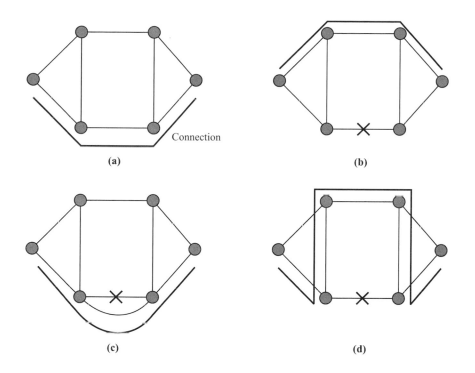

Figure 10.2   Path, span, and ring switching. (a) Working path for the connection under normal operation. (b) Path switching, where the connection is rerouted end to end on an alternate path. (c) Span switching, where the connection is rerouted on a spare link between the nodes adjacent to the failure. (d) Ring switching, where the connection is rerouted on a ring between the nodes adjacent to the failure.

schemes are used in both SONET and SDH, but their nomenclature is different. We will specify both nomenclatures but use the SONET nomenclature for the most part.

A taxonomy of the different protection schemes is given in Table 10.1. We will start by describing the different types of protection mechanisms that are used for simple point-to-point links, and then discuss how these can be applied for networks. Each protection scheme can be associated with a specific layer in the network. As we saw in Chapter 6, the SONET layer includes a *path* layer and a *line* layer. Both path layer and line layer protection schemes are used in practice. Equivalently, SDH networks use both *channel* layer and *multiplex section* (MS) layer protection schemes. A path layer protection scheme operates on individual paths or connections in the network. For example, in an OC-48 (2.5 Gb/s) ring supporting STS-1 (51 Mb/s) connections, a path layer scheme would treat each STS-1 connection independently

Table 10.1   A summary of protection schemes in SONET and SDH. N denotes the number of working interfaces that share a single protection interface. The schemes operate either in the path layer or in the SONET line layer/SDH multiplex section (MS) layer. Path layer ring schemes include unidirectional path-switched ring (UPSR) or $1 + 1$ subnetwork connection protection (SNCP). Line layer ring schemes include bidirectional line-switched ring (BLSR) or, equivalently, multiplexed section-shared protection ring (MS-SPRing).

	Protection Scheme				
SONET Term	$1 + 1$	$1:N$	UPSR		BLSR
SDH Term	$1 + 1$	$1:N$		SNCP	MS-SPRing
Type	Dedicated	Shared	Dedicated	Dedicated	Shared
Topology	Point-point	Point-point	Ring	Ring/mesh	Ring
Layer	Line/MS	Line/MS	Path/–	–/path	Line/MS

and switch them independently of each other. A line layer scheme on the other hand, operates on the entire set of connections at once and generally does not distinguish between the different connections that are part of the aggregate signal. In the former example, a line layer protection scheme in an OC-48 ring would switch all the connections within the OC-48 together. (There are some exceptions to this statement. The bidirectional line-switched rings (BLSRs) that we will study later do allow bits to be set for each connection. In the event of a failure, only those connections that are specified are switched. This is needed to ensure that some connections can be left unprotected if so desired, and also to handle node failures, as we will see in Section 10.2.4.)

10.2.1   Point-to-Point Links

Two fundamental types of protection mechanisms are used in point-to-point links: $1 + 1$ protection and 1:1 or, more generally, $1:N$ protection, as shown in Figure 10.3. Both operate in the line or multiplex section layer.

In $1 + 1$ protection, traffic is transmitted simultaneously on two separate fibers (usually over disjoint routes) from the source to the destination. Assuming unidirectional protection switching, the destination simply selects one of the two fibers for reception. If that fiber is cut, the destination simply switches over to the other fiber and continues to receive data. This form of protection is very fast and requires no signaling protocol between the two ends. Note that since connections are usually full duplex, there is actually a pair of fibers between the two nodes, say, node A and

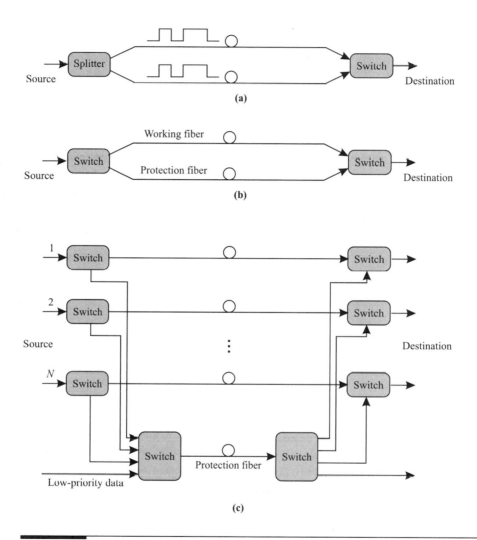

Figure 10.3   Different types of protection techniques for point-to-point links: (a) $1 + 1$ protection, where the signal is simultaneously transmitted over two paths; (b) 1:1 protection, where the signal is transmitted over a working path under normal conditions but switched to a protect path after a failure; and (c) 1:N protection, which is a more generalized form of 1:1 protection, where N working paths share a single protection path.

node B for the working traffic. One fiber carries traffic from A to B, and the other carries traffic from B to A. Likewise there is another pair of fibers for protection traffic. Node A's receiver and node B's receiver can make the switching decisions independently.

In 1:1 protection, there are still two fibers from the source to the destination. However, traffic is transmitted over only one fiber at a time, say, the working fiber. If that fiber is cut, the source and destination both switch over to the other protection fiber. As we discussed earlier, an APS protocol is required for signaling between the source and destination. For this reason, 1:1 protection is not as quick as unidirectional $1 + 1$ protection in restoring traffic because of the added communication overhead involved. However, it offers two main advantages over $1 + 1$ protection. The first is that under normal operation, the protection fiber is unused. Therefore, it can be used to transmit lower-priority traffic. This lower-priority traffic must be discarded if the working fiber is cut. SONET and SDH equipment in the field does provide support for this lower-priority or *extra traffic*. This capability is not widely used today, but carriers in the past have used this capability on occasion to carry "lower-priority" data traffic or even voice traffic, when their networks are temporarily over capacity. This is likely to change in the future with the advent of data services, as we shall see in Section 10.4. Best-effort data services, in particular, can use this capability.

Another advantage is that the 1:1 protection can be extended so as to share a single protection fiber among many working fibers. In a more general 1:N protection scheme, N working fibers share a single protection fiber. This arrangement can handle the failure of any single working fiber. Note that in the event of multiple failures, the APS protocol must ensure that only traffic on one of the failed fibers is switched over to the protection fiber.

In the previous discussion we talked about how the protection is done, but skimmed over what the triggers are for initiating protection switching. In SONET/SDH, the incoming signal is continuously monitored. Protection switching is initiated if a signal fail or a signal degrade condition is detected on the line. A signal fail represents a hard failure and is detected typically as a loss of signal or as a loss of the SONET/SDH frame. Out of the 60 ms allowed for restoration, detecting the failure and initiating protection switching must be performed within 10 ms.

10.2.2 Self-Healing Rings

Ring networks have become very popular in the carrier world as well as in enterprise networks. A ring is the simplest topology that is *2-connected*, that is, provides two separate paths between any pair of nodes that do not have any nodes or links in common except the source and destination nodes. This allows a ring network to be resilient to failures. Rings are also efficient from a fiber layout perspective—multiple sites can be interconnected with a single physical ring. In contrast, a hubbed approach would require fibers to be laid between each site and a hub node, and would require two disjoint routes between each site and the hub, which is a more expensive proposition.

Much of the carrier infrastructure today uses SONET/SDH rings. These rings are called *self-healing* since they incorporate protection mechanisms that automatically detect failures and reroute traffic away from failed links and nodes onto other routes rapidly. The rings are implemented using SONET/SDH add/drop multiplexers (ADMs), which we studied in Section 6.1. These ADMs selectively drop and add traffic from/to the ring as well as protect the traffic against failures.

The different types of ring architectures differ in two aspects: in the directionality of traffic and in the protection mechanisms used. A *unidirectional* ring carries working traffic in only one direction of the ring (say, clockwise), as shown in Figure 10.4. Working traffic from node A to node B is carried clockwise along the ring, and working traffic from B to A is also carried clockwise, on a different set of links in the ring. A *bidirectional* ring carries working traffic in both directions. Figure 10.5 shows a four-fiber bidirectional ring. Working traffic from A to B is carried clockwise, and working traffic from B to A is carried counterclockwise along the ring. Note that in both unidirectional and bidirectional SONET/SDH rings, all connections are bidirectional and use up the same amount of bandwidth in both directions. The two directions of a connection are routed differently based on the type of ring, as we discussed earlier.

The SONET/SDH standards dictate that in SONET/SDH rings, service must be restored within 60 ms after a failure. This time includes several components: the time needed to detect the failure, for which 10 ms is allocated; the time needed to signal to other nodes in the network (if needed), including the propagation delays; the actual switching time; and the time to reacquire the frame synchronization after the switch-over has occurred.

Three ring architectures have been widely deployed: two-fiber unidirectional path-switched rings (UPSR), four-fiber bidirectional line-switched rings (BLSR/4), and two-fiber bidirectional line-switched rings (BLSR/2). In SDH, the 1 + 1 path protection has been defined to operate in a more general mesh topology and is called subnetwork connection protection (SNCP). SDH multiplex section shared protection ring/4 (MS-SPRing/4) and MS-SPRing/2 are similar to BLSR/4 and BLSR/2, respectively. Table 10.2 summarizes the features of the different architectures, which we will discuss in detail in the following sections.

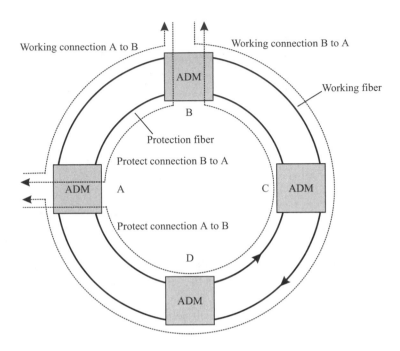

Figure 10.4   A unidirectional path-switched ring (UPSR). One of the fibers is considered the working fiber and the other the protection fiber. Traffic is transmitted simultaneously on the working fiber in the clockwise direction and on the protection fiber in the counterclockwise direction. Protection is done at the path layer.

Table 10.2   Comparison of different types of self-healing rings.

Parameter	UPSR SNCP	BLSR/4 MS-SPRing/4	BLSR/2 MS-SPRing/2
Fiber pairs	1	2	1
TX/RX pairs/node	2	4	2
Protection type	Dedicated	Shared	Shared
Protection capacity	= Working capacity	= Working capacity	= Working capacity
Link failure	Path switch	Span/ring switch	Ring switch
Node failure	Path switch	Ring switch	Ring switch
Restoration speed	Faster	Slower	Slower
Implementation	Simple	Complex	Complex

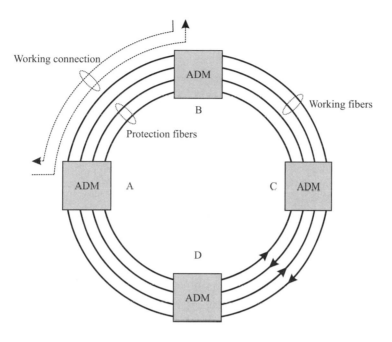

Figure 10.5   A four-fiber bidirectional line-switched ring (BLSR/4). The ring has two working fibers and two protection fibers. Traffic between two nodes is transmitted normally on the shortest path between them, and either span or ring switching is used to restore service after a failure.

10.2.3   *Unidirectional Path-Switched Rings*

Figure 10.4 shows a UPSR. One fiber is used as the working fiber and the other as the protection fiber. Traffic from node A to node B is sent simultaneously on the working fiber in the clockwise direction and on the protection fiber in the counterclockwise direction. The protection is performed at the path layer for each connection as follows. Node B continuously monitors both the working and protection fiber and selects the better signal between the two for each SONET connection. Under normal operation, suppose node B receives traffic from the working fiber. If there is a link failure, say, of link AB, then B will switch over to the protection fiber and continue to receive the data. Note that the switch-over is done on a connection-by-connection basis (see Problem 10.8). Observe that this is essentially like the $1 + 1$ scheme that we studied earlier, except that it is operating at the path layer in a ring rather than at the line layer in a point-to-point configuration.

Note that this protection scheme easily handles failures of links, transmitters/receivers, or nodes. It is simple to implement and requires no signaling protocol or communication between the nodes. The capacity required for protection purposes is equal to the working capacity. This will turn out to be the case for the other ring architectures as well.

The main drawback with the UPSR is that it does not spatially reuse the fiber capacity. This is because each (bidirectional) connection uses up capacity on every link in the ring and has dedicated protection bandwidth associated with it. Thus, there is no sharing of the protection bandwidth between connections. For example, suppose each connection requires 51 Mb/s (STS-1) of bandwidth and the ring operates at 622 Mb/s (OC-12). Then the ring could support a total of twelve 51 Mb/s connections. The BLSR architectures that we will study next do incorporate spatial reuse and can support aggregate traffic capacities higher than the transmission rate.

UPSRs are popular topologies in lower-speed local exchange and access networks, particularly where the traffic is primarily hubbed from the access nodes into a hub node in the carrier's central office. In this case, we will see that the traffic carrying capacity that a UPSR can support is the same as what the more complicated ring architectures incorporating spatial reuse can support. This makes the UPSR an attractive option for such applications due to its simplicity and, thus, lower cost. Typical ring speeds today are OC-3 (STM-1) and OC-12 (STM-4). There is no specified limit on the number of nodes in a UPSR or on the ring length. In practice, the ring length will be limited by the fact that the clockwise and counterclockwise path taken by a signal will have different delays associated with them, which in turn, will affect the restoration time in the event of a failure.

A UPSR is essentially $1 + 1$ protection implemented at the path layer in a ring.

10.2.4 *Bidirectional Line-Switched Rings*

BLSRs are much more sophisticated than UPSRs and incorporate additional protection mechanisms, as we will see below. Unlike a UPSR, they operate at the line or multiplex section layer. The BLSR equivalent in the SDH world is called a multiplex section shared protection ring (MS-SPRing).

Figure 10.5 shows a four-fiber BLSR. Two fibers are used as working fibers, and two are used for protection. Unlike a UPSR, working traffic in a BLSR can be carried on both directions along the ring. For example, on the working fiber, traffic from node A to node B is carried clockwise along the ring, whereas traffic from B to A is carried counterclockwise along the ring. Usually, traffic belonging to both directions of a connection is routed on the shortest path between the two nodes in the ring. However, in certain cases [Kha97, LC97], traffic may be routed along the longer path to reduce network congestion and make better use of the available capacity.

A BLSR can support up to 16 nodes, and this number is limited by the 4-bit addressing field used for the node identifier. The maximum ring length is limited to 1200 km (6 ms propagation delay) because of the requirements on the restoration time in the case of a failure. For longer rings, particularly for undersea applications, the 60 ms restoration time has been relaxed.

A BLSR/4 employs two types of protection mechanisms: *span switching* and *ring switching*. In span switching, if a transmitter or receiver on a working fiber fails, the traffic is routed onto the protection fiber between the two nodes on the same link, as shown in Figure 10.6. (Span switching can also be used to restore traffic in the event of a working fiber cut, provided the protection fibers on that span are routed separately from the working fibers. However, this is usually not the case.) In case of a fiber or cable cut, service is restored by ring switching, as illustrated in Figure 10.7. Suppose link AB fails. The traffic on the failed link is then rerouted by nodes A and B around the ring on the protection fibers. Ring switching is also used to protect against a node failure.

A BLSR/2, shown in Figure 10.8, can be thought of as a BLSR/4 with the protection fibers "embedded" within the working fibers. In a BLSR/2, both of the fibers are used to carry working traffic, but half the capacity on each fiber is reserved for protection purposes. Unlike a BLSR/4, span switching is not possible here, but ring switching works in much the same way as in a BLSR/4. In the event of a link failure, the traffic on the failed link is rerouted along the other part of the ring using the protection capacity available in the two fibers. As with 1:1 protection on point-to-point links, an advantage of BLSRs is that the protection bandwidth can be used to carry low-priority traffic during normal operation. This traffic is preempted if the bandwidth is needed for service restoration.

BLSRs provide spatial reuse capabilities by allowing protection bandwidth to be shared between spatially separated connections. The spatial reuse achievable in a best-case scenario is illustrated in Figure 10.9. As in the UPSR example above, consider a BLSR/2 operating at 622 Mb/s (OC-12), supporting 51 Mb/s STS-1 connections. The figure shows a ring with four nodes and STS-1 connections between each pair of adjacent nodes. Note that all four of these connections can be protected by dedicating 51 Mb/s of bandwidth around the ring that is shared by all these connections. This is because these connections do not overlap spatially and thus do not need to be restored simultaneously, as long as we are dealing with only single-failure conditions. In this example, the 622 Mb/s ring could thus support a total of 24 such

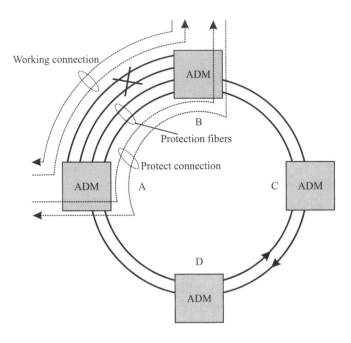

Figure 10.6 Illustrating span switching in a BLSR/4. Traffic is switched from the working fiber pair to the protection fiber pair on the same span.

51 Mb/s connections (6 connections per link; note that only half the capacity is available for working traffic, over four links), as compared to just 12 for an equivalent UPSR. This capacity increases as the number of nodes in the rings increases. An 8-node OC-12 BLSR/2 could support 48 STS-1 connections in the example above.

Thus BLSRs are more efficient than UPSRs in protecting distributed traffic patterns. Their efficiency comes from the fact that the protection capacity in the ring is shared among all the connections, as we saw above. For this reason, BLSRs are widely deployed in long-haul and interoffice networks, where the traffic pattern is more distributed than in access networks. Today, these rings operate at OC-12 (STM-16), OC-48 (STM-16), and OC-192 (STM-64) speeds. Most metro carriers have deployed BLSR/2s, while many long-haul carriers have deployed BLSR/4s. BLSR/4s can handle more failures than BLSR/2s. For example, a BLSR/4 can simultaneously handle one transmitter failure on each span in the ring. It is also easier to service than a BLSR/2 ring because multiple spans can be serviced independently without taking down the ring. However, ring management in a BLSR/4 is more complicated than in a BLSR/2 because multiple protection mechanisms have to be coordinated.

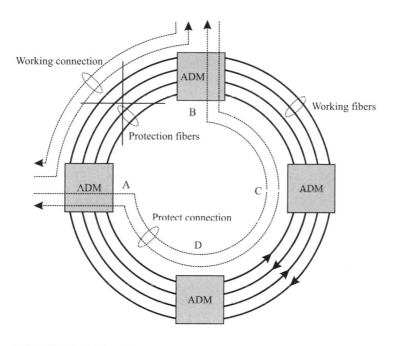

Figure 10.7   Illustrating ring switching in a BLSR/4. Traffic is rerouted around the ring by the nodes adjacent to the failure.

BLSRs are significantly more complex to implement than UPSRs. They require extensive signaling between the nodes for many reasons, as we will see below. This signaling is done using the K_1/K_2 bytes in the SONET overhead (see Chapter 6).

Handling Node Failures in BLSRs

So far, we have dealt primarily with how to handle failures of links, such as those occurring from a fiber cut. Failures of nodes are usually less likely because, in many cases, redundant configurations (such as dual power supplies and switch fabrics) are used. However, nodes may still fail because of some catastrophic events or human errors. Handling node failures complicates the BLSR restoration mechanism. The failure of a node is seen by all its adjacent nodes as failures of the links that connect them to the failed node. If each of these adjacent nodes performs restoration assuming that it is a single link failure, there can be undesirable consequences. One example is shown in Figure 10.10. Here, when node 1 fails, nodes 6 and 2 assume it is a link failure and attempt to reroute the traffic around the ring (ring switching) to restore service. This causes erroneous connections, as shown in the figure. The

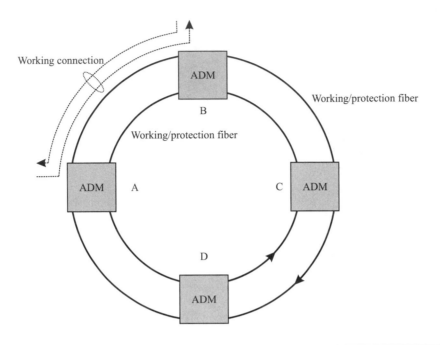

Figure 10.8   A two-fiber bidirectional line-switched ring (BLSR/2). The ring has two fibers and half the bandwidth. Ring switching is used to restore service after a failure.

only way to prevent such occurrences is to ensure that the nodes performing the restoration determine the type of failure before invoking their restoration mechanisms. This would require exchanging messages between the nodes in the network. In the preceding example, nodes 6 and 2 could first try to exchange messages around the ring to determine if they have both recorded link failures and, if so, invoke the appropriate restoration procedure. This restoration procedure can avoid these misconnections by not attempting to restore any traffic that originates or terminates at the failed node. This is called *squelching*. Thus each node in a BLSR maintains squelch tables that indicate which connections need to be squelched in the event of node failures. The price paid for this is a slower restoration time because of the coordination required between the nodes to determine the appropriate restoration mechanism to be invoked.

Low-Priority Traffic in BLSRs

Just as we saw with 1:1 protection earlier, BLSRs can use the protection bandwidth to carry low-priority or extra traffic, under normal operation. This extra traffic is lost

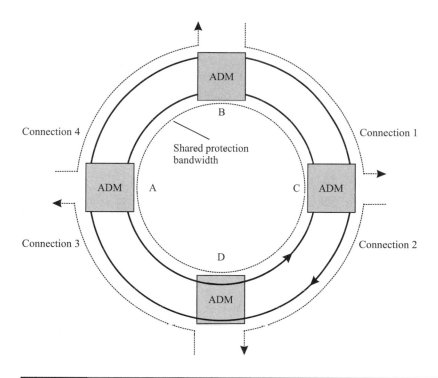

Figure 10.9   Spatial reuse in a BLSR. Multiple working connections can share protection bandwidth around the ring as long as they do not overlap on any link.

in the event of a failure. However, this feature requires additional signaling between the nodes in the event of a failure to indicate to the other nodes that they should operate in protection mode and throw away the low-priority traffic.

10.2.5   *Ring Interconnection and Dual Homing*

A single ring is only a part of the overall network. The entire network typically consists of multiple rings interconnected with each other, and a connection may have to be routed through multiple rings to get to its destination. The interconnection of these rings is thus an important aspect to be considered. The simplest way for rings to interoperate is to connect the drop sides of two ADMs on different rings back to back, as shown in Figure 10.11. The interconnection is done using signals typically at lower bit rates than the line bit rate. For instance, two OC-12 UPSRs may be interconnected by DS3 signals. In many cases, a digital crossconnect is interspersed between the two rings to provide additional grooming and multiplexing capabilities.

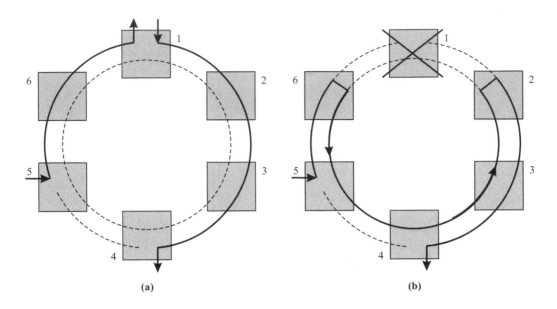

(a)   **(b)**

Figure 10.10   Erroneous connections due to the failure of a node being treated by its adjacent nodes as link failures: (a) Normal operation, with a connection from node 5 to node 1 and another connection from node 1 to node 4. (b) After node 1 fails, nodes 6 and 2 invoke ring switching independently. This causes a connection to be set up erroneously between node 5 and node 4. This problem can be prevented by first identifying the failed node and then not restoring any connections that originate or terminate at the failed node.

The problem with the approach above is that if one of the ADMs fails, or there is a problem with the cabling between the two ADMs, the interconnection is broken. A way to deal with this problem is to use *dual homing*. Dual homing makes use of two hub nodes to perform the interconnection, as shown in Figure 10.12. For traffic going between the rings, connections are set up between the originating node on one ring and both the hub nodes. Thus if one of the hub nodes fails, the other node can take over, and the end user does not see any disruption to traffic. Similarly, if there is a cable cut between the two hub nodes, alternate protection paths are now available to restore the traffic.

Rather than set up two separate connections between the originating node and the two hub nodes, the architecture uses a multicasting or *drop-and-continue* feature present in the ADMs. Consider the connection shown between an end node and the two hub nodes (hub 1 and hub 2) in Figure 10.12. In the clockwise direction of the ring, the ADM at hub 1 drops the traffic associated with the connection but

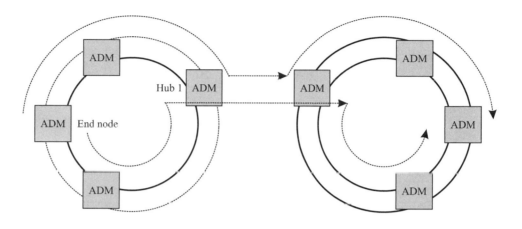

Figure 10.11   Back-to-back interconnection of SONET/SDH rings. This simple interconnection is vulnerable to the failure of one of the two nodes that form the interconnect, or of the link between these two nodes.

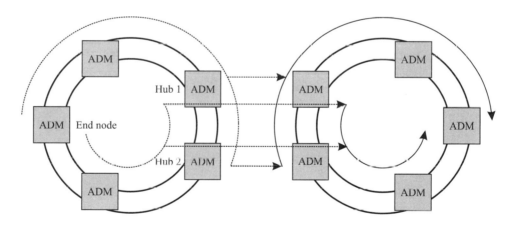

Figure 10.12   Dual homing to handle hub node failures. Each end node is connected to two hub nodes so as to be able to recover from the failure of a hub node or the failure of any interconnection between the hub nodes. The ADMs in the nodes have a "drop-and-continue" feature, which allows them to drop a traffic stream as well as have it continue onto the next ADM.

also simultaneously allows this traffic to continue along the ring, where it is again dropped at hub 2. Likewise, along the counterclockwise direction, the ADM at hub 2 uses its drop-and-continue feature to drop traffic from this connection as well as pass it through to hub 1. Note that additional bandwidth is used up between the two hub nodes on each ring to support this capability.

Dual homing is being deployed in business access networks to interconnect access UPSRs with interoffice BLSRs as well as to interconnect multiple BLSRs. It can also be applied to interconnections between two subnetworks, not necessarily two rings (although rings are the major application). In general, for dual homing to work, the dual node interconnect itself must be a protected subnetwork, so that alternate paths are available if any of the hub nodes or the links interconnecting them fails.

10.3 Protection in IP Networks

The IP layer has historically provided best-effort services. As we studied in Section 6.3, IP, by its very nature, uses dynamic, hop-by-hop routing of packets. Each router maintains a routing table of the next-hop neighbor for each destination, and incoming packets are routed based on this table. If there is a failure in the network, the intradomain routing protocol (OSPF or IS-IS) operates in a distributed manner and updates these routing tables at each router within the domain. In practice, it can take seconds after the failure is detected before the routing tables at all the routers converge and have consistent routing information. During this process, packets continue to be routed based on the current versions of the routing tables at the routers, which can be inconsistent and incorrect. This causes packets to be routed incorrectly and possibly loop within the network. Potentially, packets could therefore be lost or undergo long delays on the order of seconds after a failure is detected. Even if a router decides to route a packet along an alternate route, following the detection of a failure, packets could still loop within the network, as shown in Figure 10.13. In this example, consider packets destined for router D. Suppose link CD fails. Node C would then attempt to route packets destined for D to router B, hoping to find an alternate path to reach router D. Router B, however, still thinks that the best way to get to router D is through router C and would route that packet back to router C. This is the case until the routing tables at the routers have all converged.

The slow recovery from failures is due to the fundamental nature of IP routing—the fact that it is distributed, next-hop-based dynamic routing. Providing faster restoration times requires some way to nail down paths and have packets follow a known path through the network. This capability is provided by multi-protocol label switching (MPLS). As we studied in Section 6.3, MPLS allows label-switched

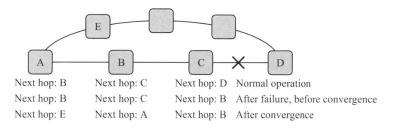

Next hop: B	Next hop: C	Next hop: D	Normal operation
Next hop: B	Next hop: C	Next hop: B	After failure, before convergence
Next hop: E	Next hop: A	Next hop: B	After convergence

Figure 10.13 An example to illustrate routing loops in an IP network after a failure. It takes many interations before the routing tables at the nodes converge to the correct routes. In the meantime, there can be routing loops.

paths (LSPs) to be set up between nodes. All packets belonging to an LSP are routed along the same path. This allows several protection schemes to be implemented within the MPLS layer (which can be viewed as a link layer under the IP network layer). For example, upon detecting a link failure, we could set up alternate LSPs for all the LSPs currently using that link, and reroute packets on the newly set up LSPs. This could be done locally to route around a failed link, or it could be done at the ends of the LSPs. A variety of protection schemes, such as $1 + 1$, ring, or shared mesh, could be implemented using this approach and are being developed currently.

The other aspect of protection in the IP layer has to do with the time taken by the IP layer to detect failures in the first place. In a typical implementation used in intradomain routing protocols [AJY00], adjacent routers exchange periodic "hello" packets between themselves. If a router misses a certain number of these packets, it declares the link to have failed and initiates rerouting. By default, the routers send hello packets every 10 seconds and declare the link down if they miss three successive hello packets. Thus it could take up to 30 seconds to detect a failure. The process can be speeded up by exchanging hello packets more frequently; however, the minimum interval is currently specified to be 1 second. More typically, core routers detect failures in about 10 seconds. Alternatively, a separate set of packets can be exchanged periodically for this purpose [HYCG00]. However, these packets can get queued up in buffers if there are a lot of other packets waiting and so may have to be processed at higher priority levels than regular packets.

Another option is to rely on the underlying SONET or optical layer to detect the failure and inform the IP layer. This can be done by having the line card inside a router look at the framing and communicate failure detection information up into the routing protocol. However, this is not usually architected into today's routers.

$\underline{10.4}$ Why Optical Layer Protection

The optical layer provides lightpaths for use by its client layers, such as the SONET, IP, or ATM layers. (Recall that the layers that use the services provided by the optical layer are called client layers of the optical layer.) We have seen that extensive protection mechanisms are available in the SONET layer, and there is some degree of protection possible in the other client layers as well. These layers were all designed to work independently of each other and not rely on protection mechanisms available in other layers. We will see below that there is a strong need for protection in the optical layer, despite the existence of protection mechanisms in the client layers.

- SONET/SDH networks incorporate extensive protection functions. However, other networks such as IP, ATM, and ESCON networks do not provide the same level of protection. As we saw in Section 10.3, IP traffic for the most part is "best-effort" traffic. However, as carrier networks become more data centric, there is an increasing expectation from both carriers and their customers that these networks will need to provide the same level of availability as SONET and SDH networks.

    One way for realizing this capability is to develop additional protection mechanisms within the IP, ATM, or other client layers, as we saw in Section 10.3. Another way to protect data networks is to rely on optical layer protection, which can be quite cost-effective and efficient.

- Significant cost savings can be realized by making use of optical layer protection instead of client layer protection. We illustrate this with two examples.

    Consider an example of a WDM ring network with lightpaths carrying higher-layer traffic. Figure 10.14 illustrates an example where there is no optical layer protection. Two SONET line terminals (LTEs) are connected to each other through lightpaths provided by the optical layer, as are two IP routers. For simplicity we look at a undirectional lightpath from LTE A to LTE B and another lightpath from router C to router D. These two lightpaths are protected by the SONET and IP layers, respectively, using $1 + 1$ protection. The working connection from LTE A to LTE B is established on wavelength λ_1 along the shortest path in the ring, and the other protection connection is established, say, on the same wavelength λ_1 around the ring. Likewise, the working connection from router C to router D may be established on λ_1 on the shortest path. However, the protection connection from router C to router D, which needs to be routed around the ring, must be allocated another wavelength, say, λ_2. Thus two wavelengths are required to support this configuration.

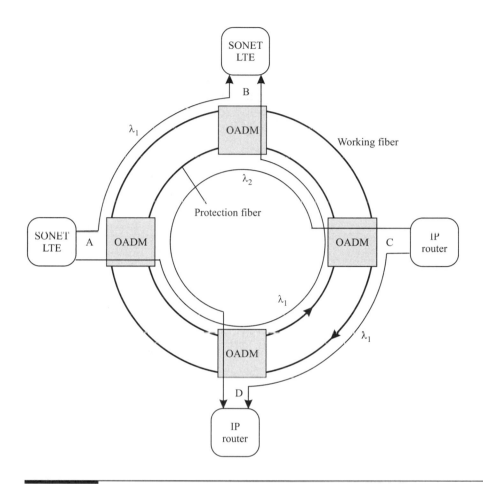

Figure 10.14   A WDM ring built using optical add/drop multiplexers (OADMs), supporting two interconnected SONET line terminals (LTEs) and two interconnected IP routers using protection provided by the SONET and IP layers, respectively. The SONET and IP boxes do not share protection bandwidth.

Figure 10.15 shows what can be gained by having the optical layer do the protection instead. Now we can eliminate the individual $1 + 1$ protection for the SONET LTEs and the IP routers and make them share a common protection wavelength around the ring. Only a single wavelength is required to support this configuration. Note, however, that only a single link cut can be handled by this arrangement, whereas the earlier arrangement of Figure 10.14 can handle some combinations of multiple fiber cuts (see Problem 10.11). Likewise, the

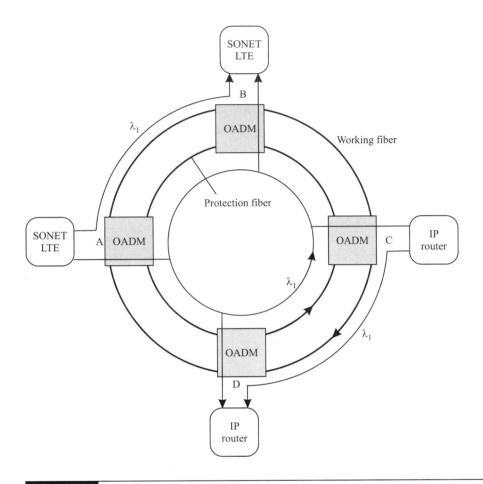

Figure 10.15 Benefit of optical layer protection. The configuration is the same as that of Figure 10.14. However, the optical layer now uses a single wavelength around the ring to protect both the SONET and IP connections.

arrangement of Figure 10.14 can support two simultaneous transmitter failures, whereas the arrangement of Figure 10.15 can support only a single such failure. Nevertheless, if we are primarily interested in handling one failure at any given time, the optical layer protection scheme of Figure 10.15 offers a clear savings in capacity.

Consider what would happen if we had to support N such pairs (N being the number of links in the ring), with each of them being adjacent on the ring. Without optical layer protection, N protection wavelengths would be required.

With optical layer protection, only one wavelength would be needed. Optical layer protection is more efficient because it shares the protection resources across multiple pairs of client layer equipment. In contrast, client layer protection mechanisms cannot share the protection resources between different or independent clients.

Another example of an IP network operating over WDM links is shown in Figure 10.16. Consider two network configuration options. Figure 10.16(a) shows the IP routers interconnected by two diversely routed WDM links. In this case, no protection is provided by the optical layer, and the protection against fiber cuts as well as equipment failures (for example, router port failure) is handled completely by the IP layer. Note that the configuration shown requires three working ports and three protect ports on each router.

Figure 10.16(b) shows a better way of realizing a network with the same capabilities, by making use of protection within the optical layer. In this case, fiber cuts are handled by the optical layer. A simple bridge-and-switch arrangement is used to connect two diversely routed fiber pairs in a single WDM system. In general, it is more efficient to have fiber cuts handled by the optical layer, since a single switch then takes care of restoring all the channels, instead of having each individual IP link take care of the restoration by itself. More importantly, this arrangement can result in a significant savings in equipment cost. In contrast with the previous configuration, this configuration requires each router to have only a single protect port instead of three. If one of the working ports in the router fails, the router directs the traffic onto the protect port. Note that this type of failure cannot be handled by the optical layer.

This example also brings out another value of optical layer protection. Generally the cost of a router port is significantly higher than the cost per port of optical layer equipment. Therefore it is cheaper to reserve protection bandwidth in the optical layer (effectively reserve ports on optical layer equipment), rather than have additional ports in IP routers for this purpose.

- The optical layer can handle some faults more efficiently than the client layers. A WDM network carries several wavelengths of traffic on a single fiber. Without optical layer protection, a fiber cut results in each traffic stream being restored independently by the client layer. In addition, the network management system is flooded with a large number of alarms for this single failure. Instead, if the optical layer were to restore this failure, fewer entities have to be rerouted (albeit larger entities), and hence the process is faster and simpler.

- Optical layer protection can be used to provide an additional degree of resilience in the network, for instance, to protect against multiple failures. An example of

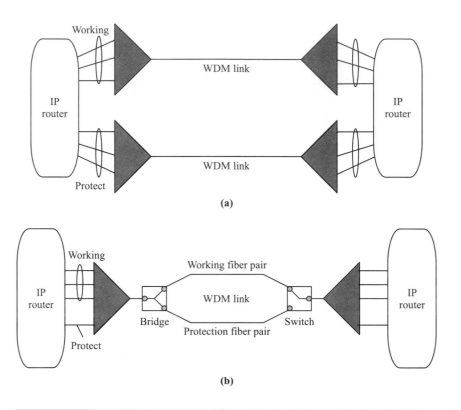

Figure 10.16   Example showing the benefit of optical layer protection compared to protecting at the IP layer. (a) All the protection is handled by the routers. Two diversely routed WDM links are used. Each IP router uses three working ports and three protect ports to protect against both fiber cuts and equipment failures. (b) A single WDM line system is deployed, with protection against fiber cuts handled by the optical layer. Equipment failures are handled by the IP layer. The IP routers now use three working ports and an additional protect port in case one of the working ports fails.

this is shown in Figure 10.17. Consider a SONET BLSR operating over lightpaths provided by the optical layer. Figure 10.17(a) shows normal operation of the network. Figure 10.17(b) shows what happens to a sample SONET connection in the event of a link failure. The BLSR does a ring switch and reroutes the connection around the ring. At this point, until the failed link is repaired, the network cannot handle another failure. Repairing a failed link can take several hours to days—a fairly long period during which the network is vulnerable to

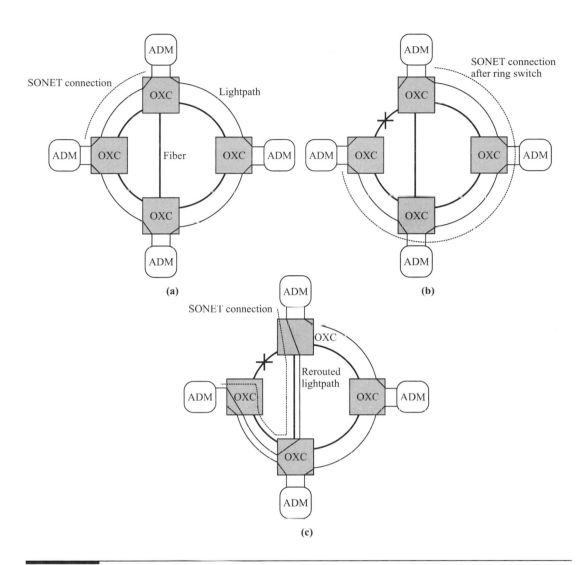

Figure 10.17   Optical layer protection used to enhance SONET protection. The thick lines indicate fiber links, the thin lines indicate lightpaths provided by the optical layer between SONET ADMs, and the dashed line indicates a SONET connection. (a) Normal operation before failure. A SONET ring is realized using lightpaths provided by the optical layer. (b) Due to a fiber failure, a lightpath connecting two adjacent SONET ADMs fails, causing the SONET ADMs to invoke ring switching to rapidly restore the SONET connection. (c) The optical crossconnects (OXCs) perform optical layer restoration and reroute the lightpath around the failure. To the SONET ring, it appears as if the failure has been restored and the ring reverts back to normal operation, ready to tackle another failure.

additional failures. Optical layer protection can be used to remove this vulnerability. In Figure 10.17(c), the optical layer reroutes the lightpath on the failed link around the failure over another optical path. At this point, as far as the BLSR is concerned, it appears as if the failed link has been restored, and the ring reverts back to normal operation. This allows the BLSR to handle additional failures while the failed link is actually being repaired.

- Finally, protection in SONET is currently based on rings (UPSR/BLSR). Ring-based schemes require that the capacity in the network reserved for protection be equal to the capacity used for working traffic. Within the optical layer, a variety of mesh-based protection schemes are being developed. These offer the promise of requiring significantly less protection capacity than ring-based schemes. Admittedly, these schemes could also be applied in the SONET layer.

However, optical layer protection does have its limitations:

- Not all failures can be handled by the optical layer. If a laser in an attached client terminal fails, the optical layer cannot do anything about it. Thus, client equipment failures need to be dealt with by the client layer.

- The optical layer may not be able to detect the appropriate conditions that would cause it to invoke protection switching. For instance, a transparent network can only monitor presence or absence of power (and in some cases, the optical signal-to-noise ratio). While it may also be able to measure power degradations, it may not know what the reasonable values for the power levels are because they vary widely depending on the type of signal being carried. Thus it can only trigger protection switching upon detecting loss of light. The bit error rate is a more precise indicator of signal quality, but a transparent network may not be able to measure bit error rate.

- The optical layer protects traffic in units of lightpaths, and it cannot protect part of the traffic within a lightpath and not protect other parts. Such functions need to be performed by the client layers.

- Protection routes in the optical layer may be longer than the primary routes, and the choice of alternate routes may be severely limited due to link budget considerations.

- We need to pay careful attention to interworking of protection schemes between the different layers. We will discuss some of these issues in Section 10.6.

10.4.1 Service Classes Based on Protection

In Chapter 9, we alluded to the fact that multiple classes of service can be provided by the optical layer based on the type of protection provided. The main differences in these classes lie in the level of connection availability provided and the restoration time for a connection. These different classes will likely be supported using different protection schemes. While no standards have been defined yet, we provide a likely set of services below:

Platinum. This provides the highest level of availability and the fastest restoration times, comparable to SONET/SDH protection schemes, typically around 60 ms. For example, a dedicated 1 + 1 protection scheme could be used to provide this class of service. This class may be viewed as a premium service and is accordingly priced.

Gold. This provides high availability and fast restoration times, typically in the range of hundreds of milliseconds. For example, a shared mesh protection scheme can provide this class of service.

Silver. This class sits below gold in terms of availability and restoration time. For example, a protection scheme that provides "best-effort" restoration may fit into this category. Another example would be a scheme wherein a connection is reattempted from scratch in case of a failure.

Bronze. Here, the optical layer provides unprotected lightpaths. In the event of a failure of the working path, the connection is lost.

Lead. This class of service would have the lowest availability and the lowest priority among all the classes. For instance, we may support this class by using protection bandwidth reserved for other classes of service. If that bandwidth is needed to protect other higher-priority traffic, connections in this class are preempted.

There is a great deal of debate about what types of applications will use these service classes and which of them will proliferate. For instance, today carriers using SONET/SDH are providing primarily platinum-type services to their customers. However, we expect that the increasing dominance of data traffic will stimulate the need for lower-priced classes of service. For example, carriers interconnecting Internet routers from Internet service providers are providing in some cases platinum services and in other cases bronze (unprotected) services. In the latter case, the IP layer handles all the restoration functions. In the former situation, it is quite possible that some of that traffic could be carried over lightpaths with a lower quality of service.

10.5 Optical Layer Protection Schemes

We next look at the different types of optical layer protection schemes. For the most part, conceptually, the schemes are similar to their SONET and SDH equivalents. However, their implementation is substantially different, for several reasons:

Table 10.3 A summary of optical protection schemes operating in the optical multiplex section (OMS) layer. Both dedicated protection rings (DPRings) and shared protection rings (SPRings) are possible.

| | Protection Scheme | | | |
	1 + 1	1:1	OMS-DPRing	OMS-SPRing
Type	Dedicated	Shared	Dedicated	Shared
Topology	Point-point	Point-point	Ring	Ring

Table 10.4 A summary of optical protection schemes operating in the optical channel layer.

| | Protection Scheme | | |
	1 + 1	OCh-SPRing	OCh-Mesh
Type	Dedicated	Shared	Shared
Topology	Mesh	Ring	Mesh

the equipment cost for WDM links grows with the number of wavelengths to be multiplexed and terminated, link budget constraints need to be taken into account when designing the protection scheme, and there may be wavelength conversion constraints to deal with.

We saw in Chapter 9 that the optical layer consists of the optical channel (OCh) layer (or path layer), the optical multiplex section (OMS) layer (or line layer), and the optical transmission section (OTS) layer. Just as SONET protection schemes fit into either the line layer (for example, BLSR) or the path layer (for example, UPSR), optical protection schemes also belong to the OCh or OMS layers. An OCh layer scheme restores one lightpath at a time, whereas an OMS layer scheme restores the entire group of lightpaths on a link and cannot restore individual lightpaths separately. Table 10.3 provides an overview of schemes operating in the optical multiplex section layer. Table 10.4 summarizes schemes operating in the optical channel layer. These schemes have not yet been standardized, and there are many variants. We have attempted to use a nomenclature that is consistent with SDH terminology.

In SONET, there is not a significant cost associated with processing each connection separately in the path layer instead of processing all the connections together in the line layer because the processing is done using application-specific integrated circuits, where the incremental cost of processing the path layer compared to the line layer is not significant. In contrast, there can be a significant difference in cost associated with OCh layer schemes relative to OMS layer schemes. An OCh layer scheme has to demultiplex all the wavelengths, whereas an OMS layer scheme operates on all the wavelengths and thus requires less equipment.

As an example, consider the two protection schemes shown in Figure 10.18. Figure 10.18(a) shows $1+1$ OMS protection, while Figure 10.18(b) shows $1+1$ OCh protection. The OMS scheme requires two WDM terminals and an additional splitter and switch. The OCh scheme, on the other hand, requires four WDM terminals and a splitter and switch per wavelength. Thus its equipment cost is higher than the cost of the OMS scheme. Indeed this is the case if all channels are to be protected. However, the cost of OCh protection can be reduced if not all channels need to be protected. Assuming multiplexers, splitters, and switches can be added on a wavelength-by-wavelength basis, the cost of OCh protection grows linearly with the number of channels that are to be protected. The cost of an OMS protection scheme, on the other hand, is independent of the number of channels to be protected. If only a small fraction of the channels are to be protected, then OCh protection is not significantly more expensive than OMS protection.

The choice of protection schemes is dictated primarily by the service classes to be supported (as discussed below) and by the type of equipment deployed. In the SONET/SDH world, protection is performed primarily by the SONET/SDH line terminals (LTEs) and add/drop multiplexers (ADMs) and not by digital crossconnects. This is the case primarily because digital crossconnects were more inefficient at performing fast protection than the LTEs and ADMs, partly because they operated on lower-speed tributaries. However, we are likely to see protection functions handled somewhat differently in the optical layer. Multiplexing equipment, such as optical line terminals and add/drop multiplexers, can provide both OCh layer and OMS layer protection in linear or ring configurations. On the other hand, optical crossconnects can provide protection in linear, ring, and mesh configurations. Unlike their digital crossconnect counterparts in the SONET/SDH world, optical crossconnects are designed to provide efficient protection. Depending on the type of crossconnect (see Section 7.4), the protection could be done either at the optical channel layer (for crossconnects that groom at the wavelength level) or at the STS-1 level (for electrical core crossconnects grooming at STS-1). Therefore one possibility is to use simple unprotected WDM point-to-point systems and rely on the optical crossconnects to perform the protection functions. Backbone networks handling large numbers of wavelengths may opt for this choice, as may operators who have already deployed a large quantity of unprotected WDM equipment in their networks. The other possibility is to rely on the WDM line terminals and add/drop multiplexers to perform this function. Metropolitan networks using small numbers of channels and not requiring the use of crossconnects may opt for this choice.

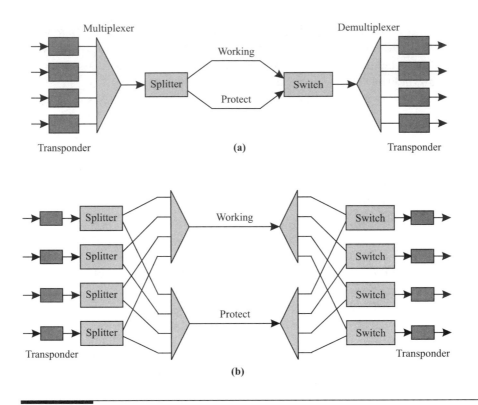

Figure 10.18   Comparison of (a) 1 + 1 OMS and (b) 1 + 1 OCh protection schemes.

10.5.1 **1 + 1 OMS Protection**

This is perhaps the simplest optical layer protection scheme and is shown in Figure 10.18(a). Because of its simplicity, it has been implemented by several vendors in their OLTs. The composite WDM signal is bridged onto two diverse paths using an optical splitter. At the other end, an optical switch is used to select the better among the two signals, based primarily on detecting the presence or absence of light signals. The split incurs an additional 3 dB loss, and the switch also adds a small amount of loss (< 1 dB). An alternative implementation uses optical amplifiers on each of the fibers and a passive combiner to combine both directions at the receiver. At any time, one amplifier is turned on and the other is turned off. This has the advantage of avoiding a single point of failure in the system (the selector switch in other implementations), but may be more expensive to implement.

10.5.2 1:1 OMS Protection

This scheme is similar to the SONET 1:1 scheme discussed in Section 10.2.1 and the benefits are similar: support for low-priority traffic and also the ability to have N working systems share a single protection system. Compared to the $1 + 1$ scheme of Figure 10.18(a), a typical implementation uses a switch at the transmitter, instead of a splitter, resulting in a somewhat lower total loss in the path. Just as in the SONET equivalent, an APS protocol is needed to provide coordination between the two ends of the link.

10.5.3 OMS-DPRing

The OMS-DPRing (dedicated protection ring) is similar to a SONET UPSR, except that it operates at the OMS (or optical line) layer, whereas the UPSR operates in the SONET path layer. It can also be thought of as an optical unidirectional line-switched ring (ULSR).

One possible implementation of an OMS-DPRing [Bat98] is shown in Figure 10.19. Signals are coupled into and out of the ring via passive couplers. Each node transmits on both directions of the ring. Note that different nodes must transmit at different wavelengths; otherwise their transmissions would collide. Under normal operation, the ring functions as a bus, with one pair of amplifiers turned off on the entire ring and all the others turned on. If there is a link failure, the amplifiers next to the failed link are turned off and the ones that were originally inactive are now turned on to restore traffic. For example, in Figure 10.19(a), the amplifier pair to the right of node A is turned off under normal operation and the other amplifiers are turned on. In Figure 10.19(b), when link CD fails, the amplifier pair at C adjacent to the failed link is turned off, and the originally inactive amplifiers at node A are turned onto create a new bus and restore traffic.

10.5.4 OMS-SPRing

The OMS-SPRing (shared protection ring) is analogous to a SONET BLSR/4 with some changes. A possible implementation of a four-fiber ring is shown in Figure 10.20. Two of the fibers have WDM equipment deployed, and the remaining two fibers around the ring are used for protection purposes and do not have attached WDM equipment. In the event of a cut, the signal is either span switched or ring switched onto the protection fibers, as shown in Figure 10.21. In both cases, not having WDM equipment on the protection fibers not only saves cost but also provides a relatively lower-loss path around the ring for the protection traffic. Optical amplifiers may be needed on the protection fibers depending on the link losses.

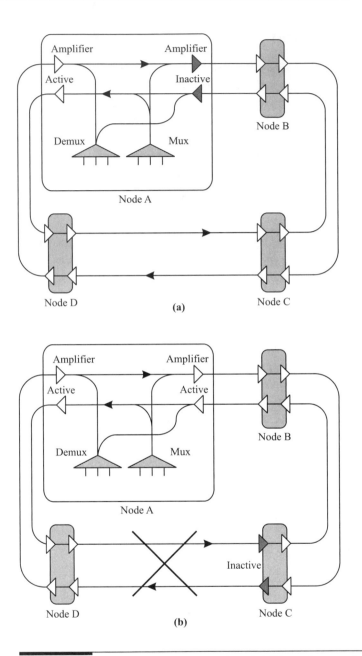

Figure 10.19   OMS-DPRing protection. (a) Normal operation. One pair of amplifiers is inactive (turned off) and the others are turned on, creating a bus. (b) After a failure, the currently inactive amplifiers are turned on and an amplifier pair adjacent to the failure is turned off to bring up the alternate path and restore traffic.

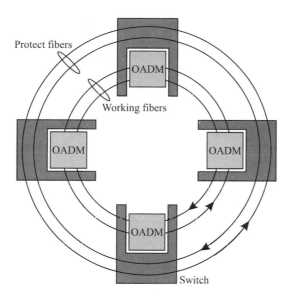

Figure 10.20   OMS-SPRing shown under normal operation. Only the working fibers are connected to optical add/drop multiplexers. The protection fibers are connected around the ring.

A two-fiber version of OMS-SPRing can also be realized by dedicating half the wavelengths on each fiber for protection purposes. By making sure that protection wavelengths on one fiber correspond to the working wavelengths on the other fiber, the signals can be rerouted without requiring wavelength conversion. This scheme, however, requires the two groups of wavelengths to be demultiplexed and multiplexed at each node, and thus is not strictly operating at the OMS layer.

10.5.5   *1:N Transponder Protection*

The OMS layer schemes that we discussed above handle link failures and node failures but do not handle failures of the end equipment, particularly the transponders. The transponders may be protected in a 1:N configuration by having a spare transponder for every N working transponders. One problem to overcome is that transponders today operate at fixed wavelengths, and so the spare transponder will operate at a different wavelength than the working transponder. When the signal is switched over to the spare transponder, we also need to set up a new lightpath on the new wavelength through the network. Alternatively, we could use a tunable laser in the spare transponder.

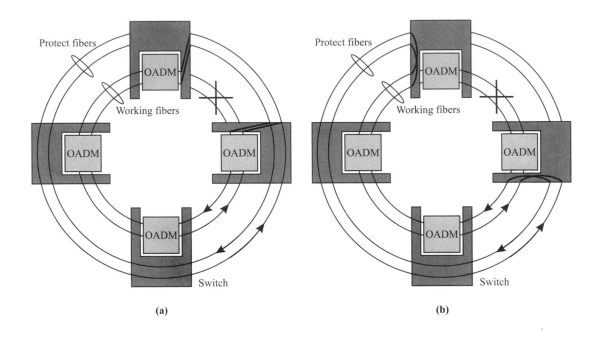

Protect fibers

Working fibers

OADM

OADM   OADM

OADM

Switch

(a)

Protect fibers

Working fibers

OADM

OADM   OADM

OADM

Switch

(b)

Figure 10.21   OMS-SPRing after a failure. (a) Span switching. (b) Ring switching.

10.5.6   **1 + 1 OCh Dedicated Protection**

In 1 + 1 OCh protection, two lightpaths on disjoint routes are set up for each client connection. As shown in Figure 10.18(b), the client signal is split at the input and the destination selects the better of the two lightpaths. As with SONET and SDH, no signaling is required. This approach works in point-to-point, ring, and mesh configurations. In the context of a ring, the scheme is also called OCh-DPRing (OCh dedicated protection ring) or optical UPSR.

Like SONET UPSRs, this approach is bandwidth inefficient in that the protection bandwidth is not shared among multiple client connections. However, it is one of the simplest protection schemes and therefore has been implemented by several vendors in optical add/drop multiplexers and crossconnects.

Figure 10.22 shows another possible implementation of the bridge and select functions within a node. Here, the signal entering the optical layer is split and sent to two transponders, and then diversely routed across the network. At the receiving end, the signal is terminated in two transponders, and the better signal is selected afterwards to be sent to the client. In Figure 10.18, the client signal is passed through

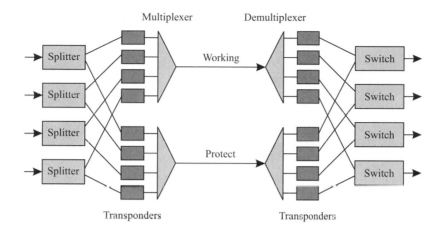

Figure 10.22   Another implementation of $1 + 1$ OCh protection. The signal from the client equipment is split and sent to two transponders for transmission over diverse paths, and at the destination the better copy is selected by an optical switch at the output of the transponders.

a transponder and split afterwards. At the receiving end, one of the two signals is selected by an optical switch before it is sent into a transponder and then onwards to the client. This uses half as many transponders as the previous option but does not protect against a transponder failure. Aside from this aspect, there are several other subtleties that affect the choice of one implementation versus the other, such as the criteria for switching from one path to another, and potential restoration time differences between the two approaches.

10.5.7   OCh-SPRing

The OCh-SPRing (shared protection ring) is somewhat similar to a SONET BLSR/4. However, the BLSR operates at the line (multiplex section) layer, whereas this scheme operates at the optical channel layer and not the optical multiplex section layer. Working lightpaths are set up on the shortest path along the ring. When a working lightpath fails, it is restored either using a span switch or a ring switch, just as in a SONET BLSR/4. Nonoverlapping lightpaths in the ring can share a single wavelength around the ring for protection, and this spatial reuse allows the OCh-SPRing to be more efficient than an OCh-DPRing for distributed traffic. The operation of the OCh-SPRing is essentially the same as that shown in Figures 10.5–10.7, where the fibers now correspond to wavelengths and the connections correspond to lightpaths.

Just as with a BLSR, fast coordination between the ring nodes is needed in order to support node failures or low-priority traffic.

10.5.8 OCh-Mesh Protection

Ring architectures are inherently suitable for sparse physical topologies and in situations where most of the traffic is confined within the ring. Many backbone networks tend to be somewhat more densely connected than rings and are essentially meshed, with traffic being fairly distributed. A typical North American long-haul carrier's backbone network may have, say, 50 nodes, with an average node having 3–4 adjacent nodes, with some nodes having as many as 5–10 adjacent nodes. For such networks, mesh protection schemes offer more bandwidth-efficient protection than rings. The bandwidth efficiency of a mesh relative to a ring depends on several factors, including the network topology, the traffic pattern, and the type of mesh protection scheme used. In general, the more dense or meshed the topology, the greater the benefit of mesh protection. Also, if traffic in the network is primarily localized, then rings can do a good job. In contrast, if traffic in the network is distributed, then rings are inefficient—many lightpaths will need to be partitioned into multiple rings, and multiple rings need to be interconnected and protected to support these lightpaths. Efficiency improvements ranging from 20% to 60% have been reported for mesh protection schemes relative to ring protection schemes [RM99a, RM99b]. Here we provide a simple example to illustrate the efficiency of mesh protection relative to ring protection. We will look at a more realistic detailed example in Section 13.2.6.

> **Example 10.1**   Consider the network shown in Figure 10.23(a), with three lightpaths to be supported. Assume that all these lightpaths need to be protected. Each lightpath uses 1 unit of capacity on each link that it traverses.
>
> First suppose we use $1 + 1$ OCh dedicated protection. We would then set up dedicated protection lightpaths as shown in Figure 10.23(b). In this case, a total of eight units of protection capacity is needed in the network.
>
> Next let us consider a configuration that uses shared ring protection (OCh-SPRing). Here we have an interesting problem of how to configure the rings themselves. One solution is to configure the rings as shown in Figure 10.23(c). In this case, lightpaths X and Y each share the same bandwidth for protection, while lightpath Z has a separate ring for protection. This configuration requires a total of eight units of capacity for protection, which is the same as for dedicated protection above. Note, however, that the protection capacity can be reduced to six units by having lightpaths X and Y share a ring but using dedicated protection for lightpath Z. Another way to look at this is that by using the eight units of

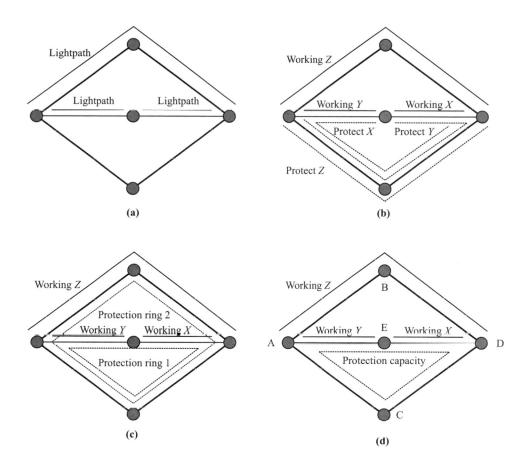

Figure 10.23   Example to illustrate the bandwidth efficiency of mesh protection relative to ring protection. (a) A mesh network with three lightpaths present. (b) Protecting the lightpaths using 1 + 1 dedicated protection. (c) Protecting the lightpaths using OCh-SPRing protection. (d) Protecting the lightpaths using OCh-mesh protection.

capacity, we can support additional lightpaths that can share the ring used to protect lightpath Z.

We now consider the case of shared mesh protection. Our mesh protection scheme works as follows. We will use the same routes used by the 1 + 1 scheme for routing the protection lightpaths. The big difference is that the protection lightpaths are not set up ahead of time, but are only set up when there is a failure.

As long as two lightpaths don't fail simultaneously, we can have them share the same protection capacity in the network. In this case, only a single lightpath fails at any given time, assuming we have to deal only with link failures. Therefore we only need to provide sufficient protection bandwidth to protect one lightpath at a time. We leave it to the reader to verify that the four units of capacity shown in Figure 10.23(d) are sufficient.

Mesh protection schemes are not new. They were used in the 1980s in networks with digital crossconnects. However, these protection schemes were centralized and operated rather slowly, taking minutes to hours to restore traffic after a failure. Also the protection was complex to manage, and there were no applicable standards. After the standardization of SONET/SDH and due to the fast 60 ms ring protection offered by SONET/SDH, these mesh-based restoration schemes were largely abandoned.

Today, we are seeing a resurrection of mesh protection schemes in the optical layer of the network for several reasons.

- The processing power available to implement mesh protection has dramatically increased over the past few decades, to the point where computationally intensive functions such as determining new routes can be performed rapidly. The communication bandwidth available for network control purposes has also gone up dramatically. To protect a network providing terabits/second of capacity, it is quite reasonable to dedicate several 2 Mb/s or 45 Mb/s lines in the network for control traffic. This was not the case earlier, where this amount of bandwidth would have been considered large, relative to the actual traffic within the network.

- Optical crossconnects and other optical layer equipment protect bandwidth at much larger granularities (lightpaths) than digital crossconnects that operate at DS1 or DS3 speeds. As a result, they have fewer entities to manage and protect. However, this situation will change as traffic grows.

- Relatively fast signaling and routing protocols have been developed for other forms of data networks, such as IP and ATM networks, and many of these protocols can be adapted for use in the optical layer.

- The 60 ms protection time requirement is not a hard number. Many carriers interested in protecting data traffic will be satisfied with protection times on the order of a few hundred milliseconds, making it easier to implement more complex protection schemes.

A variety of mesh protection schemes have been proposed, and many are currently being implemented by optical crossconnect vendors. In addition to the factors discussed above, the mesh protection schemes will have to overcome some key issues in order to facilitate widespread deployment:

- Part of the reason that SONET/SDH protection has been so successful is that the protection schemes were standardized. This is yet to happen with mesh protection schemes.

- One of the advantages of ring-based schemes is that the network is partitioned into multiple domains and each domain is protected independently. Thus one part of the network does not affect the other parts. This implies that the network can handle simultaneous multiple failures as long as they occur in different domains. Moreover, one part of the network can be serviced without impacting the protection scheme in the other parts. In order to get the full benefit of mesh protection, we will need to treat the network in its entirety as a single domain. Breaking up the network into smaller domains reduces the bandwidth efficiency unless the individual domains are reasonably large.

  Another dimension to this is the effect of software bugs or operator errors. In ring-based networks, such problems are localized, whereas in mesh networks, these problems can have a networkwide impact.

- Mesh protection schemes are considerably more complex to manage than ring protection schemes. In order to make them successful, vendors will need to provide carriers with the appropriate management tools to hide the complexity from the network operators. For instance, this could mean providing automated tools to plan and compute primary and protection routes in the network, which are otherwise fairly complex operations.

  On the plus side, however, interconnecting rings is fairly complex, and mesh protection allows for more flexible planning of capacity in the network—capacity does not have to be nailed down up front; instead it can be provisioned as needed across the network.

- The more efficient mesh protection schemes will require rapid networkwide signaling mechanisms to be implemented to propagate information related to failures and to reroute lightpaths that are affected by a failure. This in turn implies that the nodes performing the protection switching will have to be designed carefully to minimize processing latencies.

- The more efficient mesh protection schemes require that protection routing tables be maintained at the nodes. These routing tables provide information about the network topology and protection paths in the network. The tables need to be updated when lightpaths, links, or nodes are added or removed from the network. Most importantly, these tables need to be consistent across all the nodes in the network.

These protection routing tables are similar to the routing tables maintained in IP networks, which work well even in very large IP networks with thousands of nodes. However, we need to realize that routing tables in IP networks are not always consistent. If the tables are inconsistent, routing pathologies, such as looping, can be present in the network with fairly high probabilities. For example, at the end of 1995, the likelihood of encountering a major routing pathology in the Internet was 3.3% [Pax97]. These pathologies can cause packets to be forwarded incorrectly in the network, but these packets eventually find their way to their destination or are dropped by the network. In the latter event, the packets are retransmitted by a higher-layer protocol (TCP). While this approach works well in IP networks, we cannot afford to have routing pathologies in transport networks because they could prevent restoration of service after a failure. Therefore, fast and reliable topology update mechanisms need to be in place to maintain the protection routing tables.

We now look at the different variations of mesh protection. One aspect of this is whether the entire network is protected as a single domain, or whether it is broken down into multiple domains, with each domain protected independently, and the different domains then tied together. In a degenerate scenario, each domain could be a single ring, in which case we get back to the usual mode of ring-based protection.

Another important aspect that differentiates protection schemes is whether the protection routes are precomputed ahead of time (*offline*), or whether they are computed after a failure has occurred (*online*). In both cases, another dimension to consider is the degree of distributed implementation. This affects the complexity of the signaling protocols required and has a direct impact on the speed of restoration.

Let us first consider the case where the protection routes are precomputed. In this case, the protection route for a lightpath is computed at the time it is set up and stored in the network. Sufficient bandwidth is allocated on all the links so as to ensure the lightpath can be restored in the event of any possible failure. (Note that this protection bandwidth is still shared among many lightpaths and is not dedicated to a single lightpath. This is the distinction between 1 + 1 dedicated protection and shared protection.) Depending on the sophistication of the scheme used, there may be one or many possible alternate routes for a given lightpath, based on the actual failure scenario. For example, the simplest scenario is to compute a single disjoint path through the network as the protection route. Alternatively, we may use multiple protection routes, based on which link fails in the network. Clearly the amount of information needed to be stored in the network depends on the number of protection routes per lightpath.

In a centralized implementation of this scheme, a central controller in the network is notified if a failure occurs. The central controller then sets up all the alternate routes for the lightpaths by signaling to all the affected network elements to reconfigure their switches as needed. The problem with this approach is that the central controller is a single point of failure and is likely to be a significant bottleneck, both in terms of communication and processing speed.

Several variants of a distributed implementation are possible. In one variant, the failure information is flooded to all the network nodes. Each node then looks up its routing table and reconfigures its switch, based on the exact failure that occurred. Another possibility is to signal the failure to the sources/destinations of all the affected lightpaths. Each source-destination pair then sets up the alternate routing path by signaling to the nodes along the new path.

Next let us consider computing routes on the fly. In this case, new routes are computed after the failure has been discovered. One major issue that comes up in this context is whether sufficient bandwidth is available in the network to handle all the lightpaths that need to be restored. Without essentially precomputing the routes, it is not possible to determine the amount of protection bandwidth needed a priori. In this case, it is possible that some lightpaths are restored and others aren't.

Again this scheme can be implemented in a centralized or distributed manner. The distributed implementation is more complex than for the case where routes are precomputed. Here it is possible that multiple nodes acting independently may contend for the same link or wavelength resource to restore two independent lightpaths. These contentions will have to be dealt with, making the signaling scheme more complex and the recovery possibly slower. A centralized implementation would avoid such conflicts, but would suffer even worse communication and processing bottlenecks, compared to the centralized implementation for the case where the routes are precomputed.

Based on our discussions so far, we see that mesh protection requires the following functions: route computation, topology maintenance, and signaling to set up the protection routes. These functions have been implemented in IP and ATM networks. For example, in IP networks, route computation is done using a Dijkstra shortest-path-first algorithm, and the topology is maintained using a routing protocol such as OSPF (open shortest path first). Signaling has been used to establish paths in MPLS networks and ATM networks. Several signaling protocols are available for this purpose, including the resource reservation protocol (RSVP) [BZB+97], private network-network interface (PNNI) signaling protocol [ATM96], and Signaling System 7 (SS7) [ITU93]. Today, there is a significant amount of work under way to expand MPLS (called GMPLS, for generalized MPLS) [AR01] to provide similar capabilities in optical networks.

10.5.9 Choice of Protection Technique

We have explored a number of different optical layer protection options. It is still too early to determine which ones will be deployed widely. An operator wanting to offer the different types of protection on the lightpaths as discussed in Section 10.4.1 must use an OCh layer protection method. On the other hand, an operator who is satisfied with protecting all lightpaths together will likely prefer an OMS layer scheme. Many of the protection schemes discussed above are being implemented in commercial products.

10.6 Interworking between Layers

We have seen that protection functions can be done in the optical layer, SONET/SDH layers, or in the service layer (IP/ATM). How should protection in the network be coordinated between all these layers?

By default, the protection mechanisms in different layers will work independently. In fact, a single failure might trigger multiple protection mechanisms, all trying to restore service simultaneously, which would result in a large number of unnecessary alarms flooding the management center. This results in allocating protection bandwidth at each of the layers, which is inefficient.

An area of significant concern is that protection mechanisms in different layers could potentially contend with each other, preventing or delaying service restoration, although careful design can eliminate such occurrences. The following argument shows that multilayer protection schemes will eventually converge and restore traffic under the right assumptions:

Consider two network layers, a client layer operating over a server layer, each with its own protection mechanisms. If the following conditions are met, the network will always restore traffic in the event of a failure:

1. A viable protection path exists for each layer.
2. The server layer does not depend on the client layer to detect failures and invoke its protection-switching functions.
3. The client layer protection is *revertive* in the sense that it will repeatedly try switching to the other path if its current path fails.

Observe that since the server layer is independent of the client layer and does not depend on client layer indicators, in the event of a failure, the server layer will detect the failure and restore the traffic. After the failure occurs, there may be a period of time when the client layer is unable to restore service because the server layer is invoking its protection scheme. Ultimately, since the server layer converges, the client layer will see either a working path or a protection path available for it, and will therefore eventually converge.

If any of the conditions above are not met, then the protection scheme may not converge. For example, if the client layer protection is nonrevertive, it may switch over once to the protection path, discover that path is not available, and not switch back to its primary path.

While it is desirable to have some sort of coordination between protection mechanisms in different layers, this may not always be possible. For example, the protection mechanisms in different layers may actually be activated by different nodes. In some cases, it may be possible to add a priority mechanism where one layer attempts to restore service first, and only afterwards does the second layer try. One automatic way to ensure this is to have the restoration in one layer happen so quickly that the other layer doesn't even sense that a failure has occurred. For example, consider a WDM network carrying IP traffic. As we saw in Section 10.3, it can take several seconds for the IP layer to detect a failure. It is entirely feasible for the optical layer to have completed its restoration within this time scale so that the IP layer doesn't detect the failure. This may not, however, be feasible when we have SONET rings operating over a WDM network. The SONET rings detect failures very quickly and can initiate protection switching as early as 2.3 μs after a failure occurs.

Another way to implement orderly restoration would be to impose an additional *hold-off time* in the higher layer before it attempts restoration so as to provide sufficient time for the lower layer to do its restoration. However, a large hold-off time would increase the overall restoration time and is therefore not highly desirable either. In general, it would make sense to have the priorities arranged such that the layer that can provide the fastest restoration tries first.

Summary

Engineering the network for survivability plays an increasingly important role in transport networks. Protection techniques are well established in SONET and SDH and include point-to-point, dedicated protection rings, and shared protection rings. Point-to-point protection schemes work for simple systems with diverse fiber routes between node locations. Dedicated protection rings are primarily used to aggregate traffic from remote locations to one or two hub locations. Shared protection rings are used in the core parts of the network where the traffic is more distributed.

Protection in the optical layer is emerging, with several commercial products now implementing optical layer protection. Optical layer protection is needed to protect the data services that are increasingly being transported directly on the optical layer without the SONET/SDH layer being present. It can also be more efficient with respect to reducing the protection bandwidth required (by sharing the bandwidth across multiple clients) and therefore more cost-effective.

Optical channel layer protection is needed if some channels are to be protected while others are not. Optical multiplex section layer protection is more cost-effective for those cases where all the traffic needs to be protected. There is a growing trend toward the use of shared mesh protection in the optical layer, which is viewed as being more bandwidth-efficient and flexible, compared to the traditional ring-based approaches.

Further Reading

There is a vast literature on protection in SONET and SDH networks. SONET rings and protection schemes are described in ANSI T1.105.1 and Telcordia GR-253 and GR-1230. ITU G.841 describes the equivalent SDH architectures. We also refer the reader to the books by Sexton and Reid [SR97] and Wu [Wu92].

Providing reliable service in IP and MPLS networks is a topic of great interest today. Several protection schemes are being developed. See, for example, [DR00, Section 7.4], [CO99], and several Internet drafts available at *www.ietf.org*.

There is a lot of activity under way on optical layer protection schemes, with several being implemented in products today. These have not yet been standardized. [DWY99, RM99a, RM99b, Ram01, MM00, Bar00, GR00a, GR00b, Dos99, MBN99, Wu95, WO95, Tel98, GR96, GRS97] provide good coverage of the major issues. Interworking of protection schemes between different layers is covered in [Dem99, MB96].

Problems

10.1   Consider a shared protection ring with two types of restoration possible. In the first scheme, the connection is rerouted by the source and destination around the ring in the event of a failure. In the second, the connection is rerouted around the ring by the nodes adjacent to the failed link (as in a BLSR). Give an example of a traffic pattern where the first scheme uses less ring bandwidth than the second. Give another example where the two require the same amount of bandwidth.

10.2   Show that in a ring architecture if the protection capacity is less than the working capacity, then service cannot be restored under certain single failure conditions.

10.3   Compare the performance of UPSRs and BLSR/2s in cases where all the traffic is between a hub node and the other nodes. Assume the same ring speed in both

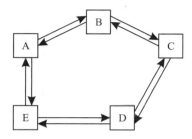

Figure 10.24 Network topology for Problem 10.6.

cases. Is a BLSR/2 any more efficient than a UPSR in traffic-carrying capacity in this scenario?

10.4 Construct a traffic distribution for which the traffic-carrying capacity of a BLSR/4 is maximized. What is this capacity as a multiple of the bit rate on the working fibers?

10.5 Assuming a uniform traffic distribution, compute the traffic-carrying capacity of a BLSR/4 as a multiple of the bit rate on the working fibers.

10.6 Consider the topology shown in Figure 10.24 over which STS-1s are to be transported as dictated by the bandwidth demands specified in the table below for each node pair. Assume all the bandwidth requirements are bidirectional.

STS-1	B	C	D	E
A	12	6	4	12
B		8	10	6
C			12	2
D				8

Given the fiber topology and the STS-1–based bandwidth requirements, we will utilize a two-fiber OC-N SONET ring architecture, but we need to determine which SONET ring architecture is the most suitable for the given network—the UPSR or the BLSR/2.

(a) Provide a detailed illustration of how the six STS-1s between nodes A and C would be transported by a UPSR and a BLSR/2. Redraw Figure 10.24 to begin each illustration.

(b) Suppose that a backhoe cuts the fiber pair between nodes B and C. Again, redrawing Figure 10.24 and referencing your illustrations above, provide a detailed illustration of how the six STS-1s between nodes A and C would be transported just after this failure for the UPSR and the BLSR/2. Use dashed lines to highlight any differences in the routing from normal operation.

(c) Using the bandwidth demands given in the table above, design best-case ring routing plans for the UPSR and the BLSR/2. Illustrate the routing on the network topology of Figure 10.24. In addition, specify the quantity of STS-1s being transported over each fiber link for both cases.

(d) Assuming that we want to use a single OC-*N* ring, what would be the minimum standard value of *N* in each case for the designed UPSR and BLSR/2?

(e) Given all of this information, which ring architecture is better suited for this application? Briefly explain your reasoning.

10.7 The UPSR, BLSR/4, and BLSR/2 are designed primarily to handle single failures. However, they can handle some cases of simultaneous multiple failures as well. Carefully characterize the types of multiple link/node failure combinations that these different architectures can handle.

10.8 The 1 + 1 protection in a SONET UPSR is not implemented at a fiber level but at an individual SONET connection level: for each connection, the receiver picks the better of the two paths. An alternative and simpler approach would be to have the receiver simply pick the better of the two fiber lines coming in, say, based on the bit error rate. In this case, the receiver would not have to look at the individual connections in order to make its decision, but rather would look at the error rate of the composite signal on the fiber. Why doesn't this work?

10.9 Suppose you had only two fibers but could use two wavelengths, say, 1.3 μm and 1.55 μm, over each fiber. This can be used to deploy a BLSR/4 ring in three different ways: (1) the two working fibers could be multiplexed over one fiber and the two protection fibers over the other, (2) a working fiber and a protection fiber in the same direction could be multiplexed over one fiber, or (3) a working fiber and a protection fiber in the opposite direction could be multiplexed over one fiber. Which option would you choose?

10.10 Consider a four-fiber BLSR that uses both span and ring switching. What are the functions required in network management to (a) coordinate span and ring switching mechanisms, and (b) allow multiple failures to be restored?

10.11 Consider the example shown in Figure 10.14. Carefully characterize the set of simultaneous multiple fiber cuts that can be handled by this arrangement.

10.12 Consider a five-node optical ring with one hub node and four access nodes. The traffic to be supported is one lightpath between each access node and the hub node. You can deploy either a two-fiber OCh-DPRing or a two-fiber OCh-SPRing in this application. No wavelength conversion is allowed inside the network, so each lightpath must use the same wavelength on every link along its path. Compare the amount of protection and working capacity needed for each case. Using a wavelength on a link counts as one unit of capacity. Would your answer change if wavelength conversion was allowed in both types of rings at any node in the ring?

10.13 Develop computer software that performs the following functions:

(a) Allows you to input a network topology graph and a set of lightpaths (source-destinations).

(b) Routes the lightpaths using a shortest-path algorithm.

(c) Computes protection bandwidth in the network for two cases: $1 + 1$ OCh protection and OCh shared mesh protection.

For $1 + 1$ OCh protection, use an algorithm to provide two disjoint shortest paths for each lightpath, such as the one in [ST84]. For shared mesh protection, use the following algorithm: for each failure i, determine the amount of protection capacity, $C_i(l)$, that would be required on each link l in the network. Prove that the total protection capacity needed on link l is then simply $\max_i C_i(l)$.

(d) Experiment with a variety of topologies, traffic patterns, and different routing/protection computation algorithms. Summarize your conclusions.

References

[AJY00] C. Alaettinoglu, V. Jacobson, and H. Yu. Towards millisecond IGP convergence. In *North American Network Operators Group Fall Meeting*, 2000. See also IETF drafts *draft-alaettinoglu-isis-convergence-00.txt* and *draft-ietf-ospf-scalability-00.txt*.

[AR01] D. Awduche and Y. Rekhter. Multiprotocol lambda switching: Combining MPLS traffic engineering control with optical crossconnects. *IEEE Communications Magazine*, 39(4):111–116, Mar. 2001.

[ATM96] ATM Forum. *Private Network-Network Interface Specification: Version 1.0*, 1996.

[Bar00] S. Baroni et al. Analysis and design of backbone architecture alternatives for IP optical networking. *IEEE Journal of Selected Areas in Communications*, 18(10):1980–1994, Oct. 2000.

[Bat98] R. Batchellor. Optical layer protection: Benefits and implementation. In *Proceedings of National Fiber Optic Engineers Conference*, 1998.

[BZB$^+$97] R. Bradon, L. Zhang, S. Berson, S. Herzog, and S. Jamin. *Resource Reservation Protocol—Version 1 Functional Specification*. Internet Engineering Task Force, Sept. 1997.

[CO99]   T. M. Chen and T. H. Oh. Reliable services in MPLS. *IEEE Communications Magazine*, 37(12):58–62, Dec. 1999.

[Dem99]   P. Demeester et al. Resilience in multilayer networks. *IEEE Communications Magazine*, 37(8):70–77, Aug. 1999.

[Dos99]   B. T. Doshi et al. Optical network design and restoration. *Bell Labs Technical Journal*, 4(1):58–84, Jan.–March 1999.

[DR00]   B. S. Davie and Y. Rekhter. *MPLS Technology and Applications*. Morgan Kaufmann, San Francisco, 2000.

[DWY99]   P. Demeester, T.-H. Wu, and N. Yoshikai, editors. *IEEE Communications Magazine: Special Issue on Survivable Communication Networks*, volume 37, Aug. 1999.

[GR96]   O. Gerstel and R. Ramaswami. Multiwavelength optical network architectures and protection schemes. In *Proceedings of Tirrenia Workshop on Optical Networks*, pages 42–51, 1996.

[GR00a]   O. Gerstel and R. Ramaswami. Optical layer survivability—a services perspective. *IEEE Communications Magazine*, 38(3):104–113, March 2000.

[GR00b]   O. Gerstel and R. Ramaswami. Optical layer survivability: An implementation perspective. *IEEE JSAC Special Issue on Optical Networks*, 18(10):1885–1899, Oct. 2000.

[GRS97]   O. Gerstel, R. Ramaswami, and G. H. Sasaki. Fault tolerant WDM rings with limited wavelength conversion. In *Proceedings of IEEE Infocom*, pages 508–516, 1997.

[HYCG00] G. Hjalmtysson, J. Yates, S. Chaudhuri, and A. Greenberg. Smart routers—simple optics: An architecture for the optical Internet. *IEEE/OSA Journal on Lightwave Technology*, 18(12):1880–1891, 2000.

[ITU93]   ITU-T. *Recommendation Q.700: Introduction to CCITT Signaling System No. 7*, 1993.

[Kha97]   S. Khanna. A polynomial time approximation scheme for the SONET ring loading problem. *Bell Labs Technical Journal*, 2(2):36–41, Spring 1997.

[LC97]   C. Y. Lee and S. G. Chang. Balancing loads on SONET rings with integer demand splitting. *Computer Operations Research*, 24(3):221–229, 1997.

[MB96]   J. Manchester and P. Bonenfant. Fiber optic network survivability: SONET/optical protection layer interworking. In *Proceedings of National Fiber Optic Engineers Conference*, pages 907–918, 1996.

[MBN99]   J. Manchester, P. Bonenfant, and C. Newton. The evolution of transport network survivability. *IEEE Communications Magazine*, 37(8):44–51, Aug. 1999.

[MM00] G. Mohan and C. S. R. Murthy. Lightpath restoration in WDM optical networks. *IEEE Network Magazine*, 14(6):24–32, Nov.–Dec. 2000.

[Pax97] V. Paxson. End-to-end routing behavior in the Internet. *IEEE/ACM Transactions on Networking*, 5(5):601–615, Oct. 1997.

[Ram01] R. Ramamurthy et al. Capacity performance of dynamic provisioning in optical networks. *IEEE/OSA Journal on Lightwave Technology*, 19(1):40–48, 2001.

[RM99a] B. Ramamurthy and B. Mukherjee. Survivable WDM mesh networks, Part I—protection. In *Proceedings of IEEE Infocom*, pages 744–751, 1999.

[RM99b] B. Ramamurthy and B. Mukherjee. Survivable WDM mesh networks, Part II—restoration. In *Proceedings of IEEE International Conference on Communication*, pages 2023–2030, 1999.

[SR97] M. Sexton and A. Reid. *Broadband Networking: ATM, SDH and SONET*. Artech House, Boston, 1997.

[ST84] J. W. Suurballe and R. E. Tarjan. A quick method for finding shortest pairs of disjoint paths. *Networks*, 14:325–336, 1984.

[Tel98] Telcordia Technologies. *Common Generic Requirements for Optical Add Drop Multiplexers (OADMs) and Optical Terminal Multiplexers (OTMs)*, Dec. 1998. GR-2979-CORE, Issue 2.

[WO95] L. Wuttisittikulkij and M. J. O'Mahony. Multiwavelength self-healing ring transparent networks. In *Proceedings of IEEE Globecom*, pages 45–49, 1995.

[Wu92] T. H. Wu. *Fiber Network Service Survivability*. Artech House, Boston, 1992.

[Wu95] T. H. Wu. Emerging techniques for fiber network survivability. *IEEE Communications Magazine*, 33(2):58–74, Feb. 1995.

11 chapter

Access Networks

IN PREVIOUS CHAPTERS, we have explored the use of optical networks for metro and long-haul network applications. The *access network* is the "last leg" of the telecommunications network that runs from the service provider's facility to the home or business. With fiber now directly available to many office buildings in metropolitan areas, networks based on SONET/SDH or Ethernet-based technologies are being used to provide high-speed access to large business users. Business users are big consumers of data services, many of which are delivered in the form of leased lines at various speeds ranging from 1.5 Mb/s to several gigabits per second. While this is happening, the telephone and cable companies are also placing a significant emphasis on the development of networks that will allow them to provide a variety of services to individual homes and small to medium businesses. This is the focus of this chapter.

Today, homes get essentially two types of services: plain old telephone service (POTS) over the telephone network and broadcast analog video over the cable network. Recently added to this mix are data services for Internet access using either digital subscriber line (DSL) technology over the telephone network or cable modem service over the cable network.

Early efforts on developing high-capacity access networks were devoted to developing networks that would accommodate various forms of video, such as video-on-demand and high-definition television. However, the range of services that users are expected to demand in the future is vast and unpredictable. Today, end users

Table 11.1   Different types of services that must be supported by an access network. The bandwidth requirements are given for each individual stream.

Service	Type	Downstream Bandwidth	Upstream Bandwidth
Telephony	Switched	4 kHz	4 kHz
ISDN	Switched	144 kb/s	144 kb/s
Broadcast video	Broadcast	6 MHz or 2–6 Mb/s	0
Interactive video	Switched	6 Mb/s	Small
Internet access	Switched	A few Mb/s	Small initially
Videoconferencing	Switched	6 Mb/s	6 Mb/s
Business services	Switched	1.5 Mb/s–10 Gb/s	1.5 Mb/s–10 Gb/s

are interested in both Internet access and other high-speed data access services, for such applications as telecommuting, distance learning, and eventually entertainment video and videoconferencing. Future, unforeseen applications are also sure to arise and make ever-increasing demands on the bandwidth available in the last mile. The term *full service* encompasses the variety of services that are expected to be delivered via access networks. A sampling of the different services and their characteristics is given in Table 11.1. Both telephone and cable companies are striving to become full-service providers.

At a broad level, these services can be classified based on three major criteria. The first is the bandwidth requirement, which can vary from a few kilohertz for telephony to several megahertz per video stream and hundreds of megabits per second for high-speed leased lines. The second is whether this requirement is *symmetric* (two way), for example, videoconferencing, or *asymmetric* (one way), for example, broadcast video. Today, while most business services are symmetric, other services tend to be asymmetric, with more bandwidth needed from the service provider to the user (the downstream direction) than from the user to the service provider (the upstream direction). The last criterion is whether the service is inherently broadcast, where every user gets the same information, for example, broadcast video, or whether the service is switched, where different users get different information, as is the case with Internet access.

In the next section, we provide an overview of the different types of existing and emerging access network architectures. We then provide a more detailed description of the two most promising access architectures—the hybrid fiber coax (HFC) network and the fiber to the curb (FTTC) approach and its variants.

11.1 Network Architecture Overview

In broad terms, an access network consists of a hub, remote nodes (RNs), and network interface units (NIUs), as shown in Figure 11.1. In the case of a telephone company, the hub is a *central office* (also called a *local exchange* in many parts of the world), and in the case of a cable company, it is called a *head end*. Each hub serves several homes or businesses via the NIUs. An NIU either may be located in a subscriber location or may itself serve several subscribers. The hub itself may be part of a larger network, but for our purposes, we can think of the hub as being the source of data to the NIUs and the sink of data from the NIUs. In many cases, rather than running cables from the hub to each individual NIU, another hierarchical level is introduced between the hub and the NIUs. Each hub may be connected to several RNs deployed in the field, with each RN in turn serving a separate set of NIUs. The network between the hub and the RN is called the *feeder* network, and the network between the RN and the NIUs is called the *distribution* network.

We saw that services could be either broadcast or switched. In the same way, the distribution network could also be either broadcast or switched. Note that in the context of services, we are using the terms *broadcast* and *switched* to denote whether all users get the same information or not. In the context of the network, we are referring to the network topology. Different combinations of services and network topologies are possible—a broadcast service may be supported by a broadcast or a switched network, and a switched service may be supported by a broadcast or a switched network. In a broadcast network, an RN broadcasts the data it receives from the feeder network to all its NIUs. In a switched network, the RN processes the data coming in and sends possibly separate data streams to different NIUs. The telephone network that we will study later is a switched network, whereas the cable television network is a broadcast network. Broadcast networks may be cheaper than switched networks, are well suited for delivering broadcast services, and have the advantage that all the NIUs are identical, making them easier to deploy. (In some switched networks that we will study, different NIUs use different wavelengths, which makes it more complicated to manage and track the inventory of NIUs in the network.) Switched networks, as their name suggests, are well suited for delivering switched services and provide more security. For example, it is not possible for one subscriber to tap into another subscriber's data, and it is more difficult for one subscriber to corrupt the entire network. Fault location is generally easier in a switched network than in a broadcast network. In broadcast networks, the "intelligence" is all at the NIUs, whereas in switched networks, it is in the network. Thus NIUs in switched networks may be simpler than in broadcast networks.

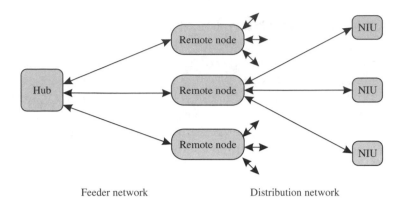

Figure 11.1 Architecture of an access network. It consists of a hub, which is a telephone company central office or cable company head end, remote nodes deployed in the field, and network interface units that serve one or more individual subscribers.

Another way of classifying access networks is based on the type of feeder network, which is the network between the hub and the RN. In one scenario, the feeder network could assign each NIU its own *dedicated* bandwidth. By dedicated bandwidth, we mean that different NIUs are assigned different frequency (or wavelength) bands in the frequency (or wavelength) domain. In another scenario, the feeder network could have a total bandwidth that is *shared* by all the NIUs. By shared bandwidth, we mean that multiple NIUs share a given bandwidth in the time domain. In this case, each NIU could potentially access the entire bandwidth for short periods. For upstream transmission from the NIUs back to the hub, we will need some form of media access control to coordinate access to the shared bandwidth by the NIUs. If the traffic from/to the NIUs is bursty, it is more efficient to share a large total amount of bandwidth among many NIUs rather than assign each NIU its own dedicated bandwidth. On the other hand, with dedicated bandwidth, each NIU can be guaranteed a certain quality of service, which is more difficult to do with shared bandwidth. A disadvantage of the shared bandwidth approach is that each NIU must have optics/electronics that operate at the total bandwidth of the network as opposed to the bandwidth needed by the NIU.

Table 11.2 classifies the different types of access networks that we will be studying in this chapter according to whether their distribution network is broadcast or switched, and whether they use dedicated or shared bandwidth in the feeder network. For example, the telephone network is a switched network with each NIU getting its own dedicated bandwidth of 4 kHz. The cable network is a broadcast network with

Table 11.2 Classification of different types of access networks, from [FRI96]. The acronyms refer to the following: HFC—hybrid fiber coax network; DSL—digital subscriber loop; and PON—passive optical network, with the T standing for telephony, W for wavelength, and WR for wavelength routed.

Distribution Network	Feeder Network	
	Shared	Dedicated
Broadcast Switched	Cable TV (HFC), TPON	WPON Telephony, DSL, WRPON

all NIUs sharing the total cable bandwidth. A broadcast star WDM passive optical network (WPON), with each NIU assigned a separate wavelength, is an example of a broadcast network but with dedicated bandwidth to each NIU. We will study this architecture in Section 11.3.

Today, two kinds of access networks reach our homes: the telephone network and the cable network. The telephone network runs over twisted-pair copper cable. It consists of point-to-point copper pairs between the telco central office and the individual home. The two wires in a pair are twisted together to reduce the crosstalk between them, hence the name *twisted pair*. This plant was designed to provide 4 kHz bandwidth to each home, although we will see that much higher bandwidths can be extracted out of it using contemporary signal-processing techniques. Wires from individual homes are aggregated as shown in Figure 11.2. The telephone network is a switched network that provides dedicated bandwidth to each user.

A typical cable network is shown in Figure 11.3. It consists of fibers between the cable company head end (analogous to a telco central office) and remote (fiber) nodes. Usually, the channels from the head end are broadcast to the remote nodes by using subcarrier multiplexing (SCM) on a laser (see Section 4.2 to understand how SCM works). From the remote node, coaxial cables go to each home. A remote node serves between 500 and 2000 homes. Such a network is called a hybrid fiber coax (HFC) network. The cable bandwidth used is between 50 and 550 MHz, and the cable carries up to 78 AM-VSB (amplitude-modulated vestigial sideband) television signals in channels placed 6 MHz apart in the American NTSC (National Television System Committee) standard. A return path in the 5 to 40 MHz window is available as well. Many cable companies have now upgraded their networks to carry the video channels in digital format. The cable network is a broadcast network where all users share a common total bandwidth. The same set of signals from the head end is delivered to all the homes.

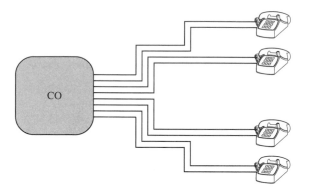

Figure 11.2   The twisted-pair telephone access network, which consists of individual twisted pairs routed from the central office (CO) to the individual subscribers.

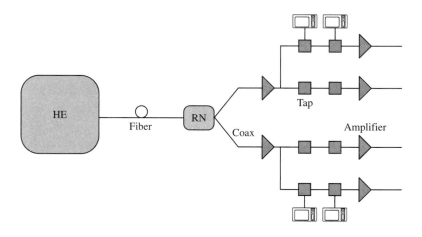

Figure 11.3   The hybrid fiber coax cable television network. The head end broadcasts signals over fiber to the remote node, which then distributes it to individual subscribers via coaxial cable drops.

The telephone and cable networks are vastly different. The telephone network provides very little bandwidth per home but incorporates sophisticated switching equipment and operations and management systems. The cable network provides a lot of bandwidth to each home, but it is all unidirectional and broadcast, with no switching and very simple management.

Several approaches have been used to upgrade the access network infrastructure to support the emerging set of new services. The *integrated services digital network* (ISDN) today provides 144 kb/s of bandwidth over the existing twisted-pair infrastructure and is available in many metropolitan areas. *Digital subscriber loop* (DSL) is another technique that works over the existing twisted-pair infrastructure but provides significantly more bandwidth than ISDN. DSL uses sophisticated modulation and coding techniques to realize a capacity of a few megabits per second over twisted pair, which is sufficient to transmit compressed video. This requires the central office (CO) and the home to each have a DSL modem. However, DSL has some limitations. The realizable bandwidth is inversely proportional to the distance between the CO and the home, and with today's technology, we can achieve several hundred kilobits per second to a few megabits per second over this infrastructure. The existing twisted-pair infrastructure incorporates several 4 kHz filters that must be removed. The bandwidth on the upstream (return) path is severely limited to a few hundred kilobits per second. Many variations and enhancements of DSL have been proposed. As in the conventional telephone network, ISDN and DSL can be classified as switched networks with dedicated bandwidth per NIU.

Satellites provide another way of delivering access services. The direct broadcast satellite system uses a geosynchronous satellite to broadcast a few hundred channels to individual homes. A satellite may provide more bandwidth than a terrestrial coaxial cable system. However, the main problem is that, unlike terrestrial systems, the amount of spatial reuse of bandwidth possible is quite limited, since a single satellite has a wide coverage area within which it broadcasts the signals. Also there is no easy way to handle the upstream traffic. Today, it is possible to have high-speed Internet access delivered via satellite, with the upstream direction carried over a regular telephone line.

Wireless access is yet another viable option. Although it suffers from limited bandwidth and range, it can be deployed rapidly and allows providers without an existing infrastructure to enter the market. Among the variants are the *multichannel multipoint distribution service* (MMDS) and the *local multipoint distribution service* (LMDS), both of which are terrestrial line-of-sight systems. MMDS provides thirty-three 6 MHz channels in the 2–3 GHz band with a range of 15 to 55 km, depending on the transmit power. LMDS operates in the 28 GHz band with 1.3 GHz of bandwidth and is suitable for short-range (3–5 km) deployment in dense metropolitan areas (the distance is also dependent on the amount of rainfall, as rain attenuates signals in this band). Optical fiberless systems using lasers transmitting over free space into the home are also being developed as an alternative approach. These systems can provide about 622 Mb/s of capacity over a line-of-sight range of 200 to 500 m.

In the context of the next-generation access network, the two main architectures being considered today are the so-called hybrid fiber coax (HFC) approach and the fiber to the curb (FTTC) approach. The HFC approach is still a broadcast architecture, whereas the FTTC approach incorporates switching.

11.2 Enhanced HFC

Although we have used the term HFC to describe the existing cable infrastructure, HFC is also the term used to describe an upgraded version of this architecture, which we will refer to as an *enhanced* HFC architecture. Since both the fiber and the coax cable carry multiple subcarrier modulated streams, and it is a broadcast network, a better term to describe the HFC architecture is *subcarrier modulated fiber coax bus* (SMFCB). The network architecture is essentially the same as that shown in Figure 11.3. In order to provide increased bandwidth per user, the network is being enhanced using a combination of several techniques. First, the transmitted frequency range can be increased, for example, up to 1 GHz from the 500 MHz in conventional HFC systems. Enhanced HFC systems being deployed today in larger metropolitan areas are already delivering up to 862 MHz of bandwidth. Within each subcarrier channel, we can use spectrally efficient digital modulation techniques, such as 256 QAM (quadrature amplitude modulation), which provides a spectral efficiency of 7 bits/Hz. In addition, we can drive fiber deeper into the network and reduce the number of homes served by a remote node down to about 50 homes, from the 500 homes typically served by an HFC network. This is being done today as well. We can also use multiple fibers and multiple wavelengths to increase the overall capacity.

In a typical enhanced HFC architecture, like the existing cable network, downstream data is broadcast from the head end to remote (fiber) nodes by using a passive optical star coupler. In recent deployments, it is common to use high-power 1.55 μm transmitters in conjunction with booster amplifiers to achieve a high split ratio. In addition, signals at 1.3 μm can be multiplexed on the same set of fibers. These 1.3 μm signals can be used in a *narrowcasting* mode. That is, these signals can be transmitted only to a selected set of users, rather than to all users. This feature can be used to provide additional bandwidth for selected groups of users.

From a remote node, several coax trees branch out to the network interface units. An NIU may serve one or more homes. Its function is to separate the signals into telephone signals and broadcast video signals, and to send the telephone signal on twisted pair and the video signal on coax to each home that it serves. Each coax leg serves about 50–500 homes. Logically, the architecture is a broadcast bus, although it is implemented as a combination of optical stars and coax trees/buses. Downstream

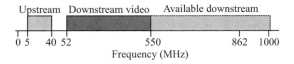

Figure 11.4   Bandwidth allocation in an enhanced HFC network.

broadcast video to the home would be sent on analog subcarrier channels. Video signals could be sent as analog AM-VSB streams, compatible with existing equipment inside homes. Digital video, as well as telephony and data services, can be carried over the same infrastructure. In addition, upstream channels can be provided in the 5–40 MHz band, which is not used for downstream traffic. Figure 11.4 shows the bandwidth usage in an enhanced HFC network.

The cable infrastructure has already been upgraded in many cities to provide Internet access services through the use of a specific modem developed for this application, called a cable modem, at the head end and at the home. The modems use a shared media Ethernet-type media access control protocol to provide this service. The peak rate of this service is on the order of a few megabits per second, but is shared among all the users in a neighborhood as the HFC network is fundamentally a broadcast network. The amount of bandwidth available per user depends on how many other users are accessing the network and the traffic generated by the other users.

Clearly, enhanced HFC is the natural evolution path for the cable service providers. It maintains compatibility with existing analog equipment and is an efficient approach to deliver broadcast services. On the other hand, it has the disadvantages of a coax-based solution, such as limited upstream bandwidth, limited reliability, and powering needed for the many amplifiers in the path.

11.3 Fiber to the Curb (FTTC)

In contrast to HFC, in FTTC, data is transmitted digitally over optical fiber from the hub, or central office, to fiber-terminating nodes called *optical network units* (ONUs). The expectation is that the fiber would get much closer to the subscriber with this architecture. Depending on how close the fiber gets to an individual subscriber, different terms are employed to describe this architecture (see Figure 11.5). In the most optimistic scenario, fiber would go to each home, in which case this architecture is called *fiber to the home* (FTTH), and the ONUs would perform the function of the NIUs. For the case where ONUs serve a few homes or buildings, say, 8–64, this can

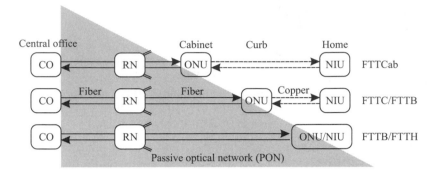

Figure 11.5 Different types of fiber access networks, based on how close the fiber gets to the end user. In many cases, the remote node may be located at the central office itself. The ONUs terminate the fiber signal, and the links between the ONUs and the NIUs are copper based.

be thought of as *fiber to the curb* (FTTC) or *fiber to the building* (FTTB). Typically, in FTTC, the fiber is within about 100 m of the end user. In this case, there is an additional distribution network from the ONUs to the NIUs. With the fiber to the cabinet (FTTCab) approach, the fiber is terminated in a cabinet in the neighborhood and is within about 1 km of the end user.

To make the FTTC architecture viable, the network from the CO to the ONU is typically a *passive optical network* (PON). The remote node is a simple passive device such as an optical star coupler, and it may sometimes be colocated in the central office itself rather than in the field. Although many different architectural alternatives can be used for FTTC, the term FTTC is usually used to describe a version where the signals are broadcast from the central office to the ONUs, and the ONUs share a common total bandwidth in time division multiplexed fashion.

In the context of FTTC, the feeder network is the portion of the network between the central office and the remote node, and the distribution network is between the remote node and the ONUs. We will see that a variety of different types of architectures can be realized by using different types of sources at the central office combined with different types of remote nodes.

Practically speaking, it is quite expensive today to transmit analog video signals over an all-fiber infrastructure; this may necessitate an analog hybrid fiber coax overlay that carries the analog video signals. The FTTC architecture is sometimes also called *baseband modulated fiber coax bus* (BMFCB) or *switched digital video* (SDV).

Table 11.3   Comparison of different PON architectures. N denotes the number of ONUs in the network. An ONU bit rate of 1 indicates that the ONU operates at the bit rate corresponding to the traffic it terminates rather than the aggregate traffic of N. Node sync refers to whether the nodes in the network must be synchronized to a common clock or not. CO sharing relates to whether the equipment is shared among multiple users or whether separate equipment is required to service each user.

Architecture	Fiber Sharing	Power Splitting	ONU Bit Rate	Node Sync	CO Sharing
All fiber	No	None	1	No	No
TPON	Yes	$1/N$	N	Yes	Yes
WPON	Yes	$1/N$	1	Yes	No
WRPON	Yes	None	1	Yes	Yes

In what follows, we shall concentrate on different alternatives for realizing the portion of the access network that is optical. Optical access network architectures must be simple, and the network must be easy to operate and service. This means that passive architectures, where the network itself does not have any switching in it and does not need to be controlled, are preferable to active ones. Passive networks also do not need to be powered, except at the end points, which provide significant cost savings to operators. Moreover, the ONU itself must be kept very simple in order to reduce cost and improve reliability. This rules out using sophisticated lasers and other optical components within the ONU. Preferably, the components used in the ONU must be capable of operating without any temperature control. The CO equipment can be somewhat more sophisticated, since it resides in a controlled environment, and its cost can be amortized over the many subscribers served out of a single CO.

The optical networks proposed for this application are commonly called PONs (passive optical networks)—all of them use passive architectures. They use some form of passive component, such as an optical star coupler or static wavelength router, as the remote node. The main advantages of using passive architectures in this case come from their reliability, ease of maintenance, and the fact that the field-deployed network does not need to be powered. Moreover, the fiber infrastructure itself is transparent to bit rates and modulation formats, and the overall network can be upgraded in the future without changing the infrastructure itself. Table 11.3 compares the different architectures.

The simplest PON architecture, shown in Figure 11.6(a), uses a separate fiber pair from the CO to each ONU. The main problem with this approach is that the cost of CO equipment scales with the number of ONUs. Moreover, the operator needs

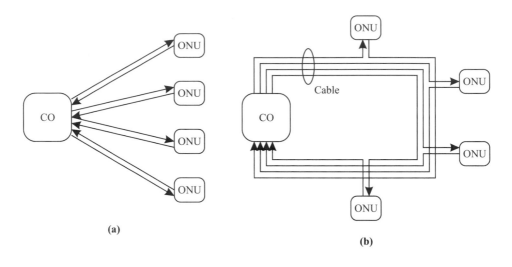

Figure 11.6   (a) The point-to-point fiber approach. (b) In practice the fibers could be laid in the form of a ring.

to install and maintain all these fiber pairs. This approach is being implemented on a limited scale today, primarily to provide high-speed services to businesses. In Japan, NTT is operating such a system at bit rates from 8 to 32 Mb/s over each fiber. Although logically there is a separate fiber pair to each ONU, physically the fibers could be laid in a ring configuration, as shown in Figure 11.6.

Instead of providing a fiber pair to each ONU, a single fiber can be used with bidirectional transmission. However, the same wavelength cannot be used to transmit data simultaneously in both directions because of uncontrolled reflections in the fiber. One way is to use time division multiplexing so that both ends don't transmit simultaneously. Another is to use different wavelengths (1.3 and 1.55 μm, for example) for the different directions.

More commonly, rather than dedicating a fiber pair per user, the fiber pair is shared by many users. The most common example of such networks are the SONET/SDH rings, which are now widely deployed to provide high-speed services to large business customers. These rings operate at speeds ranging from 155 Mb/s to 10 Gb/s. In this case, an ONU is a SONET add/drop multiplexer (ADM), and multiple ONUs can be present on the same ring. However, these rings are not considered part of the PON family. Rather they can be viewed as an alternative fiber access solution.

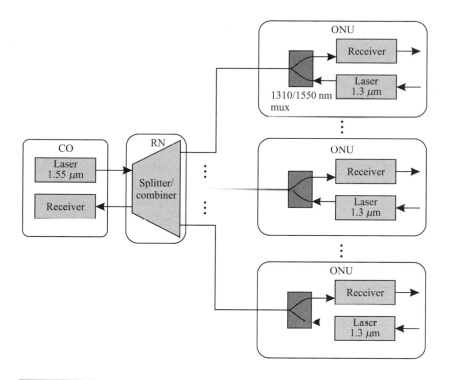

Figure 11.7 A broadcast and select TPON. The CO broadcasts its signal downstream to all the ONUs using a passive star coupler. The ONUs share an upstream channel in a time-multiplexed fashion. In this case, upstream and downstream signals are carried using different wavelengths over a single fiber.

While SONET/SDH rings are suitable for delivering the higher-speed services and addressing the needs of large business customers, the PON architectures that we will study here can provide a more cost-effective solution for addressing the needs of small- and medium-sized businesses and homes, which require a few DS1 (1.5 Mb/s) lines, DSL lines, or 10 Mb/s Ethernet connections.

The most common PON architecture is the TPON (originally called PON for telephony) architecture [Ste87], shown in Figure 11.7. The downstream traffic is broadcast by a transmitter at the CO to all the ONUs by a passive star coupler. Though the architecture is a broadcast architecture, switched services can be supported by assigning specific time slots to individual ONUs based on their bandwidth demands. For the upstream channel, the ONUs share a channel that is combined using a coupler, again via fixed time division multiplexing (TDM) or some other multiaccess protocol. In the TDM approach, the ONUs need to be synchronized to a common clock. This is done by a process called *ranging*, where each ONU measures its delay from the CO and adjusts its clock such that all the ONUs are synchronized relative to the CO. The CO then assigns time slots to each ONU as needed.

This architecture allows the relatively expensive CO equipment to be shared among all the ONUs and makes use of fairly mature low-cost optical components. The CO transmitter can be an LED or a Fabry-Perot laser, and cheap, uncooled pinFET receivers and LEDs/Fabry-Perot lasers can be used within the ONUs. The number of ONUs that can be supported is limited by the splitting loss in the star coupler. Each ONU must have electronics that run at the *aggregate* bit rate of all the ONUs. There is a trade-off between the transmit power, receiver sensitivity, bit rate, and number of ONUs (which determines the splitting loss) and the total distance covered.

As we mentioned earlier, TPONs may be more cost-effective at offering lower-speed services compared to SONET/SDH rings or Ethernet-based offerings. TPON vendors claim that it is easier to provision bandwidth in a flexible manner remotely by changing the number of time slots an individual subscriber is assigned. However, the current generation of SONET/SDH products is being enhanced to provide dynamic bandwidth provisioning as well. In a TPON, a failure of one subscriber's equipment does not affect other subscribers, whereas a SONET/SDH ring node failing affects all the nodes on the ring. However, SONET/SDH has built-in protection mechanisms to reroute traffic in the event of both equipment failures and fiber cuts and restore services rapidly. In contrast, dealing with fiber cuts is not easy in the TPON architecture, without doubling up on the fiber plant. By the same token, with the TPON architecture, additional subscribers can be added without affecting any of the other subscribers. In SONET/SDH rings, this is a more complex process.

TPON development has accelerated with the establishment of the *full service access network* standard by a large group of service providers and equipment companies. The standard specifies an ATM-based TPON architecture with a downstream bandwidth of up to 622 Mb/s and an upstream bandwidth of up to 155 Mb/s. The targeted distance is 20 km with a total fiber attenuation in the 10–30 dB range. Practical link budgets using lasers at the CO and ONUs allow a 16- to 32-way split with this approach. For example, a TPON operating at 622 Mb/s using a 32-way splitter can provide each subscriber with about 20 Mb/s of bandwidth. The TPON can operate over a single fiber pair by using different wavelengths in the upstream (1.3 μm) and downstream (1.55 μm) directions. Alternatively, it can also operate over a fiber pair using 1.3 μm transmitters. As of this writing, FSAN-based TPONs are beginning to be deployed in the field. Some carriers, notably BT, have deployed early versions of TPONs for several years, primarily for telephony. Their deployment may become more widespread as optical component costs come down.

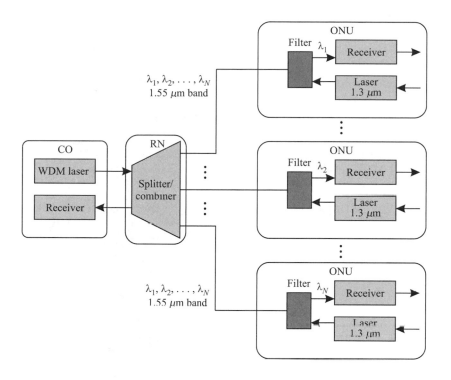

Figure 11.8   A broadcast-and-select WDM PON (WPON), which is an upgraded version of the basic PON architecture. In this case, the CO broadcasts multiple wavelengths to all the ONUs, and each ONU selects a particular wavelength. As in a conventional TPON, the ONUs time-share an upstream channel at a wavelength different from the downstream wavelengths.

Many enhancements have been proposed to the basic TPON architecture to increase its capacity and flexibility. The next step, shown in Figure 11.8, is to replace the single transceiver at the CO with a WDM array of transmitters or a single tunable transmitter to yield a WDM PON (WPON). This approach allows each ONU to have electronics running only at the rate it receives data, and not at the aggregate bit rate. However, it is still limited by the power splitting at the star coupler.

Introducing wavelength routing solves the splitting loss problem while retaining all the other advantages of the WDM PON. In addition, it allows point-to-point dedicated services to be provided to ONUs. This leads to the WRPON architecture shown in Figure 11.9.

Several types of WRPONs have been proposed and demonstrated. They all use a wavelength router, typically an arrayed waveguide grating (AWG) for the

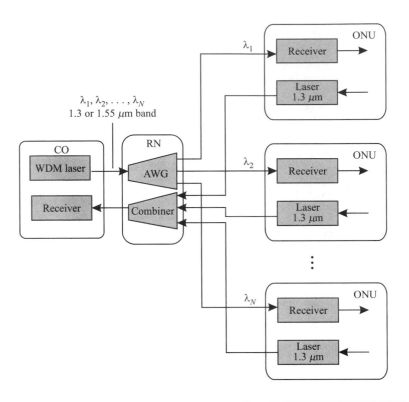

Figure 11.9   A wavelength-routing PON (WRPON). In this case, a passive arrayed waveguide grating (AWG) is used to route different wavelengths to different ONUs in the downstream direction, without incurring a splitting loss. As in the TPON and WPON architectures, the ONUs time-share a wavelength for upstream transmission.

downstream traffic, but vary in the type of equipment located at the CO and ONUs, and in how the upstream traffic is supported. The router directs different wavelengths to different ONUs. The earliest demonstration was the so-called passive photonics loop (PPL) [WKR+88, WL88]. It used 16 channels in the 1.3 μm band for downstream transmission and 16 additional channels in the 1.55 μm band for upstream transmission. However, this approach is not economical because we need two expensive lasers for each ONU—one inside the ONU and one at the central office. We describe several variants of this architecture that provide more economical sharing of resources at the CO and ONUs.

The RITENET architecture [Fri94] (see Figure 11.10) uses a tunable laser at the CO. A frame sent to each ONU from the CO consists of two parts: a *data* part, wherein data is transmitted by the CO, and a *return traffic* part, wherein no data

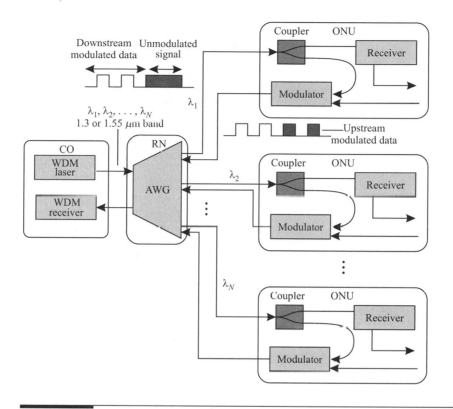

Figure 11.10   The RITENET WRPON architecture. The ONUs use an external modulator to modulate an unmodulated signal transmitted from the CO.

is transmitted but the CO laser is left turned on. Each ONU is provided with an external modulator. During the return traffic part of the frame, the ONU uses the modulator to modulate the light signal from the CO. This avoids the need for having a laser at the ONU. The upstream traffic from the ONUs is also sent to the router. The router combines all the different wavelengths and sends them out on a common port to a receiver in the CO. If a single receiver is used in the CO, then the ONUs must use time division multiplexing to get access to that receiver. Alternatively, if a separate receiver is used for each wavelength at the CO, each ONU gets a dedicated wavelength to transmit upstream back to the CO. This architecture avoids the need for having a laser at each ONU. Instead, each ONU has an external modulator.

A lower-cost alternative to RITENET is the LARNET architecture [ZJS+95] (see Figure 11.11), which uses an LED at the ONU instead of an external modulator for transmission in the upstream direction. The LED emits a broadband signal that gets "sliced" upon going through the wavelength router, as shown in Figure 11.12. Only

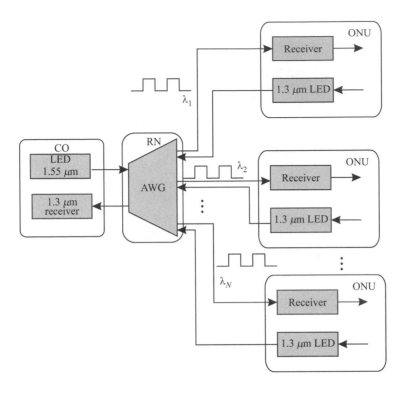

Figure 11.11   The LARNET WRPON architecture. A broadband signal from the LED at the CO is split into individual wavelength components by the AWG and broadcast to all the ONUs.

the power in the part of the LED spectrum corresponding to the passband of the wavelength router is transmitted through to the receiver at the CO. Note, however, that with N ONUs, this imposes a splitting loss of at least $1/N$—only a small fraction of the total power falls within the passband of the router.

More important, an LED can be used at the CO as well [IFD95] for downstream transmission. In this case, the signal sent by the CO LED effectively gets broadcast to all the ONUs. It is in fact possible to have two transmitters within the CO: an LED, say, at 1.3 μm, broadcasting to all the ONUs, and a tunable laser at 1.55 μm selectively transmitting to the ONUs. This is an important way to carry broadcast analog video signals over the digital switched fiber infrastructure at low cost without having to use a separate overlay network for this purpose.

WDM components for PONs are not yet mature and are more expensive than the components required for simple broadcast PONs. However, WRPONs offer much

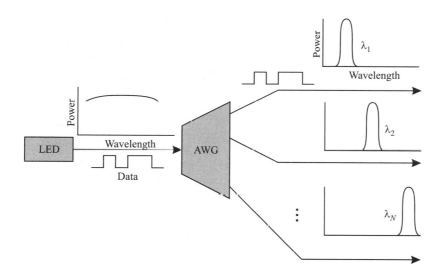

Figure 11.12 Spectral slicing: if a broadband LED signal is sent through a filter, only the portion of the LED spectrum that is passed by the filter comes out.

higher capacities than the simple broadcast PONs, and simple PONs can be upgraded to WRPONs as the need arises.

11.3.1 PON Evolution

We have studied a number of PON variants in this section. It is important to realize that there is a nice evolution path from a very simple TPON architecture to some of the more complex WRPON architectures. The evolution can be performed with minimal disruption of existing services and without wasting already-deployed equipment. In general, the terminal equipment can be upgraded as additional capacity and services are needed, without having to upgrade the outside fiber plant, which is a true long-term investment. The upgrade scenario for PONs could go as follows. The operator can start by deploying a simple broadcast TPON, which is a broadcast star network with shared bandwidth, according to the classification of Table 11.2. If more ONUs need to be supported, the operator can upgrade the network to a WDM broadcast PON, which is a broadcast network with dedicated bandwidth provided to each ONU. This can be done by upgrading the transmitters at the CO to WDM transmitters, and the operator may be able to reuse the existing ONUs. If higher capacities per ONU are needed, the operator can further upgrade the network to a wavelength-routed PON, which is a switched network with dedicated bandwidth. Moreover, this wavelength-routed PON can also support broadcast services efficiently using the spectral slicing technique described earlier. Thus there is an upgrade path starting from a broadcast network with shared bandwidth to a broadcast network with dedicated bandwidth and eventually to a switched network with dedicated bandwidth.

Summary

Service providers, both telephone operators and cable companies, are actively looking to deploy broadband access networks to provide a variety of new services. Fiber-based services are now available for many businesses in metropolitan areas. When it comes to residential access, however, fiber is yet to reach the home. SONET/SDH ring-based architectures have been deployed to support the needs of large business customers, but they are not as suited for supporting the needs of residential users and small business customers. The two main architectures for broadband access networks are the hybrid fiber coax (HFC) architecture, which is based on evolving the current plant deployed by cable television operators, and the fiber to the curb (FTTC) architecture, or equivalently a passive optical network (PON) architecture. Compared to the HFC approach, FTTC has a higher initial cost, but provides bandwidth deeper in the network and may prove to be a better longer-term solution. Although FTTC refers to a simple broadcast TDM star PON architecture, we also explored several upgrade options of the PON approach that provide higher capacities by making clever use of wavelength division multiplexing techniques. A number of major telephone carriers and manufacturers in the world have gotten together and defined the requirements for an FTTC-based architecture to enable them to deploy the full service access network [FSA98]. As of this writing, PONs based on this architecture are just beginning to be deployed.

FTTC is attractive in places where coaxial cable is not already deployed, which is the case in many countries other than the United States. FTTC also makes sense for telephone companies who lack a cable infrastructure.

Variants of FTTC have been around a long time, but deployment has been slow for several reasons. First, there is significant cost associated with building and deploying a new access network, which can take several years to pay back. Therefore there is a big barrier toward making the investment in the first place. Second, this is coupled with the uncertain outlook in terms of the revenue that can be generated from the investment. Third, optical component costs are only now starting to decline, with the development of components especially optimized for PON applications, such as low-cost, uncooled semiconductor lasers and transceivers.

The HFC approach, on the other hand, is attractive in places where coaxial cable is already deployed to the home, such as the United States. It is the logical evolution choice for cable companies who have already deployed a simpler version of the HFC architecture to provide basic cable television service.

As optical component costs come down and bandwidth needs increase, it is clear that optical fiber will play a major role in access networks—the question is how close will it get to our homes?

Further Reading

There is a vast body of literature on access networks, and several conferences have sessions devoted to it. There is an informative Web page maintained by the DSL forum (*http://www.adsl.com*). See also [Bha99] for a nice overview of the different types of DSL.

The papers in [Fra98, Aar95, Kob94, KKS00, SKY89] describe plans for deploying fiber in the access network and compare different architectural approaches. TDM PONs were first proposed in [Ste87]. At this time, a group of telephone carriers and manufacturers have gotten together and agreed upon a standard for a fiber-based access architecture, called the full service access network (FSAN) initiative. Details may be found on the Web at *www.fsanet.net* and in [FSA98, Qua98]. The International Telecommunications Union (ITU) has published a requirements standard based on FSAN [ITU98].

[FRI96, FHJ+98, VMVQ00] describe some possible evolutions of the basic TPON architecture by making clever use of WDM and optical amplifiers. A variety of WDM PONs are described in [WKR+88, WL88, Fri94, ZJS+95, IFD95, IRF96].

Problems

11.1 Do a power budget calculation for the different types of PON architectures considered in this chapter and determine the number of ONUs that can be supported in each case, assuming the following parameters:

Laser output power	−3 dBm
LED output power	−20 dBm
Transmit bit rate	155 Mb/s
Receiver sensitivity	−40 dBm
Fiber loss, including connectors	10 dB
1×8 wavelength router loss	5 dB
1×32 wavelength router loss	9 dB
1×64 wavelength router loss	12 dB
Excess splitter loss	1 dB

The normal wavelength router losses are indicated above. However, with spectral slicing, an additional loss is also incurred as only a small fraction of the spectrum is transmitted out of each port on the wavelength router. Assume that in addition to the standard loss, we get only $1/2N$ of the transmitted power in each channel, where N is the number of ONUs.

11.2   Consider the RITENET architecture shown in Figure 11.10. Suppose the laser speed at the CO is limited to 155 Mb/s. The network needs to support 20 ONUs and provide each ONU with 10 Mb/s bandwidth from the CO to the ONU and 2 Mb/s from the ONU to the CO. How could you modify the architecture to support this requirement?

References

[Aar95]   R. Aaron, editor. *IEEE Communications Magazine: Special Issue on Access to Broadband Services*, volume 33, Aug. 1995.

[Bha99]   V. K. Bhagavath. Emerging high-speed xDSL access services: Architectures, issues, insights, and implications. *IEEE Communications Magazine*, 37(11):106–114, Nov. 1999.

[FHJ$^+$98]   R. D. Feldman, E. E. Harstead, S. Jiang, T. H. Wood, and M. Zirngibl. An evaluation of architectures incorporating wavelength division multiplexing for broad-band fiber access. *IEEE/OSA Journal on Lightwave Technology*, 16(9):1546–1559, Sept. 1998.

[Fra98]   P. W. France, editor. *BT Technology Journal—Special Issue on Local Access Technologies*, volume 16, Oct. 1998.

[Fri94]   N. J. Frigo et al. A wavelength-division-multiplexed passive optical network with cost-shared components. *IEEE Photonics Technology Letters*, 6(11):1365–1367, 1994.

[FRI96]   N. J. Frigo, K. C. Reichmann, and P. P. Iannone. WDM passive optical networks: A robust and flexible infrastructure for local access. In *Proceedings of International Workshop on Photonic Networks and Technologies*, pages 201–212, 1996.

[FSA98]   *Full Services Access Network Requirements Specification*, 1998. Available on the Web at *www.fsanet.net*.

[IFD95]   P. P. Iannone, N. J. Frigo, and T. E. Darcie. WDM passive optical network architecture with bidirectional optical spectral slicing. In *OFC'95 Technical Digest*, pages 51–53, 1995. Paper TuK2.

[IRF96]   P. P. Iannone, K. C. Reichmann, and N. J. Frigo. Broadcast digital video delivered over WDM passive optical networks. *IEEE Photonics Technology Letters*, 8(7):930–932, 1996.

[ITU98]   ITU-T. *Recommendation G.983: Broadband Optical Access Systems Based on Passive Optical Networks*, 1998.

[KKS00]   D. Kettler, H. Kafka, and D. Spears. Driving fiber to the home. *IEEE Communications Magazine*, 38(11):106–110, Nov. 2000.

[Kob94]   I. Kobayashi, editor. *IEEE Communications Magazine: Special Issue on Fiber-Optic Subscriber Loops*, volume 32, Feb. 1994.

[Qua98]   J. A. Quayle et al. Achieving global consensus on the strategic broadband access network—the full service access initiative. *BT Technology Journal*, 16(4):58–70, Oct. 1998.

[SKY89]   P. W. Shumate, O. Krumpholz, and K. Yamaguchi, editors. *IEEE/OSA JLT/JSAC Special Issue on Subscriber Loop Technology*, volume 7, Nov. 1989.

[Ste87]   J. Stern et al. Passive optical local networks for telephony applications. *Electronics Letters*, 23:1255–1257, 1987.

[VMVQ00]   I. Van de Voorde, C. M. Martin, J. Vandewege, and X. Z. Qiu. The superPON demonstrator: An exploration of possible evolution paths for optical access networks. *IEEE Communications Magazine*, 38(2):74–82, Feb. 2000.

[WKR⁺88]   S. S. Wagner, H. Kobrinski, T. J. Robe, H. L. Lemberg, and L. S. Smoot. Experimental demonstration of a passive optical subscriber loop architecture. *Electronics Letters*, 24:344–346, 1988.

[WL88]   S. S. Wagner and H. L. Lemberg. Technology and system issues for the WDM-based fiber loop architecture. *IEEE/OSA Journal on Lightwave Technology*, 7(11):1759–1768, 1988.

[ZJS⁺95]   M. Zirngibl, C. H. Joyner, L. W. Stulz, C. Dragone, H. M. Presby, and I. P. Kaminow. LARnet, a local access router network. *IEEE Photonics Technology Letters*, 7(2):1041–1135, Feb. 1995.

12 chapter

Photonic Packet Switching

I N THIS CHAPTER, we study optical networks that are capable of providing packet-switched service at the optical layer. We call these networks *photonic packet-switched* (PPS) networks. Packet-switched services are provided today using electronic switches by many networks, such as IP and ATM networks. Here, we are interested in networks where the packet-switching functions are performed *optically*. The goal of PPS networks is to provide the same services that electronic packet-switched networks provide, but at much higher speeds.

The optical networks that we have studied so far provide circuit-switched services. These networks provide lightpaths, which can be established and taken down as needed. In these networks, the optical nodes do not switch signals on a packet-by-packet basis, but rather only switch at the time a circuit is established or taken down. Packet switching is done in the electronic domain by other equipment such as IP routers or ATM switches. These routers and ATM switches make use of lightpaths provided by the optical layer to establish links between themselves as needed. In addition to switching packets, routers and ATM switches make use of sophisticated software and hardware to perform the control functions needed in a packet-switched network.

We will see in this chapter that all the building blocks needed for optical packet switching are in a fairly rudimentary state today and exist only in research laboratories—they are either difficult to realize, very bulky, or very expensive, even after a decade of research in this area. Moreover, it is likely that we will need electronics to perform the intelligent control functions for the foreseeable future. Optics can be used to switch the data through, but it does not yet have the computing capabilities to perform many of the control functions required, such as processing the packet header, determining the route for the packet, prioritizing packets based on class of service, maintaining topology information, and so on.

However, there are a few motivations for researching optical packet switching. One is that optical packet switches hold the potential for realizing higher capacities than electronic routers (although this potential is yet to be demonstrated!). For instance, the capacity of the best routers today is less than 1 Tb/s, with the highest-speed interfaces being at 10 Gb/s. In contrast, optical switches are, for the most part, bit rate independent, so they can be used to switch tens to hundreds of Tb/s of traffic. At line rates of 80 Gb/s and beyond, electronic time division multiplexing (TDM) appears to be running out of gas, and optical time division multiplexing may be the way to go.

Another motivation for studying optical packet switching is that it can improve the bandwidth utilization within the optical layer. The notion is that high-speed optical links between routers are still underutilized due to the bursty nature of traffic, and using an underlying optical packet layer instead of an optical circuit layer will help improve link utilizations. The question is whether having another high-speed packet-switched layer under an already existing packet-switched layer (say, IP) will provide sufficient improvement in statistical link utilization. The answer depends on the statistical properties of the traffic. The conventionally accepted wisdom is that as many lower-speed bursty traffic streams are multiplexed through many layers, the burstiness of the aggregate stream is lower than that of the individual streams. In this case, having an optical packet layer under an electrical packet layer may not help much because the traffic entering the optical layer is already smoothed out. However, it has been shown recently that with some types of bursty traffic, notably the so-called self-similar traffic, that the burstiness of a multiplexed stream is not less than that of its constituent individual streams [PF95, ENW96]. For such traffic, using an optical packet layer provides the potential to improve the link utilization.

Figure 12.1 shows a generic example of a store-and-forward packet-switched network. In this network, the nodes $A–F$ are the switching/routing nodes; the end nodes 1–6 are the sources and sinks of packet data. We will assume that all packets are of fixed length. Packets sent by an end node will, in general, traverse multiple links and hence multiple routing nodes, before they reach their destination end node. For example, if node 1 has to send a packet to node 6, there are several possible routes that it can take, all consisting of multiple links and routing nodes. If the route chosen for this packet is $1–A–B–D–F–6$, this packet traverses the links $1–A$, $A–B$, $B–D$, $D–F$, and $F–6$. The routing nodes traversed are A, B, D, and F. Note that the route chosen may be specified by the packet itself, or the packet may simply specify only the destination node and leave the choice of route to the routing nodes in its path. In the remainder of the discussion, we will assume that the route is chosen

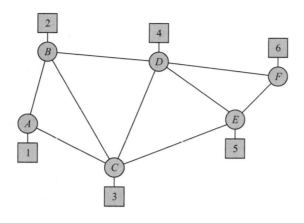

Figure 12.1 A generic store-and-forward network.

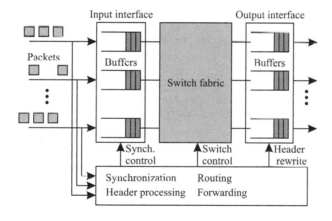

Figure 12.2 A routing node in the network of Figure 12.1.

by the routing nodes based on the packet destination that is carried in the packet header.

Figure 12.1 is also the block diagram of a PPS network. The major difference is that the links run at very high speeds (hundreds of gigabits per second) and the signals are handled mostly optically within each routing node.

Figure 12.2 shows a block diagram depicting many of the functions of a routing node, or router. In general, there is one input from, and one output to, each other routing node and end node that this routing node is connected to by a link. For example, in Figure 12.1, routing node *A* has three inputs and outputs: from/to routing node *B*, routing node *C*, and end node 1. Similarly, routing node *C* has five inputs and outputs. Routers perform the following functions (see Section 6.3 for a more detailed description of how these functions are performed by IP routers):

Routing. Routers maintain up-to-date information of the network topology. This information is maintained in the form of a routing table stored at each node.

Forwarding. For each incoming packet, a router processes the packet header and looks up its routing table to determine the output port for that packet. It may also make some changes to the header itself and reinsert the header at the output. For example, we studied in Chapter 6 that an ATM switch determines the virtual circuit identifier (VCI) in the header of the incoming cell and looks it up in its locally stored VCI table, which provides the corresponding output link and outgoing VCI. It then inserts the outgoing VCI into the cell header.

Switching. Switching is the actual process of switching the incoming packet to the appropriate output port determined by the forwarding process. In the electronic domain, the forwarding and switching functions are usually treated together as a single function, but it will prove to be useful to separate them in PPS networks.

Buffering. There are many reasons why buffering is needed in a router. Perhaps the most important one in this context is to deal with destination conflicts. Multiple packets arrive simultaneously at different inputs of a router. Several of these may have to be switched to the same output port. However, at any given time, only one packet can be switched to any given output port. Thus the router will have to buffer the other packets until they get their turn. Buffers are also used to separate packets based on their priorities or class of service.

Figure 12.2 shows buffers at the input as well as the output. We will explore the trade-offs between input and output buffering in Section 12.4. We will see that buffers are difficult to realize in the case of photonic packet switches, and most switch proposals therefore use only a small amount of buffering, usually integrated with the switch.

Multiplexing. Routers multiplex many lower-speed streams into a higher-speed stream. They also perform the reverse demultiplexing operation.

Synchronization. Synchronization can be broadly defined as the process of aligning two signal streams in time. In PPS networks, it refers either to the alignment of an incoming pulse stream and a locally available *clock* pulse stream or to the relative alignment of two incoming pulse streams. The first situation occurs during multiplexing and demultiplexing, and the second occurs at the inputs of the router where the different packet streams need to be aligned to obtain good switching performance.

PPS networks will have to perform all the functions described above. Some of these functions involve a fair amount of sophisticated logic and processing and are still best handled in the electrical domain. The routing and forwarding functions, in particular, fit into this category. Most PPS proposals to date assume that the packet header is transmitted separately from the data at a lower speed and process the header electronically. We will, however, study some of the approaches toward providing at least rudimentary header processing in the optical domain.

Due to technological constraints, it is quite difficult to perform even the remaining functions of switching, buffering, multiplexing, and synchronization in the optical domain. This will become clearer as we explore the different techniques for performing these functions. Therefore, PPS networks are at this time still in research laboratories and have not yet entered the commercial marketplace. To simplify the implementation, especially the control functions, most PPS proposals also assume the use of *fixed-size* packets, and we will make the same assumption in this chapter. Of course in reality we have to deal with varying packet sizes. If a fixed packet size is used inside the network, then the longer packets will have to be segmented at the network inputs and reassembled together at the end. Alternatively, we could design the PPS nodes to switch variable-sized packets, a more complex proposition.

The outline of this chapter is as follows. We start by describing techniques for multiplexing and demultiplexing optical signals in the time domain, followed by methods of doing synchronization in the optical domain. Synchronization requires delaying one stream with respect to the other if they are misaligned in time. In this context, we will also study how tunable optical delays can be realized. We then discuss various solutions for dealing with the buffering problem. We conclude the chapter by discussing burst switching, a variant of PPS, and some of the experimental work that has been carried out to demonstrate the various aspects of PPS.

12.1 Optical Time Division Multiplexing

At the inputs to the network, lower-speed data streams are multiplexed optically into a higher-speed stream, and at the outputs of the network, the lower-speed streams must be extracted from the higher-speed stream optically by means of a demultiplexing function. Functionally, optical TDM (OTDM) is identical to electronic TDM. The only difference is that the multiplexing and demultiplexing operations are performed entirely optically at high speeds. The typical aggregate rate in OTDM systems is on the order of 100 Gb/s, as we will see in Section 12.6.

OTDM is illustrated in Figure 12.3. Optical signals representing data streams from multiple sources are interleaved in time to produce a single data stream. The interleaving can be done on a bit-by-bit basis as shown in Figure 12.3(a). Assuming the data is sent in the form of packets, it can also be done on a packet-by-packet basis, as shown in Figure 12.3(b). If the packets are of fixed length, the recognition of packet boundaries is much simpler. In what follows, we will assume that fixed-length packets are used.

In both the bit-interleaved and the packet-interleaved case, *framing pulses* can be used. In the packet-interleaved case, framing pulses mark the boundary between packets. In the bit-interleaved case, if n input data streams are to be multiplexed, a framing pulse is used every n bits. As we will see later, these framing pulses will turn out to be very useful for demultiplexing individual packets from a multiplexed stream of packets.

Note from Figure 12.3 that very short pulses—much shorter than the bit interval of each of the multiplexed streams—must be used in OTDM systems. Given that we are interested in achieving overall bit rates of several tens to hundreds of gigabits per second, the desired pulse widths are on the order of a few picoseconds. A periodic train of such short pulses can be generated using a mode-locked laser, as described in Section 3.5.1, or by using a continuous-wave laser along with an external modulator, as described in Section 3.5.4. Since the pulses are very short, their frequency spectrum will be large. Therefore, unless some special care is taken, there will be significant pulse broadening due to the effects of chromatic dispersion. For this purpose, many OTDM experiments use suitably shaped return-to-zero (RZ) pulses, which we studied in Sections 2.5 and 4.1.

Assume that n data streams are to be multiplexed and the bit period of each of these streams is T. Also assume that framing pulses are used. Then the interpulse width is $\tau = T/(n + 1)$ because $n + 1$ pulses (including the framing pulse) must be transmitted in each bit period. Thus the temporal width τ_p of each pulse must satisfy $\tau_p \leq \tau$. Note that usually $\tau_p < \tau$ so that there is some guard time between successive pulses. One purpose of this guard time is to provide for some tolerance in the multiplexing and demultiplexing operations. Another reason is to prevent the undesirable interaction between adjacent pulses that we discussed earlier.

12.1.1 Bit Interleaving

We will first study how the bit-interleaved multiplexing illustrated in Figure 12.3(a) can be performed optically. This operation is illustrated in Figure 12.4. The periodic pulse train generated by a mode-locked laser is split, and one copy is created for each data stream to be multiplexed. The pulse train for the ith data stream, $i = 1, 2, \ldots, n$, is delayed by $i\tau$. This delay can be achieved by passing the pulse train through the

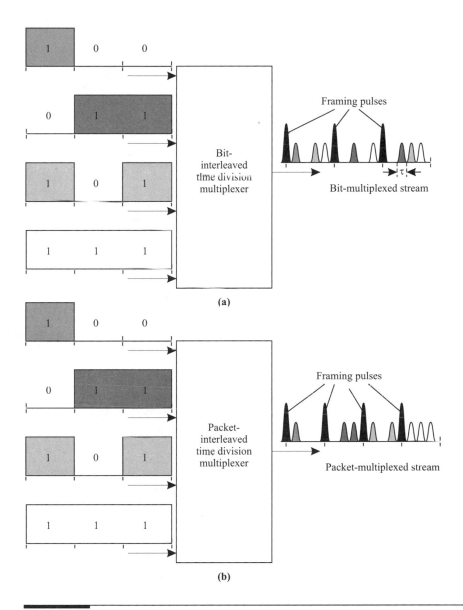

Figure 12.3 (a) Function of a bit-interleaved optical multiplexer. (b) Function of a packet-interleaved optical multiplexer. The same four data streams are multiplexed in both cases. In (b), the packet size is shown as 3 bits for illustration purposes only; in practice, packets are much larger and vary in size. Note that the data must be compressed in time in both cases.

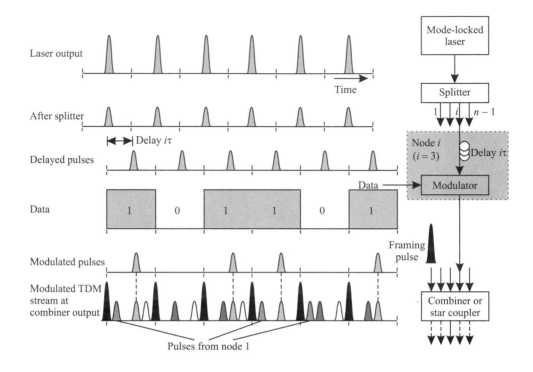

Figure 12.4 An optical multiplexer to create the bit-interleaved TDM stream shown in Figure 12.3(a). Only the operations at one node (node 3) are shown (after [Mid93, Chapter 6]).

appropriate length of optical fiber. Since the velocity of light in silica fiber is about 2×10^8 m/s, one meter of fiber provides a delay of about 5 ns. Thus the delayed pulse streams are nonoverlapping in time. The undelayed pulse stream is used for the framing pulses. Each data stream is used to externally modulate the appropriately delayed periodic pulse stream. The outputs of the external modulator and the framing pulse stream are combined to obtain the bit-interleaved optical TDM stream. The power level of the framing pulses is chosen to be distinctly higher than that of the data pulses. This will turn out to be useful in demultiplexing, as we will see. In the case of broadcast networks with a star topology, the combining operation is naturally performed by the star coupler.

The corresponding demultiplexing operation is illustrated in Figure 12.5. The multiplexed input is split into two streams using, say, a 3 dB coupler. If the jth stream from the multiplexed stream is to be extracted, one of these streams is delayed

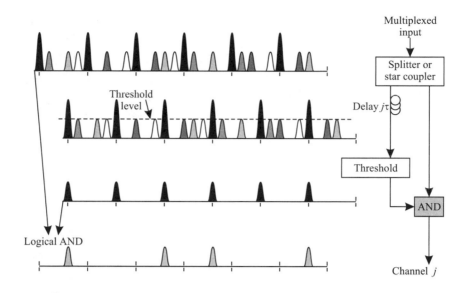

Figure 12.5   An optical demultiplexer to extract one of the multiplexed channels from a bit-interleaved TDM stream (after [Mid93, Chapter 6]).

by $j\tau$. A thresholding operation is performed on the delayed stream to extract the framing pulses. The reason the framing pulses were multiplexed with higher power than the other pulses was to facilitate this thresholding operation. Note that because of the induced delay, the extracted framing pulses coincide with the pulses in the undelayed stream that correspond to the data stream to be demultiplexed. A logical AND operation between the framing pulse stream and the multiplexed pulse stream is used to extract the jth stream. The output of the *logical AND gate* is a pulse if, during a pulse interval, both inputs have pulses; the output has no pulse otherwise. We will discuss two devices to perform the logical AND operation in Section 12.1.3: a *nonlinear optical loop mirror* and a *soliton-trapping gate*.

12.1.2   Packet Interleaving

We next consider how the packet interleaving operation shown in Figure 12.3(b) can be performed. This operation is illustrated in Figure 12.6(a). As in the case of bit interleaving, a periodic stream of narrow pulses is externally modulated by the data stream. If the bit interval is T, the separation between successive pulses is also T. We must somehow devise a scheme to reduce the interval between successive pulses to τ, corresponding to the higher-rate multiplexed signal. This can done by passing the

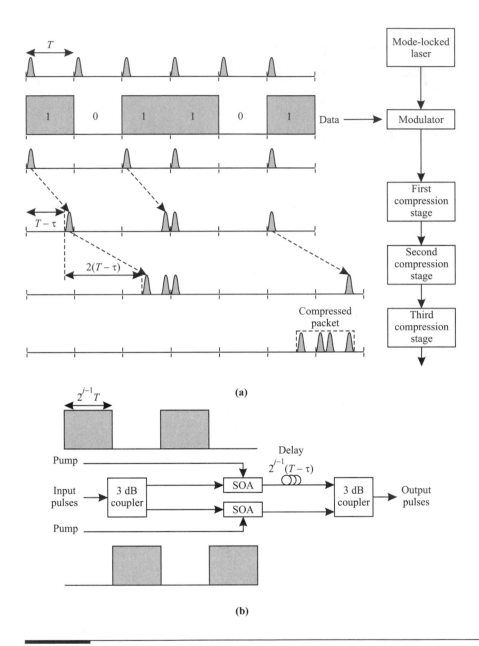

Figure 12.6   An optical multiplexer to create a packet-interleaved TDM stream. (a) The packet passes through k compression stages, where 2^k is the smallest power of two that is not smaller than the packet length l in bits. (b) Detailed view of compression stage j (after [SBP96]).

output of the external modulator through a series of compression stages. If the size of each packet is l bits, the output goes through $k = \lceil \log_2 l \rceil$ compression stages. In the first compression stage, bits $1, 3, 5, 7, \ldots$ are delayed by $T - \tau$. In the second compression stage, the pairs of bits $(1, 2), (5, 6), (9, 10), \ldots$ are delayed by $2(T - \tau)$. In the third compression stage, the bits $(1, 2, 3, 4), (9, 10, 11, 12), \ldots$ are delayed by $4(T - \tau)$. The jth compression stage is shown in Figure 12.6(b). Each compression stage consists of a pair of 3 dB couplers, two semiconductor optical amplifiers (SOAs) used as on-off switches, and a delay line. The jth compression stage has a delay line of value $2^{j-1}(T - \tau)$. It is left as an exercise (Problem 12.1) to show that the delay encountered by pulse $i, i = 1, 2, \ldots, l$, on passing through the kth compression stage is $(2^k - i)(T - \tau)$. Combined with the fact that the input pulses are separated by time T, this implies that pulse i occurs at the output at time $(2^k - 1)(T - \tau) + (i - 1)\tau$. Thus the output pulses are separated by a time interval of τ.

The demultiplexing operation is equivalent to "decompressing" the packet. In principle, this can be accomplished by passing the compressed packet through a set of decompression stages that are similar to the compression stage shown in Figure 12.6(b). This approach is discussed in Problem 12.2. Again, the number of stages required would be $k = \log\lceil l \rceil$, where l is the packet length in bits. However, the on-off switches required in this approach must have switching times on the order of the pulse width τ, making this approach impractical for the small values of τ that are of interest in photonic packet-switching networks.

A more practical approach is to use a bank of AND gates, like the one used in Figure 12.5, and convert the single (serial) high-speed data stream into multiple (parallel) lower-speed data streams that can then be processed electronically. This approach is illustrated in Figure 12.7. In this figure, a bank of five AND gates is used to break up the incoming high-speed stream into five parallel streams each with five times the pulse spacing of the multiplexed stream. This procedure is identical to what would be used to receive five bit-interleaved data streams. One input to each AND gate is the incoming data stream, and the other input is a control pulse stream where the pulses are spaced five times apart. The control pulse streams to each AND gate are appropriately offset from each other so that they select different pulses. Thus the first parallel stream would contain bits $1, 6, 11, \ldots$ of the packet, the second would contain bits $2, 7, 12, \ldots,$ and so on. This approach can also be used to demultiplex a portion of the packet, for example, the packet header, in a photonic packet switch. We will discuss this issue further in Section 12.3.

12.1.3   Optical AND Gates

The logical AND operations shown in Figures 12.5 and 12.7 are performed optically at very high speeds. A number of mechanisms have been devised for this purpose. We

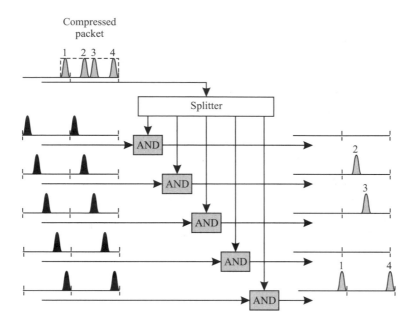

Figure 12.7 An optical demultiplexer to extract one of the multiplexed channels from a packet-interleaved TDM stream.

describe two of them. Note that the logical AND operation between two signals can be performed by an on-off switch if one of the signals is input to the switch and the other is used to control it. This viewpoint will be useful in the following discussion.

Nonlinear Optical Loop Mirror

The *nonlinear optical loop mirror* (NOLM) consists of a 3 dB directional coupler, a fiber loop connecting both outputs of the coupler, and a *nonlinear element* (NLE) located asymmetrically in the fiber loop, as shown in Figure 12.8(a). First, ignore the nonlinear element, and assume that a signal (pulse) is present at one of the inputs, shown as arm A of the directional coupler in Figure 12.8(a). Then, the two output signals are equal and undergo *exactly the same phase shift* on traversing the fiber loop. (Note that we are talking about the phase shift of the optical carrier here and not pulse delays.) We have seen in Problem 3.1 that in this case both the clockwise and the counterclockwise signals from the loop are completely reflected onto input A; specifically, no output pulse emerges from arm B in Figure 12.8(a). Hence the name fiber loop *mirror* for this configuration. However, if one of the signals were to undergo a different phase shift compared to the other, then an output pulse emerges

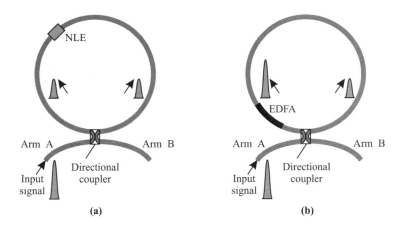

Figure 12.8   (a) A nonlinear optical loop mirror. (b) A nonlinear amplifying loop mirror.

from arm B in Figure 12.8(a). It is left as an exercise to show that the difference in the phase shifts should be π in order for all the energy to emerge from arm B (Problem 12.4).

In many early experiments with the NOLM for the purpose of switching, there was no separate NLE; rather, the intensity-dependent phase (or refractive index) change induced by the silica fiber was itself used as the nonlinearity. This intensity-dependent refractive index change is described by (2.23) and is the basis for the cancellation of group velocity dispersion effects in the case of soliton pulses. We discussed this effect in Section 2.5. An example of such a configuration is shown in Figure 12.8(b), where the pulse traversing the fiber loop clockwise is amplified by an EDFA shortly after it leaves the directional coupler. Because of the use of an amplifier within the loop, this configuration is called the *nonlinear amplifying loop mirror* (NALM). The amplified pulse has higher intensity and undergoes a larger phase shift on traversing the loop compared to the unamplified pulse.

However, these configurations are not convenient for using the NOLM as a high-speed demultiplexer. First, the intensity-dependent phase change in silica fiber is a weak nonlinearity, and typically a few hundred meters of fiber are required in the loop to exploit this effect for pulse switching. It would be desirable to use a nonlinear effect that works with shorter lengths of fiber. Second, to realize an AND gate, we require an NLE whose nonlinear properties can be conveniently controlled by the use of control pulses. The configuration shown in Figure 12.9 has both these properties and is called the *terahertz optical asymmetric demultiplexer* (TOAD).

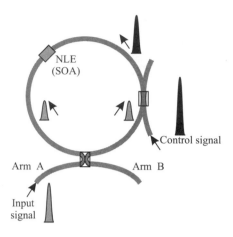

Figure 12.9   The terahertz optical asymmetric demultiplexer.

The principle of operation of the TOAD is as follows. The TOAD has another directional coupler spliced into the fiber loop for the purpose of injecting the control pulses. The control pulses carry sufficiently high power and energy so that the optical properties of the NLE are significantly altered by the control pulse for a short time interval after the control pulse passes through it. In particular, the phase shift undergone by another pulse passing through the NLE during this interval is altered. An example of a suitable NLE for this purpose is a semiconductor optical amplifier (SOA) that is driven into saturation by the control pulse. For proper operation of the TOAD as a demultiplexer, the timing between the control and signal pulses is critical. Assuming the NLE is located such that the clockwise signal pulse reaches it first, the control pulse must pass through the NLE *after* the clockwise signal pulse but *before* the counterclockwise signal pulse. If this happens, the clockwise signal pulse experiences the unsaturated gain of the amplifier, whereas the counterclockwise pulse sees the saturated gain. The latter also experiences an additional phase shift that arises due to gain saturation. Because of this asymmetry, the two halves of the signal pulse do not completely destructively interfere with each other, and a part of the signal pulse emerges from arm B of the input coupler.

Note that along with the signal pulse, the control pulse will also be present at the output. This can be eliminated by using different wavelengths for the signal and control pulses and placing an optical filter at the output to select only the signal pulse. But both wavelengths must lie within the optical bandwidth of the SOA. Another option is to use orthogonal polarization states for the signal and control pulses, and discriminate between the pulses on this basis. Whether this is done or not, the polarization state of the signal pulse must be maintained while traversing the fiber loop; otherwise, the two halves of the pulse will not interfere at the directional coupler in the desired manner after traversing the fiber loop. Another advantage of the TOAD is that because of the short length of the fiber loop, the polarization state of the pulses is maintained even if standard single-mode fiber (nonpolarization-maintaining) is used. If the fiber loop is long, it must be constructed using polarization-maintaining fiber.

Soliton-Trapping AND Gate

The soliton-trapping AND gate uses some properties of soliton pulses propagating in a birefringent fiber. In Chapter 2, we saw that in a normal fiber, the two orthogonally polarized degenerate modes propagate with the same group velocity. We also saw that in a birefringent fiber, these two modes propagate with different group velocities. As a result, if two pulses at the same wavelength but with orthogonal polarizations are launched in a birefringent fiber, they would *walk off,* or spread apart in time, because of this difference in group velocities.

However, soliton pulses are an exception to this walk-off phenomenon. Just as soliton pulses propagate in nonbirefringent silica fiber without pulse spreading due to group velocity dispersion (Section 2.5), a pair of orthogonally polarized soliton pulses propagate in birefringent fiber without walk-off. The quantitative analysis of this phenomenon is beyond the scope of this book, but qualitatively what occurs is that the two pulses undergo wavelength shifts in opposite directions so that the group velocity difference due to the wavelength shift exactly compensates the group velocity difference due to birefringence! Since the two soliton pulses travel together (don't walk off), this phenomenon is called *soliton trapping*.

The logical AND operation between two pulse streams can be achieved using this phenomenon if the two pulse streams correspond to orthogonally polarized soliton pulses. Most high-speed TDM systems use soliton pulses to minimize the effects of group velocity dispersion so that the soliton pulse shape requirement is not a problem. The orthogonal polarization of the two pulse streams can be achieved by appropriately using polarizers (see Section 3.2.1). The logical AND operation is achieved by using an optical filter at the output of the birefringent fiber.

Figure 12.10 shows the block diagram of such a soliton-trapping AND gate. It consists of a piece of birefringent fiber followed by an optical filter. Figure 12.11 illustrates the operation of this gate. When pulses of both polarizations are present at the wavelength λ, one of them gets shifted in wavelength to $\lambda + \delta\lambda$, and the other to $\lambda - \delta\lambda$. The filter is chosen so that it passes the signal at $\lambda + \delta\lambda$ and rejects the signal at λ. Thus the passband of the filter is such that one of the wavelength-shifted pulses lies within it. But the same pulse, if it does not undergo a wavelength shift,

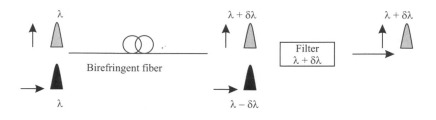

Figure 12.10   Block diagram of a soliton-trapping logical AND gate.

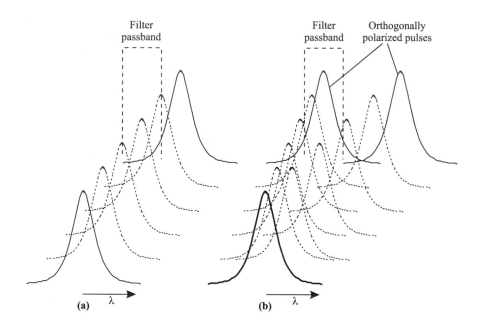

Figure 12.11   Illustration of the operation of a soliton-trapping logical AND gate. (a) Only one pulse is present, and very little energy passes through to the filter output. This state corresponds to a logical zero. (b) Both pulses are present, undergo wavelength shifts due to the soliton-trapping phenomenon, and most of the energy from one pulse passes through to the filter output. This state corresponds to a logical one.

will not be selected by the filter. Thus the filter output has a pulse (logical one) only if both pulses are present at the input, and no pulse (logical zero) otherwise.

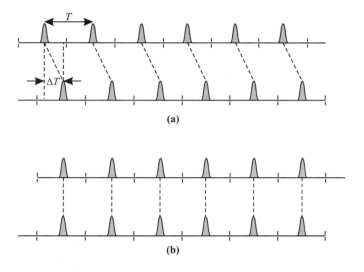

Figure 12.12 The function of a synchronizer. (a) The two periodic pulse streams with period T are out of synchronization; the top stream is ahead by ΔT. (b) The two periodic streams have been synchronized by introducing a delay ΔT in the top stream relative to the bottom stream.

12.2 Synchronization

Synchronization is the process of aligning two pulse streams in time. In PPS networks, it can refer either to the alignment of an incoming pulse stream and a locally available *clock* pulse stream or to the relative alignment of two incoming pulse streams. Recall our assumption of fixed-size packets. Thus if framing pulses are used to mark the packet boundaries, the framing pulses must occur periodically.

The function of a synchronizer can be understood from Figure 12.12. The two periodic pulse streams, with period T, shown in Figure 12.12(a) are not synchronized because the top stream is ahead in time by ΔT. In Figure 12.12(b), the two pulse streams are synchronized. Thus, to achieve synchronization, the top stream must be delayed by ΔT with respect to the bottom stream. The delays we have hitherto considered, for example, while studying optical multiplexers and demultiplexers, have been *fixed* delays. A fixed delay can be achieved by using a fiber of the appropriate length. However, in the case of a synchronizer, and in some other applications in photonic packet-switching networks, a *tunable delay* element is required since the amount of delay that has to be introduced is not known a priori. Thus we will now study how tunable optical delays can be realized.

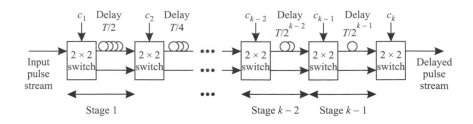

Figure 12.13   A tunable delay line capable of realizing any delay from 0 to $T - T/2^{k-1}$, in steps of $T/2^{k-1}$.

12.2.1   Tunable Delays

A tunable optical delay line capable of realizing any delay, in excess of a reference delay, from 0 to $T - T/2^{k-1}$, in steps of $T/2^{k-1}$, is shown in Figure 12.13. The parameter k controls the resolution of the delay achievable. The delay line consists of $k - 1$ fixed delays with values $T/2, T/4, \ldots, T/2^{k-1}$ interconnected by k 2×2 optical switches, as shown. By appropriately setting the switches in the cross or bar state, an input pulse stream can be made to encounter or avoid each of these fixed delays. If all the fixed delays are encountered, the total delay suffered by the input pulse stream is $T/2 + T/4 + \ldots + T/2^{k-1} = T - T/2^{k-1}$. This structure can be viewed as consisting of $k - 1$ stages followed by an output switch, as indicated in Figure 12.13. The output switch is used to ensure that the output pulse stream always exits the same output of this switch. The derivation of the control inputs $c_1, c_2, \ldots, c_k$ to the k switches is discussed in Problem 12.3.

With a tunable delay line like the one shown in Figure 12.13, two pulse streams can be synchronized to within a time interval of $T/2^k$. The value k, and thus the number of fixed delays and optical switches, must be chosen such that $2^{-k}T \ll \tau$, the pulse width. The resolution of the delay line is determined by the speed of the switches used and the precision to which the delay lines can be realized. Practically, the resolution of this approach may be on the order of 1 ns or so. We can use this approach to provide *coarse* synchronization. We will also need to perform *fine* synchronization to align bits to within a small fraction of a bit interval. One approach is to use a tunable wavelength converter followed by a highly dispersive fiber line [Bur94]. If D denotes the dispersion of the fiber used, $\Delta\lambda$ the output wavelength range, and L the length of the fiber, then we can get a relative delay variation of 0 to $D\Delta\lambda L$. If the output wavelength can be controlled in steps of $\delta\lambda$, then the delay resolution is $D\delta\lambda L$.

Given a tunable delay, the synchronization problem reduces to one of determining the relative delay, or *phase,* between two pulse streams. A straightforward approach to this problem is to compare all shifted versions of one stream with respect to the other. The comparison can be performed by means of a logical AND operation. This is a somewhat expensive approach. An alternative approach is to use an optical *phase lock loop* to sense the relative delay between the two pulse streams. Just as more than one phenomenon can be used to build an optical AND gate, different mechanisms can be used to develop an optical phase lock loop. We discuss one such mechanism that is based on the NOLM that we studied in Section 12.1.3.

12.2.2 Optical Phase Lock Loop

Consider an NOLM that does not use a separate nonlinear element but rather uses the intensity-dependent refractive index of silica fiber itself as the nonlinearity. Thus if a low-power pulse stream, say, stream 1, is injected into the loop—from arm A of the directional coupler in Figure 12.8(a)—the fiber nonlinearity is not excited, and both the clockwise and the counterclockwise propagating pulses undergo the same phase shift in traversing the loop. As a consequence, no power emerges from the output (arm B) in this case. If a high-power pulse stream, say, stream 2, is injected *in phase* (no relative delay) with, say, the clockwise propagating pulse stream, because of the intensity dependence of the refractive index of silica fiber, the refractive index seen by the clockwise pulse, and hence the phase shift undergone by it, is different from that of the counterclockwise pulse. This mismatch in the phase shift causes an output to emerge from arm B in Figure 12.8(a). Note that if the high-power pulse stream is not in phase (has a nonzero relative delay) with the clockwise propagating pulse streams, the clockwise and counterclockwise pulses undergo the same phase shift, and no output emerges from arm B of the directional coupler. To achieve synchronization between pulse streams 1 and 2, a tunable delay element can be used to adjust their relative delays till there is no output of stream 1 from the NOLM.

Note that the same problem of discriminating between the pulse streams 1 and 2 at the output of the directional coupler (arm B) as with the TOAD arises in this case as well. Since pulses from stream 2 will always be present at the output, in order to detect the absence of pulses from stream 1, the two streams must use different wavelengths or polarizations. When different wavelengths are used, because of the chromatic dispersion of the fiber, the two pulses will tend to walk away from each other, and the effect of the nonlinearity (intensity-dependent refractive index) will be reduced. To overcome this effect, the two wavelengths can be chosen to lie symmetrically on either side of the zero-dispersion wavelength of the fiber so that the group velocities of the two pulse streams are equal.

A phase lock loop can also be used to adjust the *frequency and phase* of a local clock source—a mode-locked laser—to those of an incoming periodic stream. We have seen in Section 3.5.1 that the repetition rate, or frequency, of a mode-locked laser can be determined by modulating the gain of the laser cavity. We assume that the modulation frequency of its gain medium, and hence the repetition rate of the pulses, is governed by the frequency of an electrical oscillator. The output of the NOLM can then be photodetected and used to control the frequency and phase of this electrical oscillator so that the pulses generated by the local mode-locked laser are at the same frequency and phase as that of the incoming pulse stream. We refer to [Bar96] and the references therein for the details.

Another synchronization function has to do with extracting the clock for the purposes of reading parts of the packet, such as the header, or for demultiplexing the data stream. This function can also be performed using an optical phase-locked loop. However, this function can also be performed by sending the clock along with the data in the packet. In one example [BFP93], the clock is sent at the beginning of the packet. At the switching node, the clock is separated from the rest of the packet by using a switch to read the incoming stream for a prespecified duration corresponding to the duration of the clock signal. This clock can then be used to either read parts of the packet or to demultiplex the data stream.

12.3 Header Processing

For a header of fixed size, the time taken for demultiplexing and processing the header is fixed, and the remainder of the packet is buffered optically using a delay line of appropriate length. The processing of the header bits may be done electronically or optically, depending on the kind of control input required by the switch. Electrically controlled switches employing the electro-optic effect and fabricated in lithium niobate (see Section 3.7) are most commonly used in switch-based network experiments today. In this case, the header processing can be carried out electronically (after the header bits have been demultiplexed into a parallel stream). The packet destination information from the header is used to determine the outgoing link from the switch for this packet, using a look-up table. For each input packet, the look-up table determines the correct switch setting, so that the packet is routed to the correct output port. Of course, this leads to a conflict if multiple inputs have a packet destined for the same output at the same time. This is one of the reasons for having buffers in the routing node, as explained next.

If the destination address is carried in the packet header, it can be read by demultiplexing the header bits using a bank of AND gates, for example, TOADs, as shown in Figure 12.7. However, this is a relatively expensive way of reading the header, which is a task that is easier done with electronics than with optics. Another reason for using electronics to perform this function is that the routing and forwarding functions required can be fairly complex, involving sophisticated control algorithms and look-up tables.

With this in mind, several techniques have been proposed to simplify the task of header recognition. One common technique is to transmit the header at a much lower bit rate than the packet itself, allowing the header to be received and processed relatively easily within the routing node. The packet header could also be transmitted on a wavelength that is different from the packet data. It could also be transmitted on a separate subcarrier channel on the same wavelength. All these methods allow the header to be carried at a lower bit rate than the high-speed data in the packet, allowing for easier header processing. However, given the high payload speeds involved in order to maintain reasonable bandwidth utilization without making the packet size unreasonably large, we will have to use fairly short headers and process them very quickly—this may not leave much room for sophisticated header processing. See Problem 12.5 for an example.

12.4 Buffering

In general, a routing node contains buffers to store the packets from the incoming links before they can be transmitted or forwarded on the outgoing links. Hence the name *store and forward* for these networks. In a general store-and-forward network, electronic or optical, the buffers may be present at the inputs only, at the outputs only, or at both the inputs and the outputs, as shown in Figure 12.2. The buffers may also be integrated within the switch itself in the form of random access memory and shared among all the ports. This option is used quite often in the case of electronic networks where both the memory and switch fabric are fabricated on the same substrate, say, a silicon-integrated circuit, but we will see that it is not an option for optical packet switches. We will also see that most optical switch proposals do not use input buffering for performance-related reasons.

There are at least three reasons for having to store or buffer a packet before it is forwarded on its outgoing link. First, the incoming packet must be buffered while the packet header is processed to determine how the packet must be routed. This is usually a fixed delay that can be implemented in a simple fashion. Second, the required switch input and/or output port may not be free, causing the packet to be queued at its input buffer. The switch input may not be free because other packets that arrived on the same link have to be served earlier. The switch output port may not be free because packets from other input ports are being switched to it. Third, after the packet has been switched to the required output port, the outgoing link

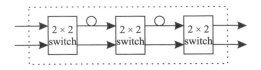

Figure 12.14   Example of a 2 × 2 routing node using a feed-forward delay line architecture.

from this port may be busy transmitting other packets, thus making this packet wait for its turn. The latter delays are variable and are implemented differently from the fixed delay required for header processing.

The lack of good buffering methods in the optical domain is a major impediment. Unlike the electronic domain, we do not have random access memory in the optical domain. Instead the only way of realizing optical buffers is to use fiber delay lines, which consist of relatively long lengths of fiber. For example, about 200 m of fiber is required for 1 μs of delay, which would be sufficient to store 10 packets, each with 1000 bits at 10 Gb/s. Thus usually very small buffers are used in photonic packet-switching networks. Note that unlike an electronic buffer, a packet cannot be accessed at an arbitrary point of time; it can exit the buffer only after a fixed time interval after entering it. This is the time taken for the packet to traverse the fiber length. This constraint must be incorporated into the design of PPS networks. Of course, by repeated traversals of the same piece of fiber, packet delays that are multiples of this basic delay can be obtained.

PPS networks typically make use of delay lines in one of two types of configurations. Figure 12.14 shows one example of a *feed-forward* architecture. In this configuration, a two-input, two-output routing node is constructed using three 2 × 2 switches interconnected by two delay lines. If each delay line can store one packet—that is, the propagation time through the delay line is equal to one slot—the routing node has a buffering capacity of two packets. If packets destined for the same output arrive simultaneously at both inputs, one packet will be routed to its correct output, and the other packet will be stored in delay line 1. This can be accomplished by setting switch 1 in the appropriate state. This packet then has the opportunity to be routed to its desired output in a subsequent slot. For example, if no packets arrive in the next slot, this stored packet can be routed to its desired output in the next slot by setting switches 2 and 3 appropriately.

The other configuration is the *feedback* configuration, where the delay lines connect the output of the switch back to its input. We will study this configuration in Section 12.4.3.

There are several options for dealing with contention resolution in an optical switch. The first option is to provide sufficient buffering in the switch to be able to handle these contentions. We will see in order to achieve reasonable packet loss probabilities, the buffers need to be able to accommodate several hundred packets. As we have seen above, this is not a trivial task in the context of optical buffers.

Another option is to drop packets whenever we have contentions. This is not attractive because such events will occur quite often unless the links are occupied by very few packets compared to their capacities. For each such event, the source must retransmit the packet, causing the effective link utilization to drop even farther.

A third option is to use the wavelength domain to help resolve conflicts. This can help reduce the amount of buffering required in a significant way.

The final option is for the packet to be *misrouted* by the switch, that is, transferred by the switch to the *wrong output*. This option, termed *deflection routing*, has received considerable study in the research literature on PPS networks.

We start by describing the various types of buffering, and the use of the wavelength domain to resolve conflicts, followed by deflection routing. The switch architectures used in the following section are idealized versions for illustration only; we will look at some actual proposals and experimental configurations in Section 12.6.

12.4.1   Output Buffering

Consider the switch with output buffering shown in Figure 12.15. Let us assume that time is divided into slots and packets arriving into the switch are aligned with respect to these time slots. In each time slot, we have packets arriving at the input ports. Of these, one or more packets may have to be switched to the same output port. In the worst case, we could have a packet arriving at each input port, with all these packets destined to a single output port. In this case, if the switch is designed to operate at N times the line rate (N being the number of ports), these packets can all be switched onto the output port. However, only one of these packets can be transmitted out during this time slot, and the other packets will have to stored in the output buffer. If the output buffer is full, then packets will have to be dropped. The packet loss probability indicates how frequently packets are dropped by the switch. For each such event, the source must retransmit the packet causing the effective link utilization to drop even farther. We can minimize the packet loss probability by increasing the buffer size. With sufficiently large output buffers, an output-buffered switch has the best possible performance with respect to packet delay and throughput, compared to other switch architectures. The throughput can be viewed as the asymptotic value

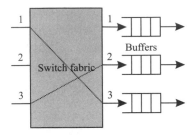

Figure 12.15 A generic switch with output buffers.

of the offered load at which the packet delay through the switch becomes very large (tends to infinity).

We can use a simple model to understand the performance of the different buffering techniques. The model assumes that in each time slot, a packet is received at the input with probability ρ. Thus ρ denotes the traffic load. It further assumes that traffic is uniformly distributed, and therefore the packet is destined to a particular output port with probability $1/N$, where N is the number of ports on the switch. While this is admittedly not a very realistic model, it gives some understanding of the trade-offs between the different buffering approaches. The parameters of interest are the desired packet loss probability, the number of packet buffers needed, and the traffic load. The number of packet buffers suggested by this model is typically smaller than what is actually required, since in reality traffic is more bursty than what is assumed by this model.

For the output-buffered switch, this simple model was analyzed in [HK88], which shows that to get a packet loss of 10^{-6} at a traffic load of 0.8, we need about 25 packet buffers per output. With sufficiently large buffers, a throughput close to 1 can be obtained.

One issue with the output-buffered switch is that the switch needs to operate at N times the line rate per port. That is, it needs to be able to switch up to N packets per time slot from different inputs onto the same output. This is quite difficult to implement with optical switches. For this reason, many optical switch proposals emulate an output-buffered switch while still operating at the line rate per port. If multiple packets arriving in a time slot are all destined to the same output port, the switch schedules different delays for each of these packets at the input so that they get switched to the output in different succeeding time slots. For example, the switch handles the first packet immediately, delays the next packet by one time slot at the input, delays the next by two slots, and so on.

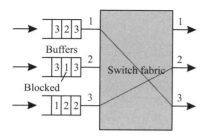

Figure 12.16 Head-of-line blocking in an input-buffered switch. Observe that the packet destined for output 1 in input buffer 2 is blocked despite the fact that the output is free.

12.4.2 Input Buffering

A switch with input buffering has buffers at the input to the switch but not at the output. These switches have relatively poor throughput due to a phenomenon called *head-of-line* (HOL) blocking, which is illustrated in Figure 12.16. When we have multiple input packets at the head of the line destined to a single output port, only one packet can be switched through. The other packets, however, may block packets behind them from being switched in the same time slot. For example, in Figure 12.16, we have packets at port 1 and port 2 at the head of their lines, both destined for port 3. Say we switch the packet at port 1 onto port 3. The second packet in line behind the head-of-line packet on port 2 is destined to output port 1 but cannot be switched to that output, even though it is free. For the traffic model considered earlier, this HOL blocking reduces the achievable throughput to 0.58 for large switch sizes [HK88]. While we can improve the throughput by selecting packets other than just the one at the head of the line, this is quite complicated and not feasible in the context of optical switches. The other problem is that the packet's delay at the input buffer cannot be determined before placing the packet in the buffer because it depends on the other inputs. In the context of optical delay lines, it means that when the packet exits the delay line, we may still not be able to switch it through as the desired output may be busy. For these reasons, optical switches with input buffers only are not a good choice.

12.4.3 Recirculation Buffering

In this approach, the buffers connect the outputs back to the inputs. Typically, some of the switch ports are reserved for buffering only, and the output of these ports is connected back to the corresponding inputs via buffers. If multiple packets destined

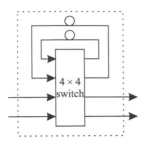

Figure 12.17   Example of a 2 × 2 routing node using a feedback delay line architecture.

for a common output port arrive simultaneously, one of them is switched to the output port while the others are switched to the recirculating buffers.

In the context of optical switches, the buffering is implemented using *feedback* delay lines. In the feedback architecture of Figure 12.17, the delay lines connect the outputs of the switch to its inputs. With two delay lines and two inputs from outside, the switch is internally a 4 × 4 switch. Again, if two packets contend for a single output, one of them can be stored in a delay line. If the delay line has length equal to one slot, the stored packet has an opportunity to be routed to its desired output in the next slot. If there is contention again, it, or the contending packet, can be stored for another slot in a delay line.

Recirculation buffering is more effective than output buffering at resolving contentions because the buffers in this case are shared among all the outputs, as opposed to having a separate buffer per output. The trade-off is that larger switch sizes are needed in this case due to the additional switch ports needed for connecting the recirculating buffers. For example, in [HK88], it is shown that a 16 × 16 switch requires a total of 112 recirculation buffers, or about 7 buffers per output, to achieve a packet loss probability of 10^{-6} at an offered load of 0.8. In contrast, we saw earlier that the output-buffered switch requires about 25 buffers per output, or a total of 400 buffers, to achieve the same packet loss probability.

In the feed-forward architecture considered earlier, a packet has a fixed number of opportunities to reach its desired output. For example, in the routing node shown in Figure 12.14, the packet has at most three opportunities to be routed to its correct destination: in its arriving slot and the next two immediate slots. On the other hand, in the feedback architecture, it appears that a packet can be stored indefinitely. This is not true in practice since photonic switches have several decibels of loss. The loss can be made up using amplifiers, but then we have to account for the cascaded amplifier noise as packets are routed through the delay line multiple times. The switch crosstalk also accumulates. Therefore, the same packet cannot be routed through the switch more than a few times. In practice, the feed-forward architecture is preferred to the feedback architecture since it attenuates the signals almost equally, regardless of the path taken through the routing node. This is because almost all the loss is in passing through the switches, and in this architecture, every packet passes through the same number of switches independent of the delay it experiences. This *low differential loss* characteristic is important in a network since it reduces the dynamic range of the signals that must be handled.

12.4.4   Using Wavelengths for Contention Resolution

One way to reduce the amount of buffering needed is to use multiple wavelengths. In the context of PPS, buffers correspond to fiber delay lines. Observe that we can store multiple packets at different wavelengths in the same delay line.

We start by looking at a baseline architecture for an output-buffered switch using delay lines that does not make use of multiple wavelengths. Figure 12.18 shows such an implementation, which is equivalent to the output-buffered switch of Figure 12.15 with B buffers per output. Up to B slots of delay are provided per output by using a set of B delay lines per output. T denotes the duration of a time slot. If multiple input packets arriving in a time slot need to go to the same output, one of them is switched out while the others are delayed by different amounts and stored in the different delay lines, so that the output contention is resolved. Note that the set of delay lines together can store more than B packets simultaneously. For instance, a single K-slot delay line can hold up to K packets simultaneously. Therefore the total number of packets that can be held by the set of delay lines in Figure 12.18 is $1 + 2 + \ldots + B = B(B+1)/2$. However, since we can have only one packet per slot transmitted out (or a total of B packets in B slots), the effective storage capacity of this set of delay lines is only B packets.

In its simplest form, we can use wavelengths *internal* to the switch to reduce the number of delay lines required. Figure 12.19 shows an example of such an output-buffered switch [ZT98]. Instead of providing a set of delay lines per output, the delay lines are shared among all the outputs. Packets entering the switch are sent through a tunable wavelength converter device. (Note that tunable wavelength converter devices are still in research laboratories today—see Section 3.8 for some of the approaches being pursued.) At the output of the switch, the packets are sent through an arrayed waveguide grating (AWG). The wavelength selected by the tunable wavelength converter and the output switch fabric port to which the packet is switched together determine the delay line to which the packet is routed by the AWG.

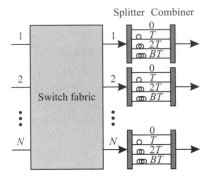

Splitter Combiner

Figure 12.18 An example of an output-buffered optical switch using fiber delay lines for buffers that does not use wavelengths for contention resolution.

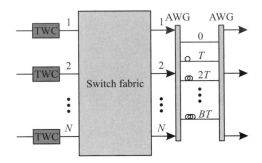

Figure 12.19 An example of an output-buffered optical switch using multiple wavelengths internal to the switch and fiber delay lines for buffers. The switch uses tunable wavelength converters and arrayed waveguide gratings.

Figure 3.25 provides a description of how the AWG works in this configuration. For example, consider the first input port on the AWG. From this port, wavelength λ_1 is routed to delay line 0, wavelength λ_2 is routed to the single-slot delay line, wavelength λ_3 is routed to the two-slot delay line, and wavelength λ_B is routed to the B-slot delay line. In order to allow a packet at each input of the AWG to be routed to each possible delay line, we need the number of wavelengths, $W = \max(N, B)$, where N is the number of inputs. Thus the delay seen by a packet can be controlled by controlling the wavelength at the output of the tunable wavelength converter device. In this case, if we have two input packets on different ports destined to the same output, their wavelengths are chosen such that one of them is delayed while the other is switched through. From a buffering perspective, this configuration is equivalent to the baseline configuration of Figure 12.18. Note that the TWCs must be on the inputs to the switch fabric (not at the outputs) since several packets may leave a switch fabric output on one time slot, on different wavelengths.

For instance, in one routing method, a packet bound for output port j is routed to output port j of the switch fabric. Its wavelength is chosen based on the delay required. With the AWG design assumed above, an incoming packet bound for output 1, requiring a single-slot delay, would be converted to wavelength λ_2 at the input, and switched to port 1 of the switch fabric.

Assuming the same traffic model as before, with $\rho = 0.8$, in order to obtain a packet loss probability of 10^{-6} for a 16×16 switch, we need a total of 25 delay lines, instead of 25 delay lines per output for the case where only a single wavelength is used inside the switch. In Section 12.6, we will study other examples of switch configurations that use wavelengths internally to perform the switching and/or buffering functions.

We next consider the situation where we have a WDM network. In this case, multiple wavelengths are used on the transmission links themselves. We can gain further reduction in the shared buffering required compared to a single-wavelength system by making use of the statistical nature of bursty traffic across multiple wavelengths. Figure 12.20 shows a possible architecture [Dan97] for such a switch, again using tunable wavelength converters and delay lines. At the inputs to the switch, the wavelengths are demultiplexed and sent through tunable wavelength converters and then into the switch fabric. The delay lines are connected to the output of the switch fabric. The W wavelengths destined for a given output port share a single set of delay lines. In this case, we have additional flexibility in dealing with contention. If two packets need to go out on the same output port, either they can be delayed in time, or they can be converted to different wavelengths and switched to the output port at the same time. The TWCs convert the input packets to the desired output wavelength, and the switch routes the packets to the correct output port and the appropriate delay line for that output.

As the number of wavelengths is increased, keeping the load per wavelength constant, the amount of buffering needed will decrease because, within any given time slot, the probability of finding another free wavelength is quite high. Basically we are sharing capacity among several wavelengths and permitting better use of that capacity. [Dan97] shows that the number of delay lines required to achieve a packet loss probability of 10^{-6} at an offered load of 0.8 per wavelength for a 16×16 switch drops from 25 per output without using multiple wavelengths to 7 per output using four wavelengths, and to 4 per output when eight wavelengths are present.

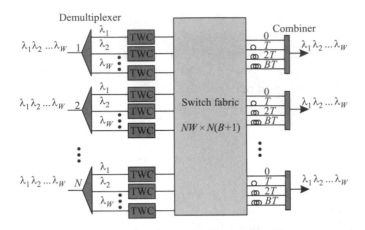

Figure 12.20 An example of an output-buffered optical switch capable of switching multiple input wavelengths. The switch uses TWCs and wavelength demultiplexers. The TWCs convert the input packets to the desired output wavelength, and the switch routes the packets to the correct output port and the appropriate delay line for that output.

Table 12.1 Number of delay lines required for different switch architectures. A uniformly distributed offered load of 0.8 per wavelength per input is assumed, with a packet loss probability of 10^{-6}. The switch size is 16×16.

Buffering Type	Input λs	Internal λs	Internal Fabric	Delay Lines per Output	Delay Lines Total
Output (Figure 12.18)	1	1	16×16	25	400
Recirculating (Figure 12.17)	1	1	23×23	7	112
Output (Figure 12.19)	1	64	16×16	Shared	26
Output (Figure 12.20)	4	4	64×128	7	112
Output (Figure 12.20)	8	8	128×80	4	64

Table 12.1 compares the number of delay lines required for the different buffering schemes that we considered in this section. Note that the number of delay lines is only one among the many parameters we must consider when designing switch architectures. The others include the switch fabric size, the number of wavelength converters required, and the number of wavelengths used internally (and the associated complexity of the multiplexers and demultiplexers). While we have illustrated a few sample architectures in Figures 12.17 through 12.20, many variants of these

architectures have been proposed that trade off these parameters against each other. See [Dan97, ZT98, Hun99, Gam98, Gui98] for more examples.

12.4.5 Deflection Routing

Deflection routing was invented by Baran in 1964 [Bar64]. It was studied and implemented in the context of processor interconnection networks in the 1980s [Hil85, Hil87, Smi81]. In these networks, just as in photonic packet-switching networks, buffers are expensive because of the high transmission speeds involved, and deflection routing is used as an alternative to buffering. Deflection routing is also sometimes called *hot-potato routing*.

Intuitively, misrouting packets rather than storing them will cause packets to take longer paths on average to get to their destinations, and thus will lead to increased delays and lesser throughput in the network. This is the price paid for not having buffers at the switches. These trade-offs have been analyzed in detail for regular network topologies such as the Manhattan Street network [GG93], an example of which is shown in Figure 12.21, or the shufflenet [KH90, AS92], another regular interconnection network, an example of which is shown in Figure 12.22, or both [Max89, FBP95]. Regular topologies are typically used for processor interconnections and may be feasible to implement in LANs. However, they are unlikely to be used in WANs, where the topologies used are usually arbitrary. Nevertheless, these analyses shed considerable light on the issues involved in the implementation of deflection routing even in wide-area photonic packet-switching networks and the resulting performance degradation, compared to buffering in the event of a destination conflict.

Before we can discuss these results, we need to slightly modify the model of the routing node shown in Figure 12.2. While discussing this figure earlier, we said that the routing node has one input link and output link from/to every other routing node and end node to which it is connected. In many cases, the end node is colocated with the routing node so that information regarding packets to be transmitted or received can be almost instantaneously exchanged between these nodes. In particular, this makes it possible for the end node to inject a new packet into its associated routing node, *only when* no other packet is intended for the same output link. Thus this new injected packet neither gets deflected nor causes deflection of other packets. This is a reasonable assumption to make in practice.

Delay

The first consequence of deflection routing is that the average delay experienced by the packets in the network is larger than in store-and-forward networks. In this

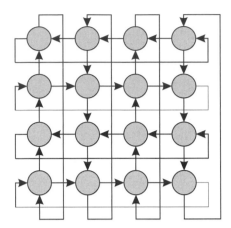

Figure 12.21   The Manhattan Street network with $4^2 = 16$ nodes. In a network with n^2 nodes, these nodes are arranged in a square grid with n rows and columns. Each node transmits to two nodes—one in the same row and another in the same column. Each node also receives from two other nodes—one in the same row and the other in the same column. Assuming n is even, the direction of transmission alternates in successive rows and columns.

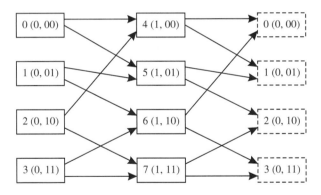

Figure 12.22   The shufflenet with eight nodes. More generally, a (Δ, k) shufflenet consists of $k\Delta^k$ nodes, arranged in k columns, each with Δ^k nodes. We can think of a (Δ, k) shufflenet in terms of the state transition diagram of a k-digit shift register, with each digit in $\{0, 1, \ldots, \Delta - 1\}$. Each node $(c, a_0 a_1 \ldots a_{k-1})$ is labeled by its column index $c \in \{0, 1, 2, \ldots, k - 1\}$ along with a k-digit string $a_0 a_1 \ldots a_{k-1}$, $a_i \in \{0, 1, \ldots, \Delta - 1\}$, $0 \le i \le k - 1$. There is an edge from a node i to another node j in the following column if node j's string can be obtained from node i's string by one shift. In other words, there is an edge from node $(c, a_0 a_1 \ldots a_{k-1})$ to a node $((c + 1) \bmod k, a_1 a_2 \ldots a_{k-1}*)$, where $* \in \{0, 1, \ldots, \Delta - 1\}$.

comparison, not only is the network topology fixed, but the statistics of the packet arrivals between each source-destination pair are also fixed. In particular, the rate of injection of new packets into the network, which is called the *arrival rate*, for each source-destination pair must be fixed. The delay experienced by a packet consists of two components. The first is the *queuing delay*—the time spent waiting in the buffers at each routing node for transmission. There is no queuing delay in the case of deflection routing. The second component of the delay experienced by a packet is the *propagation delay*—the time taken for the packet to traverse all the links from the source node to the destination node. The propagation delay is often larger for deflection routing than for routing with buffers owing to the misdirection of packets away from their destinations. As a result, in most cases, for a given arrival rate, the overall delay in deflection-routed networks is larger than the overall delay in store-and-forward networks.

Throughput

Another consequence of deflection routing is that the *throughput* of the network is decreased compared to routing with buffers. An informal definition of the throughput of these networks, which will suffice for our purposes here, is that it is the maximum rate at which *new* packets can be injected into the network from their sources. Clearly, this depends on the interconnection topology of the network and the data rates on the links. In addition, it depends on the *traffic pattern*, which must remain fixed in defining the throughput. The traffic pattern specifies the fraction of new packets for each source-destination pair. Typically, in all theoretical analyses of such networks, the throughput is evaluated for a *uniform traffic pattern*, which means that the arrival rates of new packets for all source-destination pairs in the network are equal. If all the links run at the same speed, the throughput can be conveniently expressed as a fraction of the link speed.

For Manhattan Street networks with sizes ranging from a few hundred to a few thousand nodes, deflection routing achieves 55–70% of the throughput achieved by routing with buffering [Max89]. For shufflenets in the same range of sizes, the value is only 20–30% of the throughput with buffers. However, since a shufflenet has a much higher throughput than a Manhattan Street network of the same size (for routing with buffers), the actual throughput of the Manhattan Street network in the case of deflection routing is lower than that of the shufflenet. All these results assume a uniform traffic pattern.

So what do these results imply for irregular networks? To discuss this, let us examine some of the differences in the properties of these two networks. One important property of any network is its *diameter*, which is the largest number of hops on the shortest path between any two nodes in the network. In other words, the diameter is the *maximum* number of hops between two nodes in the network. However, in most networks, the larger the diameter, the greater the number of hops that a packet has to travel even *on average* to get to its destination. The Manhattan Street network has a diameter that is proportional to $\sqrt{n}$, where n is the number of nodes in the network. On the other hand, the shufflenet has a diameter that is proportional to $\log_2 n$. (We consider shufflenets of degree 2.) Thus if we consider a Manhattan Street network and a shufflenet with the same number of nodes and edges, the Manhattan Street network will have a lower throughput for routing with buffers than the shufflenet, since each packet has to traverse more edges, on the average. For arbitrary networks, we can generalize this and say that the smaller the diameter of the network, the larger the throughput for routing with buffers.

For deflection routing, a second property of the network that we must consider is its *deflection index*. This property was introduced in [Max89], although it was not called by this name. It was formally defined and discussed in greater detail in a later paper [GG93]. The deflection index is the largest number of hops that a single deflection adds to the shortest path between some two nodes in the network. In the Manhattan Street network, a single deflection adds at most four hops to the path length, so its deflection index is four. On the other hand, the shufflenet has a deflection index of $\log_2 n$ hops. This accounts for the fact that the Manhattan Street network has a significantly larger *relative* throughput—the deflection routing throughput expressed as a fraction of the store-and-forward throughput—than the shufflenet (55–70% versus 20–30%). For arbitrary networks, we can then say that the deflection index must be kept small so that the throughput remains high in the face of deflection routing.

Combining the two observations, we can conclude that network topologies with small diameters and small deflection indices are best suited for photonic packet-switching networks. A regular topology designed by combining the Manhattan Street and shufflenet topologies and having these properties is discussed in [GG93]. In addition to choosing a good network topology (not necessarily regular), the performance of deflection-routing networks can be further improved by using appropriate *deflection rules*. A deflection rule specifies the manner in which the packets to be deflected are chosen among the packets contending for the same switch output port. The results we have quoted assume that in the event of a conflict between two packets, both packets are equally likely to be deflected. This deflection rule is termed *random*. Another possible deflection rule, called *closest-to-finish* [GG93], states that when two packets are contending for the same output port, the packet that is farther away from its destination is deflected. This has the effect of reducing the average number of deflections suffered by a packet and thus increasing the throughput.

Small Buffers

We can also consider deflection routing with a very limited number of buffers, for example, buffers of one or two packets at each input port. If this limited buffer is full, the packet is again deflected. Such limited-buffer deflection-routing strategies achieve higher throughputs compared to the purest form of deflection routing without any buffers whatsoever. We refer to [Max89, FBP95] for the quantitative details.

Livelock

When a network employs deflection routing, there is the possibility that a packet will be deflected forever and never reach its destination. This phenomenon has been called both *deadlock* [GG93] and *livelock* [LNGP96], but the term *livelock* seems to be more appropriate. Livelock is somewhat similar to routing loops encountered in store-and-forward networks (see Section 6.3), but routing loops are a transient phenomenon there, whereas livelock is an inherent characteristic of deflection routing.

Livelock can be eliminated by suitably designed deflection rules. However, proving that any particular deflection rule is livelock-free seems to be hard. We refer to [GG93, BDG95] for some further discussion of this issue (under the term *deadlock*). One way to eliminate livelocks is to simply drop packets that have exceeded a certain threshold on the hop count.

12.5   Burst Switching

Burst switching is a variant of PPS. In burst switching, a source node transmits a header followed by a packet burst. Typically the header is transmitted at a lower speed on an out-of-band control channel, although most proposals assume an out-of-band control channel. An intermediate node reads the packet header and activates its switch to connect the following burst stream to the appropriate output port if a suitable output port is available. If the output port is not available, the burst is either buffered or dropped. The main difference between burst switching and conventional photonic packet switching has to do with the fact that bursts can be fairly long compared to the packet duration in packet switching.

In burst switching, if the bursts are sufficiently long, it is possible to ask for or reserve bandwidth in the network ahead of time before sending the burst. Various protocols have been proposed for this purpose. For example, one such protocol, called Just-Enough-Time (JET), works as follows. A source node wanting to send a burst first sends out a header on the control channel alerting the nodes along the path that a burst will follow. It follows the header by transmitting the burst after a certain time period. The period is large enough to provide the nodes sufficient time to process the header and set the switches to switch the burst through when it arrives, so that additional buffering is not needed for this purpose at the nodes.

Overall, burst switching is essentially a variation of PPS where packets have variable and fairly large sizes, and little or no buffering is used at the nodes. Like packet switching, one of the main issues with burst switching is to determine the buffer sizes needed at the nodes to achieve reasonable burst drop probabilities when there is contention. The same techniques that we discussed earlier in Section 12.4 apply here as well.

12.6 Testbeds

Several PPS testbeds have been built over the last few years, and many are being built today. The main focus of most of these testbeds is the demonstration of certain key PPS functions such as multiplexing and demultiplexing, routing/switching, header recognition, optical clock recovery (synchronization or bit-phase alignment), pulse generation, pulse compression, and pulse storage. We will discuss some of these testbeds in the remainder of this section. The key features of these testbeds are summarized in Table 12.2.

12.6.1 KEOPS

KEOPS (Keys to Optical Packet Switching) [Gam98, Gui98, RMGB97] was a significant project undertaken by a group of research laboratories and universities in Europe. Its predecessor was the ATMOS (ATM optical switching) project [Mas96, RMGB97]. KEOPS demonstrated several of the building blocks for PPS and put together two separate demonstrators illustrating different switch architectures. The building blocks demonstrated include all-optical wavelength converters using cross-phase modulation in semiconductor optical amplifiers (see Section 3.8) up to 40 GHz, a packet sychronizer at 2.5 Gb/s using a tunable delay line, tunable lasers, and low-loss integrated indium phosphide Mach-Zehnder–type electro-optic switches.

The demonstrations of network functionality were performed at a data rate of 2.5 Gb/s and 10 Gb/s, with the packet header being transmitted at 622 Mb/s. The KEOPS switches used wavelengths internal to the switch as a key tool in performing the switching and buffering, instead of using large optical space switches. In this sense, the KEOPS demonstrators are variations of the architecture of Figure 12.19. The first demonstrator, shown in Figure 12.23, used a two-stage switching approach with wavelength routing. Here, the first stage routes the input signal to the appropriate delay line by converting it to a suitable wavelength and passing it through a

Table 12.2   Key features of photonic packet-switching testbeds described in Section 12.6.

Testbed	Topology	Bit Rate	Functions Demonstrated
KEOPS	Switch	2.5 Gb/s (per port)	4 × 4 switch, subnanosecond switching, all-optical wavelength conversion tunable lasers, packet synchronizer
KEOPS	Switch	10 Gb/s (per port)	16 × 16 broadcast/select, subnanosecond switching
FRONTIERNET	Switch	2.5 Gb/s (per port)	16 × 16, tunable laser
NTT	Switch	10 Gb/s (per port)	4 × 4 broadcast/select
Synchrolan (BT Labs)	Bus	40 Gb/s (aggregate)	Bit-interleaved data transmission and reception
BT Labs	Switch	100 Gb/s (per port)	Routing in a 1 × 2 switch based on optical header recognition
Princeton	Switch	100 Gb/s (per port)	Packet compression, TOAD-based demultiplexing
AON	Helix (bus)	100 Gb/s (aggregate)	Optical phase lock loop, pulse generation, compression, storage
CORD	Star	2.5 Gb/s (per port)	Contention resolution

wavelength demultiplexer. The second stage routes the packet to the correct output, again by using a tunable wavelength converter and a combination of wavelength demultiplexers and multiplexers. Each input has access to at least one delay line in each set of delay lines. Since the delay line in turn has access to all the output ports, the switch may be viewed as implementing a form of shared output buffering.

The switch controller (not shown in the figure) schedules the incoming packets onto the delay lines as follows: Each input packet is scheduled with the minimum possible delay, d, such that (1) no other packet is scheduled in the same time slot to the same output port, (2) no other packet is scheduled in the same time slot on any of the delay lines leading to the same second-stage TWC as the desired packet, and (3) in order to deliver packets in sequence of their arrival, no previous packet from the same input is scheduled to the same output port with a delay larger than d.

Another demonstrator used a broadcast-and-select approach as shown in Figure 12.24. Here packets arriving at different inputs are assigned different wavelengths. Each packet is then broadcast into an array of delay lines providing different delays. Each delay line can store multiple packets simultaneously at different

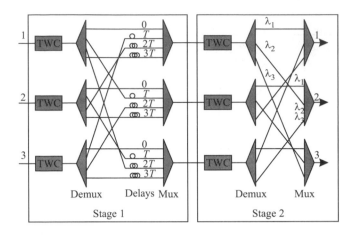

Figure 12.23   The wavelength-routing packet switch used in KEOPS.

wavelengths. Thus each input packet is made available at the output over several slots. Of these, one particular slot is selected using a combination of wavelength demultiplexers, optical switches, and wavelength multiplexers. This switch therefore emulates an output-buffered switch with a B slot buffer on each output. A 16×16 switch using this approach was demonstrated.

12.6.2   NTT's Optical ATM Switches

Researchers at NTT have demonstrated photonic packet switches using an approach somewhat similar to KEOPS [Yam98, HMY98]. Like the KEOPS switches, these switches also use wavelengths internal to the switch as a key element in performing the switching function. The FRONTIERNET switch [Yam98], shown in Figure 12.25, uses tunable wavelength converters in conjunction with an arrayed waveguide grating to perform the switching function, followed by delay line buffers. This is again an output-buffered switch, with two stages of selection. For each output, the first stage selects the time slot, and the second stage the desired wavelength within that time slot. In the experiment, the tunable converter assumes that the incoming data is electrical and uses a tunable laser and external modulator to provide a tunable optical input into an arrayed waveguide grating. A 16×16 switch operating at 2.5 Gb/s with optical delay line buffering was demonstrated.

In separate experiments [HMY98], the switching was accomplished by broadcasting a wavelength-encoded signal to a shared array of delay lines and selecting the appropriate time slot at the output, again like the KEOPS approach. A 4×4

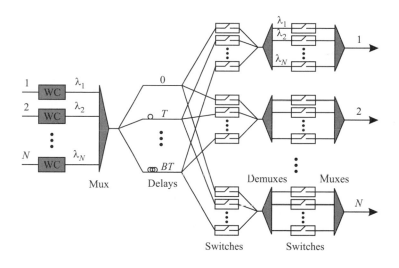

Figure 12.24   The broadcast-and-select packet switch used in KEOPS.

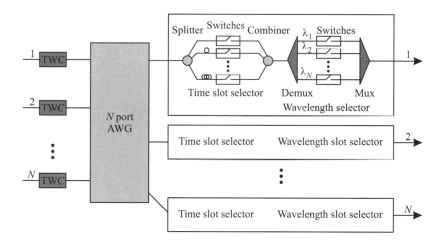

Figure 12.25   The FRONTIERNET architecture.

switch at a 10 Gb/s data rate was demonstrated. The key technologies demonstrated included tunable lasers and optical delay line buffering.

12.6.3 BT Labs Testbeds

Researchers at British Telecom (BT) Laboratories demonstrated several aspects of PPS networks [CLM97] that we discussed in this chapter. Multiplexing and de-multiplexing of high-speed signals in the optical domains was demonstrated in a prototype broadcast local-area network based on a bus topology called *Synchrolan* [LGM+97, Gun97b]. Bit interleaving was used with each of the multiplexed channels operating at a bit rate of 2.5 Gb/s. The aggregate bit rate transmitted on the bus was 40 Gb/s. The clock signal (akin to a framing pulse) was distributed along with the bit-interleaved data channels. The availability of the clock signal meant that there was no need for optical clock recovery techniques. A separate time slot was not used for the clock signal, but rather it was transmitted with a polarization orthogonal to that of the data signals. This enabled the clock signal to be separated easily from the data. In a more recent demonstration [Gun97a], the data and clock signals were transmitted over two separate standard single-mode (nonpolarization-preserving) fibers, avoiding the need for expensive polarization-maintaining components.

A PPS node was also demonstrated separately at BT Labs [Cot95]. The optical header from an incoming packet was compared with the header—local address—corresponding to the PPS node, using an optical AND gate (but of a different type than the ones we discussed). The rest of the packet was stored in a fiber delay line while the comparison was performed. The output of the AND gate was used to set a 1×2 switch so that the packet was delivered to one of two outputs based on a match, or lack of it, between the incoming packet header and the local address.

12.6.4 *Princeton University Testbed*

This testbed was developed in the Lightwave Communications Laboratory at Princeton University, funded by DARPA [Tol98, SBP96]. The goal was to demonstrate a single routing node in a network operating at a transmission rate of 100 Gb/s. Packet interleaving was used, and packets from electronic sources at 100 Mb/s were optically compressed to the 100 Gb/s rate using the techniques we described in Section 12.1. The limitations of the semiconductor optical amplifiers used in the packet compression process (Figure 12.6) require a 0.5 ns (50 bits at 100 Gb/s) guard band between successive packets. Optical demultiplexing of the compressed packet header was accomplished by a bank of AND gates, as described in Section 12.1. The TOAD architecture described in Section 12.1.3 was used for the AND gates. The number

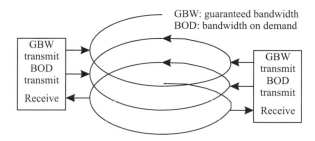

Figure 12.26   The helical LAN topology proposed to be used in the AON TDM testbed.

of TOADs to be used is equal to the length of the packet header. Thus the optically encoded *serial* packet header was converted to a parallel, electronic header by a bank of TOADs.

12.6.5   AON

This testbed was developed by the All-Optical Network (AON) consortium consisting of AT&T Bell Laboratories, Digital Equipment Corporation, and the Massachusetts Institute of Technology [Bar96]. The aim was to develop an optical TDM LAN/MAN operating at an aggregate rate of 100 Gb/s using packet interleaving. Different classes of service, specifically *guaranteed bandwidth service* and *bandwidth-on-demand service,* were proposed to be supported. The topology used is shown in Figure 12.26. This is essentially a bus topology where users transmit in the top half of the bus and receive from the bottom half. One difference, however, is that each user is attached for transmission to two points on the bus such that the guaranteed bandwidth transmissions are always upstream from the bandwidth-on-demand transmissions. Thus the topology can be viewed as having the helical shape shown in Figure 12.26; hence the name *helical LAN* (HLAN) for this network.

Experiments demonstrating an optical phase lock loop were carried out. In these experiments, the frequency and phase of a 10 Gb/s electrically controlled mode-locked laser were locked to those of an incoming 40 Gb/s stream. (Every fourth pulse in the 40 Gb/s stream coincides with a pulse from the 10 Gb/s stream.) Other demonstrated technologies include short pulse generation, pulse compression, pulse storage, and wavelength conversion.

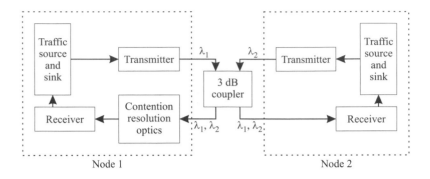

Figure 12.27   A block diagram of the CORD testbed.

12.6.6 CORD

The Contention Resolution by Delay Lines (CORD) testbed was developed by a consortium consisting of the University of Massachusetts, Stanford University, and GTE Laboratories [Chl96]. A block diagram of the testbed is shown in Figure 12.27. The testbed consisted of two nodes transmitting ATM-sized packets at 2.488 Gb/s using different transmit wavelengths (1310 nm and 1320 nm). A 3 dB coupler broadcasts all the packets to both the nodes. Each node generates packets destined to both itself and the other node. This gives rise to contentions at both the receivers. The headers of the packets from each node were carried on distinct subcarrier frequencies (3 GHz and 3.5 GHz) located outside the data bandwidth ($\approx$ 2.5 GHz). The subcarrier headers were received by tapping off a small portion of the power (10%) from the incoming signal.

Time was divided into slots, with the slot size being equal to 250 ns. Since an ATM packet is only 424/2.488 $\approx$ 170 ns long, there was a fair amount of guard band in each slot. Slot synchronization between nodes was accomplished by having nodes adjust their clocks based on their propagation delay to the hub. However, a separate synchronizer node was not used, and one of the nodes itself acted as the synchronizer (called "master" in CORD) node. The data rate on the subcarrier channels was chosen to be 80 Mb/s so that a 20-bit header can be transmitted in the 250 ns slot.

In one of the nodes, a feed-forward delay line architecture similar to that shown in Figure 12.14 was used with a WDM demux and mux surrounding it, so that signals at the two wavelengths could undergo different delays. Thus this node had greater opportunities to resolve contentions among packets destined to it. This is the origin of the name *contention resolution by delay lines* for this testbed. The current testbed is built using discrete components, including lithium niobate switches, semiconductor optical amplifiers (for loss compensation), and polarization-maintaining fiber for the delay lines. An integrated version of the *contention resolution optics* (CRO), which would integrate the three 2×2 switches and semiconductor amplifiers on a single InP substrate, is under development.

Summary

Photonic packet-switched networks offer the potential of realizing packet-switched networks with much higher capacities than may be possible with electronic packet-switched networks. However, significant advances in technology are needed to make them practical, and there are some significant roadblocks to overcome. The state of optical packet-switching technology is somewhat analogous to the state of electronic circuits before the integrated circuit was invented. All the building blocks needed for optical packet switching are in a fairly rudimentary state today and in research laboratories—they are either difficult to realize, very bulky, or very expensive. For example, optical buffering is implemented using hundreds of meters of delay lines, which are bulky and can only provide limited amounts of storage. Transmitting data at 100 Gb/s and higher line rates over any significant distances of optical fiber is still a major challenge. At this time, fast optical switches have relatively high losses, including polarization-dependent losses, and are not amenable to integration, which is essential to realize large switches. Optical wavelength converters, which have been proposed for many of the architectures, are still in their infancy today. Temperature dependence of individual components can also be a significant problem when multiplexing, demultiplexing, or synchronizing signals at such high bit rates. We also need effective ways of combatting the signal degradation through these switches. For instance, a cheap all-optical 3R regenerator along the lines of what we studied in Section 3.8 would make many of these architectures more practical. For the foreseeable future, it appears that we will continue to perform all the intelligent control functions for packet switching in the electrical domain.

In the near term, we will continue to see the optical layer being used to provide circuit-switched services, with packet-switching functions being done in the electronic domain by IP routers or ATM switches. PPS, particularly with burst switching, is being positioned as a possible future replacement for the optical circuit layer, while still retaining electronic packet switching at the higher layers. The notion is that circuit-switched links are still underutilized due to the bursty nature of traffic, and using an underlying optical packet layer instead of a circuit layer will help improve link utilizations.

Further Reading

There has been a great deal of research activity related to photonic packet switching with respect to architectures and performance evaluation, as well as experiments and testbeds. See [HA00] for a recent overview, as well as [Pru93, BPS94, Mid93]. [BIPT98, MS88] are special issues devoted to this topic.

The NOLM is described in [DW88], and its use for optical demultiplexing is described in [BDN90]. The NALM is described in [FHHH90]. The architecture of the TOAD is described in [SPGK93], and its operation is analyzed in [KGSP94]. Its use for packet header recognition is described in [GSP94]. Another nonlinear optical loop mirror structure, which uses a short loop length and an SOA within the loop, is described in [Eis92]. The soliton-trapping AND gate is described in [CHI$^+$92]. Other demultiplexing methods using high-speed modulators are described in [Mik99, MEM98]. Packet compression and decompression can also be accomplished by a technique called *rate conversion;* see [PHR97].

For a summary of optical buffering techniques, see [HCA98, Hal97]. Many of the performance results relating to buffering in packet switches may be found in [HK88]. Optical buffering at 40 Gb/s is described in [HR98]. [Dan97, Dan98] analyze the impact of using the wavelength dimension to reduce the number of buffers.

For an overview of deflection routing, see [Bor95]. For an analysis of deflection routing on the hypercube topology, see [GH92]. Some other papers on deflection routing that may be of interest are [HC93, BP96]. [BCM$^+$92] describes an early experimental demonstration of a packet-switching photonic switch using deflection routing.

Using burst switching in the context of PPS has been proposed by [QY99, Tur99, YQD01]. Similar notions have been proposed earlier in the context of electronic packet-switched networks [Ams83].

Most of the testbeds we have discussed, and some we haven't, are described in the special issues on optical networks and photonic switching [BIPT98, CHK$^+$96, FGO$^+$96]. See also [Hun99, Gui00] for another testbed architecture and demonstration using wavelength-based switching. A design for a soliton ring network operating at 100 Gb/s and using soliton logic gates such as the soliton-trapping AND gate is described in [SID93].

We have covered WDM as well as TDM techniques in the book, but haven't explored networks based on optical code division multiple access (OCDMA). Here different transmitters make use of different *codes* to *spread* their data, either in the time domain or in the frequency domain. The codes are carefully designed so that many transmitters can transmit simultaneously without interfering with one another, and the receiver can pick out a desired transmitter's signal from the others by suitably *despreading* the received signal. OCDMA networks were a popular research topic in the late 1980s and early 1990s, but they suffer from even more problems than PPS networks employing high-speed TDM. See [Sal89, SB89, PSF86, FV88] for a sampling of papers on this topic, and see [Gre93] for a good overview.

Problems

12.1 In the packet multiplexing illustrated in Figure 12.6, show that the delay encountered by pulse i, $i = 1, 2, \ldots, l$, on passing through the k compression stages is $(2^k - i)(T - \tau)$. Using the fact that the pulses are separated by time T at the input, now show that pulse i occurs at the output at time $(2^k - 1)(T - \tau) + (i - 1)\tau$. Thus the pulses are separated by a time interval of τ at the output.

12.2 Show that a compressed data packet of length l bits, obtained by the packet multiplexing technique illustrated in Figure 12.6, can be decompressed, in principle, by passing it through a series of $k = \lceil \log \rceil l$ expansion stages, where the jth expansion stage is as shown in Figure 12.28. What should be the switching time of the on-off switches used in this scheme?

12.3 Consider the tunable delay shown in Figure 12.13. Assume that a delay of $xT/2^{k-1}$ is to be realized, where x is a k-bit integer. Consider the binary representation of x,

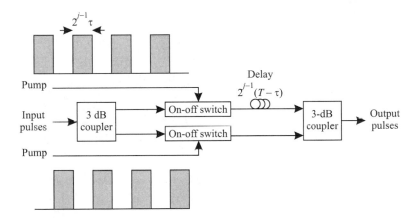

Figure 12.28 An optical packet demultiplexer can be built, in principle, by passing the compressed packet passes through k expansion stages, where 2^k is the smallest power of two that is not smaller than the packet length l in bits. The figure shows a detailed view of expansion stage j.

and find an expression for the control inputs $c_1, \ldots, c_k$. Assume that if $c_i = 1$, switch i is set in the bar state, and if $c_i = 0$, switch i is set in the cross state.

12.4   Consider the fiber loop mirror shown in Figure 12.8, and show that the nonlinear element should introduce a phase shift of π between the clockwise and counterclockwise signals in order for all the energy entering the directional coupler from arm A to be transferred to arm B.

12.5   We have seen that many photonic packet-switching proposals use a lower-rate header compared to the payload. Suppose the maximum header bit rate is 1 Gb/s and headers are 10 bytes long. The payload data rate is 100 Gb/s.

(a) We would like the duration of the payload to be 90% of the overall packet duration (including header and payload). What size does the payload need to be?

(b) If we wanted the maximum payload size to be 1000 bytes and maintain the same efficiency, at what rate would the header have to be transmitted?

(c) Suppose we need a minimum of 1 μs to process the header. This time is accounted for as an additional guard band in the overall packet, in addition to the header and payload. Again, if we want to maintain the payload at 90% of the overall packet, and the header at 10 bytes at 1 Gb/s, what size does the payload need to be?

References

[Ams83]   S. Amstutz. Burst switching—an introduction. *IEEE Communications Magazine*, 21:36–42, Nov. 1983.

[AS92]   A. S. Acampora and S. I. A. Shah. Multihop lightwave networks: A comparison of store-and-forward and hot-potato routing. *IEEE Transactions on Communications*, 40(6):1082–1090, June 1992.

[Bar64]   P. Baran. On distributed communications networks. *IEEE Transactions on Communications*, pages 1–9, March 1964.

[Bar96]   R. A. Barry et al. All-optical network consortium—ultrafast TDM networks. *IEEE JSAC/JLT Special Issue on Optical Networks*, 14(5):999–1013, June 1996.

[BCM+92]   D. J. Blumenthal, K. Y. Chen, J. Ma, R. J. Feuerstein, and J. R. Sauer. Demonstration of a deflection routing 2 × 2 photonic switch for computer interconnects. *IEEE Photonics Technology Letters*, 4(2):169–173, Feb. 1992.

[BDG95]   C. Baransel, W. Dobosiewicz, and P. Gburzynski. Routing in multihop packet switching networks: Gb/s challange. *IEEE Network*, pages 38–61, May/June 1995.

[BDN90]   K. J. Blow, N. J. Doran, and B. P. Nelson. Demonstration of the nonlinear fibre loop mirror as an ultrafast all-optical demultiplexer. *Electronics Letters*, 26(14):962–964, July 1990.

[BFP93]   A. Bononi, F. Forghieri, and P. R. Prucnal. Synchronisation in ultrafast packet switching transparent optical networks. *Electronics Letters*, 29(10):872–873, May 1993.

[BIPT98] D. J. Blumenthal, T. Ikegami, P. R. Prucnal, and L. Thylen, editors. *IEEE/OSA Journal of Lightwave Technology: Special Issue on Photonic Packet Switching Technologies, Techniques and Systems*, volume 16, Dec. 1998.

[Bor95]   F. Borgonovo. Deflection routing. In M. Steenstrup, editor, *Routing in Communication Networks*. Prentice Hall, Englewood Cliffs, NJ, 1995.

[BP96]   A. Bononi and P. R. Prucnal. Analytical evaluation of improved access techniques in deflection routing networks. *IEEE/ACM Transactions on Networking*, 4(5):726–730, Oct. 1996.

[BPS94]   D. J. Blumenthal, P. R. Prucnal, and J. R. Sauer. Photonic packet switches: Architectures and experimental implementations. *Proceedings of IEEE*, 82:1650–1667, Nov. 1994.

[Bur94]   M. Burzio et al. Optical cell synchronization in an ATM optical switch. In *Proceedings of European Conference on Optical Communication*, pages 581–584, 1994.

[CHI+92] M. W. Chbat, B. Hong, M. N. Islam, C. E. Soccolich, and P. R. Prucnal. Ultrafast soliton-trapping AND gate. *IEEE/OSA Journal on Lightwave Technology*, 10(12):2011–2016, Dec. 1992.

[CHK+96] R. L. Cruz, G. R. Hill, A. L. Kellner, R. Ramaswami, and G. H. Sasaki, editors. *IEEE JSAC/JLT Special Issue on Optical Networks*, volume 14, June 1996.

[Chl96]   I. Chlamtac et al. CORD: Contention resolution by delay lines. *IEEE JSAC/JLT Special Issue on Optical Networks*, 14(5):1014–1029, June 1996.

[CLM97]   D. Cotter, J. K. Lucek, and D. D. Marcenac. Ultra-high bit-rate networking: From the transcontinental backbone to the desktop. *IEEE Communications Magazine*, 35(4):90–95, April 1997.

[Cot95]   D. Cotter et al. Self-routing of 100 Gbit/s packets using 6 bit "keyword" address recognition. *Electronics Letters*, 31(25):2201–2202, Dec. 1995.

[Dan97]   S. L. Danielsen et al. WDM packet switch architectures and analysis of the influence of tuneable wavelength converters on the performance. *IEEE/OSA Journal on Lightwave Technology*, 15(2):219–227, Feb. 1997.

[Dan98]   S. L. Danielsen et al. Analysis of a WDM packet switch with improved performance under bursty traffic conditions due to tuneable wavelength converters. *IEEE/OSA Journal on Lightwave Technology*, 16(5):729–735, May 1998.

[DW88]   N. J. Doran and D. Wood. Nonlinear-optical loop mirror. *Optics Letters*, 13(1):56–58, Jan. 1988.

[Eis92]   M. Eiselt. Optical loop mirror with semiconductor laser amplifier. *Electronics Letters*, 28(16):1505–1506, July 1992.

[ENW96]   A. Erramilli, O. Narayan, and W. Willinger. Experimental queueing analysis with long-range dependent packet traffic. *IEEE/ACM Transactions on Networking*, 4(2):209–223, Apr. 1996.

[FBP95]   F. Forghieri, A. Bononi, and P. R. Prucnal. Analysis and comparison of hot-potato and single-buffer deflection routing in very high bit rate optical mesh networks. *IEEE Transactions on Communications*, 43(1):88–98, Jan. 1995.

[FGO+96]   M. Fujiwara, M. S. Goodman, M. J. O'Mahony, O. K. Tonguez, and A. E. Willner, editors. *IEEE/OSA JLT/JSAC Special Issue on Multiwavelength Optical Technology and Networks*, volume 14, June 1996.

[FHHH90]   M. E. Fermann, F. Haberl, M. Hofer, and H. Hochreiter. Nonlinear amplifying loop mirror. *Optics Letters*, 15(13):752–754, July 1990.

[FV88]   G. J. Foschini and G. Vannucci. Using spread spectrum in a high-capacity fiber-optic local network. *IEEE/OSA Journal on Lightwave Technology*, 6(3):370–379, March 1988.

[Gam98]   P. Gambini et al. Transparent optical packet switching: Network architecture and demonstrators in the KEOPS project. *IEEE JSAC: Special Issue on High-Capacity Optical Transport Networks*, 16(7):1245–1259, Sept. 1998.

[GG93]   A. G. Greenberg and J. Goodman. Sharp approximate models of deflection routing in mesh networks. *IEEE Transactions on Communications*, 41(1):210–223, Jan. 1993.

[GH92]   A. G. Greenberg and B. Hajek. Deflection routing in hypercube networks. *IEEE Transactions on Communications*, 40(6):1070–1081, June 1992.

[Gre93]   P. E. Green. *Fiber-Optic Networks*. Prentice Hall, Englewood Cliffs, NJ, 1993.

[GSP94]   I. Glesk, J. P. Sokoloff, and P. R. Prucnal. All-optical address recognition and self-routing in a 250 Gb/s packet-switched network. *Electronics Letters*, 30(16):1322–1323, Aug. 1994.

[Gui98]   C. Guillemot et al. Transparent optical packet switching: The European ACTS KEOPS project approach. *IEEE/OSA Journal on Lightwave Technology*, 16(12):2117–2134, Dec. 1998.

[Gui00] K. Guild et al. Cascading and routing 14 optical packet switches. In *Proceedings of European Conference on Optical Communication*, 2000.

[Gun97a] P. Gunning et al. 40 Gbit/s optical TDMA LAN over 300m of installed blown fibre. In *Proceedings of European Conference on Optical Communication*, volume 4, pages 61–64, Sept. 1997.

[Gun97b] P. Gunning et al. Optical-TDMA LAN incorporating packaged integrated Mach-Zehnder interferometer channel selector. *Electronics Letters*, 33(16):1404–1406, July 1997.

[HA00] D. K. Hunter and I. Andonovic. Approaches to optical Internet packet switching. *IEEE Communications Magazine*, 38(9).116–122, Sept. 2000.

[Hal97] K. L. Hall. All-optical buffers for high-speed slotted TDM networks. In *IEEE/LEOS Summer Topical Meeting on Advanced Semiconductor Lasers and Applications*, page 15, 1997.

[HC93] B. Hajek and R. L. Cruz. On the average delay for routing subject to independent deflections. *IEEE Transactions on Information Theory*, 39(1):84–91, Jan. 1993.

[HCA98] D. K. Hunter, M. C. Chia, and I. Andonovic. Buffering in optical packet switches. *IEEE/OSA Journal on Lightwave Technology*, 16(12):2081–2094, Dec. 1998.

[Hil85] W. D. Hillis. *The Connection Machine*. MIT Press, Cambridge, MA, 1985.

[Hil87] W. D. Hillis. The connection machine. *Scientific American*, 256(6), June 1987.

[HK88] M. G. Hluchyj and M. J. Karol. Queuing in high-performance packet switching. *IEEE JSAC*, 6(9):1587–1597, Dec. 1988.

[HMY98] K. Habara, T. Matsunaga, and K.-I. Yukimatsu. Large-scale WDM star-based photonic ATM switches. *IEEE/OSA Journal on Lightwave Technology*, 16(12):2191–2201, Dec. 1998.

[HR98] K. L. Hall and K. T. Rauschenbach. All-optical buffering of 40 Gb/s data packets. *IEEE Photonics Technology Letters*, 10(3):442–444, Mar. 1998.

[Hun99] D. K. Hunter et al. WASPNET—a wavelength switched packet network. *IEEE Communications Magazine*, 37(3):120–129, Mar. 1999.

[KGSP94] M. G. Kane, I. Glesk, J. P. Sokoloff, and P. R. Prucnal. Asymmetric loop mirror: Analysis of an all-optical switch. *Applied Optics*, 33(29):6833–6842, Oct. 1994.

[KH90] A. Krishna and B. Hajek. Performance of shuffle-like switching networks with deflection. In *Proceedings of IEEE Infocom*, pages 473–480, 1990.

[LGM+97] J. K. Lucek, P. Gunning, D. G. Moodie, K. Smith, and D. Pitcher. Synchrolan: A 40 Gbit/s optical-TDMA LAN. *Electronics Letters*, 33(10):887–888, April 1997.

[LNGP96]   E. Leonardi, F. Neri, M. Gerla, and P. Palnati. Congestion control in asynchronous high-speed wormhole routing networks. *IEEE Communications Magazine*, pages 58–69, Nov. 1996.

[Mas96]   F. Masetti et al. High speed, high capacity ATM optical switches for future telecommunication transport networks. *IEEE JSAC/JLT Special Issue on Optical Networks*, 14(5):979–998, June 1996.

[Max89]   N. F. Maxemchuck. Comparison of deflection and store-and-forward techniques in the Manhattan Street and shuffle-exchange networks. In *Proceedings of IEEE Infocom*, pages 800–809, 1989.

[MEM98]   D. D. Marcenac, A. D. Ellis, and D. G. Moodie. 80 Gbit/s OTDM using electroabsorption modulators. *Electronics Letters*, 34(1):101–103, Jan. 1998.

[Mid93]   J. E. Midwinter, editor. *Photonics in Switching, Volume II: Systems*. Academic Press, San Diego, CA, 1993.

[Mik99]   B. Mikkelsen et al. Unrepeatered transmission over 150 km of nonzero-dispersion fibre at 100 Gbit/s with semiconductor based pulse source, demultiplexer and clock recovery. *Electronics Letters*, 35(21):1866–1868, Oct. 1999.

[MS88]   J. E. Midwinter and P. W. Smith, editors. *IEEE JSAC: Special Issue on Photonic Switching*, volume 6, Aug. 1988.

[PF95]   V. Paxon and S. Floyd. Wide area traffic: The failure of Poisson modelling. *IEEE/ACM Transactions on Networking*, 3(3):226–244, June 1995.

[PHR97]   N. S. Patel, K. L. Hall, and K. A. Rauschenbach. Optical rate conversion for high-speed TDM networks. *IEEE Photonics Technology Letters*, 9(9):1277, Sept. 1997.

[Pru93]   P. R. Prucnal. Optically processed self-routing, synchronization, and contention resolution for 1-d and 2-d photonic switching architectures. *IEEE Journal of Quantum Electronics*, 29(2):600–612, Feb. 1993.

[PSF86]   P. R. Prucnal, M. A. Santoro, and T. R. Fan. Spread spectrum fiber-optic local area network using optical processing. *IEEE/OSA Journal on Lightwave Technology*, LT-4(5):547–554, May 1986.

[QY99]   C. Qiao and M. Yoo. Optical burst switching (OBS): A new paradigm for an optical Internet. *Journal of High Speed Networks*, 8(1):69–84, 1999.

[RMGB97]   M. Renaud, F. Masetti, C. Guillemot, and B. Bostica. Network and system concepts for optical packet switching. *IEEE Communications Magazine*, 35(4):96–102, Apr. 1997.

[Sal89]   J. A. Salehi. Code division multiple-access techniques in optical fiber networks—Part I: Fundamental principles. *IEEE Transactions on Communications*, 37(8):824–833, Aug. 1989.

[SB89]   J. A. Salehi and C. A. Brackett. Code division multiple-access techniques in optical fiber networks—Part II: Systems performance analysis. *IEEE Transactions on Communications*, 37(8):834–842, Aug. 1989.

[SBP96]   S.-W. Seo, K. Bergman, and P. R. Prucnal. Transparent optical networks with time-division multiplexing. *IEEE JSAC/JLT Special Issue on Optical Networks*, 14(5):1039–1051, June 1996.

[SID93]   J. R. Sauer, M. N. Islam, and S. P. Dijaili. A soliton ring network. *IEEE/OSA Journal on Lightwave Technology*, 11(12):2182–2190, Dec. 1993.

[Smi81]   B. Smith. Architecture and applications of the HEP multiprocessor system. In *Real Time Signal Processing IV, Proceedings of SPIE*, pages 241–248, 1981.

[SPGK93] J. P. Sokoloff, P. R. Prucnal, I. Glesk, and M. Kane. A terahertz optical asymmetric demultiplexcr (TOAD). *IEEE Photonics Technology Letters*, 5(7):787–790, July 1993.

[Tol98]   P. Toliver et al. Routing of 100 Gb/s words in a packet-switched optical networking demonstration (POND) node. *IEEE/OSA Journal on Lightwave Technology*, 16(12):2169–2180, Dec. 1998.

[Tur99]   J. S. Turner. Terabit burst switching. *Journal of High Speed Networks*, 8(1):3–16, 1999.

[Yam98]   Y. Yamada et al. Optical output buffered ATM switch prototype based on FRONTIERNET architecture. *IEEE JSAC: Special Issue on High-Capacity Optical Transport Networks*, 16(7):2117–2134, Sept. 1998.

[YQD01]   M. Yoo, C. Qiao, and S. Dixit. Optical burst switching for service differentiation in the next-generation optical Internet. *IEEE Communications Magazine*, 39(2):98–104, Feb. 2001.

[ZT98]   W. D. Zhong and R. S. Tucker. Wavelength routing-based photonic packet buffers and their applications in photonic switching systems. *IEEE/OSA Journal on Lightwave Technology*, 16(10):1737–1745, Oct. 1998.

13
chapter

Deployment Considerations

IN THIS CHAPTER, we will study some of the issues facing network operators as they build new networks or upgrade their networks to higher and higher capacities. We will start by understanding how the network is changing from a services perspective, and then understand the changes happening to the network infrastructure. Chapter 1 provided an overview of some of these changes, but we will examine them in detail in this chapter. We will try to understand the various architectural choices available to carriers planning their next-generation networks, in terms of the roles played by SONET/SDH, IP, and ATM. We will discuss the role played by the optical layer and the economic considerations underlying the deployment of WDM and TDM optical layer technologies in the network. We will see that long-haul networks and metro networks have different requirements that influence the choice of technology deployed. In general, it is difficult to decide between the different technologies, and network operators often employ sophisticated network design tools to help them understand the cost trade-offs between different approaches. The examples and problems in this chapter will help the reader gain a better understanding of these trade-offs.

13.1 The Evolving Telecommunications Network

The legacy transport network in place in networks run by established carriers is based on SONET and SDH. Over the past decade, we have seen the WDM optical layer play an increasing role in these networks.

Several factors are causing service providers to reexamine the way they build their transport network. The first driver is obviously the enormous growth in network traffic. Not only is the traffic doubling every year, but the traffic mix is unpredictable and changing. Another driver is the increasing dominance of data traffic, particularly Internet traffic, relative to voice traffic. Data traffic now exceeds voice traffic on the public network. This trend is likely to continue for at least the next several years. A third driver is the advent of increased competition, which is causing service providers to rethink how they deploy services. In contrast to a world where a new service request for bandwidth could take weeks to months to be fulfilled and require long-term contractual agreements, service providers are increasingly entering a world where services need to be deployed rapidly without long-term contracts at highly competitive rates.

Moreover, there is now a new generation of carriers who operate under significantly different business models than the established carriers. These different business models require different architectures. A carrier providing services to interconnect Internet service providers has very different requirements than a traditional carrier servicing voice and private circuit-switched lines. We also now have a new set of carriers' carriers. These are carriers providing bulk bandwidths (say, at 622 Mb/s and above) primarily to other carriers. These carriers' carriers have different requirements from carriers delivering low-speed services (such as 1.5 Mb/s lines) to their customers.

Before we delve into the evolution of the network, it is worth looking at what carriers look for when they deploy equipment in their network. At the end of the day, what they deploy must either enable them to reduce the cost of their network, or enable them to generate revenue from new services enabled by the equipment deployed. From a cost perspective, carriers look at *capital* cost and *operations* cost. Capital cost is the up-front cost of deploying the equipment, and operations cost represents the recurring cost of maintaining and operating the network. Capital cost includes the cost of the equipment, as well as the cost of real estate, providing for appropriate power and cooling and the fiber facilities. In the case of transmission equipment, the goal is to minimize the cost per bit transmitted per mile in the network. It is important to look at the initial entry cost, as well as the cost to add incremental capacity to already-deployed equipment.

Operations cost includes real estate rental/lease costs; recurring costs of power and cooling; labor costs to provision, maintain, and service the equipment; and costs associated with replacing failed equipment and missing service-level agreements on network availability. While most carriers will say that operations costs dominate over capital costs in their networks, capital costs are usually much easier to quantify, and hence many carriers use capital costs as the primary basis for making purchasing decisions.

Looking at the revenue side of the equation, carriers are always on the lookout for generating new revenue streams by deploying new services. These might include services tailored toward enabling new applications, for instance, providing storage networks between data centers, or modified versions of traditional services. For instance, deploying equipment that enables a carrier to set up and take down private line circuit-switched services in minutes where needed would enable a carrier to offer short-term tariffs on these services, as opposed to requiring its customers to buy the service for extended durations. Another benefit of this capability is that it reduces the time to deploy a service and extracts more revenue as a result. Yet another benefit is that it allows a carrier to better utilize its existing network resources, without having stranded bandwidth due to an inability to anticipate the traffic pattern in the network.

The factors described above are forcing carriers to deploy networks that can scale in capacity, networks that are flexible in that they are able to deliver a wide variety of services *where* needed *when* needed. The optical layer provides carriers with the ability to deliver these high-speed circuit-switched services, and also serves as the transport mechanism for carrying multiplexed low-speed packet and circuit-switched services.

13.1.1   The SONET/SDH Core Network

Figure 13.1(a) shows the core network of a typical established carrier. The network consists of interconnected SONET rings. Given today's capacity demands, many of the rings actually consist of multiple rings connecting the same set of nodes. These are called *stacked rings*. These rings operate over different fibers, or more commonly, wavelengths within the same fibers using WDM. Figure 13.1(b) shows a blowup of a large node in this network. The node has multiple WDM terminals (OLTs). Each ring passing through the node requires a SONET ADM. These ADMs are connected to the OLTs and operate at line rates of OC-48 (2.5 Gb/s) or OC-192 (10 Gb/s). The ADMs drop lower-speed traffic streams, ranging from 45 Mb/s DS3 streams to higher-speed 622 Mb/s OC-12 streams. The lower-speed traffic is handled by digital crossconnect systems (DCSs). Data traffic is brought into the network through these lower-speed signals and multiplexed to higher speeds by the SONET ADMs and the DCSs. This data enters the network typically in the form of private lines, such as DS1, DS3 or E1, E3 lines, or directly at other SONET/SDH rates. These rates are well defined and mapped into the SONET/SDH multiplexing structure. Other data traffic, such as IP traffic from routers or ATM traffic from ATM switches, can be brought into the network via DS1/DS3 lines or higher-speed optical signals such as OC-3, OC-12, and carried over the SONET/SDH infrastructure.

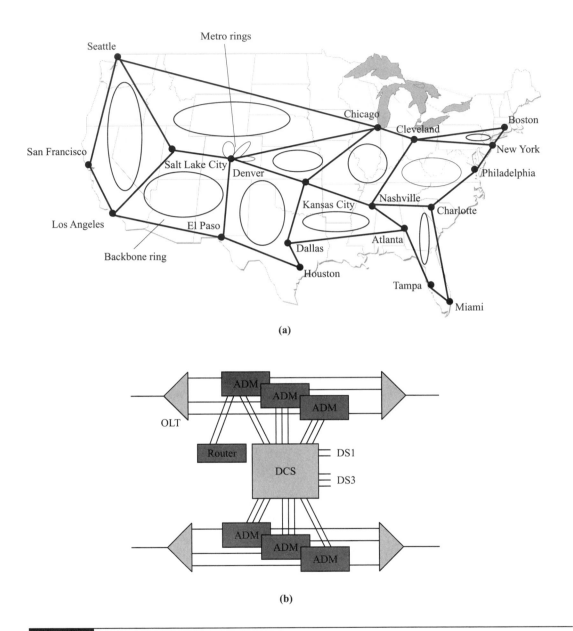

(a)

(b)

Figure 13.1 A typical carrier backbone network based on SONET/SDH, showing SONET/SDH add/drop multiplexers (ADMs) and digital crossconnects (DCSs), along with optical line terminals (OLTs) and routers. (a) The network topology, which consists of interconnected rings in the backbone, with feeder metro rings. (b) Architecture of a typical node, including OLTs, stacked up SONET ADMs, and DCSs.

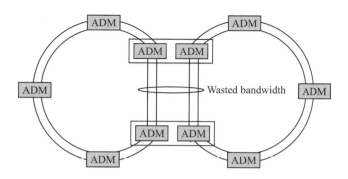

Figure 13.2 Bandwidth wasted when two rings built using ADMs share the same fiber route. Half the bandwidth on each ring along the shared route is reserved for protection.

This network was designed primarily to carry voice and private line traffic. The network provides guaranteed latency and bandwidth, and well-established protection schemes ensure high network availability. SONET/SDH also provides extensive performance monitoring and fault management capabilities. The network is mostly static, with switching provided by the DCSs in order to provision connnections. The switching is done at the time a connection is set up. Once set up, connections remain for months or years, but may have to be switched in the interim to deal with network failures or for maintenance purposes.

However, as we see the increasing dominance of data traffic and the emergence of new optical layer equipment, several deficiencies of the SONET/SDH-based network architecture become evident:

- It consists primarily of static rings where capacity is provisioned in a static manner. It does not allow the rapid provisioning of services end to end across the network on time scales of tens of milliseconds (for fast protection switching) to seconds (for rapid provisioning).

- The traffic demands themselves are more meshed, and the ring architecture is not the most efficient at supporting an inherently meshed traffic demand for several reasons. Multiple rings need to be interconnected, and the interconnection is fairly complex and done through digital crossconnects. Half the capacity on each ring in the network is reserved for protection. Moreover, if two rings share a common link, as shown in Figure 13.2, the protection capacity is reserved for each ring separately along the overlapping link. This may be useful if the network needs to protect against multiple simultaneous failures, but is otherwise wasteful.

- By default, all the traffic is protected. This does not allow carriers to offer a variety of services, some protected and others not protected. Protection is not needed for certain types of traffic, for instance, best-effort IP traffic.

- Data traffic is increasingly entering the network at higher and higher speeds. If IP routers are hooked into the tributary ports of a SONET/SDH box, the SONET/SDH equipment may need to operate at a higher line speed than the IP router ports. For instance, IP routers with OC-48c ports may need to be connected into OC-192 SONET ADMs. (Recall from Chapter 6 that the "c" stands for *concatenated*.) This is usually because the SONET equipment is designed to support tributary interfaces at lower speeds than the line interfaces. Also some variants of SONET, such as a two-fiber BLSR, reserve half the bandwidth on each fiber line for protection. For example, a two-fiber OC-48 ring carries only an OC-24 worth of working traffic on each fiber, and SONET does not provide for splitting up a concatenated SONET stream. This situation presents a problem because the IP routers are increasingly able to support ports at rates comparable to SONET/SDH line rates. Thus an IP router with OC-192c ports may need to be mapped into an OC-768 SONET ADM. The rates coming out of IP router ports may soon exceed the line rate of SONET/SDH ADMs. For this reason, it makes sense to connect IP routers with high-speed ports directly into the optical layer without going through intermediate SONET/SDH equipment.

- Some carriers are in the business of delivering high-speed, best-effort IP services. For these carriers, the SONET/SDH layer doesn't provide much of a benefit. The multiplexing and protection offered by the SONET/SDH layer is not needed. Thus significant cost can be saved by eliminating the SONET/SDH equipment for these applications. Note, however, that SONET framing still offers significant advantages: it provides a commonly used set of transport rates and provides sufficient overheads to allow detailed performance monitoring and fault management. For this reason, while SONET multiplexing and protection may not be required in IP or ATM networks, SONET framing is still widely used by IP and ATM equipment.

- SONET/SDH does not provide efficient mapping for many signals used in data networks. For example, transporting a 100 Mb/s Ethernet signal across the country requires leasing a 155 Mb/s OC-3 line.

- Finally, today, carriers lack the management and signaling systems in order to be able to provision connections end to end across their network. The current situation is that different network elements are managed by different management systems, and provisioning connections on systems already fully equipped is a time-consuming and rather manual process. For instance, each SONET ADM and DCS in the network is provisioned separately, one at a time, using element

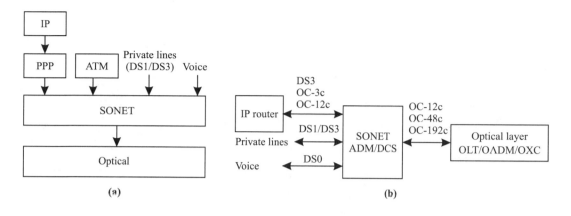

Figure 13.3 Using SONET/SDH as the common transmission layer. IP packets are encapsulated into PPP frames for link layer functions and then mapped into SONET/SDH frames for transmisison over the fiber. The bit rates indicated are for illustration purposes only. (a) The logical layered view. (b) Example of how equipment is interconnected.

management systems. While there are some umbrella network management systems that do provision end-to-end connections, these still provide limited interoperability across equipment from multiple vendors. We saw in Section 9.6.2 that signaling standards are being developed to solve this problem.

For these reasons, the network architecture is changing in some rather significant ways. The best architecture depends to a large extent on the service mix offered by the carrier, and also the legacy network that is in place in the current network. We will next describe the choices facing carriers as they plan their next-generation transport networks.

13.1.2 Architectural Choices for Next-Generation Transport Networks

The optical layer has emerged as the main transmission layer for telecommunications backbone networks. The real debate is about what set of technologies to use above the optical layer to deliver services. This in turn decides the set of boxes that will need to be deployed at the network nodes. The choices today include SONET/SDH, IP, and ATM. Figures 13.3, 13.4, and 13.5 show a variety of options available to carriers planning their next-generation networks.

Figure 13.3 shows the SONET/SDH layer as the common transmission layer above the optical layer. Other services, including ATM and IP, are carried over

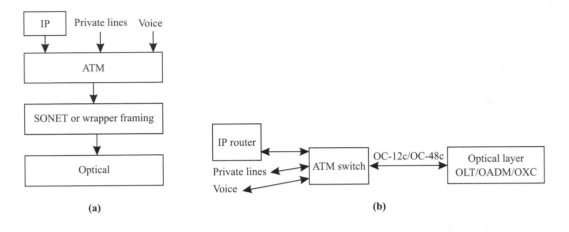

Figure 13.4 Using the ATM layer as the common service layer. IP and other services are brought into the ATM layer. The ATM switches are directly connected to optical layer equipment. The ATM switches embed their cells in frames, typically using SONET/SDH framing. The bit rates indicated are for illustration purposes only. (a) The logical layered view. (b) Example of how equipment is interconnected.

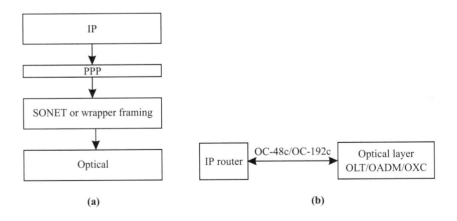

Figure 13.5 Using the IP layer as the common service layer. The routers use a framing protocol to embed the packets before they are transmitted over the optical layer. The bit rates indicated are for illustration purposes only. (a) The logical layered view. (b) Example of how equipment is interconnected.

the SONET/SDH layer. Figure 13.3(a) shows a logical view of the layers, while Figure 13.3(b) shows how the equipment is interconnected in a typical configuration. IP packets are typically carried over a link layer protocol such as PPP (point-to-point protocol), which provides link-level integrity of the frames on a link-by-link basis. These packets are then framed into SONET/SDH frames. All these functions are performed by a line card inside the router. The router is connected to a SONET/SDH box, which multiplexes this connection along with others for transmission over the optical layer.

We have already pointed out the deficiencies of this architecture. For these reasons, it is becoming clear that SONET/SDH will not remain the core transmission layer for much longer. Rather, SONET/SDH will be moved toward the edge of the network and used to multiplex lower-speed circuit-switched lines and bring them into the optical layer. Other network elements such as IP routers and ATM switches will also bring in traffic into the optical layer.

Figure 13.4 shows a model where ATM is used as the common link layer (layer 2) with all services riding above the ATM layer. In this case, the ATM switches are directly connected to optical layer equipment. The ATM switches need to use a framing protocol to embed the cells before transmitting them over the optical layer. The framing protocol allows the data to be formatted for transmission over a physical link and allows various overheads to be added for management purposes. SONET/SDH framing is widely used because of its superior management capabilities and because chipsets are widely available. The framing is done at the line cards sitting inside the ATM switch, rather than requiring a separate SONET/SDH box.

The ATM layer is relatively more mature than the IP layer in terms of providing quality-of-service (QoS) guarantees such as latency and bandwidth. For example, carriers can deliver guaranteed-bandwidth "virtual" DS1 and DS3 services using their ATM networks with a technique known as *circuit emulation*. Protection switching if needed can be provided by the SONET/SDH layer or the optical layer. Several carriers made heavy investments in ATM and are committed to that approach in the near term. In many cases, IP traffic is carried through a frame relay interface into an ATM network. This option is cheaper than having IP routers use private lines such as DS1/DS3 because the frame relay equivalents for these services are less expensive and best-effort traffic doesn't need the QoS guarantees that private lines offer.

Another reason for using the ATM layer to carry IP traffic is that ATM virtual circuits can be set up to provide virtual direct connections between routers. For example, in Figure 13.6, traffic flowing from router A to router C can be routed at intermediate node B through the ATM layer without passing through another router. This helps improve the QoS for the IP traffic. The IP layer itself is currently being enhanced to provide this capability using multi-protocol label switching (MPLS).

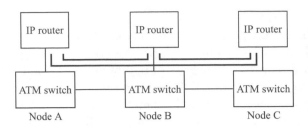

Figure 13.6 An example to illustrate the use of direct virtual connections between routers using the ATM layer. The router at node A sees a direct virtual connection with the router at node C, without having to go through the router at node B.

Thus, over time, the need for having an ATM intermediate layer would seem to disappear.

Figure 13.5 shows a model where the IP layer resides directly on top of the optical layer. The IP layer classically belongs to layer 3 of the OSI hierarchy. With the advent of MPLS, the IP layer also includes layer 2 functionality. In this case, IP routers are directly connected to optical layer equipment. In the wide-area network, SONET/SDH framing is widely used for the reasons given above, and the framing is done on line cards within the router. It is important to note that in this case, there is no need for a separate SONET/SDH box in the network, which can translate into significant cost savings. Two other framing techniques are also emerging. The first is based on Ethernet, or more specifically, Gigabit and 10-Gigabit Ethernet. The other is based on the digital wrapper standard that we studied in Chapter 9. Ethernet-based framing is likely to proliferate as high-speed Ethernet becomes widely deployed in metro networks and reaches out into long-haul networks.

The IP layer today, however, is not yet capable of providing QoS guarantees, such as guaranteed bandwidth and latencies. It also does not provide the same level of availability that SONET/SDH does with its 60 ms restoration times. For these reasons, it cannot yet support voice and private line traffic as well as the SONET/SDH network does. Such private line traffic still constitutes a significant portion of carrier revenues.

Another reason is that in the core of the network it is more efficient to switch data in larger granularities than based on individual packets. There is not as much benefit due to statistical multiplexing, because the traffic is already highly aggregated when it reaches the core. As a result, the traffic tends to be more connection oriented, which doesn't match well with the connectionless datagram approach taken by IP networks. MPLS is being developed to address this issue and will likely be widely implemented in core routers. For all these reasons, an IP over optical layer solution is employed today primarily by carriers to transport best-effort IP services and is not a universal solution. There is intense work under way to evolve the IP and optical layers to provide the same capabilities that SONET/SDH does. Adding protection functions in the optical layer and/or IP layer, as well as adding the QoS support in the IP layer, will enable the IP over optical layer architecture to migrate to a more universal solution.

IP over WDM Variants

We have talked about directly connecting IP routers to the optical layer, in the IP over WDM paradigm. In reality there are multiple ways this can be architected, as shown in Figure 13.7. The differences pertain primarily to the manner in which traffic passing through intermediate nodes is handled and the degree of agility provided in the optical layer. Before going into this in more detail, we look briefly at the capabilities of large IP routers and large optical crossconnects (OXCs). In general the trend to date has been that the total capacity that can be switched by a top-of-the-line router is much smaller than the total switching capacity of an OXC. Likewise, the OXC can be significantly denser (occupy a smaller footprint) than an equivalent router. Furthermore the cost per router port is usually much larger than the cost per equivalent OXC port. None of these are surprising, given the relative differences in functions and resulting complexity between a router and an OXC.

The simplest architecture for IP over WDM, shown in Figure 13.7(a), is to connect the IP routers directly into optical line terminals (OLTs). Passthrough traffic at intermediate nodes is handled by the routers. This, however, has the highest cost for dealing with passthrough traffic, since expensive router ports need to be used to handle all this traffic. Also a large number of router ports will be needed in this approach, requiring significant floor space and associated power and cooling issues. Unfortunately, in some carriers, the router network and the transport (optical layer) network are designed and operated by different groups independently. This often leads to a situation not unlike what we see in Figure 13.7(a).

The second approach, shown in Figure 13.7(b), is similar to the first, except that the passthrough traffic is handled by connecting patch cables between back-to-back WDM terminals within the optical layer. This approach is the lowest-cost option, as all passthrough traffic is handled without additional equipment or using up router ports. However, it is relatively inflexible in the sense that lightpaths cannot be configured dynamically in the network. Also it may be important to perform some demultiplexing and multiplexing of the lightpaths, that is, grooming, at intermediate nodes, if partial signals have to be dropped and added locally, for instance, an OC-12 signal from an OC-192 lightpath.

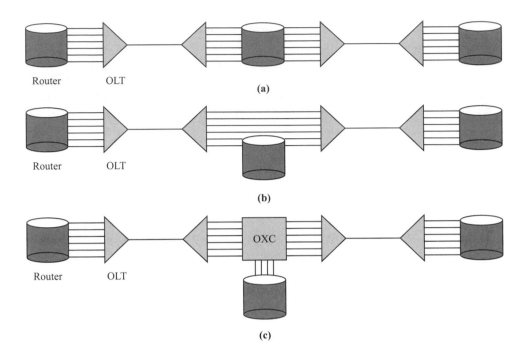

Figure 13.7 Different architectures for realizing an IP over WDM network. (a) Passthrough traffic is handled by routers. (b) Passthrough traffic is patched through in the optical layer in a static fashion. (c) Passthrough traffic is handled by an optical crossconnect (OXC) providing dynamic reconfiguration and traffic grooming.

The third approach, shown in Figure 13.7(c), uses OXCs to handle the passthrough traffic. In terms of cost, it lies between the two approaches discussed above, but provides the flexibility to set up lightpaths dynamically, as well as performs partial demultiplexing and multiplexing at intermediate nodes, if needed. As a result, this is the preferred IP over WDM architecture.

The Evolving Network

As of this writing, a variety of architectures have been implemented by carriers. One carrier has deployed an ATM over SONET/SDH over WDM network and transports lower-speed IP traffic over it. Higher-speed IP interfaces out of routers such as OC-48c interfaces are in most cases carried directly over the optical layer. Voice and private lines are carried over the SONET network. Another carrier that provides just IP services has deployed the IP over WDM architecture. Yet other carriers have deployed an ATM network operating directly over the optical layer and are using it to deliver both virtual circuit and packet-oriented services.

For these reasons, the network is migrating gradually to the architecture shown in Figure 13.8. The backbone is a mesh network made up of optical crossconnects, optical add/drop multiplexers (OADMs), and optical line terminals. The network supports a variety of traffic types, including SONET, ATM, and IP. High-speed traffic streams are directly connected into the optical layer, whereas lower-speed streams may be multiplexed and brought into the network using one of the common service layers described above. Capacity is provisioned and allocated dynamically in the network by the OXCs and the OADMs. Bandwidth-efficient protection is offered as needed on a circuit-by-circuit basis.

SONET/SDH will remain to support voice and private line traffic, as it is the best architecture for this purpose. In fact some of this multiplexing, particularly at the higher speeds, may be done by optical layer equipment, rather than separate SONET/SDH boxes. IP over the optical layer will become more ubiquitous as QoS guarantees are better implemented in the IP layer, MPLS matures to provide direct connections between routers, and protection functions are implemented well in the optical layer and/or the IP layer.

At the edges of the network, access will be provided by a new-generation network element that combines lower-rate statistical and fixed SONET-like time division multiplexing over the optical layer. We call this element a multiservice platform (MSP). By combining time division and statistical multiplexing, an MSP has the potential to deliver a variety of circuit-switched and packet-switched services to the end users of the network. The idea is to use a single box in the access part of the network to deliver a variety of services to end users, without having to deploy multiple overlay networks to support each service type.

As of this writing, a variety of MSPs are being developed by different equipment manufacturers, with a range of functionalities. At one end of the spectrum, an MSP is simply a SONET ADM, which provides data interfaces, such as Ethernet, in addition to supporting voice (DS0) and private lines (DS1/DS3, etc.). This box maps Ethernet signals into a SONET time slot and is purely a circuit-switched device, with no statistical multiplexing capabilities. Other MSPs are architected using a packet- or cell-switched internal core, which allows them to combine statistical multiplexing with time division multiplexing. These boxes perform statistical aggregation of the incoming data signals before mapping them into SONET time slots on their line sides. Finally there are MSPs that do not have any time division capabilities at all, carrying all incoming traffic over a packet-switched network such as IP or ATM. These rely on using QoS capabilities within the IP or ATM layers to provide circuit-switched–like services.

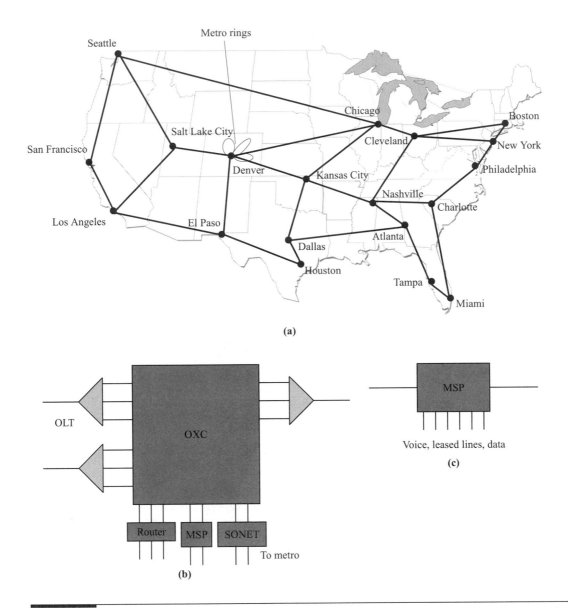

Figure 13.8 The future telecommunications network. (a) Network topology showing a meshed long-haul backbone with metro collector rings. (b) Architecture of a typical backbone node showing an OXC, OLT, IP router, SONET add/drop multiplexer, and an MSP. (c) A node on a metro ring served by an MSP. The MSP is used to deliver a variety of services including voice, private lines, and data services.

Like SONET rings, most MSPs are deployed in ring configurations and include built-in restoration capabilities, which are based on SONET mechanisms for the most part. Ring configurations work well for metro networks, as the fiber is mostly laid in rings. Laying fiber in ring configurations is economical, compared to using other configurations, such as a star (also called a hub and spoke) configuration. A star configuration requires two disjoint fiber routes to be laid between each access node and the central office. In contrast, multiple access nodes can be combined on a single fiber ring, and additional nodes can be added to the ring as needed, without having to lay new fiber routes each time a new node needs to be added. Some MSPs also include built-in WDM interfaces with optical add/drop (OADM) capabilities.

Passive optical networks (PONs) are also emerging as potential candidates to deliver services to small and medium users of bandwidth. In order to prove in, they need to be cheaper and more flexible than SONET/SDH platforms or MSPs. We studied PONs in Chapter 11.

However, as of this writing, WDM is just beginning to be used in metro networks, as its economics are not as compelling as in long-haul networks. More on this in Section 13.2.8.

13.2 Designing the Transmission Layer

We will next look at the choices that service providers have to make in choosing the right tranmission layer. The historical trend has been to increase capacity in the network and at the same time drive down the cost per bit of bandwidth. Service providers generally look for at least a fourfold increase in capacity when planning their networks. As a rule of thumb, they expect to get this fourfold increase in capacity at about 2–2.5 times the cost of current equipment.

There are fundamentally three ways of increasing transmission capacity.

1. The first approach is to light up additional fibers or to deploy additional fibers as needed. We can think of this as the *space division multiplexing* (SDM) approach: keep the bit rate the same but use more fibers.

2. The other traditional approach is to increase the transmission bit rate on the fiber. This is the TDM approach.

3. The third approach is to add additional wavelengths over the same fiber. This is the WDM approach.

Note that the three techniques are complementary to each other and are all needed in the network for a variety of reasons. For instance, using SDM, particularly when existing fibers are close to being exhausted, can be viewed as a long-term way of building up infrastructure; WDM and TDM can be viewed as providing the ability to turn up services rapidly over existing fiber infrastructure. Electronic TDM is required for grooming traffic at lower speeds in the network, where optics is not cost-effective. WDM provides the ability to scale the capacity of the infrastructure in a different dimension. Therefore, the network almost always employs a combination of these techniques in practice.

The interesting question is not whether to use SDM or TDM or WDM—all of these will be used—but to determine the right combination of these. For instance, let us look only at WDM and TDM. To get a total capacity of 80 Gb/s, should we deploy a network with 32 wavelengths at 2.5 Gb/s each, or a network with 8 wavelengths at 10 Gb/s each? This is a complicated question with many parameters affecting the right choice. When should we deploy more fibers, instead of investing in higher-capacity TDM or WDM systems? Several factors influence this decision-making process:

- Is this a new network build or an upgrade of an existing network? If it is an upgrade, we need to consider the cost of adding channels to existing systems in lieu of deploying new systems.
- The availability and cost of additional fiber.
- The type of fiber available.
- The cost of lighting up a new fiber versus adding additional capacity to an already-lit fiber.
- The relative cost of TDM and WDM equipment.

We will attempt to address some of these questions next. The problems at the end of the chapter also provide a partial insight into some of the issues.

13.2.1 Using SDM

Using additional fibers is a straightforward upgrade alternative. The viability of this approach depends on a few factors. First, are additional fibers available on the route? If so, then the next consideration is the route length. If the route length is short (typically a few tens of kilometers) and no regenerators or amplifiers are required along the route, then this is a good alternative. However, if amplifiers or regenerators are required, then this becomes an expensive proposition because each fiber requires a separate set of amplifiers or regenerators. However, it may be worth paying the price to light up a new fiber if the new equipment to be deployed over that fiber provides significantly reduced transmission costs compared to existing equipment on the already-lit fiber.

If no fibers are available on the route, then we need to look at the cost associated with laying new fiber. This varies widely. If there is space in existing conduits, fiber can be pulled through relatively inexpensively and quickly. However, if new conduits must be laid, the cost can be very expensive, even over short distances if the route is in a dense metropolitan area. If new conduits are to be laid, then the link can be populated with a large-count fiber cable. Today's fiber bundles come with hundreds of fibers.

The other aspect of this problem is the time it takes to lay new fiber. Constructing new fiber links takes months to years and requires right-of-way permits from municipalities where the new link is laid. These permits may not be easy to obtain in dense metropolitan areas, due to the widespread impact caused by digging up the streets. In contrast, upgrading an existing fiber link using either TDM or WDM can be done within days to weeks. While it is necessary in some circumstances to lay new fibers, this is not a good mechanism for rapid response to service requests.

Note that carriers are not likely to wait until the last fiber is exhausted before they consider an upgrade process. For example, an upgrade process may be triggered when it is time to light up the last few fibers on a route. This might result in installing additional fibers along the router. Alternatively, the carrier may deploy a higher-capacity TDM or WDM system on the last few fibers, and transfer the traffic from the lower-capacity fibers onto the new system deployed to free up existing fibers along the route.

13.2.2 Using TDM

Clearly, TDM is required for grooming traffic at the lower bit rates where optics is not cost-effective. The question is to what bit rate should traffic be time division multiplexed before it is transmitted over the fiber (perhaps on a wavelength over the fiber). Today's long-haul links operate mostly at rates of 2.5 Gb/s or 10 Gb/s. We will see in Section 13.2.5 that the choice of bit rate here is dictated primarily by the type of fiber available. Metropolitan interoffice links operate mostly at 2.5 Gb/s, and access links operate at even lower speeds. Here the situation is somewhat more complicated, as we will explore in Section 13.2.8.

Electronic TDM technology is already delivering the capability to reach 40 Gb/s transmission rates and may well push this out to 80 Gb/s in the future. Beyond these rates, it is likely that we will need some form of optical TDM.

At the higher bit rates, we have to deal with more severe transmission impairments over the fiber, specifically chromatic dispersion, polarization-mode dispersion (PMD), and fiber nonlinearities. With standard single-mode fiber, from Figure 5.19, the chromatic dispersion limit is about 60 km at 10 Gb/s and about 1000 km at 2.5 Gb/s, assuming transmission around 1550 nm. With practical transmitters, the distances are even smaller. The 10 Gb/s limit can be further reduced in the presence of self-phase modulation. Beyond these distances, the signal must be electronically regenerated, or some form of chromatic dispersion compensation must be employed. Practical 10 Gb/s systems being deployed today commonly use some form of chromatic dispersion compensation. This is usually cheaper than using regeneration, particularly when combined with WDM.

As we saw in Section 5.7.4, the distance limit due to PMD at 10 Gb/s is 16 times less than that at 2.5 Gb/s. On old fiber links, the PMD value can be as high as 2 ps/$\sqrt{km}$. For this value, assuming a 1 dB penalty requirement, the distance limit calculated from (5.23) is about 25 km at 10 Gb/s. Electronic regeneration or PMD compensation is required for longer distances. The PMD-induced distance limit may be even lower because of additional PMD caused by splices, connectors, and other components along the transmission path. PMD does not pose a problem in newly constructed links where the PMD value can be kept as low as 0.1 ps/$\sqrt{km}$.

Finally, nonlinear effects such as self-phase modulation limit the maximum transmission power per channel, resulting in a need for closer amplifier spacing, and thus more amplifiers in the link, leading to somewhat higher costs. At 10 Gb/s, transmission powers are usually limited to under 5 dBm per channel.

Today 10 Gb/s TDM systems are widely deployed in long-haul networks, mostly in conjunction with WDM, and 40 Gb/s TDM systems will soon become commercially available.

13.2.3 Using WDM

It may be preferable to maintain a modest transmission bit rate, say, 10 Gb/s, and have multiple wavelengths over the fiber, than to go to a higher bit rate and have fewer wavelengths. Keeping the bit rate low makes the system less vulnerable to chromatic dispersion, polarization-mode dispersion, and some types of nonlinearities, such as self-phase modulation. On the other hand, WDM systems are generally not suitable for deployment over dispersion-shifted fiber because of the limitations imposed by four-wave mixing (see Chapter 5).

WDM systems can be designed to be transparent systems. This allows different wavelengths to carry data at different bit rates and protocol formats. This can be a major advantage in some cases.

Finally, WDM provides great flexibility in building networks. For example, if there is a network node at which most of the traffic is to be passed through and a small fraction is to be dropped and added, it may be more cost-effective to use a WDM optical add/drop element than terminating all the traffic and doing the add/drop in the electrical domain.

There has been a relentless push in expanding the capacity of WDM systems over the past few years, as shown in Figure 13.9. We are now seeing systems with over 100 wavelengths becoming available. At the same time, channel spacings are being

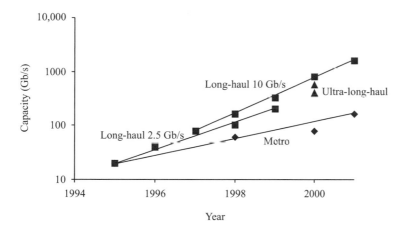

Long-haul	
Year	**Capacity**
1995	8×2.5 Gb/s
1996	16×2.5 Gb/s
1997	8×10 Gb/s
1998	40×2.5 Gb/s
1998	16×10 Gb/s
1999	80×2.5 Gb/s
1999	32×10 Gb/s
2000	80×10 Gb/s
2001	160×10 Gb/s

Ultra-long-haul	
Year	**Capacity**
2000	160×2.5 Gb/s
	56×10 Gb/s

Metro	
Year	**Capacity**
1995	20×1 Gb/s
1998	24×2.5 Gb/s
2000	32×2.5 Gb/s
2001	64×2.5 Gb/s

Figure 13.9 Trends in WDM system capacity for commercially available systems. The capacity indicated is the total capacity (bit rate × number of channels) on a fiber, between regenerators. Bit rates are usually 2.5 Gb/s or 10 Gb/s on each channel. Multiple data points for a given year indicate systems from different vendors—we can have more channels at 2.5 Gb/s or less at 10 Gb/s. Metro systems typically have regenerator spacings of about 50–75 km. Long-haul systems have regenerator spacings of about 400–600 km. Ultra-long-haul systems stretch this distance to about 2500–4000 km. Amplifier spacings in long-haul and ultra-long-haul systems are between 80 and 120 km.

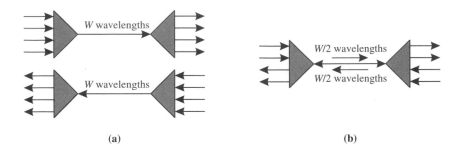

Figure 13.10   (a) Unidirectional and (b) bidirectional transmission systems.

reduced, with 50 GHz channel spacings now common and 25 GHz channel spacings being achieved in some systems. With the opening up of the new L-band window in the 1565–1625 nm range, we can expect to see further increases in the number of wavelengths. Today's state-of-the-art long-haul systems carry about 100 channels at 10 Gb/s each and have regenerator spacings of 400 to 600 km. The ultra-long-haul systems expand spacing between regenerators to about 4000 km but have somewhat lower capacities than the long-haul systems.

13.2.4 *Unidirectional versus Bidirectional WDM Systems*

A unidirectional WDM system uses two fibers, one for each direction of traffic, as shown in Figure 13.10(a). A bidirectional system, on the other hand, requires only one fiber and typically uses half the wavelengths for transmitting data in one direction and the other half for transmitting data in the opposite direction on the same fiber. Both types of systems are being deployed and have their pros and cons. We will compare the two types of systems, assuming that technology limits us to having a fixed number of wavelengths, say, W, per fiber in both cases.

1. A unidirectional system is capable of handling W full-duplex channels over two fibers. A bidirectional system handles $W/2$ full-duplex channels over one fiber. The bidirectional system, therefore, has half the total capacity, but allows a user to build capacity more gradually than a unidirectional system. Thus it may have a slightly lower initial cost. However, to go beyond $W/2$ channels, the user must buy a second bidirectional system and pay for this additional equipment at that time.

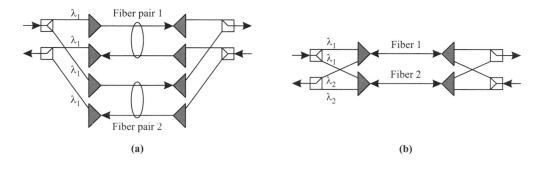

Figure 13.11   Implementing $1 + 1$ protected configurations using unidirectional and bidirectional transmission systems: (a) two unidirectional systems using four fibers, (b) two bidirectional systems using two fibers.

2. If only one fiber (not two) is available, then there is no alternative but to deploy bidirectional systems. Implementing $1 + 1$ or 1:1 configurations with unidirectional WDM systems requires a minimum of two pairs of fibers routed separately, but only requires two fibers with bidirectional systems, as shown in Figure 13.11. Note, however, that as mentioned above, the bidirectional systems provide half as much capacity.

3. Systems using distributed Raman amplification tend to be unidirectional.

4. As we saw in Chapter 10, if optical layer protection is required, unlike unidirectional systems, bidirectional systems do not require an automatic protection-switching (APS) protocol between the two ends of the link, since both ends detect a fiber cut simultaneously.

5. Consider two equivalent all-optical networks in terms of capacity. One network uses a bidirectional link between nodes with a total of W wavelengths per link. Another network uses two unidirectional links between nodes, with a total of $W/2$ wavelengths on each unidirectional link. Problem 8.10 shows that the bidirectional network is less efficient at utilizing the available capacity than the unidirectional network due to inefficiencies in wavelength assignment.

6. Bidirectional systems can potentially be configured to handle asymmetric traffic. Given a total number of wavelengths in the fiber, more wavelengths could be used in one direction compared to the other. While this may be easy to do for unamplified systems, it is more difficult to do in amplified systems because these systems typically use separate amplifiers for each direction.

7. In general, it is slightly more difficult to design the transmission system in bidirectional systems since more impairments must be taken into account, in particular, reflections, as discussed in Section 5.6.4. There are more components in the path, such as filters for separating the wavelengths in different directions, leading to higher losses. However, at high channel counts, even unidirectional systems may require these filters.

8. Although amplifiers for bidirectional systems may employ more complicated structures than unidirectional systems, they need to handle only half as many channels as unidirectional systems, which means that they can produce higher output powers per channel and provide more gain flatness. This of course assumes the use of a different amplifier for each direction, which is typically the case. However, for a given total capacity, twice as many amplifiers are required in a bidirectional system compared to a unidirectional system.

9. Bidirectional systems usually require a guard band between the two sets of wavelengths traveling in opposite directions to avoid crosstalk penalties. However, high-channel-count unidirectional systems may also require guard bands due to the hierarchical nature of the multiplexing and demultiplexing in these systems. (We studied this in Section 3.3.10.) The guard band can be eliminated by interleaving the wavelengths in opposite directions, that is, by having adjacent wavelengths travel in opposite directions on the fiber. This also has the added advantage of effectively doubling the channel spacing. For instance, if we transmit 100 channels spaced 50 GHz apart over a fiber, then we have 50 channels spaced 100 GHz apart in each direction.

13.2.5 Long-Haul Networks

The long-haul carriers in North America have links spanning several hundred to a few thousand kilometers. In Europe the links are somewhat shorter but still several hundred kilometers in length. The economics for deploying WDM on these links is quite compelling, based on the enormous savings in regenerator costs enabled by the use of optical amplifiers, as well as the time to market to deploy new services. Thus most long-haul carriers have deployed WDM extensively in their networks. The specific combination of WDM and TDM depends very much on the carrier's installed base of fiber and the type of services delivered. Among the major established carriers, AT&T and Sprint have primarily installed standard single-mode fiber. Thus WDM is an attractive option for them, and they have actively deployed WDM systems on many of their routes. Most of their links operate at 2.5 Gb/s (OC-48) rather than 10 Gb/s (OC-192). This is because of the older fiber base, with potential PMD problems as well as because of the need for a large amount of chromatic dispersion compensation on standard single-mode fiber at 10 Gb/s. In addition, these carriers for the most part provide services at relatively low bit rates, such as DS3 (45 Mb/s). The OC-192 terminals initially provided low-speed interfaces down to OC-48 rates but now provide lower-speed interfaces down to OC-3/12 rates. Thus carriers providing DS3 services need to buy additional equipment to multiplex DS3s to OC-12s or OC-48s, which adds to their equipment cost.

Another major carrier, Worldcom, has a mix of dispersion-shifted fiber and standard single-mode fiber. This made them an early adopter of 10 Gb/s systems for those routes with dispersion-shifted fiber. At the same time, they have deployed WDM systems on other routes that use standard single-mode fiber. Meanwhile, some of the new links being installed use nonzero-dispersion fiber, which allows both types of systems to be considered for deployment.

Over the past few years there have been a number of new carriers building long-haul networks worldwide. In the United States, these include Qwest and Level 3 Communications, among others. These carriers have laid new fiber routes, and many have decided to install nonzero-dispersion fiber or the large effective area fiber (LEAF). In some cases, they have hedged their bets with respect to fiber type by leaving space in the conduits to pull additional fiber through later as needed. These carriers are for the most part delivering bulk bandwidth at OC-12/48/192 rates to their customers. Thus it makes sense for them to deploy WDM at OC-192 rates, and that is what they have done.

As we have mentioned earlier, systems operating in the C-band as well as in the L-band are now available. The L-band requires a separate amplifier and is relatively more expensive than the C-band to deploy, due to the higher cost of the L-band amplifiers, compared to the C-band amplifiers (this is partially because L-band amplifiers require higher pump powers than their C-band counterparts). While most long-haul carriers have deployed C-band WDM systems, they have been slow to adopt L-band systems. This is because it is usually cheaper to deploy another C-band system over a new pair of fibers rather than add the L-band to an existing C-band system. Some of the new carriers who have recently built new fiber networks particularly have a large number of excess fibers and use this approach. Carriers who have deployed dispersion-shifted fibers are likely to be early adopters of the L-band for WDM (and other fiber bands besides the C-band) due to the difficulties associated with four-wave mixing and other nonlinearities in the C-band on this type of fiber.

13.2.6 Long-Haul Network Case Study

In this section, we look at a fairly realistic example of designing a North American long-haul backbone network. We use the network topology shown in Figure 13.8(a). We look at using conventional-reach long-haul (LH) systems as well

Table 13.1   Traffic matrix for the long-haul mesh network case study. The fiber topology is shown in Figure 13.8(a). The traffic is shown in terms of the number of 10 Gb/s wavelengths between pairs of nodes in the upper-right triangle of this matrix.

Node Number	Node Name	1	2	3	4	5	6	7	8	9	10	11	12	13	14	15	16	17	18	19	Total Traffic
1	Seattle	0	2	2	2	2	1	2	3	3	1	3	3	2	1	3	3	2	3	3	41
2	San Francisco	0	0	3	2	3	3	3	3	2	1	1	1	1	1	1	2	3	2	1	35
3	Los Angeles	0	0	0	1	1	2	3	1	1	3	1	2	1	3	1	1	3	1	3	33
4	Salt Lake City	0	0	0	0	2	1	1	3	2	1	2	3	1	2	1	2	2	3	1	32
5	El Paso	0	0	0	0	0	1	2	3	2	2	3	2	1	1	3	2	1	2	1	34
6	Denver	0	0	0	0	0	0	2	2	3	2	1	3	2	2	3	1	1	2	2	34
7	Houston	0	0	0	0	0	0	0	1	3	2	2	3	3	3	3	2	3	2	1	41
8	Dallas	0	0	0	0	0	0	0	0	1	2	1	3	1	2	1	3	1	1	1	33
9	Kansas City	0	0	0	0	0	0	0	0	0	1	2	3	3	1	1	2	3	1	1	35
10	Chicago	0	0	0	0	0	0	0	0	0	0	1	3	2	2	3	3	3	3	1	36
11	Nashville	0	0	0	0	0	0	0	0	0	0	0	2	3	3	1	2	2	3	1	34
12	Atlanta	0	0	0	0	0	0	0	0	0	0	0	0	2	1	2	3	1	1	2	40
13	Tampa	0	0	0	0	0	0	0	0	0	0	0	0	0	3	2	1	2	2	3	35
14	Miami	0	0	0	0	0	0	0	0	0	0	0	0	0	0	2	1	2	2	1	33
15	Charlotte	0	0	0	0	0	0	0	0	0	0	0	0	0	0	0	1	1	1	2	32
16	Philadelphia	0	0	0	0	0	0	0	0	0	0	0	0	0	0	0	0	2	1	1	33
17	New York	0	0	0	0	0	0	0	0	0	0	0	0	0	0	0	0	0	2	3	37
18	Boston	0	0	0	0	0	0	0	0	0	0	0	0	0	0	0	0	0	0	2	34
19	Cleveland	0	0	0	0	0	0	0	0	0	0	0	0	0	0	0	0	0	0	0	30

as ultra-long-haul (ULH) systems. We also look at the benefits of different types of protection architectures.

The network of Figure 13.8(a) has 19 nodes and 28 links interconnecting the nodes. Table 13.1 shows the assumed traffic matrix between the various nodes in terms of 10 Gb/s channels. The total end-to-end traffic amounts to 3.31 Tb/s and represents a fairly realistic network in the 2002–2003 time frame.

The first step in the design process is to route the end-to-end traffic and determine the amount of working and protection capacity required. Sophisticated algorithms are used to perform this function in practice, but we use fairly simple algorithms for this study. For 1 + 1 protection, we have to calculate a pair of working and protection paths which are node disjoint, that is, do not have any intermediate nodes (and links) in common. This ensures that the protection path will be available in case a node or link along the working path fails. We choose the working path as the shortest-length path between the end nodes. To calculate the protection path for a given pair of end nodes, we delete the intermediate nodes in the working path between those two nodes, and calculate the shortest-length path in the resulting topology.

For shared mesh protection, we use the same working and protection paths as in the 1+1 protection case. However, we do not need to allocate protection capacity for each path separately. Instead we provide only as much protect capacity as is needed to reroute the working paths affected by a single link failure. To do this, we calculate the protection capacity required on the links for every possible link failure and take the maximum over all possible link failures.

Table 13.2 shows the assumed link distances and the number of 10 Gb/s wavelengths required on each link as a result of the routing and capacity allocation discussed above. Even though the end-to-end traffic requirement between any pair of nodes is no more than 30 Gb/s (three 10 Gb/s wavelengths), there are several links that carry more than 100 wavelengths (or equivalently over 1 Tb/s of capacity). For example, the Denver–Kansas City link carries 77 working wavelengths and 78 protection wavelengths (in the case of 1 + 1 protection), or 41 protection wavelengths (in the case of shared mesh protection). In many of these links, we will end up using multiple WDM systems in parallel to meet the capacity demand.

We assume each of the 19 nodes has one or more electrical core crossconnects. The crossconnects terminate all the traffic at the node, including both traffic passing through the node as well as traffic being added/dropped at the node. Thus, there is no optical passthrough at the nodes. Table 13.3 shows the number of crossconnect ports required for the 1 + 1 and shared mesh protection cases. Each node requires a few hundred such ports. For 1 + 1 protection, the largest node is Nashville, which has 566 ports and handles 5.66 Tb/s of traffic. For shared mesh protection, the largest node is Kansas City, which has 413 ports and handles 4.13 Tb/s of traffic.

The next step in the design is to cost out the network, based on the type and quantity of equipment deployed at all the sites. Table 13.4 shows the capabilities and costs of the LH and ULH systems assumed for this study, as well as the crossconnects.

Table 13.5 shows the quantity of different types of LH and ULH equipment and crossconnects required to support the link distances and capacities shown in Table 13.2. Figure 13.12 shows the corresponding network costs in graphical form and illustrates how the network cost varies with the different options as well as the cost breakdown among the various components. Observe that both ULH and mesh protection provide cost savings. Also, with this model, the amplifier cost is relatively small compared to the cost of transponders/regenerators and crossconnects.

Note that we have assumed the use of crossconnects for both the 1+1 case and the shared mesh case. Crossconnects are essential in the shared mesh scenario, as they are the ones that provide this capability. However, 1 + 1 protection can be implemented directly by the transponders, and we do not need crossconnects for this purpose.

Table 13.2   Link distances in the network topology of Figure 13.8(a). Also shown are the number of wavelengths required on each link to support the working traffic and the protection traffic for the cases of $1 + 1$ and shared mesh protection, assuming the traffic matrix of Table 13.1.

Link	Length (km)	Working Capacity	Protection Capacity $1 + 1$	Protection Capacity Shared Mesh
Seattle–San Francisco	1235	4	43	33
San Francisco–Los Angeles	616	35	12	10
Seattle–Salt Lake City	1251	25	35	16
Los Angeles–Salt Lake City	1073	44	37	22
Seattle–Chicago	3146	12	39	39
Salt Lake City–Denver	693	91	28	14
Los Angeles–El Paso	1294	14	54	52
El Paso–Denver	1011	21	77	37
Denver–Chicago	1657	45	27	18
Denver–Kansas City	999	77	78	41
El Paso–Houston	1213	19	71	35
Houston–Dallas	409	44	46	30
Dallas–Kansas City	805	42	50	15
Dallas–Atlanta	1305	19	43	26
Kansas City–Nashville	873	68	102	80
Kansas City–Chicago	750	26	61	29
Nashville–Atlanta	388	63	47	14
Atlanta–Tampa	742	48	30	14
Tampa–Miami	370	33	45	25
Miami–Charlotte	1183	14	64	48
Nashville–Charlotte	598	30	108	47
Charlotte–Philadelphia	814	18	74	47
Nashville–Cleveland	816	27	87	35
Boston–Cleveland	1020	26	11	8
New York–Cleveland	751	50	59	22
Philadelphia–New York	165	35	57	25
New York–Boston	343	8	29	26
Chicago–Cleveland	546	89	51	42

At the intermediate nodes, passthrough connections can be patched through using manual patch panels. However, if full flexibility is desired in provisioning end-to-end connections, then crossconnects will be needed in both cases.

The outcome of the study depends critically on the relative cost and capabilities of different types of equipment, and the routing algorithm used. For instance, we have assumed that there is a small premium in cost for ULH amplifiers and transponders relative to their LH counterparts, and a small decrease in number of wavelengths

Table 13.3   Number of crossconnect ports required at each of the 19 nodes in the case of 1+1 and shared mesh protection. In 1 + 1 protection, each add/drop wavelength consumes three crossconnect ports, one for the local add/drop, one on the working path, and one on the protection path. The passthrough traffic consists of both working and protection traffic not terminating at the local node. In the shared mesh case, each add/drop wavelength consumes one port for the local add/drop and one additional port for the working path. The passthrough ports include ports to carry all the working traffic passing through the node, as well as all the ports reserved for shared protection.

Node	Add/drop λ	1 + 1 Protection		Shared Mesh Protection	
		Passthrough λ	Total Ports	Passthrough λ	Total Ports
Seattle	41	76	199	88	170
San Francisco	35	24	129	47	117
Los Angeles	33	130	229	144	210
Salt Lake City	32	196	292	180	244
El Paso	34	188	290	144	212
Denver	34	376	478	310	378
Houston	41	98	221	87	169
Dallas	33	178	277	143	209
Kansas City	35	434	539	343	413
Chicago	36	278	386	264	336
Nashville	34	464	566	330	398
Atlanta	40	170	290	144	224
Tampa	35	86	191	85	155
Miami	33	90	189	87	153
Charlotte	32	244	340	172	236
Philadelphia	33	118	217	92	158
New York	37	164	275	129	203
Boston	34	6	108	34	102
Cleveland	30	340	430	269	329

per system. If the relative cost changes, the study conclusions can change quite substantially. Figure 13.13 plots the relative cost of LH and ULH options as a function of the relative cost of transponders (and regenerators) and amplifiers.

We have only touched some of the issues affecting network design. A number of additional factors need to be taken into account while designing a more realistic network:

- We can use LH systems on shorter links and ULH systems on longer links to optimize the cost further.

- Many ULH systems include optical add/drop capability to pass through signals at intermediate nodes in the optical domain, rather than requiring all wavelengths to

Table 13.4 Characteristics of the equipment used in the backbone network study. All costs are in thousands of U.S. dollars. The ULH amplifier and transponder costs are somewhat higher compared to their LH counterparts, and the ULH system has fewer wavelengths than the LH system. For terminals (including transponders), regenerators, and crossconnects, there is a common equipment cost, and in addition a cost per port equipped. For example, an LH terminal equipped with 10 transponders would cost $800,000, and a crossconnect equipped with two ports would cost $380,000.

LH System	
Number of wavelengths per system	80
Spans between regeneration	6 × 80 km (640 km total)
Terminal common equipment cost	$200
10 Gb/s transponder cost	$60
Regenerator common equipment cost	$200
10 Gb/s regenerator cost	$100
Amplifier cost	$200
ULH System	
Number of wavelengths per system	60
Spans between regeneration	25 × 80 km (2000 km total)
Terminal common equipment cost	$200
10 Gb/s transponder cost	$75
Regenerator common equipment cost	$200
10 Gb/s regenerator cost	$125
Amplifier cost	$240
Crossconnect	
Number of 10 Gb/s ports	128
Common equipment cost	$300
Cost per 10 Gb/s port	$40

be terminated. This capability can be used to reduce the nodal costs by eliminating some of the transponders required to terminate the passthrough traffic. In this case, we also have to deal with the routing and wavelength assignment problem discussed in Chapter 8, as signals being passed through optically cannot be converted to other wavelengths.

- Using more sophisticated routing and capacity allocation algorithms will bring the cost down for both 1 + 1 and shared mesh protection.

Table 13.5   Number of amplifiers, transponders, regenerators, and crossconnects required for LH and ULH systems to realize the capacities and link distances shown in Table 13.2, for both 1 + 1 and shared mesh protection.

Part	Quantity			
	1 + 1		Shared Mesh	
	LH	ULH	LH	ULH
Amplifiers	364	487	275	413
Transponders	4984	4984	3754	3754
Terminal common equipment	75	96	59	77
Regenerators	1866	51	1432	51
Regenerator common equipment	33	1	24	1
Crossconnect ports	5646	5646	4416	4416
Crossconnect common equipment	55	55	43	43

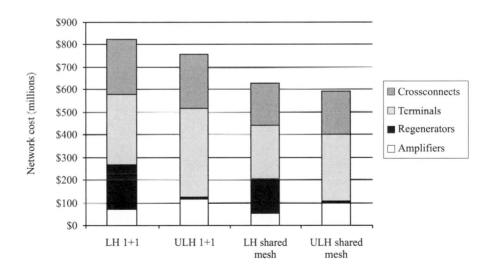

Figure 13.12   Breakdown of network costs for LH and ULH systems with 1 + 1 and shared mesh protection.

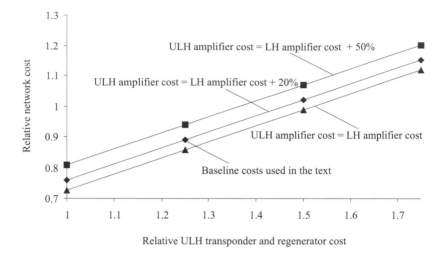

Figure 13.13   Sensitivity of study results to the relative cost of ULH and LH transponders (and regenerators) and amplifiers. The *x* axis indicates the ULH transponder and regenerator cost relative to the LH transponder and regenerator cost. The *y* axis indicates the relative network cost for ULH and LH systems assuming 1 + 1 protection.

- We have decoupled the network costing from the routing and capacity allocation. However, further cost optimization is possible by considering the two parts together. For example, in the LH case, we might choose slightly longer paths if it means using fewer regenerators on some of the links in the path.

- We have not taken into account the cost of blocking when considering crossconnects. Observe that many nodes require more than one crossconnect, given our assumption of a 1.28 Tb/s crossconnect. In this analysis, we have simply used as many crossconnects as needed to obtain the desired port counts, without considering the cost of scaling the crossconnect or the cost of blocking.

- We have implicitly assumed that there is no protection between the client equipment (for example, routers) and the optical layer equipment (such as crossconnects). In practice, we'll need to have some protection here as well and factor its cost into account.

- Traffic demands are at 10 Gb/s. We haven't dealt with aggregating and grooming lower-speed demands.

13.2.7 Long-Haul Undersea Networks

The economics of long-haul undersea links is similar to that of the long-haul terrestrial links, but with a few subtle differences. First, there are several types of undersea links commonly deployed. One type spans several thousands of kilometers across the Atlantic or Pacific oceans to interconnect North America with Europe or Asia, as shown in Figure 13.14. Another type tends to be relatively shorter haul (a few hundred kilometers), interconnecting countries either in a festoon type of arrangement or by direct links across short stretches of water. The term *festoon* means a string suspended in a loop between two points. In this context, it refers to an undersea cable used to connect two locations that are not separated by a body of water, usually neighboring countries. A trunk-and-branch configuration is also popular, where an undersea trunk cable serves several countries. Each country is connected to the trunk cable by a branching cable, with passive optical components used to perform the branching at the branching units. If a branch cable is cut, access to a particular country is lost, but other countries continue to communicate via the trunk cable. WDM is widely deployed in all these types of links.

The long-haul undersea systems tend to operate at the leading edge of technology and have to overcome significant impairments to attain the distances involved. The links use the dispersion management technique described in Section 5.8.6 by having alternating spans with positive and negative dispersion fiber to realize a total chromatic dispersion of zero but at the same time have finite chromatic dispersion at all points along the link.

The shorter-distance undersea links also stretch design objectives but in a different way. The main objective with these links is to eliminate any undersea amplifiers or repeater stations, due to their relatively higher cost of installation and maintenance. As a result, these systems use relatively high-power transmitters.

The trunk-and-branch configuration is also evolving. The early branching units contained passive splitters and combiners, but optical add/drop multiplexers are now being used to selectively drop and add specific wavelengths at different locations.

Undersea systems are designed to provide very high levels of reliability and availability due to the high cost of servicing or replacing failed parts of the network. Optical amplifiers with redundant pumping arrangements have proven to be highly reliable devices, and their failure rates are much lower than those of electronic regenerators. Likewise, optical add/drop multiplexers using passive WDM devices have been qualified for use in undersea branching configurations.

Undersea networks are very expensive to build, and the capacity on these networks is shared among a number of users. WDM allows traffic from different users to be segregated by carrying them on different channels—a useful feature.

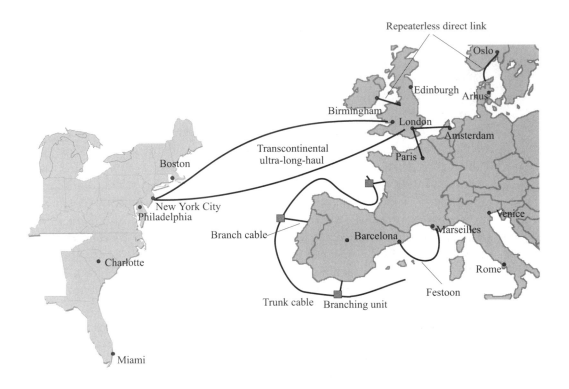

Figure 13.14   Different types of undersea networks, showing a couple of ultra-long-haul trans-Atlantic links, shorter-haul direct repeaterless links, a trunk-and-branch configuration, and a festoon.

One key difference between undersea links and terrestrial links is that, in most cases, undersea links are deployed from scratch with new fibers rather than over existing fiber plant. It is rare to upgrade an existing long-haul amplified undersea link, as the cost of laying a new link is not significantly higher than the cost of upgrading an existing link. This provides more flexibility in design choices.

13.2.8   Metro Networks

The metro network can be broken up into two parts. The first part is the metro access network and extends from the carrier's central office to the carrier's customer locations, serving to collect traffic from them into the carrier's network. The second part of this network is the metro interoffice network—the part of the network between carrier central offices. The access network today typically consists of rings a few kilometers to a few tens of kilometers in diameter, and traffic is primarily hubbed into the central office. The interoffice network tends to be several to a few tens of kilometers between sites, and traffic tends to be more distributed.

Because of the shorter spans involved, the case for WDM links is less compelling in metro networks. The other alternatives, namely, using multiple fibers or using higher-speed TDM, are quite viable in many situations. Despite this, however, there hasn't been widespread deployment of OC-192 in the metro network. One reason is that OC-192 interfaces have only recently appeared on metro systems. Another reason is that carriers in this part of the network are interested in delivering low-speed services at DS1 (1.5 Mb/s) or DS3(45 Mb/s) rates and OC-192 equipment is only now becoming a cost-effective alternative for this application.

On the other hand, reasons other than pure capacity growth are driving the deployment of WDM in these networks. Metro carriers need to provide a variety of different types of connections to their customers. The service mix includes leased private line services; statistical multiplexing types of services such as frame relay, ATM, and IP; Gigabit Ethernet; ESCON; and Fibre Channel. In many cases this service mix is supported by having a set of overlay networks, each dedicated to supporting a different service. These overlay networks are ideally realized using a single infrastructure. Due to its transparent nature, a WDM network provides a better infrastructure than most others, such as SONET/SDH, for this purpose.

Another factor is that the traffic distribution changes much more rapidly in metro networks than in long-haul networks. This drives the need to be able to rearrange network capacity quickly and efficiently as needed. Reconfigurable WDM networks allow capacity to be provided as needed in an efficient manner.

A big driver for WDM deployment in metro networks has been the need for large enterprises to interconnect their data centers. These data centers are separated by several kilometers to a few tens of kilometers. All transactions are mirrored at both sites. This allows the enterprise to recover quickly from a disaster when one of the centers fails. There may be other reasons for this as well, such as lower real estate costs at one location than at the other. Peripheral equipment such as disk farms can be placed at the cheaper site. The bandwidth requirement for such applications is large. The large mainframes at these data centers need to be interconnected by several hundred channels, each at up to 1 Gb/s. For example, IBM mainframes communicate using hundreds of ESCON channels, discussed in Chapter 6, running at 200 Mb/s each, or Fibre Channel at 1 Gb/s, as discussed in Chapter 6. Typically, these data centers tend to be located in dense metropolitan areas where most of the installed fiber is already in use. Moreover, these networks use a large variety of protocols and bit rates. These two factors make WDM an attractive option for these types of networks.

These networks are sometimes called *storage-area networks*. This is the primary application for most of the WDM networks deployed in metro networks today.

Because of the nature of the traffic and a large amount of passthrough traffic in these networks, a strong case can be made for deploying WDM rings with optical add/drop multiplexers instead of higher-speed TDM rings. We present a detailed case study of a metro access network in Section 13.2.9.

It is important to realize that despite the shorter spans for metro networks, optical amplifiers may still be needed for several reasons:

1. Although spans are short, in many cases the fiber in the ground is old, has many connectors in its path, and thus has relatively high loss. For example, a 10 km metro link may have a loss as high as 10 dB.

2. The loss is not just due to spans—a large component of the loss comes from the loss of optical add/drop multiplexers, each of which can add several decibels of loss.

3. Finally, protection requirements drive the need for alternate spans that may be much longer (for example, around a ring) than the working spans.

As of this writing, there has been widespread deployment of private WDM links for enterprise applications in the metro network. Several carriers in the United States have deployed WDM in their metro networks, but many are still considering the relative benefits of WDM versus other alternatives in this part of the network. As such the deployment is not yet as ubiquitous as it is in the long-haul network.

13.2.9 Metro Ring Case Study

We now look at a detailed example of upgrading a metro ring, based on a study done in [GR99]. Consider a four-node access ring with three remote nodes homing into a hub node. Assume for simplicity that all traffic is between the hub node and the remote nodes, with no traffic between the remote nodes themselves. Initially we have a SONET ring operating at OC-3 (155 Mb/s) capacity. Suppose the capacity on this ring is exhausted and that no spare fibers are available along the ring. We now have a couple of different options for upgrading the ring. The first option is to upgrade the ring to the next higher speed—OC-12 (622 Mb/s). This requires replacing or upgrading the SONET add/drop multiplexers (ADMs) at all the nodes.

This is the TDM upgrade path. The other alternative is to introduce WDM and build multiple "virtual" rings at different wavelengths over the same fiber pair. We can do this in incremental steps, one additional ring at a time. For example, as shown in Figure 13.15, we can start by adding another ring at a different wavelength connecting one of the remote nodes (the one that needs more capacity, say, node 1)

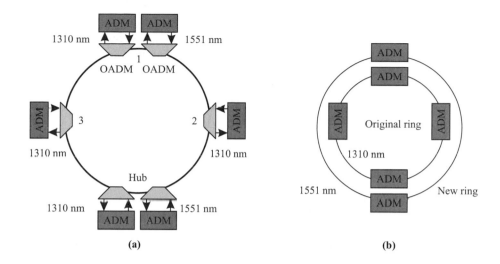

Figure 13.15   Using WDM to upgrade a four-node access ring. One additional ring is added at a different wavelength. (a) The physical topology and (b) the lightpath topology showing the connectivity between the SONET ADMs.

to the hub. In order to do this, we would need to introduce wavelength add/drop multiplexers (OADMs) at each node to drop the appropriate wavelengths. These can be "coarse" OADMs, since it is likely that the original ring is operating at 1310 nm, and we would add new rings in the 1550 nm WDM window. We would also need to add additional SONET ADMs at node 1 and at the hub, say, at OC-3 rates, if node 1 desires another OC-3 of capacity into the hub. Note that only two SONET OC-3 ADMs need to be added in this scenario. We can continue this upgrade path by adding additional rings, as shown in Figure 13.16. As we add additional rings, we will need to deploy additional "dense" OADMs at the nodes to separate out the different wavelengths used inside the 1550 nm wavelength window.

The key point to note in the WDM scenario is that, compared to the TDM scenario, the existing SONET equipment is preserved, and additional (SONET) hardware is only added at nodes that need additional capacity, requiring a potentially smaller up-front capital expense.

Note that a similar upgrade process can be used to upgrade an OC-12 ring to an OC-48 ring, or an OC-48 ring to an OC-192 ring. Moreover the WDM approach allows flexibility in dealing with non-SONET protocols on the different wavelengths and can provide future scalability. For example, once an OC-3 ring is upgraded to an OC-12 ring using TDM, what happens if the OC-12 ring runs out of capacity? The

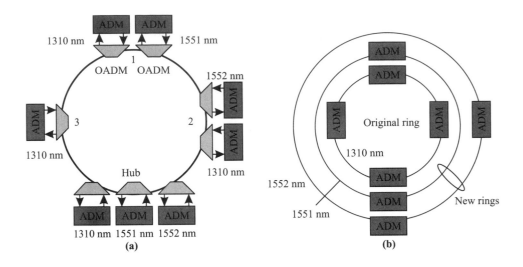

Figure 13.16   Continuing the upgrade process from Figure 13.15. Two additional rings are added at different wavelengths to the base configuration. (a) The physical topology and (b) the lightpath topology showing the connectivity between the SONET ADMs.

OC-12 hardware would then have to be replaced by OC-48 hardware. In contrast, with the WDM upgrade path, additional wavelengths can be added at higher bit rates (for example, we could retain, say, two existing OC-12 rings, and add a third ring at OC-48 or OC-192). Therefore the WDM solution is more "future-proof," compared to the TDM solution.

The key question we've left unanswered is how the two approaches compare from a cost perspective. This depends to a large extent on the cost of the OADMs relative to the SONET ADMs. Figure 13.17 shows the network cost for the upgrades described in this example, assuming the equipment costs shown in Table 13.6. The three sets of lines correspond to an upgrade from an OC-3 to an OC-12 ring, an OC-12 to an OC-48 ring, and an OC-48 to an OC-192 ring. For the cost numbers we have assumed, it appears that at low bit rates (OC-3 to OC-12) it is more cost-effective to do a TDM upgrade, whereas at the higher bit rates (OC-48 to OC-192), it may be cheaper to do a WDM upgrade. These numbers clearly will change over time based on the relative costs of the OADMs and the SONET ADMs, but the trends shown in Figure 13.17 will still be the same. We leave it to the reader to generate the numbers used to plot the lines in Figure 13.17.

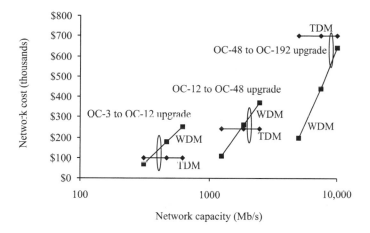

Figure 13.17 Relative network cost for the TDM and WDM upgrades. Three sets of upgrades are shown: OC-3 to OC-12, OC-12 to OC-48, and OC-48 to OC-192. The horizontal lines indicate the TDM upgrade path, and the slanted lines indicate the WDM upgrade path.

Table 13.6 Equipment cost assumptions for Figure 13.17. Prices of all the equipment listed here are coming down, due to competition and improvements in technology. These numbers are reasonably indicative of prices in 1999/2000.

Equipment	Cost without WDM Interfaces (U.S. $)	Cost with WDM Interfaces (U.S. $)
OC-3 ADM		15,000
OC-12 ADM	25,000	35,000
OC-48 ADM	60,000	80,000
OC-192 ADM	175,000	
Coarse OADM		10,000
Dense OADM		20,000

13.2.10 From Opaque Links to Agile All-Optical Networks

The optical layer itself is evolving, not just in terms of raw capacity, but also in terms of functionality. The optical network originally consisted of WDM links, with all the functions at the end of the link performed in the electrical domain. These networks are sometimes called *opaque* networks. Due to the high cost of optical-to-electrical (O/E) conversions, particularly at the higher bit rates, it makes sense to minimize the number of these converters in the network. The first step in this direction was the development of ultra-long-haul systems, which provided longer reach between regenerators. The second step is to handle as much of the traffic passing through a node in the optical domain as possible. An all-optical OADM or OXC performs this function. Having optical passthrough instead of electrical processing can lead to an order of magnitude savings in the cost, given that the cost of O/E conversions dominates the cost of the node itself. There are associated savings in power and floor space as well, given that the O/E devices consume most of the power and occupy most of the floor space in WDM equipment. Even further cost savings can be realized by passing signals through in bands of wavelengths, instead of individual wavelengths. These networks are called *all-optical* or *transparent* networks.

The next step in the evolution of the optical layer is to add agility. An agile network provides the ability to set up and take down lightpaths as needed and allows carriers to provision and deploy services rapidly. With the introduction of optical crossconnects and reconfigurable optical add/drop multiplexers, *agile opaque* networks are becoming a reality today. It is only a matter of time before *agile all-optical* networks arrive. Adding agility to an all-optical network results in even further complexities to be tackled at the physical layer, such as adaptive power and dispersion management. These problems have already been tackled at least partially in the context of ultra-long-haul systems.

While an all-optical network provides significant advantages, it also has its limitations. Certain functions, such as wavelength conversion, regeneration, and traffic grooming at fine granularities (for example, at STS-1 or 51 Mb/s) will need to be done in the electrical domain. As we saw in Chapter 8, we may not be able to completely handle all the passthrough traffic in the optical domain, due to inefficiencies in how traffic is groomed in the network. For these reasons, a practical node will end up using a combination of all-optical and electrical crossconnects. The all-optical crossconnects can be used to switch signals through in the optical domain as much as possible, and signals needing to be regenerated, converted from one wavelength to another, or groomed will be handed down to the electrical layer.

Another subtle aspect of the all-optical network is related to interoperability between systems from multiple vendors. As we saw in Chapter 9, it is difficult for equipment from different vendors to interoperate at the wavelength layer. Interoperability between vendors needs to be done through regenerators/transponders. This implies that the all-optical network by itself is a single-vendor network. Transponders are needed at the edges of this network to provide interoperability with other all-optical networks. A realistic network will therefore consist of all-optical islands or subnets, interconnected with other such subnets through transponders at the boundaries.

Summary

This chapter addressed architectural alternatives for the new generation of carrier networks. These networks are different from the established legacy networks based on SONET/SDH. This is driven by the increasing dominance of data over voice and the emergence of new carriers with vastly different business models offering different types of services. An established carrier offering a mix of services may choose to overlay SONET/SDH and IP or ATM over the optical layer. New carriers offering predominantly data-oriented services may opt to deploy IP or ATM directly over the optical layer, not deploying any SONET/SDH at all. The optical layer is becoming ubiquitous in both long-haul and metro networks. The optical layer here provides circuit-switched lightpaths to the higher layers. Note that the optical layer is not performing any packet-switching functions. These functions are best left to the electronic layers. Optical packet-switching technology is still in research laboratories.

The next-generation metro access network will likely use a hybrid packet-circuit network element as the key element to deliver services. The core of the network is migrating away from a SONET ring-based architecture to a meshed optical-layer-based architecture, with protection functions implemented in the optical layer.

Within the optical layer, TDM, WDM, and SDM are all used to provide capacity. The right combination of these techniques is not an easy choice and depends on a variety of factors including the length of the link, the availability of spare fibers, the type of fiber and its dispersion and nonlinear characteristics, and the type of services to be deployed using the network. The problems at the end of this chapter will give the reader an inkling of what such a comparison might involve. Network planners need to make their own analysis of the different alternatives, perhaps with the aid of some network planning and design tools, to decide which way to go.

The optical layer itself is migrating from an opaque network, consisting of WDM links with electrical processing at the ends of the link, to an all-optical network, where traffic is passed through in the optical domain at intermediate nodes. At the same time, the optical network is moving from a static network to an agile network, where lightpaths can be set up and taken down as needed.

Further Reading

The subject matter in this chapter is widely covered in the business press and by investment houses. Several market research firms publish reports on various segments of the optical networking industry. These include Communications Industry Researchers (*www.cir-inc.com*), Electronicast (*www.electronicast.com*), KMI (*www.kmicorp.com*), Ovum (*www.ovum.com*), Pioneer Consulting (*www.pioneerconsulting.com*), Ryan, Hankin and Kent (*www.rhk.com*), Strategies Unlimited (*www.strategies-u.com*), and Yankee Group (*www.yankeegroup.com*). There have been many studies published about the relative economics of various architectural options. Be warned these are rather biased views, as the assumptions made significantly impact the outcome, and these assumptions are usually biased toward supporting the products offered by the vendor doing the study. The various options for supporting IP over WDM have been explored in many papers; for instance, [Mae00] provides a relative cost analysis. See [PCW+00, Coo00, OSF00, PCK00] for a recent sampling of papers related to metro WDM economics. [DSGW00, Dos01] explore the value proposition behind ultra-long-haul WDM systems. The National Fiber Optic Engineers' Conference usually has many papers on these topics.

Problems

13.1 Imagine that you are a planner for a long-haul carrier planning to deploy an IP over WDM network. Your job is to make the right technology and vendor choice for your network. You are given the following information. The initial requirement is to deploy 20 Gb/s of capacity between two nodes. You anticipate that this capacity will grow to 80 Gb/s in a year and over a few years grow to 320 Gb/s. You have a choice of several WDM systems from different vendors with the following prices and capabilities:

Vendor	A	B	C
Number of channels	80	128	32
Bit rate per channel	OC-192	OC-48	OC-192
Distance between regenerators	640 km	1920 km	1920 km
Amplifier spacing	80 km	80 km	80 km
OLT common equipment	$200,000	$275,000	$300,000
Transponder	$50,000	$25,000	$80,000
Amplifier	$150,000	$100,000	$125,000

Assume that the common equipment prices for the optical line terminals include any amplifiers if needed. One transponder is needed for each channel at each end of the link. Once the distance between regenerators is exceeded, the signals need to be regenerated by using two terminals back to back with transponders.

Compute the cost of each solution for a 640 km link, a 1280 km link, and a 1920 km link. Draw a diagram of each configuration. What are your conclusions? Other than the costs computed above, what other factors might influence your choice?

13.2 Consider the same problem as in Problem 13.1 with one difference. For the 1280 and 1920 km cases, between the two nodes is a third node spaced 600 km from the first node, where half the capacity needs to be dropped and added. For this case, assume that vendor B and vendor C offer systems where you can use back-to-back terminals at this intermediate node without requiring transponders for the passthrough channels. (Transponders are still needed for the channels dropped and added.) Repeat your analysis. What are your conclusions?

13.3 Imagine that you are a planner for a metro carrier. The links in your network are fairly short, with a maximum span length of 40 km. You want to compare SDM, TDM, and WDM options for realizing a two-node link. Assume the following costs.

Equipment	Cost (U.S. $)
Pulling fiber through existing conduit (per km)	300
Laying new conduit, including fiber (per km)	20,000
OC-48 BLSR/2 ADM	
Common equipment	40,000
Additional per OC-12 drop	5,000
Additional per STS-1 drop	750
OC-192 BLSR/2 ADM	
Common equipment	125,000
Additional per OC-12 drop	5000
OC-12 BLSR/2 ADM	15,000
Additional per STS-1 drop	750
Metro WDM terminal (OLT)	
Common equipment	30,000
Additional per transponder	10,000

You need to deliver 10 Gb/s of capacity in the form of OC-12s (622 Mb/s) to your customers. Compare the cost of the following options for the scenario where fibers are available versus fiber needs to be pulled through existing conduit versus new conduit needs to be laid: (a) OC-48 ADMs over separate fibers, (b) OC-48 ADMs in conjunction with WDM terminals over a single fiber pair, (c) OC-192 ADMs, and (d) WDM terminals with no SONET equipment.

Factor in the cost of protection as well. Assume that two diversely routed fiber pairs are available between the two sites. Whenever SONET is used, protection is done in the SONET boxes, and no protection is done in the OLTs. For the case with no SONET equipment, the protection is done at the optical multiplex section by the OLTs—assume that the cost is already factored into the OLTs.

Repeat this problem for the case where the capacity needs to be delivered in the form of STS-1s to your customers. In this case, the options available to you are: (a) OC-48 ADMs over separate fibers, (b) OC-48 ADMs in conjunction with WDM terminals over a single fiber pair, (c) OC-192 ADMs back-ended by OC-12 ADMs, and (d) WDM terminals back-ended by OC-12 ADMs.

Draw a diagram of the different configurations. What are your conclusions? Other than the costs computed above, what other factors might influence your choice?

13.4   This is an extension of the previous problem related to planning a metro network. We will explore the use of optical add/drops in this problem. You now have to create a linear network of three nodes A, B, and C. The link between node A and node B is 40 km, and the link between node B and node C is also 40 km. You need 5 Gb/s of capacity between A and B, 5 Gb/s between B and C, and another 5 Gb/s between A and C. All capacity is to be delivered as OC-12s. In addition to the equipment available above, you also have the option of using an OADM at node B that works with OLTs at node A and node C. The WDM system has a reach of 80 km with an intermediate OADM. The OADM has a common equipment cost, including any needed amplifiers, of $50,000 and can drop as many wavelengths as needed. Transponders are needed for the added and dropped channels.

In addition, assume that SONET ADMs have a maximum reach of 40 km. Signals need to be regenerated after this, and the regenerator costs are as follows: OC-48 regenerator, $10,000; OC-192 regenerator, $30,000.

Now consider the following solutions: (a) Fibers are available, and you use OC-48 ADMs over them. In this case you need to use a regenerator at node B for passthrough traffic or another OC-48 ADM for multiplexing and demultiplexing local traffic. Consider also the cases where fiber needs to be pulled through existing conduit and also of conduit exhaust. (b) OC-48 ADMs along with OLTs and OADMs.

(c) OC-12 delivery directly using OLTs and OADMs, no SONET. (d) OC-192 ADM with another ADM at node B to demultiplex and multiplex local traffic.

For this problem, ignore any protection needed. Note that this could result in cheaper equipment, but for our purposes, assume that the equipment costs don't change.

Compare the costs of these alternatives. What do you conclude?

13.5 You are looking at deploying an optical crossconnect at a large node in a carrier network. The crossconnect is connected to OLTs and drops traffic down to IP routers. You have three options to consider: (1) an electrical crossconnect (EXC) solution, where the crossconnect uses short-reach interfaces connected to transponders in the OLTs and to short-reach interfaces in the routers; (2) an opaque photonic crossconnect solution, where the photonic crossconnect (PXC) is connected to transponders in the OLTs and to short-reach interfaces in the routers; (3) a transparent photonic crossconnect solution, where the photonic crossconnect is connected to the OLTs directly without transponders, but transponders are used between the routers and the crossconnect.

Assume the following:

Item	Cost (U.S. $)	Power	Footprint
WDM OC-48 transponder	$25,000	75 W	64 ports/rack
WDM OC-192 transponder	$50,000	150 W	32 ports/rack
EXC switch fabric		10,000 W	1 rack
EXC OC-48 port	$15,000	50 W	256 ports/rack
EXC OC-192 port	$30,000	100 W	64 ports/rack
PXC port	$25,000	2 W	256 ports/rack

Assume that the EXC supports a maximum of 512 OC-48 ports or 128 OC-192 ports and that the PXC supports 1024 ports.

Compare the cost and floor space taken up for the three options above for the following situations. (Include any transponders used, but neglect the routers as they are common to all the scenarios.) Summarize your findings.

(a) The node is switching 256 OC-48 wavelengths coming in from the WDM systems, of which 25%, 50%, or 75% of the traffic may be dropped locally into router ports. (For example, with a 25% drop, you would need a total of 320 ports on the crossconnect.)

(b) The node is switching 256 OC-192 wavelengths coming in from the WDM systems, of which 25%, 50%, or 75% of the traffic may be dropped locally into router ports.

References

[Coo00]   H. K. Cook. The economics of metro DWDM deployment. In *Proceedings of National Fiber Optic Engineers Conference*, 2000.

[Dos01]   B. Doshi et al. Ultra-long-reach systems, optical transparency and networks. In *OFC 2001 Technical Digest*, 2001. Paper TuG4.

[DSGW00]   A. Dwiwedi, M. Sharma, J. M. Grochocinski, and R. E. Wagner. Value of reach extension in long-distance networks. In *Proceedings of National Fiber Optic Engineers Conference*, 2000.

[GR99]   O. Gerstel and R. Ramaswami. Upgrading SONET rings with WDM instead of TDM: An economic analysis. In *OFC'99 Technical Digest*, pages 75–77, 1999.

[Mae00]   Y. Maeno et al. Cost comparison of an IP/OTN integrated node against a pure IP routing node. In *Proceedings of National Fiber Optic Engineers Conference*, 2000.

[OSF00]   G. Ocakoglu, K. Struyve, and P. Falcao. The business case for DWDM metro systems in a pan-European carrier environment. In *Proceedings of National Fiber Optic Engineers Conference*, 2000.

[PCK00]   G. N. S. Prasanna, E. A. Caridi, and R. M. Krishnaswamy. Metropolitan IP-optical networks: A systematic case study. In *Proceedings of National Fiber Optic Engineers Conference*, 2000.

[PCW⁺00]   V. Poudyal, R. H. Cardwell, O. J. Wasem, J. E. Baran, and A. Rajan. Comparison of network alternatives for transporting high capacity tributaries for IP router interconnection. In *Proceedings of National Fiber Optic Engineers Conference*, 2000.

appendix

Acronyms

Acronym	Expansion
1R	Regeneration without reshaping or retiming
2R	Regeneration with reshaping but no retiming
3R	Regeneration with retiming and reshaping
AAL	ATM adaptation layer
ACTS	Advanced communications technologies and services
ADM	Add/drop multiplexer
AGC	Automatic gain control
AIS	Alarm indication signal
ANSI	American National Standards Institute
AOTF	Acousto-optic tunable filter
APD	Avalanche photodetector
APS	Automatic protection switching
AR	Anti-reflective
ASE	Amplified spontaneous emission
ATM	Asynchronous transfer mode
ATMOS	ATM optical switching
AWG	Arrayed waveguide grating
BCH	Bose-Chaudhuri-Hochquenghem (code)
BDI	Backward defect indicator
BER	Bit error rate
BLSR	Bidirectional line-switched ring

Acronym	Expansion
CBR	Constant bit rate
CGM	Cross-gain modulation
CLP	Cell loss priority
CMIP	Common management information protocol
CO	Central office
CORBA	Common object request broker
CPM	Cross-phase modulation
CRC	Cyclic redundancy check
CW	Continuous wave
DARPA	Defense Advanced Research Projects Agency
DBR	Distributed Bragg reflector
DCF	Dispersion compensating fiber
DCN	Data communications network
DCS	Digital crossconnect
DFB	Distributed feedback
DGD	Differential group delay
DM	Direct modulation
DPRing	Dedicated protection ring
DSB	Double sideband
DSF	Dispersion-shifted fiber
DSL	Digital subscriber loop
DTMF	Dielectric thin film multicavity filter
DWDM	Dense wavelength division multiplexing
EA	Electro absorption
EDFA	Erbium-doped fiber amplifier
EDFFA	Erbium-doped fluoride fiber amplifier
EMS	Element management system
ESCON	Enterprise serial connection
FCC	Federal Communications Commission
FDDI	Fiber distributed data interface
FDI	Forward defect indicator
FDM	Frequency division multiplexing
FEC	Forward error correction
FET	Field effect transistor
FOM	Figure of merit
FP	Fabry-Perot
FSAN	Full service access network
FSR	Free spectral range

Acronym	Expansion
FTP	File transfer protocol
FTTC	Fiber to the curb
FWHM	Full-width half maximum
FWM	Four-wave mixing
GCSR	Grating coupled sampled reflector
GDMO	Guidelines for definition of managed objects
GFC	Generic flow control
GMPLS	Generalized multiprotocol label switching
GVD	Group velocity dispersion
HDLC	High-level data link control
HEC	Header error control
HFC	Hybrid fiber coax
HIPPI	High performance parallel interface
HOL	Head of line
IBM	International Business Machines
IETF	Internet Engineering Task Force
ILP	Integer linear program
IP	Internet Protocol
IR	Intermediate reach
ISI	Intersymbol interference
IS-IS	Intermediate system–intermediate system
ISO	International Standards Organization
ISP	Internet service provider
ITU	International Telecommunications Union
KEOPS	Keys to optical packet switching
LAN	Local-area network
LDP	Label distribution protocol
LEAF	Large effective area fiber
LED	Light emitting diode
LH	Long-haul
LLN	Linear lightwave network
LMDS	Local multipoint distribution service
LR	Long reach
LSP	Label-switched path
LSR	Label-switched router
LT	Line terminal
LTD	Lightpath topology design
LTE	Line terminating equipment

Acronym	Expansion
MEMS	Micro-electro-mechanical systems
MIB	Management information base
MILP	Mixed integer linear program
MLM	Multilongitudinal mode
MMDS	Multichannel multipoint distribution service
MMF	Multimode fiber
MPLS	Multiprotocol label switching
MS	Multiplex section
MSP	Multiservice platform
MZI	Mach-Zehnder interferometer
NA	Numerical aperture
NALM	Nonlinear amplifying loop mirror
NEBS	Network equipment building system
NLSE	Nonlinear Schrödinger equation
NNI	Network-to-network interface
NOLM	Nonlinear optical loop mirror
NP	Nonpolynomial
NRZ	Non-return-to-zero
NSIF	Network and Services Interoperability Forum
NTSC	National Television Standards Committee
NZ-DSF	Nonzero-dispersion-shifted fiber
OADM	Optical add/drop multiplexer
OBLSR	Optical bidirectional line-switched ring
OBPSR	Optical bidirectional path-switched ring
OCDMA	Optical code division multiple access
OCh	Optical channel
OC-x	Optical carrier-x ($x = 1, 3, 12, 48, 192, 768, \ldots$)
O/E/O	Optical-to-electrical-to-optical
OFC	Optical Fiber Communications Conference
OIF	Optical Internetworking Forum
OLT	Optical line terminal
OMS	Optical multiplex section
ONU	Optical network unit
OOK	On-off keying
OSC	Optical supervisory channel
OSI	Open systems interconnection
OSPF	Open shortest path first
OTDM	Optical time division multiplexing
OTN	Optical transport network

Acronym	Expansion
OUPSR	Optical unidirectional path-switched ring
OXC	Optical crossconnect
PC	Personal computer
PDH	Plesiochronous digital hierarchy
PDL	Polarization-dependent loss
PLL	Phase-locked loop
PMD	Polarization-mode dispersion
PNNI	Private network-to-network interface
PON	Passive optical network
PPP	Point-to-point protocol
PPS	Photonic packet switching
PSK	Phase-shift keying
PT	Payload type
PWDM	Point-to-point WDM
QAM	Quadrature amplitude modulation
QOS	Quality of service
RF	Radio frequency
RN	Remote node
RS	Reduced slope
RSVP	Resource reservation protocol
RWA	Routing and wavelength assignment
RZ	Return-to-zero
SAN	Storage-area network
SBCON	Single byte command code sets connection architecture
SBS	Stimulated Brillouin scattering
SCM	Subcarrier multiplexing
SDH	Synchronous digital hierarchy
SDM	Space division multiplexing
SLM	Single longitudinal mode
SMF	Single-mode fiber
SN	Sequence number
SNMP	Simple network management protocol
SNR	Signal-to-noise ratio
SOA	Semiconductor optical amplifier
SONET	Synchronous optical network
SOP	State of polarization
SPE	Synchronous payload envelope
SPM	Self-phase modulation
SPRing	Shared protection ring

Acronym	Expansion
SR	Short reach
SRS	Stimulated Raman scattering
SSB	Single sideband
STM-x	Synchronous transport module-x ($x = 1, 4, 16, 64, 256, \ldots$)
STS-x	Synchronous transport signal-x ($x = 1, 3, 12, 48, 192, \ldots$)
SWP	Spatial walk-off polarizer
TCP	Transmission control protocol
TDM	Time division multiplexing
TE	Transverse electric
TFMF	Thin-film multicavity filter
TL-1	Transaction Language-1
TM	Transverse magnetic
TMN	Telecommunications management network
TOAD	Terahertz optical asymmetric demultiplexer
TPON	PON for telephony
TWC	Tunable wavelength converter
UBR	Unspecified bit rate
UDP	User datagram protocol
ULH	Ultra-long-haul
UNEQ	Unequipped
UNI	User network interface
UPSR	Unidirectional path-switched ring
UV	Ultraviolet
VC	Virtual circuit
VC	Virtual container
VCI	Virtual circuit identifier
VCO	Voltage control oscillator
VCSEL	Vertical cavity surface emitting laser
VLSI	Very large scale integrated circuits
VOA	Variable optical attenuator
VP	Virtual path
VPI	Virtual path identifier
VT	Virtual tributary
WA	Wavelength assignment
WDM	Wavelength division multiplexing
WPON	Wavelength PON
WRPON	Wavelength-routed PON

appendix B

Symbols and Parameters

Table B.1 Parameters and symbols used in Part I (dimensionless unless otherwise indicated).

Parameter	Symbol	Typical Value/Units
Effective area	A_e	50 μm^2
Pulse envelope	$A(z, t)$	
Fiber core radius	a	4 μm (SMF)
Bit rate	B	Mb/s or Gb/s
Electrical bandwidth	B_e	GHz
Optical bandwidth	B_o	GHz
Bit error rate	BER	10^9–10^{-15}
Normalized effective index	b	
Capacitance	C	μF (microfarad)
Speed of light in vacuum	c	3×10^8 m/s
Dispersion parameter	D	ps/nm-km
Electric flux density	$\mathbf{D}$	coulombs/m^2
Material dispersion	D_M	ps/nm-km
Polarization-mode dispersion	D_{PMD}	ps/$\sqrt{\text{km}}$
Waveguide dispersion	D_W	ps/nm-km
Dispersion-shifted fiber	DSF	$D = 0$ (1.55 μm)
Electric field	$\mathbf{E}$	V/m
Energy level	E	differences, ΔE, expressed in nm using $\Delta E = hc/\lambda$
Electronic charge	e	1.6×10^{-19} coulombs
Amplifier noise figure	F	dB
Finesse	F	
Optical carrier frequency	f_c	THz
Pump frequency	f_p	THz

Table B.1   Parameters and symbols used in Part I (dimensionless unless otherwise indicated) *(continued)*.

Parameter	Symbol	Typical Value/Units
Signal frequency	f_s	THz
Amplifier gain	G	
Amplifier unsaturated gain	G_{max}	
Brillouin gain coefficient	g_B	4×10^{-11} m/W
Raman gain coefficient	g_R	6×10^{-14} m/W
Magnetic field	$\mathbf{H}$	A/m
Planck's constant	h	6.63×10^{-34} J/Hz
Photocurrent	I_p	μA or nA
Thermal noise current	I_{th}	3 pA/$\sqrt{\text{Hz}}$
Boltzmann's constant	k_B	1.38×10^{-23} J/°K
Dispersion length	L_D	km
Effective length	L_e	km
Link length	L	km
Nonlinear length	L_{NL}	km
Coupling length	l	μm
Distance between amplifiers	l	km
Average number of photons per 1 bit	M	
Nonzero-dispersion-shifted-fiber	NZ-DSF	$-6 \le D \le 6$ ps/nm-km (1.55 μm)
Effective index	n_{eff}	
Refractive index	n	
Spontaneous emission factor	n_{sp}	
Core refractive index	n_1	
Cladding refractive index	n_2	
Nonlinear index coefficient	$\bar{n}$	2.2–3.4 $\times 10^{-8}$ μm^2/W
Amplifier output saturation power	P_{out}^{sat}	mW
Amplifier saturation power	P^{sat}	mW
Electric polarization	$\mathbf{P}$	coulombs/m^2
Linear polarization	$\mathcal{P}_L$	coulombs/m^2
Local-oscillator power	P_{LO}	dBm
Nonlinear polarization	$\mathcal{P}_{NL}$	coulombs/m^2
Power	P	W or mW
Power penalty	PP	dB
Penalty (signal-dependent noise)	PP$_{sig-dep}$	dB
Penalty (signal-independent noise)	PP$_{sig-indep}$	dB
Receiver sensitivity	$\bar{P}_{sens}$	dBm
Load resistance	R_L	Ω or kΩ
Photodetector responsivity	$\mathcal{R}$	A/W
Reflectivity	R	
Real part of x	$\Re[x]$	
Extinction ratio	r	

Table B.1 Parameters and symbols used in Part I (dimensionless unless otherwise indicated) *(continued)*.

Parameter	Symbol	Typical Value/Units
Standard single-mode fiber	SMF	$D = 17$ ps/nm-km (1.55 μm), $D = 0$ (1.3 μm)
Signal-to-noise ratio	SNR	dB or no units
Bit period	T	ns
Decision threshold	T_d	
V-number	V	
Optical frequency	ν	Hz
Number of wavelengths	W	
Absorption coefficient	α	1/cm
Fiber attenuation	α	0.22 dB/km at 1.55 μm
Propagation constant	β	1/μm
Group velocity	$1/\beta_1$	m/s
GVD parameter	β_2	s^2/m (or in terms of D)
Coupling ratio	γ	0–1
Nonlinear propagation coefficient	γ	2.6 /W-km
Fractional core-cladding refractive index difference	Δ	
Brillouin gain bandwidth	Δf_B	20 MHz at 1.55 μm
Interchannel spacing	$\Delta\lambda$	nm
Permittivity of vacuum	ϵ_0	8.854×10^{-12} F/m
Detector quantum efficiency	η	1 for pinFETs
Four-wave mixing efficiency	η	
Input coupling efficiency	η_i	
Output coupling efficiency	η_o	
Chirp factor	κ	
Coupling coefficient	κ	1/μm
Grating period	Λ	μm
Filter center wavelength	λ_0	μm
Wavelength	λ	μm or nm
Permeability of vacuum	μ_0	$4\pi \times 10^{-7}$ H/m
Shot noise power	σ_{shot}^2	
Thermal noise power	σ_{th}^2	
Signal-spontaneous noise power	$\sigma_{\text{sig-spont}}^2$	
Spontaneous-spontaneous noise power	$\sigma_{\text{spont-spont}}^2$	
Phase	ϕ	radians
Susceptibility	χ	
Third-order susceptibility	$\chi^{(3)}$	6×10^{-15} cm^3/erg
Angular frequency	ω, ω_0	

Standards

C.1   International Telecommunications Union (ITU-T)

These standards can be ordered through *www.itu.ch*.

C.1.1   Fiber

G.652. Characteristics of a single-mode optical fiber cable.

G.653. Characteristics of a dispersion-shifted single-mode optical fiber cable.

G.655. Characteristics of a nonzero-dispersion-shifted single-mode optical fiber cable.

C.1.2   SDH (Synchronous Digital Hierarchy)

G.691. Optical interfaces for single-channel STM-64, STM-256 systems, and other SDH systems with optical amplifiers.

G.707. Network node interface for the synchronous digital hierarchy (SDH).

G.708. Sub STM-0 network node interface for the synchronous digital hierarchy (SDH).

G.774. Synchronous digital hierarchy (SDH) management information model for the network element view. Several addendums exist.

G.780. Vocabulary of terms for synchronous digital hierarchy (SDH) networks and equipment.

G.781. Synchronization layer functions.

G.783. Characteristics of synchronous digital hierarchy (SDH) equipment functional blocks.

G.784. Synchronous digital hierarchy (SDH) management.

G.803. Architecture of transport networks based on the synchronous digital hierarchy (SDH).

G.805. Generic functional architecture of transport networks.

G.831. Management capabilities of transport networks based on the synchronous digital hierarchy (SDH).

G.841. Types and characteristics of SDH network protection architectures.

G.842. Interworking of SDH network protection architectures.

G.957. Optical interfaces for equipments and systems relating to the synchronous digital hierarchy.

C.1.3   Optical Networking

G.692. Optical interfaces for multichannel systems with optical amplifiers.

G.709. Interface for the optical transport network (OTN).

G.798. Characteristics for the OTN equipment functional blocks.

G.871. Framework for recommendations.

G.872. Architecture for optical transport networks (OTN).

G.874. Management aspect of optical transport network elements.

G.875. OTN management information model for the network element view.

G.957. Optical interfaces for equipment and systems related to SDH.

G.959. Optical networking physical layer interfaces.

G.983. Broadband optical access systems based on passive optical networks (PON).

G.astn. Automatic switched networks.

G.vsr. Optical interfaces for intraoffice systems.

C.1.4   Management

M.3000. Overview of TMN recommendations.

M.3010. Principles for a telecommunications management network.

M.3100. Generic network information model.

Q.822. Stage 1, stage 2, and stage 3 description for the Q3 interface—performance management.

X.744. Information technology—open systems interconnection—systems management: Software management function.

C.2 Telcordia

These standards can be ordered through *www.telcordia.com*.

C.2.1 Physical and Environmental

FR-2063. Network Equipment-Building System (NEBS) family of requirements (NEBSFR).

C.2.2 SONET

GR-253. Synchronous optical network (SONET) transport systems: Common generic criteria.

GR-496. SONET add-drop multiplexer (SONET ADM) generic criteria.

GR-1230. SONET Bi-directional line-switched ring equipment generic criteria.

GR-1244. Clocks for the synchronized network: Common generic criteria.

GR-1250. Generic requirements for synchronous optical network (SONET) file transfer.

GR-1365. SONET private line service interface generic criteria for end users.

GR-1374. SONET inter-carrier interface physical layer generic criteria for carriers.

GR-1377. SONET OC-192 transport system generic criteria.

GR-1400. SONET dual-fed unidirectional path switched ring (UPSR) equipment generic criteria.

GR-2875. Generic requirements for digital interface systems.

GR-2899. Generic criteria for SONET two-channel (1310/1550-nm) wavelength division multiplexed systems.

GR-2900. SONET asymmetric multiplex functional criteria.

GR-2950. Information model for SONET digital cross-connect systems (DCSs).

GR-2954. Transport performance management based on the TMN architecture.

GR-2996. Generic criteria for SONET digital cross-connect systems.

GR-3000. Generic requirements for SONET element management systems (EMSs).

GR-3001. Generic requirements for SONET network management systems (NMSs).

C.2.3 Optical Networking

GR-1209. Generic requirements for fiber optic branching components.

GR-1377. SONET OC-192 transport system generic criteria.

GR-2918. DWDM network transport systems with digital tributaries for use in metropolitan area applications: Common generic criteria.

GR-2979. Common generic requirements for optical add-drop multiplexers (OADMs) and optical terminal multiplexers (OTMs).

GR-2998. Generic requirements for wavelength division multiplexing (WDM) element management systems (EMSs).

GR-2999. Generic requirements for wavelength division multiplexing (WDM) network management systems (NMSs).

GR-3009. Optical cross-connect generic requirements.

C.3 American National Standards Institute (ANSI)

These can be ordered from *www.ansi.org*.

C.3.1 SONET

T1.105. Telecommunications—synchronous optical network (SONET)—basic description including multiplex structures, rates, and formats.

T1.105.01. Telecommunications—synchronous optical network (SONET)—automatic protection switching. See also all the other T1.105.* documents.

C.3.2 ESCON and Fibre Channel

X3.289. Information technology—Fibre Channel—fabric generic requirements (FC-FG).

X3.296. Information technology—single byte command code connection (SBCON) architecture. (This is the ANSI version of IBM's ESCON).

X3.303. Fibre Channel physical and signaling interface-3 (FC-PH-3).

appendix D

Wave Equations

THE PROPAGATION of electromagnetic waves is governed by the following *Maxwell's equations:*

$$\nabla \cdot \mathbf{D} = \rho \tag{D.1}$$

$$\nabla \cdot \mathbf{B} = 0 \tag{D.2}$$

$$\nabla \times \mathbf{E} = -\frac{\partial \mathbf{B}}{\partial t} \tag{D.3}$$

$$\nabla \times \mathbf{H} = \mathbf{J} + \frac{\partial \mathbf{D}}{\partial t} \tag{D.4}$$

Here, ρ is the charge density, and $\mathbf{J}$ is the current density. We assume that there are no free charges in the medium so that $\rho = 0$. For such a medium, $\mathbf{J} = \sigma \mathbf{E}$, where σ is the conductivity of the medium. Since the conductivity of silica is extremely low ($\sigma \approx 0$), we assume that $\mathbf{J} = 0$; this amounts to assuming a lossless medium.

In any medium, we also have, from (2.5) and (2.6),

$$\mathbf{D} = \epsilon_0 \mathbf{E} + \mathbf{P},$$

where $\mathbf{P}$ is the electric polarization of the medium and

$$\mathbf{B} = \mu_0 (\mathbf{H} + \mathbf{M}),$$

where $\mathbf{M}$ is the magnetic polarization of the medium. Since silica is a nonmagnetic material, we set $\mathbf{M} = 0$.

Using these relations, we can eliminate the flux densities from Maxwell's curl equations (D.3) and (D.4) and write them only in terms of the field vectors **E** and **H**, and the electric polarization **P**. For example,

$$\nabla \times \nabla \times \mathbf{E} = -\mu_0 \epsilon_0 \frac{\partial^2 \mathbf{E}}{\partial t^2} - \mu_0 \frac{\partial^2 \mathbf{P}}{\partial t^2}. \tag{D.5}$$

To solve this equation for **E**, we have to relate **P** to **E**. If we neglect nonlinear effects, we can assume the linear relation between **P** and **E** given by (2.7) and further, because of the homogeneity assumption, we can write $\chi(t)$ for $\chi(\mathbf{r}, t)$. We relax this assumption when we discuss nonlinear effects in Section 2.4.

We can solve (D.5) for **E** most conveniently by using Fourier transforms. The Fourier transform $\tilde{\mathbf{E}}$ of **E** is defined by (2.4); $\tilde{\mathbf{P}}$ and $\tilde{\mathbf{H}}$ are defined similarly. It follows from the properties of Fourier transforms that

$$\mathbf{E}(\mathbf{r}, t) = \frac{1}{2\pi} \int_{-\infty}^{\infty} \tilde{\mathbf{E}}(\mathbf{r}, \omega) \exp(-i\omega t)\, d\omega.$$

By differentiating this equation with respect to t, we obtain the Fourier transform of $\partial \mathbf{E}/\partial t$ as $-i\omega \tilde{\mathbf{E}}$.

Taking the Fourier transform of (D.5), we get

$$\nabla \times \nabla \times \tilde{\mathbf{E}} = \mu_0 \epsilon_0 \omega^2 \tilde{\mathbf{E}} + \mu_0 \omega^2 \tilde{\mathbf{P}}.$$

Using (2.8) to express $\tilde{\mathbf{P}}$ in terms of $\tilde{\mathbf{E}}$, this reduces to

$$\nabla \times \nabla \times \tilde{\mathbf{E}} = \mu_0 \epsilon_0 \omega^2 \tilde{\mathbf{E}} + \mu_0 \epsilon_0 \omega^2 \tilde{\chi} \tilde{\mathbf{E}}.$$

We denote $c = 1/\sqrt{\mu_0 \epsilon_0}$; c is the speed of light in a vacuum. When losses are neglected, as we have neglected them, $\tilde{\chi}$ is real, and we can write $n(\omega) = \sqrt{1 + \tilde{\chi}(\omega)}$, where n is the refractive index. Note that this is the same as (2.9), which we used as the definition for the refractive index. With this notation,

$$\nabla \times \nabla \times \tilde{\mathbf{E}} = \frac{\omega^2 n^2}{c^2} \tilde{\mathbf{E}}. \tag{D.6}$$

By using the identity,

$$\nabla \times \nabla \times \tilde{\mathbf{E}} = \nabla(\nabla \cdot \tilde{\mathbf{E}}) - \nabla^2 \tilde{\mathbf{E}},$$

(D.6) can be rewritten as

$$\nabla^2 \tilde{\mathbf{E}} + \frac{\omega^2 n^2}{c^2} \tilde{\mathbf{E}} = \nabla(\nabla \cdot \tilde{\mathbf{E}}). \tag{D.7}$$

Because of our assumption of a homogeneous medium (χ independent of $\mathbf{r}$) and using (D.1) and (2.9), we get

$$0 = \nabla \cdot \tilde{\mathbf{D}} = \epsilon_0 \nabla \cdot (1 + \tilde{\chi}) \tilde{\mathbf{E}} = \epsilon_0 n^2 \nabla \cdot \tilde{\mathbf{E}}. \tag{D.8}$$

This enables us to simplify (D.7) and obtain the wave equation (2.10) for $\tilde{\mathbf{E}}$. Following similar steps, the wave equation (2.11) can be derived for $\tilde{\mathbf{H}}$.

appendix E

Pulse Propagation in Optical Fiber

IN MATHEMATICAL TERMS, chromatic dispersion arises because the propagation constant β is not proportional to the angular frequency ω, that is, $d\beta/d\omega \neq$ constant (independent of ω). $d\beta/d\omega$ is denoted by β_1, and β_1^{-1} is called the *group velocity*. As we will see, this is the velocity with which a pulse propagates through the fiber (in the absence of chromatic dispersion). Chromatic dispersion is also called *group velocity dispersion*.

If we were to launch a pure monochromatic wave at frequency ω_0 into a length of optical fiber, the magnitude of the (real) electric field vector associated with the wave would be given by

$$|\mathbf{E}(\mathbf{r}, t)| = J(x, y) \cos(\omega_0 t - \beta(\omega_0)z). \tag{E.1}$$

Here the z coordinate is taken to be along the fiber axis, and $J(x, y)$ is the distribution of the electric field along the fiber cross section and is determined by solving the wave equation. This equation can be derived as follows.

For the fundamental mode, the longitudinal component is of the form $E_z = 2\pi J_l(x, y) \exp(i\beta z)$. Here $J_l(x, y)$ is a function only of $\rho = \sqrt{x^2 + y^2}$ due to the cylindrical symmetry of the fiber and is expressible in terms of Bessel functions. The transverse component of the fundamental mode is of the form $E_x(E_y) = 2\pi J_t(x, y) \exp(i\beta z)$, where again $J_t(x, y)$ depends only on $\sqrt{x^2 + y^2}$ and can be expressed in terms of Bessel functions. Thus, for each of the solutions corresponding to the fundamental mode, we can write

$$\tilde{\mathbf{E}}(\mathbf{r}, \omega) = 2\pi J(x, y) e^{i\beta(\omega)z} \hat{e}(x, y), \tag{E.2}$$

where $J(x, y) = \sqrt{J_l(x, y)^2 + J_t(x, y)^2}$ and the $\hat{e}$ is the unit vector along the direction of $\tilde{E}(\mathbf{r}, \omega)$. In this equation, we have explicitly written β as a function of ω to emphasize this dependence. In general, $J()$ and $\hat{e}()$ are also functions of ω, but this dependence can be neglected for pulses whose spectral width is much smaller than their center frequency. This condition is satisfied by pulses used in optical communication systems. Equation (E.1) now follows from (E.2) by taking the inverse Fourier transform.

This pure monochromatic wave propagates at a velocity $\omega_0/\beta(\omega_0)$. This is called the *phase velocity of the wave*. In practice, signals used for optical communication are not monochromatic waves but pulses having a nonzero spectral width. To understand how such pulses propagate, consider a pulse consisting of just two spectral components: one at $\omega_0 + \Delta\omega$ and the other at $\omega_0 - \Delta\omega$. Further assume that $\Delta\omega$ is small so that we may approximate

$$\beta(\omega_0 \pm \Delta\omega) \approx \beta_0 \pm \beta_1 \Delta\omega,$$

where $\beta_0 = \beta(\omega_0)$ and

$$\beta_1 = \frac{d\beta}{d\omega}\bigg|_{\omega=\omega_0}.$$

The magnitude of the electric field vector associated with such a pulse would be given by

$$
\begin{aligned}
|E(\mathbf{r}, t)| &= J(x, y) \left[\cos\left((\omega_0 + \Delta\omega)t - \beta(\omega_0 + \Delta\omega)z\right) + \right. \\
&\quad \left. \cos\left((\omega_0 - \Delta\omega)t - \beta(\omega_0 - \Delta\omega)z\right) \right] \\
&\approx 2J(x, y) \cos(\Delta\omega t - \beta_1 \Delta\omega z) \cos(\omega_0 t - \beta_0 z).
\end{aligned}
$$

This pulse can be viewed in time t and space z as the product of a very rapidly varying sinusoid, namely, $\cos(\omega_0 t - \beta_0 z)$, which is also called the *phase* of the pulse, and a much more slowly varying *envelope*, namely, $\cos(\Delta\omega t - \beta_1 \Delta\omega z)$. Note that in this case the phase of the pulse travels at a velocity of ω_0/β_0, whereas the envelope of the pulse travels at a velocity of $1/\beta_1$. The quantity ω_0/β_0 is called the *phase velocity* of the pulse, and $1/\beta_1$ is called the *group velocity*.

In general, pulses used for optical communication can be represented in this manner as the product of a slowly varying envelope function (of z and t), which is usually not a sinusoid, and a sinusoid of the form $\cos(\omega_0 t - \beta_0 z)$, where ω_0 is termed the *center frequency* of the pulse. And just as in the preceding case, the envelope of the pulse propagates at the group velocity, $1/\beta_1$. This concept can be stated more precisely as follows.

Consider a pulse whose shape, or envelope, is described by $A(z, t)$ and whose center frequency is ω_0. Assume that the pulses have *narrow spectral width*. By this we mean that most of the energy of the pulse is concentrated in a frequency band whose width is negligible compared to the center frequency ω_0 of the pulse. This assumption is usually satisfied for most pulses used in optical communication systems. With this assumption, it can be shown that the magnitude of the (real) electric field vector associated with such a pulse is

$$|\mathbf{E}(\mathbf{r}, t)| = J(x, y)\Re[A(z, t)e^{-i(\omega_0 t - \beta_0 z)}], \tag{E.3}$$

where $\Re[q]$ denotes the real part of q (see, for example, [Agr97]). Here β_0 is the value of the propagation constant β at the frequency ω_0. $J(x, y)$ has the same significance as before. It is mathematically convenient to allow the pulse envelope $A(z, t)$ to be complex valued so that it captures not only the change in the pulse shape during propagation but also any induced phase shifts. Thus if $A(z, t) = |A(z, t)| \exp(i\phi_A(z, t))$, the phase of the pulse is given by

$$\phi(t) = \omega_0 t - \beta_0 z - \phi_A(z, t). \tag{E.4}$$

To get the description of the actual pulse, we must multiply $A(z, t)$ by $\exp(-i(\omega_0 t - \beta_0 z))$ and take the real part. We will illustrate this in (E.6).

Here we have also assumed that the pulse is obtained by modulating a nearly monochromatic source at frequency ω_0. This means that the frequency spectrum of the optical source has negligible width compared to the frequency spectrum of the pulse. We will consider the effect of relaxing this assumption later in this section.

By assuming that the higher derivatives of β with respect to ω are negligible, we can derive the following partial differential equation for the evolution of the pulse shape $A(z, t)$ [Agr97]:

$$\frac{\partial A}{\partial z} + \beta_1 \frac{\partial A}{\partial t} + \frac{i}{2}\beta_2 \frac{\partial^2 A}{\partial t^2} = 0. \tag{E.5}$$

Here,

$$\beta_2 = \left.\frac{d^2\beta}{d\omega^2}\right|_{\omega=\omega_0}.$$

Note that if β were a linear function of ω, that is, $\beta_2 = 0$, then $A(z, t) = F(t - \beta_1 z)$, where F is an arbitrary function that satisfies (E.5). Then $A(z, t) = A(0, t - \beta_1 z)$ for all z and t, and arbitrary pulse shapes propagate without change in shape (and at velocity $1/\beta_1$). In other words, *if the group velocity is independent of ω, no broadening of the pulse occurs.* Thus β_2 is the key parameter governing group velocity or chromatic dispersion. It is termed the *group velocity dispersion parameter* or, simply, *GVD parameter*.

E.1   Propagation of Chirped Gaussian Pulses

Mathematically, a chirped Gaussian pulse at $z = 0$ is described by the equation

$$
\begin{aligned}
G(t) &= \Re\left[A_0 e^{-\frac{1+i\kappa}{2}\left(\frac{t}{T_0}\right)^2} e^{-i\omega_0 t} \right] \\
&= A_0 e^{-\frac{1}{2}\left(\frac{t}{T_0}\right)^2} \cos\left(\omega_0 t + \frac{\kappa}{2}\left(\frac{t}{T_0}\right)^2 \right).
\end{aligned}
\tag{E.6}
$$

The peak amplitude of the pulse is A_0. The parameter T_0 determines the width of the pulse. It has the interpretation that it is the half-width of the pulse at the $1/e$-intensity point. (The intensity of a pulse is the square of its amplitude.) The *chirp factor* κ determines the degree of chirp of the pulse. From (E.4), the phase of this pulse is

$$
\phi(t) = \omega_0 t + \frac{\kappa t^2}{2 T_0^2}.
$$

The instantaneous angular frequency of the pulse is the derivative of the phase and is given by

$$
\frac{d}{dt}\left(\omega_0 t + \frac{\kappa}{2}\frac{t^2}{T_0^2} \right) = \omega_0 + \frac{\kappa}{T_0^2} t.
$$

We define the *chirp factor* of a Gaussian pulse as T_0^2 times the derivative of its instantaneous angular frequency. Thus the chirp factor of the pulse described by (E.6) is κ. This pulse is said to be *linearly chirped* since the instantaneous angular frequency of the pulse increases or decreases *linearly* with time t, depending on the sign of the chirp factor κ. In other words, the chirp factor κ is a constant, independent of time t, for linearly chirped pulses.

Let $A(z, t)$ denote a chirped Gaussian pulse as a function of time and distance. At $z = 0$,

$$
A(0, t) = A_0 e^{-\frac{1+i\kappa}{2}\left(\frac{t}{T_0}\right)^2}.
\tag{E.7}
$$

If we solve (E.5) for a chirped Gaussian pulse (so the initial condition for this differential equation is that $A(0, t)$ is given by (E.7)), we get

$$
A(z, t) = \frac{A_0 T_0}{\sqrt{T_0^2 - i\beta_2 z(1 + i\kappa)}} \exp\left(-\frac{(1 + i\kappa)(t - \beta_1 z)^2}{2\left(T_0^2 - i\beta_2 z(1 + i\kappa)\right)} \right).
\tag{E.8}
$$

This can be rewritten in the form

$$
A(z,t) = \left[A_z e^{-\frac{1+i\kappa}{2}\left(\frac{t-\beta_1 z}{T_z}\right)^2} e^{i\phi_z} \right]
\tag{E.9}
$$

Comparing with (E.6), we see that $A(z,t)$ is also the envelope of a chirped Gaussian pulse for all $z > 0$, and the chirp factor κ remains unchanged. However, the width of this pulse increases as z increases if $\beta_2 \kappa > 0$. This happens because the parameter governing the pulse width is now

$$
\begin{aligned}
T_z^2 &= \left(\Re \left[\frac{1+i\kappa}{T_0^2 - i\beta_2 z(1+i\kappa)} \right] \right)^{-1} \\
&= T_0^2 \left[\left(1 + \frac{\beta_2 z \kappa}{T_0^2} \right)^2 + \left(\frac{\beta_2 z}{T_0^2} \right)^2 \right],
\end{aligned}
\tag{E.10}
$$

which monotonically increases with increasing z if $\beta_2 \kappa > 0$. A measure of the pulse broadening at distance z is the ratio T_z/T_0. The analytical expression (2.13) for this ratio follows from (E.10).

E.2   Nonlinear Effects on Pulse Propagation

So far, we have understood the origins of SPM and CPM and the fact that these effects result in changing the phase of the pulse as a function of its intensity (and the intensity of other pulses at different wavelengths in the case of CPM). To understand the magnitude of this phase change or chirping and how it interacts with chromatic dispersion, we will need to go back and look at the differential equation governing the evolution of the pulse shape as it propagates in the fiber. We will also find that this relationship is important in understanding the fundamentals of solitons in Section 2.5.

We will consider pulses for which the magnitude of the associated (real) electric field vector is given by (E.3), which is

$$
|\mathbf{E}(\mathbf{r},t)| = J(x,y)\Re[A(z,t)e^{-i(\omega_0 t - \beta_0 z)}].
$$

Recall that $J(x,y)$ is the transverse distribution of the electric field of the fundamental mode dictated by the geometry of the fiber, $A(z,t)$ is the complex envelope of the pulse, ω_0 is its center frequency, and $\Re[\cdot]$ denotes the real part of its argument. Let A_0 denote the peak amplitude of the pulse, and $P_0 = A_0^2$ its peak power.

We have seen that the refractive index becomes intensity dependent in the presence of SPM and is given by (2.23) for a plane monochromatic wave. For non-monochromatic pulses with envelope A propagating in optical fiber, this relation must be modified so that the frequency *and* intensity-dependent refractive index is now given by

$$\hat{n}(\omega, E) = n(\omega) + \bar{n}|A|^2/A_e. \tag{E.11}$$

Here, $n(\omega)$ is the linear refractive index, which is frequency dependent because of chromatic dispersion, but also intensity independent, and A_e is the effective cross-sectional area of the fiber, typically 50 μm^2 (see Figure 2.15 and the accompanying explanation). The expression for the propagation constant (2.22) must also be similarly modified, and the frequency and intensity-dependent propagation constant is now given by

$$\hat{\beta}(\omega, E) = \beta(\omega) + \frac{\omega}{c}\frac{\bar{n}|A|^2}{A_e}. \tag{E.12}$$

Note that in (E.11) and (E.12) when we use the value $\bar{n} = 3.2 \times 10^{-8}$ μm^2/W, the intensity of the pulse $|A|^2$ must be expressed in watts (W). We assume this is the case in what follows and will refer to $|A|^2$ as the power of the pulse (though, strictly speaking, it is only proportional to the power).

For convenience, we denote

$$\gamma = \frac{\omega}{c}\frac{\bar{n}}{A_e} = \frac{2\pi}{\lambda}\frac{\bar{n}}{A_e}$$

and thus $\hat{\beta} = \beta + \gamma|A|^2$. Comparing this with (E.11), we see that γ bears the same relationship to the propagation constant β as the nonlinear index coefficient $\bar{n}$ does to the refractive index n. Hence, we call γ the *nonlinear propagation coefficient*. At a wavelength $\lambda = 1.55$ μm and taking $A_e = 50$ μm^2, $\gamma = 2.6$ /W-km.

To take into account the intensity dependence of the propagation constant, (E.5) must be modified to read

$$\frac{\partial A}{\partial z} + \beta_1\frac{\partial A}{\partial t} + \frac{i}{2}\beta_2\frac{\partial^2 A}{\partial t^2} = i\gamma|A|^2 A. \tag{E.13}$$

In this equation, the term $\frac{i}{2}\beta_2\frac{\partial^2 A}{\partial t^2}$ incorporates the effect of chromatic dispersion, as discussed in Section 2.3, and the term $i\gamma|A|^2 A$ incorporates the intensity-dependent phase shift.

Since this equation incorporates the effect of chromatic dispersion also, the combined effects of chromatic dispersion and SPM on pulse propagation can be analyzed using this equation as the starting point. These effects are qualitatively different from that of chromatic dispersion or SPM acting alone.

In order to understand the relative effects of chromatic dispersion and SPM, it is convenient to introduce the following change of variables:

$$\tau = \frac{t - \beta_1 z}{T_0}, \qquad \xi = \frac{z}{L_D} = \frac{z|\beta_2|}{T_0^2}, \qquad \text{and} \qquad U = \frac{A}{\sqrt{P_0}}. \tag{E.14}$$

In these new variables, (E.13) can be written as

$$i\frac{\partial U}{\partial \xi} - \frac{\text{sgn}(\beta_2)}{2}\frac{\partial^2 U}{\partial \tau^2} + N^2|U|^2 U = 0, \tag{E.15}$$

where

$$N^2 = \gamma P_0 L_D = \frac{\gamma P_0}{|\beta_2|/T_0^2}.$$

Equation (E.15) is called the nonlinear Schrödinger equation (NLSE).

The change of variables introduced by (E.14) has the following interpretation. Since the pulse propagates with velocity β_1 (in the absence of chromatic dispersion), $t - \beta_1 z$ is the time axis in a reference frame moving with the pulse. The variable τ is the time in this reference frame but in units of T_0, which is a measure of the pulse width. The variable ξ measures distance in units of the *chromatic dispersion length* $L_D = T_0^2/|\beta_2|$, which we already encountered in Section 2.3. The quantity P_0 represents the peak power of the pulse, and thus U is the envelope of the pulse normalized to have unit peak power.

Note that the quantity $1/\gamma P_0$ also has the dimensions of length; we call it the *nonlinear length* and denote it by L_{NL}. Using $\gamma = 2.6$ /W-km and $P_0 = 1$ mW, we get $L_{NL} = 384$ km. If the pulse power P_0 is increased to 10 mW, the nonlinear length decreases to 38 km. The nonlinear length serves as a convenient normalizing measure for the distance z in discussing nonlinear effects, just as the chromatic dispersion length does for the effects of chromatic dispersion. Thus the effect of SPM on pulses can be neglected for pulses propagating over distances $z \ll L_{NL}$. Then we can write the quantity N introduced in the NLSE as $N^2 = L_D/L_{NL}$. Thus it is the ratio of the chromatic dispersion and nonlinear lengths. When $N \ll 1$, the nonlinear length is much larger than the chromatic dispersion length so that the nonlinear effects can be neglected compared to those of chromatic dispersion. This amounts to saying that the third term (the one involving N) in the NLSE can be neglected. In this case, the NLSE reduces to (E.5) for the evolution of pulses in the presence of chromatic dispersion alone, with the change of variables given by (E.14).

The NLSE serves as the starting point for the discussion of the combined effects of GVD and SPM. For arbitrary values of N, the NLSE has to be solved numerically. These numerical solutions are important tools for the understanding of the combined effects of chromatic dispersion and nonlinearities on pulses and are discussed extensively in [Agr95]. The qualitative description of these solutions in both the normal and anomalous chromatic dispersion regimes is discussed in Section 2.4.5.

We can use (E.13) to estimate the SPM-induced chirp for Gaussian pulses. To do this, we neglect the chromatic dispersion term and consider the equation

$$\frac{\partial A}{\partial z} + \beta_1 \frac{\partial A}{\partial t} = i\gamma |A|^2 A. \tag{E.16}$$

By using the variables τ and U introduced in (E.14) instead of t and A, and $L_{NL} = (\gamma P_0)^{-1}$, this reduces to

$$\frac{\partial U}{\partial z} = \frac{i}{L_{NL}} |U|^2 U. \tag{E.17}$$

Note that we have not used the change of variable ξ for z since L_D is infinite when chromatic dispersion is neglected. This equation has the solution

$$U(z, \tau) = U(0, \tau) e^{iz|U(0,\tau)|^2/L_{NL}}. \tag{E.18}$$

Thus the SPM causes a phase change but no change in the envelope of the pulse. Note that the initial pulse envelope $U(0, \tau)$ is arbitrary; so this is true for all pulse shapes. Thus *SPM by itself leads only to chirping, regardless of the pulse shape*; it is chromatic dispersion that is responsible for pulse broadening. The SPM-induced chirp, however, modifies the pulse-broadening effects of chromatic dispersion.

E.3   Soliton Pulse Propagation

In the anomalous chromatic dispersion regime (1.55 μm band for standard single-mode fiber and most dispersion-shifted fibers), the GVD parameter β_2 is negative. Thus sgn(β_2) = -1, and the NLSE of (E.15) can be written as

$$i\frac{\partial U}{\partial \xi} + \frac{1}{2}\frac{\partial^2 U}{\partial \tau^2} + N^2 |U|^2 U = 0. \tag{E.19}$$

An interesting phenomenon occurs in this anomalous chromatic dispersion regime when N is an integer. In this case, the modified NLSE (E.19) can be solved analytically, and the resulting pulse envelope has an amplitude that is independent of ξ (for $N = 1$) or periodic in ξ (for $N \geq 2$). This implies that these pulses propagate with no change in their widths or with a periodic change in their widths. The solutions of this equation are termed *solitons,* and N is called the *order* of the soliton.

It can be verified that the solution of (E.19) corresponding to $N = 1$ is

$$U(\xi, \tau) = e^{i\xi/2}\mathrm{sech}\tau. \tag{E.20}$$

The pulse corresponding to this envelope is called the *fundamental soliton*. The fundamental soliton pulse and its envelope are sketched in Figure 2.25(a) and (b), respectively. (As in the case of chirped Gaussian pulses in Section 2.3, the frequency of the pulse is shown vastly diminished for the purposes of illustration.)

Note that (in a reference frame moving with the pulse) the magnitude of the fundamental soliton pulse envelope, or the pulse shape, does not change with the distance coordinate z. However, the pulse acquires a phase shift that is linear in z as it propagates.

Recall that the order of the soliton, N, is defined by

$$N^2 = \gamma P_0 L_D = \frac{\gamma P_0}{|\beta_2|/T_0^2}.$$

Since γ and β_2 are fixed for a given fiber and operating wavelength, for a fixed soliton order, the peak power P_0 of the pulse increases as the pulse width T_0 decreases. Since operation at very high bit rates requires narrow pulses, this also implies that large peak powers are necessary in soliton communication systems.

It can also be verified that the solution of (E.19) corresponding to $N = 2$ is

$$U(\xi, \tau) = 4e^{i\xi/2}\frac{\cosh 3\tau + 3\cosh \tau e^{i4\xi}}{\cosh 4\tau + 4\cosh 2\tau + 3\cos 4\xi}. \tag{E.21}$$

The magnitude of this normalized pulse envelope is sketched in Figure E.1 as a function of ξ and τ. The periodicity of the pulse envelope with respect to ξ can be clearly seen from this plot. In each period, the pulse envelope first undergoes compression due to the positive chirping induced by SPM and then undergoes broadening, finally regaining its original shape.

Further Reading

Pulse propagation is covered in detail in [Agr95]. The classic papers by Marcuse [Mar80, Mar81] are a must-read for anyone wishing to dig deeper into the mathematics of Gaussian and chirped Gaussian pulse propagation.

References

[Agr95] G. P. Agrawal. *Nonlinear Fiber Optics*, 2nd edition. Academic Press, San Diego, CA, 1995.

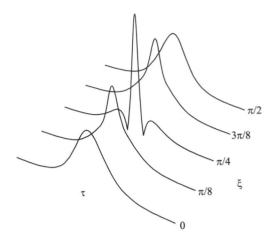

Figure E.1   The magnitude of the pulse envelope of the second-order soliton.

[Agr97]   G. P. Agrawal. *Fiber-Optic Communication Systems*. John Wiley, New York, 1997.

[Mar80]   D. Marcuse. Pulse distortion in single-mode fibers. *Applied Optics*, 19:1653–1660, 1980.

[Mar81]   D. Marcuse. Pulse distortion in single-mode fibers. 3: Chirped pulses. *Applied Optics*, 20:3573–3579, 1981.

appendix F

Nonlinear Polarization

THE LINEAR EQUATION (2.7) for the relationship between the induced polarization P and the applied electric field E holds when the power levels and/or bit rates are moderate. When this is not the case, this must be generalized to include higher powers of $E(\mathbf{r}, t)$. For an isotropic medium and an electric field polarized along one direction so that it has a single component $E(\mathbf{r}, t)$, this relationship can be written as follows:

$$
\begin{aligned}
\mathcal{P}(\mathbf{r}, t) \;=\; & \epsilon_0 \int_{-\infty}^{t} \chi^{(1)}(\mathbf{r}, t - t_1) E(\mathbf{r}, t_1) \, dt_1 \\
& + \epsilon_0 \int_{-\infty}^{t} \int_{-\infty}^{t} \chi^{(2)}(t - t_1, t - t_2) E(\mathbf{r}, t_1) E(\mathbf{r}, t_2) \, dt_1 \, dt_2 \\
& + \epsilon_0 \int_{-\infty}^{t} \int_{-\infty}^{t} \int_{-\infty}^{t} \chi^{(3)}(t - t_1, t - t_2, t - t_3) E(\mathbf{r}, t_1) E(\mathbf{r}, t_2) E(\mathbf{r}, t_3) \, dt_1 \, dt_2 \, dt_3 \\
& + \cdots .
\end{aligned}
\tag{F.1}
$$

Now $\chi^{(1)}(\mathbf{r}, t)$ is called the *linear susceptibility* to distinguish it from $\chi^{(i)}(\mathbf{r}, t)$, $i = 2, 3, \ldots$, which are termed the *higher-order nonlinear susceptibilities*. Owing to certain symmetry properties of the silica molecule, $\chi^{(2)}(\mathbf{r}, t) = 0$. The effect of the higher-order susceptibilities $\chi^{(4)}(,)$, $\chi^{(5)}(,)$, $\ldots$, is negligible in comparison with that of $\chi^{(3)}(,)$. Thus we can write (F.1) as

$$
\mathcal{P}(\mathbf{r}, t) = \mathcal{P}_L(\mathbf{r}, t) + \mathcal{P}_{NL}(\mathbf{r}, t).
$$

Here $\mathcal{P}_L(\mathbf{r}, t)$ is the *linear polarization* given by (2.18). The *nonlinear polarization* $\mathcal{P}_{NL}(\mathbf{r}, t)$ is given by

$$\mathcal{P}_{NL}(\mathbf{r}, t) = \epsilon_0 \int_{-\infty}^{t} \int_{-\infty}^{t} \int_{-\infty}^{t} \chi^{(3)}(t - t_1, t - t_2, t - t_3)$$

$$E(\mathbf{r}, t_1) E(\mathbf{r}, t_2) E(\mathbf{r}, t_3) \, dt_1 \, dt_2 \, dt_3. \qquad \text{(F.2)}$$

The nonlinear response of the medium occurs on a very narrow time scale of less than 100 fs—much smaller than the time scale of the linear response—and thus can be assumed to be instantaneous for pulse widths greater than 1 ps. Note that even if the pulse occupies only a tenth of the bit interval, this assumption is satisfied for bit rates greater than 100 Gb/s. We will consider only this instantaneous nonlinear response case in this book. When this assumption is satisfied,

$$\chi^{(3)}(t - t_1, t - t_2, t - t_3) = \chi^{(3)} \delta(t - t_1) \delta(t - t_2) \delta(t - t_3),$$

where $\chi^{(3)}$ on the right-hand side is now a constant, independent of t. This assumption enables us to simplify (F.2) considerably. It now reduces to

$$\mathcal{P}_{NL}(\mathbf{r}, t) = \epsilon_0 \chi^{(3)} E^3(\mathbf{r}, t),$$

which is equation (2.19).

G appendix

Multilayer Thin-Film Filters

To understand the principle of operation of dielectric thin-film multicavity filters, we need to digress and discuss some results from electromagnetic theory.

G.1 Wave Propagation at Dielectric Interfaces

A plane electromagnetic wave is one whose electric and magnetic fields vary only in the spatial coordinate along the direction of propagation. In other words, along any plane perpendicular to the direction of propagation, the electric and magnetic fields are constant. The ratio of the amplitude of the electric field to that of the magnetic field at any such plane is called the *impedance* at that plane. In a medium that supports only one propagating wave (so there is no reflected wave), this impedance is called the *intrinsic impedance* of the medium and is denoted by η. If ϵ is the dielectric permittivity of the medium and μ is its magnetic permeability, $\eta = \sqrt{\mu/\epsilon}$. If we denote the *intrinsic impedance of vacuum* by η_0, for a nonmagnetic dielectric medium with refractive index n, the intrinsic impedance $\eta = \eta_0/n$. (A nonmagnetic dielectric material has the same permeability as that of a vacuum. Since most commonly used dielectrics are nonmagnetic, in the rest of the discussion, we assume that the dielectrics considered are nonmagnetic.)

Consider the interface between two dielectrics with refractive indices n_1 and n_2, illustrated in Figure G.1(a). Assume that a plane electromagnetic wave is incident normal to this interface. The reflection coefficient at this interface is the ratio of the amplitude of the electric field in the reflected wave to that in the incident wave. From

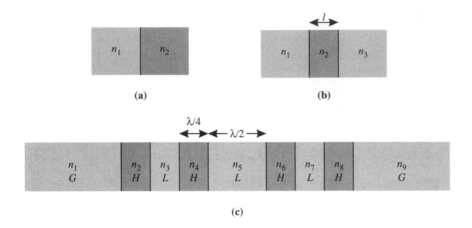

Figure G.1   (a) The interface between two dielectric media. (b) A dielectric slab or film placed between two other dielectric media. (c) Multiple dielectric slabs or films stacked together.

the principles of electromagnetics [RWv93, Section 6.7], it can be shown that the *reflection coefficient* at this interface (for normal incidence) is

$$\rho = \frac{\eta_2 - \eta_1}{\eta_2 + \eta_1} = \frac{n_1 - n_2}{n_1 + n_2}. \tag{G.1}$$

Thus the fraction of power transmitted through this interface is

$$1 - |\rho|^2 = 1 - \left| \frac{n_1 - n_2}{n_1 + n_2} \right|^2.$$

Here, as in the rest of the discussion, we assume that the dielectrics are lossless so that no power is absorbed by them.

Now consider a slab of a dielectric material of thickness l and refractive index n_2 (dielectric 2) that is placed between two dielectrics with refractive indices n_1 and n_3 (dielectrics 1 and 3, respectively). Assume that dielectrics 1 and 3 have very large, essentially infinite, thicknesses. This is illustrated in Figure G.1(b). A part of any signal incident from dielectric 1 will be reflected at the 1-2 interface and a part transmitted. Of the transmitted part, a fraction will be reflected at the 2-3 interface. Of this reflected signal, another fraction will be reflected at the 2-1 interface and the remainder transmitted to dielectric 1 and added to the first reflected signal, and so on. In principle, the *net* signal reflected at the 1-2 interface can be calculated by adding all the reflected signals calculated using the reflection coefficients given by

(G.1), with the proper phases. But the whole process can be simplified by using the concept of impedances and the following result concerning them.

If the impedance at some plane in a dielectric is Z_L, called the *load impedance*, the impedance at distance l in front of it, called the *input impedance*, is given, as a function of the wavelength λ, by

$$Z_i = \eta \left(\frac{Z_L \cos(2\pi n l/\lambda) + i\eta \sin(2\pi n l/\lambda)}{\eta \cos(2\pi n l/\lambda) + i Z_L \sin(2\pi n l/\lambda)} \right). \tag{G.2}$$

Here, η is the intrinsic impedance of the dielectric, and n is its refractive index. Note that in a single dielectric medium, $Z_L = \eta$, and (G.2) yields $Z_i = \eta$ as well. This agrees with our earlier statement that the impedance at all planes in a single dielectric medium is η.

The reason that the concept of impedance is useful for us is that the reflection and transmission coefficients may be expressed in terms of impedances. Specifically, the reflection coefficient at an interface with load impedance Z_L, in a dielectric with intrinsic impedance η, is given by

$$\rho = \frac{Z_L - \eta}{Z_L + \eta}. \tag{G.3}$$

The transmission coefficient at the same interface is given by

$$\tau = 1 - \rho = \frac{2Z_L}{Z_L + \eta}. \tag{G.4}$$

Note that (G.1) is a special case of (G.3) obtained by setting $\eta = \eta_1$ and $Z_L = \eta_2$.

Now consider again the case of a single dielectric slab, placed between two other dielectrics, illustrated in Figure G.1(b). The impedance at the 2-3 interface is η_3. Thus the impedance at the 1-2 interface may be calculated using (G.2) as

$$Z_{12} = \eta_2 \left(\frac{\eta_3 \cos(2\pi n l/\lambda) + i\eta_2 \sin(2\pi n l/\lambda)}{\eta_2 \cos(2\pi n l/\lambda) + i\eta_3 \sin(2\pi n l/\lambda)} \right).$$

Using this, the reflection coefficient at the 1-2 interface can be obtained from (G.3) as

$$\rho = \frac{Z_{12} - \eta_1}{Z_{12} + \eta_1}.$$

If the slab of a dielectric of thickness l shown in Figure G.1(b) is viewed as a filter, its power transfer function—the fraction of power transmitted by it—is given by

$$T(\lambda) = 1 - |\rho|^2.$$

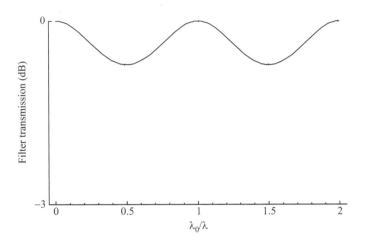

Figure G.2   Transfer function of the filter shown in Figure G.1(b) for $n_1 = n_3 = 1.52$, $n_2 = 2.3$, and $l = \lambda_0/2n_2$.

Let $\lambda_0 = 2nl$ so that the optical path length in the slab is a half wavelength. Note that $T(\lambda_0) = 1$. In Figure G.2, $T(\lambda)$ is plotted as a function of λ_0/λ, assuming $n_1 = n_3 = 1.5$ and $n_2 = 2.3$.

Note that for the case $n_1 = n_3$, this filter becomes a Fabry-Perot filter (see Problem 3.12).

This result can be generalized to an arbitrary number of dielectric slabs as follows. Consider a series of k dielectrics with refractive indices $n_1, n_2, \ldots, n_k$ (not necessarily distinct) and thicknesses $l_1, l_2, \ldots, l_k$, which are stacked together as shown in Figure G.1. We also assume that l_1 and l_k are very large, essentially infinite. This can be viewed as a filter of which a special case is the DTMF. We assume that the input signal is incident normal to the 1-2 interface. If we find the reflection coefficient, ρ, at the 1-2 interface, we can determine the power transfer function, $T(\lambda)$, of the filter, using $T(\lambda) = 1 - |\rho|^2$.

Using the impedance machinery, this is quite easy to do. If η_i is the intrinsic impedance of dielectric i, $i = 1, \ldots, k$, $\eta_i = \eta_0/n_i$. We start at the right end of the filter, at the $(k-1)$-(k) interface. The impedance at this plane is just the intrinsic impedance of medium k, namely, η_k. The intrinsic impedance at the $(k-2)$-$(k-1)$ interface can be calculated using (G.2) with $Z_L = \eta_k$, $\eta = \eta_{k-1}$, $n = n_{k-1}$, and $l = l_{k-1}$. Continuing in the same manner, we can recursively calculate the input impedances at the interfaces $(k - 3)$-$(k - 2), \ldots, 1$-2. From this, the reflection co-efficient at the 1-2 interface can be calculated using (G.3), and the power transfer function of the filter can be determined.

G.2   Filter Design

Although the power transfer function of any given stack of dielectrics can be determined using the preceding procedure, designing a filter of this type to meet a given filter requirement is a more typical problem encountered in practice. The multiple dielectric slab structure exemplified by Figure G.1(c) is quite versatile, and a number of well-known filter transfer functions, such as the Butterworth and the Chebyshev, may be synthesized using it [Kni76]. However, the synthesis of these filters calls for a variety of dielectric materials with different refractive indices. This may be a difficult requirement to meet in practice.

It turns out, however, that very useful filter transfer functions can be synthesized using just two different dielectric materials, a low-index dielectric with refractive index n_L and a high-index dielectric with refractive index n_H [Kni76]. Assume we want to synthesize a bandpass filter with center wavelength λ_0. Then, a general structure for doing this is to use alternate layers of high-index and low-index dielectrics with thicknesses equivalent to a quarter or a half wavelength at λ_0. (A quarter-wavelength slab of the dielectric with refractive index n_L would have a thickness $\lambda_0/4n_L$.) Since these thicknesses at optical wavelengths are quite small, the term *thin film* is more appropriately used instead of *slab*. The dielectric thin films that are a half-wavelength thick at λ_0 are called the *cavities* of the filter. A particularly useful filter structure consists of a few cavities separated by several quarter-wavelength films. If H and L denote quarter-wavelength films (at λ_0) of the high- and low-index dielectrics, respectively, then we can represent any such filter by a sequence of Hs and Ls. Two Ls or two Hs in succession would represent a half-wavelength film. For example, if the lightly shaded dielectrics are of low index and the darker shaded are of high index, the filter consisting of the multiple dielectric films 2-8 shown in Figure G.1(c) can be represented by the sequence $HLHLLHLH$. If the surrounding dielectrics, 1 and 9, are denoted by G (for glass), the entire structure in Figure G.1(c) can be represented by the sequence $GHLHLLHLHG$. If we know the refractive indices n_G, n_L, and n_H of the G, L, and H dielectrics, respectively, the transfer function of the filter can be calculated using the procedure outlined. For $n_G = 1.52$, a typical value for the cover glass, $n_L = 1.46$, which is the refractive index of SiO_2 (a low-index dielectric), and $n_H = 2.3$, which is the refractive index of TiO_2 (a high-index dielectric), this transfer function is plotted in Figure G.3. From this figure, we see that the main lobe is quite

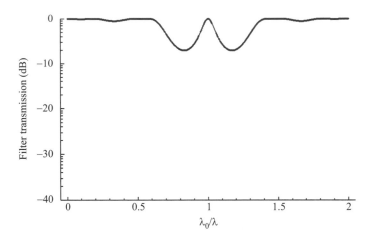

Figure G.3   Transfer function of the filter shown in Figure G.1(c) for $n_G = 1.52$, $n_L = 1.46$, and $n_H = 2.3$.

wide compared to the center wavelength, and the side lobe suppression is less than 10 dB. Clearly, a better transfer function is needed if the filter is to be useful.

A narrower passband and greater side lobe suppression can be achieved by the use of more quarter-wavelength films than just three. For example, the filter described by the sequence

$$G(HL)^9HLL(HL)^9HG$$

has the transfer function shown in Figure G.4. The notation $(HL)^k$ denotes the sequence $HL \cdot HL \cdot \ldots \cdot HL$ (k times). Note that this filter is a single-cavity filter since it uses just one half-wave film. However, it uses 38 quarter-wave films, 19 on each side of the cavity.

The transfer function of a dielectric thin-film filter is periodic in frequency or in λ_0/λ, just like the Fabry-Perot filter. In Figure G.4(a), the transfer function of the filter for one complete period is shown. However, this figure hides the passband structure of the filter. Therefore, the transfer function of the filter is shown in Figure G.4(b) for a narrow spectral range around the center wavelength λ_0. The passband structure of the filter can now be clearly seen. The resemblance to the Fabry-Perot filter transfer function (Figure 3.17) is no accident (see Problem 3.12).

The use of multiple cavities leads to a flatter passband and a sharper transition from the passband to the stop band. Both effects are illustrated in Figure 3.19, where the filter transfer function, around the center wavelength λ_0, is plotted for a

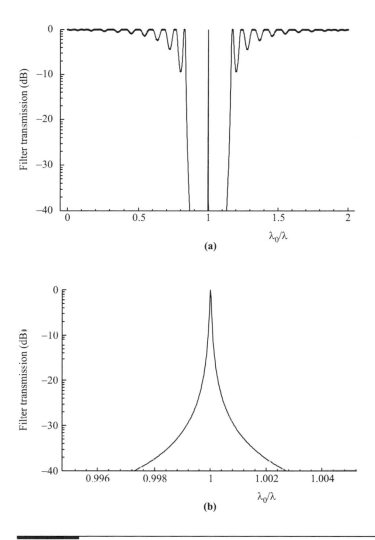

Figure G.4   Transfer function of a single-cavity dielectric thin-film filter. The sequence structure is $G(HL)^9 HLL(HL)^9 HG$. $n_G = 1.52$, $n_L = 1.46$, and $n_H = 2.3$.

single-cavity, two-cavity, and three-cavity dielectric thin-film filter. The single-cavity filter is the same as the one considered here. The two-cavity filter is described by the sequence

$$G(HL)^6 HLL(HL)^{12} HLL(HL)^6 HG.$$

The three-cavity filter is described by the sequence

$$G(HL)^5 HLL(HL)^{11} HLL(HL)^{11} HLL(HL)^5 HG.$$

Again, the values $n_G = 1.52$, $n_L = 1.46$, and $n_H = 2.3$ were used.

References

[Kni76] Z. Knittl. *Optics of Thin Films*. John Wiley, New York, 1976.

[RWv93] S. Ramo, J. R. Whinnery, and T. van Duzer. *Fields and Waves in Communication Electronics*. John Wiley, New York, 1993.

appendix

Random Variables and Processes

IN MANY PLACES in the book, we use random variables and random processes to model noise, polarization, and network traffic. Understanding the statistical nature of these parameters is essential in predicting the performance of communication systems.

H.1 Random Variables

A random variable X is characterized by a probability distribution function

$$F_X(x) = P\{X \le x\}.$$

The derivative of $F_X(x)$ is the probability density function

$$f_X(x) = \frac{dF_X(x)}{dx}.$$

Note that

$$\int_{-\infty}^{\infty} f_X(x)dx = 1.$$

In many cases, we will be interested in obtaining the expectation, or ensemble average, associated with this probability function. The expectation of a function $g(x)$ is defined as

$$E[g(X)] = \int_{-\infty}^{\infty} f_X(x)g(x)dx.$$

751

The mean of X is defined to be

$$E[X] = \int_{-\infty}^{\infty} x f_X(x) dx,$$

and the mean square (second moment) of X is

$$E[X^2] = \int_{-\infty}^{\infty} x^2 f_X(x) dx.$$

The variance of X is defined as

$$\sigma_X^2 = E[X^2] - (E[X])^2.$$

In many cases, we are interested in determining the statistical properties of two or more random variables that are not independent of each other. The joint probability distribution function of two random variables X and Y is defined as

$$F_{X,Y}(x, y) = P\{X \leq x, Y \leq y\}.$$

Sometimes we are given some information about one of the random variables and must estimate the distribution of the other. The conditional distribution of X given Y is denoted as

$$F_{X|Y}(x|y) = P\{X \leq x | Y \leq y\}.$$

An important relation between these distributions is given by Bayes' theorem:

$$F_{X|Y}(x|y) = \frac{F_{X,Y}(x, y)}{F_Y(y)}.$$

H.1.1   Gaussian Distribution

A random variable X is said to follow a Gaussian distribution if its probability density function

$$f_X(x) = \frac{1}{\sqrt{2\pi}\sigma} e^{(x-\mu)^2/\sigma^2}, \quad -\infty \leq x \leq \infty.$$

Here, μ is the mean and σ^2 the variance of X. In order to compute bit error rates, we will need to compute the probability that $X \geq v$, which is defined as the function

$$Q(v) = \int_v^{\infty} f_X(x) dx.$$

This function can be numerically evaluated. For example, $Q(v) = 10^{-9}$ if $v = 6$, and $Q(v) = 10^{-15}$ if $v = 8$.

Also if X and Y are jointly distributed Gaussian random variables, then it can be proved that

$$E[X^2Y^2] = E[X^2]E[Y^2] + 2(E[XY])^2. \tag{H.1}$$

H.1.2   Maxwell Distribution

The Maxwellian probability density function is useful to calculate penalties due to polarization-mode dispersion. A random variable X is said to follow a Maxwellian distribution if its probability density function

$$f_X(x) = \frac{\sqrt{2}}{\alpha^3\sqrt{\pi}}x^2 e^{-x^2/2\alpha^2}, \qquad x \geq 0,$$

where α is a parameter associated with the distribution. The mean and mean-square value of X can be computed as

$$E[X] = 2\alpha\sqrt{\frac{2}{\pi}}$$

and

$$E[X^2] = 3\alpha^2 = \frac{3}{8}\pi(E[X])^2.$$

Therefore, the variance

$$\sigma_X^2 = E[X^2] - (E[X])^2 = \alpha^2\left(3 - \frac{8}{\pi}\right).$$

It can also be shown that

$$P(X > 3E[X]) \approx 4 \times 10^{-5}.$$

H.1.3   Poisson Distribution

A discrete random variable X takes on values from a discrete but possibly infinite set $S = \{x_1, x_2, x_3, \ldots\}$. It is characterized by a probability mass function $P(x)$, which is the probability that X takes on a value x. The expectation of a function $g(X)$ is defined as

$$E[g(X)] = \sum_{i|x_i \in S} g(x_i)P(x_i).$$

X is a Poisson random variable if

$$P(i) = \frac{e^{-r}r^i}{i!}, \quad i = 0, 1, 2, \ldots,$$

where r is a parameter associated with the distribution. It is easily verified that $E[X] = r$ and $\sigma_X^2 = r$.

H.2 Random Processes

Random processes are useful to model time-varying stochastic events. A random process $X(t)$ is simply a sequence of random variables $X(t_1)$, $X(t_2)$, ..., one for each instant of time. The first-order probability distribution function is given by

$$F(x,t) = P\{X(t) \leq x\},$$

and the first-order density function by

$$f(x,t) = \frac{\partial F(x,t)}{\partial x}.$$

The second-order distribution function is the joint distribution function

$$F(x_1, x_2, t_1, t_2) = P\{X(t_1) \leq x_1, X(t_2) \leq x_2\},$$

and the corresponding second-order density function is defined as

$$f(x_1, x_2, t_1, t_2) = \frac{\partial^2 F(x_1, x_2, t_1, t_2)}{\partial x_1 \partial x_2}.$$

The mean of the process is

$$\mu(t) = E[X(t)] = \int_{-\infty}^{\infty} x f(x,t) dx.$$

The autocorrelation of the process is

$$R_X(t_1, t_2) = E[X(t_1)X(t_2)] = \int_{-\infty}^{\infty} \int_{-\infty}^{\infty} x_1 x_2 f(x_1, x_2, t_1, t_2) dx_1 dx_2.$$

The autocovariance of the process is defined as

$$L_X(t_1, t_2) = R_X(t_1, t_2) - E[X(t_1)]E[X(t_2)].$$

The random process is *wide-sense stationary* if it has a constant mean

$$E[X(t)] = \mu,$$

and the autocorrelation (and autocovariance) depends only on $\tau = t_1 - t_2$, that is, $R_X(\tau) = E[X(t)X(t + \tau)]$ and $L_X(\tau) = R_X(\tau) - \mu^2$. For a wide-sense stationary random process, the *power spectral density* is the Fourier transform of the autoco-variance and is given by

$$S_X(f) = \int_{-\infty}^{\infty} L_X(\tau)e^{-i2\pi f\tau}d\tau.$$

Note that the variance of the random process is given by

$$\sigma_X^2 = L_X(0) = \frac{1}{2\pi}\int_{-\infty}^{\infty} S_X(f)df.$$

In many cases, we will represent noise introduced in the system as a stationary random process. In this case, the spectral density is useful to represent the spectral distribution of the noise. For example, in a receiver, the noise $X(t)$ and signal are sent through a low pass filter with impulse response $h(t)$. The transfer function of the filter $H(f)$ is the Fourier transform of its impulse response. In this case, the spectral density of the output noise process $Y(t)$ can be expressed as

$$S_Y(f) = S_X(f)|H(f)|^2.$$

Suppose the filter is an ideal low pass filter with bandwidth B_e; that is, $H(f) = 1, -B_e \leq f \leq B_e$ and 0 otherwise. The variance of the noise process at its output is simply

$$\sigma_Y^2 = L_Y(0) = \frac{1}{2\pi}\int_{-B_e}^{B_e} S_X(f)df.$$

H.2.1   Poisson Random Process

Poisson random processes are used to model the arrival of photons in an optical communication system. They are also used widely to model the arrival of traffic in a communication network. The model is accurate primarily for voice calls, but it is used for other applications as well, without much real justification.

A Poisson process $X(t)$ is characterized by a rate parameter λ. For any two time instants t_1 and $t_2 > t_1$, $X(t_2) - X(t_1)$ is the number of arrivals during the time interval $(t_1, t_2]$. The number of arrivals during this interval follows a Poisson distribution; that is,

$$P\left(X(t_2) - X(t_1) = n\right) = e^{-\lambda(t_2-t_1)}\frac{(\lambda(t_2 - t_1))^n}{n!},$$

where n is a nonnegative integer. Therefore, the mean number of arrivals during this time interval is

$$E[X(t_2) - X(t_1)] = \lambda(t_2 - t_1).$$

A Poisson process has many important properties that make it easier to analyze systems with Poisson traffic than other forms of traffic. See [BG92] for a good summary.

H.2.2   Gaussian Random Process

In many cases, we model noise as a wide-sense stationary Gaussian random process $X(t)$. It is also common to assume that at any two instants of time $t_1 \neq t_2$ the random variables $X(t_1)$ and $X(t_2)$ are independent Gaussian variables with mean μ. For such a process, we can use (H.1) and write

$$E[X^2(t)X^2(t+\tau)] = (E[X^2(t)])^2 + 2(E[X(t)]E[X(t+\tau)])^2,$$

that is,

$$E[X^2(t)X^2(t+\tau)] = R_X^2(0) + 2R_X^2(\tau).$$

▬ Further Reading

There are several good books on probability and random processes. See, for example, [Pap91, Gal99].

▬ References

[BG92]   D. Bertsekas and R. G. Gallager. *Data Networks*. Prentice Hall, Englewood Cliffs, NJ, 1992.

[Gal99]   R. G. Gallager. *Discrete Stochastic Processes*. Kluwer, Boston, 1999.

[Pap91]   A. Papoulis. *Probability, Random Variables, and Stochastic Processes*, 3rd edition. McGraw-Hill, New York, 1991.

I
appendix

Receiver Noise Statistics

W̲E START OUT BY DERIVING an expression for the statistics of the photocurrent in the *pin* receiver, along the lines of [BL90, RH90]. It is useful to think of the photodetection process in the following way. Each time a photon hits the receiver, the receiver generates a small current pulse. Let t_k denote the arrival times of photons at the receiver. Then the photocurrent generated can be expressed as

$$I(t) = \sum_{k=-\infty}^{\infty} eh(t - t_k), \tag{I.1}$$

where e is the electronic charge and $eh(t - t_k)$ denotes the current impulse due to a photon arriving at time t_k. Note that since $eh(t - t_k)$ is the current due to a single electron, we must have

$$\int_{-\infty}^{\infty} eh(t - t_k)dt = e.$$

The arrival of photons may be described by a Poisson process, whose rate is given by $P(t)/hf_c$. Here, $P(t)$ is the instantaneous optical power, and hf_c is the photon energy. The rate of generation of electrons may then also be considered to be a Poisson process, with rate

$$\lambda(t) = \frac{\mathcal{R}}{e} P(t),$$

where $\mathcal{R} = \eta e / hf_c$ is the responsivity of the photodetector, η being the quantum efficiency.

757

To evaluate (I.1), let us break up the time axis into small intervals of length δt, with the kth interval being $[(k - 1/2)\delta t, (k + 1/2)\delta t)$. Let N_k denote the number of electrons generated during the kth interval. Using these notations, we can rewrite (I.1) as

$$I(t) = \sum_{k=-\infty}^{\infty} e N_k h(t - k\delta t).$$

Note that since the intervals are nonoverlapping, the N_k are independent Poisson random variables, with rate $\lambda(k\delta t)\delta t$.

We will first compute the mean value and autocorrelation functions of the photocurrent for a given optical power $P(.)$. The mean value of the photocurrent is

$$E[I(t)|P(.)] = \sum_{k=-\infty}^{\infty} e E[N_k] h(t - k\delta t) = \sum_{k=-\infty}^{\infty} e\lambda(k\delta t)\delta t\ h(t - k\delta t).$$

In the limit when $\delta t \to 0$, this can be rewritten as

$$E[I(t)|P(.)] = \int_{-\infty}^{\infty} e\lambda(\tau)h(t - \tau)d\tau = \mathcal{R} \int_{-\infty}^{\infty} P(\tau)h(t - \tau)d\tau.$$

Likewise, the autocorrelation of the photocurrent can be written as

$$
\begin{aligned}
E[I(t_1)I(t_2)|P(.)] &= \int_{-\infty}^{\infty} e^2\lambda(\tau)h(t_1 - \tau)h(t_2 - \tau)d\tau \\
&\quad + E[I(t_1)|P(.)]E[I(t_2)|P(.)] \\
&= e\mathcal{R} \int_{-\infty}^{\infty} P(\tau)h(t_1 - \tau)h(t_2 - \tau)d\tau \\
&\quad + \mathcal{R}^2 \int_{-\infty}^{\infty} P(\tau)h(t_1 - \tau)d\tau \int_{-\infty}^{\infty} P(\tau)h(t_2 - \tau)d\tau.
\end{aligned}
$$

An ideal photodetector generates pure current impulses for each received photon. For such a detector $h(t) = \delta(t)$, where $\delta(t)$ is the impulse function with the properties that $\delta(t) = 0, t \neq 0$ and $\int_{-\infty}^{\infty} \delta(t)dt = 1$. For this case, the mean photocurrent becomes

$$E[I(t)|P(.)] = \mathcal{R}P(t),$$

and its autocorrelation is

$$E[I(t_1)I(t_2)|P(.)] = e\mathcal{R}P(t_1)\delta(t_2 - t_1) + \mathcal{R}^2 P(t_1)P(t_2).$$

Removing the conditioning over $P(.)$ yields

$$E[I(t)] = \mathcal{R}E[P(t)], \tag{I.2}$$

and

$$E[I(t_1)I(t_2)] = e\mathcal{R}E[P(t_1)]\delta(t_2 - t_1) + \mathcal{R}^2 E[P(t_1)P(t_2)].$$

The autocovariance of $I(t)$ is then given as

$$
\begin{aligned}
L_I(t_1, t_2) &= E[I(t_1)I(t_2)] - E[I(t_1)]E[I(t_2)] \\
&= e\mathcal{R}E[P(t_1)]\delta(t_2 - t_1) + \mathcal{R}^2 L_P(t_1, t_2),
\end{aligned} \tag{I.3}
$$

where L_P denotes the autocovariance of $P(t)$.

I.1   Shot Noise

First let us consider the simple case when there is a constant power P incident on the receiver. For this case, $E[P(t)] = P$ and $L_P(\tau) = 0$, and (I.2) and (I.3) can be written as

$$E[I(t)] = \mathcal{R}P$$

and

$$L_I(\tau) = e\mathcal{R}P\delta(\tau),$$

where $\tau = t_2 - t_1$. The power spectral density of the photocurrent is the Fourier transform of the autocovariance and is given by

$$S_I(f) = \int_{-\infty}^{\infty} L_I(\tau)e^{-i2\pi f\tau}d\tau = e\mathcal{R}P.$$

Thus the shot noise current can be thought of as being a white noise process with a flat spectral density as given here. Within a receiver bandwidth of B_e, the shot noise power is given by

$$\sigma_{\text{shot}}^2 = \int_{-B_e}^{B_e} S_I(f)df = 2e\mathcal{R}PB_e.$$

Therefore, the photocurrent can be written as

$$I = \overline{I} + i_s,$$

where $\overline{I} = \mathcal{R}P$ and i_s is the shot noise current with zero mean and variance $e\mathcal{R}PB_e$.

I.2   Amplifier Noise

An optical amplifier introduces spontaneous emission noise to the signal in addition to providing gain. Consider a system with an optical preamplifier shown in Figure 4.7. The electric field at the input to the receiver may be written as

$$E(t) = \sqrt{2P} \cos(2\pi f_c t + \Phi) + N(t).$$

Here, P is the signal power, f_c is the carrier frequency, and Φ is a random phase uniformly distributed in $[0, 2\pi]$. $N(t)$ represents the amplifier spontaneous emission noise. For our purposes, we will assume that this is a zero-mean Gaussian noise process with autocorrelation $R_N(\tau)$.

The received power is given by

$$P(t) = E^2(t) = 2P \cos^2(2\pi f_c t + \Phi) + 2\sqrt{2P} N(t) \cos(2\pi f_c t + \Phi) + N^2(t).$$

The mean power is

$$E[P(t)] = P + R_N(0). \tag{I.4}$$

To calculate the autocovariance, note that since $N(t)$ is a Gaussian process,

$$E[N^2(t)N^2(t + \tau)] = R_N^2(0) + 2R_N^2(\tau)$$

using the moment formula (H.1). Using this fact, the autocovariance of $P(.)$ can be calculated to be

$$L_P(\tau) = 2R_N^2(\tau) + 4P R_N(\tau) \cos(2\pi f_c \tau) + \frac{P^2}{2} \cos(4\pi f_c \tau). \tag{I.5}$$

The corresponding spectral density is given by

$$
\begin{aligned}
S_P(f) &= \int_{-\infty}^{\infty} L_P(\tau) e^{-i2\pi f \tau} d\tau \\
&= 2S_N(f) * S_N(f) + 2P[S_N(f - f_c) + S_N(f + f_c)] \\
&\quad + \frac{P^2}{4}[\delta(f - 2f_c) + \delta(f + 2f_c)].
\end{aligned} \tag{I.6}
$$

The $*$ denotes the convolution operator, where $f(x) * g(x) = \int_{-\infty}^{\infty} f(u)g(x - u)du$.

After photodetection, the last term in (I.5) and (I.6) can be omitted because the $2f_c$ components will be filtered out.

In order to derive the noise powers, we return to (I.3) and substitute for $E[P(.)]$ and $L_P(.)$ from (I.4) and (I.6), respectively, to obtain

$$L_I(\tau) = e\mathcal{R}[P + R_N(0)]\delta(\tau) + \mathcal{R}^2[4P R_N(\tau) \cos(2\pi f_c \tau)] + \mathcal{R}^2[2R_N^2(\tau)].$$

We also have

$$S_I(f) = eR[P + R_N(0)] + R^2 2P[S_N(f - f_c) + S_N(f + f_c)]$$
$$+ R^2[2S_N(f) * S_N(f)]. \quad (I.7)$$

The first term on the right-hand side represents the shot noise terms due to the signal and the amplifier noise. The second term represents the signal-spontaneous beat noise, and the last term is the spontaneous-spontaneous beat noise. Note that we have so far assumed that the amplifier noise is Gaussian but with an arbitrary spectral shape $S_N(f)$. In practice, it is appropriate to assume that the amplifier noise is centered at f_c and is white over an optical bandwidth $B_o < 2f_c$, with

$$S_N(f) = \begin{cases} \frac{P_n(G-1)}{2}, & |f \pm f_c| < \frac{B_o}{2} \\ 0, & \text{otherwise.} \end{cases}$$

Here, P_n is given by $n_{sp}hf_c$, where n_{sp} is the spontaneous emission factor. Correspondingly, we have

$$R_N(0) = \int_{-\infty}^{\infty} S_N(f)df = P_n(G-1)B_o.$$

The spectral density of the photocurrent $S_I(f)$ from (I.7) is plotted in Figure I.1, assuming the preceding value for $S_N(f)$. Note that, as before, the shot noise is white, but the signal-spontaneous beat noise spectrum has a rectangular shape, and the spontaneous-spontaneous beat noise a triangular shape. Moreover, the incident optical power P is given by GP_i, where P_i is the input power to the amplifier.

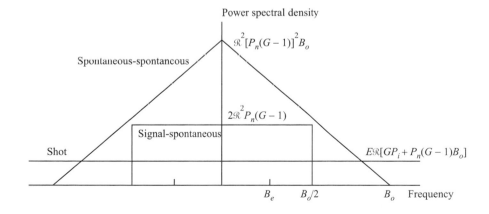

Figure I.1 Photocurrent spectral density.

Note that the photocurrent is passed through a low pass filter with bandwidth B_e. The noise power at the output of the filter is given by

$$\sigma^2 = \int_{-B_e}^{B_e} S_I(f)df = \sigma_{\text{shot}}^2 + \sigma_{\text{sig-spont}}^2 + \sigma_{\text{spont-spont}}^2,$$

where

$$\sigma_{\text{shot}}^2 = 2e\mathcal{R}[GP_i + P_n(G-1)B_o]B_e,$$

$$\sigma_{\text{sig-spont}}^2 = 4\mathcal{R}^2 GP_i P_n(G-1)B_e,$$

and

$$\sigma_{\text{spont-spont}}^2 = \mathcal{R}^2[P_n(G-1)]^2(2B_o - B_e)B_e.$$

▬ References

[BL90]   J. R. Barry and E. A. Lee. Performance of coherent optical receivers. *Proceedings of IEEE*, 78(8):1369–1394, Aug. 1990.

[RH90]   R. Ramaswami and P. A. Humblet. Amplifier induced crosstalk in multi-channel optical networks. *IEEE/OSA Journal on Lightwave Technology*, 8(12):1882–1896, Dec. 1990.

Bibliography

[Aar95] R. Aaron, editor. *IEEE Communications Magazine: Special Issue on Access to Broadband Services*, volume 33, Aug. 1995.

[AB98] M.-C. Amann and J. Buus. *Tunable Laser Diodes*. Artech House, Boston, 1998.

[ABC+94] A. Aggarwal, A. Bar-Noy, D. Coppersmith, R. Ramaswami, B. Schieber, and M. Sudan. Efficient routing and scheduling algorithms for optical networks. In *Proceedings of 5th Annual ACM-SIAM Symposium on Discrete Algorithms*, pages 412–423, Jan. 1994.

[Abo01] O. Aboul-Magd et al. *Constraint-Based LSP Setup Using LDP*. Internet Engineering Task Force, 2001. *draft-ietf-mpls-cr-ldp-05.txt*.

[ACKP97] V. Auletta, I. Caragiannis, C. Kaklamanis, and P. Persiano. Bandwidth allocation algorithms on tree-shaped all-optical networks with wavelength converters. In *Proceedings of the 4th International Colloquium on Structural Information and Communication Complexity*, 1997.

[AD93] G. P. Agrawal and N. K. Dutta. *Semiconductor Lasers*. Kluwer Academic Press, Boston, 1993.

[Agr95] G. P. Agrawal. *Nonlinear Fiber Optics*, 2nd edition. Academic Press, San Diego, CA, 1995.

[Agr97] G. P. Agrawal. *Fiber-Optic Communication Systems*. John Wiley, New York, 1997.

[AI93] M.-C. Amann and S. Illek. Tunable laser diodes utilising transverse tuning scheme. *IEEE/OSA Journal on Lightwave Technology*, 11(7):1168–1182, July 1993.

763

[AJY00]   C. Alaettinoglu, V. Jacobson, and H. Yu. Towards millisecond IGP convergence. In *North American Network Operators Group Fall Meeting*, 2000. See also IETF drafts *draft-alaettinoglu-isis-convergence-00.txt* and *draft-ietf-ospf-scalability-00.txt*.

[AKB+92]   R. C. Alferness, U. Koren, L. L. Buhl, B. I. Miller, M. G. Young, T. L. Koch, G. Raybon, and C. A. Burrus. Widely tunable InGaAsP/InP laser based on a vertical coupler filter with 57-nm tuning range. *Applied Physics Letters*, 60:3209–3211, 1992.

[AKW00]   R. C. Alferness, H. Kogelnik, and T. H. Wood. The evolution of optical systems: Optics everywhere. *Bell Labs Technical Journal*, 5(1):188–202, Jan.–March 2000.

[Ale93]   S. B. Alexander et al. A precompetitive consortium on wide-band all-optical networks. *IEEE/OSA Journal on Lightwave Technology*, 11:714–735, May/June 1993.

[Alf99]   R. Alferness, editor. *Bell Labs Technical Journal: Optical Networking*, volume 4, Jan.–Mar. 1999.

[Ame88]   American National Standards Institute. Z136.2. *Safe Use of Optical Fiber Communication Systems Utilizing Laser Diodes and LED Sources*, 1988.

[Ame97]   American National Standards Institute. X3.296. *Single-Byte Command Code Sets CONnection (SBCON) Architecture*, 1997.

[Ame98]   American National Standards Institute. X3.303. *Fibre Channel Physical and Signalling Interface-3 (FC-3)*, 1998.

[AMO93]   R. K. Ahuja, T. L. Magnanti, and J. B. Orlin. *Network Flows: Theory, Algorithms, and Applications*. Prentice Hall, Englewood Cliffs, NJ, 1993.

[Ams83]   S. Amstutz. Burst switching—an introduction. *IEEE Communications Magazine*, 21:36–42, Nov. 1983.

[And00]   P. A. Andrekson. High speed soliton transmission on installed fibers. In *OFC 2000 Technical Digest*, pages TuP2–1/229–231, 2000.

[AP94]   S. Aidarus and T. Plevyak, editors. *Telecommunications Network Management into the 21st Century*. IEEE Press, Los Alamitos, CA, 1994.

[AR01]   D. Awduche and Y. Rekhter. Multiprotocol lambda switching: Combining MPLS traffic engineering control with optical crossconnects. *IEEE Communications Magazine*, 39(4):111–116, Mar. 2001.

[AS92]   A. S. Acampora and S. I. A. Shah. Multihop lightwave networks: A comparison of store-and-forward and hot-potato routing. *IEEE Transactions on Communications*, 40(6):1082–1090, June 1992.

[ATM96]   ATM Forum. *Private Network-Network Interface Specification: Version 1.0*, 1996.

[AY86] Y. Arakawa and A. Yariv. Quantum well lasers—gain, spectra, dynamics. *IEEE Journal of Quantum Electronics*, 22(9):1887–1899, Sept. 1986.

[BA94] F. Bruyère and O. Audouin. Assessment of system penalties induced by polarization mode dispersion in a 5 Gb/s optically amplified transoceanic link. *IEEE Photonics Technology Letters*, 6(3):443–445, March 1994.

[Bac96] E.-J. Bachus et al. Coherent optical systems implemented for business traffic routing and access: The RACE COBRA project. *IEEE/OSA JLT/JSAC Special Issue on Multiwavelength Optical Technology and Networks*, 14(6):1309–1319, June 1996.

[Bak01] B. Bakhshi et al. 1 Tb/s (101 × 10 Gb/s) transmission over transpacific distance using 28 nm C-band EDFAs. In *OFC 2001 Technical Digest*, pages PD21/1–3, 2001.

[Bal96] K. Bala et al. WDM network economics. In *Proceedings of National Fiber Optic Engineers Conference*, pages 163–174, 1996.

[Bar64] P. Baran. On distributed communications networks. *IEEE Transactions on Communications*, pages 1–9, March 1964.

[Bar96a] R. A. Barry, editor. *IEEE Network: Special Issue on Optical Networks*, volume 10, Nov. 1996.

[Bar96b] R. A. Barry et al. All-optical network consortium—ultrafast TDM networks. *IEEE JSAC/JLT Special Issue on Optical Networks*, 14(5):999–1013, June 1996.

[Bar00] S. Baroni et al. Analysis and design of backbone architecture alternatives for IP optical networking. *IEEE Journal of Selected Areas in Communications*, 18(10):1980–1994, Oct. 2000.

[Bat98] R. Batchellor. Optical layer protection: Benefits and implementation. In *Proceedings of National Fiber Optic Engineers Conference*, 1998.

[BC90] P. N. Butcher and D. Cotter. *The Elements of Nonlinear Optics*, volume 9 of *Cambridge Studies in Modern Optics*. Cambridge University Press, Cambridge, 1990.

[BCM+92] D. J. Blumenthal, K. Y. Chen, J. Ma, R. J. Feuerstein, and J. R. Sauer. Demonstration of a deflection routing 2 × 2 photonic switch for computer interconnects. *IEEE Photonics Technology Letters*, 4(2):169–173, Feb. 1992.

[BDG95] C. Baransel, W. Dobosiewicz, and P. Gburzynski. Routing in multihop packet switching networks: Gb/s challange. *IEEE Network*, pages 38–61, May/June 1995.

[BDN90] K. J. Blow, N. J. Doran, and B. P. Nelson. Demonstration of the nonlinear fibre loop mirror as an ultrafast all-optical demultiplexer. *Electronics Letters*, 26(14):962–964, July 1990.

[Ben65]   V. E. Beneš. *Mathematical Theory of Connecting Networks and Telephone Traffic.* Academic Press, New York, 1965.

[Ben96a]   A. F. Benner. *Fibre Channel.* McGraw-Hill, New York, 1996.

[Ben96b]   I. Bennion et al. UV-written in-fibre Bragg gratings. *Optical Quantum Electronics,* 28(2):93–135, Feb. 1996.

[Ber76]   C. Berge. *Graphs and Hypergraphs.* North Holland, Amsterdam, 1976.

[Ber96a]   J.-C. Bermond et al. Efficient collective communication in optical networks. In *23rd International Colloquium on Automata, Languages and Programming—ICALP '96,* Paderborn, Germany, pages 574–585, 1996.

[Ber96b]   L. Berthelon et al. Experimental assessment of node cascadability in a reconfigurable survivable WDM ring network. In *Proceedings of Topical Meeting on Broadband Optical Networks,* 1996.

[Ber96c]   L. Berthelon et al. Over 40,000 km across a layered network by recirculation through an experimental WDM ring network. In *Proceedings of European Conference on Optical Communication,* 1996.

[BFP93]   A. Bononi, F. Forghieri, and P. R. Prucnal. Synchronisation in ultrafast packet switching transparent optical networks. *Electronics Letters,* 29(10):872–873, May 1993.

[BG92]   D. Bertsekas and R. G. Gallager. *Data Networks.* Prentice Hall, Englewood Cliffs, NJ, 1992.

[BG95]   D. Bienstock and O. Gunluk. Computational experience with a difficult mixed-integer multicommodity flow problem. *Mathematical Programming,* 68:213–237, 1995.

[BH96]   R. A. Barry and P. A. Humblet. Models of blocking probability in all-optical networks with and without wavelength changers. *IEEE JSAC/JLT Special Issue on Optical Networks,* 14(5):858–867, June 1996.

[Bha99]   V. K. Bhagavath. Emerging high-speed xDSL access services: Architectures, issues, insights, and implications. *IEEE Communications Magazine,* 37(11):106–114, Nov. 1999.

[Big90]   N. Biggs. Some heuristics for graph colouring. In R. Nelson and R. J. Wilson, editors, *Graph Colourings,* Pitman Research Notes in Mathematics Series, pages 87–96. Longman Scientific & Technical, Burnt Mill, Harlow, Essex, UK, 1990.

[Big01]   S. Bigo et al. 10.2 Tb/s (256×42.7 Gbit/s PDM/WDM) transmission over 100 km TeraLight fiber with 1.28bit/s/Hz spectral efficiency. In *OFC 2001 Technical Digest,* pages PD25/1–3, 2001.

[BIPT98]   D. J. Blumenthal, T. Ikegami, P. R. Prucnal, and L. Thylen, editors. *IEEE/OSA Journal of Lightwave Technology: Special Issue on Photonic Packet Switching Technologies, Techniques and Systems*, volume 16, Dec. 1998.

[Bir96]   A. Birman. Computing approximate blocking probabilities for a class of optical networks. *IEEE JSAC/JLT Special Issue on Optical Networks*, 14(5):852–857, June 1996.

[BK95]   A. Birman and A. Kershenbaum. Routing and wavelength assignment methods in single-hop all-optical networks with blocking. In *Proceedings of IEEE Infocom*, pages 431–438, 1995.

[BKLW00]   W. F. Brinkman, T. L. Koch, D. V. Lang, and D. W. Wilt. The lasers behind the communications revolution. *Bell Labs Technical Journal*, 5(1):150–167, Jan.–March 2000.

[BL90]   J. R. Barry and E. A. Lee. Performance of coherent optical receivers. *Proceedings of IEEE*, 78(8):1369–1394, Aug. 1990.

[Bla95]   U. Black. *Network Management Standards*. McGraw-Hill, New York, 1995.

[BM00]   R. Berry and E. Modiano. Reducing electronic multiplexing costs in SONET/WDM rings with dynamically changing traffic. *IEEE Journal of Selected Areas in Communications*, 18:1961–1971, 2000.

[Bor95]   F. Borgonovo. Deflection routing. In M. Steenstrup, editor, *Routing in Communication Networks*. Prentice Hall, Englewood Cliffs, NJ, 1995.

[BOS99]   P. C. Becker, N. A. Olsson, and J. R. Simpson. *Erbium-Doped Fiber Amplifiers: Fundamentals and Technology*. Academic Press, San Diego, CA, 1999.

[BP96]   A. Bononi and P. R. Prucnal. Analytical evaluation of improved access techniques in deflection routing networks. *IEEE/ACM Transactions on Networking*, 4(5):726–730, Oct. 1996.

[BPS94]   D. J. Blumenthal, P. R. Prucnal, and J. R. Sauer. Photonic packet switches: Architectures and experimental implementations. *Proceedings of IEEE*, 82:1650–1667, Nov. 1994.

[Bra89]   C. A. Brackett, editor. *IEEE Communications Magazine: Special Issue on Lightwave Systems and Components*, volume 27, Oct. 1989.

[Bre01]   J. F. Brennan III et al. Dispersion and dispersion-slope correction with a fiber Bragg grating over the full C-band. In *OFC 2001 Technical Digest*, pages PD12/1–3, 2001.

[Buc95]   J. A. Buck. *Fundamentals of Optical Fibers*. John Wiley, New York, 1995.

[Bur86]   W. E. Burr. The FDDI optical data link. *IEEE Communications Magazine*, 24(5):18–23, May 1986.

[Bur94]   M. Burzio et al. Optical cell synchronization in an ATM optical switch. In *Proceedings of European Conference on Optical Communication*, pages 581–584, 1994.

[BW99]   M. Born and E. Wolf. *Principles of Optics: Electromagnetic Theory of Propagation, Diffraction and Interference of Light*. Cambridge University Press, 1999.

[BZB⁺97] R. Bradon, L. Zhang, S. Berson, S. Herzog, and S. Jamin. *Resource Reservation Protocol—Version 1 Functional Specification*. Internet Engineering Task Force, Sept. 1997.

[Cah98]   R. Cahn. *Wide Area Network Design: Concepts and Tools for Optimization*. Morgan Kaufmann, San Francisco, 1998.

[Cai01]   J.-X. Cai et al. 2.4 Tb/s (120 × 20 Gb/s) transmission over transoceanic distance with optimum FEC overhead and 48% spectral efficiency. In *OFC 2001 Technical Digest*, pages PD20/1–3, 2001.

[CDdM90] F. Curti, B. Daino, G. de Marchis, and F. Matera. Statistical treatment of the evolution of the principal states of polarization in single-mode fibers. *IEEE/OSA Journal on Lightwave Technology*, 8(8):1162–1166, Aug. 1990.

[CdLS92] S. A. Calta, S. A. deVeer, E. Loizides, and R. N. Strangewayes. Enterprise systems connection (ESCON) architecture—system overview. *IBM Journal of Research and Development*, 36(4):535–551, July 1992.

[CEG⁺96] G. K. Chang, G. Ellinas, J. K. Gamelin, M. Z. Iqbal, and C. A. Brackett. Multiwavelength reconfigurable WDM/ATM/SONET network testbed. *IEEE/OSA JLT/JSAC Special Issue on Multiwavelength Optical Technology and Networks*, 14(6):1320–1340, June 1996.

[CGK92]   I. Chlamtac, A. Ganz, and G. Karmi. Lightpath communications: An approach to high-bandwidth optical WAN's. *IEEE Transactions on Communications*, 40(7):1171–1182, July 1992.

[CGK93]   I. Chlamtac, A. Ganz, and G. Karmi. Lightnets: Topologies for high-speed optical networks. *IEEE/OSA Journal on Lightwave Technology*, 11(5/6):951–961, May/June 1993.

[CGS93]   I. Cidon, I. S. Gopal, and A. Segall. Connection establishment in high-speed networks. *IEEE/ACM Transactions on Networking*, 1(4):469–482, Aug. 1993.

[CH00]   C. J. Chang-Hasnain. Tunable VCSEL. *IEEE Journal of Selected Topics in Quantum Electronics*, 6(6):978–987, Nov./Dec. 2000.

[Cha94]   G. K. Chang et al. Experimental demonstration of a reconfigurable WDM/ATM/SONET multiwavelength network testbed. In *OFC'94 Technical Digest*, 1994. Postdeadline paper PD9.

[Chb98] M. W. Chbat et al. Towards wide-scale all-optical networking: The ACTS optical pan-European network (OPEN) project. *IEEE JSAC: Special Issue on High-Capacity Optical Transport Networks*, 16(7):1226–1244, Sept. 1998.

[Che90] K.-W. Cheung. Acoustooptic tunable filters in narrowband WDM networks: System issues and network applications. *IEEE Journal of Selected Areas in Communications*, 8(6):1015–1025, Aug. 1990.

[CHI+92] M. W. Chbat, B. Hong, M. N. Islam, C. E. Soccolich, and P. R. Prucnal. Ultrafast soliton-trapping AND gate. *IEEE/OSA Journal on Lightwave Technology*, 10(12):2011–2016, Dec. 1992.

[Chi97] D. Chiaroni et al. New 10 Gb/s 3R NRZ optical regenerative interface based on semiconductor optical amplifiers for all-optical networks. In *Proceedings of European Conference on Optical Communication*, pages 41–43, 1997. Postdeadline paper.

[CHK+96] R. L. Cruz, G. R. Hill, A. L. Kellner, R. Ramaswami, and G. H. Sasaki, editors. *IEEE JSAC/JLT Special Issue on Optical Networks*, volume 14, June 1996.

[Chl96] I. Chlamtac et al. CORD: Contention resolution by delay lines. *IEEE JSAC/JLT Special Issue on Optical Networks*, 14(5):1014–1029, June 1996.

[Chr84] A. R. Chraplyvy. Optical power limits in multichannel wavelength-division-multiplexed systems due to stimulated Raman scattering. *Electronics Letters*, 20:58, 1984.

[Chr90] A. R. Chraplyvy. Limitations on lightwave communications imposed by optical-fiber nonlinearities. *IEEE/OSA Journal on Lightwave Technology*, 8(10):1548–1557, Oct. 1990.

[CK94] R. J. Campbell and R. Kashyap. The properties and applications of photosensitive germanosilicate fibre. *International Journal of Optoelectronics*, 9(1):33–57, 1994.

[Cla99] T. Clark. *Designing Storage-Area Networks*. Addison-Wesley, Reading, MA, 1999.

[Clc94] B. Clesca et al. Gain flatness comparison between erbium-doped fluoride and silica fiber amplifiers with wavelength-multiplexed signals. *IEEE Photonics Technology Letters*, 6(4):509–512, April 1994.

[CLM97] D. Cotter, J. K. Lucek, and D. D. Marcenac. Ultra-high bit-rate networking: From the transcontinental backbone to the desktop. *IEEE Communications Magazine*, 35(4):90–95, April 1997.

[Clo53] C. Clos. A study of nonblocking switching networks. *Bell System Technical Journal*, 32:406–424, March 1953.

[CM00] A. L. Chiu and E. H. Modiano. Traffic grooming algorithms for reducing electronic multiplexing costs in WDM ring networks. *IEEE/OSA Journal on Lightwave Technology*, 18:2–12, 2000.

[CMLF00] T. Cinkler, D. Marx, C. P. Larsen, and D. Fogaras. Heuristic algorithms for joint configuration of the optical and electrical layer in multi-hop wavelength routing networks. In *Proceedings of IEEE Infocom*, 2000.

[CNW90] N. K. Cheung, G. Nosu, and G. Winzer, editors. *IEEE JSAC: Special Issue on Dense WDM Networks*, volume 8, Aug. 1990.

[CO99] T. M. Chen and T. H. Oh. Reliable services in MPLS. *IEEE Communications Magazine*, 37(12):58–62, Dec. 1999.

[Col00] L. A. Coldren. Monolithic tunable diode lasers. *IEEE Journal of Selected Topics in Quantum Electronics*, 6(6):988–999, Nov./Dec. 2000.

[Com00] D. E. Comer. *Internetworking with TCP/IP: Vol. I: Principles, Protocols and Architecture*. Prentice Hall, Englewood Cliffs, NJ, 2000.

[Coo00] H. K. Cook. The economics of metro DWDM deployment. In *Proceedings of National Fiber Optic Engineers Conference*, 2000.

[Cot95] D. Cotter et al. Self-routing of 100 Gbit/s packets using 6 bit "keyword" address recognition. *Electronics Letters*, 31(25):2201–2202, Dec. 1995.

[CS83] J. S. Cook and O. I. Szentisi. North American field trials and early applications in telephony. *IEEE JSAC*, 1:393–397, 1983.

[CSH00] G. K. Chang, K. I. Sato, and D. K. Hunter, editors. *IEEE/OSA Journal of Lightwave Technology: Special Issue on Optical Networks*, volume 18, 2000.

[CT91] T. M. Cover and J. A. Thomas. *Elements of Information Theory*. Wiley, New York, 1991.

[CT98] A. R. Chraplyvy and R. W. Tkach. Terabit/second transmission experiments. *IEEE Journal of Quantum Electronics*, 34(11):2103–2108, 1998.

[Dan95] S. L. Danielsen et al. Detailed noise statistics for an optically preamplified direct detection receiver. *IEEE/OSA Journal on Lightwave Technology*, 13(5):977–981, 1995.

[Dan97] S. L. Danielsen et al. WDM packet switch architectures and analysis of the influence of tuneable wavelength converters on the performance. *IEEE/OSA Journal on Lightwave Technology*, 15(2):219–227, Feb. 1997.

[Dan98] S. L. Danielsen et al. Analysis of a WDM packet switch with improved performance under bursty traffic conditions due to tuneable wavelength converters. *IEEE/OSA Journal on Lightwave Technology*, 16(5):729–735, May 1998.

[Dar87] T. E. Darcie. Subcarrier multiplexing for multiple-access lightwave networks. *IEEE/OSA Journal on Lightwave Technology*, LT-5:1103–1110, 1987.

[DEK91] C. Dragone, C. A. Edwards, and R. C. Kistler. Integrated optics $N \times N$ multiplexer on silicon. *IEEE Photonics Technology Letters*, 3:896–899, Oct. 1991.

[Dem99]   P. Demeester et al. Resilience in multilayer networks. *IEEE Communications Magazine*, 37(8):70–77, Aug. 1999.

[Der95]   F. Derr. Design of an 8 × 8 optical cross-connect switch: Results on subsystems and first measurements. In *ECOC'95 Optical Networking Workshop*, 1995. Paper S2.2.

[Des94]   E. Desurvire. *Erbium-Doped Fiber Amplifiers: Principles and Applications*. John Wiley, New York, 1994.

[Dij59]   E. W. Dijkstra. A note on two problems in connexion with graphs. *Numerical Mathematics*, pages 269–271, 1959.

[DL00]   S. S. Dixit and P. J. Lin, editors. *IEEE Communications Magazine: Optical Networks Come of Age*, volume 38, Feb. 2000.

[DMJ+96]   T. Durhuus, B. Mikkelsen, C. Joergensen, S. Lykke Danielsen, and K. E. Stubkjaer. All optical wavelength conversion by semiconductor optical amplifiers. *IEEE/OSA JLT/JSAC Special Issue on Multiwavelength Optical Technology and Networks*, 14(6):942–954, June 1996.

[Dos99]   B. T. Doshi et al. Optical network design and restoration. *Bell Labs Technical Journal*, 4(1):58–84, Jan. March 1999.

[Dos01]   B. Doshi et al. Ultra-long-reach systems, optical transparency and networks. In *OFC 2001 Technical Digest*, 2001. Paper TuG4.

[dP95]   M. de Prycker. *Asynchronous Transfer Mode: Solution for Broadband ISDN*. Prentice Hall, London, 1995.

[DR00]   B. S. Davie and Y. Rekhter. *MPLS Technology and Applications*. Morgan Kaufmann, San Francisco, 2000.

[Dra89]   C. Dragone. Efficient $n \times n$ star couplers using Fourier optics. *IEEE/OSA Journal on Lightwave Technology*, 7(3):479–489, March 1989.

[DSGW00]   A. Dwiwedi, M. Sharma, J. M. Grochocinski, and R. E. Wagner. Value of reach extension in long-distance networks. In *Proceedings of National Fiber Optic Engineers Conference*, 2000.

[DW88]   N. J. Doran and D. Wood. Nonlinear-optical loop mirror. *Optics Letters*, 13(1):56–58, Jan. 1988.

[dW90]   D. de Werra. Heuristics for graph coloring. In G. Tinhofer, E. Mayr, and H. Noltemeier, editors, *Computational Graph Theory*, volume 7 of *Computing, Supplement*, pages 191–208. Springer-Verlag, Berlin, 1990.

[DWY99]   P. Demeester, T.-H. Wu, and N. Yoshikai, editors. *IEEE Communications Magazine: Special Issue on Survivable Communication Networks*, volume 37, Aug. 1999.

[DYJ00]   S. S. Dixit and A. Yla-Jaaski, editors. *IEEE Communications Magazine: WDM Optical Networks: A Reality Check*, volume 38, Mar. 2000.

[Eis92]   M. Eiselt. Optical loop mirror with semiconductor laser amplifier. *Electronics Letters*, 28(16):1505–1506, July 1992.

[EM00]   J. M. H. Elmirghani and H. T. Mouftah. All-optical wavelength conversion technologies and applications in DWDM networks. *IEEE Communications Magazine*, 38(3):86–92, Mar. 2000.

[ENW96]   A. Erramilli, O. Narayan, and W. Willinger. Experimental queueing analysis with long-range dependent packet traffic. *IEEE/ACM Transactions on Networking*, 4(2):209–223, Apr. 1996.

[Epw95]   R. E. Epworth. Optical transmission system. U.S. Patent 5463487, 1995.

[ES92]   J. C. Elliott and M. W. Sachs. The IBM enterprise systems connection architecture. *IBM Journal of Research and Development*, 36(4):577–591, July 1992.

[FBP95]   F. Forghieri, A. Bononi, and P. R. Prucnal. Analysis and comparison of hot-potato and single-buffer deflection routing in very high bit rate optical mesh networks. *IEEE Transactions on Communications*, 43(1):88–98, Jan. 1995.

[FDW01]   D. A. Francis, S. P. Dijaili, and J. D. Walker. A single-chip linear optical amplifier. In *OFC 2001 Technical Digest*, pages PD13/1–3, 2001.

[FGO+96]   M. Fujiwara, M. S. Goodman, M. J. O'Mahony, O. K. Tonguez, and A. E. Willner, editors. *IEEE/OSA JLT/JSAC Special Issue on Multiwavelength Optical Technology and Networks*, volume 14, June 1996.

[FHHH90]   M. E. Fermann, F. Haberl, M. Hofer, and H. Hochreiter. Nonlinear amplifying loop mirror. *Optics Letters*, 15(13):752–754, July 1990.

[FHJ+98]   R. D. Feldman, E. E. Harstead, S. Jiang, T. H. Wood, and M. Zirngibl. An evaluation of architectures incorporating wavelength division multiplexing for broad-band fiber access. *IEEE/OSA Journal on Lightwave Technology*, 16(9):1546–1559, Sept. 1998.

[Flo00]   F. A. Flood. L-band erbium-doped fiber amplifiers. In *OFC 2000 Technical Digest*, pages WG1-1–WG1-4, 2000.

[FNS+92]   A. Frank, T. Nishizeki, N. Saito, H. Suzuki, and E. Tardos. Algorithms for routing around a rectangle. *Discrete Applied Mathematics*, 40:363–378, 1992.

[Fou00]   J. E. Fouquet. Compact optical cross-connect switch based on total internal reflection in a fluid-containing planar lightwave circuit. In *OFC 2000 Technical Digest*, pages TuM1-1–TuM1-4, 2000.

[Fra93]   A. G. Fraser. Banquet speech. In *Proceedings of Workshop on High-Performance Communication Subsystems*, Williamsburg, VA, Sept. 1993.

[Fra98a]   P. W. France, editor. *BT Technology Journal—Special Issue on Local Access Technologies*, volume 16, Oct. 1998.

[Fra98b]   T. Franck et al. Duobinary transmitter with low intersymbol interference. *IEEE Photonics Technology Letters*, 10:597–599, 1998.

[Fri94]   N. J. Frigo et al. A wavelength-division-multiplexed passive optical network with cost-shared components. *IEEE Photonics Technology Letters*, 6(11):1365–1367, 1994.

[FRI96]   N. J. Frigo, K. C. Reichmann, and P. P. Iannone. WDM passive optical networks: A robust and flexible infrastructure for local access. In *Proceedings of International Workshop on Photonic Networks and Technologies*, pages 201–212, 1996.

[FSA98]   *Full Services Access Network Requirements Specification*, 1998. Available on the Web at *www.fsanet.net*.

[FTC95]   F. Forghieri, R. W. Tkach, and A. R. Chraplyvy. WDM systems with unequally spaced channels. *IEEE/OSA Journal on Lightwave Technology*, 13(5):889–897, May 1995.

[FTCV96]   F. Forghieri, R. W. Tkach, A. R. Chraplyvy, and A. M. Vengsarkar. Dispersion compensating fiber: Is there merit in the figure of merit? In *OFC'96 Technical Digest*, pages 255–257, 1996.

[Fuk01]   K. Fukuchi et al. 10.92 Tb/s (273 × 40 Gb/s) triple-band/ultra-dense WDM optical-repeatered transmission experiment. In *OFC 2001 Technical Digest*, pages PD24/1–3, 2001.

[FV88]   G. J. Foschini and G. Vannucci. Using spread spectrum in a high-capacity fiber-optic local network. *IEEE/OSA Journal on Lightwave Technology*, 6(3):370–379, March 1988.

[Gal99]   R. G. Gallager. *Discrete Stochastic Processes*. Kluwer, Boston, 1999.

[Gam98]   P. Gambini et al. Transparent optical packet switching: Network architecture and demonstrators in the KEOPS project. *IEEE JSAC: Special Issue on High-Capacity Optical Transport Networks*, 16(7):1245–1259, Sept. 1998.

[Gar98]   L. D. Garrett et al. The MONET New Jersey demonstration network. *IEEE JSAC: Special Issue on High-Capacity Optical Transport Networks*, 16(7):1199–1219, Sept. 1998.

[GEE94]   E. L. Goldstein, L. Eskildsen, and A. F. Elrefaie. Performance implications of component crosstalk in transparent lightwave networks. *IEEE Photonics Technology Letters*, 6(5):657–670, May 1994.

[GG93]   A. G. Greenberg and J. Goodman. Sharp approximate models of deflection routing in mesh networks. *IEEE Transactions on Communications*, 41(1):210–223, Jan. 1993.

[GH92]   A. G. Greenberg and B. Hajek. Deflection routing in hypercube networks. *IEEE Transactions on Communications*, 40(6):1070–1081, June 1992.

[GJ79]   M. R. Garey and D. S. Johnson. *Computers and Intractability—A Guide to the Theory of NP Completeness*. W. H. Freeman, San Francisco, 1979.

[GJR96] P. E. Green, F. J. Janniello, and R. Ramaswami. Multichannel protocol-transparent WDM distance extension using remodulation. *IEEE JSAC/JLT Special Issue on Optical Networks*, 14(6):962–967, June 1996.

[GK97]   O. Gerstel and S. Kutten. Dynamic wavelengh allocation in all-optical ring networks. In *Proceedings of IEEE International Conference on Communication*, 1997.

[Gla00] A. M. Glass et al. Advances in fiber optics. *Bell Labs Technical Journal*, 5(1):168–187, Jan.–March 2000.

[GLM$^+$00] O. Gerstel, B. Li, A. McGuire, G. Rouskas, K. Sivalingam, and Z. Zhang, editors. *IEEE JSAC: Special Issue on Protocols and Architectures for Next-Generation Optical Networks*, Oct. 2000.

[Glo71] D. Gloge. Weakly guiding fibers. *Applied Optics*, 10:2252–2258, 1971.

[GLS99] O. Gerstel, P. Lin, and G. Sasaki. Combined WDM and SONET network design. In *Proceedings of IEEE Infocom*, 1999.

[Gna96] A. H. Gnauck et al. One terabit/s transmission experiment. In *OFC'96 Technical Digest*, 1996. Postdeadline paper PD20.

[Gol00] E. A. Golovchenko et al. Modeling of transoceanic fiber-optic WDM communication systems. *IEEE Journal of Selected Topics in Quantum Electronics*, 6:337–347, 2000.

[Gor00] W. J. Goralski. *SONET*. McGraw-Hill, New York, 2000.

[GR96]   O. Gerstel and R. Ramaswami. Multiwavelength optical network architectures and protection schemes. In *Proceedings of Tirrenia Workshop on Optical Networks*, pages 42–51, 1996.

[GR99]   O. Gerstel and R. Ramaswami. Upgrading SONET rings with WDM instead of TDM: An economic analysis. In *OFC'99 Technical Digest*, pages 75–77, 1999.

[GR00a] O. Gerstel and R. Ramaswami. Optical layer survivability—a services perspective. *IEEE Communications Magazine*, 38(3):104–113, March 2000.

[GR00b] O. Gerstel and R. Ramaswami. Optical layer survivability: An implementation perspective. *IEEE JSAC Special Issue on Optical Networks*, 18(10):1885–1899, Oct. 2000.

[GR00c] J. Gruber and R. Ramaswami. Moving towards all-optical networks. *Lightwave*, 34(8):40–49, Dec. 2000.

[GR00d]   J. Gruber and R. Ramaswami. Towards agile all-optical networks. *Lightwave*, Dec. 2000.

[Gre93]   P. E. Green. *Fiber-Optic Networks*. Prentice Hall, Englewood Cliffs, NJ, 1993.

[GRL00]   J. Gruber, P. Roorda, and F. Lalonde. The photonic switch crossconnect (PSX)—its role in evolving optical networks. In *Proceedings of National Fiber Optic Engineers Conference*, pages 678–689, 2000.

[GRS97a] O. Gerstel, R. Ramaswami, and G. H. Sasaki. Benefits of limited wavelength conversion in WDM ring networks. In *OFC'97 Technical Digest*, pages 119–120, 1997.

[GRS97b] O. Gerstel, R. Ramaswami, and G. H. Sasaki. Fault tolerant WDM rings with limited wavelength conversion. In *Proceedings of IEEE Infocom*, pages 508–516, 1997.

[GRS98]   O. Gerstel, R. Ramaswami, and G. H. Sasaki. Cost effective traffic grooming in WDM rings. In *Proceedings of IEEE Infocom*, 1998.

[Gru95]   S. G. Grubb et al. High power 1.48 μm cascaded Raman laser in germanosilicate fibers. In *Optical Amplifiers and Applications*, page 197, 1995.

[GRW00]   O. Gerstel, R. Ramaswami, and W-K. Wang. Making use of a two stage multiplexing scheme in a WDM network. In *OFC 2000 Technical Digest*, pages ThD1-1–ThD1-3, 2000.

[GSKR99] O. Gerstel, G. H. Sasaki, S. Kutten, and R. Ramaswami. Worst-case analysis of dynamic wavelength allocation in optical networks. *IEEE/ACM Transactions on Networking*, 7(6):833–846, Dec. 1999.

[GSP94]   I. Glesk, J. P. Sokoloff, and P. R. Prucnal. All-optical address recognition and self-routing in a 250 Gb/s packet-switched network. *Electronics Letters*, 30(16):1322–1323, Aug. 1994.

[Gui98]   C. Guillemot et al. Transparent optical packet switching: The European ACTS KEOPS project approach. *IEEE/OSA Journal on Lightwave Technology*, 16(12):2117–2134, Dec. 1998.

[Gui00]   K. Guild et al. Cascading and routing 14 optical packet switches. In *Proceedings of European Conference on Optical Communication*, 2000.

[Gun97a] P. Gunning et al. 40 Gbit/s optical TDMA LAN over 300m of installed blown fibre. In *Proceedings of European Conference on Optical Communication*, volume 4, pages 61–64, Sept. 1997.

[Gun97b] P. Gunning et al. Optical-TDMA LAN incorporating packaged integrated Mach-Zehnder interferometer channel selector. *Electronics Letters*, 33(16):1404–1406, July 1997.

[GW94]   A. Ganz and X. Wang. Efficient algorithm for virtual topology design in multihop lightwave networks. *IEEE/ACM Transactions on Networking*, 2(3):217–225, June 1994.

[HA00]   D. K. Hunter and I. Andonovic. Approaches to optical Internet packet switching. *IEEE Communications Magazine*, 38(9):116–122, Sept. 2000.

[Hal97]   K. L. Hall. All-optical buffers for high-speed slotted TDM networks. In *IEEE/LEOS Summer Topical Meeting on Advanced Semiconductor Lasers and Applications*, page 15, 1997.

[Ham50]   R. W. Hamming. Error detecting and error correcting codes. *Bell System Technical Journal*, 29, 1950.

[Har00]   J. S. Harris. Tunable long-wavelength vertical-cavity lasers: The engine of next generation optical networks? *IEEE Journal of Selected Topics in Quantum Electronics*, 6(6):1145–1160, Nov./Dec. 2000.

[HC93]   B. Hajek and R. L. Cruz. On the average delay for routing subject to independent deflections. *IEEE Transactions on Information Theory*, 39(1):84–91, Jan. 1993.

[HCA98]   D. K. Hunter, M. C. Chia, and I. Andonovic. Buffering in optical packet switches. *IEEE/OSA Journal on Lightwave Technology*, 16(12):2081–2094, Dec. 1998.

[HD97]   G. R. Hill and P. Demeester, editors. *IEEE Communications Magazine: Special Issue on Photonic Networks in Europe*, volume 35, April 1997.

[Hec98]   J. Hecht. *Understanding Fiber Optics*. Prentice Hall, Englewood Cliffs, NJ, 1998.

[Hec99]   J. Hecht. *City of Light: The Story of Fiber Optics*. Oxford University Press, New York, 1999.

[HFKV96]   F. Heismann, M. T. Fatehi, S. K. Korotky, and J. J. Veselka. Signal tracking and performance monitoring in multi-wavelength optical networks. In *Proceedings of European Conference on Optical Communication*, pages 3.47–3.50, 1996.

[HH90]   P. A. Humblet and W. M. Hamdy. Crosstalk analysis and filter optimization of single- and double-cavity Fabry-Perot filters. *IEEE Journal of Selected Areas in Communications*, 8(6):1095–1107, Aug. 1990.

[HH96]   A. M. Hill and A. J. N. Houghton. Optical networking in the European ACTS programme. In *OFC'96 Technical Digest*, pages 238–239, San Jose, CA, Feb. 1996.

[Hil85]   W. D. Hillis. *The Connection Machine*. MIT Press, Cambridge, MA, 1985.

[Hil87]   W. D. Hillis. The connection machine. *Scientific American*, 256(6), June 1987.

[Hil93]   G. R. Hill et al. A transport network layer based on optical network elements. *IEEE/OSA Journal on Lightwave Technology*, 11:667–679, May/June 1993.

[HJKM78] K. O. Hill, D. C. Johnson, B. S. Kawasaki, and R. I. MacDonald. CW three-wave mixing in single-mode optical fibers. *Journal of Applied Physics*, 49(10):5098–5106, Oct. 1978.

[HK88] M. G. Hluchyj and M. J. Karol. Queuing in high-performance packet switching. *IEEE JSAC*, 6(9):1587–1597, Dec. 1988.

[HK97] Y. Hamazumi and M. Koga. Transmission capacity of optical path overhead transfer scheme using pilot tone for optical path networks. *IEEE/OSA Journal on Lightwave Technology*, 15(12):2197–2205, Dec. 1997.

[HMY98] K. Habara, T. Matsunaga, and K.-I. Yukimatsu. Large-scale WDM star-based photonic ATM switches. *IEEE/OSA Journal on Lightwave Technology*, 16(12):2191–2201, Dec. 1998.

[HR98] K. L. Hall and K. T. Rauschenbach. All-optical buffering of 40 Gb/s data packets. *IEEE Photonics Technology Letters*, 10(3):442–444, Mar. 1998.

[HSS98] A. M. Hill, A. A. M. Saleh, and K. Sato, editors. *IEEE JSAC: Special Issue on High-Capacity Optical Transport Networks*, volume 16, Sept. 1998.

[Hui01] R. Hui et al. 10 Gb/s SCM system using optical single side-band modulation. In *OFC 2001 Technical Digest*, pages MM4/1–4, 2001.

[Hun99] D. K. Hunter et al. WASPNET—a wavelength switched packet network. *IEEE Communications Magazine*, 37(3):120–129, Mar. 1999.

[HYCG00] G. Hjalmtysson, J. Yates, S. Chaudhuri, and A. Greenberg. Smart routers—simple optics: An architecture for the optical Internet. *IEEE/OSA Journal on Lightwave Technology*, 18(12):1880–1891, 2000.

[IFD95] P. P. Iannone, N. J. Frigo, and T. E. Darcie. WDM passive optical network architecture with bidirectional optical spectral slicing. In *OFC'95 Technical Digest*, pages 51–53, 1995. Paper TuK2.

[Int93] International Electrotechnical Commission. *60825-1: Safety of Laser Products—Part 1: Equipment Classification, Requirements and User's Guide*, 1993.

[Int00] International Electrotechnical Commission. *60825-2: Safety of Laser Products—Part 2: Safety of Optical Fiber Communication Systems*, 2000.

[IRF96] P. P. Iannone, K. C. Reichmann, and N. J. Frigo. Broadcast digital video delivered over WDM passive optical networks. *IEEE Photonics Technology Letters*, 8(7):930–932, 1996.

[Ish83] H. Ishio. Japanese field trials and applications in telephony. *IEEE JSAC*, 1:404–412, 1983.

[ISSV96] E. Iannone, R. Sabella, L. De Stefano, and F. Valeri. All optical wavelength conversion in optical multicarrier networks. *IEEE Transactions on Communications*, 44(6):716–724, June 1996.

[ITU93] ITU-T. *Recommendation Q.700: Introduction to CCITT Signaling System No. 7*, 1993.

[ITU96] ITU-T SG15/WP 4. *Rec. G.681: Functional Characteristics of Interoffice and Long-Haul Line Systems Using Optical Amplifiers, Including Optical Multiplexing*, 1996.

[ITU98] ITU-T. *Recommendation G.983: Broadband Optical Access Systems Based on Passive Optical Networks*, 1998.

[ITU99] ITU-T. *Rec. G.664: Optical Safety Procedures and Requirements for Optical Transport Systems*, 1999.

[Jac96] J. L. Jackel et al. Acousto-optic tunable filters (AOTFs) for multiwavelength optical cross-connects: Crosstalk considerations. *IEEE/OSA JLT/JSAC Special Issue on Multiwavelength Optical Technology and Networks*, 14(6):1056–1066, June 1996.

[Jai96] M. Jain. Topology designs for wavelength routed optical networks. Technical report, Indian Institute of Science, Bangalore, Jan. 1996.

[JBM95] S. V. Jagannath, K. Bala, and M. Mihail. Hierarchical design of WDM optical networks for ATM transport. In *Proceedings of IEEE Globecom*, pages 2188–2194, 1995.

[JCC93] V. Jayaraman, Z.-M. Chuang, and L. A. Coldren. Theory, design and performance of extended tuning range semiconductor lasers with sampled gratings. *IEEE Journal of Quantum Electronics*, 29:1824–1834, July 1993.

[Jeu90] L. B. Jeunhomme. *Single-Mode Fiber Optics*. Marcel Dekker, New York, 1990.

[JQE91] *IEEE Journal of Quantum Electronics*, June 1991.

[KA96] M. Kovacevic and A. S. Acampora. On the benefits of wavelength translation in all optical clear-channel networks. *IEEE JSAC/JLT Special Issue on Optical Networks*, 14(6):868–880, June 1996.

[Kam96] I. P. Kaminow et al. A wideband all-optical WDM network. *IEEE JSAC/JLT Special Issue on Optical Networks*, 14(5):780–799, June 1996.

[Kan99] J. Kani et al. Interwavelength-band nonlinear interactions and their suppression in multiwavelength-band WDM transmission systems. *IEEE/OSA Journal on Lightwave Technology*, 17:2249–2260, 1999.

[Kar01] M. Karlsson et al. Higher order polarization mode dispersion compensator with three degrees of freedom. In *OFC 2001 Technical Digest*, pages MO1/1–3, 2001.

[Kas95]   N. Kashima. *Passive Optical Components for Optical Fiber Transmission*. Artech House, Boston, 1995.

[Kas99]   R. Kashyap. *Fibre Bragg Gratings*. Academic Press, San Diego, CA, 1999.

[KBW96]   L. G. Kazovsky, S. Benedetto, and A. E. Willner. *Optical Fiber Communication Systems*. Artech House, Boston, 1996.

[Ker93]   A. Kershenbaum. *Telecommunications Network Design Algorithms*. McGraw-Hill, New York, 1993.

[KF86]   M. V. Klein and T. E. Furtak. *Optics,* 2nd edition. John Wiley, New York, 1986.

[KGSP94] M. G. Kane, I. Glesk, J. P. Sokoloff, and P. R. Prucnal. Asymmetric loop mirror: Analysis of an all-optical switch. *Applied Optics*, 33(29):6833–6842, Oct. 1994.

[KI166]   K. C. Kao and G. A. Hockham. Dielectric-fiber surface waveguides for optical frequencies. *Proceedings of IEE*, 133(3):1151–1158, July 1966.

[KH90]   A. Krishna and B. Hajek. Performance of shuffle-like switching networks with deflection. In *Proceedings of IEEE Infocom*, pages 473–480, 1990.

[Kha97]   S. Khanna. A polynomial time approximation scheme for the SONET ring loading problem. *Bell Labs Technical Journal*, 2(2):36–41, Spring 1997.

[KK90]   T. L. Koch and U. Koren. Semiconductor lasers for coherent optical fiber communications. *IEEE/OSA Journal on Lightwave Technology*, 8(3):274–293, 1990.

[KK97a]   I. P. Kaminow and T. L. Koch, editors. *Optical Fiber Telecommunications IIIA*. Academic Press, San Diego, CA, 1997.

[KK97b]   I. P. Kaminow and T. L. Koch, editors. *Optical Fiber Telecommunications IIIB*. Academic Press, San Diego, CA, 1997.

[KKM70]   F. P. Kapron, D. B. Keck, and R. D. Maurer. Radiation losses in glass optical waveguides. *Applied Physics Letters*, 17(10):423–425, Nov. 1970.

[KKS00]   D. Kettler, H. Kafka, and D. Spears. Driving fiber to the home. *IEEE Communications Magazine*, 38(11):106–110, Nov. 2000.

[KLHN93] M. J. Karol, C. Lin, G. Hill, and K. Nosu, editors. *IEEE/OSA Journal of Lightwave Technology: Special Issue on Broadband Optical Networks*, May/June 1993.

[KM88]   K. Kobayashi and I. Mito. Single frequency and tunable laser diodes. *IEEE/OSA Journal on Lightwave Technology*, 6(11):1623–1633, November 1988.

[KM98]   F. W. Kerfoot and W. C. Marra. Undersea fiber optic networks: Past, present and future. *IEEE JSAC: Special Issue on High-Capacity Optical Transport Networks*, 16(7):1220–1225, Sept. 1998.

[Kni76]   Z. Knittl. *Optics of Thin Films*. John Wiley, New York, 1976.

[Kob94]   I. Kobayashi, editor. *IEEE Communications Magazine: Special Issue on Fiber-Optic Subscriber Loops*, volume 32, Feb. 1994.

[KPEJ97]   C. Kaklamanis, P. Persiano, T. Erlebach, and K. Jansen. Constrained bipartite edge coloring with applications to wavelength routing in all-optical networks. In *International Colloquium on Automata, Languages, and Programming*, 1997.

[Kra99]   J. M. Kraushaar. *Fiber Deployment Update: End of Year 1998*. Federal Communications Commission, Sept. 1999. Available from *http://www.fcc.gov*.

[KS97]   V. Kumar and E. Schwabe. Improved access to optical bandwidth in trees. In *Proceedings of the ACM Symposium on Distributed Algorithms*, 1997.

[KS98]   R. M. Krishnaswamy and K. N. Sivarajan. Design of logical topologies: A linear formulation for wavelength routed optical networks with no wavelength changers. In *Proceedings of IEEE Infocom*, 1998.

[KSHS01]   A. M. J. Koonen, M. K. Smit, H. Herrmann, and W. Sohler. Wavelength selective devices. In H. Venghaus and N. Grote, editors, *Devices for Optical Communication Systems*. Springer-Verlag, Heidelberg, 2001.

[KWK+98]   M. Koga, A. Watanabe, T. Kawai, K. Sato, and Y. Ohmori. Large-capacity optical path cross-connect system for WDM photonic transport network. *IEEE JSAC: Special Issue on High-Capacity Optical Transport Networks*, 16(7):1260–1269, Sept. 1998.

[LA91]   J.-F. P. Labourdette and A. S. Acampora. Logically rearrangeable multihop lightwave networks. *IEEE Transactions on Communications*, 39(8):1223–1230, Aug. 1991.

[Lao99]   H. Laor. 576 × 576 optical cross connect for single-mode fiber. In *Proceedings of Annual Multiplexed Telephony Conference*, 1999.

[LC82]   S. Lin and D. J. Costello. *Error Correcting Codes*. Prentice Hall, Englewood Cliffs, NJ, 1982.

[LC97]   C. Y. Lee and S. G. Chang. Balancing loads on SONET rings with integer demand splitting. *Computer Operations Research*, 24(3):221–229, 1997.

[Lee91]   T. P. Lee. Recent advances in long-wavelength semiconductor lasers for optical fiber communication. *Proceedings of IEEE*, 79(3):253–276, March 1991.

[LGM+97]   J. K. Lucek, P. Gunning, D. G. Moodie, K. Smith, and D. Pitcher. Synchrolan: A 40 Gbit/s optical-TDMA LAN. *Electronics Letters*, 33(10):887–888, April 1997.

[LGT98]   L. Y. Lin, E. L. Goldstein, and R. W. Tkach. Free-space micromachined optical switches with submillisecond switching time for large-scale optical crossconnects. *IEEE Photonics Technology Letters*, 10(4):525–528, Apr. 1998.

[Lin89]   C. Lin, editor. *Optoelectronic Technology and Lightwave Communications Systems*. Van Nostrand Reinhold, New York, 1989.

[Liu98]   Y. Liu et al. Advanced fiber designs for high capacity DWDM systems. In *Proceedings of National Fiber Optic Engineers Conference*, 1998.

[LL84]   J. P. Laude and J. M. Lerner. Wavelength division multiplexing/demultiplexing (WDM) using diffraction gratings. *SPIE-Application, Theory and Fabrication of Periodic Structures*, 503:22–28, 1984.

[LM93]   E. A. Lee and D. G. Messerschmitt. *Digital Communication*, 2nd edition. Kluwer, Boston, 1993.

[LNGP96]   E. Leonardi, F. Neri, M. Gerla, and P. Palnati. Congestion control in asynchronous high-speed wormhole routing networks. *IEEE Communications Magazine*, pages 58–69, Nov. 1996.

[LS00]   G. Li and R. Simha. On the wavelength assignment problem in multifiber WDM star and ring networks. In *Proceedings of IEEE Infocom*, 2000.

[LZ89]   T. P. Lee and C-N. Zah. Wavelength-tunable and single-frequency lasers for photonic communication networks. *IEEE Communications Magazine*, 27(10):42–52, Oct. 1989.

[LZNA98]   G. Luo, J. L. Zyskind, J. A. Nagel, and M. A. Ali. Experimental and theoretical analysis of relaxation-oscillations and spectral hole burning effects in all-optical gain-clamped EDFA's for WDM networks. *IEEE/OSA Journal on Lightwave Technology*, 16:527–533, 1998.

[Mac74]   J. B. MacChesney et al. Preparation of low-loss optical fibers using simultaneous vapor deposition and fusion. In *Proceedings of 10th International Congress on Glass*, volume 6, pages 40–44, Kyoto, Japan, 1974.

[Mae98]   M. Maeda. Management and control of optical networks. *IEEE Journal of Selected Areas in Communications*, 16(6):1008–1023, Sept. 1998.

[Mae00]   Y. Maeno et al. Cost comparison of an IP/OTN integrated node against a pure IP routing node. In *Proceedings of National Fiber Optic Engineers Conference*, 2000.

[Mar74]   D. Marcuse. *Theory of Dielectric Optical Waveguides*. Academic Press, New York, 1974.

[Mar80]   D. Marcuse. Pulse distortion in single-mode fibers. *Applied Optics*, 19:1653–1660, 1980.

[Mar81]   D. Marcuse. Pulse distortion in single-mode fibers. 3: Chirped pulses. *Applied Optics*, 20:3573–3579, 1981.

[Mas96]   F. Masetti et al. High speed, high capacity ATM optical switches for future telecommunication transport networks. *IEEE JSAC/JLT Special Issue on Optical Networks*, 14(5):979–998, June 1996.

[Max89]   N. F. Maxemchuck. Comparison of deflection and store-and-forward techniques in the Manhattan Street and shuffle-exchange networks. In *Proceedings of IEEE Infocom*, pages 800–809, 1989.

[MB96]   J. Manchester and P. Bonenfant. Fiber optic network survivability: SONET/optical protection layer interworking. In *Proceedings of National Fiber Optic Engineers Conference*, pages 907–918, 1996.

[MBN99]   J. Manchester, P. Bonenfant, and C. Newton. The evolution of transport network survivability. *IEEE Communications Magazine*, 37(8):44–51, Aug. 1999.

[MBRM96]   B. Mukherjee, D. Banerjee, S. Ramamurthy, and A. Mukherjee. Some principles for designing a wide-area optical network. *IEEE/ACM Transactions on Networking*, 4(5):684–696, 1996.

[McE77]   R. J. McEliece. *The Theory of Information and Coding: A Mathematical Framework for Communication*. Addison-Wesley, Reading, MA, 1977.

[McG98]   K. A. McGreer. Arrayed waveguide gratings for wavelength routing. *IEEE Communications Magazine*, 36(12):62–68, Dec. 1998.

[McG99]   A. McGuire. Management of optical transport networks. *IEE Electronics and Communication Engineering Journal*, 11(3):155–163, June 1999.

[MEM98]   D. D. Marcenac, A. D. Ellis, and D. G. Moodie. 80 Gbit/s OTDM using electroabsorption modulators. *Electronics Letters*, 34(1):101–103, Jan. 1998.

[Mid93]   J. E. Midwinter, editor. *Photonics in Switching, Volume II: Systems*. Academic Press, San Diego, CA, 1993.

[Mik99]   B. Mikkelsen et al. Unrepeatered transmission over 150 km of nonzero-dispersion fibre at 100 Gbit/s with semiconductor based pulse source, demultiplexer and clock recovery. *Electronics Letters*, 35(21):1866–1868, Oct. 1999.

[MK88]   S. D. Miller and I. P. Kaminow, editors. *Optical Fiber Telecommunications II*. Academic Press, San Diego, CA, 1988.

[MKR95]   M. Mihail, C. Kaklamanis, and S. Rao. Efficient access to optical bandwidth. In *IEEE Symposium on Foundations of Computer Science*, pages 548–557, 1995.

[MM98]   A. Mecozzi and D. Marcenac. Theory of optical amplifier chains. *IEEE/OSA Journal on Lightwave Technology*, 16:745–756, 1998.

[MM00]   G. Mohan and C. S. R. Murthy. Lightpath restoration in WDM optical networks. *IEEE Network Magazine*, 14(6):24–32, Nov.–Dec. 2000.

[Mor96]   T. Morioka et al. 100 Gb/s × 10 channel OTDM/WDM transmission using a single supercontinuum WDM source. In *OFC'96 Technical Digest*, 1996. Postdeadline paper PD21.

[MS88]   J. E. Midwinter and P. W. Smith, editors. *IEEE JSAC: Special Issue on Photonic Switching*, volume 6, Aug. 1988.

[MS96]   W. C. Marra and J. Schesser. Africa ONE: The Africa optical network. *IEEE Communications Magazine*, 34(2):50–57, Feb. 1996.

[MS98]   D. E. McDysan and D. L. Spohn. *ATM: Theory and Application*. McGraw-Hill, New York, 1998.

[MS00]   P. P. Mitra and J. B. Stark. Nonlinear limits to the information capacity of optical fibre communications. *Nature*, pages 1027–1030, 2000.

[MT83]   A. Moncalvo and F. Tosco. European field trials and early applications in telephony. *IEEE JSAC*, 1:398–403, 1983.

[MYK82]   T. Mukai, Y. Yamamoto, and T. Kimura. S/N and error-rate performance of AlGaAs semiconductor laser preamplifier and linear repeater systems. *IEEE Transactions on Microwave Theory and Techniques*, 30(10):1548–1554, 1982.

[MZB97]   N. M. Margalit, S. Z. Zhang, and J. E. Bowers. Vertical cavity lasers for telecom applications. *IEEE Communications Magazine*, 35(5):164–170, May 1997.

[Nak00]   M. Nakazawa et al. Ultrahigh-speed long-distance TDM and WDM soliton transmission technologies. *IEEE Journal of Selected Topics in Quantum Electronics*, 6:363–396, 2000.

[NE00]   S. Namiki and Y. Emori. Recent advances in ultra-wideband Raman amplifiers. In *OFC 2000 Technical Digest*, pages FF-1–FF-2, 2000.

[NE01]   S. Namiki and Y. Emori. Ultra-broadband Raman amplifiers pumped and gain-equalized by wavelength-division-multiplexed high-power laser diodes. *IEEE Journal of Selected Topics in Quantum Electronics*, 7(1):3–16, Jan./Feb. 2001.

[Nei00]   D. T. Neilson et al. Fully provisioned 112 × 112 micro-mechanical optical crossconnect with 35.8 Tb/s demonstrated capacity. In *OFC 2000 Technical Digest*, pages 204–206, 2000. Postdeadline paper PD-12.

[Nel01]   Lynn E. Nelson. Challenges of 40 Gb/s WDM transmission. In *OFC 2001 Technical Digest*, pages ThF1/1–3, 2001.

[Neu88]   E.-G. Neumann. *Single-Mode Fibers*. Springer-Verlag, Berlin, 1988.

[NKM98]   D. Nesset, T. Kelly, and D. Marcenac. All-optical wavelength conversion using SOA nonlinearities. *IEEE Communications Magazine*, 36(12):56–61, Dec. 1998.

[NO94]   K. Nosu and M. J. O'Mahony, editors. *IEEE Communications Magazine: Special Issue on Optically Multiplexed Networks*, volume 32, Dec. 1994.

[NR01] A. Neukermans and R. Ramaswami. MEMS technology for optical networking applications. *IEEE Communications Magazine*, 39(1):62–69, Jan. 2001.

[NS02] T. K. Nayak and K. N. Sivarajan. A new approach to dimensioning optical networks. *IEEE Journal of Selected Areas in Communications*, to appear, 2002.

[NSK99] M. Nakazawa, K. Suzuki, and H. Kubota. Single-channel 80 Gbit/s soliton transmission over 10000 km using in-line synchronous modulation. *Electronics Letters*, 35:1358–1359, 1999.

[NTM00] A. Narula-Tam and E. Modiano. Dynamic load balancing for WDM-based packet networks. In *Proceedings of IEEE Infocom*, 2000.

[OLH89] R. Olshanksy, V. A. Lanzisera, and P. M. Hill. Subcarrier multiplexed lightwave systems for broadband distribution. *IEEE/OSA Journal on Lightwave Technology*, 7(9):1329–1342, Sept. 1989.

[Ols89] N. A. Olsson. Lightwave systems with optical amplifiers. *IEEE/OSA Journal on Lightwave Technology*, 7(7):1071–1082, July 1989.

[O'M88] M. J. O'Mahony. Semiconductor laser amplifiers for future fiber systems. *IEEE/OSA Journal on Lightwave Technology*, 6(4):531–544, April 1988.

[Ona96] H. Onaka et al. 1.1 Tb/s WDM transmission over a 150 km 1.3 μm zero-dispersion single-mode fiber. In *OFC'96 Technical Digest*, 1996. Postdeadline paper PD19.

[Ono98] T. Ono et al. Characteristics of optical duobinary signals in terabit/s capacity, high spectral efficiency WDM systems. *IEEE/OSA Journal on Lightwave Technology*, 16:788–797, 1998.

[OSF00] G. Ocakoglu, K. Struyve, and P. Falcao. The business case for DWDM metro systems in a pan-European carrier environment. In *Proceedings of National Fiber Optic Engineers Conference*, 2000.

[OSYZ95] M. J. O'Mahony, D. Simeonidou, A. Yu, and J. Zhou. The design of a European optical network. *IEEE/OSA Journal on Lightwave Technology*, 13(5):817–828, May 1995.

[OWS96] S. Okamoto, A. Watanabe, and K.-I. Sato. Optical path cross-connect node architectures for photonic transport network. *IEEE/OSA JLT/JSAC Special Issue on Multiwavelength Optical Technology and Networks*, 14(6):1410–1422, June 1996.

[OY98] T. Ono and Y. Yano. Key technologies for terabit/second WDM systems high spectral efficiency of over 1 bit/s/hz. *IEEE Journal of Quantum Electronics*, 34:2080–2088, 1998.

[PAP86] M. J. Potasek, G. P. Agrawal, and S. C. Pinault. Analytic and numerical study of pulse broadening in nonlinear dispersive optical fibers. *Journal of Optical Society of America B*, 3(2):205–211, Feb. 1986.

[Pap91]   A. Papoulis. *Probability, Random Variables, and Stochastic Processes*, 3rd edition. McGraw-Hill, New York, 1991.

[Pax97]   V. Paxson. End-to-end routing behavior in the Internet. *IEEE/ACM Transactions on Networking*, 5(5):601–615, Oct. 1997.

[PCK00]   G. N. S. Prasanna, E. A. Caridi, and R. M. Krishnaswamy. Metropolitan IP-optical networks: A systematic case study. In *Proceedings of National Fiber Optic Engineers Conference*, 2000.

[PCW+00] V. Poudyal, R. H. Cardwell, O. J. Wasem, J. E. Baran, and A. Rajan. Comparison of network alternatives for transporting high capacity tributaries for IP router interconnection. In *Proceedings of National Fiber Optic Engineers Conference*, 2000.

[PD99]   L. L. Peterson and B. S. Davie. *Computer Networks: A Systems Approach*. Morgan Kaufmann, San Francisco, 1999.

[Per73]   S. D. Personick. Applications for quantum amplifiers in simple digital optical communication systems. *Bell System Technical Journal*, 52(1):117–133, Jan. 1973.

[Per99]   R. Perlman. *Interconnections: Bridges, Routers, Switches, and Internetworking Protocols*. Addison-Wesley, Reading, MA, 1999.

[PF95]   V. Paxon and S. Floyd. Wide area traffic: The failure of Poisson modelling. *IEEE/ACM Transactions on Networking*, 3(3):226–244, June 1995.

[PHR97]   N. S. Patel, K. L. Hall, and K. A. Rauschenbach. Optical rate conversion for high-speed TDM networks. *IEEE Photonics Technology Letters*, 9(9):1277, Sept. 1997.

[PL01]   D. Penninckx and S. Lanne. Reducing PMD impairments. In *OFC 2001 Technical Digest*, pages TuP1/1–3, 2001.

[PN87]   K. Padmanabhan and A. N. Netravali. Dilated networks for photonic switching. *IEEE Transactions on Communications*, 35:1357–1365, 1987.

[Pro00]   J. G. Proakis. *Digital Communications*, 4th edition. McGraw-Hill, New York, 2000.

[Pru89]   P. R. Prucnal, editor. *IEEE Network: Special Issue on Optical Multiaccess Networks*, volume 3, March 1989.

[Pru93]   P. R. Prucnal. Optically processed self-routing, synchronization, and contention resolution for 1-d and 2-d photonic switching architectures. *IEEE Journal of Quantum Electronics*, 29(2):600–612, Feb. 1993.

[PS95]   J. S. Patel and Y. Silberberg. Liquid crystal and grating-based multiple-wavelength cross-connect switch. *IEEE Photonics Technology Letters*, 7(5):514–516, May 1995.

[PSF86] P. R. Prucnal, M. A. Santoro, and T. R. Fan. Spread spectrum fiber-optic local area network using optical processing. *IEEE/OSA Journal on Lightwave Technology*, LT-4(5):547–554, May 1986.

[PTCF91] C. D. Poole, R. W. Tkach, A. R. Chraplyvy, and D. A. Fishman. Fading in lightwave systems due to polarization-mode dispersion. *IEEE Photonics Technology Letters*, 3(1):68–70, Jan. 1991.

[Qua98] J. A. Quayle et al. Achieving global consensus on the strategic broadband access network—the full service access initiative. *BT Technology Journal*, 16(4):58–70, Oct. 1998.

[QY99] C. Qiao and M. Yoo. Optical burst switching (OBS): A new paradigm for an optical Internet. *Journal of High Speed Networks*, 8(1):69–84, 1999.

[Ram01] R. Ramamurthy et al. Capacity performance of dynamic provisioning in optical networks. *IEEE/OSA Journal on Lightwave Technology*, 19(1):40–48, 2001.

[RH90] R. Ramaswami and P. A. Humblet. Amplifier induced crosstalk in multi-channel optical networks. *IEEE/OSA Journal on Lightwave Technology*, 8(12):1882–1896, Dec. 1990.

[Rig95] P.-J. Rigole et al. 114-nm wavelength tuning range of a vertical grating assisted codirectional coupler laser with a super structure grating distributed Bragg reflector. *IEEE Photonics Technology Letters*, 7(7):697–699, July 1995.

[RL93] R. Ramaswami and K. Liu. Analysis of effective power budget in optical bus and star networks using erbium-doped fiber amplifiers. *IEEE/OSA Journal on Lightwave Technology*, 11(11):1863–1871, Nov. 1993.

[RLB95] P. Roorda, C.-Y. Lu, and T. Boutlier. Benefits of all-optical routing in transport networks. In *OFC'95 Technical Digest*, pages 164–165, 1995.

[RM99a] B. Ramamurthy and B. Mukherjee. Survivable WDM mesh networks, Part I—protection. In *Proceedings of IEEE Infocom*, pages 744–751, 1999.

[RM99b] B. Ramamurthy and B. Mukherjee. Survivable WDM mesh networks, Part II—restoration. In *Proceedings of IEEE International Conference on Communication*, pages 2023–2030, 1999.

[RMGB97] M. Renaud, F. Masetti, C. Guillemot, and B. Bostica. Network and system concepts for optical packet switching. *IEEE Communications Magazine*, 35(4):96–102, Apr. 1997.

[RN76] H.-D. Rudolph and E.-G. Neumann. Approximations for the eigenvalues of the fundamental mode of a step-index glass fiber waveguide. *Nachrichtentechnische Zeitschrift*, 29(14):328–329, 1976.

[Ros86] F. E. Ross. FDDI—a tutorial. *IEEE Communications Magazine*, 24(5):10–17, May 1986.

[RS95]   R. Ramaswami and K. N. Sivarajan. Routing and wavelength assignment in all-optical networks. *IEEE/ACM Transactions on Networking*, pages 489–500, Oct. 1995. An earlier version appeared in *Proceedings of IEEE Infocom'94*.

[RS96]   R. Ramaswami and K. N. Sivarajan. Design of logical topologies for wavelength-routed optical networks. *IEEE JSAC/JLT Special Issue on Optical Networks*, 14(5):840–851, June 1996.

[RS97a]   R. Ramaswami and G. H. Sasaki. Multiwavelength optical networks with limited wavelength conversion. In *Proceedings of IEEE Infocom*, pages 490–499, 1997.

[RS97b]   R. Ramaswami and A. Segall. Distributed network control for optical networks. *IEEE/ACM Transactions on Networking*, Dec. 1997.

[RT84]   P. K. Runge and P. R. Trischitta. The SL undersea lightwave system. *IEEE/OSA Journal on Lightwave Technology*, 2:744–753, 1984.

[RU94]   P. Raghavan and E. Upfal. Efficient routing in all-optical networks. In *Proceedings of 26th ACM Symposium on Theory of Computing*, pages 134–143, May 1994.

[RWv93]   S. Ramo, J. R. Whinnery, and T. van Duzer. *Fields and Waves in Communication Electronics*. John Wiley, New York, 1993.

[Ryf01]   R. Ryf et al. 1296-port MEMS transparent optical crossconnect with 2.07 Petabit/s switch capacity. In *OFC 2001 Technical Digest*, 2001. Postdeadline paper PD28.

[Sab01]   O. A. Sab. FEC techniques in submarine transmission systems. In *OFC 2001 Technical Digest*, pages TuF1/1–3, 2001.

[Sal89]   J. A. Salehi. Code division multiple-access techniques in optical fiber networks— Part I: Fundamental principles. *IEEE Transactions on Communications*, 37(8):824–833, Aug. 1989.

[SAS96]   S. Subramaniam, M. Azizoglu, and A. K. Somani. Connectivity and sparse wavelength conversion in wavelength-routing networks. In *Proceedings of IEEE Infocom*, pages 148–155, 1996.

[SB87]   R. A. Spanke and V. E. Beneš. An *n*-stage planar optical permutation network. *Applied Optics*, 26, April 1987.

[SB89]   J. A. Salehi and C. A. Brackett. Code division multiple-access techniques in optical fiber networks—Part II: Systems performance analysis. *IEEE Transactions on Communications*, 37(8):834–842, Aug. 1989.

[SBJC90] D. A. Smith, J. E. Baran, J. J. Johnson, and K.-W. Cheung. Integrated-optic acoustically-tunable filters for WDM networks. *IEEE Journal of Selected Areas in Communications*, 8(6):1151–1159, Aug. 1990.

[SBJS93]   T. E. Stern, K. Bala, S. Jiang, and J. Sharony. Linear lightwave networks: Performance issues. *IEEE/OSA Journal on Lightwave Technology*, 11:937–950, May/June 1993.

[SBP96]   S.-W. Seo, K. Bergman, and P. R. Prucnal. Transparent optical networks with time-division multiplexing. *IEEE JSAC/JLT Special Issue on Optical Networks*, 14(5):1039–1051, June 1996.

[SBW87]   N. Shibata, R. P. Braun, and R. G. Waarts. Phase-mismatch dependence of efficiency of wave generation through four-wave mixing in a single-mode optical fiber. *IEEE Journal of Quantum Electronics*, 23:1205–1210, 1987.

[SGS99]   J. M. Simmons, E. L. Goldstein, and A. A. M. Saleh. Quantifying the benefit of wavelength add-drop in WDM rings with distance-independent and dependent traffic. *IEEE/OSA Journal on Lightwave Technology*, 17:48–57, 1999.

[Sha48]   C E. Shannon. A mathematical theory of communication. *Bell System Technical Journal*, 27(3):379–423, July 1948.

[SIA92]   Y. Suematsu, K. Iga, and S. Arai. Advanced semiconductor lasers. *Proceedings of IEEE*, 80:383–397, 1992.

[SID93]   J. R. Sauer, M. N. Islam, and S. P. Dijaili. A soliton ring network. *IEEE/OSA Journal on Lightwave Technology*, 11(12):2182–2190, Dec. 1993.

[SKA00]   H. Sunnerud, M. Karlsson, and P. A. Andrekson. Analytical theory for PMD-compensation. *IEEE Photonics Technology Letters*, 12:50–52, 2000.

[SKA01]   H. Sunnerud, M. Karlsson, and P. A. Andrekson. A comparison between NRZ and RZ data formats with respect to PMD-induced system degradation. *IEEE Photonics Technology Letters*, 13:448–450, 2001.

[SKN01]   K. Suzuki, H. Kubota, and M. Nakazawa. 1 Tb/s (40 Gb/s x 25 channel) DWDM quasi-DM soliton transmission over 1,500 km using dispersion-managed single-mode fiber and conventional C-band EDFAs. In *OFC 2001 Technical Digest*, pages TuN7/1–3, 2001.

[SKY89]   P. W. Shumate, O. Krumpholz, and K. Yamaguchi, editors. *IEEE/OSA JLT/JSAC Special Issue on Subscriber Loop Technology*, volume 7, Nov. 1989.

[SM00]   B. Schein and E. Modiano. Quantifying the benefit of configurability in circuit-switched WDM ring networks. In *Proceedings of IEEE Infocom*, 2000.

[SMB00]   D. T. Schaafsma, E. Miles, and E. M. Bradley. Comparison of conventional and gain-clamped semiconductor optical amplifiers for wavelength-division-multiplexed transmission systems. *IEEE/OSA Journal on Lightwave Technology*, 18(7):922–925, July 2000.

[Smi72] R. G. Smith. Optical power handling capacity of low loss optical fibers as determined by stimulated Raman and Brillouin scattering. *Applied Optics*, 11(11):2489–2160, Nov. 1972.

[Smi81] B. Smith. Architecture and applications of the HEP multiprocessor system. In *Real Time Signal Processing IV, Proceedings of SPIE*, pages 241–248, 1981.

[SNA97] G. H. Smith, D. Novak, and Z. Ahmed. Technique for optical SSB generation to overcome dispersion penalties in fibre-radio systems. *Electronics Letters*, 33:74–75, 1997.

[SNIA90] N. Shibata, K. Nosu, K. Iwashita, and Y. Azuma. Transmission limitations due to fiber nonlinearities in optical FDM systems. *IEEE Journal of Selected Areas in Communications*, 8(6):1068–1077, Aug. 1990.

[Son95] G. H. Song. Toward the ideal codirectional Bragg filter with an acousto-optic-filter design. *IEEE/OSA Journal on Lightwave Technology*, 13(3):470–480, March 1995.

[Soo92] J. B. D. Soole et al. Wavelength selectable laser emission from a multistripe array grating integrated cavity laser. *Applied Physics Letters*, 61:2750–2752, 1992.

[SOW95] K. I. Sato, S. Okamoto, and A. Watanabe. Photonic transport networks based on optical paths. *International Journal of Communication Systems (UK)*, 8(6):377–389, Nov./Dec. 1995.

[Spa87] R. A. Spanke. Architectures for guided-wave optical space switching systems. *IEEE Communications Magazine*, 25(5):42–48, May 1987.

[SPGK93] J. P. Sokoloff, P. R. Prucnal, I. Glesk, and M. Kane. A terahertz optical asymmetric demultiplexer (TOAD). *IEEE Photonics Technology Letters*, 5(7):787–790, July 1993.

[SR97] M. Sexton and A. Reid. *Broadband Networking: ATM, SDH and SONET*. Artech House, Boston, 1997.

[SS96a] M. A. Scobey and D. E. Spock. Passive DWDM components using microplasma optical interference filters. In *OFC'96 Technical Digest*, pages 242–243, San Jose, Feb. 1996.

[SS96b] C. A. Siller and M. Shafi, editors. *SONET/SDH: A Sourcebook of Synchronous Networking*. IEEE Press, Los Alamitos, CA, 1996.

[SS99] A. A. M. Saleh and J. M. Simmons. Architectural principles for optical regional and metropolitan access networks. *IEEE/OSA Journal on Lightwave Technology*, 17(12), Dec. 1999.

[SS00] A. Sridharan and K. N. Sivarajan. Blocking in all-optical networks. In *Proceedings of IEEE Infocom*, 2000.

[ST84]   J. W. Suurballe and R. E. Tarjan. A quick method for finding shortest pairs of disjoint paths. *Networks*, 14:325–336, 1984.

[ST91]   B. E. A. Saleh and M. C. Teich. *Fundamentals of Photonics*. Wiley, New York, 1991.

[Sta83] J. R. Stauffer. FT3C—a lightwave system for metropolitan and intercity applications. *IEEE JSAC*, 1:413–419, 1983.

[Sta99] J. B. Stark. Fundamental limits of information capacity for optical communications channels. In *Proceedings of European Conference on Optical Communication*, pages 1–28, Nice, France, Sept. 1999.

[Ste87] J. Stern et al. Passive optical local networks for telephony applications. *Electronics Letters*, 23:1255–1257, 1987.

[Ste90] T. E. Stern. Linear lightwave networks: How far can they go? In *Proceedings of IEEE Globecom*, pages 1866–1872, 1990.

[Ste94] W. R. Stevens. *TCP/IP Illustrated, Volume 1*. Addison-Wesley, Reading, MA, 1994.

[Stu00] K. E. Stubkjaer. Semiconductor optical amplifier-based all-optical gates for high-speed optical processing. *IEEE Journal of Selected Topics in Quantum Electronics*, 6(6):1428–1435, Nov./Dec. 2000.

[Sub00] M. Subramanian. *Network Management: Principles and Practice*. Addison-Wesley, Reading, MA, 2000.

[SV96]   M. W. Sachs and A. Varma. Fibre channel and related standards. *IEEE Communications Magazine*, 34(8):40–49, Aug. 1996.

[TCF+95] R. W. Tkach, A. R. Chraplyvy, F. Forghieri, A. H. Gnauck, and R. M. Derosier. Four-photon mixing and high-speed WDM systems. *IEEE/OSA Journal on Lightwave Technology*, 13(5):841–849, May 1995.

[Tel98] Telcordia Technologies. *Common Generic Requirements for Optical Add-Drop Multiplexers (OADMs) and Optical Terminal Multiplexers (OTMs)*, Dec. 1998. GR-2979-CORE, Issue 2.

[Tel99] Telcordia Technologies. *SONET Transport Systems: Common Generic Criteria*, 1999. GR-253-CORE Issue 2, Revision 2.

[Tie95] L. F. Tiemeijer et al. Reduced intermodulation distortion in 1300 nm gain-clamped MQW laser amplifiers. *IEEE Photonics Technology Letters*, 7(3):284–286, Mar. 1995.

[Toh93] Y. Tohmori et al. Over 100 nm wavelength tuning in superstructure grating (SSG) DBR lasers. *Electronics Letters*, 29:352–354, 1993.

[Tol98] P. Toliver et al. Routing of 100 Gb/s words in a packet-switched optical networking demonstration (POND) node. *IEEE/OSA Journal on Lightwave Technology*, 16(12):2169–2180, Dec. 1998.

[TOT96] H. Takahashi, K. Oda, and H. Toba. Impact of crosstalk in an arrayed-waveguide multiplexer on $n \times n$ optical interconnection. *IEEE/OSA JLT/JSAC Special Issue on Multiwavelength Optical Technology and Networks*, 14(6):1097–1105, June 1996.

[TOTI95] H. Takahashi, K. Oda, H. Toba, and Y. Inoue. Transmission characteristics of arrayed $n \times n$ wavelength multiplexer. *IEEE/OSA Journal on Lightwave Technology*, 13(3):447–455, March 1995.

[TS00] R. H. Thornburg and B. J. Schoenborn. *Storage Area Networks: Designing and Implementing a Mass Storage System*. Prentice Hall, Englewood Cliffs, NJ, 2000.

[TSN94] H. Takahashi, S. Suzuki, and I. Nishi. Wavelength multiplexer based on SiO_2–Ta_2O_5 arrayed-waveguide grating. *IEEE/OSA Journal on Lightwave Technology*, 12(6):989–995, June 1994.

[Tuc75] A. Tucker. Coloring a family of circular arcs. *SIAM Journal on Applied Mathematics*, 29(3):493–502, 1975.

[Tur99] J. S. Turner. Terabit burst switching. *Journal of High Speed Networks*, 8(1):3–16, 1999.

[Udu99] D. K. Udupa. *TMN Telecommunications Management Network*. McGraw-Hill, New York, 1999.

[US86] U. S. Food and Drug Administration, Department of Radiological Health. *Requirements of 21 CFR Chapter J for Class 1 Laser Products*, Jan. 1986.

[Vak99] D. Vakhshoori et al. 2 mW CW singlemode operation of a tunable 1550 nm vertical cavity surface emitting laser. *Electronics Letters*, 35(11):900–901, May 1999.

[Ven96a] A. M. Vengsarkar et al. Long-period fiber-grating-based gain equalizers. *Optics Letters*, 21(5):336–338, 1996.

[Ven96b] A. M. Vengsarkar et al. Long-period gratings as band-rejection filters. *IEEE/OSA Journal on Lightwave Technology*, 14(1):58–64, Jan. 1996.

[VMVQ00] I. Van de Voorde, C. M. Martin, J. Vandewege, and X. Z. Qiu. The superPON demonstrator: An exploration of possible evolution paths for optical access networks. *IEEE Communications Magazine*, 38(2):74–82, Feb. 2000.

[VPM01] G. Vareille, F. Pitel, and J. F. Marccrou. 3 Tb/s (300 × 11.6 Gbit/s) transmission over 7380 km using 28 nm C+L-band with 25 GHz channel spacing and NRZ format. In *OFC 2001 Technical Digest*, pages PD22/1–3, 2001.

[VS91]   A. R. Vellekoop and M. K. Smit. Four-channel integrated-optic wavelength demultiplexer with weak polarization dependence. *IEEE/OSA Journal on Lightwave Technology*, 9:310–314, 1991.

[vT68]   H. L. van Trees. *Detection, Estimation, and Modulation Theory, Part I.* John Wiley, New York, 1968.

[WASG96]   R. E. Wagner, R. C. Alferness, A. A. M. Saleh, and M. S. Goodman. MONET: Multiwavelength optical networking. *IEEE/OSA JLT/JSAC Special Issue on Multiwavelength Optical Technology and Networks*, 14(6):1349–1355, June 1996.

[WD96]   N. Wauters and P. Demeester. Design of the optical path layer in multiwavelength cross-connected networks. *IEEE JSAC/JLT Special Issue on Optical Networks*, 14(6):881–892, June 1996.

[Wei98]   Y. Wei et al. Connection management for multiwavelength optical networking. *IEEE Journal of Selected Areas in Communications*, 16(6):1097–1108, Sept. 1998.

[WF83]   A. X. Widmer and P. A. Franaszek. A DC-balanced, partitioned-block, 8B-10B transmission code. *IBM Journal of Research and Development*, 27(5):440–451, Sept. 1983.

[Wil96]   G. Wilfong. Minimizing wavelengths in an all-optical ring network. In *7th International Symposium on Algorithms and Computation*, pages 346–355, 1996.

[Wil00a]   A. E. Willner, editor. *IEEE Journal of Selected Topics in Quantum Electronics: Millennium Issue*, volume 6, Nov./Dec. 2000.

[Wil00b]   B. J. Wilson et al. Multiwavelength optical networking management and control. *IEEE/OSA Journal on Lightwave Technology*, 18(12):2038–2057, 2000.

[WJ90]   J. M. Wozencraft and I. M. Jacobs. *Principles of Communication Engineering.* Waveland Press, Prospect Heights, IL, 1990. Reprint of the originial 1965 edition.

[WK92]   J. H. Winters and S. Kasturia. Adaptive nonlinear cancellation for high-speed fiber-optic systems. *IEEE/OSA Journal on Lightwave Technology*, 10:971–977, 1992.

[WKR+88]   S. S. Wagner, H. Kobrinski, T. J. Robe, H. L. Lemberg, and L. S. Smoot. Experimental demonstration of a passive optical subscriber loop architecture. *Electronics Letters*, 24:344–346, 1988.

[WL88]   S. S. Wagner and H. L. Lemberg. Technology and system issues for the WDM-based fiber loop architecture. *IEEE/OSA Journal on Lightwave Technology*, 7(11):1759–1768, 1988.

[WL96]   K.-Y. Wu and J.-Y. Liu. Liquid-crystal space and wavelength routing switches. In *Proceedings of Lasers and Electro-Optics Society Annual Meeting*, pages 28–29, 1996.

[WMB92] J. Willems, G. Morthier, and R. Baets. Novel widely tunable integrated optical filter with high spectral selectivity. In *Proceedings of European Conference on Optical Communication*, pages 413–416, 1992.

[WO95] L. Wuttisittikulkij and M. J. O'Mahony. Multiwavelength self-healing ring transparent networks. In *Proceedings of IEEE Globecom*, pages 45–49, 1995.

[Woo00] E. L. Wooten et al. A review of lithium niobate modulators for fiber-optic communication systems. *IEEE Journal of Selected Topics in Quantum Electronics*, 6(1):69–82, Jan./Feb. 2000.

[WOS90] W. I. Way, R. Olshansky, and K. Sato, editors. Special issue on applications of RF and microwave subcarriers to optical fiber transmission in present and future broadband networks. *IEEE Journal of Selected Areas in Communications*, 8(7), Sept. 1990.

[Wu92] T. H. Wu. *Fiber Network Service Survivability*. Artech House, Boston, 1992.

[Wu95] T. H. Wu. Emerging techniques for fiber network survivability. *IEEE Communications Magazine*, 33(2):58–74, Feb. 1995.

[WV00] J. Walrand and P. Varaiya. *High-Performance Communication Networks*. Morgan Kaufmann, San Francisco, 2000.

[WW98] G. Wilfong and P. Winkler. Ring routing and wavelength translation. In *Proceedings of the Symposium on Discrete Algorithms (SODA)*, pages 334–341, 1998.

[Yam80] Y. Yamamoto. Noise and error-rate performance of semiconductor laser amplifiers in PCM-IM transmission systems. *IEEE Journal of Quantum Electronics*, 16:1073–1081, 1980.

[Yam98] Y. Yamada et al. Optical output buffered ATM switch prototype based on FRONTIERNET architecture. *IEEE JSAC: Special Issue on High-Capacity Optical Transport Networks*, 16(7):2117–2134, Sept. 1998.

[Yan96] Y. Yano et al. 2.6 Tb/s WDM transmission experiment using optical duobinary coding. In *Proceedings of European Conference on Optical Communication*, 1996. Postdeadline paper Th.B.3.1.

[Yar65] A. Yariv. Internal modulation in multimode laser oscillators. *Journal of Applied Physics*, 36:388, 1965.

[Yar89] A. Yariv. *Quantum Electronics*, 3rd edition. John Wiley, New York, 1989.

[Yar97] A. Yariv. *Optical Electronics in Modern Communications*. Oxford University Press, 1997.

[YLES96] J. Yates, J. Lacey, D. Everitt, and M. Summerfield. Limited-range wavelength translation in all-optical networks. In *Proceedings of IEEE Infocom*, pages 954–961, 1996.

[Yoo96] S. J. B. Yoo. Wavelength conversion techniques for WDM network applications. *IEEE/OSA JLT/JSAC Special Issue on Multiwavelength Optical Technology and Networks*, 14(6):955–966, June 1996.

[You95] M. G. Young et al. Six-channel WDM transmitter module with ultra-low chirp and stable λ selection. In *Proceedings of European Conference on Optical Communication*, pages 1019–1022, 1995.

[YQD01] M. Yoo, C. Qiao, and S. Dixit. Optical burst switching for service differentiation in the next-generation optical Internet. *IEEE Communications Magazine*, 39(2):98–104, Feb. 2001.

[ZA95] Z. Zhang and A. S. Acampora. A heuristic wavelength assignment algorithm for multihop WDM networks with wavelength routing and wavelength reuse. *IEEE/ACM Transactions on Networking*, 3(3):281–288, June 1995.

[Zah92] C. E. Zah et al. Monolithic integration of multiwavelength compressive strained multiquantum-well distributed-feedback laser array with star coupler and optical amplifiers. *Electronics Letters*, 28:2361–2362, 1992.

[ZCC+96] J. Zhou, R. Cadeddu, E. Casaccia, C. Cavazzoni, and M. J. O'Mahony. Crosstalk in multiwavelength optical cross-connect networks. *IEEE/OSA JLT/JSAC Special Issue on Multiwavelength Optical Technology and Networks*, 14(6):1423–1435, June 1996.

[Zhu01] B. Zhu et al. 3.08 Tb/s (77 × 42.7 Gb/s) transmission over 1200 km of non-zero dispersion-shifted fiber with 100-km spans using C- L-band distributed Raman amplification. In *OFC 2001 Technical Digest*, pages PD23/1–3, 2001.

[Zir91] M. Zirngibl. Gain control in erbium-doped fiber amplifiers by an all-optical feedback loop. *Electronics Letters*, 27:560, 1991.

[Zir96] M. Zirngibl et al. An 18-channel multifrequency laser. *IEEE Photonics Technology Letters*, 8:870–872, 1996.

[Zir98] M. Zirngibl. Analytical model of Raman gain effects in massive wavelength division multiplexed transmission systems. *Electronics Letters*, 34:789, 1998.

[ZJ94] M. Zirngibl and C. H. Joyner. A 12-frequency WDM laser source based on a transmissive waveguide grating router. *Electronics Letters*, 30:700–701, 1994.

[ZJS+95] M. Zirngibl, C. H. Joyner, L. W. Stulz, C. Dragone, H. M. Presby, and I. P. Kaminow. LARnet, a local access router network. *IEEE Photonics Technology Letters*, 7(2):1041–1135, Feb. 1995.

[ZO94]   J. Zhou and M. J. O'Mahony. Optical transmission system penalties due to fiber polarization mode dispersion. *IEEE Photonics Technology Letters*, 6(10):1265–1267, Oct. 1994.

[ZT98]   W. D. Zhong and R. S. Tucker. Wavelength routing-based photonic packet buffers and their applications in photonic switching systems. *IEEE/OSA Journal on Lightwave Technology*, 16(10):1737–1745, Oct. 1998.

[Zys96] J. L. Zyskind et al. Fast power transients in optically amplified multiwavelength optical networks. In *OFC'96 Technical Digest*, 1996. Postdeadline paper PD31.

Index

Bold page numbers indicate definition.